Advances in Intelligent Systems and Computing

Volume 510

Series editor

Janusz Kacprzyk, Polish Academy of Sciences, Warsaw, Poland
e-mail: kacprzyk@ibspan.waw.pl

About this Series

The series "Advances in Intelligent Systems and Computing" contains publications on theory, applications, and design methods of Intelligent Systems and Intelligent Computing. Virtually all disciplines such as engineering, natural sciences, computer and information science, ICT, economics, business, e-commerce, environment, healthcare, life science are covered. The list of topics spans all the areas of modern intelligent systems and computing.

The publications within "Advances in Intelligent Systems and Computing" are primarily textbooks and proceedings of important conferences, symposia and congresses. They cover significant recent developments in the field, both of a foundational and applicable character. An important characteristic feature of the series is the short publication time and world-wide distribution. This permits a rapid and broad dissemination of research results.

More information about this series at http://www.springer.com/series/11156

Tai-He Fan · Shui-Li Chen
San-Min Wang · Yong-Ming Li
Editors

Quantitative Logic and Soft Computing 2016

Proceedings of the 4th International Conference on Quantitative Logic and Soft Computing (QLSC2016) held at Hangzhou, China, 14–17 October, 2016

Springer

Editors
Tai-He Fan
Zhejiang Sci-Tech University
Hangzhou
China

San-Min Wang
Zhejiang Sci-Tech University
Hangzhou
China

Shui-Li Chen
Jimei University
Xiamen
China

Yong-Ming Li
Shaanxi Normal University
Xi'an
China

ISSN 2194-5357 ISSN 2194-5365 (electronic)
Advances in Intelligent Systems and Computing
ISBN 978-3-319-46205-9 ISBN 978-3-319-46206-6 (eBook)
DOI 10.1007/978-3-319-46206-6

Library of Congress Control Number: 2016951970

Printed on acid-free paper

This Springer imprint is published by Springer Nature
The registered company is Springer International Publishing AG
The registered company address is: Gewerbestrasse 11, 6330 Cham, Switzerland

Preface

The Fourth International Conference on Quantitative Logic and Soft Computing (QLSC2016) was held on October 14–17, 2016, in Zhejiang Sci-Tech University, Hangzhou, China. QLSC2016 was the fourth in a series of conferences on quantitative logic and soft computing. It follows the successful QLSC2012 in Xi'an, China, QLSC2010 in Xiamen, China, and QLSC2009 in Shanghai, China. It was a major symposium for scientists, engineers, and practitioners to present their updated results, ideas, developments, and applications in all areas of quantitative logic and soft computing. It aimed to strengthen relations between industry research laboratories and universities, and to create a primary symposium for scientists in fields related to quantitative logic and soft computing worldwide as follows:

1. Quantitative logic and uncertainty logic;
2. Automata and quantification of software;
3. Fuzzy connectives and fuzzy reasoning;
4. Fuzzy logical algebras;
5. Artificial intelligence and soft computing; and
6. Fuzzy sets theory and applications.

Early in 2009, at the closing ceremony of the first QLSC conference in Shanghai, when congratulating the success of the conference our beloved Prof. Guo-Jun Wang said that QLSC is a vey good and helpful opportunity for researchers in the related fields and hope that QLSC should go on continuously. It was very sad that Prof. Guo-Jun Wang left us forever in the winter of 2013; this conference was a memorial to him.

Over viewing the QLSC2016 proceedings, we have put a step forward but still have a long way to go. QLSC2016 received more than 70 submissions. Each paper has undergone a rigorous review process. Only high-quality papers are included. The proceeding contains 61 papers, including 5 invited papers or abstracts from keynote speakers. It consists of 7 parts: (1) keynote speakers; (2) quantitative logic and uncertainty logic; (3) automata and quantification of software; (4) fuzzy connectives and fuzzy reasoning; (5) fuzzy logical algebras; (6) artificial intelligence and soft computing; and (7) fuzzy sets theory and applications.

Putting together the conference proceedings was a team effort. Special thanks were due to the authors for providing all materials; to all keynote speakers for their kindness to present excellent speeches to the conference; to the program committee and external reviewers for peer-reviewing papers and providing valuable suggestions; to the organizing committee and all the local volunteers, especially for their excellent management work for QLSC2016; to Prof. Janusz Kacprzyk for his warmheartedness to suggest that the proceedings to be included in the "Advances in Intelligent Systems and Computing" series; to Dr. Thomas Ditzinger (Editor at Springer Publishing Co.) for his excellent work on the final version of this volume; to three graduate students Wen-Wen Zhang, Hong-Mei Wang, and Hao-Yue Liu for compiling work on the papers before submitted to the publisher; and to our main organizer—Zhejiang Sci-Tech University in China for the great support.

Hangzhou, China
October 2016

Tai-He Fan
Shui-Li Chen
San-Min Wang
Yong-Ming Li

Conference Organizations

Honorary Chair
Ru-Qian Lu, Academician, Chinese Academy of Sciences, China

Chairs
Song-Liang Qiu, President of Zhejiang Sci-Tech University, China
Bin Zhao, Shaanxi Normal University, China
Dao-Wu Pei, Zhejiang Sci-Tech University, China

Program Committee
Chair
Paul P. Wang, Duke University, USA

Co-chairs
Yong-Ming Li, Shaanxi Normal University, China
San-Min Wang, Zhejiang Sci-Tech University, China

Members
Michał Baczyński, University of Silesia, Poland
Kai-Yuan Cai, Beihang University, China
Fei-Long Cao, China Jiliang University, China
Yong-Zhi Cao, Peking University, China
Guo-Qing Chen, Tsinghua University, China
Shui-Li Chen, Jimei University, China
Yi-Xiang Chen, East China Normal University, China
Jiu-Lun Fan, Xi'an University of Posts and Telecommunications, China
Tai-He Fan, Zhejiang Sci-Tech University, China
Janos C. Fodor, Obuda University, Hungary
Li Fu, Qinghai Nationalities University, China
Xin-Bo Gao, Xidian University, China
Lluis Godo, Spanish National Research Council, Spain
Long Hong, Nanjing University of Posts and Telecommunications
Bao-Qing Hu, Wuhan University, China

Xiao-Jing Hui, Yan'an University, China
Hui Kou, Sichuan University, China
Hong-Xing Li, Dalian Science and Technology University, China
Jin-Jin Li, Minnan Normal University, China
Jun Li, Lanzhou University of Technology, China
Qing-Guo Li, Hunan University, China
Sheng-Gang Li, Shaanxi Normal University, China
Yong-Ming Li, Shaanxi Normal University, China
Bao-Ding Liu, Tsinghua University, China
Hua-Wen Liu, Shandong University, China
Guang-Wu Meng, Liaocheng University, China
Zhi-Wen Mo, Sichuan Normal University, China
Vilem Novak, University of Ostrava at Ostrava, Czech Republic
Dao-Wu Pei, Zhejiang Sci-Tech University, China
Irina Perfilieva, University of Ostrava at Ostrava, Czech Republic
Witold Pedrycz, University of Alberta, Canada
Feng Qin, Jiangxi Normal University, China
Ke-Yun Qin, Southwest Jiaotong University, China
Yan-Hong She, Xi'an Petroleum University, China
Paul P. Wang, Duke University, USA
San-Min Wang, Zhejiang Sci-Tech University, China
Xue-Ping Wang, Sichuan Normal University, China
Xu-Zhu Wang, Taiyuan University of Science and Technology, China
Zhu-Deng Wang, Yancheng Teachers College, China
Zhen-Yuan Wang, University of Texas, USA
Hong-Bo Wu, Shaanxi Normal University, China
Wei-Zhi Wu, Zhejiang Ocean University, China
Luo-Shan Xu, Yangzhou University, China
Yang Xu, Southwest Jiaotong University, China
Zhong-Qiang Yang, Shantou University, China
Yi-Yu Yao, University of Regina, Canada
Jian-Min Zhan, Hubei University for Nationalities, China
Ming-Sheng Ying, UTS, Australia
De-Xue Zhang, Sichuan University, China
Guo-Qiang Zhang, CWRU, USA
Xiao-Hong Zhang, Shanghai Maritime University, China
Guang-Quan Zhang, UTS, Australia
Xing-Fang Zhang, Liaocheng University, China
Bin Zhao, Shaanxi Normal University, China
Dong-Sheng Zhao, Nanyang Technological University, Singapore
Hong-Jun Zhou, Shaanxi Normal University, China

Organizing Committee Chairs

Jue-Liang Hu, Zhejiang Sci-Tech University, China
Tai-He Fan, Zhejiang Sci-Tech University, China

Organizing Committee Co-chairs
Yi-Xiang Chen, East China Normal University, China
Shui-Li Chen, Jimei University, China

Organizing Committee Members
Zuo-Hua Ding, Zhejiang Sci-Tech University, China
Tai-He Fan, Zhejiang Sci-Tech University, China
Shu-Guang Han, Zhejiang Sci-Tech University, China
Zhong Li, Zhejiang Sci-Tech University, China
Hai-Yu Pan, Taizhou College, China
Dao-Wu Pei, Zhejiang Sci-Tech University, China
Yan-Hong She, Xi'an Petroleum University, China
Hui-Xian Shi, Shaanxi Normal University, China
San-Min Wang, Zhejiang Sci-Tech University, China
Rui Yang, Zhejiang Sci-Tech University, China
Wei-Ping Wang, Zhejiang Sci-Tech University, China

Publication Committee Chairs
Tai-He Fan, Zhejiang Sci-Tech University, China, E-mail: taihefan@163.com
Shui-Li Chen, Jimei University, China, E-mail: sgzx@jmu.edu.cn

Sponsored by
Zhejiang Sci-Tech University, China

Financially Supported by
National Natural Science Foundation of China Key Project (No. 11531009)

Contents

Part IV Fuzzy Connectives and Fuzzy Reasoning

Part V Fuzzy Logical Algebras

Part I
Keynote Speeches

Information Processing with Information Granules of Higher Type and Higher Order

Witold Pedrycz

Abstract The apparent challenges in data analytics calls for new advanced technologies. Granular Computing along with a diversity of its formal settings offers a badly needed conceptual and algorithmic environment that becomes instrumental in this setting. In virtue of the key facets of data analytics, there is a genuine quest to foster new development avenues of Granular Computing by bringing concepts of information granules of higher type and higher order.

In essence, information granules of higher type, say type-2 are information granules whose description is provided in terms of information granules rather than in a numeric fashion. Commonly encountered examples of type-2 information granules are fuzzy sets of type-2. Information granules of higher order, especially information granules of order-2 are those construct defined over a collection of information granules forming a universe of discourse. We elaborate on selected ways in which information granules of higher type and higher order are constructed, especially when using a principle of justifiable granularity.

Discussed are fundamental constructs in which such information granules of higher types play a pivotal role. Those are the architectures realizing data and knowledge fusion (aggregation). Here we demonstrate that a sound aggregation of pieces of data (knowledge) leads to information granules of higher type whereas the elevated type of information granularity is inherently reflective of the diversity of sources of knowledge being encountered in this aggregation process. It is also shown that the development of information granules of higher type is advocated in the realization of a hierarchy of processing and coping with a distributed nature of data. The emergence of information granules of higher order is demonstrated in the characterization of evolution of concepts (so-called generic information granules) when coping with the dynamics of data streams.

W. Pedrycz (✉)
Department of Electrical and Computer Engineering, University of Alberta, Edmonton, Canada
e-mail: wpedrycz@ualberta.ca

T.-H. Fan et al. (eds.), *Quantitative Logic and Soft Computing 2016*,
Advances in Intelligent Systems and Computing 510,
DOI 10.1007/978-3-319-46206-6_1

The talk is made self-contained and covers all required prerequisites about fundamentals of Granular Computing.

Similarity-Based Logics for Approximate Entailments

Lluís Godo

Abstract Reasoning under practical circumstances is often inexact. Assumptions might be fulfilled only in an approximate way but conclusions are drawn anyway. Different epistemic aspects may be involved, like uncertainty, preference or similarity. In order to formalise such kind of reasoning we need to go beyond classical propositional logic. In this presentation we will deal with logics for similarity-based reasoning. This kind of reasoning can be cast in the more general framework of reasoning by analogy and has applications, for example, in classification, case-based reasoning, or interpolation.

In his seminal work on similarity-based reasoning, Ruspini proposes [5] the interpretation of fuzzy sets in terms of (crisp) sets and fuzzy similarity relations. To this end, he builds up a framework for approximate inference that is based on the mutual similarity of the propositions involved. Following these lines, a number of approaches have dealt with fuzzy similarity-based reasoning from a logical perspective. In particular, the so-called logic of *Approximate Entailment* (LAE), and the logic of a dual notion of *Strong Entailment* (LSE) have been studied [2, 7]. These logics formalise the effect of small changes on the validity of logical relationships. For instance, consider a pair of propositions such that none is implied by the other one; we may then still ask if one proposition is a consequence of the other one by means of a slight change. Conversely, for a pair of propositions one of which is a consequence of the other one, we may ask if this consequence relation is stable under small changes.

To formalise these kind of inferences, one needs to specify what is meant by "*approximate*". Several approaches dealing with statements interpreted in metric spaces or logics on comparative similarity have also been considered in the literature, see e.g. [1, 4, 6]. Here we will follow a quantitative approach and use fuzzy similarity relations to model the notion of approximate entailment [3]. In this talk we will present the main notions and properties of LAE, a propositional graded modal logic where propositions are interpreted, as in classical logic, by subsets of a fixed set,

L. Godo (✉)
Artificial Intelligence Research Institute, IIIA - CSIC, 08913 Bellaterra, Spain
e-mail: godo@iiia.csic.es

T.-H. Fan et al. (eds.), *Quantitative Logic and Soft Computing 2016*,
Advances in Intelligent Systems and Computing 510,
DOI 10.1007/978-3-319-46206-6_2

called the set of worlds, that in addition it is assumed to be endowed with a fuzzy similarity relation, which associates with each pair of two worlds their degree of resemblance. The basic semantic structures are hence *fuzzy similarity spaces*, which consist of a set of worlds and a fuzzy similarity relation, and the core syntactic objects of LAE are implications between propositions endowed with a degree. The intended meaning of a statement of the form $A >_c B$ is that B is an approximate consequence of A to the degree c, where c is a real number between 0 and 1. We will present a complete axiomatization, and moreover we will show some extensions as well as how the framework can be enhanced in case of dealing with similarities on linearly ordered scales [8].

Acknowledgments The author acknowledges partial support by the Spanish MINECO/FEDER project RASO (TIN2015-71799-C2-1-P)

References

1. Alenda, R., Olivetti, N., Schwind, C.: Comparative concept similarity over minspaces: axiomatisation and tableaux calculus. In: Giese, M., et al. (eds.) Automated Reasoning with Analytic Tableaux and Related Methods. Proceedings of the 18th International Conference TABLEAUX 2009, pp. 17–31. Springer, Berlin (2009)
2. Esteva, F., Godo, L., Rodríguez, R.O., Vetterlein, T.: Logics for approximate and strong entailments. Fuzzy Sets Syst. **197**, 59–70 (2012)
3. Godo, L., Rodríguez, R.O.: Logical approaches to fuzzy similarity-based reasoning: an overview. In: Della Riccia, G., et al. (eds.) Preferences and Similarities. CISM Courses and Lectures, vol. 504, pp. 75–128. Springer, Berlin (2008)
4. Kutz, O., Sturm, H., Suzuki, N.-Y., Wolter, F., Zakharyaschev, M.: Logics of metric spaces. ACM Trans. Comput. Log. **4**(2), 260–294 (2003)
5. Ruspini, E.H.: On the semantics of fuzzy logic. Int. J. Approx. Reason. **5**, 45–88 (1991)
6. Sheremet, M., Tishkovsky, D., Wolter, F., Zakharyaschev, M.: A logic for concepts and similarity. J. Log. Comput. **17**, 415–452 (2007)
7. Vetterlein, T.: Logic of approximate entailment in quasimetric spaces. Int. J. Approx. Reason. **64**, 39–53 (2015)
8. Vetterlein, T., Esteva, F., Godo, L.: Logics for approximate entailment in ordered universes of discourse. Int. J. Approx. Reason. **71**, 50–63 (2016)

Theoretical and Applicational Aspects of Fuzzy Implication Functions

Michał Baczyński

Abstract Fuzzy implication functions are one of the main mathematical operations in fuzzy logic. They generalize the classical two-valued implication, which takes values in $\{0, 1\}$, to fuzzy logic, where the truth values belong to the unit interval $[0, 1]$. This family of functions plays a significant role in the development of fuzzy systems. The study of this class of operations has been extensively developed in the last 40 years from both theoretical and applicational points of view. In our presentation we will concentrate on both streams. Firstly, we present the mathematical aspects of fuzzy implications, namely, analytical and algebraic.

We show basic facts and we try to find concise answers for the following questions:

(i) How can we get the implications from fuzzy logic connectives?
(ii) How can we get the implications from unary functions?
(iii) What are the properties, characterizations and representations for different families of fuzzy implications?
(iv) Which functional equations are investigated for fuzzy implications?

Secondary, we present different applications of fuzzy implication functions. We mainly discuss their role in approximate reasoning, fuzzy control, mathematical morphology and image processing.

M. Baczyński (✉)
Institute of Mathematics, University of Silesia, ul. Bankowa 14,
40-007 Katowice, Poland
e-mail: michal.baczynski@us.edu.pl

T.-H. Fan et al. (eds.), *Quantitative Logic and Soft Computing 2016*,
Advances in Intelligent Systems and Computing 510,
DOI 10.1007/978-3-319-46206-6_3

References

1. Baczyński, M., Beliakov, G., Bustince, H., Pradera, A. (eds.): Advances in Fuzzy Implication Functions. Studies in Fuzziness and Soft Computing, vol. 300. Springer, Berlin (2013)
2. Baczyński, M., Jayaram, B.: Fuzzy Implications. Studies in Fuzziness and Soft Computing, vol. 231. Springer, Berlin (2008)
3. Baczyński, M., Jayaram, B., Massanet, S., Torrens, J.: Fuzzy implications: past, present, and future. In: Kacprzyk, J., Pedrycz, W. (eds.) Springer Handbook of Computational Intelligence, pp. 183–202. Springer, Berlin (2015)
4. Combs, W.E., Andrews, J.E.: Combinatorial rule explosion eliminated by a fuzzy rule configuration. IEEE Trans. Fuzzy Syst. **6**, 1–11 (1998)
5. Mas, M., Monserrat, M., Torrens, J., Trillas, E.: A survey on fuzzy implication functions. IEEE Trans. Fuzzy Syst. **15**, 1107–1121 (2007)

A Novel Description of Factor Logic

Hai-Tao Liu, Yi-Xiang Chen, Pei-Zhuang Wang and Hua-Can He

Abstract This paper presents a novel description for factor logic, which puts state description into a factor space, and sets up truth sets for formulae. Along this way, factor space may provide a suitable platform for the development of quantitative logic. Factor logic can be also described as a derivative system of Boolean logic with added hypotheses Γ, which can be used in data mining and function-structure analysis in switch systems.

Keywords Factor logic · Quantitative logic · Factor space · Decision tree · Fault analysis

1 Introduction

Quantitative logic [1–3] was initiated by Professor G.J. Wang, a great mathematician in China. Recently, Dr. Zhou uses probability quantitative logic [4] as a state description space for a platform to set up the truth degree for formulae. Factor space was originally developed by Professor P.Z. Wang in [5–9]. This paper merges those two areas: quantitative logic and factor space into a new framework and develop a novel description of factor logic following Professor Wang's approach in quantitative logic. This paper puts state description on factor space and presents a new method about factor logic without strict logical statements and proofs. It is developed as a kind of qualified quantitative logic. Factor logic can be also built as a derivative system of Boolean logic with added hypotheses, which can be used in data mining

H.-T. Liu · P.-Z. Wang
Liaoning Technical University, Fuxin, Liaoning, China
e-mail: peizhuangw@126.com

Y.-X. Chen (✉)
Software Engineering Institute, East China Normal University, Shanghai, China
e-mail: yxchen@sei.ecnu.edu.cn

H.-C. He
NorthWestern Polytechnical University, Xi'an, China

T.-H. Fan et al. (eds.), *Quantitative Logic and Soft Computing 2016*,
Advances in Intelligent Systems and Computing 510,
DOI 10.1007/978-3-319-46206-6_4

and function-structure analysis, some examples in its applications can be found in the paper. The authors of this paper express the feeling of commemorating Professor G.J. Wang.

The organization of this paper is as follows: Preliminary is stated in Sect. 2, A novel description on factor space is given in Sect. 3, Sect. 4 introduces some applications of factor logic. Section 5 is a brief conclusion.

2 Preliminary

Factor space [5] was initiated by Professor P.Z. Wang in 1982. We simply restate some of its basic results here.

Definition 1 (*Factor Space*) A factor space [6, 7], defined on universe of discussion U, is a family of sets $\psi = (\{X(f)\}_{f\in F}; U)$ satisfying:

1 $F = (F, \vee, \wedge, c, 1, 0)$ is a complete Boolean algebra;
2 $X(0) = \{\emptyset\}$;
3 For any $T \subseteq F$, if $\{f \mid f \in T\}$ is irreducible (i.e., $s \neq t \Rightarrow s \wedge t = 0 (s, t \in T)$), then $X(\{f \mid f \in T\}) = \prod_{f\in T} X(f)$, where $\prod$ stands for Cartesian product;
4 $\forall f \in T$, there is a mapping with the same symbol $f : f \in X(f)$.

F is called the set of factors, $f \in F$ a factor on U and $X(f)$ the phase space of factor f.

A factor f is said to be simpler than a factor g or g is said to be more complex than f if $f \leq g$. For any $F' \subseteq F$, let $F' = \vee\{f \mid f \in F'\}$. F' is joint mapping of mappings: $F'(u) = \{f(u) \mid f \in F'\}$.

Definition 2 (*Background Relation* [8]) For given n factors $f_1, f_2, \ldots, f_n$, we call

$$R = R_F = X(F) = \{a = (a_1, \ldots, a_n) \in X(f_1) \times \cdots \times X(f_n) \mid \exists u \in U; f_1(u) = a_1, \ldots, f_n(u) = a_n\} \tag{1}$$

to be the background relation between $f_1, \ldots, f_n$, or the background set of F.

Denote $X = X(f_1) \times \cdots \times X(f_n)$. Then $X(F) = X$ if $f_1, \ldots, f_n$ are independent; if $X(F) \neq X$, then the phase configuration could not be generated freely, the Cartesian product space includes some virtual configurations. The real configurations consists the background set R, in which R is the real product phase space of $f_1 \ldots f_n$.

Background set R is a very important terminology in factor space.

1. R generates all concepts on factor space ψ: For any $a = (a_1, \ldots, a_n) \in R$, denote $[a] = F^{-1}(a) = \{u \in U | f_1(u) = a_1, \ldots,$
   $f_n(u) = a_n\}$. Denote $U^* = \{[a] \mid a \in R\}$, which is the quotient space of U with respect to F. It is obvious that F is an isomorphism between U^* and R. Denote

$A = \{\alpha = (a, [a]) \mid a \in R\}$, which is called the atom concept set on ψ, α is called an atom concept and a, $[a]$ are called intension and extension of α respectively. The Boolean algebra C generated by A is called the concept algebra on ψ.

2. R determines all tautologies on factor space ψ:
For simplicity, suppose that $F = \{f, g\}$. For any $A \subseteq X(f)$,

$$A^* = \{y \in X(g) \mid \exists x \in X(f); (x, y) \in R\}. \tag{2}$$

Theorem 1 *$A \to B$ is a tautology if and only if $A^* \subseteq B$.*

3 Description on Factor Logic

A factor space $\psi = (\{x(f)\}_{f \in F}; U)$ with background set $R = F(U) \subseteq X(F) = X(f_1) \times \cdots \times X(f_n)$ determines a logic system L_{factor} as follows:

1. A symbol set $S = X(f_1) + \cdots + X(f_n)$ (plus bracket pair and $1, 0$), where $X(f_j) = \{x_{1j}, \ldots, x_{n(j)j}\}$ is the phase space of factor f_j and x_{ij} is the i-th phase of factor f_j. Each element in S is called a letter.
2. Formulae set $F(S)$, which is the Boolean algebra $(F(S), \vee, \wedge, \neg)$ generated from S.
All letters are called atom formulae; each conjunction of letters $x_{i(1)j(1)} \wedge \cdots \wedge x_{i(k)j(k)}$ (abridged $x_{i(1)j(1)} \ldots x_{i(k)j(k)}$) is called a ($k$-length) block; each disjunction of blocks $b_1 \vee \cdots \vee b_t$ (abridged $b_1 + \cdots + b_t$) is called a disjunction normal form. A formula p is called a tautology if $p = \mathbf{1}$ (i.e., $p \to \mathbf{1}$ and $\mathbf{1} \to p$); p is called a contradiction if $p = \mathbf{0}$.
3. Axioms: Boolean logic axioms added Assumptions:

   Γ_1 Family axiom: $X(f_i) = \{x_{1j}, \ldots, x_{m(i)j}\}$ is called j-th family, letter of same family must obey that

$$x_{1j} \vee \cdots \vee x_{m(i)j} = \mathbf{1}; x_{ij} \wedge x_{kj} = \mathbf{0}(i \neq k); x_{ij} = \vee\{x_{ij} \mid i \neq k\}; \tag{3}$$

   Γ_2 Background axiom: Background set R can be represented as a disjunction normal form $r \in F(S)$ ($r = \mathbf{1}$, without indicate r). For any formula $p \in F(S)$, we have that

$$p \to p \wedge r; p \wedge r \to p (p \in F(S)). \tag{4}$$

4. Valuation field $W_2 = \{0, 1\} = \{\{0, 1\}, \vee, \wedge, \neg\}$.
5. Modus Ponens: $\{p, p \to q\} \vdash q$.

which is a derivative system of Boolean logic by adding hypothesis $\Gamma = \Gamma_1 + \Gamma_2$. This paper has not given the proof of the strongly complete theorem: $\Gamma \vdash p$ iff $\Gamma \models p$.

Family axiom Γ_1 emphasizes that one family has only one letter appears in a disjunction: $x_{ij} \wedge x_{kj} = \mathbf{0}(i \neq k)$. For example, High and Short are two letters in a same family, then $High' \wedge Short'$ is a contradiction. A letter is not a minimal phase, but an n-length block $\mathbf{x} = x_{1(1)1} \ldots x_{i(n)n}$ is a minimal phase in phase space.

Background axiom emphasizes that the truth-degree of a formula is independent to those phases outside the background set R. In other words, the meaningful set of formulae is $F_r(S) = \{p \wedge r \mid p \in F(S)\}$.

Denote $F^+(S) = \{(p, x) \mid p \in (S), x \in R\}$. A proposition is a pair $(p, \mathbf{x}) \in F^+(S)$ and a predicate is a function $p(\mathbf{x}) = (p, \mathbf{x})$.

Suppose that $\mathbf{x} = (x_{1(1)1}, \ldots, x_{i(n)n})$ is a phase of R, letter x_{ij} appears in block x if and only if the name of x in family j is the very x_{ij}, i.e., $x_{i(j)j} = x_{ij}$ or $i(j) = i$. Denote mapping $t : S \in 2^R : t(x_{ij}) = \{\mathbf{x} = (x_{i(1)1} \cdots x_{i(n)n}) \mid i(j) = i\} = \mathbf{b}_{ij}$, which transfers x_{ij} to a subset $\mathbf{x}_{ij}$ in R. Extending the mapping on to $F_R(S)$, we have that

$$t(p \vee q) = t(p) \cup t(q), t(p \wedge q) = t(p) \cap t(q), t(\neg p) = (t(p))^c.$$

Definition 3 (*Truth Set*) For any $p \in F(S)$, we call $P = t(p)$ to be the truth set of formula p.

Without declaring, formula and its truth set are written by lower-case and capital letters respectively.

Definition 4 (*Domain of Interpretation*) R is called the domain of interpretation of L_{factor}. Mapping v: $F^+(S) \in W_2$ is called a assignment, if $v(p, \mathbf{x}) = 1$ $(\mathbf{x} \in P)$, $v(p, \mathbf{x}) = 0(\mathbf{x} \notin P)$.

Proposition 1 *Formula $p \to q$ is a tautology if and only if $P \subseteq Q$.*

Proof $p \to q$ is a tautology if and only if $p \to q = \mathbf{1}$, i.e., $\neg p \vee q = \mathbf{1}$, if and only if $t(\neg p \vee q) = R$, i.e., $P^c \cup Q = R$. Since $p = p \wedge r$, $q = q \wedge r$, since both P and Q are subsets of R, so we have $p \to q$ is a tautology if and only if $P \subseteq Q$. □

Factor logic has its own characteristics: A formula p in $F(S)$ is not a proposition but a concept; While the pair $(p, \mathbf{x}) \in F^+(S)$ is a proposition and $p(\mathbf{x})$ is a predicate. An assignment v does not evaluate concepts but propositions.

Proposition 2 *The length of blocks must not be greater than n.*

Proof Each non-zero block includes at most one letter from a family, and there are n families, thus the Proposition holds. □

It is not difficult to prove that the truth set of letter x_{ij} is the cylinder extension of $\{x_{ij}\}$ within R:

$$\underset{\sim}{x}_{ij} = [\{x_{ij}\}X(f_1) \times \cdots \times X(f_{i-1}) \times X(f_{i+1}) \times \cdots \times X(f_n)] \cap R. \tag{5}$$

Similarly, the truth set of block with 2-letter $x_{ij} \wedge x_{kl}$ is the cylinder extension of $x_{i(1)1} \times \cdots \times x_{i(n)}$ within R:

$$\underset{\sim}{x}_{ij} \wedge \underset{\sim}{x}_{kl} = [\{x_{ij}\} \times \{x_{kl}\} \times \{X(f_j) \mid j \neq i, k\}] \cap R. \quad (6)$$

The longer block, the smaller truth set, for n-length block $x_{i(1)1} \times \cdots \times x_{i(n)}$, its truth set becomes a point in R or empty:

$$x_{i(1)1} \times \cdots \times x_{i(n)} = x_{i(1)1\times\cdots\times i(n)n} \cap R.$$

Definition 5 (*Prime Implicants*) If p implicates q and there is no $p' \neq p$ such that p implicates p' and p' implicates q, then p is called the prime implicant of q. A disjunction normal form is called a minimal disjunction normal form if all blocks are prime implicants.

Proposition 3 *If letter x_{ij} does not appear in any block of disjunction normal form $(\neg q) \wedge r$, then x_{ij} is a prime implicant of q.*

Proof Let Q and Q^c be the truth sets of q and $\neg q$ respectively. We have that $Q \cup Q^c = U^*$. The truth set of $(\neg q) \wedge r$ is $Q^c \cap R$, and the truth set of letter x_{ij} is $\underset{\sim}{x}_{ij}$. If letter x_{ij} does not appear in any block of disjunction normal form $(\neg q) \wedge r$, then $\underset{\sim}{x}_{ijj} \cap (Q^c \cap R) = \emptyset$. Since truth sets can be freely changed outside the background set R (See Hypothesis Γ_2), we can write that $\underset{\sim}{x}_{ijj} \cap (Q^c) = \phi$. It means that $\underset{\sim}{x}_{ijj} \subseteq Q$. Therefore the letter x_{ij} implicate q. Since a letter has the maximal truth set with respect to a block, it must be prime. □

4 Application of Factor Logic in Data Analysis

Could quantitative logic be applied in data analysis directly? Yes, it is. There is an example cited from online representation of decision tree. Decision tree is a public method in data mining. For easy statement, we use factor spaces language to explain the table, which has a little change and inverted table axis.

Example 1 There is a universe $U = \{u1, \ldots, u14\}$, consisting of 14 groups of persons, the number of persons in each group is shown as frequent in the lowest row. There are four conditional factors and a resulted factor. They are:

1. $f_1 =$ Age, $X(f_1) = \{Old, Middle, Young\} = \{O, M, Y\}$;
2. $f_2 =$ Income, $X(f_2) = \{High, Average, Low\} = \{H, A, L\}$;
3. $f_3 =$ Student?, $X(f_3) = \{Student, Non-student\} = \{S, N\}$;
4. $f_4 =$ Reputation, $X(f_4) = \{Good, Fair\} = \{G, F\}$.

The resulted factor is $g = Purchase, X(g) = \{Buy, Empty\} = \{B, E\}$ (Table 1).

Table 1 shows peoples phases under different factors. Each row stands of a factor, each column stands for a group of persons. The task of decision tree is rule-extracting by means of learning from the table.

Table 1 Customers purchase statistics

U	u01	u02	u03	u04	u05	u06	u07	u08	u09	u10	u11	u12	u13	u14
Age	M	O	O	M	Y	O	Y	M	M	Y	Y	O	Y	O
Income	H	A	L	L	L	A	A	A	H	H	H	L	A	A
Student?	N	N	S	S	S	S	S	N	S	N	N	S	N	N
Reputation	F	F	F	G	F	F	G	G	F	F	G	G	F	G
Purchase	B	B	B	B	B	B	B	B	B	E	E	E	E	E
Frequent	128	80	64	64	64	132	64	32	32	64	64	64	128	64

From the view of the factor logic, each column is a rule. For example, the first column shows that: If somebody's age is middle and income is high and is not a student and reputation is fair, then he/she is a buyer.

There are 14 columns, we can extract 14 rules and the rule-extraction can be written as two disjunction normal forms:

$$\begin{aligned} &MHNF + OANF + OLSF + MLSG + YLSF \\ &+ OASF + YASG + MANG + MHSF = B; \end{aligned} \tag{7}$$

$$YHNF + YHNG + OLSG + YANF + OANG = E. \tag{8}$$

where $MHNF = M \wedge H \wedge N \wedge F$ and $\wedge$ is written as $+$.

Without any algorithm, the task of decision tree can be resolved by logic immediately! But when the number of blocks is enough large, the extraction is difficult to be understood in use. In logic, there needs to find out the minimal disjunction normal form.

The symbol set S consists of ten letters belonging to four conditional factors:

$$S = X(f_1) + X(f_2) + X(f_3) + X(f_4) = \{O, M, Y; H, A, L; S, N; G, F\}.$$

According to Axiom Γ_1, there are 36 4-length blocks, they indeed forms the Cartesian product space X:

$$\begin{aligned} X &= X(f_1) \times X(f_2) \times X(f_3) \times X(f_4) \\ &= \{OHSG, OHSF, OHNG, OHNF, OASG, OASF, OANG, OANF, OLSG, \\ &\quad OLSF, OLNG, OLNF, MHSG, MHSF, MHNG, MHNF, MASG, \\ &\quad MASF, MANG, MANF, MLSG, MLSF, MLNG, MLNF, YHSG, \\ &\quad YHSF, YHNG, YHNF, YASG, YASF, YANG, YANF, YLSG, YLSF, \\ &\quad YLNG, YLNF\} \end{aligned}$$

Having 1024 persons take part in the statistics to form the table, the table does completely reflect the populations information. According to Axiom Γ_2, there are only 14 real phase-configurations from the 4 conditional factors. Deleting all virtual configurations from X, we get the background set R:

$$R = \{MHNF, OANF, OLSF, MLSG, YLSF, OASF, YASG, MANG, MHSF, YHNF, YHNG, OLSG, YANF, OANG\}$$

From (7), we get that

$$\begin{aligned} B &= MHNF + OANF + OLSF + MLSG + YLSF \\ &\quad + OASF + YASG + MANG + MHSF \\ &= (MHNF + MLSG + MANG + MHSF) \\ &\quad + (OANF + OLSF + OASF) + (YASG + YLSF) \\ &= M(HNF + LSG + ANG + HSF) \\ &\quad + OF(AN + LS + AS) + YS(AG + LF). \end{aligned}$$

According to Proposition 3, M is a prime implicant form of B if M does not occur in E. Let $\underline{E}$ be the set of letters occurring in one of blocks in E. We have that $\underline{E} = \{O, Y; H, A, L; S, N; G, F\}$, then $R - \underline{E} = \{M\}$. It means that M does not occur in E. So that M is a prime implicant form of B. Delete all implicants of M, we have that

$$B = M + OF(AN + LS + AS) + YS(AG + LF).$$

Similarly, since OF does not occur in E (See (2)), OF is a prime implicant form of B. Delete all implicants of OF, we have that

$$B = M + OF + YS(AG + LF).$$

Similarly, since YS does not occur in E (See (8)), YS is a prime implicant form of B. Delete all implicants of YS, we have one minimal disjunction normal form of B:

$$B = M + OF + YS$$

From (8), we get that

$$\begin{aligned} E &= YHNF + YHNG + OLSG + YANF + OANG \\ &= (YHNF + YHNG + YANF) + (OANG + OLSG) \\ &= YN(HF + HG + AF) + OG(AN + LS). \end{aligned}$$

Since YN does not occur in any block of B, according to Proposition 3, YN is a prime implicant form of E. Delete all implicants of YN, we have that

$$E = YN + OG(AN + LS).$$

Fig. 1 Purchase decision tree

/Good ⟶ **No**
/Old-(Reputation)-Fair ⟶ **Buyer**
(Age)-Middle ⟶ **Buyer**
\ Young -(Student?)-Student ⟶ **Buyer**
\ Non-student ⟶ **No**

Similarly, since OG does not occur in B, OG is a prime implicant form of E. Delete all implicants of OF, we have one minimal disjunction normal form of E.

$$E = YS + OG.$$

Now, we can get the decision tree shown in Fig. 1.

Wang HD has gotten more simple method on decision tree by means of factor space theory [8]. Now, factor logic presents another simple method. The open problem is: How to select prime implicants with 1-length, 2-length or short length blocks?

Factor logic has also been applied in the structure-function analysis [9], shown in following two examples.

Example 2 Given a electronic system having five switches $x_1, \ldots, x_5$, they can be described by five factors $F = (f_1, f_2, f_3, f_4, f_5)$ respectively. Each one has its own phase space $X(f_j) = \{x_{1j}, x_{0j}\} = \{x_j, \underline{x}_j\}$ (x_{1j} stands for that the switch x_j is connected and x_{0j} stands for that x_j is disconnected). The resulted factor g has phase space $X(g) = \{T, F\}$ (T stand for the system being transformation, and F stands for fault). The table is given in the next paper.

There are 32 objects in the universe $U = \{u_1, \ldots, u_{32}\}$, and $X = X(f_1) \times \cdots \times X(f_5)$ contains $2^5 = 32$ objects, so that the background set is the same as the Cartesian product space: $R = T + F$;

$$S = \{x_1, x_1; x_2, x_2; x_3, x_3; x_4, x_4; x_5, x_5\}$$

$$\begin{aligned}\underline{T} = \{&x_1\underline{x}_2\underline{x}_3x_4\underline{x}_5, \underline{x}_1\underline{x}_2x_3\underline{x}_4x_5, \underline{x}_1\underline{x}_2x_3\underline{x}_4x_5, x_1x_2x_3\underline{x}_4\underline{x}_5, x_1\underline{x}_2x_3x_4\underline{x}_5,\\ &x_1\underline{x}_2\underline{x}_3x_4x_5, \underline{x}_1x_2x_3x_4x_5, x_1x_2\underline{x}_3x_4\underline{x}_5, x_1\underline{x}_2x_3\underline{x}_4x_5, \underline{x}_1x_2x_3\underline{x}_4x_5,\\ &x_1x_2x_3x_4\underline{x}_5, x_1\underline{x}_2x_3x_4x_5, \underline{x}_1x_2x_3x_4x_5, x_1x_2\underline{x}_3x_4x_5, x_1x_2x_3\underline{x}_4x_5,\\ &x_1x_2x_3x_4x_5\}\end{aligned}$$

$$\begin{aligned}F = \{&\underline{x}_1\underline{x}_2\underline{x}_3\underline{x}_4\underline{x}_5, x_1\underline{x}_2\underline{x}_3\underline{x}_4\underline{x}_5, \underline{x}_1x_2\underline{x}_3\underline{x}_4\underline{x}_5, \underline{x}_1\underline{x}_2x_3\underline{x}_4\underline{x}_5, \underline{x}_1\underline{x}_2\underline{x}_3x_4\underline{x}_5, \underline{x}_1\underline{x}_2\underline{x}_3\underline{x}_4x_5,\\ &x_1x_2\underline{x}_3\underline{x}_4\underline{x}_5, x_1\underline{x}_2x_3\underline{x}_4\underline{x}_5, x_1\underline{x}_2\underline{x}_3\underline{x}_4x_5, \underline{x}_1x_2x_3\underline{x}_4\underline{x}_5, \underline{x}_1x_2\underline{x}_3x_4\underline{x}_5, \underline{x}_1x_2\underline{x}_3\underline{x}_4x_5,\\ &\underline{x}_1\underline{x}_2x_3x_4\underline{x}_5, \underline{x}_1x_2x_3x_4\underline{x}_5, \underline{x}_1x_2\underline{x}_3x_4x_5, x_1x_2\underline{x}_3\underline{x}_4x_5\}\end{aligned}$$

$$\begin{aligned}T = {}&x_1\underline{x}_2\underline{x}_3x_4\underline{x}_5 + \underline{x}_1\underline{x}_2x_3\underline{x}_4x_5 + \underline{x}_1\underline{x}_2x_3\underline{x}_4x_5 + x_1x_2x_3\underline{x}_4\underline{x}_5 + x_1\underline{x}_2x_3x_4\underline{x}_5 +\\ &x_1\underline{x}_2\underline{x}_3x_4x_5 + \underline{x}_1x_2x_3x_4x_5 + x_1x_2\underline{x}_3x_4\underline{x}_5 + x_1\underline{x}_2x_3\underline{x}_4x_5 + \underline{x}_1x_2x_3\underline{x}_4x_5 +\\ &x_1x_2x_3x_4\underline{x}_5 + x_1\underline{x}_2x_3x_4x_5 + \underline{x}_1x_2x_3x_4x_5 + x_1x_2\underline{x}_3x_4x_5 + x_1x_2x_3\underline{x}_4x_5 +\\ &x_1x_2x_3x_4x_5\end{aligned}$$

$$= x_1(\underline{x}_2\underline{x}_3x_4\underline{x}_5 + x_2x_3\underline{x}_4\underline{x}_5 + \underline{x}_2x_3x_4\underline{x}_5 + \underline{x}_2\underline{x}_3x_4x_5 + x_2\underline{x}_3x_4\underline{x}_5 + \underline{x}_2x_3\underline{x}_4x_5 + x_2x_3x_4\underline{x}_5 + \underline{x}_2x_3x_4x_5 + x_2\underline{x}_3x_4x_5 + x_2x_3\underline{x}_4x_5 + x_2x_3x_4x_5) + \underline{x}_1\underline{x}_2x_3\underline{x}_4x_5 + \underline{x}_1\underline{x}_2x_3\underline{x}_4x_5 + \underline{x}_1x_2x_3x_4x_5 + \underline{x}_1x_2x_3\underline{x}_4x_5 + \underline{x}_1x_2x_3x_4x_5.$$

Unfortunately, letter x_1 can be found in F, so that x_1 is not a prime implicant of T. In this case, do not be disappointed, continue search two letter pairs:

$$T = x_1x_4(\underline{x}_2\underline{x}_3\underline{x}_5 + \underline{x}_2x_3\underline{x}_5 + \underline{x}_2\underline{x}_3x_5 + x_2\underline{x}_3\underline{x}_5 + x_2x_3\underline{x}_5 + \underline{x}_2x_3x_5 + x_2\underline{x}_3x_5 + x_2x_3x_5) + x_3x_5(\underline{x}_1\underline{x}_2\underline{x}_4 + \underline{x}_1x_2x_4 + \underline{x}_1\underline{x}_2x_4 + x_1\underline{x}_2\underline{x}_4 + \underline{x}_1x_2\underline{x}_4 + \underline{x}_1x_2x_4 + x_1x_2\underline{x}_4) + x_1x_2x_3\underline{x}_4\underline{x}_5.$$

Since x_1 and x_4 are not in any block of F simultaneously, according to Proposition 3, x_1x_4 is a prime disjunction normal form of T. Deleting all implicants of x_1x_4, we have that

$$T = x_1x_4 + x_3x_5(\underline{x}_1\underline{x}_2\underline{x}_4 + \underline{x}_1x_2x_4 + \underline{x}_1\underline{x}_2x_4 + x_1\underline{x}_2\underline{x}_4 + \underline{x}_1x_2\underline{x}_4 + \underline{x}_1x_2x_4 + x_1x_2\underline{x}_4) + x_1x_2x_3\underline{x}_4\underline{x}_5.$$

Since x_3 and x_5 are not in any block of F simultaneously, according to Proposition 3, x_3x_5 is a prime disjunction normal form of T. Deleting all implicants of x_3x_5, we have that

$$T = x_1x_4 + x_3x_5 + x_1x_2x_3\underline{x}_4\underline{x}_5.$$

Since x_1, x_2 and x_3 are not in any block of F simultaneously, according to Proposition 3, $x_1x_2x_3$ is a prime disjunction normal form of T. Deleting all implicants of $x_1x_2x_3$, we have that

$$T = x_1x_4 + x_3x_5 + x_1x_2x_3. \tag{9}$$

The target of a switch system is not a decision making or a classification, which is a function-structure analysis. The table tells us the systems' function, we need to find out what is the structure of the switches. The formula (9) is the structure, and it is easily to draw a picture in Fig. 2.

In this example the background set $R = X$, the Hypothesis Γ is redundant, and the Proposition 3 is the same as classical method of prime implication extraction in Boolean logic. Factor logic can extend the method to constrained switch systems. If we set a constraint, $x_3 = x_1 + x_2$ for example, then the configuration of switch is not free, 16 columns in Table 2 are violates the constraint. Table 2 becomes to Table 3 in the following example, How to extract prime implication?

Example 3 Based on the Table 3, we define these two formulae T and F as follows.

$$T = x_1x_4(\underline{x}_2x_3\underline{x}_5 + x_2x_3\underline{x}_5 + \underline{x}_2 + x_3x_5 + x_2x_3x_5) + \underline{x}_1\underline{x}_2\underline{x}_3x_4x_5 + x_1\underline{x}_2x_3\underline{x}_4x_5 + \underline{x}_1x_2x_3\underline{x}_4x_5 + \underline{x}_1x_2x_3x_4x_5 + x_1x_2x_3\underline{x}_4x_5 + x_1x_2x_3\underline{x}_4\underline{x}_5.$$

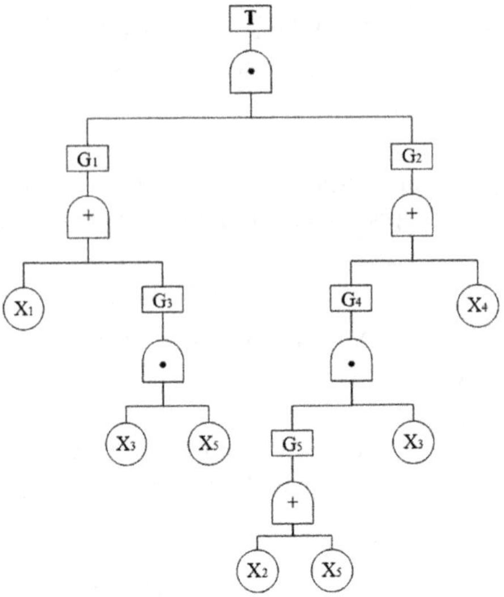

Fig. 2 Structure of five switches

Table 2 Function of a switch system

U	1	2	3	4	5	6	7	8	9	10	11	12	13	14	15	16
f_1^*	$\underline{x}_1$	x_1	$\underline{x}_1$	$\underline{x}_1$	$\underline{x}_1$	$\underline{x}_1$	x_1	x_1	x_1	x_1	$\underline{x}_1$	$\underline{x}_1$	$\underline{x}_1$	$\underline{x}_1$	$\underline{x}_1$	$\underline{x}_1$
f_2	$\underline{x}_2$	$\underline{x}_2$	x_2	$\underline{x}_2$	$\underline{x}_2$	$\underline{x}_2$	x_2	$\underline{x}_2$	$\underline{x}_2$	$\underline{x}_2$	x_2	x_2	x_2	$\underline{x}_2$	$\underline{x}_2$	$\underline{x}_2$
f_3	$\underline{x}_3$	$\underline{x}_3$	$\underline{x}_3$	x_3	$\underline{x}_3$	$\underline{x}_3$	$\underline{x}_3$	x_3	$\underline{x}_3$	$\underline{x}_3$	x_3	$\underline{x}_3$	$\underline{x}_3$	x_3	x_3	$\underline{x}_3$
f_4	$\underline{x}_4$	$\underline{x}_4$	$\underline{x}_4$	$\underline{x}_4$	x_4	$\underline{x}_4$	$\underline{x}_4$	$\underline{x}_4$	x_4	$\underline{x}_4$	$\underline{x}_4$	x_4	$\underline{x}_4$	x_4	$\underline{x}_4$	x_4
f_5	$\underline{x}_5$	$\underline{x}_5$	$\underline{x}_5$	$\underline{x}_5$	$\underline{x}_5$	x_5	$\underline{x}_5$	$\underline{x}_5$	$\underline{x}_5$	x_5	$\underline{x}_5$	$\underline{x}_5$	x_5	$\underline{x}_5$	x_5	x_5
g	F	F	F	F	F	F	F	F	T	F	F	F	F	F	T	T
U	17	18	19	20	21	22	23	24	25	26	27	28	29	30	31	32
f_1^*	x_1	x_1	x_1	$\underline{x}_1$	$\underline{x}_1$	x_1	$\underline{x}_1$	x_1	x_1	$\underline{x}_1$	x_1	x_1	$\underline{x}_1$	x_1	x_1	x_1
f_2	x_2	$\underline{x}_2$	$\underline{x}_2$	x_2	x_2	x_2	$\underline{x}_2$	$\underline{x}_2$	x_2	x_2	x_2	$\underline{x}_2$	x_2	x_2	x_2	x_2
f_3	x_3	$\underline{x}_3$	x_3	x_3	$\underline{x}_3$	$\underline{x}_3$	x_3	x_3	$\underline{x}_3$	x_3	x_3	x_3	x_3	$\underline{x}_3$	x_3	x_3
f_4	$\underline{x}_4$	x_4	x_4	x_4	x_4	$\underline{x}_4$	x_4	$\underline{x}_4$	x_4	$\underline{x}_4$	x_4	x_4	x_4	x_4	x_4	x_4
f_5	$\underline{x}_5$	$\underline{x}_5$	x_5	$\underline{x}_5$	x_5	x_5	x_5	x_5	$\underline{x}_5$	x_5	$\underline{x}_5$	x_5	x_5	x_5	x_5	x_5
g	T	T	T	F	F	F	T	T	T	T	T	T	T	T	T	T

$$F = \underline{x}_1\underline{x}_2\underline{x}_3\underline{x}_4\underline{x}_5 + \underline{x}_1\underline{x}_2\underline{x}_3x_4\underline{x}_5 + \underline{x}_1\underline{x}_2\underline{x}_3\underline{x}_4x_5 + x_1\underline{x}_2x_3\underline{x}_4\underline{x}_5 + \underline{x}_1x_2x_3\underline{x}_4\underline{x}_5 + \underline{x}_1\underline{x}_2\underline{x}_3x_4x_5.$$

Since x_1 and x_4 are not in any block of F simultaneously, according to Proposition 3, x_1x_4 is a prime disjunction normal form of T. Deleting all implicants of x_1x_4, we have that

$$T = x_1x_4 + x_3x_5(x_1\underline{x}_2\underline{x}_4 + \underline{x}_1x_2\underline{x}_4 + \underline{x}_1x_2x_4 + x_1x_2\underline{x}_4) + x_1x_2x_3\underline{x}_4\underline{x}_5 + \underline{x}_1\underline{x}_2\underline{x}_3x_4x_5.$$

Table 3 Function of constrained switch system

U	1	5	6	8	11	16	17	18	20	24	26	27	28	29	31	32
f_1	$\underline{x}_1$	$\underline{x}_1$	$\underline{x}_1$	x_1	$\underline{x}_1$	$\underline{x}_1$	x_1	x_1	$\underline{x}_1$	x_1	$\underline{x}_1$	x_1	x_1	$\underline{x}_1$	x_1	x_1
f_2	$\underline{x}_2$	$\underline{x}_2$	$\underline{x}_2$	$\underline{x}_2$	x_2	$\underline{x}_2$	x_2	$\underline{x}_2$	x_2	$\underline{x}_2$	x_2	x_2	$\underline{x}_2$	x_2	x_2	x_2
f_3	$\underline{x}_3$	$\underline{x}_3$	$\underline{x}_3$	x_3	x_3	$\underline{x}_3$	x_3	x_3	x_3	x_3	x_3	x_3	x_3	x_3	x_3	x_3
f_4	$\underline{x}_4$	x_4	$\underline{x}_4$	$\underline{x}_4$	$\underline{x}_4$	x_4	$\underline{x}_4$	x_4	x_4	$\underline{x}_4$	$\underline{x}_4$	x_4	x_4	x_4	$\underline{x}_4$	x_4
f_5	$\underline{x}_5$	$\underline{x}_5$	x_5	$\underline{x}_5$	$\underline{x}_5$	x_5	$\underline{x}_5$	$\underline{x}_5$	$\underline{x}_5$	x_5	x_5	$\underline{x}_5$	x_5	x_5	x_5	x_5
g	F	F	F	F	F	T	T	T	F	T	T	T	T	T	T	T

Since x_3 and x_5 are not in any block of F simultaneously, according to Proposition 3, x_3x_5 is a prime disjunction normal form of T. Deleting all implicants of x_3x_5, we have that

$$T = x_1x_4 + x_3x_5 + x_1x_2x_3\underline{x}_4\underline{x}_5 + \underline{x}_1\underline{x}_2\underline{x}_3x_4x_5.$$

Since x_4 and x_5 are not in any block of F simultaneously, according to Proposition 3, x_4x_5 is a prime disjunction normal form of T. Deleting all implicants of x_4x_5, we have that

$$T = x_1x_4 + x_3x_5 + x_1x_2x_3\underline{x}_4\underline{x}_5 + x_4x_5.$$

Since x_1, x_2 and x_3 are not in any block of F simultaneously, according to Proposition 3, $x_1x_2x_3$ is a prime disjunction normal form of T. Deleting all implicants of $x_1x_2x_3$, we have that

$$T = x_1x_4 + x_3x_5 + x_1x_2x_3 + x_4x_5.$$

It is the structure of the switch system, which is the same structure of Fig. 2, but each switch x3 is binding as the parallel connection of x_1 and x_2. The 5 switch system indeed degenerated into a 4 switch system. We omit the structure graph here.

5 Conclusions

Factor logic is a derivative system of Boolean logic with added hypothesis. Even though it is not a strict theory, the examples in application show that it is worthy to be developed in the future. The motive of this paper is to commemorate Prof. Wang GJ. Factor logic takes state description into factor space and set up truth set for formulae, it may provide a suitable platform for the development of quantitative logic.

Acknowledgments The second author acknowledges the support form the National Natural Science Foundation of China (Grant No. 61370100) and the third author acknowledges that this work is supported by the National Natural Science Foundation of China (Grant No. 61350003).

References

1. Wang, G.J.: Introduction of Mathematical Logic and Resolution Principle (in Chinese). Science Press, Beijing (2000)
2. Wang, G.J.: Non-classical Logic and Approximate Reasoning (in Chinese). Science Press, Beijing (2008)
3. Pei, D.W., Wang, G.J.: The completeness of formal system and its applications. Sci. China, Ser. E. **32**(1), 56–64 (2002)
4. Zhou, H.J.: Probabilistic Quantitative Logic and Its Applications (in Chinese). Science Press, Beijing (2015)
5. Wang, P.Z., Sugeno, M.: The factors field and background structure for fuzzy subsets. Fuzzy Math. **2**, 45–54 (1992)
6. Wang, P.Z., Li, H.X.: Fuzzy Systems and Fuzzy Computer (in Chinese). Scientific Press, Beijing (1995)
7. Wang, P.Z., Liu, Z.L., Shi, Y., Guo, S.C.: Factor Space, the Theoretical Base of Data Science. Ann. Data. Science **1**(2), 233–251 (2014)
8. Wang, H.D., Wang, P.Z., Shi, Y., Liu, H.T.: Improved factorial analysis algorithm in factor spaces. International Conference on Informatics, pp. 201–204 (2014)
9. Cui, T.J., Wang, P.Z., Li, S.S., Wang, L.G.: Function-structure analysis in factor space and invert analysis in SFT systems [J] (Accepted)

A Brief Introduction to Probabilistically Quantitative Logic with Its Applications

Hong-Jun Zhou

Abstract Uncertainty is a fundamental and unavoidable feature of our real life, which has many distinct representation forms such as randomness, fuzziness, ambiguity, inaccuracy, incompleteness and roughness. Accordingly, many different mathematical models for dealing with these uncertainties, like probability, fuzzy set theory, Dempster-Shafer theory of evidence and rough set theory, have been introduced and also applied with great success in many fields. In order to construct uncertainty models with more powerful abilities of linguistic expression and logical inference, researchers have been devoting to intersections of different branches of mathematics of uncertainty, among which the resulting interdiscipline by combining probability and many-valued propositional logics is called probabilistically quantitative logic. This paper presents a brief introduction to probabilistically quantitative logic from four viewpoints of its research approaches and applications. Main contents include probabilistic truth degrees of propositions, logic systems for reasoning about probabilities of many-valued events, generalized state theory on residuated lattices, consistency degrees of formal theories, characterizations of maximally consistent theories, Stone representations of R_0-algebras, and generalized state based similarity convergence in residuated lattices with its Cauchy completion.

Keywords Probabilistically quantitative logic · Truth degree · Bosbach state · Riečan state · Consistency degree

1 Introduction

Nowhere in our real life there does not exist uncertainty, which has many distinct representation forms such as randomness, fuzziness, ambiguity, inaccuracy, incompleteness and roughness [1–4]. On the other hand, modern mathematics based on two-

H.-J. Zhou (✉)
College of Mathematics and Information Science, Shaanxi Normal University, Xi'an 710119, China
e-mail: hjzhou@snnu.edu.cn

T.-H. Fan et al. (eds.), *Quantitative Logic and Soft Computing 2016*, Advances in Intelligent Systems and Computing 510,
DOI 10.1007/978-3-319-46206-6_5

valued logic and modern techniques developed thereafter have made great achievements, but classical mathematics, after all, is an idealized model of the complicated real world, and hence, modern mathematics cannot, or at least only with low efficiency, deal with uncertainties existing in our real life [5, 6]. Indeed, it can be traced back to the 17th century that people began to study random phenomena, and established probability theory [7] in the 1930s. Meanwhile, other mathematical theories modelling respective forms of uncertainty, e.g., many-valued logics [8] born in the 1920s; fuzzy set theory [9] born in the 1960s and diverse branches of fuzzy mathematics [10] developed soon after; Demspter-Shafer theory of evidence [11] built in the 1970s; rough set theory [12] introduced in the 1980s; and credibility measure [4] in recent decades, have also emerged in succession and grown gradually in strength. These mathematical models of uncertainty have also obtained remarkable achievements in such fields as information processing, computational intelligence, automated control, multi-criteria decision-making and data mining. However, the mathematics of uncertainty has not been constructed as perfect as the magnificent building of modern mathematics, because distinct forms of uncertainty in our real life interweave with each other, and are so complicated that each mathematical theory of uncertainty as mentioned above is insufficient to represent other forms of uncertainty. For example, probability theory is good at representing and reasoning about randomness of events but not at fuzziness of the events, and fuzzy logic is just to the contrary [13, 14]. In order to make up their limitations and build a connecting bridge so as to improve their power of linguistic expression and logical inference, researchers tend to combining two or more related disciplines into a new interdiscipline. In fact, interdisciplinarity has become an inevitable trend in the development of science today. Mathematics of uncertainty, information science and computational intelligence all belong to such category. The purpose of the interdisciplinarity is to further simulate human intelligence, while human brain is the intelligent source of all modern civilization, and hence simulation of human intelligence is and will forever be a challenging task and is, of course, very difficult because there are no mature patterns to follow.

The present paper concerns combination of probability and many-valued propositional logics. Such an idea can be traced back to Ramsey [15] from Cambridge University in the 1930s. Main approaches appearing in this community can be classified into three categories: semantic quantification, algebraic axiomatization and modal formalization. Firstly, in the 1980s, Adam [16] from Stanford University and Hailperin [17] from Lehigh University introduced probability logic by using probability distributions on the set of state descriptions of propositions in two-valued propositional logic, and in recent two decades, Wang et. al. [18–26] proposed quantitative logic by integral of truth-functions with respect to countably infinite product measures of uniform probability on the truth value sets of many-valued propositional logics. The above approaches to measure the extent to which propositions are true by virtue of probability measures on valuation domains outside the underlying propositional logic systems are called semantic quantifications. Secondly, with almost the same intent of semantic quantification but using algebraic axiomatization approach, Mundici [27] introduced in 1995 the notion of state on MV-algebras by extending the Kolmogorov axioms of probability measures on Boolean algebras. Since its birth, state has attracted great interests of researchers and has been generalized to other logical algebras such as BL-algebras, MTL-algebras, residuated lattices and Hilbert algebras [28–38]. Considering that the underlying logical algebras with a state are not universal algebras and hence they do not provide an algebraizable logic in the sense of Blok and Pigozzi [39] for reasoning about probabilities of many-valued events, Flaminio and Montagna [40] enlarged the language of MV-algebras by adding a unary internal operator equationally described so as to preserve the basic properties of a state in its original meaning. The resulting algebras are called MV-algebras with internal states or state MV-algebras. This topic has aroused considerable interests of researchers in the fields of non-classical logics and many-valued probability theory. Internal states have also been extended to more general logical algebras [41–46]. Recently, a common generalization of the approaches for states and internal states was introduced by Georgescu and Mureşan [47] by replacing the codomain [0, 1] of Bosbach and Riečan states on residuated lattices with an arbitrary residuated lattice. The resulting notions are called generalized Bosbach state (of type I, or of type II) and generalized Riečan state, respectively. So far the whole framework of generalized states has been established [48–51]. Lastly, completely independent of the abovementioned two main directions of semantic quantification and algebraic axiomatization, and particularly for providing a logical foundation for reasoning about probabilities of many-valued events inside Łukasiewicz infinite-valued propositional logic, Hájek et. al. [52, 53] proposed a modal-like logic system by adding a modality interpreted as the probability of the underlying proposition and by adding additional axioms for modal formulas which are copies of axioms of states on MV-algebras, and finally proved its Kripke type completeness with respect to states on MV-algebras.

The purpose of the present paper is to provide a brief state of the art survey of main achievements about combinations of probability measures and many-valued propositional logics from the above mentioned three viewpoints of semantic quantification, modal formalization and algebraic axiomatization, and also their applications in related fields. These results create a new branch of mathematics of uncertainty which we call probabilistically quantitative logic. The rest of the paper is structured as follows: Sect. 2 reviews semantic quantification approach to many-valued propositional logics by integrals of truth functions with respect to uncertainty measures on the space of valuations. In this section we will focus on the theory of probabilistic truth degrees of propositions in Łukasiewicz finite or infinite-valued propositional logic and in the formal deductive system $\mathscr{L}^*$ (equivalently, NM-logic), and the theory of truth degree of Choquet integral type in Łukasiewicz infinite-valued propositional logic. Section 3 deals with Hájek's modal formalization towards the probabilistic truth degree functions studied in Sect. 2 and provides one modal logic system for reasoning about probabilities of many-valued events. Section 4 reviews main results about generalized states on residuated lattices. Section 5 recalls some applications of probabilistic truth degrees of propositions in consistency degrees of logic theories, characterizations of maximally consistent theories, three-valued Stone representations of R_0-algebras and generalized-state-based similarity convergence with its Cauchy completion. Section 6 concerns concluding remarks and possible further studies.

2 Semantic Quantification

2.1 *Probabilistic Truth Degrees of Propositions*

In probability logic, the probability of a proposition in two-valued propositional logic is defined as the probability of the set of state-descriptions at which the underlying proposition is true given a probability distribution on the set of all state-descriptions of the proposition. Note that probability logic was established only inside the framework of two-valued propositional logic. Moreover, such a way is suitable to define probabilities of finitely many propositions under one probability distribution while we have countably infinite many propositions. Quantitative logic defined the notion of truth degree for all propositions in diverse infinite or finite-valued propositional logics by using the countably infinite product of uniform probability measures on the set of truth values, but the usage of product measures determines the independence of atomic propositions. Considering the problems existing as above we [54–57] introduced the notion of probabilistic truth degree for propositions by means of the integral of truth functions with respect to Borel probability measures on the space of valuations.

In this section we review the notion of probabilistic truth degree of propositions in, as examples, many-valued Łukasiewicz propositional logics Ł_n and Ł, and many-valued R_0-type propositional logics $\mathscr{L}_n^*$ and $\mathscr{L}^*$. Note that, in n-valued propositional logics Ł_n and $\mathscr{L}_n^*$, the set of truth values is $W_n = \left\{0, \frac{1}{n-1}, \ldots, \frac{n-2}{n-1}, 1\right\}$, and in Ł and $\mathscr{L}^*$, the set of truth values is $W = [0, 1]$. Of course, the logical operations in different truth value sets are determined by the underlying logic systems. Accordingly, denote the set of valuations Ω_n by W_n^ω, and denote Ω by W^ω. Let $F(S)$ be the set of all well-formed formulae generated by the set $S = \{p_1, p_2, \ldots\}$ of atomic propositions and the logic connectives $\neg$, $\vee$ and $\rightarrow$. For more details on Ł_n, Ł, $\mathscr{L}_n^*$ and $\mathscr{L}^*$, we refer to [58, 59].

We first review the notion of probabilistic truth degree of propositions in n-valued propositional logics. To do so, view $X_k = W_n$ as a discrete topological space ($k = 1, 2, \ldots$), and $\Omega_n = X = \prod_{k=1}^{\infty} X_k$ as a product space, called *valuation space*. Let $\mathscr{B}(X_k)$ and $\mathscr{B}(\Omega)$ be the set of all Borel subsets of X_k and of Ω, respectively, μ a Borel probability measure defined on $\mathscr{B}(\Omega)$.

Definition 1 In Ł_n and $\mathscr{L}_n^*$, let $\varphi \in F(S)$, and define

$$\tau_{n,\mu}(\varphi) = \sum_{i=0}^{n-1} \frac{i}{n-1} \mu \left(\varphi^{-1} \left(\frac{i}{n-1} \right) \right). \tag{1}$$

Then $\tau_{n,\mu}(\varphi)$ is called *μ-truth degree* of φ, where φ, at the right hand of (1), is viewed as a function $\varphi : \Omega \rightarrow W_n$, $\varphi(v) = v(\varphi)$, $v \in \Omega$. Unless confusion arises, we shall drop the subscript n from $\tau_{n,\mu}(\varphi)$.

Remark 1 (i) Take proposition $\varphi = \varphi(p_1, \ldots, p_m) \in F(S)$, then $\forall i = 0, \ldots, n-1$, $\varphi^{-1}(\frac{i}{n-1}) = \overline{\varphi}^{-1}(\frac{i}{n-1}) \times \prod_{k=m+1}^{\infty} X_k$, where $\overline{\varphi}$ is the truth-function induced by φ. Recall that in $Ł_n$, $\overline{\varphi}$ is just a McNaughton function. Since $\forall E \subseteq W_n^m, E \times \prod_{k=m+1}^{\infty} X_k \in \mathscr{B}(\Omega)$, we have that in $Ł_n$ or in $\mathscr{L}_n^*$, φ is continuous from the product space Ω into the discrete space W_n, and so Borel measurable. Therefore, φ can be viewed as a random variable function on $(\Omega, \mathscr{B}(\Omega), \mu)$, and in this case

$$\tau_\mu(\varphi) = \sum_{i=0}^{n-1} \frac{i}{n-1} \mu\left(\varphi^{-1}\left(\frac{i}{n-1}\right)\right) = \int_{\mathscr{B}(\Omega)} \varphi(v) \mathrm{d}\mu$$

is the mathematical expectation of φ.

(ii) Let μ be a Borel probability measure on Ω, $m \in \mathbb{N}$, and define $\mu(m) : \mathscr{B}(W_n^m) \to [0, 1]$:

$$\mu(m)(E) = \mu\left(E \times \prod_{k=m+1}^{\infty} X_k\right), \quad E \in \mathscr{B}(W_n^m).$$

Then $\mu(m)$ is a Borel probability measure on W_n^m. For $\varphi = \varphi(p_1, \ldots, p_m) \in F(S)$, we have

$$\begin{aligned} \tau_\mu(\varphi) &= \sum_{i=0}^{n-1} \tfrac{i}{n-1} \mu\left(\varphi^{-1}\left(\tfrac{i}{n-1}\right)\right) \\ &= \sum_{i=0}^{n-1} \tfrac{i}{n-1} \mu\left(\overline{\varphi}^{-1}\left(\tfrac{i}{n-1}\right) \times \prod_{k=m+1}^{\infty} X_k\right) \\ &= \sum_{i=0}^{n-1} \tfrac{i}{n-1} \mu(m)\left(\overline{\varphi}^{-1}\left(\tfrac{i}{n-1}\right)\right). \end{aligned} \tag{2}$$

(iii) $\tau_\mu(\varphi)$ can be further written as

$$\begin{aligned} \tau_\mu(\varphi) &= \sum_{i=0}^{n-1} \tfrac{i}{n-1} \mu(m)\left(\overline{\varphi}^{-1}\left(\tfrac{i}{n-1}\right)\right) \\ &= \sum_{i=0}^{n-1} \tfrac{i}{n-1}\left(\sum\left\{\mu(m)(\{(x_1, \ldots, x_m)\}) \mid (x_1, \ldots, x_m) \in \overline{\varphi}^{-1}\left(\tfrac{i}{n-1}\right)\right\}\right) \\ &= \sum_{i=0}^{n-1} \left(\sum\left\{\tfrac{i}{n-1} \mu(m)(\{(x_1, \ldots, x_m)\}) \mid (x_1, \ldots, x_m) \in \overline{\varphi}^{-1}\left(\tfrac{i}{n-1}\right)\right\}\right) \\ &= \sum_{i=0}^{n-1} \left(\sum\left\{\overline{\varphi}(x_1, \ldots, x_m) \mu(m)(\{(x_1, \ldots, x_m)\}) \mid (x_1, \ldots, x_m) \in \overline{\varphi}^{-1}\left(\tfrac{i}{n-1}\right)\right\}\right) \\ &= \sum\{\overline{\varphi}(x_1, \ldots, x_m) \mu(m)(\{(x_1, \ldots, x_m)\}) \mid (x_1, \ldots, x_m) \in W_n^m\}. \end{aligned} \tag{3}$$

(iv) Let $n = 2$ and $\varphi = \varphi(p_1, \ldots, p_m) \in F(S)$. Then $\overline{\varphi}$ is a Boolean function, and (3) reduces to

$$\tau_{2,\mu}(\varphi) = \sum\{\mu(m)(x_1, \dots, x_m) \mid (x_1, \dots, x_m) \in \overline{\varphi}^{-1}(1)\}. \tag{4}$$

Recall that each disjunctive component of the disjunctive normal form of φ is called a *state-description* of φ, which is one-to-one corresponding to a vector in $\overline{\varphi}^{-1}(1)$. Thus $\mu(m)$ can be viewed as a probability distribution on the set of all state-descriptions of φ and $\tau_{2,\mu}(\varphi)$ in (4) is the probability of φ as defined in [16, 17]. What is important here is that, by virtue of μ, one can define the probabilities of all propositions in $F(S)$.

(v) Let μ be the product probability measure on Ω generated by the uniform probability measures on W_n, then $\mu(m)$ is the uniform probability measure on W_n^m, and hence,

$$\begin{aligned}\tau_\mu(\varphi) &= \sum_{i=0}^{n-1} \tfrac{i}{n-1}\mu(m)\left(\overline{\varphi}^{-1}\left(\tfrac{i}{n-1}\right)\right)\\ &= \sum_{i=0}^{n-1} \tfrac{i}{n-1} \cdot \tfrac{|\overline{\varphi}^{-1}(\frac{i}{n-1})|}{n^m} \\ &= \tfrac{1}{n^m}\sum_{i=0}^{n-1} \tfrac{i}{n-1}\left|\overline{\varphi}^{-1}\left(\tfrac{i}{n-1}\right)\right|.\end{aligned} \tag{5}$$

This is just the computation formula for truth degrees of propositions given in [18, 19]. Different from [18, 19, 21, 23, 25], μ in Definition 1 is not necessarily a product probability measure.

(vi) For each valuation $v = (v_1, v_2, \dots) \in \Omega = W_n^\omega$, take the Borel probability measure μ_k on $X_k = W_n$ satisfying $\mu_k(\emptyset) = 0$, $\mu_k(X_k) = 1$, and

$$\mu_k(\{x_k\}) = \begin{cases} 1, & x_k = v_k, \\ 0, & x_k \neq v_k, \end{cases} \quad k = 1, 2, \dots.$$

Let μ be the product measure generated by $\mu_1, \mu_2, \dots$, then it is routine to verify that $\tau_\mu = v$. This shows that each valuation $v \in \Omega$ is a truth degree function in the sense of Definition 1.

Example 1 Let $\varphi_1 = p_1, \varphi_2 = p_2 \to p_3, \varphi_3 = p_1 \vee p_2 \vee p_3, \varphi_4 = p_1 \wedge p_2 \wedge p_3$. Compute $\tau_\mu(\varphi_i)$ $(i = 1, \dots, 4)$ in $Ł_3$ and $\mathscr{L}_3^*$, respectively.

Solution Since $\varphi_i (i = 1, \dots, 4)$ involves just p_1, p_2 and p_3, it is enough to consider Borel probability measures $\mu(3)$ on $W_3^3 = \{0, \frac{1}{2}, 1\}^3$. Note that, $\mathscr{L}_3^*$ is equivalent to $Ł_3$, i.e., they have the same provable theorems, hence each proposition above has the same truth degrees in these two logic systems.

(i) Suppose that $\mu(3)\{(1, 1, 1)\} = 0.3$, $\mu(3)(\{(\frac{1}{2}, \frac{1}{2}, \frac{1}{2})\}) = 0.2$, $\mu(3)(\{(x_1, x_2, x_3)\}) = 0.02$ iff $(x_1, x_2, x_3) \in W_3^3 - \{(1, 1, 1), (\frac{1}{2}, \frac{1}{2}, \frac{1}{2})\}$, then:

$\overline{p_1}^{-1}(1) = \{(1, x_2, x_3) \mid x_2, x_3 \in W_3\}$, $\overline{p_1}^{-1}(\frac{1}{2}) = \{(\frac{1}{2}, x_2, x_3) \mid x_2, x_3 \in W_3\}$, so one has

$$\begin{aligned}\tau_\mu(p_1) &= \mu(3)(\overline{p_1}^{-1}(1)) + \tfrac{1}{2}\mu(3)\left(\overline{p_1}^{-1}\left(\tfrac{1}{2}\right)\right)\\ &= 0.3 + 0.02 \times 8 + \tfrac{1}{2} \times (0.2 + 0.02 \times 8)\\ &= 0.64;\end{aligned}$$

$\overline{\varphi_2}^{-1}(1) = \{(x_1, x_2, x_3) \mid x_1, x_2, x_3 \in W_3, x_2 \leq x_3\}$, $\overline{\varphi_2}^{-1}(\frac{1}{2}) = \{(x_1, x_2, x_3) \mid x_2 = x_3 + \frac{1}{2}\}$, thus

$$\begin{aligned}\tau_\mu(\varphi_2) &= \mu(3)(\overline{\varphi_2}^{-1}(1)) + \tfrac{1}{2}\mu(3)\left(\overline{\varphi_2}^{-1}\left(\tfrac{1}{2}\right)\right)\\ &= (0.3 + 0.2 + 0.02 \times 16) + \tfrac{1}{2} \times (0.02 \times 6)\\ &= 0.88;\end{aligned}$$

$\overline{\varphi_3}(1) = \{(x_1, x_2, x_3) \mid \max\{x_1, x_2, x_3\} = 1\}$, $\overline{\varphi_3}^{-1}(\frac{1}{2}) = \{(x_1, x_2, x_3) \mid \max\{x_1, x_2, x_3\} = \frac{1}{2}\} = \{0, \frac{1}{2}\}^3 - \{(0, 0, 0)\}$, then

$$\begin{aligned}\tau_\mu(\varphi_3) &= \mu(3)(\overline{\varphi_3}^{-1}(1)) + \tfrac{1}{2}\mu(3)\left(\overline{\varphi_3}^{-1}\left(\tfrac{1}{2}\right)\right)\\ &= (0.3 + 0.02 \times 18) + \tfrac{1}{2} \times (0.2 + 0.02 \times 6)\\ &= 0.82;\end{aligned}$$

$\overline{\varphi_4}^{-1}(1) = \{(1, 1, 1)\}$, $\overline{\varphi_4}^{-1}(\frac{1}{2}) = \{\frac{1}{2}, 1\}^3 - \{(1, 1, 1)\}$, and thus

$$\begin{aligned}\tau_\mu(\varphi_4) &= \mu(3)(\overline{\varphi_4}^{-1}(1)) + \tfrac{1}{2}\mu(3)\left(\overline{\varphi_4}^{-1}\left(\tfrac{1}{2}\right)\right)\\ &= 0.3 + \tfrac{1}{2} \times (0.2 + 0.02 \times 6)\\ &= 0.46.\end{aligned}$$

(ii) Let $\mu(3)$ be the uniform probability distribution on $\{0, \frac{1}{2}, 1\}^3$, then $\tau_\mu(\varphi_1) = \frac{1}{2}$, $\tau_\mu(\varphi_2) = \frac{7}{9}$, $\tau_\mu(\varphi_3) = \frac{5}{6}$, $\tau_\mu(\varphi_4) = \frac{1}{6}$.

Example 2 Let μ be the product probability measure on Ω generated by the uniform probability measures on W_n, see (5), calculate the μ-truth degrees of p_1, $p_2 \to p_3$, $p_1 \vee p_2$ and $p_1 \wedge \cdots \wedge p_m$ in $Ł_n$ and in $\mathscr{L}_n^*$, respectively.

(i) In $Ł_n$,

$$\begin{aligned}\tau_\mu(p_1) &= \tfrac{1}{n^1}\sum_{i=0}^{n-1}\tfrac{i}{n-1}\left|\overline{p_1}^{-1}\left(\tfrac{i}{n-1}\right)\right|\\ &= \tfrac{1}{n}\sum_{i=1}^{n-1}\tfrac{i}{n-1}\\ &= \tfrac{1}{n(n-1)}\sum_{i=1}^{n-1} i\\ &= \tfrac{1}{n(n-1)}\cdot\tfrac{n(n-1)}{2} = \tfrac{1}{2};\end{aligned}$$

$$\begin{aligned}\tau_\mu(p_2 \to p_3) &= \tfrac{1}{n^2}\sum_{i=0}^{n-1}\tfrac{i}{n-1}\left|\overline{p_2 \to p_3}^{-1}\left(\tfrac{i}{n-1}\right)\right|\\ &= \tfrac{1}{n^2(n-1)}\left(\tfrac{n(n+1)(n-1)}{2} + \sum_{i=1}^{n-2} i(i+1)\right)\\ &= \tfrac{1}{6n^2(n-1)}(5n^2 - n)(n-1)\\ &= \tfrac{5n-1}{6n};\end{aligned}$$

$$\begin{aligned}\tau_\mu(p_1 \vee p_2) &= \tfrac{1}{n^2}\sum_{i=0}^{n-1}\tfrac{i}{n-1}(2i+1)\\ &= \tfrac{4n+1}{6n};\end{aligned}$$

$$\begin{aligned}\tau_\mu(p_1 \wedge \cdots \wedge p_m) &= \tfrac{1}{n^m}\sum_{i=0}^{n-1}\tfrac{i}{n-1}\mid \overline{p_1 \wedge \cdots \wedge p_m}^{-1}(\tfrac{i}{n-1})\mid\\ &= \tfrac{1}{n^m}\sum_{i=1}^{n-1}\tfrac{i}{n-1}[(n-i)^m-(n-i-1)^m]\\ &= \tfrac{1}{n^m(n-1)}[(n-1)^m+(n-2)^m+\cdots+1^m]\\ &= \tfrac{1}{n^m(n-1)}\sum_{k=1}^{n-1}k^m.\end{aligned}$$

(ii) In $\mathscr{L}_n^*$, since $p_1,\ p_1 \vee p_2,\ p_1 \wedge \cdots \wedge p_m$ contains no implications, they each have the same μ-truth degrees as in $Ł_n$, i.e., $\tau_\mu(p_1) = \frac{1}{2}$, $\tau_\mu(p_1 \vee p_2) = \frac{4n+1}{6n}$, $\tau_\mu(p_1 \wedge \cdots \wedge p_m) = \frac{1}{n^m(n-1)}\sum_{k=1}^{n-1}k^m$. It now remains to find $\tau_\mu(p_2 \to p_3)$. Since $\overline{p_2 \to p_3}(x_2, x_3) = 1$ iff $x_2 \leq x_3$, we have $|\overline{p_2 \to p_3}^{-1}(1)| = n + (n-1) + \cdots + 1 = \frac{n(n+1)}{2}$. Let $\overline{p_2 \to p_3}(x_2, x_3) = \frac{i}{n-1} (1 \leq i \leq n-2)$. It is not difficult to verify that

$$|\overline{p_2 \to p_3}^{-1}\left(\frac{i}{n-1}\right)| = \begin{cases} n-1, & i = \frac{n-1}{2} \text{ and } n \text{ is odd},\\ 2i+1, & i < \frac{n-1}{2} \text{ and } n \text{ is odd},\\ 2(n-1-i), & i > \frac{n-1}{2} \text{ and } n \text{ is odd},\\ 2i+1, & i \leq \frac{n-1}{2} \text{ and } n \text{ is even},\\ 2(n-1-i), & i > \frac{n-1}{2} \text{ and } n \text{ is even}.\end{cases}$$

Thus, in the case when n is odd,

$$\begin{aligned}\tau_\mu(p_2 \to p_3) &= \tfrac{1}{n^2}\times\tfrac{n(n+1)}{2} + \tfrac{1}{2}\times\tfrac{n-1}{n^2} + \sum_{i=1}^{\frac{n-1}{2}-1}\tfrac{i}{n-1}\cdot\tfrac{2i+1}{n^2}\\ &\quad + \sum_{i=\frac{n-1}{2}+1}^{n-2}\tfrac{i}{n-1}\cdot\tfrac{2(n-1-i)}{n^2}\\ &= \tfrac{6n^2+n-1}{8n^2};\end{aligned}$$

and in the case when n is even,

$$\begin{aligned}\tau_\mu(p_2 \to p_3) &= \tfrac{1}{n^2}\times\tfrac{n(n+1)}{2} + \sum_{i=1}^{\frac{n}{2}-1}\tfrac{i}{n-1}\cdot\tfrac{2i+1}{n^2}\\ &\quad + \sum_{i=\frac{n}{2}}^{n-2}\tfrac{i}{n-1}\cdot\tfrac{2(n-1-i)}{n^2}\\ &= \tfrac{6n^2-5n-2}{8n(n-1)}.\end{aligned}$$

Therefore

$$\tau_\mu(p_2 \to p_3) = \begin{cases} \frac{6n^2+n-1}{8n^2}, & n \text{ is odd}, \\ \frac{6n^2-5n-2}{8n(n-1)}, & n \text{ is even}. \end{cases}$$

$\tau_{n,\mu}$ has the following basic properties.

Proposition 1 *Let μ be a Borel probability measure on $\Omega_n = W_n^\omega$, then in $Ł_n$ and $\mathscr{L}_n^*$:*
(i) $0 \leq \tau_\mu(\varphi) \leq 1$;
(ii) *If φ is a tautology (contradiction), then $\tau_\mu(\varphi) = 1(\tau_\mu(\varphi) = 0)$;*
(iii) *If φ and ψ are logically equivalent, then $\tau_\mu(\varphi) = \tau_\mu(\psi)$;*
(iv) $\tau_\mu(\varphi) + \tau_\mu(\neg\varphi) = 1$;
(v) $\tau_\mu(\varphi) + \tau_\mu(\psi) = \tau_\mu(\varphi \vee \psi) + \tau_\mu(\varphi \wedge \psi)$
(vi) *If $\vdash \varphi \to \psi$, then $\tau_\mu(\varphi) \leq \tau_\mu(\psi)$*
(vii) $\tau_\mu(\varphi \wedge \psi) \geq \tau_\mu(\varphi) + \tau(\psi) - 1$;
(viii) $\tau_\mu(\psi) \leq \tau_\mu(\varphi \to \psi)$.

Proposition 2 *Let μ be a Borel probability measure on Ω, then in $Ł_n$:*
(i) $\tau_\mu(\varphi) + \tau_\mu(\varphi \to \psi) = \tau_\mu(\psi) + \tau_\mu(\psi \to \varphi)$;
(ii) $\tau_\mu(\varphi) + \tau_\mu(\psi) = \tau_\mu(\varphi \oplus \psi) + \tau_\mu(\varphi \& \psi)$.

Proposition 3 ([55, 57]) *In $Ł_n$, we have*
(i) *There is a one-to-one correspondence between deductively closed logic theories and topologically closed subsets of the valuation space Ω;*
(ii) *Deductively closed logic theories each have the form of Ker(τ_μ), where Ker$(\tau_\mu) = \{\varphi \in F(S) \mid \tau_\mu(\varphi) = 1\}$, for some Borel probability measure μ on Ω.*

Proposition 4 *Let μ be a non-atomic (i.e., $\forall v \in \Omega$, $\mu(\{v\}) = 0$) Borel probability measure on Ω_n, then:*
(i) *$H_\mu = \{\tau_\mu(\varphi) \mid \varphi \in F(S)\}$ is a countable subset of* $[0, 1]$; (ii) *for μ the product measure generated by uniform probability measures, $\{\frac{k}{n^m} \mid k = 0, \ldots, n^m; m \in \mathbb{N}\} \subseteq H_\mu$;*
(iii) *the μ-logic metric space $(F(S), \rho_\mu)$ has no isolated points, where $\rho_\mu(\varphi, \psi) = \tau_\mu((\varphi \to \psi) \wedge (\psi \to \varphi))$.*

In the following we review the integrated truth degree functions $\tau_{\infty,\mu}$ in Ł and in $\mathscr{L}^*$, which are connected to $\tau_{n,\mu}$ in Definition 1 by a limit theorem.

Definition 2 Let $\varphi = \varphi(p_1, \ldots, p_m) \in F(S)$, μ a Borel probability measure on $\Omega = [0, 1]^\omega$, and define

$$\begin{aligned} \tau_{\infty,\mu}(\varphi) &= \int_{[0,1]^\omega} \varphi(v) \mathrm{d}\mu \\ &= \int_{[0,1]^m} \overline{\varphi_\infty}(x_1, \ldots, x_m) \mathrm{d}\mu(m), \end{aligned} \tag{6}$$

where $\overline{\varphi_\infty} : [0, 1]^m \to [0, 1]$ is the truth-function induced by φ in Ł (or in $\mathscr{L}^*$). $\tau_{\infty,\mu}(\varphi)$ is called the *μ-integrated truth degree* of φ.

Theorem 1 *Let $\varphi = \varphi(p_1, \ldots, p_m) \in F(S)$, μ a Borel probability measure on $\Omega_n = W_n^\omega$, then μ can be extended, in a certain way [55, 57], to a Borel probability measure (still denoted by μ) on $\Omega = [0, 1]^\omega$ such that*
(i) *in Łukasiewicz many-valued propositional logics, $\lim_{n\to\infty} \tau_{n,\mu}(\varphi) = \tau_{\infty,\mu}(\varphi)$;*
(ii) *in R_0-type propositional logics, $\lim_{n\to\infty} \tau_{n,\mu}(\varphi) = \tau_{\infty,\mu}(\varphi)$ whenever $\mu(\{(x_1, \ldots, x_m) \mid (x_1, \ldots, x_m)$ is a discontinuity point of $\overline{\varphi_\infty}\}) = 0$.*

2.2 Axiomatic Definition of Probabilistic Truth Degree Function and Its Representation

In this subsection we recall the axiomatic definition of the probabilistic truth degree functions in $Ł_n$ and in Ł and their representation by using (1). For the axiomatic definition of the probabilistic truth degree function in $\mathscr{L}^*$ we refer to [60].

Definition 3 In $Ł_n$ and Ł, a function $\tau : F(S) \to [0, 1]$ is called a *probabilistic truth degree function* if τ satisfies:
(ŁK1) $0 \leq \tau(\varphi) \leq 1$;
(ŁK2) If φ is an axiom, then $\tau(\varphi) = 1$;
(ŁK3) If $\vdash \varphi \to \psi$, then $\tau(\varphi) \leq \tau(\psi)$;
(ŁK4) If $\varphi \& \psi$ is a refutable formula, then $\tau(\varphi \oplus \psi) = \tau(\varphi) + \tau(\psi)$.
We will add the subscript n or ∞ to τ to indicate precisely the underlying logic $Ł_n$ or Ł when necessary.

For properties of τ and its characterizations by other axioms, please see [57]. By Kroupa-Panti's integral representations of states on MV-algebras [29, 30] one can get the representation for probabilistic truth degree function.

Theorem 2 (i) *For a probabilistic truth degree function τ in $Ł_n$, there exists a unique Borel probability measure μ on $\Omega = W_n^\omega$ such that τ can be represented by* (1), *i.e.,*
$\forall \varphi \in F(S)$, $\tau(\varphi) = \sum_{i=0}^{n-1} \frac{i}{n-1}\mu(\varphi^{-1}(\frac{i}{n-1}))$;
(ii) *For a probabilistic truth degree function τ in Ł, there exists a unique Borel probability measure on $\Omega = [0, 1]^\omega$ such that τ can be represented by* (6), *i.e.,*

$$\begin{aligned}\tau(\varphi) &= \int_\Omega \varphi(v)\mathrm{d}\mu \\ &= \int_{[0,1]^m} \overline{\varphi}(x_1, \ldots, x_m)\mathrm{d}\mu(m),\end{aligned}$$

$\varphi = \varphi(p_1, \ldots, p_m) \in F(S)$.

2.3 Choquet Type Truth Degree of Propositions

Logic propositions are all proper symbol strings of finite length, and any logic inference does not involve infinitely many propositions, hence the requirement on the countable additivity of probability measures on the valuation space seems too strong to be widely accepted by logicians. Finite additivity of measures can even be released because the dependence between information in our real life is not linear. We introduced in [56] the notion of Choquet type truth degrees of propositions in Ł by using the Choquet integral of McNaughton functions with respect to general fuzzy measures on the valuation space and proved that a truth degree function has good properties when the fuzzy measure has just finite additivity and particularly, the truth degree function will reduce to the probabilistic truth degree function in (6) when the fuzzy measure is a Borel probability measure.

Definition 4 Let $\Omega = [0, 1]^\omega$, $\mathscr{B}$ the set of all Borel subsets of Ω, and μ a fuzzy measure on $(\Omega, \mathscr{B})$. Then, $\forall \varphi \in F(S)$, define

$$\tau_{\mu,C}(\varphi) = (C)\int_\Omega \varphi \mathrm{d}\mu = \int_0^1 \mu(\varphi^{-1}([\alpha, 1]))\mathrm{d}\alpha \tag{7}$$

as the *μ-Choquet type truth degree* of φ. For simplicity, denote $\tau_{\mu,C}$ by τ if no confusion arises.

Remark 2 (i) Let μ be a fuzzy measure on $(\Omega, \mathscr{B})$, $\varphi \in F(S)$, $\alpha \in [0, 1]$. Then, by the continuity of $\varphi : \Omega \to [0, 1]$, we have $\varphi^{-1}([\alpha, 1]) \in \mathscr{B}$, hence $\mu(\varphi^{-1}([\alpha, 1]))$ is well-defined and is non-increasing with respect to α, so $\mu(\varphi^{-1}([\alpha, 1]))$ is Rieman integrable on $[0, 1]$. This shows that τ in (7) is well defined.

(ii) Let $\varphi = \varphi(p_1, \ldots, p_m) \in F(S), \alpha \in [0, 1]$, then $\varphi^{-1}([\alpha, 1]) = \overline{\varphi}^{-1}([\alpha, 1]) \times \prod_{k=m+1}^{\infty} X_k$. Hence

$$\begin{aligned}\mu(\varphi^{-1}([\alpha, 1])) &= \mu\left(\overline{\varphi}^{-1}([\alpha, 1]) \times \prod_{k=m+1}^{\infty} X_k\right)\\ &= \mu(m)(\overline{\varphi}^{-1}([\alpha, 1])),\end{aligned}$$

So $\tau(\varphi) = (C)\int_{[0,1]^m} \overline{\varphi}\mathrm{d}\mu(m) = \int_0^1 \mu(m)(\overline{\varphi}^{-1}([\alpha, 1]))\mathrm{d}\alpha$. This shows that given a proposition, its μ-Choquet type truth degree is determined by the fuzzy measure $\mu(m)$ on $[0, 1]^m$. Thus it suffices to consider fuzzy measures on $[0, 1]^m$, $\mu(m)$ is sometimes denoted by μ too.

(iii) By Theorem 11.1 in [61], $\tau(\varphi) = (C)\int_\Omega \varphi\mathrm{d}\mu = \int_0^1 \mu(\varphi^{-1}((\alpha, 1]))\mathrm{d}\alpha$.

Example 3 Let λ be the Lebesgue measure on $[0, 1]^2$, then $\mu(2) = \lambda^2$ is a fuzzy measure on $[0, 1]^2$, which has no additivity. Calculate $\tau(p_1)$, $\tau(p_1 \wedge p_2)$, $\tau(p_1 \to p_2)$ and $\tau((p_1 \to p_2) \to \neg p_1 \vee p_2)$.

Solution Note that $\overline{p_1}$ can be extended to a binary function $\overline{p_1}(x_1, x_2) = \overline{p_1}(x_1)$, $(x_1, x_2) \in [0, 1]^2$.

$$\begin{aligned}\tau(p_1) &= (C)\int_{[0,1]^2} \overline{p_1}\mathrm{d}\mu = \int_0^1 \mu(\{(x_1, x_2) \mid x_1 \geq \alpha\})\mathrm{d}\alpha \\ &= \int_0^1 (\lambda(\{(x_1, x_2) \mid x_1 \geq \alpha\}))^2 \mathrm{d}\alpha \\ &= \int_0^1 (1-\alpha)^2 \mathrm{d}\alpha \\ &= \tfrac{1}{3}.\end{aligned}$$

$$\begin{aligned}\tau(p_1 \vee p_2) &= (C)\int_{[0,1]^2} \overline{p_1 \vee p_2}\mathrm{d}\alpha \\ &= \int_0^1 \mu(\{(x_1, x_2) \mid x_1 \geq \alpha x_2 \geq \alpha\})\mathrm{d}\alpha \\ &= \int_0^1 (\lambda(\{(x_1, x_2) \mid x_2 \geq x_1, x_2 \geq \alpha\} \\ &\qquad \cup\{(x_1, x_2) \mid x_1 > x_2, x_1 \geq \alpha\}))^2 \mathrm{d}\alpha \\ &= \int_0^1 (\lambda(\{(x_1, x_2) \mid x_2 \geq x_1, x_2 \geq \alpha\}) \\ &\qquad +\lambda(\{(x_1, x_2) \mid x_1 > x_2, x_1 \geq \alpha\}))^2 \mathrm{d}\alpha \\ &= \int_0^1 (1-\alpha^2)^2 \mathrm{d}\alpha \\ &= \tfrac{8}{15}.\end{aligned}$$

The remaining calculations are left to the readers: $\tau(p_1 \wedge p_2) = \frac{1}{5}$, $\tau((p_1 \to p_2) \to \neg p_2 \vee p_2) = \frac{4}{15}$, $\tau(p_1 \to p_2) = \frac{43}{60}$.

For the properties of τ corresponding to that of μ we refer to [56].

Theorem 3 *Let μ be a Borel probability measure on Ω, then* $\tau(\varphi) = \int_0^1 \mu(\varphi^{-1}([\alpha, 1]))\mathrm{d}\alpha = \int_\Omega \varphi \mathrm{d}\mu$.

3 Modal Formalization

Following Hájek's modal formalization approach [52, 53], one can construct modal-like logic systems for reasoning about truth degrees of propositions, by abstracting as a modality P the probabilistic truth degree function τ in Ł$_n$ defined by (1) (equivalently by Definition 3), and as axioms of modal formulas some basic identities of τ, like $\tau(\neg\varphi) = 1 - \tau(\varphi)$, which provide a logic foundation for semantic quantification. We discussed in [57] three modal-like logic systems, but due to limit of the length of the paper we review only the simplest one, denoted PQ(Ł$_n$, Ł).

Definition 5 The alphabet of PQ(Ł$_n$, Ł) consists of: propositional variables $p_1, p_2, \ldots$, denote by S is the set of all propositional variables, i.e., $S = \{p_1, p_2, \ldots\}$; the primitive logic connectives $\neg$ and $\to$ of Łukasiewicz propositional logic; a modality P and necessary punctuations , ,(,). The set of well-formed formulas of PQ(Ł$_n$, Ł) is divided into two parts: $F(S)$ and $MF(S)$, where

(i) $F(S)$ is just the set of all well-formed propositions of $Ł_n$ generated by S by using the logic connectives $(\neg, \rightarrow)$. Members of $F(S)$ are denoted by $\varphi, \psi, \ldots$.
(ii) $MF(S)$ is generated by $\{P(\varphi) \mid \varphi \in F(S)\}$ by using the logic connectives $(\neg, \rightarrow)$, more precisely, $P(\varphi) \in MF(S)$ $(\varphi \in F(S))$; if $P(\varphi), P(\psi) \in MF(S)$, then $\neg P(\varphi)$, $P(\varphi) \rightarrow P(\psi) \in MF(S)$. $P(\varphi)$ is called an *atomic modal formula*, and members of $MF(S)$ are called *modal formulas*, denoted by $\Phi, \Psi, \ldots$.

Definition 6 The axiom schemes and inference rules of PQ($Ł_n$, Ł) are:
(i) substitution instances of axioms (Ł1)–(Ł4) of Ł by modal formulas;
(ii) axioms (Ł5) and (Ł6) of $Ł_n$;
(iii) three axiom schemes about P:
(PQ1) $P(\neg\varphi) \equiv \neg P(\varphi)$;
(PQ2) $P(\varphi \rightarrow \psi) \rightarrow (P(\varphi) \rightarrow P(\psi))$;
(PQ3) $P(\varphi \oplus \psi) \equiv (P(\varphi) \rightarrow P(\varphi \& \psi)) \rightarrow P(\psi)$.
Inference rules are modus ponens for modal and non-modal formulas, and generalization: from φ infer $P(\varphi)$.

Definition 7 A semantic model of PQ($Ł_n$, Ł) is a quadruple $(\Omega, v, \tau, \|\cdot\|_{\tau,v})$, where
(i) Ω is the set of all valuations of $Ł_n$, and $v \in \Omega$ is a valuation;
(ii) τ is a truth degree function as defined in Definitions 1, 3,
(iii) $\|\cdot\|_{\tau,v}$ is a mapping $\|\cdot\|_{\tau,v} : F(S) \cup MF(S) \rightarrow [0, 1]$ such that

- $\|\varphi\|_{\tau,v} = v(\varphi)$,
- $\|P(\varphi)\|_{\tau,v} = \tau(\varphi)$,
- $\|\neg\Phi\|_{\tau,v} = 1 - \|\Phi\|_{\tau,v}$,
- $\|\Phi \rightarrow \Psi\|_{\tau,v} = \|\Phi\|_{\tau,v} \rightarrow \|\Psi\|_{\tau,v} = (1 - \|\Phi\|_{\tau,v} + \|\Psi\|_{\tau,v}) \wedge 1$.

Theorem 4 ([53]) *Let Γ be a finite modal theory of PQ($Ł_n$, Ł), $\Phi \in MF(S)$, then*

$$\Gamma \vdash \Phi \text{ iff } \Gamma \models \Phi.$$

4 Algebraic Axiomatization

Mundici [27] introduced the notion of state on MV-algebras by extending the Kolmogorov axioms of probability theory. The state theory is closely related to the theory of probabilistic truth degree of propositions, and it shows that there is a one-to-one correspondence between state operators on the Łukasiewicz Lindenbaum algebra and the probabilistic truth degree functions in Łukasiewicz propositional logic. In short, the former is a generalized and axiomatized version of the latter, while the latter is a semantically analyzed version. In general, they are different from each other, because, for example, the above correspondence does not hold in $\mathscr{L}^*$. The state theory has received rapid development in recent decades, and many profound results were obtained [28–30, 33, 51]. From the side of generalizing domains, states have been extended to different logical algebras like BL-algebras, MTL-algebras and residuated lattices [31–38]; from the side of generalizing codmains, internal states [40–46] and generalized states [47–51] have also been established. Limited to the length of the paper, this section reviews only the theory of generalized states which has the most general framework.

Throughout this section, let M and L be (bounded, integral and commutative) residuated lattices, and $s : M \to L$ a mapping. Different generalizing approaches produced in all three kinds of generalized states are as follows.

Definition 8 Suppose that $s(0) = 0$, $s(1) = 1$, then the following statements are equivalent:
(i) $\forall x, y \in M, s(d_M(x, y)) = s(x \vee y) \to s(x \wedge y)$;
(ii) $\forall x, y \in M$ with $y \leq x, s(x \to y) = s(x) \to s(y)$;
(iii) $\forall x, y \in M, s(x \to y) = s(x) \to s(x \wedge y)$;
(iv) $\forall x, y \in M, s(x \to y) = s(x \vee y) \to s(y)$.
Such s is called a *generalized Bosbach state of type I*, or *type I state* for short.

Definition 9 Suppose that $s(0) = 0$, $s(1) = 1$, then the following statements are equivalent:
(i) $\forall x, y \in M, s(x \vee y) = s(d_M(x, y)) \to s(x \wedge y)$;
(ii) $\forall x, y \in M, s(x) = s(x \to y) \to s(x \wedge y)$;
(iii) $\forall x, y \in M$ with $y \leq x, s(x) = s(x \to y) \to s(y)$;
(iv) $\forall x, y \in M, s(x \vee y) = s(x \to y) \to s(y)$;
(v) $\forall x, y \in M, s(x \to y) \to s(y) = s(y \to x) \to s(x)$.
Such s is called a *generalized Bosbach state of type II*, or *type II state* for short.

Definition 10 A mapping $m : M \to L$ is called a *generalized Riečan state*, if $\forall x, y \in M$
(i) $m(1) = 1$;
(ii) if $x \perp y$, then $m(x) \perp m(y)$, and $m(x \oplus y) = m(x) \oplus m(y)$.

Example 4 (i) Each residuated lattice homomorphism $s : M \to L$ is a type I state. The identity mapping $\mathrm{id}_M : M \to M$ is a type II state iff M is an MV-algebra.
(ii) Let $L = [0, 1]_{\mathrm{MV}}$, then each Bosbach state on M is both a type I state and a type II state. Conversely all order-preserving type I states and type II states on M reduce to Bosbach states in sense of [32, 35].
(iii) Let M be a Heyting algebra, $a \in M$. Then $s_a : M \to M$, defined by $s_a(x) = a \to x$ for $x \in M$, is an order-preserving type I state on M.
(iv) Let $M = \{0, a, b, c, d, 1\}$, where $0 < b < 1$; $0 < d < c < a < 1$, and binary operations $\otimes$, $\to$ are given by the following tables:

$\to$	0	a	b	c	d	1
0	1	1	1	1	1	1
a	0	1	b	c	c	1
b	c	1	1	c	c	1
c	b	1	b	1	a	1
d	b	1	b	1	1	1
1	0	a	b	c	d	1

$\odot$	0	a	b	c	d	1
0	0	0	0	0	0	0
a	0	a	b	d	d	a
b	0	b	b	0	0	b
c	0	d	0	d	d	c
d	0	d	0	d	d	d
1	0	a	b	c	d	1

Then $(M, \wedge, \vee, \otimes, \rightarrow, 0, 1)$ is a residuated lattice. The endofunctions of M given in the next table are all generalized Riečan states, in which s_1 to s_6 are type I states, and s_3 to s_6 are type II states:

x	0	a	b	c	d	1
$s_1(x)$	0	a	0	1	a	1
$m_1(x)$	0	a	0	1	1	1
$m_2(x)$	0	a	b	c	c	1
$s_2(x)$	0	a	b	c	d	1
$m_3(x)$	0	a	c	b	b	1
$m_4(x)$	0	a	1	0	0	1
$m_5(x)$	0	1	0	1	a	1
$s_3(x)$	0	1	0	1	1	1
$s_4(x)$	0	1	b	c	c	1
$m_6(x)$	0	1	b	c	d	1
$s_5(x)$	0	1	c	b	b	1
$s_6(x)$	0	1	1	0	0	1

In the following propositions we list some basic properties of each type of generalized states.

Proposition 5 *For a type I state $s : M \to L$, we have for all $x, y \in M$,*
(i) $s(\neg x) = \neg s(x)$;
(ii) $s(x \vee y) \to s(x) = s(y) \to s(x \wedge y)$;
(iii) $s((x \to y) \to y) = s(x \to y) \to s(y)$;
(iv) $s((x \to y) \to y) = (s(x \vee y) \to s(y)) \to s(y)$;
(v) $s(x \vee y) \to s(x) \wedge s(y) = s(x) \vee s(y) \to s(x \wedge y)$;
(vi) $s(x) \otimes s(x \to x \otimes y) \leq s(x \otimes y)$.

Proposition 6 *For an order-preserving type I state $s : M \to L$, we have for all $x, y, a, b \in M$,*
(i) $s(x) \otimes s(y) \leq s(x \otimes y)$;
(ii) $s(x) \ominus s(y) \leq s(x \ominus y)$;
(iii) $s(x \to y) \leq s(x) \to s(y)$;
(iv) $s(x \to y) \otimes s(y \to x) \leq d_L(s(x), s(y))$;
(v) $s(d_M(x, y)) \leq d_L(s(x), s(y))$;
(vi) $s(d_M(a, x)) \otimes s(d_M(b, y)) \leq d_L(s(d_M(a, b)), s(d_M(x, y)))$.

Proposition 7 *For a type II state $s : M \to L$, we have for all $x, y \in M$,*
(i) $s(\neg\neg x) = s(x)$;
(ii) $s(\neg\neg x \to x) = 1$;
(iii) $s((x \to y) \to y) = s((y \to x) \to x) = s(x \vee y)$;
(iv) $s(x \to \neg\neg y) = s(x \to y) = s(\neg\neg x \to y)$;
(v) $s(x \to y) = s(\neg y \to \neg x) = s(\neg\neg x \to \neg\neg y)$;
(vi) $s(x \to x \otimes y) = s(\neg x \vee y)$;
(vii) $s(\neg\neg x \wedge y \to z) = s(x \wedge y \to z)$;

(viii) $s((x \to y) \vee (y \to x)) = 1$;
(ix) $s(x \to y \vee z) = s((x \to y) \vee (x \to z))$;
(x) $s(x \wedge y \to z) = s((y \to z) \vee (x \to z))$.

Proposition 8 *For a generalized Riečan state $m : M \to L$, we have for all $x \in M$,*
(i) $m(\neg\neg x) = \neg\neg m(x)$;
(ii) $\neg\neg m(\neg x) = \neg m(x)$;
(iii) $m(\neg x) = \neg m(x)$.

Proposition 9 *Let M be an MV-algebra, L a residuated lattice, $s : M \to L$ a mapping satisfying $s(0) = 1$ and $s(1) = 1$. Then the following two statements are equivalent:*
(i) *s is an order-preserving type I state;*
(ii) *For all $x, y \in M$;*
(a) $s(\neg x) = \neg s(x)$;
(b) $s(x \to y) \to (s(x) \to s(y)) = 1$;
(c) $s(x \oplus y) = (s(x) \to s(x \otimes y)) \to s(y)$.

A mapping $s : M \to L$ satisfying $s(0) = 0$ and $s(1) = 1$ is called a *state-morphism* if it is a residuated lattice homomorphism.

Proposition 10 (i) *Every state morphism is an order-preserving type I state.*
(ii) *For an order-preserving type I state $s : M \to L$, if the quotient $M/Ker(s)$ is a linearly ordered residuated lattice, then s is a state-morphism.*
(iii) *Let L be a linearly ordered residuated lattice, and $s : M \to L$ an order-preserving type I state, then s is a state-morphism iff the quotient $M/Ker(s)$ is linearly ordered.*
(iv) *Let M be an MTL-algebra, L a linearly ordered residuated lattice, and $s : M \to L$ an order-preserving type I state, then s is a state-morphism iff $Ker(s)$ is a prime filter.*

We use the following two theorems to summarize the main results.

Theorem 5 (i) *For a type II state $s : M \to L$, $Ker(s)$ is an MV-filter of M.*
(ii) *Let M be a residuated lattice and L an MV-algebra, then $s : M \to L$ is an order-preserving type I state iff s is a type II state.*
(iii) *Let L be an involutive residuated lattice, and $s : M \to L$ an order-preserving type I state, then the quotient $M/Ker(s)$ is an involutive residuated lattice.*
(iv) *Let M be a divisible residuated lattice, L an involutive residuated lattice, and $s : M \to L$ an order-preserving type I state, then $M/Ker(s)$ is an MV-algebra.*
(v) *Let M be an MTL-algebra, L an MV-algebra, and $s : M \to L$ an order-preserving type I state, then the quotient $M/Ker(s)$ is an MV-algebra.*
(vi) *For an order-preserving type I state $s : M \to L$, s preserves the implication operation whenever it preserves either join or meet operation.*
(vii) *Let M be a divisible residuated lattice, L an MV-algebra, and $s : M \to L$ an order-preserving type I state, then the properties of join-preserving, meet-preserving and implication-preserving of s are mutually equivalent.*

Theorem 6 (i) *Every type II state* $s : M \to L$ *is an order-preserving type I state, but not vice versa.*
(ii) *Every order-preserving type I state* $s : M \to L$ *is a generalized Riečan state, but not vice versa.*
(iii) *Let* M *be a Glivenko residuated lattice, and* L *an involutive residuated lattice, then each generalized Riečan state* $m : M \to L$ *is an order-preserving type I state.*
(iv) *If* M *is an involutive residuated lattice, then each generalized Riečan state* $s : M \to L$ *is an order-preserving type I state.*

5 Applications and Extensions

In this section we will briefly review the applications of probabilistically quantitative logic in the following five aspects. For more details we refer to [48, 57, 62–69].

5.1 Consistency Degrees of Logic Theories

A logic theory is said to be *inconsistent* if a contradiction is a consequence of it. Inconsistency theories have the same deductive closure $F(S)$, while consistent theories have different structures and also different deductive closures. Hence it is an interesting topic to study the extent to which a theory is consistent.

We discuss the theory of consistency degree of theories in the logic $\mathscr{L}^*$ as examples.

Definition 11 Let Γ be a theory, $2^{(\Gamma)}$ the set of all finite subtheories of Γ, $\Sigma = \{\varphi_1, \ldots, \varphi_m\} \in 2^{(\Gamma)}$, $|\Sigma| = m$, $\omega(m) = (k_1, \ldots, k_m) \in \mathbb{N}^m$ with assumption that $\omega(0) = \emptyset$, and $\bar{0}$ a contradiction. Denote

$$\Sigma(\omega(m)) \to \bar{0} = \begin{cases} \varphi_1^{k_1} \& \cdots \& \varphi_m^{k_m} \to \bar{0}, & m > 0, \\ \bar{0}, & m = 0, \end{cases} \tag{8}$$

and define

$$Entail(\Gamma) = \sup\{\tau(\Sigma(\omega(m)) \to \bar{0}) \mid \Sigma \in 2^{(\Gamma)}, |\Sigma| = m, \omega(m) \in \mathbb{N}^m, m \in \mathbb{N}\}, \tag{9}$$

where τ is as defined by (6) and μ is the Lebesgue measure. Then $\mathrm{Entail}(\Gamma)$ is called the *entailment degree of* Γ.

Example 5 Find $\mathrm{Entail}(\Gamma)$ in $\mathscr{L}^*$, where (i) $\Gamma = \emptyset$, (ii) $\Gamma = \{p\}$, (iii) $\Gamma = \{p \wedge \neg p\}$, (iv) $\Gamma = \{p \to \neg p\}$.

Solution
(i) Let $\Gamma = \emptyset$, then $\forall \Sigma \in 2^{(\Gamma)}$, $\Sigma = \emptyset$, and thus $\Sigma(\omega(m)) \to \bar{0} = \bar{0}$. Therefore, $\tau(\Sigma(\omega(m)) \to \bar{0}) = 0$, $\text{Entail}(\Gamma) = 0$.
(ii) $\tau(p^2 \to \bar{0}) = 1 - \tau(p^2) = 1 - \int_0^1 x^2 \mathrm{d}x = 1 - \int_{\frac{1}{2}}^1 x\mathrm{d}x = \frac{5}{8}$, and hence $\text{Entail}(\Gamma) = \frac{5}{8}$.
(iii) $\tau((p \wedge \neg p)^2 \to \bar{0}) = 1 - \int_0^1 (x \wedge (1-x))^2 \mathrm{d}x = 1$, and so $\text{Entail}(\Gamma) = 1$.
(iv) $\tau((p \to \neg p)^2 \to \bar{0}) = 1 - \int_0^1 (x \to 1-x)^2 \mathrm{d}x = 1 - \int_0^{\frac{1}{2}} \mathrm{d}x - \int_{\frac{1}{2}}^1 (1-x)^2 \mathrm{d}x = 1 - \frac{1}{2} - 0 = \frac{1}{2}$, we have $\text{Entail}(\Gamma) = \frac{1}{2}$.

Definition 12 For a theory Γ, define

$$Consist(\Gamma) = 1 - \frac{1}{2} Entail(\Gamma)(1 + i(\Gamma)), \tag{10}$$

where $i(\Gamma) = \max\{[1 - \sup_{v \in \Omega} v(\varphi)] \mid \varphi \in D(\Gamma)\}$. Then $\text{Consist}(\Gamma)$ is called the *consistency degree* of Γ.

Theorem 7 *For a theory Γ, we have:*
(i) *Γ is inconsistent iff $Consist(\Gamma) = 0$;*
(ii) *Γ is consistent iff $\frac{1}{2} \le Consist(\Gamma) \le 1$;*
(iii) *Γ is consistent and $Entail(\Gamma) = 1$ iff $Consist(\Gamma) = \frac{1}{2}$;*
(iv) *Γ is quasi-consistent iff $Consist(\Gamma) = 1$.*

5.2 *Three Methods of Graded Reasoning*

The idea behind Eqs. (8) and (9) suggests a method for graded reasoning by replacing contradiction with a general proposition.

Definition 13 Let Γ be a theory, $2^{(\Gamma)}$ the set of all finite subtheories of Γ, $\Sigma = \{\varphi_1, \ldots, \varphi_m\} \in 2^{(\Gamma)}$, $|\Sigma| = m$, $\omega(m) = (k_1, \ldots, k_m) \in \mathbb{N}^m$, $m \in \mathbb{N}$ with assumption that $\omega(0) = \emptyset$, and $\varphi \in F(S)$. Put

$$\Sigma(\omega(m)) \to \varphi = \begin{cases} \varphi_1^{k_1} \& \cdots \& \varphi_m^{k_m} \to \varphi, & m > 0, \\ \varphi, & m = 0, \end{cases}$$

and define

$$Entail(\Gamma, \varphi) = \sup\{\tau(\Sigma(\omega(m)) \to \varphi) \mid \Sigma \in 2^{(\Gamma)}, |\Sigma| = m, \omega(m) \in \mathbb{N}^m, m \in \mathbb{N}\}.$$

Then *Entail*(Γ, φ) is called the *entailment degree* of Γ w.r.t. φ.

Using the notion of Hausdorff metric one can construct the second method of graded reasoning.

Definition 14 Let Γ be a theory, $\varphi \in F(S)$, and define

$$H(\Gamma, \varphi) = 1 - \inf\{H(D(\Gamma), D(\Sigma)) \mid \Sigma \subseteq F(S), \Sigma \vdash \varphi\},$$

where H is the Hausdorff metric [6] on $\mathscr{P}(F(S)) - \{\emptyset\}$. Then φ is called a *consequence of Γ in the degree $H(\Gamma, \varphi)$*.

Thirdly, the pseudo metric on $F(S)$ defined by $\rho(\varphi, \psi) = 1 - \tau((\varphi \to \psi) \wedge (\psi \to \varphi))$ provides us one more method for graded reasoning. Let Γ be a theory, $\varphi \in F(S)$, then $\rho(\varphi, D(\Gamma)) = \inf\{\rho(\varphi, \psi) \mid \psi \in D(\Gamma)\}$ represents the distance of φ to $D(\Gamma)$, and so $1 - \rho(\varphi, D(\Gamma))$ measures the extent to which φ is a consequence of Γ.

Theorem 8 *Let Γ be a theory, $\varphi \in F(S)$, then*

$$H(\Gamma, \varphi) \leq \mathit{Entail}(\Gamma, \varphi) = 1 - \rho(\varphi, D(\Gamma)).$$

5.3 *Characterizations of Maximally Consistent Theories*

The indices introduced in Sect. 5.1 as well as those in the literature cannot distinguish different maximally consistent theories, so we need to find other ways to study maximally consistent theories. Luckily, we can get clear descriptions of the structures of maximally consistent theories in several logic systems such as $Ł_n$, Ł, $\mathscr{L}^*$ and NMG-logic [57]. Here we review only the characterizations of maximally consistent theories in $\mathscr{L}^*$.

Theorem 9 *Let $Q = \{(\alpha_1, \alpha_2, \ldots) \mid \alpha_m \in \{0, \frac{1}{2}, 1\}, m \in \mathbb{N}\}$. $\forall \alpha = (\alpha_1, \alpha_2, \ldots) \in Q$, define $S(\alpha) = \{\varphi_1, \varphi_2, \ldots\}$, where φ_m satisfies:*

$$\varphi_m = \begin{cases} p_m, & \alpha_m = 1, \\ (\neg p_m^2)\&(\neg(\neg p_m)^2), & \alpha_m = \frac{1}{2}, \\ \neg p_m, & \alpha_m = 0, \end{cases}$$

$m = 1, 2, \ldots$. *Then $M = \{D(S(\alpha)) \mid \alpha \in Q\}$ is the set of all maximally consistent theories in $\mathscr{L}^*$.*

We can further introduce Stone topology $\mathscr{T}$ on M, and prove that $(M, \mathscr{T})$ is a Cantor space, of which topologically closed subsets are one-to-one corresponding to Łukasiewicz theories [67, 68]. These results are also special cases of the results in R_0-algebras we will review in the following subsection.

5.4 Three-Valued Stone Representations in R_0-Algebras

The results about structural and topological characterizations of maximally consistent theories in $\mathscr{L}^*$ can be further extended to R_0-algebras. Throughout this subsection, let M be an R_0-algebra.

Theorem 10 *For a filter F of M, the following are equivalent:*
(i) *F is maximal;*
(ii) *$\forall x \in M$, exactly one of $x \in F$ and $\neg x^2 \in F$ holds;*
(iii) *$\forall x \in M$, exactly one of $x \in F$, $\neg x \in F$ and $(\neg x^2) \otimes (\neg(\neg x)^2) \in F$ holds.*

Theorem 11 *The mapping $F \mapsto h$ is a one-to-one correspondence between maximal filters F and R_0-homomorphisms $h : M \to W_3$, where*

$$h(x) = \begin{cases} 1, & x \in F, \\ \frac{1}{2}, & (\neg x^2) \otimes (\neg(\neg x)^2) \in F, \\ 0, & \neg x \in F. \end{cases}$$

Denote

$$\mathrm{Max}(M) = \{F \mid F \text{ is a maximal filter of } M\}.$$

Theorem 12 *Define $s : M \to \{0, \frac{1}{2}, 1\}^{Max(M)}$ as follows:*

$$s(x)(F) = F(x), \ x \in M.$$

and let

$$\mathscr{S} = \{s(x) \mid x \in M\}.$$

Then
(i) *$\mathscr{S}$ forms a subbasis for some three-valued topology, denoted by δ, on $Max(M)$;*
(ii) *$(Max(M), \delta)$ is zero-dimensional and covering compact, and $(M, \iota_{0.5}(\delta))$ is a Stone space.*

Theorem 13 *Let $MV(M)$ be the MV-skeleton of M, $B(M)$ the Boolean skeleton of M, $(Max(M), \delta)$ the three-valued Stone space of M, and $(Max(M), \iota_{0.5}(\delta))$ the Stone space of M. Then*
(i) *The set $Clop(M, \delta)$ of all clopen subsets of $(Max(M), \delta)$ forms an MV-algebra under pointwise operations, and is isomorphic to $MV(M)$;*
(ii) *The set $Clop(M, \mathscr{T})$ of all clopen subsets of $(Max(M), \iota_{0.5}(\delta))$ forms a Boolean algebra under the inclusion order, and is isomorphic to $B(M)$;*
(iii) *The mapping $F \mapsto C(F) = \{F^* \in Max(M) \mid F^* \supseteq F\}$ is a one-to-one correspondence between MV-filters F of M and topologically closed subsets $C(F)$ of $(Max(M), \iota_{0.5}(\delta))$.*

5.5 Similarity Cauchy Completion of Residuated Lattices W.r.t. Order-Preserving Type I States

Let M and L be residuated lattices as in Sect. 4. An order-preserving type I state $s : M \to L$ induces a similarity relation $\rho_s(x, y) = s(d_M(x, y))$ on M, which together with the biresiduation d_L of L can define ρ_s-similarity convergence on M, and then the problem of corresponding Cauchy completion will arise.

Definition 15 (i) We say that a sequence $\{x_n\} \subseteq L$ *d-converges* at $x \in L$, denoted by $x_n \overset{d}{\to} x$ or $\lim\limits_{n\to\infty} x_n = x$, if there exists a sequence $\{z_n\} \subseteq L$ such that $\{z_n\} \uparrow 1$, and $\forall n \in \mathbb{N}$, $z_n \leq d_L(x_n, x)$. In this case x is called a *limit* of $\{x_n\}$.

(ii) We say that a sequence $\{x_n\}$ of M *ρ_s-converges* at x, denoted by $x_n \overset{\rho_s}{\to} x$, if $\lim\limits_{n\to\infty} \rho_s(x_n, x) = 1$.

Theorem 14 *Every order-preserving type I state $s : M \to L$ is ρ_s-continuous, i.e., $\lim\limits_{n\to\infty} s(x_n) = s(x)$ whenever $x_n \overset{\rho_s}{\to} x$.*

Let M and L be residuated lattices such that L is d-Cauchy complete, $s : M \to L$ an order-preserving type I state, and $\rho_s = s \circ d_M$. In the following we construct the s-Cauchy completion of M.

Denote by $\mathscr{C}_s(M)$ the set of all ρ_s-Cauchy sequences of M, and define on $\mathscr{C}_s(M)$ the operations $\circ \in \{\wedge, \vee, \otimes, \to, \leftrightarrow\}$ as follows:

$$\underline{x} \circ \underline{y} = \{x_n \circ y_n\},$$

where $\underline{x} = \{x_n\}$, $\underline{y} = \{y_n\} \in \mathscr{C}_s(M)$, then $(\mathscr{C}_s(M), \wedge, \vee, \otimes, \to, \underline{0}, \underline{1})$ forms a residuated lattice, where $\underline{0} = \{0\}_{n\in\mathbb{N}}$, $\underline{1} = \{1\}_{n\in\mathbb{N}}$ are constant sequences.

Define on $\mathscr{C}_s(M)$ a binary relation $\sim$ by

$$\underline{x} \sim \underline{y} \text{ iff } \lim_{n,m\to\infty} \rho_s(x_n, y_n) = 1,$$

where $\underline{x} = \{x_n\}$, $\underline{y} = \{y_n\} \in \mathscr{C}_s(M)$, then $\sim$ is a congruence relation. Let $\widetilde{M}_s = \mathscr{C}_s(M)/\sim$, denote by $\widetilde{\underline{x}}$ the class of $\underline{x} = \{x_n\}$ under $\sim$, and define

$$\widetilde{\underline{x}} \circ \widetilde{\underline{y}} = \widetilde{\underline{x} \circ \underline{y}}$$

$\underline{x} = \{x_n\}$, $\underline{y} = \{y_n\} \in \mathscr{C}_s(M)$, $\circ \in \{\wedge, \vee, \otimes, \to, \leftrightarrow\}$, then $(\widetilde{M}_s, \wedge, \vee, \otimes, \to, \widetilde{\underline{0}}, \widetilde{\underline{1}})$ is also a residuated lattice.

$\forall x \in M$, denote $\underline{x} = \{x\}_{n\in\mathbb{N}}$ being a constant consequence, then $\underline{x} \in \mathscr{C}_s(M)$. Define $\varphi_s : M \to \widetilde{M}_s$ by

$$\varphi_s(x) = \widetilde{\underline{x}}, \ x \in M.$$

Then φ_s is a residuated lattice homomorphism.

Theorem 15 *Let M and L be residuated lattices with L d-Cauchy complete, and $s : M \to L$ an order-preserving type I state, then:*
(i) $\widetilde{M}_s$ is a residuated lattice, and is involutive if L is;
(ii) $\widetilde{s}$ is a faithful order-preserving type I state;
(iii) φ_s is a residuated lattice homomorphism, and $\widetilde{s} \circ \varphi_s = s$;
(iv) φ_s is injective iff s is faithful;
(v) $\widetilde{\rho}_s = \rho_{\widetilde{s}}$;
(vi) given $\{x_n\} \subseteq M$ and $x \in M$, $\varphi_s(x_n) \xrightarrow{\widetilde{\rho}_s} \varphi_s(x)$ whenever $x_n \xrightarrow{\rho_s} x$;
(vii) for any residuated lattice C, any faithful order-preserving type I state $m : C \to L$ such that C is ρ_m-Cauchy complete, and any residuated lattice homomorphism $f : M \to C$ such that $m \circ f = s$, then there exists a unique residuated lattice homomorphism $\widetilde{f} : \widetilde{M}_s \to C$ such that $m \circ \widetilde{f} = \widetilde{s}$ and $\widetilde{f} \circ \varphi_s = f$.

$\widetilde{M}_s$ is called the *s-Cauchy similarity completion* of M.

Lastly let us consider the case when $L = [0, 1]_{\mathrm{MV}}$. Then each order-preserving type I state $s : M \to L$ reduces to a Bosbach state on M [32, 33, 35]. Put $\delta_s : M^2 \to [0, 1]$ by $\delta_s(x, y) = 1 - \rho_s(x, y)$, then δ_s is a pseudo-metric on M, and δ_s is a metric iff s is faithful, hence (M, δ_s) is a pseudo-metric space. Following the standard construction procedure of metric completion of a metric space one can construct the metric completion of (M, δ_s). Here we are interested in the relationship between $\widetilde{M}_s$ and the metric completion of M constructed in the standard way. A routine verification shows that the $\widetilde{M}_s$ is actually the metric completion of (M, δ_s), where the metric $\widetilde{\delta}_s$ on $\widetilde{M}_s$ is defined by $\widetilde{\delta}_s(\widetilde{\underline{x}}, \widetilde{\underline{y}}) = \lim_{n\to\infty} \delta_s(x_n, y_n) = 1 - \lim_{n\to\infty} \rho_s(x_n, y_n) = 1 - \widetilde{\rho}_s(\widetilde{\underline{x}}, \widetilde{\underline{y}})$.

6 Further Studies

(i) Further improvements of probabilistically quantitative logic, such as topological structures of logic metric spaces, syntactic version of semantic quantification and its completeness.
(ii) Further studies on states, internal states and generalized states on general residuated lattices (not necessarily bounded, or not integral, or not commutative).
(iii) Study on logic foundation of generalized states by Hájek's modal formalization, in which semantic theory and its completeness are challenging work.
(iv) Structural characterizations of nuclei on Glivenko logical algebras, like MV-algebras, Heyting algebras, BL-algebras, MTL-algebras.
(v) This paper does not refer to quantification of first-order predicate logic and of its fragments [70–74]. There are wide research spaces waiting in this line.
(vi) ...

Acknowledgments This work was partially supported by the National Natural Science Foundation of China (61473336), the Youth Science and Technology Program of Shaanxi Province (2016KJXX-24) and the Fundamental Research Funds for the Central Universities (201403001).

References

1. Halpern, J.Y.: Reasoning about Uncertainty. The MIT Press, London (2003)
2. Lu, R.Q.: Artificial Intelligence (in Chinese). Science Press, Beijing (2003)
3. Shi, Z.Z., Wang, W.J.: Artificial Intelligence (in Chinese). Science Press, Beijing (2007)
4. Liu, B.D.: Uncertainty Theory. Springer, Berlin (2011)
5. Xu, L.Z.: Handbook of Modern Mathematics (in Chinese). Huazhong University of Science and Technology Press, Wuhan (2011)
6. Wang, G.J.: Computational Intelligence-Word Computing and Fuzzy Sets (in Chinese). Higer Education Press, Beijing (2005)
7. Kolmogorov, A.N.: Foundations of the Theory of Probability. Chelsea Publishing Company, New York (1956)
8. Łukasiewicz, J.: On three-valued logic. Ruch Filozoficzny. **5**, 170–171 (1920)
9. Zadeh, L.A.: Fuzzy sets. Inf. Control **8**(3), 338–353 (1965)
10. Bede, B.: Mathematics of Fuzzy Sets and Fuzzy Logic. Springer, New York (2013)
11. Shafer, G.: A Mathematical Theory of Evidence. Princeton University Press, Princeton (1976)
12. Pawlak, Z.: Rough Sets: Theoretical Aspects of Reasoning about Data. Kluwer, Boston (1991)
13. Buckley, J.J.: Fuzzy Probability and Statistics. Springer, New York (2006)
14. Ross, T.J., Booker, J.M., Parkinson, W.J.: Fuzzy Logic and Probability Applications: Building the Gap. SIAM, Philadelphia, ASA, Alexandria, VA (2002)
15. Ramsey, F.P.: Foundations of Mathematics and Other Logical Essays. Harcout Brace and Company, New York (1931)
16. Adam, E.W.: A Primer of Probability Logic. CSLI Publications, Stanford (1998)
17. Hailperin, T.: Sentential Probabilistic Logic. Associated University Press, London (1996)
18. Wang, G.J., Zhou, H.J.: Quantitative logic. Inf. Sci. **179**(3), 226–247 (2009)
19. Wang, G.J., Zhou, H.J.: Introduction to Mathematical Logic and Resolution Principle. Science Press, Beijing (2009)
20. Wang, G.J., Wang, W.: Logic metric space (in Chinese). Acta Math. Sinica **44**(1), 159–168 (2001)
21. Wang, G.J., Fu, L.: Theory of truth degrees of propositions in two-valued logic. Sci. China A **31**(11), 998–1008 (2001)
22. Wang, G.J., Leung, Y.: Integrated semantics and logic metric spaces. Fuzzy Sets Syst. **136**, 71–91 (2003)
23. Wang, G.J., Li, B.J.: Theory of truth degrees of formulas in Łukasiewicz n-valued proportional logic and a limit theorem. Sci. China E **35**(6), 561–569 (2005)
24. Zhou, H.J., Wang, G.J., Zhou, W.: Consistency degrees of theories and methods of graded reasoning in n-valued R_0-logic (NM-logic). Int. J. Approx. Reas. **43**, 117–132 (2006)
25. Wang, G.J., Hui, X.J.: Randomization of classical inference patterns and its application. Sci. China Ser. F. **50**(6), 867–877 (2007)
26. Wu, H.B.: The generalized truth degree of quantitative logic in the logic system $\mathscr{L}_n^*$ (n-valued NM-logic system). Comput. Math. Appl. **59**(8), 2587–2596 (2010)
27. Mundici, D.: Averaging the truth-value in Łukasiewicz sentential logic. Studia Logica **55**(1), 113–127 (1995)
28. Kroupa, T.: Every state on semisimple MV-algebras is integral. Fuzzy Sets Syst. **157**(20), 2771–2782 (2006)
29. Panti, G.: Invariant measures in free MV-algebras. Commun. Algebra **36**(8), 2849–2861 (2008)
30. Mundici, D.: Advanced Łukasiewicz Calculus and MV-algebras. Springer, New York (2011)
31. Riečan, B.: On the probability on BL-algebras. Acta Mathematica Nitra **4**, 3–13 (2000)
32. Georgescu, G.: Bosbach states on fuzzy structures. Soft Comput. **8**(3), 217–230 (2004)
33. Dvurečenskij, A., Rachůnek, J.: Probabilistic averaging in bounded commutative residuated ℓ-monoids. Discret. Math. **306**(13), 1317–1326 (2006)
34. Liu, L.Z.: States on finite monoidal t-norm based algebras. Inf. Sci. **181**(7), 1369–1383 (2011)
35. Ciungu, L.C.: Bosbach and Riečan states on residuated lattices. J. Appl. Funct. Anal. **3**(2), 175–188 (2008)

36. Turunen, E., Mertanen, J.: States on semi-divisible residuated lattices. Soft Comput. **12**(4), 353–357 (2008)
37. Mertanen, J., Turunen, E.: States on semi-divisible generalized residuated lattices reduce to states on MV-algebras. Fuzzy Sets Syst. **159**(22), 3051–3064 (2008)
38. Buşneag, C.: States on Hilbert algebras. Studia Logica **94**(2), 177–188 (2010)
39. Blok, W., Pigozzi, D.: Algebraizable logics. Memoirs Am. Math. Soc. **77** (1989)
40. Flaminio, T., Montagna, F.: MV-algebras with internal states and probabilistic fuzzy logic. Int. J. Approx. Reas. **50**(1), 138–152 (2009)
41. Di Nola, A., Dvurečenskij, A., Lettieri, A.: On varieties of MV-algebras with internal states. Int. J. Approx. Reas. **51**(6), 680–694 (2010)
42. Di Nola, A., Dvurečenskij, A.: State-morphism MV-algebras. Ann. Pure Appl. Logic **161**(2), 161–173 (2009)
43. Ciungu, L.C., Dvurečenskij, A., Hyčko, M.: State BL-algebra. Soft Comput. **15**(4), 619–634 (2010)
44. Dvurečenskij, A., Rachůnek, J., Šalounová, D.: State operators on generalizations of fuzzy structures. Fuzzy Sets Syst. **187**(1), 58–76 (2012)
45. Botur, M., Dvurečenskij, A.: State-morphism algebras-general approach. Fuzzy Sets Syst. **218**, 90–102 (2013)
46. Borzooei, R.A., Dvurečenskij, A., Zahiri, O.: State BCK-algebras and state-morphism BCK-algebras. Fuzzy Sets Syst. **244**, 86–105 (2014)
47. Georgescu, G., Mureşan, C.: Generalized Bosbach states. http://ariv.org/abs/1007.2575 (2010)
48. Zhou, H.J., Zhao, B.: Generalized Bosbach and Riečan states based on relative negations in residuated lattices. Fuzzy Sets Syst. **187**(1), 33–57 (2012)
49. Zhao, B., Zhou, H.J.: Generalized Bosbach and Riečan states on nucleus-based-Glivenko residuated lattices. Archive Math. Logic. **52**(7–8), 689–706 (2013)
50. Ciungu L.C., Georgescu G., Mureşan C.: Generalized Bosbach states: part I, II. Archive Math. Logic **52**(3–4), 335–376; 52(7–8), 707–732 (2013)
51. Ciungu, L.C.: Non-commutative Multiple-Valued Logic Algebras. Springer, New York (2014)
52. Hájek, P.: Metamathematics of Fuzzy Logic. Kluwer, Dordrecht (1998)
53. Flaminio, T., Godo, L.: A logic for reasoning about the probability of fuzzy events. Fuzzy Sets Syst. **158**(6), 625–638 (2007)
54. Zhou, H.J., Wang, G.J.: Borel probabilistic and quantitative logic. Sci. China Inf. Sci. **54**(9), 1843–1854 (2011)
55. Zhou, H.J.: Theory of Borel probabilistic truth degrees of propositions in Łukasiewicz propositional logics and a limit theorem (in Chinese). J. Softw. **23**(9), 2235–2247 (2012)
56. Zhou, H.J., She, Y.H.: Theory of Choquet integral truth degrees of propositions in Łukasiewicz propositional logic (in Chinese). Acta Electronica Sinica **41**(12), 2327–2333 (2013)
57. Zhou, H.J.: Probabilistically Quantitative Logic with its Applications (in Chinese). Science Press, Beijing (2015)
58. Gottwald, S.: A Treatise on Many-Valued Logics. Research Studies Press, Baldock (2001)
59. Wang, G.J.: Non-classical Mathematical Logic and Approximate Reasoning (in Chinese). Sience Press, Beijing (2000)
60. Aguzzoli, S., Gerla, B.: Probability measures in the logic of nilpotent minimum. Studia Logica **94**(2), 151–176 (2010)
61. Deneberg, D.: Non-additive Measure and Integral. Kluwer, Dordrecht (1994)
62. Wang, G.J., Zhang, W.X.: Consistency degrees of finite theories in Łukasiewicz propositional logic. Fuzzy Sets Syst. **149**(2), 275–284 (2005)
63. Zhou, X.N., Wang, G.J.: Consistency degrees of theories in some systems of propositional fuzzy logic. Fuzzy Sets Syst. **152**(3), 321–331 (2005)
64. Wang, G.J.: Comparison of deduction theorems in diverse logic systems. New Math. Natl. Comput. **1**(1), 65–77 (2005)
65. Zhou, H.J., Wang, G.J.: A new theory consistency index based on deduction theorems in several logic systems. Fuzzy Sets Syst. **157**(3), 427–443 (2006)

66. Zhou, H.J., Wang, G.J.: Generalized consistency degrees of theories w. r. t. formulas in several standard complete logic systems. Fuzzy Sets Syst. **157**(15), 2058–2073 (2006)
67. Zhou, H.J., Wang, G.J.: Characterizations of maximal consistent theories in the formal deductive system $\mathscr{L}^*$ (NM-logic) and Cantor Space. Fuzzy Sets Syst. **158**(23), 2591–2604 (2007)
68. Zhou, H.J., Wang, G.J.: Three and two-valued Łukasiewicz theories in the formal deductive system $\mathscr{L}^*$ (NM-logic). Fuzzy Sets Syst. **159**(22), 2970–2982 (2008)
69. Zhou, H.J., Zhao, B.: Stone-like representation theorems and three-valued filters in R_0-algebras (nilpotent minimum algebras). Fuzzy Sets Syst. **162**(1), 1–26 (2011)
70. Halpern, J.Y.: An analysis of first-order logics of probability. Artif. Intell. **46**(3), 311–350 (1990)
71. Lynch, J.: Probabilities of first-order sentences about unary functions. Trans. Am. Math. Soc. **287**(2), 543–568 (1985)
72. Wang, G.J., Qin, X.Y., Zhou, X.N.: An intrinsic fuzzy set on the universe of discourse of predicate formulas. Fuzzy Sets Syst. **157**(24), 3145–3158 (2006)
73. Wang, G.J.: Axiomatic theory of truth degree for a class of first-order formulas and its applications (in Chinese). Sci. China Inf. Sci. **42**(5), 648–662 (2012)
74. Wang, G.J., Duan, Q.L.: Theory of (n) truth degrees of formulas in modal logic and a consistency theorem. Sci. China F. **52**(1), 70–83 (2009)

Part II
Quantitative Logic and Uncertainty Logic

A Quantitative Approach for Linear Temporal Logic

Hui-Xian Shi

Abstract The present paper aims to construct a quantitative approach for Linear Temporal Logic. Based on a certain kind of probabilistic measure with respect to the Kripke structure DTMC, we define the satisfaction degrees for LTL formulae, as a quantitative notion extending the classical case in model checking. Meanwhile, the concept of similarity degree between LTL formulae is presented, and a corresponding pseudo-metric on the set of all LTL formulae is induced, which enables the LTL logic metric space constructible.

Keywords Linear temporal logic · DTMC · Satisfaction degree · Quantitative logic · Logic metric space

1 Introduction

Quantitative Logic [1, 2] focuses on the combination of formal inference and numerical calculation methods in the area of mathematical logic theory. It was initially formed mainly in propositional logics by grading basic logical notions and constructing truth degree theory for formulae. Later on, such research rapidly developed, not only limited within the scope of propositional logics, but also covering more kinds of expressive logics, such as predicate logics, modal logics as well as temporal logics [3–5].

Temporal logics have been studied in ancient times in different areas such as philosophy. Their application to verifying complex computer systems was proposed by Pnueli and greatly developed recently [6–8]. As an expansion of propositional logics, temporal logics possess stronger expression ability resulting from various possible combination of distinguished types of modalities.

H.-X. Shi (✉)
School of Mathematics and Information Science, Shaanxi Normal University,
Xi'an 710062, China
e-mail: rubyshi@163.com

T.-H. Fan et al. (eds.), *Quantitative Logic and Soft Computing 2016*,
Advances in Intelligent Systems and Computing 510,
DOI 10.1007/978-3-319-46206-6_6

In this paper, we will focus our attention on Linear Temporal Logic (LTL for short), a temporal logic that is based on a linear-time perspective [9–11], and aim to construct a quantitative approach for LTL. Based on a certain kind of probabilistic measure with respect to the Kripke structure DTMC, we define the satisfaction degrees for LTL formulae, as a quantitative notion extending the classical case in model checking. Meanwhile, the concept of similarity degree between LTL formulae is presented, and a corresponding pseudo-metric on the set of all LTL formulae is induced, which enables the LTL logic metric space constructible.

2 Preliminaries

In this section, we briefly overview some preliminaries that we need. For more details, please see [9–11].

Definition 1 A DTMC is a tuple $\mathcal{D}= (S, \mathbf{P}, \ell_{\text{init}}, AP, L)$, where S is a nonempty set of states; $\mathbf{P} : S \times S \longrightarrow [0, 1]$ is a *transition mapping* satisfying

$$\sum_{s' \in S} \mathbf{P}(s, s') = 1 \tag{1}$$

for every $s \in S$; $\ell_{\text{init}} : S \longrightarrow [0, 1]$ is an *initial mapping* satisfying

$$\sum_{s \in S} \ell_{\text{init}}(s) = 1; \tag{2}$$

AP is a set of atomic propositions; and $L : S \rightarrow 2^{AP}$ is a *labeling function*. We call $\mathcal{D}$ a *finite* DTMC whenever S and AP are finite sets.

For a finite DTMC $\mathcal{D}$ with $S = \{s_1, s_2, \ldots, s_n\}$, the transition mapping $\mathbf{P} : S \times S \longrightarrow [0, 1]$ can be viewed as a matrix $(\mathbf{P}_{ij})_{n\times n}$ where $\mathbf{P}_{ij}$=$\mathbf{P}(s_i, s_j)$. In this case, we also call $\mathbf{P}$ a *transition matrix*. Similarly, the initial mapping $\ell_{\text{init}} : S \longrightarrow [0, 1]$ can be viewed as an n-dimensional vector $(\ell_{\text{init}}(s))_{s\in S}$.

In the following, we only consider finite DTMCs.

Definition 2 Let $\mathcal{D}= (S, \mathbf{P}, \ell_{\text{init}}, AP, L)$ be a DTMC, and $\pi = s_0 s_1 s_2 \ldots$ be a non-empty sequence of states in $\mathcal{D}$, where $\mathbf{P}(s_i, s_{i+1}) > 0$ holds for every $i \in \mathbf{N}$. We call π a *path* in $\mathcal{D}$, and denote $\pi[i] = s_i$, $\pi[i...] = s_i s_{i+1} s_{i+2} \ldots$ $(i \in \mathbf{N})$.

For convenience, we write π_{fin} when π is a finite path, and we denote the set of all infinite paths initialed at s as $\text{Paths}^{\mathcal{D}}(s)$, also $\text{Paths}^{\mathcal{D}}_{\text{fin}}(s)$ in the finite case. Similarly, $\text{Paths}(\mathcal{D})$ represents the set of all infinite paths initialed at some state in I, where

$$I = \{s \in S \mid \ell_{\text{init}}(s) > 0\}. \tag{3}$$

Definition 3 Let $\pi = s_0 s_1 s_2 \ldots$ be a path in the DTMC $\mathcal{D}$. Then trace $(\pi) = L(s_0)L(s_1)L(s_2)\ldots \in (2^{AP})^\omega$ is called the *trace* of π, and

$$\text{Traces } (\mathcal{D}) = \{\text{trace } (\pi) | \pi \in \text{Paths}(\mathcal{D})\}. \tag{4}$$

For a sequence $\sigma = A_0 A_1 A_2 \ldots \in (2^{AP})^\omega$, we can similarly define $\sigma[i]$, $\sigma[i...]$ ($i \in \mathbf{N}$) as in Definition 2.

Definition 4 LTL formulae can be constructed in BNF as follows.

$$\Phi := \text{true} \mid a \mid \neg\varphi \mid \varphi_1 \wedge \varphi_2 \mid \bigcirc\varphi \mid \varphi_1 \sqcup \varphi_2, \quad a \in AP.$$

In addition, false $= \neg$true, $\varphi_1 \vee \varphi_2 = \neg(\neg\varphi_1 \wedge \neg\varphi_2)$, $\varphi_1 \to \varphi_2 = \neg\varphi_1 \vee \varphi_2$, $\Diamond\varphi =$ true $\sqcup\, \varphi$, $\Box\varphi = \neg\Diamond\neg\varphi$, $\bigcirc^k\varphi = \bigcirc^{k-1}(\bigcirc\varphi)$ $(k \geq 1)$, and $\bigcirc^0\varphi = \varphi$.

Definition 5 Let φ be an LTL formula. Then

$$\text{Words } (\varphi) = \{\sigma \in (2^{AP})^\omega | \sigma \models \varphi\} \tag{5}$$

is called the *satisfaction set* of φ, where $\sigma \models \varphi$ is defined by induction as follows:

(i) $\sigma \models$ true always holds;
(ii) $\sigma \models a$ if and only if $a \in \sigma[0]$, where $a \in AP$;
(iii) $\sigma \models \neg\varphi$ if and only if $\sigma \models \varphi$ does not hold;
(iv) $\sigma \models \varphi_1 \wedge \varphi_2$ if and only if $\sigma \models \varphi_1$ and $\sigma \models \varphi_2$;
(v) $\sigma \models \bigcirc\varphi$ if and only if $\sigma[1...] \models \varphi$; and
(vi) $\sigma \models \varphi_1 \sqcup \varphi_2$ if and only if $\exists j \in \mathbf{N}$ such that $\sigma[j...] \models \varphi_2$ and $\sigma[i...] \models \varphi_1$ holds for every i with $0 \leq i < j$.

Definition 6 Let $\pi = s_0 s_1 s_2 \ldots$ be a path in the DTMC $\mathcal{D}$. Define

$$\pi \models \varphi \text{ if and only if trace } (\pi) \models \varphi, \tag{6}$$

$$\mathcal{D} \models \varphi \text{ if and only if Traces } (\mathcal{D}) \subseteq \text{Words } (\varphi). \tag{7}$$

Corollary 1 *Let $\pi = s_0 s_1 s_2 \ldots$ be a path in the DTMC $\mathcal{D}$, and $\varphi, \varphi_1, \varphi_2$ be LTL formulae. Then*

(i) $\pi \models$ true always holds, whereas $\pi \models$ false never;
(ii) $\pi \models a$ if and only if $a \in L(s_0)$, where $a \in AP$;
(iii) $\pi \models \neg\varphi$ if and only if $\pi \models \varphi$ does not hold;
(iv) $\pi \models \varphi_1 \wedge \varphi_2$ if and only if $\pi \models \varphi_1$ and $\pi \models \varphi_2$; $\pi \models \varphi_1 \vee \varphi_2$ if and only if $\pi \models \varphi_1$ or $\pi \models \varphi_2$; $\pi \models \varphi_1 \to \varphi_2$ if and only if $\pi \models \varphi_1$ does not hold or $\pi \models \varphi_2$;
(v) $\pi \models \bigcirc\varphi$ if and only if $\pi[1...] \models \varphi$;
(vi) $\pi \models \varphi_1 \sqcup \varphi_2$ if and only if $\exists j \in \mathbf{N}$ such that $\pi[j...] \models \varphi_2$ and $\pi[i...] \models \varphi_1$ holds for every i with $0 \leq i < j$; and

(vii) $\pi \models \Diamond\varphi$ *if and only if* $\exists j \in \boldsymbol{N}$ *such that* $\pi[j...] \models \varphi$; $\pi \models \Box\varphi$ *if and only if* $\pi[i...] \models \varphi$ *holds for every* $i \in \boldsymbol{N}$.

Definition 7 Let φ, ψ be LTL formulae. If for every DTMC $\mathcal{D}$, $\mathcal{D} \models \varphi$ if and only if $\mathcal{D} \models \psi$, then we say that φ and ψ are *equivalent*, denoted as $\varphi \equiv \psi$.

An LTL formula φ is called a *tautology* if $\mathcal{D} \models \varphi$ holds for every DTMC $\mathcal{D}$, whereas φ is called a *contradiction* whenever $\neg\varphi$ is a tautology.

Corollary 2 *Let* φ *be an LTL formulae. Then*

(i) φ *is a tautology if and only if* $\neg\varphi$ *is a contradiction;*
(ii) φ *is a tautology if and only if* $\varphi \equiv$ *true; and*
(iii) φ *is a contradiction if and only if* $\varphi \equiv$ *flase.*

3 Satisfaction Degrees for LTL Formulae

Let $\mathcal{D}$ be a DTMC and φ an LTL formula. If $\mathcal{D} \models \varphi$ holds, then every infinite initial path definitely satisfies φ. In this case, we can quantitatively consider the notion $\mathcal{D} \models \varphi$ with degree 1, as the maximum value in [0, 1]. Similarly, if $\mathcal{D} \models \neg\varphi$ holds, then none of the infinite initial paths satisfy φ, and we can quantitatively consider the notion $\mathcal{D} \models \varphi$ with degree 0, as the minimum value in [0, 1]. However in general case, neither $\mathcal{D} \models \varphi$ nor $\mathcal{D} \models \neg\varphi$ would hold even for the same formula φ. Under this circumstance, how should we quantitatively consider the notion $\mathcal{D} \models \varphi$? In order to solve this issue, we firstly need to find some probabilistic measure for $\mathcal{D}$.

3.1 Probabilistic Measure in DTMC

Let $\mathcal{D}= (S, \mathbf{P}, \ell_{\text{init}}, AP, L)$ be a DTMC, and $s \in S$, $\pi_{\text{fin}} = s_0 s_1 s_2 \ldots s_n \in \text{Paths}_{\text{fin}}(s)$, where $s_0 = s$. Define

$$C(\pi_{\text{fin}}) = \{\pi \in \text{Paths}(s) \mid \pi_{\text{fin}} \ll \pi\}, \tag{8}$$

where $\pi_{\text{fin}} \ll \pi$ states that π_{fin} is a prefix of π. Also, define the mapping $Pr^{\mathcal{D}}$: $\{C(\pi_{\text{fin}}) \mid \pi_{\text{fin}} \in \text{Paths}_{\text{fin}}(s), s \in S\} \longrightarrow [0, 1]$ as

$$Pr^{\mathcal{D}}(C(\pi_{\text{fin}})) = \ell_{\text{init}}(s) \times \mathbf{P}(\pi_{\text{fin}}), \tag{9}$$

where

$$\mathbf{P}(\pi_{\text{fin}}) = \mathbf{P}(s_0 s_1 \ldots s_n) = \begin{cases} 1, & n = 0, \\ \prod_{i=0}^{n-1} \mathbf{P}(s_i, s_{i+1}), & n > 0. \end{cases} \tag{10}$$

In the following, we will prove that $Pr^{\mathcal{D}}$ is actually a probabilistic measure on Paths($\mathcal{D}$) that we need.

Definition 8 ([12]) Let X be a nonempty set, and $\mathcal{F} \subseteq 2^X$. Then $\mathcal{F}$ is called a *semi-ring* on X, if

(i) $\emptyset \in \mathcal{F}$;
(ii) $A, B \in \mathcal{F}$ implies $A \cap B \in \mathcal{F}$; and
(iii) If $A, B \in \mathcal{F}$ and $A \subseteq B$, then there exists a finite family of pairwise disjoint sets $C_1, \dots, C_k \in \mathcal{F}$ such that $B - A = \cup_{i=1}^k C_i$.

Lemma 1 ([12]) *Let $\mathcal{F}$ be a semi-ring on X, and $\mu : \mathcal{F} \longrightarrow [0, 1]$ satisfy*

(i) $\mu(\emptyset) = 0$;
(ii) For every finite family of pairwise disjoint sets $A_1, \dots, A_k \in \mathcal{F}$, $\cup_{i=1}^k A_i \in \mathcal{F}$ implies $\mu(\cup_{i=1}^k A_i) = \sum_{i=1}^k \mu(A_i)$; and
(iii) For every countably infinite family of sets $A_1, A_2, \dots \in \mathcal{F}$, $\cup_i A_i \in \mathcal{F}$ implies $\mu(\cup_i A_i) \leq \sum_i \mu(A_i)$.
Then the mapping μ can be uniquely extended as a measure on $\sigma(\mathcal{F})$, where $\sigma(\mathcal{F})$ is the σ-algebra on X generated by $\mathcal{F}$.

Remark 1 For a given finite DTMC $\mathcal{D}= (S, \mathbf{P}, \ell_{\text{init}}, AP, L)$, let $X = \text{Paths}(\mathcal{D})$, $\mathcal{F} = \{C(\pi_{\text{fin}}) \mid \pi_{\text{fin}} \in \text{Paths}_{\text{fin}}(s), s \in I\} \cup \{\emptyset\}$. Then $\mathcal{F} \subseteq 2^X$ together with the three conditions in Definition 8 hold for $\mathcal{F}$ (see [5] for details), and $\mathcal{F}$ becomes a semi-ring on X as a result. In this case, considering the mapping $Pr^{\mathcal{D}}$ defined by Eq. (9), its restriction on $\mathcal{F}$ (still written as $Pr^{\mathcal{D}}$ for convenience) actually satisfy the three conditions in Lemma 1 (see [5] for details). Thus $Pr^{\mathcal{D}}$ can be uniquely extended as a measure, even a probabilistic measure, on $\Sigma_{\text{Paths}(\mathcal{D})}$, the σ-algebra on Paths($\mathcal{D}$) generated by $\mathcal{F}$, which makes (Paths($\mathcal{D}$), $\Sigma_{\text{Paths}(\mathcal{D})}$, $Pr^{\mathcal{D}}$) a probabilistic measure space.

3.2 *Satisfaction Degrees for LTL Formulae*

Definition 9 Let $\mathcal{D}= (S, \mathbf{P}, \ell_{\text{init}}, AP, L)$ be a finite DTMC, and φ an LTL formula. Define

$$\tau^{\mathcal{D}}(\varphi) = Pr^{\mathcal{D}}(\{\pi \in \text{Paths}(\mathcal{D}) \mid \pi \models \varphi\}). \quad (11)$$

We call $\tau^{\mathcal{D}}(\varphi)$ the *satisfaction degree* of φ with respect to $\mathcal{D}$.

Remark 2 In Definition 9, measurability of the set $\{\pi \in \text{Paths}(\mathcal{D}) \mid \pi \models \varphi\}$ should be guaranteed first. In fact, for every ω-regular property $\mathcal{L} \subseteq (2^{AP})^\omega$, Ref. [10] has proved that $\{\pi \in \text{Paths}(\mathcal{D}) \mid \text{trace}(\pi) \in \mathcal{L}\}$ is measurable with respect to $Pr^{\mathcal{D}}$. Thus $\{\pi \in \text{Paths}(\mathcal{D}) \mid \pi \models \varphi\}$ $=\{\pi \in \text{Paths}(\mathcal{D}) \mid \text{trace}(\pi) \models \varphi\}$ $=\{\pi \in \text{Paths}(\mathcal{D}) \mid \text{trace}(\pi) \in \text{Words}(\varphi)\}$ is also measurable with respect to $Pr^{\mathcal{D}}$, since Words $(\varphi) \subseteq (2^{AP})^\omega$ is an ω-regular property. As a result, Eq. (11) is well-defined.

Remark 3 The satisfaction degree $\tau^{\mathcal{D}}(\varphi)$ defined in Eq. (11) compares the set of paths satisfying φ with Paths($\mathcal{D}$), and quantitatively considers the measure as their ratio. Under this circumstance, the notion of $\tau^{\mathcal{D}}(\varphi)$ can be seen as a quantitative extension of the classical case of satisfaction, with $\mathcal{D} \models \varphi$ and $\mathcal{D} \models \neg\varphi$ being two extreme cases when $\tau^{\mathcal{D}}(\varphi)$ equals 1 or 0, respectively.

Example 1 Let $\mathcal{D}= (S, \mathbf{P}, \ell_{\text{init}}, AP, L)$ be a finite DTMC with $S = \{s_0, s_1, s_2, s_3\}$ $AP = \{a, b, c\}$, $L(s_0) = \{b\}$, $L(s_1) = \{a, b\}$, $L(s_2) = \{c\}$, $L(s_3) = \{a, c\}$, and

$$\mathbf{P} = \begin{pmatrix} 0 & 1 & 0 & 0 \\ 0 & 0 & 0.01 & 0.99 \\ 1 & 0 & 0 & 0 \\ 0 & 0 & 0 & 1 \end{pmatrix} \qquad \ell_{\text{init}} = \begin{pmatrix} 1 \\ 0 \\ 0 \\ 0 \end{pmatrix}$$

Compute the satisfaction degrees of $\varphi_1 = \neg a \rightarrow b \wedge c$, $\varphi_2 = \bigcirc a \wedge \bigcirc^2 a$ and $\varphi_3 = b \sqcup (\Diamond c)$, respectively.

Solution 1 (i) For every $\pi \in$ Paths($\mathcal{D}$), $\pi \models \varphi_1$ holds if and only if $\pi \models a$ or $\pi \models b \wedge c$, if and only if $a \in L(s_0)$ or $b, c \in L(s_0)$. Thus $\tau^{\mathcal{D}}(\varphi_1) = Pr^{\mathcal{D}}(\{\pi \in \text{Paths}(\mathcal{D}) \mid \pi \models \varphi_1\}) = Pr^{\mathcal{D}}(\emptyset) = 0$.

(ii) For every $\pi \in$ Paths($\mathcal{D}$), $\pi \models \varphi_2$ holds if and only if $\pi \models \bigcirc a$ and $\pi \models \bigcirc^2 a$, if and only if $a \in L(\pi[1])$ and $a \in L(\pi[2])$. Thus $\tau^{\mathcal{D}}(\varphi_2) = Pr^{\mathcal{D}}(\{\pi \in \text{Paths}(\mathcal{D}) \mid \pi \models \varphi_2\}) = Pr^{\mathcal{D}}(C(s_0 s_1 s_3)) = \ell_{\text{init}}(s_0) \times \mathbf{P}(s_0 s_1 s_3) = 1 \times 1 \times 0.99 = 0.99$.

(iii) For every $\pi \in$ Paths($\mathcal{D}$), $\pi \models \varphi_3$ holds if and only if $\exists j \in \mathbf{N}$ such that $\pi[j...] \models \Diamond c$ and $\pi[i...] \models b$ holds for every i with $0 \leq i < j$, if and only if $\exists j, k \in \mathbf{N}$, $k \geq j$ such that $c \in L(\pi[k])$ and $b \in L(\pi[i])$ holds for every i with $0 \leq i < j$. Thus $\tau^{\mathcal{D}}$ $(\varphi_3) = Pr^{\mathcal{D}}(\{\pi \in \text{Paths}(\mathcal{D}) \mid \pi \models \varphi_3\}) = Pr^{\mathcal{D}}(\text{Paths}(\mathcal{D})) = Pr^{\mathcal{D}}(C(s_0)) = \ell_{\text{init}}(s_0) \times \mathbf{P}(s_0) = 1 \times 1 = 1$.

Lemma 2 *Let φ, ψ be LTL formulae. If $\varphi \equiv \psi$, then Words (φ) = Words (ψ).*

Proposition 1 *Let $\mathcal{D}= (S, \boldsymbol{P}, \ell_{init}, AP, L)$ be a finite DTMC, and φ, ψ be LTL formulae.*

(i) If φ is a tautology, then $\tau^{\mathcal{D}}(\varphi) = 1$; Especially, $\tau^{\mathcal{D}}(true) = 1$.
(ii) If φ is a contradiction, then $\tau^{\mathcal{D}}(\varphi) = 0$; Especially, $\tau^{\mathcal{D}}(false) = 0$.
(iii) $\tau^{\mathcal{D}}(\varphi) + \tau^{\mathcal{D}}(\neg\varphi) = 1$.
(iv) $\tau^{\mathcal{D}}(\varphi \vee \psi) + \tau^{\mathcal{D}}(\varphi \wedge \psi) = \tau^{\mathcal{D}}(\varphi) + \tau^{\mathcal{D}}(\psi)$.
(v) If $\varphi \equiv \psi$, then $\tau^{\mathcal{D}}(\varphi) = \tau^{\mathcal{D}}(\psi)$.

Proof (i) If φ is a tautology, then $\mathcal{D} \models \varphi$, and $\tau^{\mathcal{D}}(\varphi) = Pr^{\mathcal{D}}(\{\pi \in \text{Paths}(\mathcal{D}) \mid \pi \models \varphi\}) = Pr^{\mathcal{D}}(\text{Paths}(\mathcal{D})) = Pr^{\mathcal{D}}(\cup_{s \in I}\text{Paths}(s)) = \sum_{s \in I} Pr^{\mathcal{D}}(\text{Paths}(s)) = \sum_{s \in I} Pr^{\mathcal{D}}(C(s)) = \sum_{s \in I} \ell_{\text{init}}(s) \times \mathbf{P}(s) = \sum_{s \in I} \ell_{\text{init}}(s) = \sum_{s \in S} \ell_{\text{init}}(s) = 1$.

(ii) If φ is a contradiction, then $\mathcal{D} \models \neg\varphi$, and $\tau^{\mathcal{D}}(\varphi) = Pr^{\mathcal{D}}(\{\pi \in \text{Paths}(\mathcal{D}) \mid \pi \models \varphi\}) = Pr^{\mathcal{D}}(\emptyset)$=0.

(iii) $\tau^{\mathcal{D}}(\varphi) + \tau^{\mathcal{D}}(\neg\varphi) = Pr^{\mathcal{D}}(\{\pi \in \text{Paths}(\mathcal{D}) \mid \pi \models \varphi\}) + Pr^{\mathcal{D}}(\{\pi \in \text{Paths}(\mathcal{D}) \mid \pi \models \neg\varphi\}) = Pr^{\mathcal{D}}(\{\pi \in \text{Paths}(\mathcal{D}) \mid \pi \models \varphi\}) + Pr^{\mathcal{D}}(\text{Paths}(\mathcal{D}) - \{\pi \in \text{Paths}(\mathcal{D}) \mid \pi \models \varphi\}) = 1$.

(iv) $\tau^{\mathcal{D}}(\varphi \vee \psi) + \tau^{\mathcal{D}}(\varphi \wedge \psi) = Pr^{\mathcal{D}}(\{\pi \in \text{Paths}(\mathcal{D}) \mid \pi \models \varphi \vee \psi\}) + Pr^{\mathcal{D}}(\{\pi \in \text{Paths}(\mathcal{D}) \mid \pi \models \varphi \wedge \psi\}) = Pr^{\mathcal{D}}(\{\pi \in \text{Paths}(\mathcal{D}) \mid \pi \models \varphi\} \cup \{\pi \in \text{Paths}(\mathcal{D}) \mid \pi \models \psi\}) + Pr^{\mathcal{D}}(\{\pi \in \text{Paths}(\mathcal{D}) \mid \pi \models \varphi\} \cap \{\pi \in \text{Paths}(\mathcal{D}) \mid \pi \models \psi\}) = Pr^{\mathcal{D}}(\{\pi \in \text{Paths}(\mathcal{D}) \mid \pi \models \varphi\}) + Pr^{\mathcal{D}}(\{\pi \in \text{Paths}(\mathcal{D}) \mid \pi \models \psi\}) = \tau^{\mathcal{D}}(\varphi) + \tau^{\mathcal{D}}(\psi)$.

(v) If $\varphi \equiv \psi$, then $\tau^{\mathcal{D}}(\varphi) = Pr^{\mathcal{D}}(\{\pi \in \text{Paths}(\mathcal{D}) \mid \pi \models \varphi\}) = Pr^{\mathcal{D}}(\{\pi \in \text{Paths}(\mathcal{D}) \mid \text{trace}(\pi) \models \varphi\}) = Pr^{\mathcal{D}}(\{\pi \in \text{Paths}(\mathcal{D}) \mid \text{trace}(\pi) \models \psi\}) = Pr^{\mathcal{D}}(\{\pi \in \text{Paths}(\mathcal{D}) \mid \pi \models \psi\}) = \tau^{\mathcal{D}}(\psi)$. □

Proposition 2 *Let $\mathcal{D} = (S, \mathbf{P}, \ell_{init}, AP, L)$ be a finite DTMC, φ, ψ be LTL formulae, and α, $\beta \in [0, 1]$.*

(i) If $\tau^{\mathcal{D}}(\varphi) \geq \alpha$, $\tau^{\mathcal{D}}(\varphi \to \psi) \geq \beta$, then $\tau^{\mathcal{D}}(\psi) \geq \alpha + \beta - 1$.
(ii) If $\tau^{\mathcal{D}}(\varphi \to \psi) \geq \alpha$, $\tau^{\mathcal{D}}(\psi \to \gamma) \geq \beta$, then $\tau^{\mathcal{D}}(\varphi \to \gamma) \geq \alpha + \beta - 1$.
(iii) If $\varphi \to \psi$ is a tautology, then $\tau^{\mathcal{D}}(\varphi) \leq \tau^{\mathcal{D}}(\psi)$.

Proof (i) By Proposition 1, $\tau^{\mathcal{D}}(\varphi \to \psi) = \tau^{\mathcal{D}}(\neg\varphi \vee \psi) = (1 - \tau^{\mathcal{D}}(\varphi)) + \tau^{\mathcal{D}}(\psi) - \tau^{\mathcal{D}}(\neg\varphi \wedge \psi) \geq \beta$. Thus $1 + \tau^{\mathcal{D}}(\psi) - \tau^{\mathcal{D}}(\neg\varphi \wedge \psi) \geq \beta + \tau^{\mathcal{D}}(\varphi) \geq \alpha + \beta$, and $\tau^{\mathcal{D}}(\psi) \geq \alpha + \beta - 1 + \tau^{\mathcal{D}}(\neg\varphi \wedge \psi) \geq \alpha + \beta - 1$.

(ii) Since $(\psi \to \gamma) \to ((\varphi \to \psi) \to (\varphi \to \gamma))$ is a tautology in LTL, we have $\tau^{\mathcal{D}}((\psi \to \gamma) \to ((\varphi \to \psi) \to (\varphi \to \gamma))) = 1$. By (i) and $\tau^{\mathcal{D}}(\psi \to \gamma) \geq \beta$, we also have $\tau^{\mathcal{D}}((\varphi \to \psi) \to (\varphi \to \gamma)) \geq \beta + 1 - 1 = \beta$. Again by (i) and $\tau^{\mathcal{D}}(\varphi \to \psi) \geq \alpha$, we obtain $\tau^{\mathcal{D}}(\varphi \to \gamma) \geq \alpha + \beta - 1$.

(iii) Since $\varphi \to \psi$ is a tautology in LTL, we have $\tau^{\mathcal{D}}(\varphi \to \psi) = 1$. Thus $\tau^{\mathcal{D}}(\psi) \geq \tau^{\mathcal{D}}(\varphi) + 1 - 1 = \tau^{\mathcal{D}}(\varphi)$ by (i). □

Proposition 3 *Let $\mathcal{D} = (S, \mathbf{P}, \ell_{init}, AP, L)$ be a finite DTMC, and φ, ψ be LTL formulae. Then*

(i) $\tau^{\mathcal{D}}(\varphi \sqcup \psi) \geq \tau^{\mathcal{D}}(\psi)$, $\tau^{\mathcal{D}}(\varphi \sqcup \psi) \geq \tau^{\mathcal{D}}(\varphi \wedge \bigcirc\psi)$,
(ii) $\tau^{\mathcal{D}}(\Box\varphi) \leq \tau^{\mathcal{D}}(\varphi) \leq \tau^{\mathcal{D}}(\Diamond\varphi)$,
(iii) $\tau^{\mathcal{D}}(\Box\varphi) \leq \tau^{\mathcal{D}}(\bigcirc^k\varphi) \leq \tau^{\mathcal{D}}(\Diamond\varphi)$, $k \in \mathbf{N}$.

4 LTL Logic Metric Space

Definition 10 Let $\mathcal{D} = (S, \mathbf{P}, \ell_{\text{init}}, AP, L)$ be a finite DTMC, and φ, ψ be LTL formulae. Define

$$\xi^{\mathcal{D}}(\varphi, \psi) = \tau^{\mathcal{D}}((\varphi \to \psi) \wedge (\psi \to \varphi)). \tag{12}$$

We call $\xi^{\mathcal{D}}(\varphi, \psi)$ the *similarity degree* between φ and ψ with respect to $\mathcal{D}$.

Definition 11 Let $\mathcal{D} = (S, \mathbf{P}, \ell_{\text{init}}, AP, L)$ be a finite DTMC, and φ, ψ, γ be LTL formulae. Then

(i) $\xi^{\mathcal{D}}(\varphi, \varphi) = 1$,
(ii) $\xi^{\mathcal{D}}(\varphi, \psi) = \xi^{\mathcal{D}}(\psi, \varphi)$,
(iii) $\xi^{\mathcal{D}}(\varphi, \psi) = 1 + \tau^{\mathcal{D}}(\varphi \wedge \psi) - \tau^{\mathcal{D}}(\varphi \vee \psi)$,
(iv) $\xi^{\mathcal{D}}(\varphi, \psi) + \xi^{\mathcal{D}}(\varphi, \neg\psi) = 1$,
(v) $\xi^{\mathcal{D}}(\varphi, \psi) + \xi^{\mathcal{D}}(\psi, \gamma) \leq 1 + \xi^{\mathcal{D}}(\varphi, \gamma)$,
(vi) $\varphi \equiv \psi$ implies $\xi^{\mathcal{D}}(\varphi, \psi) = 1$.

Proof (i) and (ii) are trivial.

(iii) By Proposition 1 and Eq. (9), $\xi^{\mathcal{D}}(\varphi, \psi) = \tau^{\mathcal{D}}((\varphi \to \psi) \wedge (\psi \to \varphi)) = \tau^{\mathcal{D}}((\neg\varphi \vee \psi) \wedge (\neg\psi \vee \varphi)) = \tau^{\mathcal{D}}((\neg\varphi \wedge \neg\psi) \vee (\varphi \wedge \psi)) = \tau^{\mathcal{D}}(\neg\varphi \wedge \neg\psi) + \tau^{\mathcal{D}}(\varphi \wedge \psi) - \tau^{\mathcal{D}}((\neg\varphi \wedge \neg\psi) \wedge (\varphi \wedge \psi)) = \tau^{\mathcal{D}}(\neg\varphi \wedge \neg\psi) + \tau^{\mathcal{D}}(\varphi \wedge \psi) = 1 - \tau^{\mathcal{D}}(\varphi \vee \psi) + \tau^{\mathcal{D}}(\varphi \wedge \psi)$.

(iv) By Proposition 1 and (iii), $\xi^{\mathcal{D}}(\varphi, \psi) + \xi^{\mathcal{D}}(\varphi, \neg\psi) = [\tau^{\mathcal{D}}(\neg\varphi \wedge \neg\psi) + \tau^{\mathcal{D}}(\varphi \wedge \psi)] + [\tau^{\mathcal{D}}(\neg\varphi \wedge \psi) + \tau^{\mathcal{D}}(\varphi \wedge \neg\psi)] = [\tau^{\mathcal{D}}(\neg\varphi \wedge \neg\psi) + \tau^{\mathcal{D}}(\neg\varphi \wedge \psi)] + [\tau^{\mathcal{D}}(\varphi \wedge \psi) + \tau^{\mathcal{D}}(\varphi \wedge \neg\psi)] = [\tau^{\mathcal{D}}((\neg\varphi \wedge \neg\psi) \vee (\neg\varphi \wedge \psi)) + \tau^{\mathcal{D}}((\neg\varphi \wedge \neg\psi) \wedge (\neg\varphi \wedge \psi))] + [\tau^{\mathcal{D}}((\varphi \wedge \psi) \vee (\varphi \wedge \neg\psi)) + \tau^{\mathcal{D}}((\varphi \wedge \psi) \wedge (\varphi \wedge \neg\psi))] = \tau^{\mathcal{D}}(\neg\varphi) + \tau^{\mathcal{D}}(\varphi) = 1$.

(v) By Proposition 1 and (iii), $\xi^{\mathcal{D}}(\varphi, \psi) + \xi^{\mathcal{D}}(\psi, \gamma) - \xi^{\mathcal{D}}(\varphi, \gamma) = [1 + \tau^{\mathcal{D}}(\varphi \wedge \psi) - \tau^{\mathcal{D}}(\varphi \vee \psi)] + [1 + \tau^{\mathcal{D}}(\psi \wedge \gamma) - \tau^{\mathcal{D}}(\psi \vee \gamma)] - [1 + \tau^{\mathcal{D}}(\varphi \wedge \gamma) - \tau^{\mathcal{D}}(\varphi \vee \gamma)] = 1 + \tau^{\mathcal{D}}(\varphi \wedge \psi) + \tau^{\mathcal{D}}(\psi \wedge \gamma) - \tau^{\mathcal{D}}(\varphi \wedge \gamma) - \tau^{\mathcal{D}}(\varphi \vee \psi) - \tau^{\mathcal{D}}(\psi \vee \gamma) + \tau^{\mathcal{D}}(\varphi \vee \gamma) = 1 - 2\tau^{\mathcal{D}}(\psi) + 2\tau^{\mathcal{D}}(\varphi \wedge \psi) + 2\tau^{\mathcal{D}}(\psi \wedge \gamma) - 2\tau^{\mathcal{D}}(\varphi \wedge \gamma) \leq 1$.

(vi) If $\varphi \equiv \psi$, then $(\varphi \to \psi) \wedge (\psi \to \varphi)$ is a tautology in LTL. Thus $\xi^{\mathcal{D}}(\varphi, \psi) = 1$ by Eq. (9). □

Definition 12 Let $\mathcal{D} = (S, \mathbf{P}, \ell_{\text{init}}, AP, L)$ be a finite DTMC, and φ, ψ be LTL formulae. Then

(i) $\xi^{\mathcal{D}}(\psi, \varphi \sqcup \psi) = 1 + \tau^{\mathcal{D}}(\psi) - \tau^{\mathcal{D}}(\varphi \sqcup \psi)$,
(ii) $\xi^{\mathcal{D}}(\bigcirc\varphi, \Diamond\varphi) = 1 + \tau^{\mathcal{D}}(\bigcirc\varphi) - \tau^{\mathcal{D}}(\Diamond\varphi)$,
(iii) $\xi^{\mathcal{D}}(\bigcirc\varphi, \Box\varphi) = 1 + \tau^{\mathcal{D}}(\Box\varphi) - \tau^{\mathcal{D}}(\bigcirc\varphi)$.

Definition 13 Let $\mathcal{D} = (S, \mathbf{P}, \ell_{\text{init}}, AP, L)$ be a finite DTMC, and φ, ψ be LTL formulae. Define

$$\rho^{\mathcal{D}}(\varphi, \psi) = 1 - \xi^{\mathcal{D}}(\varphi, \psi), \tag{13}$$

We call $\rho^{\mathcal{D}}(\varphi, \psi)$ the *pseudo-metric* between φ and ψ with respect to $\mathcal{D}$.

Proposition 4 *Let $\mathcal{D} = (S, \mathbf{P}, \ell_{init}, AP, L)$ be a finite DTMC, and φ, ψ, γ be LTL formulae. Then*

(i) $\rho^{\mathcal{D}}(\varphi, \varphi) = 0$.
(ii) $\rho^{\mathcal{D}}(\varphi, \psi) = \rho^{\mathcal{D}}(\psi, \varphi)$.
(iii) $\rho^{\mathcal{D}}(\varphi, \psi) + \rho^{\mathcal{D}}(\varphi, \neg\psi) = 1$.
(iv) $\rho^{\mathcal{D}}(\varphi, \psi) + \rho^{\mathcal{D}}(\psi, \gamma) \geq \rho^{\mathcal{D}}(\varphi, \gamma)$.
(v) $\rho^{\mathcal{D}}(\varphi, \psi) = \tau^{\mathcal{D}}(\varphi \vee \psi) - \tau^{\mathcal{D}}(\varphi \wedge \psi)$.
(vi) $\varphi \equiv \psi$ *implies* $\rho^{\mathcal{D}}(\varphi, \psi) = 0$.
(vii) $\rho^{\mathcal{D}}(\psi, \varphi \sqcup \psi) = \tau^{\mathcal{D}}(\varphi \sqcup \psi) - \tau^{\mathcal{D}}(\psi)$.

(viii) $\rho^{\mathcal{D}}(\bigcirc\varphi, \Diamond\varphi) = \tau^{\mathcal{D}}(\Diamond\varphi) - \tau^{\mathcal{D}}(\bigcirc\varphi)$.
(ix) $\rho^{\mathcal{D}}(\bigcirc\varphi, \Box\varphi) = \tau^{\mathcal{D}}(\bigcirc\varphi) - \tau^{\mathcal{D}}(\Box\varphi)$.

Remark 4 Let Form (LTL) denote the set of all LTL formulae. Then by Proposition 4 (i)(ii)(iv), we can conclude that the mapping $\rho^{\mathcal{D}}$: Form(LTL) $^2 \longrightarrow [0, 1]$ is actually a pseudo-metric [13] on Form(LTL), which makes (Form(LTL), $\rho^{\mathcal{D}}$) a logic metric space, called the *LTL logic metric space*.

Acknowledgments Project was supported by the National Natural Science Foundation of China (11501343).

References

1. Wang, G.J., Zhou, H.J.: Introduction to Mathematical Logic and Resolution Principle. Science Press, Beijing and U.K. Alpha Science International Limited, Oxford (2009)
2. Wang, G.J., Zhou, H.J.: Quantitative logic. Info. Sci. **179**, 226–247 (2009)
3. Wang, G.J., Shi, H.X.: Lattice-valued modal propositional logic and its completeness. Sci. China Info. Sci. **53**(11), 2230–2239 (2010)
4. Wang, G.J.: Axiomatic theory of truth degree in a class of first-order logics with its application. Sci. China Info. Sci. **42**(5), 648–662 (2012) (in Chinese)
5. Shi, H.X.: A Quantitative Approach for Modal Logic and its Application in Model Checking. SNNU, Xi'an (2013) (in Chinese)
6. Kroger, F., Merz, S.: Temporal Logic and State Systems. Springer, Berlin (2008)
7. Huth, M., Ryan, M.: Logic in Computer Science-Modelling and Reasoning about Systems, 2nd edn. Cambridge University Press, Cambridge (2004)
8. Manna, Z., Pnueli, A.: Temporal Verification of Reactive Systems. Springer, New York (1995)
9. Clarke, E.M., Emerson, E.A.: Design and synthesis of synchronisation skeletons using branching time temporal logic. In: Logic of Programs, LNCS, pp. 52–71 (1981)
10. Baier, C., Katoen, J.P.: Principles of Model Checking. MIT Press, London (2008)
11. Clarke, E.M., Grumberg, O., Pilid, D.: Model Checking. MIT Press, London (2000)
12. Billingsley, P.: Probability and Measure. Wiley, New York (1995)
13. Kelley, J.L.: General Topology. Springer, New York (1975)

A New Theory of T Truth Degree on Gödel n-Valued Propositional Logic System

Nai-Diao Zhu, Xiao-Jing Hui and Xiao-Li Gao

Abstract By adding new operators Δ and $\sim$, axiomatic expansion of *Gödel* n-valued propositional logic system is introduced, which is denoted by $Gödel_{\sim}$. In this paper, the concept of t truth degree of propositional formula is put forward in $Gödel_{\sim}$ (t take Δ, $\sim$), and the MP rule, HS rule and some related properties are studied; the concepts of t similarity degree, t pseudo-metric between propositional formulas, and t divergent degree and t consistent degree of theory Γ in $Gödel_{\sim}$ are obtained, and their correlation properties are discussed.

Keywords T truth degree · T similarity degree · T pseudo-metric · T divergent degree · T consistent degree

1 Introduction

In the literature [1, 2], the concept of truth degree is given by professor Wang in both the classical propositional logic system and multi-valued propositional logic system. On this basis, some related researches on truth degree and random truth degree are carried out by a large number of scholars in different logic systems (see [3–15]). The basic connectives $\sim$ and Δ are introduced by Esteva et al., axiomatic expansion BL_{Δ} system of basic logic system *BL* is proposed and the system $SBL_{\sim}$ is established, in which Δ deductive theorem and the strong complete theorem are both tenable (see [16–19]). So some related research on *Gödel* propositional logic system can be carried out smoothly.

N.-D. Zhu (✉) · X.-J. Hui · X.-L. Gao
College of Mathematics and Computer Science,
Yan'an University, Yan'an 716000, China
e-mail: 792266404@qq.com

X.-J. Hui
e-mail: xhmxiaojing@163.com

X.-L. Gao
e-mail: 951402445@qq.com

T.-H. Fan et al. (eds.), *Quantitative Logic and Soft Computing 2016*,
Advances in Intelligent Systems and Computing 510,
DOI 10.1007/978-3-319-46206-6_7

In $SBL_\sim$ fuzzy logic system, by virtue of Δ deductive theorem and strong completeness theorem, it is possible to establish quantitative logic theory in $Gödel_\sim$. In this paper, by adding new operators $\Delta, \sim$, axiomatic expansion of *Gödel* n-valued propositional logic system is introduced, the concepts of t truth degree of propositional formula, t similarity degree, t pseudo-metric between propositional formulas, t divergent degree and t consistent degree of theory Γ in $Gödel_\sim$ are given. Their related properties are discussed.

2 Preliminaries

Definition 1 ([18]) The axiom system BL_Δ is as follows:
(BL) The axioms system of BL;
($\Delta 1$) $\Delta A \vee \neg \Delta A$;
($\Delta 2$) $\Delta(A \vee B) \rightarrow (\Delta A \vee \Delta B)$;
($\Delta 3$) $\Delta A \rightarrow B$;
($\Delta 4$) $\Delta A \rightarrow \Delta\Delta A$;
($\Delta 5$) $\Delta(A \rightarrow B) \rightarrow (\Delta A \rightarrow \Delta B)$.

The inference rules of BL_Δ are the MP rule and Δ rules, the MP rule is that from $A, A \rightarrow B$ one can infer B, while the Δ rule is that from A one can infer ΔA.

Definition 2 ([19]) As the axiomatic expansion of *SBL*, the axiom system $SBL_\sim$ is as follows:
(SBL) The axioms system of SBL;
(~ 1) $\sim\sim A \rightarrow A$;
(~ 2) $\neg A \rightarrow \sim A$;
(~ 3) $\Delta(A \rightarrow B) \rightarrow \Delta(\sim B \rightarrow \sim B)$.

The inference rules of $SBL_\sim$ are also MP rules and Δ rules. If $\mathcal{L}$ is the axiomatic expansion of *SBL*, so $\mathcal{L}_\sim$ notes for the expansion of $\mathcal{L}$, it is just as *SBL* expansion for $SBL_\sim$, $Gödel_\sim$ and $\Pi_\sim$ are the two basic types of axiomatic expansion of $SBL_\sim$. If not otherwise specified, in the following all discussions are carried out in $Gödel_\sim$ n-valued propositional logic system.

Theorem 1 ([15] Δ Deduction Theorem) *Let $\mathcal{L}$ be an axiomatic expansion of BL_Δ. Then for each theory Γ and for all formulas A and B, we have:*

$$\Gamma, A \vdash B \text{ if and only if } \Gamma \vdash \Delta A \rightarrow B.$$

Theorem 2 ([19] Strong Completeness Theorem) *Let $\mathcal{L}$ be an axiomatic expansion of $SBL_\sim$. Then for theory Γ and formula A, the following conditions are equivalent:*
(i) $\Gamma \vdash A$;
(ii) $e(A) = 1$ *for each $\mathcal{L}$-algebra B and each model e of the theory Γ.*

3 T Truth Degree of Propositional Formula and Its Related Properties

Definition 3 Let $S = \{p_1, p_2, \ldots\}$ be a countable set, $\sim, \Delta, \vee, \wedge, \rightarrow$ respectively are unary, unary, binary, binary and binary operation in S, $F(s)$ is the type of $(1, 1, 2, 2, 2)$ free algebra generated by S, then we say that elements of $F(S)$ are proposition formulas or formulas. Notes: $\sim x = 1 - x$, $\Delta x = \{^{1,x=1}_{0,x<1}$, $x \vee y = \max\{x, y\}$, $x \wedge y = \min\{x, y\}$, $x \rightarrow y = \{^{1,x\leq y}_{y,x>y}$.

Definition 4 Let $A = A(p_1, p_2, \ldots, p_m) \in F(S)$, then A corresponds to an element function $\overline{A}$. In *Gödel*$_\sim$ n-valued propositional logic system, $\overline{A} : G_n^m \rightarrow G_n, \overline{A}(x_1, x_2, \ldots, x_m)$ is by operational signs $\sim, \Delta, \vee, \wedge, \rightarrow$ connected to $x_1, x_2, \ldots, x_m$, the way just as $A = A(p_1, p_2, \ldots, p_m)$ is by conjunction $\sim, \Delta, \vee, \wedge, \rightarrow$ connected to the atomic formula $p_1, p_2, \ldots, p_m$. Then we say that $\overline{A}$ is the function induced by A.

Definition 5 Let $A = A(p_1, p_2, \ldots, p_m) \in F(S)$ be a propositional formula containing m atomic formulas $p_1, p_2, \ldots, p_m$, $\overline{A}(x_1, x_2, \ldots, x_m)$ is the function induced by A, define

$$\tau_n(tA) = \frac{1}{n^m} \sum_{(x_1,x_2,\ldots,x_m)\in G_n^m} \overline{tA}(x_1, x_2, \ldots, x_m).$$

Then we say that $\tau_n(tA)$ is the t truth degree of formula A.

Theorem 3 *Let* $A, B \in F(S)$, *then*
(i) *A is tautology if and only if* $\tau_n(\Delta A) = 1, \tau_n(\sim A) = 0$; *A is contradiction if and only if* $\tau_n(\sim A) = 1$;
(ii) $A \approx B$ *if and only if* $\tau_n(tA) = \tau_n(tB)$;
(iii) $\tau_n(\sim A) = 1 - \tau_n(A), \tau_n(\sim \Delta A) = 1 - \tau_n(\Delta A)$;
(iv) *If* $\vdash A \rightarrow B$, *then* $\tau_n(\Delta A) \leq \tau_n(\Delta B), \tau_n(\sim A) \geq \tau_n(\sim B)$.

Proof (i) For $\forall(x_1, x_2, \ldots, x_m) \in G_n^m$, since A is tautology if and only if $\overline{A}(x_1, x_2, \ldots, x_m) = 1, \overline{\Delta A}(x_1, x_2, \ldots, x_m) = 1, \overline{\sim A}(x_1, x_2, \ldots, x_m) = 0$ if and only if

$$\frac{1}{n^m} \sum_{(x_1,x_2,\ldots,x_m)\in G_n^m} \overline{\Delta A}(x_1, x_2, \ldots, x_m) = 1,$$

$$\frac{1}{n^m} \sum_{(x_1,x_2,\ldots,x_m)\in G_n^m} \overline{\sim A}(x_1, x_2, \ldots, x_m) = 0,$$

so by Definition 5, we get $\tau_n(\Delta A) = 1, \tau_n(\sim A) = 0$.
In the same way, since A is contradiction if and only if $\overline{A}(x_1, x_2, \ldots, x_m) = 0$ if and only if $\overline{\sim A}(x_1, x_2, \ldots, x_m) = 1$ if and only if

$$\frac{1}{n^m}\sum_{(x_1,x_2,\ldots,x_m)\in G_n^m}\overline{\sim A}(x_1,x_2,\ldots,x_m)=1,$$

so by Definition 5, we get $\tau_n(\sim A)=1$.

(ii) Let $A\in F(S)$, $p_i\in F(S)$, since by [4], $A\otimes(P_i\to P_i)\approx A$, so when $A\approx B$, let A, B contains the same atomic formula $p_1,p_2,\ldots,p_m$. For $(x_1,x_2,\ldots,x_m)\in G_n^m$, since $A\approx B$ if and only if $\overline{A}(x_1,x_2,\ldots,x_m)=\overline{B}(x_1,x_2,\ldots,x_m)$ if and only if $\overline{tA}(x_1,x_2,\ldots,x_m)=\overline{tB}(x_1,x_2,\ldots,x_m)$ if and only if

$$\frac{1}{n^m}\sum_{(x_1,x_2,\ldots,x_m)\in G_n^m}\overline{tA}(x_1,x_2,\ldots,x_m)=\frac{1}{n^m}\sum_{(x_1,x_2,\ldots,x_m)\in G_n^m}\overline{tB}(x_1,x_2,\ldots,x_m),$$

so by Definition 5, we get $\tau_n(tA)=\tau_n(tB)$.

(iii) For $(x_1,x_2,\ldots,x_m)\in G_n^m$, since $\overline{\sim A}(x_1,x_2,\ldots,x_m)=1-\overline{A}(x_1,x_2,\ldots,x_m)$, $\overline{\sim\Delta A}(x_1,x_2,\ldots,x_m)=1-\overline{\Delta A}(x_1,x_2,\ldots,x_m)$, so we have

$$\begin{aligned}&\frac{1}{n^m}\sum_{(x_1,x_2,\ldots,x_m)\in G_n^m}\overline{\sim A}(x_1,x_2,\ldots,x_m)\\&=\frac{1}{n^m}\sum_{(x_1,x_2,\ldots,x_m)\in G_n^m}1-\overline{A}(x_1,x_2,\ldots,x_m),\end{aligned}$$

$$\begin{aligned}&\frac{1}{n^m}\sum_{(x_1,x_2,\ldots,x_m)\in G_n^m}\overline{\sim\Delta A}(x_1,x_2,\ldots,x_m)\\&=\frac{1}{n^m}\sum_{(x_1,x_2,\ldots,x_m)\in G_n^m}1-\overline{\Delta A}(x_1,x_2,\ldots,x_m),\end{aligned}$$

so by Definition 5, we get

$$\tau_n(\sim A)=1-\tau_n(A),\ \tau_n(\sim\Delta A)=1-\tau_n(\Delta A).$$

(iv) Let A, B contains the same atomic formula $p_1,p_2,\ldots,p_m$. For $\forall(x_1,x_2,\ldots,x_m)\in G_n^m$, since $\vdash A\to B$, so we have $\overline{A\to B}(x_1,x_2,\ldots,x_m)=1$. Again $\overline{A\to B}(x_1,x_2,\ldots,x_m)=\overline{A}(x_1,x_2,\ldots,x_m)\to\overline{B}(x_1,x_2,\ldots,x_m)=1$, so we have $\overline{A}(x_1,x_2,\ldots,x_m)\le\overline{B}(x_1,x_2,\ldots,x_m)$. Combined with Δ, $\sim$, we have

$$\overline{\Delta A}(x_1,x_2,\ldots,x_m)\le\overline{\Delta B}(x_1,x_2,\ldots,x_m),$$

$$\overline{\sim A}(x_1,x_2,\ldots,x_m)\ge\overline{\sim B}(x_1,x_2,\ldots,x_m),$$

so we have

$$\frac{1}{n^m}\sum_{(x_1,x_2,\ldots,x_m)\in G_n^m}\overline{\Delta A}(x_1,x_2,\ldots,x_m)$$
$$\leq \frac{1}{n^m}\sum_{(x_1,x_2,\ldots,x_m)\in G_n^m}\overline{\Delta B}(x_1,x_2,\ldots,x_m),$$
$$\frac{1}{n^m}\sum_{(x_1,x_2,\ldots,x_m)\in G_n^m}\overline{\sim A}(x_1,x_2,\ldots,x_m)$$
$$\geq \frac{1}{n^m}\sum_{(x_1,x_2,\ldots,x_m)\in G_n^m}\overline{\sim B}(x_1,x_2,\ldots,x_m),$$

so by Definition 5, we get

$$\tau_n(\Delta A) \leq \tau_n(\Delta B),\ \tau_n(\sim A) \geq \tau_n(\sim B).$$ □

Theorem 4 *$A, B \in F(S)$, then $\tau_n(mA \vee nB) = \tau_n(mA) + \tau_n(nB) - \tau_n(mA \wedge nB)$.*

Proof Suppose that A, B contain the same atomic formulas $p_1, p_2, \ldots, p_m$. For $a, b \in G_n$, we have $ma \vee nb = ma + nb - ma \wedge nb$, so for $x_1, x_2, \ldots, x_m \in G_n^m$, we have

$$\overline{mA \vee nB}(x_1,x_2,\ldots,x_m) = \overline{mA}(x_1,x_2,\ldots,x_m) + \overline{nB}(x_1,x_2,\ldots,x_m)$$
$$-\overline{mA \wedge nB}(x_1,x_2,\ldots,x_m).$$

Therefore,

$$\frac{1}{n^m}\sum_{(x_1,x_2,\ldots,x_m)\in G_n^m}\overline{mA \vee nB}(x_1,x_2,\ldots,x_m)$$
$$= \frac{1}{n^m}\sum_{(x_1,x_2,\ldots,x_m)\in G_n^m}\overline{mA}(x_1,x_2,\ldots,x_m)$$
$$+ \frac{1}{n^m}\sum_{(x_1,x_2,\ldots,x_m)\in G_n^m}\overline{nB}(x_1,x_2,\ldots,x_m)$$
$$- \frac{1}{n^m}\sum_{(x_1,x_2,\ldots,x_m)\in G_n^m}\overline{mA \wedge nB}(x_1,x_2,\ldots,x_m).$$

so by Definition 5, we get

$$\tau_n(mA \vee nB) = \tau_n(mA) + \tau_n(nB) - \tau_n(mA \wedge nB).$$

□

Theorem 5 *Let* $A, B \in F(S)$, *then*
(i) $\tau_n(\Delta A \to \sim B) = \tau_n(\Delta A \wedge \sim B) - \tau_n(\Delta A) + 1$;
(ii) $\tau_n(\Delta A \to \Delta B) = \tau_n(\Delta A \wedge \Delta B) - \tau_n(\Delta A) + 1$;
(iii) $\tau_n(\Delta A \to B) = \tau_n(\Delta A \wedge B) - \tau_n(\Delta A) + 1$.

Proof Suppose that A, B contains the same atomic formulas $p_1, p_2, \ldots, p_m$.

(i) For $a, b \in G_n$, we have $\Delta a \to \sim b = \Delta a \wedge \sim b - \Delta a + 1$. So $\forall x_1, x_2, \ldots, x_m \in G_n^m$, we have

$$\overline{\Delta A \to \sim B}(x_1, x_2, \ldots, x_m) = \overline{\Delta A \wedge \sim B}(x_1, x_2, \ldots, x_m) - \overline{\Delta A}(x_1, x_2, \ldots, x_m) + 1.$$

Therefore,

$$\sum_{(x_1, x_2, \ldots, x_m) \in G_n^m} \overline{\Delta A \to \sim B}(x_1, x_2, \ldots, x_m) = \sum_{(x_1, x_2, \ldots, x_m) \in G_n^m} \overline{\Delta A \wedge \sim B}(x_1, x_2, \ldots, x_m) - \sum_{(x_1, x_2, \ldots, x_m) \in G_n^m} \overline{\Delta A}(x_1, x_2, \ldots, x_m) + 1.$$

Dividing both sides $\frac{1}{n^m}$, by Definition 5, we get

$$\tau_n(\Delta A \to \sim B) = \tau_n(\Delta A \wedge \sim B) - \tau_n(\Delta A) + 1.$$

(ii) $\forall a, b \in G_n$, we have $\Delta a \to \Delta b = \Delta a \wedge \Delta b - \Delta a + 1$. So $\forall x_1, x_2, \ldots, x_m \in G_n^m$, we have

$$\overline{\Delta A \to \Delta B}(x_1, x_2, \ldots, x_m) = \overline{\Delta A \wedge \Delta B}(x_1, x_2, \ldots, x_m) - \overline{\Delta A}(x_1, x_2, \ldots, x_m) + 1.$$

Therefore,

$$\sum_{(x_1, x_2, \ldots, x_m) \in G_n^m} \overline{\Delta A \to \Delta B}(x_1, x_2, \ldots, x_m) = \sum_{(x_1, x_2, \ldots, x_m) \in G_n^m} \overline{\Delta A \wedge \Delta B}(x_1, x_2, \ldots, x_m) - \sum_{(x_1, x_2, \ldots, x_m) \in G_n^m} \overline{\Delta A}(x_1, x_2, \ldots, x_m) + 1.$$

Dividing both sides $\frac{1}{n^m}$, by Definition 5, we get

$$\tau_n(\Delta A \to \Delta B) = \tau_n(\Delta A \wedge \Delta B) - \tau_n(\Delta A) + 1.$$

(iii) $\forall a, b \in G_n$, we have $\Delta a \to b = \Delta a \wedge b - \Delta a + 1$. So $\forall x_1, x_2, \ldots, x_m \in G_n^m$, we have

$$\overline{\Delta A \to B}(x_1, x_2, \ldots, x_m) = \overline{\Delta A \wedge B}(x_1, x_2, \ldots, x_m) - \overline{\Delta A}(x_1, x_2, \ldots, x_m) + 1.$$

Therefore,

$$\begin{aligned} &\sum_{(x_1, x_2, \ldots, x_m) \subset G_n^m} \overline{\Delta A \to B}(x_1, x_2, \ldots, x_m) \\ = &\sum_{(x_1, x_2, \ldots, x_m) \in G_n^m} \overline{\Delta A \wedge B}(x_1, x_2, \ldots, x_m) \\ - &\sum_{(x_1, x_2, \ldots, x_m) \in G_n^m} \overline{\Delta A}(x_1, x_2, \ldots, x_m) + 1. \end{aligned}$$

Dividing both sides $\frac{1}{n^m}$, by Definition 5, we get

$$\tau_n(\Delta A \to B) = \tau_n(\Delta A \wedge B) - \tau_n(\Delta A) + 1.$$

□

Corollary 1 (MP rules of t truth degree) *Let* $A, B \in F(S)$, *then*
(i) $\tau_n(\Delta A \to \sim B) + \tau_n(\Delta A) \leq \tau_n(\sim B) + 1$;
(ii) *If* $\tau_n(\Delta A) \geq \alpha$, $\tau_n(\Delta A \to \sim B) \geq \beta$, *then* $\tau_n(\sim B) \geq \alpha + \beta + 1$;
(iii) $\tau_n(\Delta A \to \Delta B) + \tau_n(\Delta A) \leq \tau_n(\Delta B) + 1$;
(iv) *If* $\tau_n(\Delta A) \geq \alpha$, $\tau_n(\Delta A \to \Delta B) \geq \beta$, *then* $\tau_n(\Delta B) \geq \alpha + \beta + 1$.

Theorem 6 *Let* $A, B, C \in F(S)$, *then*
(i) $\tau_n(\Delta A \to \sim B) + \tau_n(\sim B \to \sim C) \leq \tau_n(\Delta A \to \sim C) + 1$;
(ii) $\tau_n(\Delta A \to \Delta B) + \tau_n(\Delta B \to \Delta C) \leq \tau_n(\Delta A \to \Delta C) + 1$.

Proof (i) By Corollary 1, we have

$$\tau_n(\Delta A \to \sim B) + \tau_n((\Delta A \to \sim B) \to (\Delta A \to \sim C)) \leq \tau_n(\Delta A \to \sim C) + 1.$$

Suppose that A, B, C contain the same atomic formulas $p_1, p_2, \ldots, p_m$,
$\forall a, b, c \in G_n$, $(\sim b \to \sim c) \to ((\Delta a \to \sim b) \to (\Delta a \to \sim c)) = 1$,
then we have $\vdash (\sim B \to \sim C) \to ((\Delta A \to \sim B) \to (\Delta A \to \sim C))$,
by Theorem 3(iv), we have

$$\tau_n(\sim B \to \sim C) \leq \tau_n((\Delta A \to \sim B) \to (\Delta A \to \sim C)).$$

Integrated the above two sides, we get

$$\tau_n(\Delta A \to \sim B) + \tau_n(\sim B \to \sim C) \leq \tau_n(\Delta A \to \sim C) + 1.$$

(ii) By Corollary 1, we get

$$\tau_n(\Delta A \to \Delta B) + \tau_n((\Delta A \to \Delta B) \to (\Delta A \to \Delta C)) \leq \tau_n(\Delta A \to \Delta C) + 1$$

Suppose that A, B, C contains the same atomic formulas $p_1, p_2, \ldots, p_m$,
for $a, b, c \in G_n$, $(\Delta b \to \Delta c) \to ((\Delta a \to \Delta b) \to (\Delta a \to \Delta c)) = 1$,
then we have $\vdash (\Delta B \to \Delta C) \to ((\Delta A \to \Delta B) \to (\Delta A \to \Delta C))$, by Theorem 3 (iv), we get $\tau_n(\Delta B \to \Delta C) \leq \tau_n((\Delta A \to \Delta B) \to (\Delta A \to \Delta C))$,
Integrated the above two sides, we get

$$\tau_n(\Delta A \to \Delta B) + \tau_n(\Delta B \to \Delta C) \leq \tau_n(\Delta A \to \Delta C) + 1.$$

□

Corollary 2 (HS rules of t truth degree) *Let $A, B, C \in F(S)$, then*
(i) *If $\tau_n(\Delta A \to \sim B) \geq \alpha$, $\tau_n(\sim B \to \sim C) \geq \beta$, then $\tau_n(\Delta A \to \sim C) \geq \alpha + \beta + 1$;*
(ii) *If $\tau_n(\Delta A \to \Delta B) \geq \alpha$, $\tau_n(\Delta B \to \Delta C) \geq \beta$, then $\tau_n(\Delta A \to \Delta C) \geq \alpha + \beta + 1$.*

4 T Similarity Degree and T Pseudo-Metric of Propositional Formulas

Definition 6 Let $A, B \in F(S)$, define $\xi_n(mA, nB) = \tau_n((mA \to nB) \wedge (nB \to mA))$, then it's said that $\xi_n(mA, nB)$ is the t similarity degree of proposition formulas A and B.

Theorem 7 *Let $A, B \in F(S)$, then*
(i) *If one of A and B is a tautology, another is a contradiction, then $\xi_n(mA, nB) = 0$;*
(ii) *If A and B are all tautologies or are all contradictions, then $\xi_n(mA, nB) = 1$.*

Proof Suppose that A, B contain the same atomic formulas $p_1, p_2, \ldots, p_m$.

(i) When one of A and B is a tautology, another is a contradiction, $\forall x_1, x_2, \ldots, x_m \in G_n^m$, then we have $\overline{(mA \to nB) \wedge (nB \to mA)}(x_1, x_2, \ldots, x_m) = 0$. So we have $\sum_{(x_1, x_2, \ldots, x_m) \in G_n^m} \overline{(mA \to nB) \wedge (nB \to mA)}(x_1, x_2, \ldots, x_m) = 0$.
Dividing both sides $\frac{1}{n^m}$, by Definition 5, we get

$$\tau_n((mA \to nB) \wedge (nB \to mA)) = 0.$$

According to Definition 6, we have $\xi_n(mA, nB) = 0$.

(ii) When A and B are tautologies or contradictions, for $x_1, x_2, \ldots, x_m \in G_n^m$, we have $\overline{(mA \to nB) \wedge (nB \to mA)}(x_1, x_2, \ldots, x_m) = 1$.
$\sum_{(x_1, x_2, \ldots, x_m) \in G_n^m} \overline{(mA \to nB) \wedge (nB \to mA)}(x_1, x_2, \ldots, x_m) = 1$.
Dividing both sides $\frac{1}{n^m}$, by Definition 5,we get

$$\tau_n((mA \to nB) \wedge (nB \to mA)) = 1.$$

According to the Definition 6, we get $\xi_n(mA, nB) = 1$. □

Theorem 8 *Let* $A, B \in F(S)$, *then*
(i) $\xi_n(\Delta A, \sim B) = \tau_n(\Delta A \to \sim B) + \tau_n(\sim B \to \Delta A) - 1 = 1 - \tau_n(\Delta A \vee \sim B) + \tau_n(\Delta A \wedge \sim B)$;
(ii) $\xi_n(\Delta A \to \sim B, \sim B \to \Delta A) = \xi_n(\Delta A \vee \sim B, \Delta A \wedge \sim B,) = \xi_n(\Delta A, \sim B)$.

Proof Suppose that A, B contain the same atomic formulas $p_1, p_2, \ldots, p_m$.

(i) By Theorem 4 and Definition 6, we get $\xi_n(\Delta A, \sim B) = \tau_n(\Delta A \to \sim B) + \tau_n(\sim B \to \Delta A) - \tau_n((\Delta A \to \sim B) \vee (\sim B \to \Delta A))$. Since $\vdash (\Delta A \to \sim B) \vee (\sim B \to \Delta A)$, so by Theorem 3(i), we have

$$\tau_n((\Delta A \to \sim B) \vee (\sim B \to \Delta A)) = 1.$$

So we have $\xi_n(\Delta A, \sim B) = \tau_n(\Delta A \to \sim B) + \tau_n(\sim B \to \Delta A) - 1$.
Since

$$\begin{aligned}
&\tau_n(\Delta A \to \sim B) + \tau_n(\sim B \to \Delta A) - 1 \\
&= \tau_n(\Delta A \wedge \sim B) - \tau_n(\Delta A) + 1 + \tau_n(\sim B \wedge \Delta A) - \tau_n(\sim B) + 1 - 1 \\
&= 2\tau_n(\Delta A \wedge \sim B) - \tau_n(\Delta A) - \tau_n(\sim B) + 1 \\
&= 1 - \tau_n(\Delta A \vee \sim B) + \tau_n(\Delta A \wedge \sim B).
\end{aligned}$$

In summary, we get

$$\begin{aligned}
&\xi_n(\Delta A, \sim B) = \tau_n(\Delta A \to \sim B) + \tau_n(\sim B \to \Delta A) - 1 \\
&= 1 - \tau_n(\Delta A \vee \sim B) + \tau_n(\Delta A \wedge \sim B).
\end{aligned}$$

(ii) Since (i), we get

$$\begin{aligned}
&\xi_n(\Delta A \to \sim B, \sim B \to \Delta A) \\
&= 1 - \tau_n((\Delta A \to \sim B) \vee (\sim B \to \Delta A)) + \tau_n((\Delta A \to \sim B) \wedge (\sim B \to \Delta A)) \\
&= \tau_n((\Delta A \to \sim B) \wedge (\sim B \to \Delta A)) = \xi_n(\Delta A, \sim B). \\
&\xi_n(\Delta A \vee \sim B, \sim B \wedge \Delta A) \\
&= 1 - \tau_n((\Delta A \vee \sim B) \vee (\Delta A \wedge \sim B)) + \tau_n((\Delta A \vee \sim B) \wedge (\Delta A \wedge \sim B)) \\
&= 1 - \tau_n(\Delta A \vee \sim B) + \tau_n(\Delta A \wedge \sim B) = \xi_n(\Delta A, \sim B).
\end{aligned}$$

In summary, we get

$$\xi_n(\Delta A \to \sim B, \sim B \to \Delta A) = \xi_n(\Delta A \vee \sim B, \Delta A \wedge \sim B) = \xi_n(\Delta A, \sim B).$$

□

Definition 7 Define binary nonnegative functions $\rho_n : F(S) \times F(S) \to [0, 1]$, as follows

$$\rho_n(mA, nB) = 1 - \xi_n(mA, nB), A, B \in F(S),$$

then we say that $\rho_n(mA, nB)$ is a pseudo-metric on $F(S)$.

Theorem 9 *Let $A, B \in F(S)$, then*
(i) *If one of A and B is a tautology, another is a contradiction, then $\rho_n(mA, nB) = 1$;*
(ii) *If both A and B are tautologies or both are contradictions, then $\rho_n(mA, nB) = 0$.*

Proof The prove of (i), (ii) can be obtained by Theorem 7 and Definition 7. □

5 T Divergent Degree and T Consistency Degree of Theory Γ of Propositional Formulas

Definition 8 Let Γ be a theory of $F(S)$, define

$$\mathrm{div}(\Gamma) = \sup\{\rho_n(mA, nB)|A, B \in D(\Gamma)\}.$$

Then it's said that $\mathrm{div}(\Gamma)$ is the t divergent degree of theory Γ. When $\mathrm{div}(\Gamma) = 1$, then it's said that Γ is a theory of full divergence.

Example 1 Calculating the divergence degree of $\Gamma = \{p, \sim p\}$.
Solution: Since $\vdash p \to (\sim p \to tA)$ is set up for each $A \in F(S)$, so $D(\Gamma) = F(S)$, namely $\mathrm{div}(\Gamma) = 1$, therefore Γ is fully divergent.

Definition 9 Let Γ be a theory of $F(S)$, define

$$i(\Gamma) = 1 - \min\{\lceil 1 - \rho_n(mA, nB)\rceil |A, B \in D(\Gamma)\}.$$

Then it's said that $i(\Gamma)$ is the pole index of Γ.

Proposition 1 *Let Γ be a theory of $F(S)$, define Γ is inconsistent if and only if $i(\Gamma) = 1$, Γ is consistent if and only if $i(\Gamma) = 0$.*

Proof Γ is inconsistent if and only if $\rho_n(mA, nB) = 1$ if and only if $i(\Gamma) = 1$, Γ is consistent if and only if $\rho_n(mA, nB) = 0$ if and only if $i(\Gamma) = 0$. □

Definition 10 Let Γ be a theory of $F(S)$, define

$$\eta(\Gamma) = 1 - \frac{1}{2}\mathrm{div}(\Gamma)(1 + i(\Gamma)).$$

Then it's said that $\eta(\Gamma)$ is the η-t consistent degree of theory Γ.

Proposition 2 *Let Γ be a theory of $F(S)$,then*
(i) *Γ is fully consistent, namely Γ is a full theorem if and only if $\eta(\Gamma) = 1$;*
(ii) *Γ is consistent if and only if $\frac{1}{2} \leq \eta(\Gamma) \leq 1$, Γ is consistent and fully divergent if and only if $\eta(\Gamma) = \frac{1}{2}$;*
(iii) *Γ is inconsistent if and only if $\eta(\Gamma) = 0$.*

Proof (i) Let Γ is fully consistent, then $\forall A, B \in D(\Gamma)$, we have $\rho_n(mA, nB) = 0$, thus $\mathrm{div}(\Gamma) = 0$, so by Definition 10, we get $\eta(\Gamma) = 1$. On the other hand, let $\eta(\Gamma) = 1$, then by Definition 10, we get $\frac{1}{2}\mathrm{div}(\Gamma)(1 + i(\Gamma)) = 0$, and $\frac{1}{2}(1 + i(\Gamma)) \neq 0$, so $\mathrm{div}(\Gamma) = 0$, thus the formulas of $D(\Gamma)$ are all theorems, therefore Γ is fully consistent.

(ii) By Proposition 1, when Γ is consistent, then $i(\Gamma) = 0$, thus by Definition 10, we have $\frac{1}{2} \leq \eta(\Gamma) \leq 1$. On the other hand, when Γ is inconsistent, then $i(\Gamma) = 1$ and $\mathrm{div}(\Gamma) = 1$, thus $\eta(\Gamma) = 0$, $\frac{1}{2} \leq \eta(\Gamma) \leq 1$ is not set up.

When $\eta(\Gamma) = \frac{1}{2}$, then by Definition 10, we get $\mathrm{div}(\Gamma)(1 + (i(\Gamma)) = 1$, and $\mathrm{div}(\Gamma) \leq 1$, so $\mathrm{div}(\Gamma) = 1, i(\Gamma) = 0$, therefore by Definition 8 and the Proposition 1, we get that Γ is consistent and fully divergent. On the other hand, since Γ is consistent and fully divergent, so we get $\mathrm{div}(\Gamma) = 1, i(\Gamma) = 0$, so by Definition 10, we get $\eta(\Gamma) = \frac{1}{2}$.

(iii) by (ii), we have that when Γ is inconsistent, then $\eta(\Gamma) = 0$. On the other hand, let $\eta(\Gamma) = 0$, then by Definition 5.10, we get $\mathrm{div}(\Gamma)(1 + (i(\Gamma)) = 2$. Since $\mathrm{div}(\Gamma) \leq 1$, so $i(\Gamma) = 1$, therefore by Proposition 1, we get that Γ is inconsistent. □

6 Conclusions

In this paper, by adding new operators Δ, $\sim$, axiomatic expansion of *Gödel* n-valued propositional logic system is introduced, the concepts of t truth degree of propositional formula, and t similarity degree, t pseudo-metric between propositional formulas, and t divergent degree and t consistent degree of theory Γ in $Gödel_\sim$ are given. And their correlation properties are discussed. The research of approximate reasoning about $Gödel_\sim$ n-valued propositional logic system, we will be discussed in another paper.

Acknowledgments Project Supported by the National Natural Science Foundation of China under Grant (11471007); the Natural Science Foundation of Shaanxi Province under Grant (2014JM1020); Graduate Innovation Fund of Yan'an University (YCX201612).

References

1. Wang, G.J.: Introduction to quantitative logic. Fuzzy Syst. Math. **26**(4), 1–11 (2012)
2. Wang, G.J., Liu, B.C.: The theory of relative Γ-tautology degree of formulas in four propositional logic. Chin. J. Eng. Math. **24**(4), 598–610 (2007)
3. Zhou, J.R., Wu, H.B.: An equivalent definition and some properties of truth degree in Łukasiewicz proposition logic system. Chin. J. Eng. Math. **30**(4), 580–590 (2013)
4. Hui, X.J., Wang, G.J.: Randomized studies and applications of classical reasoning model. Sci. China: E Ser. **37**(6), 801–812 (2007)
5. Hui, X.J., Wang, G.J.: Randomized studies and applications of classical reasoning model (II). Fuzzy Syst. Math. **22**(3), 21–26 (2008)
6. Zuo, W.B.: The theory of μ-truth degree of formula in many-valued propositional logic. Chin. J. Syst. Sci. Math. **31**(7), 879–892 (2011)
7. Li, J., Wang, G.J.: The theory of α-truth degree of *Gödel* n-valued propositional logic system. Chin. J. Softw. **18**(1), 33–39 (2007)
8. Wu, H.B., Zhou, J.R.: The form of mean representation of truth degree with applications in quantitative logic. Acta Electron. Sin. **40**(9), 1821–1828 (2012)
9. Han, B.H., Li, Y.M.: The approximate reasoning of quantitative logic. Fuzzy Syst. Math. **24**(5), 1–7 (2010)
10. Wang, G.J., Gao, X.N.: The concept and its application of the truth degree of the theory in proposition logic system. J. Shaanxi Normal Univ. (Natural science edition). **37**(5), 1–6 (2009)
11. Yu, H., Zhan, W.R., Wang, G.J.: The distribution of the truth degree. divergent degree and consistent degree in logic system L_n. J. Shaanxi Normal Univ. (Natural science edition). **36**(5), 6–9 (2008)
12. Wei, H.X.: The consistent degree of the theory in n-valued Łukasiewicz proposition logic system. Comput. Eng. Appl. 1–5
13. Cui, Y.L., Wu, H.B.: The relative divergent degree and relative consistent of the theory in L_n^* system. Fuzzy Syst. Math. **25**(6), 53–59 (2011)
14. Hui, X.J.: Quantified axiomatic extension systems of $SBL_{\sim}$ based on truth value. Sci. Sin. Inf. **44**(7), 900–911 (2014)
15. Cintula, P.: Weakly implicative (fuzzy) logics I: basic properties. Arch. Math. Log. **45**, 673–704 (2006)
16. Esteva, F., Godo, L., Hjek, P., et. al.: Residuated fuzzy logics with an involutive negation. Arch. Math. Log. **39**, 103–124 (2000)
17. Flaminio, T., Marchioni, E.: T-norm based logics with an independent involutive negation. Fuzzy Set Syst. **157**, 3125–3144 (2006)
18. Baaz, M.: Infinite-valued *Gödel* logic with 0–1 projections and relativisations. Comput. Sci. Phys. Lect. Notes Log. **6**, 23–33 (1996)
19. Cintula, P., Klement, E.P., Mesiar, R., et. al.: Fuzzy logics with an additional involutive negation. Fuzzy Set Syst. **161**, 390–411 (2010)

Soundness and Completeness of Fuzzy Propositional Logic with Three Kinds of Negation

Zheng-Hua Pan

Abstract In order to distinguish and deal with different negations of fuzzy knowledge, we have presented a fuzzy propositional logic with contradictory negation, opposite negation, medium negation, FSCOM for short. In this paper, we further study semantics of FLCOM. Based on the three-value interpretation of FLCOM, the soundness theorem and completeness theorem for FLCOM are proved.

Keywords Fuzzy propositional logic · Negation · Semantic interpretation · Soundness · Completeness

1 Introduction

The concept of negation plays a special role in knowledge representation and knowledge reasoning, especially in fuzzy knowledge representation and knowledge reasoning. In known fuzzy sets and fuzzy logic theories, such as FS (fuzzy sets), IFS (intuitionistic fuzzy sets), RS (rough sets) and Hájek's BL (basic logic) [1–4], fuzzy negation N is a generalization of the classical complement or negation $\neg(\neg 0 = 1, \neg 1 = 0)$, N is defined as $N(x) = 1 - x$ or $N(x) = x \quad \to 0$. Among these theories, there has only one sort of negation, which is just distinction of expressions, they have no essential difference with understanding the concept of negation. However, some scholars have cognized that negation is not a clean concept from a logical point of view, there are different negations in Database Query Languages (such as SQL), Production Rule Systems (such as CLIPS and Jess), Natural Language Processing and Semantic Web and so on [5–12]. Wagner et al. proposed that there are (at least) two kinds of negations in the above domains, a *weak negation* expressing non-truth (in the sense of "she doesn't like snow" or "he doesn't trust you"), and a strong negation expressing explicit falsity (in the sense of "she dislikes snow" or "he distrusts you") [5, 6]; Ferré introduced an epistemic extension of the concept of

Z.-H. Pan (✉)
School of Science, Jiangnan University, Wuxi 214122, China
e-mail: panzh@jiangnan.edu.cn; pan_zhenghua@163.com

T.-H. Fan et al. (eds.), *Quantitative Logic and Soft Computing 2016*,
Advances in Intelligent Systems and Computing 510,
DOI 10.1007/978-3-319-46206-6_8

negation, considered that there are extensional negation and intentional negation in logical concept analysis and natural language [8]; Kaneiwa proposed a "description logic $ALC_{\sim}$ with classical negation and strong negation, the classical negation $\neg$ represents the negation of a statement, the strong negation $\sim$ may be more suitable for expressing explicit negative information (or negative facts), in other words, $\sim$ indicates information that is directly opposite and exclusive to a statement rather than its complementary negation [9]; Pan et al. introduced an epistemic extension for the concept of negation, the negative relations in fuzzy concepts should distinguish contradictory relation, opposite relation and medium negative relation [10–12], and then presented the fuzzy sets with contradictory negation, opposite negation and medium negation (FSCOM for short) [13], the fuzzy propositional logic system with contradictory negation, opposite negation and medium negation (FLCOM for short) [14].

This paper study semantics of FLCOM, proving the soundness theorem and completeness theorem for FLCOM based on a three-valued interpretation of FLCOM.

2 Preliminaries

In [14], the negation of a fuzzy concept P was distinguished as the contradictory negation $\neg P$, the opposite negation $\urcorner P$ and the medium negation $\sim P$.

(i) The contradictory negative relation between the fuzzy concepts P and $\neg P$

Characteristics of relation: the boundary between the extensions of $\neg P$ and P is uncertain, and the extensions relation must be "either this or that".

For example, the relation between "young people"(P) and "non-young people" ($\neg P$) within the genus concept "*people*", the relation between "quick velocity"(P) and "non-quick velocity"($\neg P$) within the genus concept "velocity". The characteristic of these relations can be expressed as in the following figure (Fig. 1).

(ii) The opposite negative relation between fuzzy concepts P and $\urcorner P$

Characteristics of relation: the boundaries between the extensions of $\urcorner P$ and P are uncertain, and the extensions relation is not "either this or that".

For example, the relation between "young people"(P) and "old people"($\urcorner P$) within the genus concept "people", the relation between "quick velocity"(P) and "*slow velocity*"($\urcorner P$) within the genus concept "*velocity*". The characteristic of these relations can be expressed as the following figure (Fig. 2).

(iii) The medium negative relations between P (or $\urcorner P$) and $\sim P$

Characteristics of relation: the boundaries between the extensions of P (or $\urcorner$ P) and $\sim P$ are uncertain, t and the extensions relation are not "either this or that".

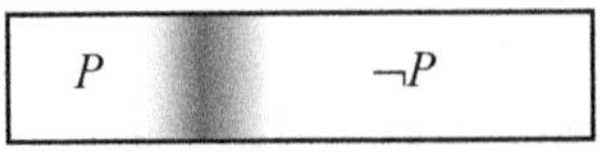

Fig. 1 The boundary between the extensions of $\neg P$ and P is uncertain, and the extensions relation must be "either this or that"

Fig. 2 The boundaries between the extensions of $\urcorner P$ and P are uncertain, and the extensions relation is not "either this or that"

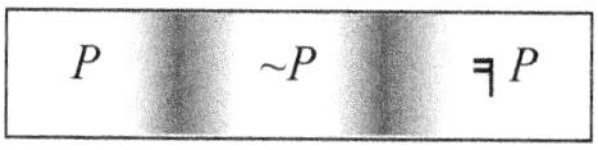

Fig. 3 The boundaries between the extensions of P (or $\urcorner P$) and $\sim P$ are uncertain, and the extensions relations are not "either this or that"

For example, the relation between "young people"(P) (or "old people"($\urcorner P$)) and "middle-aged people"($\sim P$) within the genus concept "people", the relation between "night"(P) (or "daylight"($\urcorner P$)) and "dawn"($\sim P$), the relation between "conductor"(P) (or "dielectric"($\urcorner P$)) and "semiconductor"($\sim P$). The characteristic of these relations can be expressed as the following figure (Fig. 3).

In [13], the contradictory negation $A^{\neg}$ of a fuzzy set A in FSCOM have relations with the opposite negation $A^{\urcorner}$ and the medium negation $A^{\sim}$: $A^{\neg} = A^{\urcorner} \cup A^{\sim}$. Accordingly, the above contradictory negation $P^{\neg}$ of a fuzzy concept P has the following relation with the opposite negation $P^{\urcorner}$ and the medium negation $P^{\sim}$:

$$\neg P = \urcorner P \cup \sim P \tag{1}$$

Based on the above cognition, [14] presented the following fuzzy propositional logic system with contradictory negation, opposite negation and medium negation (FLCOM for short):

Definition 1 (i) Let $\Im$ be the set of proposition variables, $\neg$ (contradictory negation), $\urcorner$ (opposite negation), $\sim$ (medium negation) and $\wedge$ (conjunction) are propositions connectives.

(ii) The following formulas are axioms:

(A1) $A \to (B \to A)$;
(A2) $(A \to (A \to B)) \to (A \to B)$;
(A3) $(A \to B) \to ((B \to C) \to (A \to C))$;
(M_1) $(A \to \neg B) \to (B \to \neg A)$;
(M_2) $(A \to \urcorner B) \to (B \to \urcorner A)$;
(H) $\neg A \to (A \to B), \urcorner A \to (A \to B)$;
($\wedge$) $A \wedge B \to A, A \wedge B \to B$;
($Y_{\urcorner}$) $\urcorner A \to \neg A \wedge \neg \sim A$;
($Y_{\sim}$) $\sim A \to \neg A \wedge \neg \urcorner A$;

(iii) The following deduction forms are deduction rules:

(Dr1) $A_1, A_2, \ldots, A_n \vdash A_i (1 \le i \le \text{n})$;
(Dr2) $A \rightarrow B, A \vdash B$.

In [14], a lot of particular properties of FLCOM have been proven. We can also prove the following deduction forms in FLCOM:

Proposition 1 *In FLCOM,*

$$\urcorner \sim A \vdash B \tag{2}$$

$$\sim \urcorner A \vdash \sim A; \sim A \vdash \sim \urcorner A \tag{3}$$

$$\urcorner\urcorner A \vdash A; A \vdash \urcorner\urcorner A \tag{4}$$

$$B \vdash A \rightarrow B; \urcorner A \vdash A \rightarrow B \tag{5}$$

$$A, \urcorner B \vdash \urcorner (A \rightarrow B); \urcorner (A \rightarrow B) \vdash A, \urcorner B \tag{6}$$

Now that FLCOM is a formalized system, the deduction form $\Sigma \vdash A$ (Σ may be empty set) is provable in FLCOM, which can be defined as follows:

Definition 2 The deduction form $\Gamma \vdash A$ (Γ may be empty set) is provable in FLCOM, if there exists a finite sequence of deduction forms $E_1, E_2, \ldots, E_n$ such that $E_n = \Gamma \vdash A$ and for each $E_n (1 \le k \le n)$, E_k is an axiom of FLCOM or E_k follows from E_i and $E_j (i < k, j < k)$ using the deduction rule of FLCOM, then $E_1, E_2, \ldots, E_n$ is called a proof of $\Gamma \vdash A$, n length of proof. $\Gamma \vdash A$ is denoted by $\vdash A$ when Γ is the empty set, A is provable in FLCOM when $\vdash A$ is provable in FLCOM . If Σ denotes an infinite set of formulas and there is a finite set $\Gamma \subset \Sigma$ such that $\Gamma \vdash A$ is provable in FLCOM, then $\Sigma \vdash A$ is provable in FLCOM.

3 The Soundness and Completeness of FLCOM

We can prove the soundness theorem and completeness theorem of FLCOM based on a three-valued interpretation.

Definition 3 (*three-valued interpretation*) Let Σ be a set of formulas of FLCOM. A Mapping $\partial : \Sigma \rightarrow \{0, 1/2, 1\}$ is called a three-valued assignment for all $A, B \in \Sigma$, if

(a) $\partial(A) + \partial(\urcorner A) = 1$;

(b) $\partial(\sim A) = \begin{cases} 1/2, & \text{when } \partial(A) = 1 \\ 1, & \text{when } \partial(A) = 1/2 \\ 1/2, & \text{when } \partial(A) = 0; \end{cases}$

(c) $\partial(A \rightarrow B)$ is a binary function $\Re(\partial(A), \partial(B))$;

$\Re$	1	1/2	0
1	1	1/2	0
1/2	1	1	1/2
0	1	1	1

(d) $\partial(A \wedge B) = min(\partial(A), \partial(B))$;
(e) $\partial(\neg A) = max(\partial(\urcorner A), \partial(\sim A))$.

Definition 4 (*three-valued valid formula*) (i) If $\partial(A) = 1$ for each three-valued assignment ∂ then A is called a three-valued valid formula, which is denoted by $\models A$. (ii) For each $B \in \Sigma$, if there is a three-valued assignment ∂ such that $\partial(B) = 1$, then Σ is called three-valued satiable. (iii) For each three-valued assignment ∂, if ∂ satisfies $\{A\}$ when ∂ satisfies Σ then $\Sigma \vdash A$ is called a three-valued valid deduction form, which is denoted by $\Sigma \models A$.

Theorem 1 (Soundness of FLCOM) *(a) If* $\vdash A$*, then* $\models A$*. (b) If* $\Sigma \vdash A$*, then* $\Sigma \models A$*.*

Proof (a) Suppose that $\vdash A$. According to Definition 2, there exists a proof of $\vdash A$: $E_1, E_2, \ldots, E_n$, in which $E_n = \vdash A$. We induct on the length n as follows:

(i) If $n = 1$, then $E_1 = \vdash A$ by Definition 2, namely, A is an axiom of FLCOM. $\vdash A$ is a three-valued valid deduction form, since the axioms of FLCOM are three-valued valid formulas. $\models A$ by Definition 4.

(ii) Suppose that (a) holds if $n < k$. When $n = k$, according to Definition 2, E_n is an axiom of FLCOM or E_n follows from E_i and $E_j (i < n, j < n)$ by using the deduction rule of FLCOM. If E_n is an axiom of FLCOM, then (a) holds as (I). If E_n follows from A_i and $A_j (i < n, j < n)$ using the deduction rule, then E_i and E_j are three-valued valid deduction forms by the induction hypothesis, so E_n is three-valued valid deduction form, namely, $\vdash A$ is a three-valued valid deduction form. So $\models A$ by Definition 4.

Therefore, (a) holds by the (I) and (II). (b) Can be proved similarly. □

Definition 5 (*Maximally consistent set*) Let Σ be a set of formulas of FLCOM (Σ may be finite or infinite). If $\Sigma \vdash A$ is provable in FLCOM for each formula A, then Σ is called an inconsistent set. If Σ is an consistent set, and that for each consistent set $\Sigma' \supseteq \Sigma$ sure that $\Sigma' = \Sigma$, then Σ is called a maximally consistent set.

Proposition 2 *For all consistent set* Σ*, there is a maximally consistent set* Σ^* *and* $\Sigma \subseteq \Sigma^*$.

Proof Let $A_1, A_2, \ldots, A_n$, be formulas of FLCOM. Defining $\Sigma_n : \Sigma_1 = \Sigma$, if $\Sigma_n \cup \{A_n\}$ is consistent then $\Sigma_{n+1} = \Sigma_n \cup \{A_n\}$, or else $\Sigma_{n+1} = \Sigma_n$. Let $\Sigma^* = \bigcup_{n=1}^{\infty} \Sigma_n$, Σ^* is proved to be a maximally consistent set and $\Sigma^* \supseteq \Sigma$ by Definition 5. □

Lemma 1 *Let A be a formula of FLCOM,* $A_1, A_2 \in \{A, \urcorner A, \sim A\}$ *and* $A_1 \neq A_2$, Σ *a set of formulas of FLCOM. The following sentences are equivalent:*

(i) Σ *is an inconsistent set;*
(ii) *there exists a* $A \in \Sigma$ *such that* $\Sigma \vdash A_1, A_2$ *is provable in FL*COM;
(iii) *there exists a* $A \in \Sigma$ *such that* $\Sigma \vdash \rceil \sim A$ *is provable in FL*COM.

Proof By Theorem 4 in [14] any two formulas in $\{A, \rceil A, \sim A\}$ are contradictory formulas thus they are inconsistent formulas, so A_1 and A_2 are inconsistent formulas. If Σ is an inconsistent set, then there exists a formula A in Σ such that $\Sigma \vdash A_1, A_2$ is provable in FLCOM according to Definition 5, so (i) $\Rightarrow$ (ii). If there exists a formula A in Σ such that $\Sigma \vdash A_1, A_2$ is provable in FLCOM, on account of A_1 and A_2 are inconsistent formulas, so $A_1, A_2 \vdash B$ (B is a formula of FLCOM), that is $\Sigma \vdash A_1, A_2 \vdash B$. Let $B = \rceil \sim A$ by the arbitrariness of the formula B of FLCOM, so (ii) $\Rightarrow$ (iii). If there exists a formula A in Σ such that $\Sigma \vdash A_1, A_2$ is provable in FLCOM, then $\Sigma \vdash \rceil \sim A \vdash B$ by (2), thus Σ is an inconsistent set, so (iii) $\Rightarrow$ (i). □

Lemma 2 *Let* Σ *be a maximally consistent set. Then*
(i) $\Sigma \vdash A$ *is provable in FL*COM*, if and only if* $A \in \Sigma$.
(ii) If $A \vdash B$ *and* $B \vdash A$ *are provable in FL*COM*, then* $A \in \Sigma$ *if and only if* $B \in \Sigma$.

Proof (i) Suppose that $A \notin \Sigma$, thus $\Sigma \cup \{A\}$ is an inconsistent set since Σ is a maximally consistent set. So, $\rceil A \in \Sigma$ or $\sim A \in \Sigma$. If $\rceil A \in \Sigma$, $\Sigma \vdash \rceil A$ is provable in FLCOM by (Dr1), $\Sigma \vdash A$ is unprovable in FLCOM by Lemma 1. Similarly, if $\sim A \in \Sigma$, $\Sigma \vdash \sim A$ is provable in FLCOM by (Dr1), $\Sigma \vdash A$ is unprovable in FLCOM by Lemma 1. Suppose that $\Sigma \vdash A$ is unprovable in FLCOM, thus $A \notin \Sigma$ by (Dr1).
(ii) can be proved similarly. □

Lemma 3 *Let* Σ *be a maximally consistent set, A and B are formulas of FL*COM. *Then,*

(i) *One and only holds in* $A \in \Sigma$, $\rceil A \in \Sigma$ *and* $\sim A \in \Sigma$.
(ii) $A \to B \in \Sigma$ *if and only if* $\rceil A \in \Sigma$ *or* $B \in \Sigma$.
(iii) $\rceil (A \to B) \in \Sigma$ *if and only if both* $A \in \Sigma$ *and* $\rceil B \in \Sigma$.

Proof (i) If there are two of them in $A \in \Sigma$, $\rceil A \in \Sigma$ and $\sim A \in \Sigma$ that hold, then Σ is an inconsistent set by Lemma 1, this is in contradiction with the hypothesis.

(ii) If $A \to B \in \Sigma$, then $\Sigma \vdash A \to B$ by Lemma 2. If $\rceil A \in \Sigma$, then $\Sigma \vdash \rceil A$ by Lemma 2. For $\vdash \rceil A \to (A \to B)$ by (H), so $\vdash (A \to B)$ according (Dr2), that is $A \to B \in \Sigma$ by Lemma 2. If $B \in \Sigma$, $\Sigma \vdash B$ by Lemma 2. For $\vdash B \to (A \to B)$ by (A), so $\vdash (A \to B)$ according (Dr2), that is $A \to B \in \Sigma$ by Lemma 2.

(iii) It is obvious by Lemma 2 and (6). □

Lemma 4 *Let* Σ *be a maximally consistent set,* (A_1, A_2, A_3) *a permutation of* $(A, \rceil A, \sim A)$. *If* $\Sigma \cup \{A_1\}$ *and* $\Sigma \cup \{A_2\}$ *are inconsistent sets, then* $\Sigma \vdash A_3$ *is provable in FL*COM.

Proof Since Σ is a maximally consistent set, and (A_1, A_2, A_3) is a permutation of $(A, \rceil A \sim A)$, then there exist $A_i \in \{A, \rceil A, \sim A\}$ and $A_i \in \Sigma (1 \le i \le 3)$. Suppose that $\Sigma \vdash A_3$ is unprovable in FLCOM, then $A_3 \notin \Sigma$, and $A_1 \notin \Sigma$, $A_2 \notin \Sigma$ for $\Sigma \cup \{A_1\}$ and $\Sigma \cup \{A_2\}$ are inconsistent sets. Thus $A_1, A_2, A_3 \notin \Sigma$ is in contradiction with $A_i \in \Sigma (1 \le i \le 3)$. □

Lemma 5 *Let Σ be a consistent set, Σ^* a maximally consistent set, A a formula of FLCOM. Then there exists a three-valued assignment ∂ such that*

(i) $\partial(A) = 1$ *if and only if* $A \in \Sigma^*$;
(ii) $\partial(A) = 1/2$ *if and only if* $\sim A \in \Sigma^*$;
(iii) $\partial(A) = 0$ *if and only if* $\urcorner A \in \Sigma^*$.

Proof If p is an atomic formula of FLCOM, by Definition 3, we may define ∂ as follows:

$$\partial(p) = \begin{cases} 1, & \text{if } p \in \Sigma^* \\ 1/2, & \text{if } \sim p \in \Sigma^* \\ 0, & \text{if } \urcorner\, p \in \Sigma^* \end{cases}$$

We induct to the number n of connective in A as follows:
(1) When $n = 0$, then A is an atomic formula of FLCOM, (i), (ii) and (iii) hold by (3).
(2) Suppose that (i), (ii) and (iii) hold if $n < k$. When $n = k$, A has the following three cases: $A = \urcorner\, B$, or $A =\sim B$, or $A = B \to C$.

Let $A = \urcorner\, B$. According to Definition 3, there are the following three cases:

(a) $\partial(\urcorner\, B) = 1$ if and only if $\partial(B) = 0$, so $\urcorner\, B \in \Sigma^*$ by (3), that is $A \in \Sigma^*$.

(b) $\partial(\urcorner\, B) = 1/2$ if and only if $\partial(B) = 1/2$, so $\sim B \in \Sigma^*$ by (3), thus $\sim \urcorner\, B \in \Sigma^*$ by (4), that is $\sim A \in \Sigma^*$.

(c) $\partial(\urcorner\, B) = 0$ if and only if $\partial(B) = 1$, so $B \in \Sigma^*$ by (3), thus $\urcorner\urcorner\, B \in \Sigma^*$ by (4), that is $\urcorner\, A \in \Sigma^*$.

Let $A =\sim B$. It is divided into three parts to prove:

(a) According to Definition 3, $\partial(\sim B) = 1$ if and only if $\partial(B) = 1/2$, so $\sim B \in \Sigma^*$ by (3), that is $A \in \Sigma^*$.

(b) In order to prove that $\partial(\sim B) = 1/2$ if and only if $\sim\sim B \in \Sigma^*$ (i.e. $\sim A \in \Sigma^*$), we may prove that $\partial(\sim B) \neq 1/2$ if and only if $\sim\sim B \notin \Sigma^*$ as follows: suppose that $\partial(\sim B) \neq 1/2$, according to Definition 3, $\partial(B) = 1/2$, and then $\sim B \in \Sigma^*$ by (3), so $\sim\sim B \notin \Sigma^*$ by (i) in Lemma 3; suppose that $\sim\sim B \notin \Sigma^*$, according to (i) in Lemma 3, $\sim B \in \Sigma^*$ or $\urcorner \sim B \in \Sigma^*$, since $\urcorner \sim B \in \Sigma^*$ does not hold (or else $\Sigma^* \vdash \urcorner \sim B \vdash C$ is provable in FLCOM by Lemma 2 and (2), that is, Σ^* is inconsistent), so $\sim B \in \Sigma^*$, and then $\partial(B) = 1/2$ by (3), $\partial(\sim B) = 1$ by Definition 3, that is $(\sim B) \neq 1/2$.

(c) In order to prove that $\partial(\sim B) = 0$ if and only if $\urcorner \sim B \in \Sigma^*$ (i.e. $\urcorner\, A \in \Sigma^*$), we may prove that $\partial(\sim B) \neq 0$ if and only if $\urcorner \sim B \notin \Sigma^*$ as follows: suppose that $\urcorner \sim B \notin \Sigma^*$, according to (i) in Lemma 3, $\sim B \in \Sigma^*$ or $\sim\sim B \in \Sigma^*$. If $\sim B \in \Sigma^*$, then $\partial(B) = 1/2$ by (3), and then $\partial(\sim B) = 1 (\neq 0)$ by Definition 3; if $\sim\sim B \in \Sigma^*$, then $\sim B \notin \Sigma^*$ by (i) in Lemma 3, so $B \in \Sigma^*$ or $\urcorner\, B \in \Sigma^*$, so $\partial(B) = 1$ or $\partial(B) = 0$ by (3), thus $\partial(\sim B) = 1/2$ by Definition 3, that is $\partial(\sim B) \neq 0$. Contrarily, suppose that $\partial(\sim B) \neq 0$, according to Definition 3, $\partial(B) = 1$, or $\partial(B) = 1/2$ or $\partial(B) = 0$, by Lemma 1 and Σ^* is consistent set, there is $\urcorner \sim B \notin \Sigma^*$ when $\partial(B) = 1$, or $\partial(B) = 1/2$ or $\partial(B) = 0$.

Let $A = B \to C$. It is divided into three parts to prove:

(a) According to Definition 3, $\partial(B \to C) = 1$ if and only if $\partial(B) = 0$ or $\partial(C) = 1$, which hold if and only if $\urcorner B \in \Sigma^*$ or $C \in \Sigma^*$ by (3), so $B \to C \in \Sigma^*$ by (ii) in Lemma 3.

(b) According to Definition 3, $\partial(B \to C) = 0$ if and only if $\partial(B) = 1$ or $\partial(C) = 0$, which implies $B \in \Sigma^*$ or $\urcorner C \in \Sigma^*$ by (3), so $\urcorner (B \to C) \in \Sigma^*$ by (iii) in Lemma 3.

(c) In order to prove that $\partial(B \to C) = 1/2$ if and only if $\sim (B \to C) \in \Sigma^*$, we may prove that $\partial(B \to C) \neq 1/2$ if and only if $\sim (B \to C) \notin \Sigma^*$ as follows: according to Definition 3, $\partial(B \to C) \neq 1/2$ if and only if $\partial(B \to C) = 1$ or $\partial(B \to C) = 0$, which holds if and only if $B \to C \in \Sigma^*$ or $\urcorner (B \to C) \in \Sigma^*$ by above (a) and (b), so $\sim (B \to C) \notin \Sigma^*$ by (i) in Lemma 3. □

Theorem 2 (completeness of FLCOM)
(a) *If* $\Sigma \models A$*, then* $\Sigma \vdash A$ *is provable in FL*COM;
(b) *If* $\models A$*, then* $\vdash A$ *is provable in FL*COM.

Proof (a) if $\Sigma \vdash A$ is unprovable in FLCOM, according to Lemma 4, $\Sigma \cup \{\urcorner A\}$ and $\Sigma \cup \{\sim A\}$ are consistent sets, so there exists a three-valued assignment ∂ such that $\partial(A) \leq 1/2$ if ∂ satisfied Σ. Thus $\Sigma \models A$ does not hold, by Definition 4.
(b) This can be proved similarly. □

Theorem 3 (compactness of FLCOM) *Let* Σ *be an infinite set of formulas of FL*COM. *For each finite set* $\Gamma \subset \Sigma$*, if* Γ *is a consistent set then* Σ *is also consistent set.*

Proof According to Lemma 1, if Σ is an inconsistent set then $\Sigma \vdash \urcorner \sim A$ is provable in FLCOM. By Definition 2, there exist a finite set $\Gamma \subset \Sigma$ such that $\Gamma \vdash \urcorner \sim A$ is provable in FLCOM, and then $\Sigma \vdash B$ by (2), namely, Γ is an inconsistent set. □

4 Conclusions

This paper presents a fuzzy propositional logic system with contradictory negation, opposite negation and medium negation (FLCOM) and study semantics of FLCOM, and introduce a three-value interpretation of FLCOM, the soundness theorem and the completeness theorem of FLCOM are proved.

Acknowledgments This work is supported by the National Natural Science Foundation of China (61375004) and the Fundamental Research Funds for the Central Universities (JUSRP51317B).

References

1. Zadeh, L.A.: Fuzzy sets. Inf. Control **8**(3), 338–353 (1965)
2. Pawlak, Z.: Rough sets. Int. J. Comput. Inf. Sci. **11**, 341–356 (1982)
3. Atanassov, K.: Intuitionistic fuzzy sets. Fuzzy Sets Syst. **20**(1), 87–96 (1996)

4. Hájek, P.: Metamathematics of Fuzzy Logic. Kluwer Academic Publishers, Dordrecht (1998)
5. Wagner, G.: Web Rules Need Two Kinds of Negation. Lecture Notes in Computer Science, vol. 2901. Springer, Berlin (2003)
6. Analyti, A., Antoniou, G., Wagner, G., et al.: Negation and negative information in the W3C resource description framework. Ann. Math., Comput. Teleinf. **1**(2), 25–34 (2004)
7. Dung, P.M., Mancarella, P.: Production systems need negation as failure. IEEE Trans. Knowl. Data Eng. **14**(2), 336–353 (2002)
8. Ferré, S.: Negation, Opposition, and Possibility in Logical Concept Analysis. Lecture Notes in Artificial intelligence, vol. 3874, pp. 130–145. Springer, Berlin (2006)
9. Kaneiwa, K.: Negations in description logic with contraries, contradictories, and subcontraries. New Gener. Comput. **25**(4): 443–468
10. Pan, Z.H., Zhu, W.J.: A new cognition and processing on contradictory knowledge. In: Proceedings of the IEEE-International Conference on Machine Learning and Cybernetics, vol. 3, pp. 1532–1537. IEEE-Computer Society, Washington (2006)
11. Pan, Z.H., Zhang, S.L.: Differentiation and processing on contradictory relation and opposite relation in knowledge. In: Proceedings of the IEEE-Fourth International Conference on Fuzzy Systems and Knowledge Discovery (FSKD), vol. 4, pp. 334–3389. IEEE-Computer Society, Washington (2007)
12. Pan, Z.H., Wang, C., Zhang, L.J.: Three kinds of negations of fuzzy knowledge and applications to decision making in financial investment. Lecture Notes in Artificial Intelligence, vol. 6422, pp. 391–401. Springer, Berlin (2010)
13. Pan, Z.H.: Fuzzy set with three kinds of negation and its applications in fuzzy decision making. Lecture Notes in Artificial Intelligence, vol. 7002, pp. 533–542. Springer, Berlin (2011)
14. Pan, Z.H.: Fuzzy propositional logic system of distinguish between three kinds of negation and its application. J. Softw. **25**(6), 1255–1272 (2014)

Normal Form of n-Valued Lukasiewicz Logic Formulas

Qing-Ping Wang

Abstract The disjunctive normal form and conjunctive normal form of Boolean functions are very important to construct logic formulas in classical logic system. Shannon expansion in symbolic computation tree logic is generalized to prove the normal form of Boolean functions. In n-valued Lukasiewicz logic system L_n, the expansion of n-valued McNaughton functions which are induced by logic formulas is studied. The quasi disjunctive normal form and quasi conjunctive normal form of m-ary n-valued McNaughton functions are given.

Keywords Shannon expansion · Boolean function · n-valued McNaughton function · Normal form

1 Introduction

Since the technology of model checking was proposed, it has been successful applied in various fields including aerospace, electronic commerce, communication technology and medical systems. Model checking is carried out on the basis of classical logic [1–11]. In classical logic system, the normal form is an important method to structural logic formulas. The structure of logic formulas can be clearly understood by means of the normal form. As everyone knows, the number of classical propositional logic formulas which include with m atomic formulas is equal to m-ary Boolean functions,

---

alfred.hofmann, ursula.barth, ingrid.haas, frank.holzwarth, anna.kramer, leonie.kunz, christine.reiss, nicole.sator, erika.siebert-cole, peter.strasser, lncs@springer.com

---

Q.-P. Wang (✉)
School of Statistics, Jiangxi University of Finance and Economics,
Nanchang, People's Republic of China
e-mail: lc_wqp@163.com
URL: http://www.springer.com/lncs

Q.-P. Wang
Research Center of Applied Statistics, Jiangxi University of Finance and Economics,
Nanchang, People's Republic of China

T.-H. Fan et al. (eds.), *Quantitative Logic and Soft Computing 2016*,
Advances in Intelligent Systems and Computing 510,
DOI 10.1007/978-3-319-46206-6_9

that is, 2^{2^m}. But for many-valued logic, the problem is much more complicated. For example, in three-value Lukasiewicz logic, the number of three-value propositional logic formulas which incl with m atomic formulas is not equal to 3^{3^m}. Because expansions of Boolean functions are given by Shannon expansion, it is not complicated, but the idea is clever. In this paper, firstly, the disjunctive normal form and conjunctive normal form of Boolean functions are proved by Shannon expansion. Then the quasi normal form of n-valued logic formulas is proposed, and the expansion n-value McNaughton function is given. Finally, in n-value Lukasiewicz logic system L_n, the quasi normal form and construction method of logical formulas are obtained.

2 Preliminaries

Definition 1 ([12]) Given $S = \{p_1, p_2, \ldots\}$. Let us define the set $F(S)$ inductively as follows.

(1) $p_1, p_2, \ldots \in F(S)$.

(2) If $A, B \in F(S)$, then $\neg A, A \to B \in F(S)$.

(3) Every element of $F(S)$ is generated by (1) and (2).

Then $F(S)$ is a free algebra of type $(\neg, \to)$ generated by S. Each member of $F(S)$ is called a formula (also called proposition) of classical logic L, and each member of S is called an atomic formula (atomic proposition) of L.

In Boolean algebra $\{0, 1\}$, operations $\neg, \to, \vee, \wedge$ are defined as follows.

$$\neg 0 = 1, \neg 1 = 0, a \to b = 0 \text{ iff } a = 1 \text{ and } b = 0, \tag{1}$$

$$a \vee b = \max\{a, b\}, a \wedge b = \min\{a, b\}. \tag{2}$$

Definition 2 ([12]) Let $\upsilon : F(S) \to \{0, 1\}$ be a mapping. υ is called a valuation of $F(S)$ if υ is a homomorphism of type $(\neg, \to)$, i.e.

$$\upsilon(\neg A) = \neg\upsilon(A), \upsilon(A \to B) = \upsilon(A) \to \upsilon(B).$$

$\upsilon(A)$ is also called the valuation of A. The set of all valuations of $F(S)$ is denoted by Ω.

Let $A, B \in F(S)$. A is called a tautology, if $\upsilon(A) = 1$ holds for every $\upsilon \in \Omega$. A is called a contradiction, if $\upsilon(A) = 0$ for every $\upsilon \in \Omega$. A and B are said to be logically equivalent, in symbols, $A \approx B$, if $\upsilon(A) = \upsilon(B)$ holds for every $\upsilon \in \Omega$.

Definition 3 ([12]) A function $f : \{0, 1\}^m \to 0, 1$ is called an m-ary Boolean function.

Proposition 1 ([12]) *Every Boolean function can be induced by some formula.*

Definition 4 ([12]) Let $A(p_1, \ldots, p_m) \in F(S)$. A is said to be in disjunctive normal form and in conjunctive normal form, respectively, if A has the respective forms

$$(Q_{11} \wedge \cdots \wedge Q_{1s}) \vee \cdots \vee (Q_{t1} \wedge \cdots \wedge Q_{ts}), \quad (3)$$

and

$$(Q_{11} \vee \cdots \vee Q_{1s}) \wedge \cdots \wedge (Q_{t1} \vee \cdots \vee Q_{ts}), \quad (4)$$

where each $Q_{ij} = p_j$ or $Q_{ij} = \neg p_j (j = 1, \ldots, s; i = 1, \ldots, t)$.

Remark 1 A formula built up from either atomic propositions or the negations of atomic formulas only by disjunctive (conjunctive) connectives is said to be in simple disjunctive (conjunctive) normal form. For example, $p_1 \vee p_2 \vee p_3$ is in simple disjunctive normal form, and $\neg p_1 \wedge p_2$ is in simple conjunctive normal form. A simple disjunctive (conjunctive) normal form can be viewed as either a disjunctive normal form or a conjunctive normal form.

Definition 5 ([1]) Let $f(x_1, \ldots, x_m)$ be a m-ary Boolean function, then

$$f(x_1, \ldots, x_m) = (\neg x_1 \wedge f(0, x_2, \ldots, x_m)) \vee (x_1 \wedge f(1, x_2, \ldots, x_m)). \quad (5)$$

is called Shannon expansion.

In the following, we give the generalization of Shannon expansion.

Proposition 2 *Let $f(x_1, \ldots, x_m)$ be a m-ary Boolaen function, denote $x_k^0 = \neg x_k, x_k^1 = x_k, (k = 1, 2)$, then*

$$f(x_1, x_2, x_3, \ldots, x_m) = \vee\{x_1^{\alpha_1} \wedge x_2^{\alpha_2} \wedge f(\alpha_1, \alpha_2, x_3, \ldots, x_m) : (\alpha_1, \alpha_2) \in \{0, 1\}^2\}. \quad (6)$$

Proof From Shannon expansion, we have

$$f(x_1, x_2, x_3, \ldots, x_m) = (\neg x_1 \wedge f(0, x_2, x_3, \ldots, x_m)) \vee (x_1 \wedge f(1, x_2, x_3, \ldots, x_m)). \quad (7)$$

and
$f(0, x_2, x_3, \ldots, x_m) = (\neg x_2 \wedge f(0, 0, x_3, \ldots, x_m)) \vee (x_2 \wedge f(0, 1, x_3, \ldots, x_m))$,
$f(1, x_2, x_3, \ldots, x_m) = (\neg x_2 \wedge f(1, 0, x_3, \ldots, x_m)) \vee (x_2 \wedge f(1, 1, x_3, \ldots, x_m))$,
substitute in Eq. (7)
$f(x_1, x_2, x_3, \ldots, x_m) = \vee\{x_1^{\alpha_1} \wedge x_2^{\alpha_2} \wedge f(\alpha_1, \alpha_2, x_3, \ldots, x_m) : (\alpha_1, \alpha_2) \in \{0, 1\}^2\}$.
More generally, the disjunctive normal form of Boolean functions can be obtained by Shannon expansion and its extended form.

Corollary 1 *Let $f(x_1, \ldots, x_m)$ be a m-ary Boolaen function, then $f(x_1, \ldots, x_m) = \vee\{x_1^{\alpha_1} \wedge \cdots \wedge x_m^{\alpha_m} \wedge f(\alpha_1, \ldots, \alpha_m) : (\alpha_1, \ldots, \alpha_m) \in \{0, 1\}^m\} = \vee\{x_1^{\alpha_1} \wedge \cdots \wedge x_m^{\alpha_m} : (\alpha_1, \ldots, \alpha_m) \in f^{-1}(1)\}$.*

Remark 2 If $f^{-1}(1) = \emptyset$, i.e. $f(x_1, \ldots, x_m) = 0$, we can denote $f(x_1, \ldots, x_m) = x_1 \wedge \neg x_1$.

In the following, we give the Dual form of Shannon expansion and its extended form.

Proposition 3 *Let $f(x_1, \ldots, x_m)$ be a m-ary Boolean function, denote $x_k^0 = x_k$, $x_k^1 = \neg x_k$, $(k = 1, 2)$, then*

$$f(x_1, x_2, x_3, \ldots, x_m) = \wedge\{x_1^{\beta_1} \vee x_2^{\beta_2} \vee f(\beta_1, \beta_2, x_3, \ldots, x_m) : (\beta_1, \beta_2) \in \{0, 1\}^2\}. \quad (8)$$

Corollary 2 *Let $f(x_1, \ldots, x_m)$ be a m-ary Boolaen function, then $f(x_1, \ldots, x_m) = \wedge\{x_1^{\beta_1} \vee \cdots \vee x_m^{\beta_m} \vee f(\beta_1, \ldots, \beta_m) : (\beta_1, \ldots, \beta_m) \in \{0, 1\}^m\} = \wedge\{x_1^{\beta_1} \vee \cdots \vee x_m^{\beta_m} : (\beta_1, \ldots, \beta_m) \in f^{-1}(0)\}$.*

Remark 3 If $f^{-1}(0) = \emptyset$, i.e. $f(x_1, \ldots, x_m) = 1$, we can denote $f(x_1, \ldots, x_m) = x_1 \vee \neg x_1$.

Theorem 1 ([12]) *Every formula is logically equivalent to a formula in disjunctive (conjunctive) normal form.*

3 The Normal Form of n-Valued McNaughton Function

In this paper, denote $L(n) = \{0, \frac{1}{n-1}, \frac{2}{n-1}, \ldots, \frac{n-2}{n-1}, 1\}$, $h_0(x) = \neg(\neg x \to x)$, $h_{\frac{1}{2}}(x) = (x \to \neg x) \wedge (\neg x \to x)$, $h_1(x) = \neg(x \to \neg x)$.

In n-valued Lukasiewicz logic system L_n, the definitions of valuation, tautology, contradiction, function induced by formula A are similar to the definitions in classical logic system L. However, in L_n,

$$x \to y = min\{1 - x + y, 1\}, x, y \in L(n).$$

Definition 6 ([13]) If function $g : (L(n))^m \to L(n)$ can be induced by any logic formula in L_n, then $g(x_1, \ldots, x_m)$ is called a m-ary n-valued McNaughton function.

Remark 4 In classical logic system, every m-ary Boolean function $f : \{0, 1\}^m \to \{0, 1\}$ can be induced by a formula. However, in n-value logic system, every m-ary n-value function can not be necessarily induced by a formula. Therefore, in n-value logic system, the structure of m-ary n-value functions which are induced by logic formulas is more complicated than Boolean functions. In the following, we will generate Shannon expansion techniques to solve this problem. Firstly, the normal form of three-valued McNaughton function is researched.

Proposition 4 *Let $g(x_1, \ldots, x_m)$ be three-valued McNaughton function, then*

$$\begin{aligned} g(x_1, x_2, \ldots, x_m) = &(h_0(x_1) \wedge g(0, x_2, \ldots, x_m)) \\ &\vee(h_{\frac{1}{2}}(x_1) \wedge g(\tfrac{1}{2}, x_2, \ldots, x_m)) \\ &\vee(h_1(x_1) \wedge g(1, x_2, \ldots, x_m)). \end{aligned}$$

Proposition 5 *Let $g(x_1, \ldots, x_m)$ be three-valued McNaughton function, denote $x_k^0 = h_0(x_k), x_k^{\frac{1}{2}} = h_{\frac{1}{2}}(x_k), x_k^1 = h_1(x_k), (k = 1, 2)$, then*

$$g(x_1, x_2, x_3, \ldots, x_m) = \vee \left\{ x_1^{\alpha_1} \wedge x_2^{\alpha_2} \wedge g(\alpha_1, \alpha_2, x_3, \ldots, x_m) : (\alpha_1, \alpha_2) \in \left\{0, \frac{1}{2}, 1\right\}^2 \right\}.$$

Corollary 3 *Let $g(x_1, \ldots, x_m)$ be three-valued McNaughton function, then* $g(x_1, x_2, x_3, \ldots, x_m) = \vee\{x_1^{\alpha_1} \wedge \cdots \wedge x_m^{\alpha_m} \wedge g(\alpha_1, \ldots, \alpha_m) : (\alpha_1, \ldots, \alpha_m) \in \{0, \frac{1}{2}, 1\}^m\} = (\vee\{x_1^{\alpha_1} \wedge \cdots \wedge x_m^{\alpha_m} : (\alpha_1, \ldots, \alpha_m) \in g^{-1}(1)\}) \vee (\vee\{x_1^{\alpha_1} \wedge \cdots \wedge x_m^{\alpha_m} \wedge \frac{1}{2} : (\alpha_1, \ldots, \alpha_m) \in g^{-1}(\frac{1}{2})\})$.

Remark 5 (i) $(\vee\{x_1^{\alpha_1} \wedge \cdots \wedge x_m^{\alpha_m} : (\alpha_1, \ldots, \alpha_m) \in g^{-1}(1)\}) \vee (\vee\{x_1^{\alpha_1} \wedge \cdots \wedge x_m^{\alpha_m} \wedge \frac{1}{2} : (\alpha_1, \ldots, \alpha_m) \in g^{-1}(\frac{1}{2})\})$ is called the quasi disjunctive normal form of m-ary three-valued McNaughton function. Note that in the disjunctive normal form, $x_i^{\alpha_i} (i = 1, 2, \ldots, m)$ is x_i or $\neg x_i$. However, in quasi disjunctive normal form, $x_i^{\alpha_i}$ is an one-ary function which is obtained by x_i through calculating $\neg, \rightarrow, \vee, \wedge$.

(ii) If $g^{-1}(1) = g^{-1}(\frac{1}{2}) = \emptyset$, i.e. $g(x_1, \ldots, x_m) = 0$, then we denote $g(x_1, \ldots, x_m) = x_1^0 \wedge x_1^1$, is an one-ary function, it can be as a simple quasi disjunctive normal form.

In the following, we give the Dual form of Proposition 4.

Proposition 6 *Let $g(x_1, \ldots, x_m)$ be three-valued McNaughton function, then*

$$\begin{aligned} g(x_1, x_2, \ldots, x_m) = &(\neg h_0(x_1) \vee g(0, x_2, \ldots, x_m)) \\ &\wedge(\neg h_{\frac{1}{2}}(x_1) \vee g(\tfrac{1}{2}, x_2, \ldots, x_m)) \\ &\wedge(\neg h_1(x_1) \vee g(1, x_2, \ldots, x_m)). \end{aligned}$$

Proposition 7 *Let $g(x_1, \ldots, x_m)$ be three-valued McNaughton function, denote $x_k^0 = \neg h_0(x_k), x_k^{\frac{1}{2}} = \neg h_{\frac{1}{2}}(x_k), x_k^1 = \neg h_1(x_k), (k = 1, 2)$, then*

$$g(x_1, x_2, x_3, \ldots, x_m) = \wedge \left\{ x_1^{\beta_1} \vee x_2^{\beta_2} \vee g(\beta_1, \beta_2, x_3, \ldots, x_m) : (\beta_1, \beta_2) \in \left\{0, \frac{1}{2}, 1\right\}^2 \right\}.$$

Corollary 4 *Let $g(x_1, \ldots, x_m)$ be three-valued McNaughton function, then*

$g(x_1, x_2, x_3, \ldots, x_m) = \wedge\{x_1^{\beta_1} \vee \cdots \vee x_m^{\beta_m} \vee g(\beta_1, \ldots, \beta_m) : (\beta_1, \ldots, \beta_m) \in \{0, \frac{1}{2}, 1\}^m\} = (\wedge\{x_1^{\beta_1} \vee \cdots \vee x_m^{\beta_m} : (\beta_1, \ldots, \beta_m) \in g^{-1}(0)\}) \wedge (\wedge\{x_1^{\beta_1} \wedge \cdots \vee x_m^{\beta_m} \vee \frac{1}{2} : (\beta_1, \ldots, \beta_m) \in g^{-1}(\frac{1}{2})\})$.

Remark 6 (i) $(\wedge\{x_1^{\beta_1} \vee \cdots \vee x_m^{\beta_m} : (\beta_1, \ldots, \beta_m) \in g^{-1}(0)\}) \wedge (\wedge\{x_1^{\beta_1} \vee \cdots \vee x_m^{\beta_m} \vee \frac{1}{2} : (\beta_1, \ldots, \beta_m) \in g^{-1}(\frac{1}{2})\})$ is called the quasi conjunctive normal form of m-ary three-valued McNaughton function. Note that in the conjunctive normal form, $x_i^{\beta_i}(i = 1, 2, \ldots, m)$ is x_i or $\neg x_i$. However, in quasi conjunctive normal form, $x_i^{\beta_i}$ is an one-ary function which is obtained by x_i through calculating $\neg, \rightarrow, \vee, \wedge$.

(ii) If $g^{-1}(0) = g^{-1}(\frac{1}{2}) = \emptyset$, i.e. $g(x_1, \ldots, x_m) = 1$, then we denote $g(x_1, \ldots, x_m) = x_1^0 \vee x_1^1$, is an one-ary function, it can be as a simple quasi disjunctive normal form.

Now, we research the n-value McNaughton function. Then, the range of McNaughton function is $L(n) = \{0, \frac{1}{n-1}, \frac{2}{n-1}, \ldots, \frac{n-2}{n-1}, 1\}$. In order to give the expansion of n-value McNaughton function, first, we give the following Lemma.

Lemma 1 ([14]) *In n-valued Lukasiewicz logic system L_n, $\forall k \in \{0, 1, 2, \ldots, n-2, n-1\}$, there is an one-ary logic formula $A_k(p)$, such that $\overline{A_k}(x)$ which is induced by $A_k(p)$ satisfies* $\overline{A_k}(x) = \begin{cases} 1, x = \frac{k}{n-1} \\ 0, x \neq \frac{k}{n-1} \end{cases}$, $(x \in Ł(n))$.

Proposition 8 *Let $g(x_1, \ldots, x_m)$ be n-valued McNaughton function, then*

$$\begin{aligned} g(x_1, x_2, \ldots, x_m) = &(\overline{A_0}(x) \wedge g(0, x_2, \ldots, x_m)) \\ &\vee(\overline{A_1}(x) \wedge g(\tfrac{1}{n-1}, x_2, \ldots, x_m)) \\ &\vee \cdots \\ &\vee(\overline{A_{n-1}}(x) \wedge g(1, x_2, \ldots, x_m)). \end{aligned}$$

Proposition 9 *Let $g(x_1, \ldots, x_m)$ be n-valued McNaughton function, denote $x_l^{\frac{k}{n-1}} = \overline{A_k}(x_l)$, $(l = 1, 2; k = 0, 1, 2, \ldots, n-2, n-1)$, then*
$g(x_1, x_2, x_3, \ldots, x_m) = \vee\{x_1^{\alpha_1} \wedge x_2^{\alpha_2} \wedge g(\alpha_1, \alpha_2, x_3, \ldots, x_m) : (\alpha_1, \alpha_2) \in \{0, \frac{1}{n-1}, \frac{2}{n-1}, \ldots, \frac{n-2}{n-1}, 1\}^2\}$.

Corollary 5 *Let $g(x_1, \ldots, x_m)$ be n-valued McNaughton function, then*
$g(x_1, \ldots, x_m) = \vee\{x_1^{\alpha_1} \wedge \cdots \wedge x_m^{\alpha_m} \wedge g(\alpha_1, \ldots, \alpha_m) : (\alpha_1, \ldots, \alpha_m) \in \{0, \frac{1}{n-1}, \frac{2}{n-1}, \ldots, \frac{n-2}{n-1}, 1\}^m\} = \vee_{k=1}^{n-1}\{x_1^{\alpha_1} \wedge \cdots \wedge x_m^{\alpha_m} \wedge \frac{k}{n-1} : (\alpha_1, \ldots, \alpha_m) \in g^{-1}(\frac{k}{n-1})\}$.

Remark 7 (i) $\vee_{k=1}^{n-1}\{x_1^{\alpha_1} \wedge \cdots \wedge x_m^{\alpha_m} \wedge \frac{k}{n-1} : (\alpha_1, \ldots, \alpha_m) \in g^{-1}(\frac{k}{n-1})\}$ is called the quasi disjunctive normal form of m-ary n-valued McNaughton function. $\forall k = 1, 2, \ldots, n-1$, if $g^{-1}(\frac{k}{n-1}) = \emptyset$, i.e. $g(x_1, \ldots, x_m) = 0$, then we denote $g(x_1, \ldots, x_m) = \overline{A_0}(x_1) \wedge \neg\overline{A_0}(x_1)$, is an one-ary function, it can be as a simple quasi disjunctive normal form.

In the following, we give the Dual form of Proposition 8.

Proposition 10 *Let $g(x_1, \ldots, x_m)$ be n-valued McNaughton function, then*

$$\begin{aligned} g(x_1, x_2, \ldots, x_m) = & (\neg \overline{A_0}(x) \vee g(0, x_2, \ldots, x_m)) \\ & \wedge(\neg \overline{A_1}(x) \vee g(\tfrac{1}{n-1}, x_2, \ldots, x_m)) \\ & \wedge \cdots \\ & \wedge(\neg \overline{A_{n-1}}(x) \vee g(1, x_2, \ldots, x_m)). \end{aligned}$$

Proposition 11 *Let $g(x_1, \ldots, x_m)$ be n-valued McNaughton function, denote $x_l^{\frac{k}{n-1}} = \neg\overline{A_k}(x_l)$, $(l = 1, 2; k = 0, 1, 2, \ldots, n-2, n-1)$, then $g(x_1, x_2, x_3, \ldots, x_m) = \wedge\{x_1^{\beta_1} \vee x_2^{\beta_2} \vee g(\beta_1, \beta_2, x_3, \ldots, x_m) : (\beta_1, \beta_2) \in \{0, \frac{1}{n-1}, \frac{2}{n-1}, \ldots, \frac{n-2}{n-1}, 1\}^2\}$.*

Corollary 6 *Let $g(x_1, \ldots, x_m)$ be n-valued McNaughton function, then $g(x_1, \ldots, x_m) = \wedge\{x_1^{\beta_1} \vee \cdots \vee x_m^{\beta_m} \vee g(\beta_1, \ldots, \beta_m) : (\beta_1, \ldots, \beta_m) \in \{0, \frac{1}{n-1}, \frac{2}{n-1}, \ldots, \frac{n-2}{n-1}, 1\}^m\} = \wedge_{k=0}^{n-2}\{x_2^{\beta_1} \vee \cdots \vee x_m^{\beta_m} \vee \frac{k}{n-1} : (\beta_1, \ldots, \beta_m) \in g^{-1}(\frac{k}{n-1})\}$.*

Remark 8 (i) $\wedge_{k=0}^{n-2}\{x_2^{\beta_1} \vee \cdots \vee x_m^{\beta_m} \vee \frac{k}{n-1} : (\beta_1, \ldots, \beta_m) \in g^{-1}(\frac{k}{n-1})\}$ is called the quasi conjunctive normal form of m-ary n-valued McNaughton function. $\forall k = 0, 1, 2, \ldots, n-2$, if $g^{-1}(\frac{k}{n-1}) = \emptyset$, i.e. $g(x_1, \ldots, x_m) = 1$, then we denote $g(x_1, \ldots, x_m) = \overline{A_0}(x_1) \vee \neg\overline{A_0}(x_1)$, is an one-ary function, it can be as a simple quasi conjunctive normal form.

4 Conclusions

In this paper, the expansions of the n-value McNaughton function are researched by the method of Shannon expansions of Boolean function. Then the quasi disjunctive normal form and quasi conjunctive normal form of m-ary n-valued McNaughton function are gained. In this way, we can get the constructing methods of logic formulas, then we can continue research counting problems of logic formulas.

Acknowledgments I would like to express my gratitude to all those who helped me during the writing of this thesis. This work is supported by the National Natural Science Foundation of China (Grant Nos. 61562030), Youth Natural Science Foundation of Jiangxi Province (Grant Nos. 20144BAB2020002).

References

1. Christel, B., Joost-Pieter, K.: Principles of Model Checking. The MIT Press, Massachusetts (2008)
2. Pallab, D., Chakrabarti, P.P., Jatindra, K.D., Sriram, S.: Min-max computation tree logic. Artif. Intell. **127**(1), 137–162 (2001)
3. Franjo, I., Zijiang, Y., Malay, K.G., Aarti, G., Pranav, A.: Efficient SAT-based bounded model checking for software verification. Theor. Comput. Sci. **404**(3), 256–274 (2008)

4. Stylianos, B., Panagiotis, K., Andrew, P.: An intruder model with message inspection for model checking security protocols. Comput. Secur. **29**(1), 16–34 (2010)
5. Olaf, B., Bernhard, S.: Model checking the full modal mu-calculus for infinite sequential processes. Theor. Comput. Sci. **221**(1–2), 251–270 (1999)
6. Alessio, L., Wojciech, P., Bozena, W.: Bounded model checking for knowledge and real time. Artif. Intell. **171**(16–17), 1011–1038 (2007)
7. Franco, R., Alessio, L.: Automatic verification of multi-agent systems by model checking via ordered binary decision diagrams. J. Appl. Log. **5**(2), 235–251 (2007)
8. Pablo, M.-G., Rubn, F.-F., Jos-Luis, S.-R., Baltasar, F.-M.: Model-checking for adventure videogames. Inf. Softw. Technol. **51**(3), 564–580 (2009)
9. Luciano, B., Vahid, R., Adel, T.R., Paola, S.: An efficient solution for model checking graph transformation systems. Electron. Notes Theor. Comput. Sci. **213**(1), 3–21 (2008)
10. Hanifa, B., Rachid, H.: CTL* model checking for time Petri nets. Theor. Comput. Sci. **353**(1–3), 208–227 (2006)
11. Mohammad, A., Fabio, S.: Termination criteria for bounded model checking: extensions and comparison. Electron. Notes Theor. Comput. Sci. **144**(1), 51–66 (2006)
12. Wang, G.J.: Introduction to Mathematical Logic and Resolution Principle, 2nd edn, pp. 16–40. Science Press, Beijing (2006)
13. Cignoli, R., D'Ottaviano, I.M.L., Mundici, D.: Algebraic Foundations of Many-Valued Reasoning, pp. 7–30. Kluwer Academic Publishers, Dordrecht (2000)
14. Wang, Q.P., Wang, G.J.: Normal form of Lukasiewicz logic formulae and related counting problems. J. Softw. **24**(3), 433–453 (2013)

Relative Divergence Degree and Relative Consistency Degree of Theories in a Kind of Goguen Propositional Logic System

Xiao-Li Gao, Xiao-Jing Hui and Nai-Diao Zhu

Abstract Using induced function of formula, the paper gave definition of Γ-k truth degree, Γ-k similarity degree and Γ-k pseudo-metric of formula relative to local finite theory Γ under k conjunction in Goguen$_{\sim,\Delta}$ propositional logic system, and proved the MP rule, HS rule, and some basic properties of Γ-k truth degree. At the meantime, the concept of relative divergence degree and relative consistency degree of any theory Γ relative to the fixed theory Γ_0 in Goguen$_{\sim,\Delta}$ system is introduced. Important relations between relative divergence degree and relative consistency degree are obtained.

Keywords Goguen$_{\sim,\Delta}$ system · Γ-k truth degree · Relative divergence degree · Relative consistency degree

1 Introduction

A large number of scholars had devoted themselves to the research on degree of propositional logic conclusion and have achieved a lot of achievements since the idea was put forward by Pavelka in the 1970s [1]. Starting from the degree of logical concept, the theory of truth degree of formulas in propositional logic system was given in [2–5]. The concept of random truth degree of the formula in propositional logic system was given by using the method of random assignment in paper [6–8]. Using Borel probability measure, the theory of Borel probability truth degree was proposed in [9, 10] in propositional logic system. The relative Γ-truth degrees

X.-L. Gao (✉) · X.-J. Hui · N.-D. Zhu
College of Mathematics and Computer Science, Yan'an University,
Yan'an 716000, China
e-mail: 951402445@qq.com

X.-J. Hui
e-mail: xhmxiaojing@163.com

N.-D. Zhu
e-mail: 792266404@qq.com

T.-H. Fan et al. (eds.), *Quantitative Logic and Soft Computing 2016*,
Advances in Intelligent Systems and Computing 510,
DOI 10.1007/978-3-319-46206-6_10

of formula in logic system relative to the locally finite theory was proposed and application scope of the theory of truth degrees was broadened in [11–13].

It is known that the research on Gödel logic system and Goguen logic system has been hindered due to strong negativity. By introducing basic connectives $\sim$ and Δ, the system $SBL_{\sim}$, in which the deductive theorem and the strong complete theorem are both tenable is established in [14–17]. So, the research on Gödel propositional logic system and Goguen proposition logic system can be carried out smoothly.

Axiomatic extensions of n-valued Goguen propositional logic system is first studied in this paper. Then, the definition of Γ-k truth degree, Γ-k similarity degree and Γ-k pseudo-metric of formula relative to local finite theory Γ under k conjunction and its related properties are given by using induced function. Finally, the concept of relative divergence degree and relative compatibility of arbitrary theory Γ relative to a particular theory are introduced in the $Goguen_{\sim,\Delta}$ propositional logic system and important relationship between relative divergence and relative compatibility degree is obtained.

2 Preliminaries

Definition 1 ([16]) The axiom system BL_{Δ} is as follows:
(BL) The axioms system of BL;
($\Delta1$) $\Delta A \vee \rightharpoondown \Delta A$;
($\Delta2$) $\Delta(A \vee B) \to (\Delta A \vee \Delta B)$;
($\Delta3$) $\Delta A \to B$;
($\Delta4$) $\Delta A \to \Delta\Delta A$;
($\Delta5$) $\Delta(A \to B) \to (\Delta A \to \Delta B)$.

The inference rules of BL_{Δ} are MP rules and Δ rules. MP rules is that B is inferred from A and $A \to B$. Δ rules is $A \to \Delta A$.

If $\mathcal{L}$ is axiomatic extension of BL, the $\mathcal{L}_{\Delta}$ is marked as extension of $\mathcal{L}$, which is the same as BL_{Δ} extension of BL. Δ Deduction Theorem are established in the system BL_{Δ}.

Theorem 1 [18] (Δ Deduction Theorem) *Let $\mathcal{L}$ be an axiomatic extension of BL_{Δ}. Then, for any theory Γ, formulas A and B, we have:*

$$\Gamma, A \vdash B \textit{ if and only if } \Gamma, \Delta A \to B.$$

SBL is an axiomatic extension when axiom $\neg\neg A \vee \neg A$ is add to BL. SBL_{Δ} is also an axiomatic extension of SBL. The system of $SBL_{\sim}$ is an logic system when the connective $\sim$ is added in the SBL system.

Definition 2 ([17]) As an extension of axiomatic of SBL, the axiom system $SBL_{\sim}$ is as follows:
(SBL) The axioms system of SBL;

$(\sim 1) \sim\sim A \to A$;
$(\sim 2)\ \neg A \to \sim A$;
$(\sim 3)\ \Delta(A \to B) \to \Delta(\sim B \to \sim B)$.

If ΔA is $\neg \sim A$ in system $SBL_\sim$, the relationship between systems SBL_Δ and $SBL_\sim$ can be established. An equivalent axiom system $SBL_\sim$ is as follows:
(SBL_Δ) The axioms system of SBL_Δ;
$(\sim 1) \sim\sim A \to A$;
$(\sim 3)\ \Delta(A \to B) \to \Delta(\sim B \to \sim B)$.

The inference rules of $SBL_\sim$ are also MP rules and Δ rules. If $\mathcal{L}$ is an extension of axiomatic of SBL, $\mathcal{L}_\sim$ can be noted as an extension of $\mathcal{L}$, which is just as SBL expansed to $SBL_\sim$. $Gödel_\sim$ and $\Pi_\sim$ are the two basic types of axiomatic extension of $SBL_\sim$. Δ Deduction Theorem established in the system $SBL_\sim$ because $SBL_\sim$ is also an axiomatic extension of BL_Δ.

Theorem 2 [17] (Strong Completeness Theorem) *Let $\mathcal{L}$ be an axiomatic extension of $SBL_\sim$. Then, for theory Γ and formula A, the following conditions are equivalent:*
(i) $\Gamma \vdash A$;
(ii) $e(A) = 1$ *for each model of each $\mathcal{L}$-algebra and theory Γ.*

3 Definitions and Properties of Γ-k Truth Degree, Γ-k Similarity Degree and Γ-k Pseudo-Metric

Definition 3 Let $S=\{p_1, p_2, \ldots\}$ be a countable set. $\sim$, Δ, $\vee$, $\wedge$ and $\to$ are operations on S, in which $\sim$ and Δ are unary operation; $\vee$, $\wedge$ and $\to$ are binary operation. F(S) is free algebra of type $(\sim, \Delta, \vee, \wedge, \to)$ generated by S. Then, an element in F(S) is said be a propositional formula or formula, and an element in S is said to be an atomic formula.

Definition 4 Goguen propositional logic system is also called Product system, denoted by Π. Let $\Pi_{\sim,\Delta} = \{0, \frac{1}{n-1}, \ldots, \frac{n-2}{n-1}, 1\}$, and qualify any $x, y \in \Pi_{\sim,\Delta}$, $\sim x = 1 - x$, $\Delta x = \{ {}^{1, x=1}_{0 x<1}$, $x \vee y = max\{x, y\}$, $x \wedge y = min\{x, y\}$, $x \to y = \{ {}^{1, x \leq y}_{\frac{y}{x}, x>y}$. Then, $Goguen_{\sim,\Delta}$ system is an extension of n-valued Product propositional logic system, which is denoted by $\Pi_{\sim,\Delta}$.

Definition 5 Let $A = A(p_1, p_2, \ldots, p_m) \in F(S)$. Then, A corresponds to a function $\overline{A}$ of n-valued and m-element. In $\Pi_{\sim,\Delta}$, $\overline{A} : \Pi_{\sim,\Delta} \to [0, 1]$, and $\overline{A}(x_1, x_2, \ldots, x_m)$ is connect to $x_1, x_2, \ldots, x_m$ by operational sign $\sim$, Δ, $\vee$, $\wedge$, $\to$, which is the same as $A = A(p_1, p_2, \ldots, p_m)$ is connected to the atomic formula $p_1, p_2, \ldots, p_m$ by conjunction $\sim$, Δ, $\vee$, $\wedge$, $\to$. Then it's said that $\overline{A}$ is a function induced by formula A.

Definition 6 In $\Pi_{\sim,\Delta}$. Let $\overline{A}(x_1, x_2, \ldots, x_m)$ be a function induced by propositional formula $A = A(p_1, p_2, \ldots, p_m)$ in F(S), $l \geqslant 0$. Let us define: $\forall (x_1, \ldots, x_m, \ldots, x_{m+l}) \in \Pi_{\sim,\Delta}$, $\overline{A}^l : \Pi_{\sim,\Delta} \to [0, 1]$, $\overline{A}^l(x_1, \ldots, x_m, \ldots, x_{m+l}) = \overline{A}(x_1, x_2, \ldots, x_m)$, and mark $\overline{A}^l$ as the extension of function $\overline{A}$ to l-element.

Let $\Gamma \subseteq F(S)$, $A \in F(S)$, and qualify $S_\Gamma = \{p \in S | \exists B \in \Gamma$, in which B is made up of atomic propositional $p\}$, $S_A = \{p \in S | p$ appears in $A\}$. If S_Γ is finite, we say Γ a local finite theory of Goguen$_{\sim,\Delta}$ propositional logic system.

Definition 7 In $\Pi_{\sim,\Delta}$, let $\Gamma \subseteq F(S)$, and S_Γ is finite, $A \in F(S)$, $S = S_\Gamma \cup S_A = \{p_1, p_2, \ldots, p_m\}$, then

$$\tau_{n,\Gamma}(kA) = \begin{cases} 1 & , N(\Gamma) = \emptyset \\ \frac{1}{N(\Gamma)} \sum\limits_{(x_1,x_2,\ldots,x_m)\in N(\Gamma)} k\overline{A}(x_1, x_2, \ldots, x_m), & N(\Gamma) \neq \emptyset \end{cases}.$$

Thereinto, $N(\Gamma) = \{(x_1, x_2, \ldots, x_m) \in \Pi_{\sim,\Delta} | \forall B \in \Gamma, \overline{B}(x_1, x_2, \ldots, x_m) = 1\}$. $\tau_{n,\Gamma}(kA)$ is said to be Γ-k truth degree of formula A relative to local finite theory Γ under k conjunction, denoted as Γ-k truth degree.

Unless otherwise specified, in the following we assume that: 1. Discussion in $\Pi_{\sim,\Delta}$; 2. κ, λ, μ *and* η take any $\sim$, Δ; 3. Basic grammar, semantic concepts such as theorem, logic equivalence, tautology, and contradiction are the same as classical propositional logic.

Theorem 3 *Let* $\Gamma \subseteq F(S)$, $A \in F(S)$, *and* S_Γ *be finite,* $S = S_\Gamma \cup S_A = \{p_1, p_2, \ldots, p_m\}$, *and* $S^* = \{p_1, p_2, \ldots, p_m, p_{m+1}, \ldots, p_{m+l}\} \subseteq F(S)$. *Then,*

$$\tau_{n,\Gamma}(kA) = \begin{cases} 1 & , N^*(\Gamma) = \emptyset \\ \frac{1}{N^*(\Gamma)} \sum\limits_{(x_1,\ldots,x_m,x_{m+1},\ldots,x_{m+l})\in N^*(\Gamma)} k\overline{A}(x_1, \ldots, x_m, x_{m+1}, \ldots, x_{m+l}), & N^*(\Gamma) \neq \emptyset \end{cases}.$$

Thereinto, $N^*(\Gamma) = \{(x_1, \ldots, x_m, x_{m+1}, \ldots, x_{m+l}) \in \Pi_{\sim,\Delta} | \forall B \in \Gamma, \overline{B}^l(x_1, \ldots, x_m, x_{m+1}, \ldots, x_{m+l}) = 1\}$.

Proof Since $N(\Gamma) = \{(x_1, x_2, \ldots, x_m) \in \Pi_{\sim,\Delta} | \forall B \in \Gamma, \overline{B}(x_1, x_2, \ldots, x_m) = 1\}$, $N^*(\Gamma) = \{(x_1, \ldots, x_m, x_{m+1}, \ldots, x_{m+l}) \in \Pi_{\sim,\Delta} | \forall B \in \Gamma, \overline{B}^l(x_1, \ldots, x_m, x_{m+1}, \ldots, x_{m+l}) = 1\}$. By Definition 6, we know for each $(x_1, \ldots, x_m, x_{m+1}, \ldots, x_{m+l}) \in \Pi_{\sim,\Delta}$, $\overline{B}^l(x_1, \ldots, x_m, x_{m+1}, \ldots, x_{m+l}) = \overline{B}(x_1, x_2, \ldots, x_m)$. So $|N^*(\Gamma)| = |N(\Gamma)| \times n^l$. Thus, if $|N^*(\Gamma)| = \emptyset$, $|N(\Gamma)| = 0$, we have $\tau_{n,\Gamma}(kA) = 1$. If $|N^*(\Gamma)| \neq \emptyset$, $|N(\Gamma)| \neq \emptyset$, since $\overline{A}^l(x_1, \ldots, x_m, \ldots, x_{m+l}) = \overline{A}(x_1, x_2, \ldots, x_m)$, so $\sum\limits_{(x_1,\ldots,x_m,\ldots,x_{m+l})\in\Pi_{\sim,\Delta}} k\overline{A}(x_1, \ldots, x_m, \ldots, x_{m+l}) = \sum\limits_{(x_1,x_2,\ldots,x_m)\in\Pi_{\sim,\Delta}\times n^l} k\overline{A}(x_1, x_2, \ldots, x_m) = \sum\limits_{(x_1,x_2,\ldots,x_m)\in\Pi_{\sim,\Delta}} k\overline{A}(x_1, x_2, \ldots, x_m) \times n^l$.

Also $\frac{1}{N^*(\Gamma)}\sum_{(x_1,\ldots,x_m,\ldots,x_{m+l})\in N^*(\Gamma)} k\overline{A}(x_1,\ldots,x_m,\ldots,x_{m+l})$

$= \frac{1}{N(\Gamma)\times n^l}\sum_{(x_1,x_2,\ldots,x_m)\in N(\Gamma)} k\overline{A}(x_1,x_2,\ldots,x_m)\times n^l$

$= \frac{1}{N(\Gamma)}\sum_{(x_1,x_2,\ldots,x_m)\in N(\Gamma)} k\overline{A}(x_1,x_2,\ldots,x_m).$

Thus we have $\tau_{n,\Gamma}(kA) = \frac{1}{N^*(\Gamma)}\sum_{(x_1,\ldots,x_m,\ldots,x_{m+l})\in N^*(\Gamma)} k\overline{A}(x_1,\ldots,x_m,\ldots,x_{m+l})$. □

For convenient, in this paper $N^*(\Gamma)$ and $\sum_{(x_1,\ldots,x_m,\ldots,x_{m+l})\in N^*(\Gamma)} k\overline{A}(x_1,\ldots,x_m,\ldots,x_{m+l})$ will be denoted by $N(\Gamma)$ and $\sum_{(x_1,x_2,\ldots,x_m)\in N(\Gamma)} k\overline{A}(x_1,x_2,\ldots,x_m)$.

Theorem 4 *Let $\Gamma_1 \subseteq \Gamma_2 \subseteq F(S)$, $A \in F(S)$, and S_Γ be finite. If $\tau_{n,\Gamma_1}(kA) = 1$, $\tau_{n,\Gamma_2}(kA) = 1$.*

Proof Since $\Gamma_1 \subseteq \Gamma_2$, $N(\Gamma_2) \subseteq N(\Gamma_1)$. If $N(\Gamma_2) = \varnothing$, $\tau_{n,\Gamma_2}(kA) = 1$. If $N(\Gamma_2) \neq \varnothing$, we can obtain $N(\Gamma_1) \neq \varnothing$. Since $\tau_{n,\Gamma_1}(kA) = 1$, and $\frac{1}{N(\Gamma_1)}\sum_{(x_1,x_2,\ldots,x_m)\in N(\Gamma_1)} k\overline{A}(x_1,x_2,\ldots,x_m) = 1$, $|N(\Gamma_1)| = \sum_{(x_1,x_2,\ldots,x_m)\in N(\Gamma_1)} k\overline{A}(x_1,x_2,\ldots,x_m)$. For any $(x_1,x_2,\ldots,x_m)\in N(\Gamma_1)$, we know $k\overline{A}(x_1,x_2,\ldots,x_m) = 1$. For any $(x_1,x_2,\ldots,x_m)\in N(\Gamma_2)$, we know $k\overline{A}(x_1,x_2,\ldots,x_m) = 1$. Thus, we get $|N(\Gamma_2)| = \sum_{(x_1,x_2,\ldots,x_m)\in N(\Gamma_2)} k\overline{A}(x_1,x_2,\ldots,x_m)$. So we get $\tau_{n,\Gamma_2}(kA) = 1$. □

Theorem 5 *Let $\Gamma \subseteq F(S)$, $A \in F(S)$, and S_Γ be finite.*
(i) *If $\vDash \lambda A \rightarrow \mu B$, $\tau_{n,\Gamma}(\lambda A) \le \tau_{n,\Gamma}(\mu B)$;*
(ii) *If $\vDash \lambda A \approx \mu B$, $\tau_{n,\Gamma}(\lambda A) = \tau_{n,\Gamma}(\mu B)$;*
(iii) *If $N(\Gamma) \neq \varnothing$, $\tau_{n,\Gamma}(\sim kA) = 1 - \tau_{n,\Gamma}(kA)$.*

Proof Suppose that A, B contains the same atomic formula $p_1, p_2, \ldots, p_m$, for any $(x_1,x_2,\ldots,x_m)\in N(\Gamma)$

(i) If $\vDash \lambda A \rightarrow \mu B$, $(\lambda\overline{A} \rightarrow \mu\overline{B})(x_1,x_2,\ldots,x_m) = 1$, $\lambda\overline{A}(x_1,x_2,\ldots,x_m) \rightarrow \mu\overline{B}(x_1,x_2,\ldots,x_m) = 1$. Thus we have $\lambda\overline{A}(x_1,x_2,\ldots,x_m) \le \mu\overline{B}(x_1,x_2,\ldots,x_m)$. Further we get $\sum_{(x_1,x_2,\ldots,x_m)\in N(\Gamma)} \lambda\overline{A}(x_1,x_2,\ldots,x_m) \le \sum_{(x_1,x_2,\ldots,x_m)\in N(\Gamma)} \mu\overline{B}(x_1,x_2,\ldots,x_m)$. So we obtain $\frac{1}{N(\Gamma)}\sum_{(x_1,x_2,\ldots,x_m)\in N(\Gamma)} \lambda\overline{A}(x_1,x_2,\ldots,x_m) \le \frac{1}{N(\Gamma)}\sum_{(x_1,x_2,\ldots,x_m)\in N(\Gamma)} \mu\overline{B}(x_1,x_2,\ldots,x_m)$. According to Definition 7, we can get $\tau_{n,\Gamma}(\lambda A) \le \tau_{n,\Gamma}(\mu B)$.

(ii) If $\vDash \lambda A \approx \mu B$, $\lambda\overline{A}(x_1,x_2,\ldots,x_m) = \mu\overline{B}(x_1,x_2,\ldots,x_m)$. Thus we have $\sum_{(x_1,x_2,\ldots,x_m)\in N(\Gamma)} \lambda\overline{A}(x_1,x_2,\ldots,x_m) = \sum_{(x_1,x_2,\ldots,x_m)\in N(\Gamma)} \mu\overline{B}(x_1,x_2,\ldots,x_m)$. So we obtain $\frac{1}{N(\Gamma)}\sum_{(x_1,x_2,\ldots,x_m)\in N(\Gamma)} \lambda\overline{A}(x_1,x_2,\ldots,x_m) = \frac{1}{N(\Gamma)}\sum_{(x_1,x_2,\ldots,x_m)\in N(\Gamma)} \mu\overline{B}(x_1,x_2,\ldots,x_m)$. According to Definition 7, we can get $\tau_{n,\Gamma}(\lambda A) = \tau_{n,\Gamma}(\mu B)$.

(iii) If $N(\Gamma) \neq \varnothing$, we have $\tau_{n,\Gamma}(\sim kA) = \frac{1}{N(\Gamma)}\sum_{(x_1,x_2,\ldots,x_m)\in N(\Gamma)} \sim k\overline{A}(x_1,x_2,\ldots,x_m) = \frac{1}{N(\Gamma)}\sum_{(x_1,x_2,\ldots,x_m)\in N(\Gamma)} (1 - k\overline{A}(x_1,x_2,\ldots,x_m)) = 1 - \tau_{n,\Gamma}(kA)$. □

Lemma 1 *Let* $\forall a, b \in \Pi_{\sim,\Delta}$. *Then* $\lambda a \vee \mu b = \lambda a + \mu b - (\lambda a \wedge \mu b)$.

Proof Firstly, let $*_1 = (\lambda a \vee \mu b) - \lambda a - \mu b + (\lambda a \wedge \mu b)$. Then, two cases are discussed. (1) If $\lambda a \geq \mu b$, $*_1 = \lambda a - \lambda a - \mu b + \mu b = 0$. (2) If $\lambda a < \mu b$, $*_1 = \mu b - \lambda a - \mu b + \lambda a = 0$. □

Theorem 6 *Let* $\Gamma \subseteq F(S)$, $A \in F(S)$, *and* S_Γ *be finite. Then*

$$\tau_{n,\Gamma}(\lambda A \vee \mu B) = \tau_{n,\Gamma}(\lambda A) + \tau_{n,\Gamma}(\mu B) - \tau_{n,\Gamma}(\lambda A \wedge \mu B).$$

Proof Suppose that A, B contains the same atomic formula $p_1, p_2, \ldots, p_m$, for any $(x_1, x_2, \ldots, x_m) \in N(\Gamma)$. By Lemma 1 we know $\lambda\overline{A}(x_1, x_2, \ldots, x_m) \vee \mu\overline{B}(x_1, x_2, \ldots, x_m) = \lambda\overline{A}(x_1, x_2, \ldots, x_m) + \mu\overline{B}(x_1, x_2, \ldots, x_m) - (\lambda\overline{A}(x_1, x_2, \ldots, x_m) \wedge \mu\overline{B}(x_1, x_2, \ldots, x_m))$, and

$\lambda\overline{A}(x_1, x_2, \ldots, x_m) \vee \mu\overline{B}(x_1, x_2, \ldots, x_m) = (\lambda\overline{A} \vee \mu\overline{B})(x_1, x_2, \ldots, x_m)$,

$\lambda\overline{A}(x_1, x_2, \ldots, x_m) \wedge \mu\overline{B}(x_1, x_2, \ldots, x_m) = (\lambda\overline{A} \wedge \mu\overline{B})(x_1, x_2, \ldots, x_m)$.

Then we have $(\lambda\overline{A} \vee \mu\overline{B})(x_1, x_2, \ldots, x_m) = \lambda\overline{A}(x_1, x_2, \ldots, x_m) + \mu\overline{B}(x_1, x_2, \ldots, x_m) - (\lambda\overline{A} \wedge \mu\overline{B})(x_1, x_2, \ldots, x_m)$.

So we get

$$\sum_{(x_1,x_2,\ldots,x_m)\in N(\Gamma)} (\lambda\overline{A} \vee \mu\overline{B})(x_1, x_2, \ldots, x_m) = \sum_{(x_1,x_2,\ldots,x_m)\in N(\Gamma)} \lambda\overline{A}(x_1, x_2, \ldots, x_m) + \sum_{(x_1,x_2,\ldots,x_m)\in N(\Gamma)} \mu\overline{B}(x_1, x_2, \ldots, x_m) - \sum_{(x_1,x_2,\ldots,x_m)\in N(\Gamma)} (\lambda\overline{A} \wedge \mu\overline{B})(x_1, x_2, \ldots, x_m).$$

Thus $\frac{1}{|N(\Gamma)|} \sum_{(x_1,x_2,\ldots,x_m)\in N(\Gamma)} (\lambda\overline{A} \vee \mu\overline{B})(x_1, x_2, \ldots, x_m)$

$$= \frac{1}{|N(\Gamma)|} \sum_{(x_1,x_2,\ldots,x_m)\in N(\Gamma)} \lambda\overline{A}(x_1, x_2, \ldots, x_m) + \frac{1}{|N(\Gamma)|} \sum_{(x_1,x_2,\ldots,x_m)\in N(\Gamma)} \mu\overline{B}(x_1, x_2, \ldots, x_m) - \frac{1}{|N(\Gamma)|} \sum_{(x_1,x_2,\ldots,x_m)\in N(\Gamma)} (\lambda\overline{A} \wedge \mu\overline{B})(x_1, x_2, \ldots, x_m).$$

By Definition 7, we know $\tau_{n,\Gamma}(\lambda A \vee \mu B) = \tau_{n,\Gamma}(\lambda A) + \tau_{n,\Gamma}(\mu B) - \tau_{n,\Gamma}(\lambda A \wedge \mu B)$. □

Lemma 2 *Let* $\forall a, b \in \Pi_{\sim,\Delta}$. *Then,* $\mu b \geq \lambda a + (\lambda a \rightarrow \mu b)$.

Proof Firstly, let $*_2 = \mu b - \lambda a - (\lambda a \rightarrow \mu b) + 1$. Then, two cases are discussed. (1) If $\lambda a \leq \mu b$, $*_2 = \mu b - \lambda a \geq 0$. (2) If $\lambda a > \mu b$, $*_2 = \mu b - \lambda a - \frac{\mu b}{\lambda a} + 1 = \frac{\mu b(\lambda a - 1)}{\lambda a} - \frac{\lambda a(\lambda a - 1)}{\lambda a} = \frac{(\mu b - \lambda a)(\lambda a - 1)}{\lambda a} \geq 0$. In summary, we can get $\mu b \geq \lambda a + (\lambda a \rightarrow \mu b)$. □

Theorem 7 (MP rules of Γ-k truth degree) *Let $\Gamma \subseteq F(S)$, $A, B \in F(S)$, and S_Γ be finite. If $\tau_{n,\Gamma}(\lambda A) \geq \alpha$, $\tau_{n,\Gamma}(\lambda A \to \mu B) \geq \beta$, then $\tau_{n,\Gamma}(\mu B) \geq \alpha + \beta - 1$.*

Proof Suppose that A, B contains the same atomic formula $p_1, p_2, \ldots, p_m$, for any $(x_1, x_2, \ldots, x_m) \in N(\Gamma)$, by Lemma 2, we get $\mu\overline{B}(x_1, x_2, \ldots, x_m) \geq \lambda\overline{A}(x_1, x_2, \ldots, x_m) + (\lambda\overline{A}(x_1, x_2, \ldots, x_m) \to \mu\overline{B}(x_1, x_2, \ldots, x_m)) - 1$.

So, $\frac{1}{|N(\Gamma)|} \sum\limits_{(x_1,x_2,\ldots,x_m)\in N(\Gamma)} \mu\overline{B}(x_1, x_2, \ldots, x_m)$

$\geq \frac{1}{|N(\Gamma)|} \sum\limits_{(x_1,x_2,\ldots,x_m)\in N(\Gamma)} \lambda\overline{A}(x_1, x_2, \ldots, x_m) + \frac{1}{|N(\Gamma)|} \sum\limits_{(x_1,x_2,\ldots,x_m)\in N(\Gamma)} (\lambda\overline{A} \to \mu\overline{B})$ $(x_1, x_2, \ldots, x_m) - \frac{1}{|N(\Gamma)|} \sum\limits_{(x_1,x_2,\ldots,x_m)\in N(\Gamma)} 1.$

By Definition 7, we get $\tau_{n,\Gamma}(\mu B) \geq \alpha + \beta - 1$. □

Theorem 8 (HS rules of Γ-k truth degree) *Let $\Gamma \subseteq F(S)$, $A, B, C \in F(S)$, and S_Γ be finite. If $\tau_{n,\Gamma}(\lambda A \to \mu B) \geq \alpha$, $\tau_{n,\Gamma}(\mu B \to \eta C) \geq \beta$, then $\tau_{n,\Gamma}(\lambda A \to \eta C) \geq \alpha + \beta - 1$.*

Proof Suppose that A, B, C contains the same atomic formula $p_1, p_2, \ldots, p_m$. It is obvious that for all $a, b, c \in \Pi_{\sim,\Delta}$, $(\lambda a \to \mu b) \to ((\mu b \to \eta c) \to (\lambda a \to \eta c)) = 1$.

So, $(\lambda A \to \mu B) \to ((\mu B \to \eta C) \to (\lambda A \to \eta C))$ is tautology. By Theorems 5(i) and 7, we get $\tau_{n,\Gamma}((\mu B \to \eta C) \to (\lambda A \to \eta C)) \geq \tau_{n,\Gamma}(\lambda A \to \mu B) \geq \alpha$, then $\tau_{n,\Gamma}(\lambda A \to \eta C) \geq \tau_{n,\Gamma}(\mu B \to \eta C) + \tau_{n,\Gamma}((\mu B \to \eta C) \to (\lambda A \to \eta C)) - 1 \geq \alpha + \beta - 1$. □

Definition 8 Let $\Gamma \subseteq F(S)$, $A, B \in F(S)$, and S_Γ be finite. Then

$$\xi_{n,\Gamma}(\lambda A, \mu B) = \tau_{n,\Gamma}((\lambda A \to \mu B) \wedge (\mu B \to \lambda A)).$$

$\xi_{n,\Gamma}(\lambda A, \mu B)$ is called the Γ-k similarity degree of formule A, B relative to local finite theory Γ under the λ, μ conjunction.

Theorem 9 *Let $\Gamma \subseteq F(S)$, $A, B \in F(S)$, and S_Γ be finite. Then*

$$\xi_{n,\Gamma}(\lambda A, \mu B) = \tau_{n,\Gamma}(\lambda A \to \mu B) + \tau_{n,\Gamma}(\mu B \to \lambda A) - 1.$$

Proof Suppose that A, B contains the same atomic formula $p_1, p_2, \ldots, p_m$, by Theorem 6 and the Definition 8, we have $\xi_{n,\Gamma}(\lambda A, \mu B) = \tau_{n,\Gamma}((\lambda A \to \mu B) \wedge (\mu B \to \lambda A)) = \tau_{n,\Gamma}(\lambda A \to \mu B) + \tau_{n,\Gamma}(\mu B \to \lambda A) - \tau_{n,\Gamma}((\lambda A \to \mu B) \vee (\mu B \to \lambda A)) = \tau_{n,\Gamma}(\lambda A \to \mu B) + \tau_{n,\Gamma}(\mu B \to \lambda A) - 1$. □

Definition 9 Let $\Gamma \subseteq F(S)$, $A, B \in F(S)$, S_Γ be finite, and qualify $\rho_{n,\Gamma} : F(S) \times F(S) \to [0, 1]$. Then,

$$\rho_{n,\Gamma}(\lambda A, \mu B) = 1 - \xi_{n,\Gamma}(\lambda A, \mu B).$$

$\rho_{n,\Gamma}(\lambda A, \mu B)$ is called the Γ-k pseudo-metric between formule A, B relative to local finite theory Γ under the λ, μ conjunction, abbreviated as Γ-k pseudo-metric. $(F(S), \rho_{n,\Gamma})$ is called a logic pseudo-metric space.

Theorem 10 *Let $\Gamma \subseteq F(S)$, $A, B \in F(S)$, and S_Γ be finite. Then,*

$$\rho_{n,\Gamma}(\lambda A, \mu B) = 1 - \tau_{n,\Gamma}((\lambda A \to \mu B) \wedge (\mu B \to \lambda A)).$$

Proof It is easy to prove it by Definitions 8 and 9. □

4 The Divergence Degree and the Consistency Degree of Any Theory Γ Relative to the Fixed Theory Γ_0

Definition 10 Let $\Gamma \subset F(S)$, and $div(\Gamma) = \sup\{\rho(\lambda A, \mu B) | \lambda A, \mu B \in D(\Gamma)\}$. $div(\Gamma)$ is called the divergence degree of theory Γ. If $div(\Gamma) = 1$, Γ is called a fully divergent theory.

Definition 11 Let $\Gamma, \Gamma_0 \subset F(S)$. A deduction starting from Γ which is a finite formule sequence $\lambda A_1, \lambda A_2, \ldots, \lambda A_n$ relative to the theory Γ_0. For any $i \leq n$, λA_i is axiom, or $\lambda A_i \in \Gamma_0$, or $\lambda A_i \in \Gamma$. For $j < k \leq i$, λA_i is obtained from λA_j, and λA_k using the MP rule. Then λA_n is called a conclusion of Γ relative to theory Γ_0, which is denoted by $\Gamma \vdash_{\Gamma_0} \lambda A_n$ or $\lambda A_n \in D_{\Gamma_0}(\Gamma)$.

Definition 12 Let $\Gamma_0 \subseteq F(S)$, S_{Γ_0} be finite. $\forall \Gamma \subseteq F(S)$, define

$$div_{\Gamma_0}(\Gamma) = \sup\{\rho_{n,\Gamma_0}(\lambda A, \mu B) | \lambda A, \mu B \in D_{\Gamma_0}(\Gamma)\}.$$

$div_{\Gamma_0}(\Gamma)$ is called the divergence degree of theory Γ relative to the fixed theory Γ_0.

Theorem 11 *Let $\Gamma_0 \subseteq F(S)$, S_{Γ_0} be finite. $\forall \Gamma \subseteq F(S)$, then*
(i) $0 \leq div_{\Gamma_0}(\Gamma) \leq 1$;
(ii) *If $\Gamma_0 = \varnothing$ or Γ_0 is a theorem set, then $div_{\Gamma_0}(\Gamma) = div(\Gamma)$;*
(iii) *If $\overline{0} \in D(\Gamma_0)$, then $div_{\Gamma_0}(\Gamma) = 0$.*

Proof (i) Obvious.

(ii) If $\Gamma_0 = \varnothing$ or Γ_0 is a theorem set, $D(\Gamma_0 \cup \Gamma) = D(\Gamma)$. Furthermore $\lambda A, \mu B \in D_{\Gamma_0}(\Gamma)$. According to Theorem 10 we have $\rho_{n,\Gamma_0}(\lambda A, \mu B) = 1 - \tau_{n,\Gamma_0}((\lambda A \to \mu B) \wedge (\mu B \to \lambda A)) = 1 - \tau((\lambda A \to \mu B) \wedge (\mu B \to \lambda A)) = \rho(\lambda A, \mu B)$. So we have $div_{\Gamma_0}(\Gamma) = div(\Gamma)$.

(iii) If $\overline{0} \in D(\Gamma_0)$, by Definition 12 and $\lambda A, \mu B \in D_{\Gamma_0}(\Gamma)$, we have $\tau_{n,\Gamma_0}((\lambda A \to \mu B) \wedge (\mu B \to \lambda A)) = 1$. That is, $\rho_{n,\Gamma_0}(\lambda A, \mu B) = 1 - \tau_{n,\Gamma_0}((\lambda A \to \mu B) \wedge (\mu B \to \lambda A)) = 0$. Thus we have $div_{\Gamma_0}(\Gamma) = 0$. □

Theorem 12 *Let $\Gamma_0 \subseteq F(S)$, S_{Γ_0} be finite, $\forall \Gamma \subseteq F(S)$. Then,*

$$div_{\Gamma_0}(\Gamma - (\Gamma_0 \cap \Gamma)) = div_{\Gamma_0}(\Gamma).$$

Proof From Definition 12 we know, $div_{\Gamma_0}(\Gamma - (\Gamma_0 \cap \Gamma)) = \sup\{\rho_{n,\Gamma_0}(\lambda A, \mu B) | \lambda A, \mu B \in D((\Gamma - (\Gamma_0 \cap \Gamma)) \cup \Gamma_0)\} = \sup\{\rho_{n,\Gamma_0}(\lambda A, \mu B) | \lambda A, \mu B \in D(\Gamma_0 \cup \Gamma)\} = div_{\Gamma_0}(\Gamma)$. □

Corollary 1 *Let $\Gamma_0 \subseteq F(S)$, S_{Γ_0} be finite, $\forall \Gamma \subset F(S)$. If $\Gamma \subseteq \Gamma_0$, then $div_{\Gamma_0}(\Gamma) = 0$.*

Definition 13 Let $\Gamma, \Gamma_0 \subset F(S)$. If $\overline{0} \in D(\Gamma_0) \cup \Gamma$, and $N(\Gamma) \neq \varnothing$, then we say that the theory Γ is not compatible with the specific theory Γ_0. Otherwise, we say that the theory Γ is compatible with the specific theory Γ_0.

Specifically. If $\Gamma \subseteq D(\Gamma_0)$, then we say that the theory Γ is completely compatible with the specific theory Γ_0.

Theorem 13 *Let $\Gamma_0 \subseteq F(S)$, S_{Γ_0} be finite, $\forall \Gamma \subseteq F(S)$. Then*
(i) *Theory Γ is completely compatible with the specific theory Γ_0, if and only if $div_{\Gamma_0}(\Gamma) = 0$;*
(ii) *Theory Γ is not compatible with the specific theory Γ_0, if and only if $\exists \lambda A, \mu B \in D(\Gamma_0 \cup \Gamma)$, s.t. $\rho_{n,\Gamma_0}(\lambda A, \mu B) = 1$;*
(iii) *Theory Γ is compatible with the specific theory Γ_0, if and only if $\forall \lambda A, \mu B \in D(\Gamma_0 \cup \Gamma)$, s.t. $\rho_{n,\Gamma_0}(\lambda A, \mu B) < 1$.*

Proof (i) Necessity: If theory Γ is completely compatible with the specific theory Γ_0, $\Gamma \subseteq D(\Gamma)$, $D(\Gamma_0 \cup \Gamma) = D(\Gamma_0)$. So for any $\lambda A, \mu B \in D(\Gamma_0 \cup \Gamma)$, we have $\lambda A, \mu B \in D(\Gamma_0)$. Thus $\rho_{n,\Gamma_0}(\lambda A, \mu B) = 1 - \tau_{n,\Gamma_0}((\lambda A \to \mu B) \wedge (\mu B \to \lambda A)) = 0$. According to Definition 12 we can obtain $div_{\Gamma_0}(\Gamma) = 0$.

Sufficiency: If $div_{\Gamma_0}(\Gamma) = 0$. By Definition 12 we know $\lambda A \in \Gamma$, $\rho_{n,\Gamma_0}(\lambda A, T) = 0$(T is theorem in $\Pi_{\sim,\Delta}$). $\lambda A \approx ((\lambda A \to T) \wedge (T \to \lambda A))$is obvious. According to Theorem 5(ii) we have $\tau_{n,\Gamma_0}(\lambda A) = \tau_{n,\Gamma_0}(\lambda A \to T) \wedge (T \to \lambda A)) = 1 - \rho_{n,\Gamma_0}(\lambda A, T) = 1$, then $\lambda A \in D(\Gamma_0)$. Thus we get $\Gamma \subseteq D(\Gamma_0)$. So we can get that theory Γ is completely compatible with the specific theory Γ_0.

(ii) Necessity: If theory Γ is not compatible with the specific theory Γ_0, $\overline{0} \in D(\Gamma_0) \cup \Gamma$, and $N(\Gamma) \neq \varnothing$. According to $N(\Gamma) \neq \varnothing$ and Definition 7 we know $\tau_{n,\Gamma_0}(\overline{0}) = 0$. $(\overline{0} \to T) \wedge (T \to \overline{0}) \approx \overline{0}$ is obvious. Due to Theorem 5(ii), we get $\tau_{n,\Gamma_0}((\overline{0} \to T) \wedge (T \to \overline{0})) = \tau_{n,\Gamma_0}(\overline{0}) = 0$. Thus there exists $\overline{0}$,T in $D(\Gamma_0 \cup \Gamma)$, we have $\rho_{n,\Gamma_0}(\overline{0}, T) = 1 - \tau_{n,\Gamma_0}((\overline{0} \to T) \wedge (T \to \overline{0})) = 1$.

Sufficiency: If exist $\lambda A, \mu B \in D(\Gamma_0 \cup \Gamma)$, such that $\rho_{n,\Gamma_0}(\lambda A, \mu B) = 1$, then $\tau_{n,\Gamma_0}((\lambda A \to \mu B) \wedge (\mu B \to \lambda A)) = 1 - \rho_{n,\Gamma_0}(\lambda A, \mu B) = 0$. According to Definition 13, we know $N(\Gamma) \neq \varnothing$.

Let's prove $\overline{0} \in D(\Gamma_0 \cup \Gamma)$.

On the one hand, according to $\lambda A, \mu B \in D(\Gamma_0 \cup \Gamma)$ and $\vdash \lambda A \to (\mu B \to \lambda A)$, $\vdash \mu B \to (\lambda A \to \mu B)$, we can use MP rule to get $\Gamma_0 \cup \Gamma \vdash (\lambda A \to \mu B) \wedge (\mu B \to \lambda A)$. On the other hand, according to Theorem 5(iii), we can obtain $\tau_{n,\Gamma_0}(\sim ((\lambda A \to \mu B) \wedge$

$(\mu B \to \lambda A))) = 1 - \tau_{n,\Gamma_0}((\lambda A \to \mu B) \wedge (\mu B \to \lambda A)) = \rho_{n,\Gamma_0}(\lambda A, \mu B) = 1$. Thus, $\Gamma_0 \cup \Gamma \vdash \sim ((\lambda A \to \mu B) \wedge (\mu B \to \lambda A))$. For $\vdash ((\lambda A \to \mu B) \wedge (\mu B \to \lambda A)) \to (\sim ((\lambda A \to \mu B) \wedge (\mu B \to \lambda A)) \to ((\lambda A \to \mu B) \wedge (\mu B \to \lambda A)) \wedge (\sim ((\lambda A \to \mu B) \wedge (\mu B \to \lambda A))))$, we can get $\Gamma_0 \cup \Gamma \vdash ((\lambda A \to \mu B) \wedge (\mu B \to \lambda A)) \wedge (\sim ((\lambda A \to \mu B) \wedge (\mu B \to \lambda A)))$ using MP rule. Since $((\lambda A \to \mu B) \wedge (\mu B \to \lambda A)) \wedge (\sim ((\lambda A \to \mu B) \wedge (\mu B \to \lambda A))) \approx \bar{0}$. We obtain $\Gamma_0 \cup \Gamma \vdash \bar{0}$. That is, $\bar{0} \in D(\Gamma_0 \cup \Gamma)$. Theory Γ is not compatible with the specific theory Γ_0.

(iii) The conclusion can be proved by using (ii). □

Definition 14 Let $\Gamma_0 \subseteq F(S)$, S_{Γ_0} be finite, $\forall \Gamma \subseteq F(S)$,

$$i_{\Gamma_0}(\Gamma) = \sup\{[\rho_{n,\Gamma_0}(\lambda A, \mu B)] | \forall \lambda A, \mu B \in D_{\Gamma_0}(\Gamma)\}.$$

$i_{\Gamma_0}(\Gamma)$ is called the polar index of theory Γ relative to the fixed theory Γ_0. Thereinto, $[\rho_{n,\Gamma_0}(\lambda A, \mu B)]$ is the largest integer in$\rho_{n,\Gamma_0}(\lambda A, \mu B)$.

Theorem 14 *Let $\Gamma_0 \subseteq F(S)$, S_{Γ_0} be finite, $\forall \Gamma \subseteq F(S)$.*
(i) $i_{\Gamma_0}(\Gamma) = 1$, *if and only if theory Γ is not compatible with the specific theory Γ_0;*
(ii) $i_{\Gamma_0}(\Gamma) = 0$, *if and only if theory Γ is compatible with the specific theory Γ_0.*

Proof (i) According to Definition 14 and Theorem 13(ii), we know $i_{\Gamma_0}(\Gamma) = 1$. If and only if, there exists $\lambda A, \mu B \in D_{\Gamma_0}(\Gamma)$, $\rho_{n,\Gamma_0}(\lambda A, \mu B) = 1$ is tenable. If and only if theory of Γ is not compatible with the specific theory of Γ_0.

(ii) According to Definition 14 and Theorem 13(iii), we know $i_{\Gamma_0}(\Gamma) = 0$. If and only if for any $\lambda A, \mu B \in D_{\Gamma_0}(\Gamma)$. $\rho_{n,\Gamma_0}(\lambda A, \mu B) < 1$ is tenable. If and only if theory of Γ is compatible with the specific theory of Γ_0. □

Definition 15 Let $\Gamma_0 \subseteq F(S)$, S_{Γ_0} be finite, $\forall \Gamma \subseteq F(S)$,

$$\eta_{\Gamma_0}(\Gamma) = 1 - \frac{1}{2} div_{\Gamma_0}(\Gamma)(1 + i_{\Gamma_0}(\Gamma)).$$

$\eta_{\Gamma_0}(\Gamma)$ is called as η_{Γ_0}-consistency degrees of theory Γ relative to the fixed theory Γ_0.

Theorem 15 *Let $\Gamma_0 \subseteq F(S)$, S_{Γ_0} be finite, $\forall \Gamma \subseteq F(S)$ (i) Γ is completely compatible with the specific theory Γ_0, if and only if $\eta_{\Gamma_0}(\Gamma) = 1$;*
(ii) *Γ is compatible with the specific theory Γ_0, if and only if $\frac{1}{2} \le \eta_{\Gamma_0}(\Gamma) \le 1$;*
(iii) *Γ is compatible with the specific theory Γ_0 and fully divergent, if and only if $\eta_{\Gamma_0}(\Gamma) = \frac{1}{2}$;*
(iv) *Γ is not compatible with the specific theory Γ_0, if and only if $\eta_{\Gamma_0}(\Gamma) = 0$.*

Proof (i) According to Theorem 13(i) and Definition 15, we know Γ is completely compatible with the specific theory Γ_0, if and only if $div_{\Gamma_0}(\Gamma) = 0$, if and only if $\eta_{\Gamma_0}(\Gamma) = 1$.

(ii) According to Theorem 11(i) and 13(ii) and Definition 15, we know Γ is compatible with the specific theory Γ_0,if and only if $i_{\Gamma_0}(\Gamma) = 0$, if and only if $\eta_{\Gamma_0}(\Gamma) = 1 - \frac{1}{2} div_{\Gamma_0}(\Gamma)$, if and only if $\frac{1}{2} \le \eta_{\Gamma_0}(\Gamma) \le 1$.

(iii) Necessity: it is easy to prove it by Definition 12 and Theorem 14(ii).

Sufficiency: If $\eta_{\Gamma_0}(\Gamma) = \frac{1}{2}$, $div_{\Gamma_0}(\Gamma)(1 + i_{\Gamma_0}(\Gamma)) = 1$. If $i_{\Gamma_0}(\Gamma) = 1$, by Theorem 14(i), we know Γ is not compatible with the specific theory Γ_0. According to Theorem 13(ii) and Definition 14, we can get $div_{\Gamma_0}(\Gamma) = 1$. So, $\eta_{\Gamma_0}(\Gamma) = 1 - \frac{1}{2}div_{\Gamma_0}(\Gamma)(1 + i_{\Gamma_0}(\Gamma)) = 0$. This conflicts with the conditions. Thus, we can get $i_{\Gamma_0}(\Gamma) = 0$, $div_{\Gamma_0}(\Gamma) = 1$. Γ is compatible with the specific theory Γ_0 and fully divergent due to Definition 12 and Theorem 14(ii).

(iv) Necessity: If Γ is not compatible with the specific theory Γ_0, by Theorems 13(ii) and 14(i), we know $i_{\Gamma_0}(\Gamma) = 1$, $div_{\Gamma_0}(\Gamma) = 1$. Then we can obtain $\eta_{\Gamma_0}(\Gamma) = 1 - \frac{1}{2}div_{\Gamma_0}(\Gamma)(1 + i_{\Gamma_0}(\Gamma)) = 0$.

Sufficiency: If there is $\eta_{\Gamma_0}(\Gamma) = 0$, $div_{\Gamma_0}(\Gamma)(1 + i_{\Gamma_0}(\Gamma)) = 2$. Since $i_{\Gamma_0}(\Gamma) \in \{0, 1\}$ and $0 \leq div_{\Gamma_0}(\Gamma) \leq 1$, we get $i_{\Gamma_0}(\Gamma) = 1$ and $div_{\Gamma_0}(\Gamma) = 1$. According to Theorem 14(i), we know Γ is not compatible with the specific theory Γ_0. □

5 Conclusions

The definition of Γ-k truth degree, Γ-k similarity degree and Γ-k pseudo-metric of formula relative to local finite theory Γ under the k conjunction and its related properties are also given by using induced function. The concept of relative divergence degree and relative compatibility of arbitrary theory Γ relative to a particular theory Γ_0 are introduced in the $Goguen_{\sim,\Delta}$ propositional logic system and the important relationship between them is also obtained. We will continue discussing how these good properties in other logical system are in the future.

Acknowledgments The work is supported by National Natural Science Foundation of China under Grant No.11471007; the work is supported by Natural Science Foundation of Shaanxi Province of China under Grant No.2014JM1020; The work is supported by Graduate Innovation Found of Yan'an University No.YCX201612.

References

1. Pavelka, J.: On fuzzy logic I: Many-valued rules of inference. II: Enriched lattice and semantics of propositional calculi. III: Semantical completeness of some many-valued propositional calculi. Zeitschrf Math Logik und Grundlagender Math. 25, 45–52, 119–134, 447–464 (1979)
2. Wang, G.J.: Quantitative logic(I). Chin. J. Eng. Math. **23**(2), 191–215 (2006)
3. Pei, D.W.: Fuzzy Logic Theory and Its Application Based on t-Norm. Science Press, Beijing (2013)
4. Wang, G.J., Liu, B.C.: The theory of relative Γ-tautology degree of formulas in four propositional logics. Chin. J. Eng. Math. **24**(4), 598–610 (2007)
5. Zhou, J.R., Wu, H.B.: An equivalent definition and some properties of truth degrees in Łukasiewicz propositions logic system. Chin. J. Eng. Math. **30**(4), 580–590 (2013)
6. Hui, X.J., Wang, G.J.: Randomization of classical inference patterns and its application. Sci. China(Series E) **37**(6), 801–812 (2007)

7. Hui, X.J.: Randomization of 3-valued propositional logic system. Acta Mathematicae Applicatae Sinica **32**(1), 19–27 (2009)
8. Wang, G.J., Hui, X.J.: Generalization of fundamental theorem of probability logic. Acta Electronica Sinica **35**(7), 1333–1340 (2007)
9. Zhou, H.J.: Theory of Borel probability truth degrees of propositions in Łukasiewicz propositional logics and a limit theorem. J. Softw. **23**(9), 2235–2247 (2012)
10. She, Y.H., He, X.L.: Borel probabilistic rough truth degree of formulas in rough logic. J. Softw. **25**(5), 970–983 (2014)
11. Wu, H.B.: The theory of Γ-truth degrees of formulas and limit theorem in Łukasiewicz propositional logic. Sci. China Inf. Sci. **44**(12), 1542–1559 (2014)
12. Wu, H.B., Zhou, J.R.: The Γ-truth degree of formulas in propositional logic system $R_0Ł_{3n+1}$ with properties. Chin. J. Comput. **38**(8), 1672–1679 (2015)
13. Cui, Y.L., Wu, H.B.: The relative divergence degree and relative consistency degree of theories in the propositional logic system L_n^*. Fuzzy Syst. Math. **25**(6), 53–59 (2011)
14. Esteva, F., Godo, L., Hjek, P., et al.: Residuated fuzzy logics with an involutive negation. Arch. Math. Logic **39**, 103–124 (2000)
15. Flaminio, T., Marchioni, E.: T-norm based logics with an independent an involutive negation. Fuzzy Set Syst. **157**, 3125–3144 (2006)
16. Baaz, M.: Infinite-valued Gödel logic with 0–1 projections and relativisations. Comput. Sci. Phys. Lect. Notes Logic **6**, 23–33 (1996)
17. Cintula, P., Klement, E.P., Mesiar, R., et al.: Fuzzy logics with an additional involutive negation. Fuzzy Set Syst. **161**, 390–411 (2010)
18. Hui, X.J.: Quantified axiomatic extension systems of $SBL_{\sim}$ based on truth value. Sci. China Inf. Sci. **44**(7), 900–911 (2014)

A Class of Fuzzy Modal Propositional Logic Systems with Three Kinds of Negation

Cheng Chen, Li-Juan Zhang and Zheng-Hua Pan

Abstract Distinguishing and dealing with different negations is a basic in fuzzy knowledge representation and reasoning, fuzzy propositional logic formal system with contradictory negation, opposition negation and medium negation (FLCOM) is capable of describing various negative relations in fuzzy knowledge. Based on FLCOM, the fuzzy modal propositional logic system with three kinds of negations MKCOM, as well as MKCOM's expansion systems MTCOM, MS_4COM and MS_5COM are proposed in this paper, the semantics of MKCOM is discussed, the soundness and completeness theorems of MKCOM are proved.

Keywords Fuzzy propositional logic · Negation · Medium modal propositional logic · Fuzzy modal propositional logic

1 Introduction

Modal Logic, as a research concerning inevitability and possibility of non-classical logical theory of propositional calculus, gives a very appropriate non-true value system. In recent years, in order to extend the value of modal logic, some scholars remain committed to the theoretical study of modal logic, and made many new modal logic systems. In 1985, Wu-Jia Zhu and Xi-An Xiao put forward the proposition logic system MP and its expansion MP* [1, 2]. In 1989, Jing Zou gave a new modal logic system and its semantics based on intermediary logic, and expanded the modal concept of medium logic [3]. On the basis of medium logic, Dong-Mo Zhang and Wu-Jia Zhu established a kind of modal logic system MK, MT, MS4 and MS5 with different structures and semantic forms in 1995 [4, 5].

Along with the development of knowledge research, the negation of knowledge plays a more and more important role in the knowledge processing.

C. Chen · L.-J. Zhang · Z.-H. Pan (✉)
School of Science, Jiangnan University, Wuxi 214122, China
e-mail: panzh@jiangnan.edu.cn; pan_zhenghua@163.com

T.-H. Fan et al. (eds.), *Quantitative Logic and Soft Computing 2016*,
Advances in Intelligent Systems and Computing 510,
DOI 10.1007/978-3-319-46206-6_11

In recent years some scholars proposed that uncertain knowledge processing needs different kinds of negations in many fields. Pan proposed that there are five kinds of negative relations in crisp knowledge and fuzzy knowledge in 2006, and Pan introduced three kinds of negations in fuzzy knowledge, namely contradictory negation, opposition negation and medium negation in 2008 [6], thereby introduced a novel fuzzy set called fuzzy set with contradictory, negation, opposite negation and medium negation, FScom for short in 2012 [7], the fuzzy propositional logic system with contradictory negation, opposite negation and medium negation, FLCOM for short in 2013 [8].

This paper attempts to construct a kind of Fuzzy Modal Propositional Logic System with contradictory negation, opposite negation and medium negation based on the formal system of the medium modal logic MK and f FLcom.

2 Preliminaries

FLCOM (Fuzzy Propositional Logic with Contradictory negation, Opposite negation and Medium negation) was introduced in [8], it is a formal logic system which differentiates contradiction, opposition and medium negations.

Definition 1 (i) Let S be a nonempty set, and its elements are called atomic propositions or atomic formula, "$\neg$" (Contradictory negation), "$\urcorner$" (Opposite negation), "$\sim$" (Medium negation), "$\rightarrow$" (Implication), "$\wedge$" (Conjunctive) and "$\vee$" (Disjunctive) is a conjunction," (" and ")" is bracket. The fuzzy proposition, which is composed of the connection words and brackets, can also be called fuzzy formula. Let $\Im$ ibe the set which is composed of all fuzzy formulae.

(ii) The following formulae in $\Im$ are called axioms:
(A_1): $A \rightarrow (B \rightarrow A)$;
(A_2): $(A \rightarrow (A \rightarrow B)) \rightarrow (A \rightarrow B)$;
(A_3): $(A \rightarrow B) \rightarrow ((B \rightarrow C) \rightarrow (A \rightarrow C))$;
(M_1): $(A \rightarrow \neg B) \rightarrow (B \rightarrow \neg A)$;
(M_2): $(A \rightarrow \urcorner B) \rightarrow (B \rightarrow \urcorner A)$;
(H): $\neg A \rightarrow (B \rightarrow A)$;
(C): $((A \rightarrow \neg A) \rightarrow B) \rightarrow ((A \rightarrow B) \rightarrow B)$;
($\vee_1$): $A \rightarrow A \vee B$;
($\vee_2$): $B \rightarrow A \vee B$;
($\wedge_1$): $A \wedge B \rightarrow A$;
($\wedge_2$): $A \wedge B \rightarrow B$;
($Y\urcorner$): $\urcorner A \rightarrow \neg A \wedge \neg \sim A$;
($Y\sim$): $\sim A \rightarrow \neg A \wedge \neg\urcorner A(\neg A \wedge \neg \sim A \rightarrow \urcorner A)$.

(iii) The following deduction forms are the deduction rules:
(MP) $A \rightarrow B, A \vdash B$.

The formal system is composed of (i), (ii) and (iii), is called a fuzzy propositional logic system with Contradictory negation, Opposite negation and Medium negation, FLCOM for short.

Definition 2 In FLCOM, the formula fuzzy A with ¬ A,╕A and ~ A have the following relationship

$$\neg A = ╕A \vee \sim A.$$

The following is a semantic interpretation of FLCOM:

Definition 3 (*λ-assignment*) Let ∂: $\Im \to [0, 1]$ be a mapping and is called a **λ**-assignment of $\Im$ for $\lambda \in (0, 1)$, if $\partial(A) + \partial(╕A) = 1$

$$If\, \lambda \in [1/2, 1)\; and\; \partial(A) \in (\lambda, 1],\; then\; \partial(\sim A) = \lambda - \frac{2\lambda - 1}{1 - \lambda}(\partial(A) - \lambda) \quad (2)$$

$$If\, \lambda \in [1/2, 1)\; and\; \partial(A) \in [0, 1 - \lambda),\; then\; \partial(\sim A) = \lambda - \frac{2\lambda - 1}{1 - \lambda}\partial(A) \quad (3)$$

$$If\, \lambda \in (0, 1/2]\; and\; \partial(A) \in (1 - \lambda, 1],\; then\; \partial(\sim A) = 1 - \frac{1 - 2\lambda}{\lambda}(\partial(A) + \lambda - 1) - \lambda \quad (4)$$

$$If\, \lambda \in (0, 1/2]\; and\; \partial(A) \in [0, \lambda),\; then\; \partial(\sim A) = 1 - \frac{1 - 2\lambda}{\lambda}\partial(A) - \lambda \quad (5)$$

$$Else, \partial(\sim A) = max(\partial(A), 1 - \partial(A)) \quad (6)$$

$$\partial(A \vee B) = \max(\partial(A), \partial(B)), \partial(A \wedge B) = \max(\partial(A), \partial(B)),$$

$$\partial(A \to B) = \Re(\partial(A), \partial(B)),\; here\; \Re : [0, 1]^2 \to [0, 1]\; is\; a\; function\; of\; two\; variables.$$

Theorem 1 *In* FLCOM, $\vdash A \to A$.

Proof
(a) $A \to (A \to A)$ (A_1)
(b) $(A \to (A \to A)) \to (A \to A)$ (A_2)
(c) $A \to A$ (a) (b) □

Theorem 2 *In* FLCOM, $\vdash A \to (B \to (A \wedge B))$.

Proof $\vdash A \wedge B \to A \wedge B$ by Theorem 1, because $\vdash A \to (B \to$ C), so $\vdash A \to (B \to (A \wedge B))$. □

3 MKCOM: Fuzzy Modal Propositional Logic with Three Kinds of Negation

Classical modal logic is a logic system with modal words "inevitable" ($\Box$) and "may" ($\Diamond$). Based on Medium Logic System, Wu-Jia Zhu and Dong-Mo Zhang established a series of MK, MT, MS_4 and MS_5 of the medium modal logic system with different structures and semantic forms. Among them, MK is the basic system. The main idea of MK is "inevitable A" ($\Box A$) is interpreted as A in all possible worlds as true, "inevitable A" false as in all possible world in which A is false, the opposite negation of $\Box A$ is true (denoted as $\urcorner \Box A$). And "inevitable a" in the intermediate state understood as the existence of a through the possible world, the medium negation of $\Box A$ is true (denoted as $\sim \Box A$), where a intermediate state, but in all through the possible world A is not false. Similarly, "may A" ($\Diamond A$) is true can be understood as the existence of a through the possible world in which a is true, "may$A$" false can be understood as in any through the world in A are false, the opposite negation of A as true (denoted as $\urcorner \Diamond A$), "may A" in the intermediate state to show that there is a up to the world, the medium negation of $\Diamond A$ is true (denoted by $\sim \Diamond A$), where A in the intermediate state, but in all up to the world A are not really. In the above ideas, it is proved that MK has the following properties [4]:

$$\Diamond A \dashv\vdash \urcorner \Box \urcorner A, \sim \Box A \dashv\vdash \Box \neg \urcorner A \wedge \Diamond \sim A, \sim \Diamond A \dashv\vdash \Box \neg A \wedge \Diamond \sim A$$

For the distinction between three kinds of negative fuzzy propositional logic FLCOM, how to establish the modal propositional logic based on FLCOM? To this end, based on the medium modal logic formal system MK, we give a kind of fuzzy modal propositional logic system with three kinds of negative MKCOM as follows.

Definition 4 (i) The basic set of symbols and the formation rules of the FLCOM in the form of the modal word $\Box$, $\Diamond$ constitute the form of language as $\mathfrak{R}$.

(ii) The following formulas in $\mathfrak{R}$ are axioms:
(A_1) $A \to (B \to A)$;
(A_2) $(A \to (A \to B)) \to (A \to B)$;
(A_3) $(A \to B) \to ((B \to C) \to (A \to C)$;
(M_1) $(A \to \neg B) \to (B \to \neg A)$;
(M_2) $(A \to \urcorner B) \to (B \to \urcorner A)$;
(H) $\neg A \to (B \to A)$;
(C) $((A \to \neg A) \to B) \to ((A \to B) \to B)$;
($\vee_1$) $A \to A \vee B$;
($\vee_2$) $B \to A \vee B$;
($\wedge_1$) $A \wedge B \to A$;
($\wedge_2$) $A \wedge B \to B$;
($Y\urcorner$) $\urcorner A \to \neg A \wedge \neg \sim A$;
($Y\sim$) $\sim A \to \neg A \wedge \neg \urcorner A(\neg A \wedge \neg \sim A \to \urcorner A)$.

(iii) Inference rule:
(R_1) If $\vdash A$, then $\vdash \Box A$;
(R_2) $\Box(A \rightarrow B) \vdash \Box A \rightarrow \Box B$;
(R_3) $\Diamond A \dashv\vdash \urcorner\Box\urcorner A$;
(R_4) $\sim \Box A \dashv\vdash \Box\neg\urcorner A \wedge \Diamond \sim A$;
(R_5) $\sim \Diamond A \dashv\vdash \Box\neg A \wedge \Diamond \sim A$;

From the above (i), (ii) and (iii), FLCOM forms a system, it is called a fuzzy propositional logic system with three kinds of negations.
Note. in the above inference rules, (R_2) corresponding to the classical modal formal system K rules. (R_3), (R_4), and (R_5) reflects the relationship between the word $\Box$ and $\Diamond$.
Intuitively speaking, R_3 represents "*A* may be true if and only if not inevitable*A*false"; R_4 represents" inevitable*A* really taking the mid-value if and only if *A* does not necessarily take false value but may take the mid-value"; R_5 represents "*A* may really taking the mid-value if and only if *A* not necessarily true but may take the mid-value".

In MKCOM, the following conclusions can be proved:

Lemma 1 *If* $\vdash A \leftrightarrow B$, *then* $\vdash \Box A \leftrightarrow \Box B$.

Proof

(a) $\vdash A \leftrightarrow B$	
(b) $\vdash \Box A \leftrightarrow B)$	a) (R_1)
(c) $\vdash \Box(A \leftrightarrow B) \rightarrow (\Box A \leftrightarrow \Box B)$	(b) (R_2)
(d) $\vdash \Box A \leftrightarrow \Box B$	(b) (c)

$\Box$

Lemma 2 $\sim \Diamond A \dashv\vdash \sim \urcorner A \Box \urcorner A$.

Proof Since $A \dashv\vDash \urcorner\urcorner A$, and $\neg A \dashv\vDash \neg\urcorner\urcorner A$, so $\Box\neg A \dashv\vDash \Box\neg\urcorner\urcorner A$. Because of $\sim \urcorner A \dashv\vDash \sim A$, we can get $\urcorner \sim \urcorner A \dashv\vDash \urcorner \sim A$, so $\Box\urcorner \sim \urcorner A \dashv\vDash \Box\urcorner \sim A$, therefore $\urcorner\Box\urcorner \sim \urcorner A \dashv\vDash \urcorner\Box \urcorner \sim A$. From the above two results available $\sim \Diamond A \dashv\vdash \Box\neg A \wedge \Diamond \sim A \dashv\vdash \Box\neg A \wedge \urcorner\Box\urcorner \sim A \dashv\vdash \Box\neg A \wedge \urcorner\Box\urcorner \sim \urcorner A \dashv\vdash \Box\neg A \wedge \Diamond \sim \urcorner A \dashv\vdash \Box\neg\urcorner\urcorner A \wedge \Diamond \sim \urcorner A \dashv\vdash \sim \Box\urcorner A \vdash \dashv \sim \urcorner\Box\urcorner A$. $\Box$

Theorem 3 **(Substitution theorem).** *If* $A \dashv\vDash B$, $\partial(A)$ *is an arbitrary formula in* $\Re$, *then* $\partial(A) \dashv\vDash \partial(B)$.

Theorem 4 *In* MKCOM,
(1) $\Box(A \wedge B) \vdash \Box A \wedge \Box B$;
(2) $\Box A \wedge \Box B \vdash \Box(A \wedge B)$;
(3) $\Box A \vee \Box B \vdash \Box(A \vee B)$;
(4) $\Diamond(A \vee B) \vdash \Diamond A \vee \Diamond B$;
(5) $\Diamond A \vee \Diamond B \vdash \Diamond(A \vee B)$;
(6) $\Diamond A \wedge B) \vdash \Diamond A \wedge \Diamond B$;

Proof We choose to prove (1), (2) and (3) are similar.

(1)

(a) $A \wedge B \to A$ ($\wedge_1$)

(b) $\Box(A \wedge B \to A)$ (a) (R_1)

(c) $\Box(A \wedge B) \to \Box A$ (b)(R_2)

(d) $\Box(A \wedge B) \to \Box B$ (c)

(e) $\Box(A \wedge B) \to (\Box A \wedge \Box B)$ (c) (d) ($\wedge_2$)

(2)

(a) $A \to (B \to (A \wedge B))$ (A_2)

(b) $\Box A \to \Box(B \to A \wedge B))$ a)(R_1)

(c) $\Box A$ assumption

(d) $\Box B \to A \wedge B))$ (b) (c)

(e) $\Box B \to \Box(A \wedge B)$ (d) (R_2)

(f) $\Box B$ assumption

(g) $\Box(A \wedge B))$ (e) (f)

(h) $(\Box A \wedge \Box B) \to \Box(A \wedge B))$ d) f)

(3)

(a) $A \to A \vee B$ ($\vee_1$)

(b) $\Box(A \to A \vee B)$ (R_1)

(c) $\Box A \to \Box(A \vee B))$ (b) (R_2)

(d) $\Box B \to \Box(A \vee B))$ (c)

(e) $\Box A \vee \Box B$ assumption

(f) $\Box(A \vee B)$ (c) (d) (e) □

We can prove the following result in MKCOM.

Theorem 5 *In* MKCOM,

(1) $\neg\Box A \vdash \Diamond\neg A$, $\Diamond\neg A \vdash \neg\Box A$;

(2) $\neg\Diamond A \vdash \Box\neg A$, $\Box\neg A + \neg\Diamond A$.

4 Semantic Interpretation of MKCOM

The semantic interpretation of the intermediate Fuzzy Modal Logic System MKCOM consists of the following formal structure of quad $<W, R, V, T>$, where W is called possible set, R is a binary relation on W, V is a mapping from $\{P_1, P_2, \ldots\} \times W$ to the true values set T, $T = \{0, 1/2, 1\}$, for arbitrary A and $w \in W$, $V(A, w)$ represents the value of the formula A in the possible world, which can be recursively defined as follows:

(1) $V(P_n, w) \in \{0, 1/2, 1\}, n = 1, 2, 3, \ldots$.

(2) $V(\urcorner A, w) = \begin{cases} 0, & \text{if } V(A, w) = 1 \\ 1/2, & \text{if } V(A, w) = 1/2 \\ 1, & \text{if } V(A, w) = 0 \end{cases}$

(3) $V(\sim A, w) = \begin{cases} 1, & \text{if } V(A, w) = 1/2 \\ 1/2, & \text{else} \end{cases}$

(4) $V(\neg A, w) = \max(V(\urcorner A, w), V(\sim A, w)) = \begin{cases} 1/2, & \text{if } V(A, w) = 1 \\ 1, & \text{if } V(A, w) = 1/2 \text{ or } V(A, w) = 0 \end{cases}$

(5) $V(A \wedge B, w) = \min(V(A, w), V(B, w))$

(6) $V(A \vee B, w) = \min(V(A, w), V(B, w))$

(7) $V(A \to B, w) = \begin{cases} 1, & \text{if } V(A, w) = 0 \text{ or } V(B, w) = 1, \\ & \text{or } V(A, w) = 1/2 \text{ and } V(B, w) = 1/2 \\ 0, & \text{if } V(A, w) = 1 \text{ and } V(B, w) = 0 \\ 1/2, & \text{else} \end{cases}$

(9) $V(\Box A, w) = \begin{cases} 1, & \forall w' \in R[w] \text{ bring } (V(A, w') = 1) \\ 0, & \forall w' \notin R[w] \text{ bring } (V(A, w') = 0) \\ 1/2, & \text{else} \end{cases}$

(10) $V(\Diamond A, w) = \begin{cases} 1, & \forall w' \in R[w] \text{ bring } (V(A, w') = 1) \\ 0, & \forall w' \notin R[w] \text{ bring } (V(A, w') = 0) \\ 1/2, & \text{else} \end{cases}$

Definition 5 The deduction form $\Gamma \vdash A$ (Γ may be empty set) is provable in MKCOM, if there exists a finite sequence of deduction forms $E_1, E_2, \ldots, E_n$ such that $E_n = \Gamma \vdash A$ and for each $E_n (1 \le k \le n)$, either E_k is an axiom of MKCOM or E_k follows from E_i *and* $E_j (i < k, j < k)$ using the deduction rule of MKCOM, then $E_1, E_2, \ldots, E_n$ is called a proof of $\Gamma \vdash A$, n length of proof. $\Gamma \vdash A$ is writes as $\vdash A$ when Γ is empty set, A is provable in FLCOM when $\vdash$ A is provable in FLCOM. If Σ denote an infinite set of formulas and there is finite set $\Gamma \subset \Sigma$ such that $\Gamma \vdash A$ is provable in MKCOM, then $\Sigma \vdash A$ is called provable in MKCOM.

Definition 6 (*Three-valued valid formula*) For any formal system $<W, R, V, T>$, *if* $V(A, w) = 1$ for any three-valued assignment V, then A is called three-valued valid formula, which is denoted by $\models$ A. For any $A \in \Gamma$ if there is $\models A$, then it is denoted by $\Gamma \models A$.

Definition 7 The formula set Γ is a consistent set if and only if $\Gamma \vdash A$ and $\neg A$ is established, when there is no formula A.

Definition 8 If Γ is consistent set and no coordination for any A and Γ W {A} in Γ, then Γ is called a maximal consistent set.

Theorem 6 *(Soundness of MK* COM*).*
(a) If $\vdash A$*, then* $\models A$.
(b) If $\Gamma \vdash A$*, then* $\Gamma \models A$.

Proof (a). Assume that $\vdash A$, one has to prove the existence of $\vdash A$: $E_1, E_2 \ldots E_n$, by Definition 5, where $E_n = \vdash A$. The following induction is about the length n of $E_1, E_2 \ldots E_n$, where $E_1, E_2 \ldots E_n$, is a proof of $\vdash A$.

(i). If n =1, according to Definition 5, we can get $E_1 \vdash A$, then A is called an axiom in MKCOM.Because axiom of MKCOM is tautology, so $\vdash A$ is just an eternal truth reasoning. According to Definition 6, we can get $\models A$.

(ii). Assuming that when $n < k$, (a) was established. When $n = k$, E_n is an axiomatic reasoning of MKCOM, or E_n can be obtained by E_i and E_j $(i < n, j < n)$ using inference rules of MKCOM. If E_n is an axiomatic reasoning of MKCOM, say, (I) is used to get E_n; Soppoe that E_n is obtained by E_i and E_j $(i < n, j < n)$ using inference rules of MKCOM, E_i and E_j is eternal truth reasoning by the induction hypothesis, so E_n is an eternal truth reasoning, that is, $\vdash A$ is an eternal truth reasoning. According to Definition 6, we can get $\models A$.

Therefore, (a) holds by the (i) and (ii). Similarly we can prove (b). □

Lemma 3 *Let A be a formula of* MKCOM, A_1 *and* A_2 *are two formulae which does not repeated in* $\{A, \neg A, \sim A\}$, Γ *is a formula set of* MKCOM. *Then the following propositions are equivalent:*
(a) Γ *is no ta consistent set;*
(b) there exists $A \in \Gamma$*, so that* $\Gamma \vdash A_1$ *and* A_2 *can be proved in* MKCOM;
(c) there exists $A \in \Gamma$ *that* $\Gamma \vdash \neg \sim A$ *can be proved in* MKCOM.

Proof Arbitrary two formula are contradictory (i.e. not coordinated) in $\{A, \neg A, \sim A\}$, so A_1 and A_2 are not coordinate formulae. If Γ is inconsistent, there exists $A \in \Gamma$, then $\Gamma \vdash A_1$ and A_2 in MKCOM can permit by Definition 7, so we can get (a) $\Rightarrow$ (b). If there exists $A \in \Gamma$, then $\Gamma \vdash A_1$ and A_2in MKCOM can permit. And because A_1 and A_2 are uncoordinated formulae, we can get $\Gamma \vdash A_1$ and $A_2 \vdash B$. For arbitrary formula B, we can make $B = \neg \sim A$, so we get (b) $\Rightarrow$ (c). If there exists $A \in \Gamma$, then $\Gamma \vdash \neg \sim A$ in MKCOM can permit. And because of the presence of $\Gamma \vdash \neg \sim A \vdash B$, Γ is not coordinated, so we get (c) $\Rightarrow$ (a). □

Lemma 4 *If* Γ *is a maximal consistent set in* MKCOM, *and A is an arbitrary formula, then* $A \in \Gamma$ *when and only when* $\Gamma \vdash A$.

Proof assume that $A \notin \Gamma$, because Γ is a maximal consistent set, H is not a coordinated set. Thus, there are two kinds of situations: (1) $\neg A \in \Gamma$, or (2) $\sim A \in \Gamma$. If $\neg A \in \Gamma$ and $\Gamma \vdash \neg A$ in MKCOM had proved, $\Gamma \vdash A$ can't permit by Lemma 3. If $\sim A \in \Gamma$ and $\Gamma \vdash \sim A$ in MKCOM had proved, $\Gamma \vdash A$ can't permit by Lemma 3. Assume that $\Gamma \vdash A$ does not permit, namely $A \notin \Gamma$. □

Lemma 5 *If Γ is a maximal consistent set in* MKCOM, *and A is an arbitrary formula, then $A \in \Gamma$ and $\neg A \in \Gamma$ have one and only one established.*

Proof If $A \in \Gamma$ and $\neg A \in \Gamma$ are both established, since $\neg A \dashv\vdash \daleth A \vee \sim A$, namely $A \in \Gamma$, $\daleth A \in \Gamma$ and $\sim A \in \Gamma$ were founded at the same time, Γ is not coordinated by the Lemma 3, they are relative to the Γ are the maximal consistent set contradiction. □

Lemma 6 *If Γ is a maximal consistent set in* MKCOM, *and A is an arbitrary formula, then $A \in \Gamma$, $\sim A \in \Gamma$ and $\daleth A \in \Gamma$ have one and only one established.*

Proof If there are two established among $A \in \Gamma$, $\sim A \in \Gamma$ and $\daleth A \in \Gamma$, then Γ is not coordinated by Lemma 3, and they are relative to Γ are the maximal consistent set contradiction. □

Theorem 7 *Any consistent set of formulas can be extended to a maximal consistent set in* MKCOM.

Proof All formulas of MKCOM are enumerated as: $A_1, A_2, \ldots, A_n$, Inductively define Γ_n: $\Gamma_1 = \Gamma$, if $\Gamma_n \cup \{A_n\}$ is consistent, define $\Gamma_{n+1} = \Gamma_n \cup \{A_n\}$, otherwise get $\Gamma_{n+1} = \Gamma_n$. We set up $\Gamma^* = \Gamma_n$, and can verify that Γ^* is a maximal consistent set that contains Γ by Definition 8. □

Theorem 8 *Let Γ and Δ in* MKCOM *be a maximal consistent set, then $\{A : \Box A \in \Gamma\} \subset \Delta$ if and only if $\{\Diamond A : A \in \Delta\} \subset \Gamma$.*

Proof If there is $\{A : \Box A \in \Gamma\} \subset \Delta$, and assuming that $A \in \Delta$, we only need to prove $\Diamond A \in \Gamma$. If $\sim \Diamond A \in \Gamma$, then $\Gamma \vdash \sim \Diamond A$, we can get $\Gamma \vdash \Box \neg A$ by rule R_5. So $\neg A \in \Delta$ can be obtained, which is contradictory with $A \in \Delta$. If $\daleth \Diamond A \in \Gamma$, then we have $\Gamma \vdash \daleth \Diamond A$, and because of $\Gamma \vdash \Box \daleth A$, so there is $\daleth A \in \Delta$, which is also contradictory with $A \in \Delta$, thus $\Diamond A \in \Gamma$ by Lemma 6.

If $\{\Diamond A : A \in \Delta\} \subset \Gamma$, and assuming that $\Box A \in \Gamma$ also holds, we only need to verify $A \in \Delta$. If $\sim A \in \Delta$, then $\Diamond \sim A \in \Gamma$, that is $\Gamma \vdash \Diamond \sim A$. On the other hand, it can be easy to prove $\vdash \daleth \Box A \vee \Box \neg A$, then we can get $\Gamma \vdash (\Diamond \sim A \wedge \daleth \Box A) \vee (\Diamond \sim A \wedge \Box \neg \daleth A)$, so we have $\Gamma \vdash \daleth \Box A \vee \sim \Box A$, which is contradictory with $\Box A \in \Gamma$. If $\daleth A \in \Delta$, then $\Diamond \daleth A \in \Gamma$, we can get $\daleth \daleth \Diamond \daleth A \in \Gamma$, namely $\daleth \Box A \in \Gamma$ can be obtained, which is contradictory with $\Box A \in \Gamma$. So it is concluded that $A \in \Delta$. □

Lemma 9 *If Γ is a maximal consistent set of* MKCOM *system, then*
(1) $\Box A \in \Gamma \Leftrightarrow A \in \Delta$ makes Δ the maximal consistent set $\{A : \Box A \in \Gamma\} \subset \Delta$.
(2) $\Diamond A \in \Gamma \Leftrightarrow A \in \Delta$ makes Δ the maximal consistent set $\{\Diamond A : A \in \Delta\} \subset \Gamma$.

Proof (1) "$\Rightarrow$": Obviously.

"$\Leftarrow$": If any maximal consistent set Δ of $\{A : \Box A \in \Gamma\} \subset \Delta$ make $A \in \Delta$, it is not hard to prove $\{A : \Box A \in \Gamma \vdash \} A$, so statement $A_1, A_2, \ldots, A_n \vdash A$ is valid in $\{A : \Box A \in \Gamma\}$ and $\vdash (A_1 \wedge A_2 \wedge \ldots \wedge A_n) \Rightarrow A$, $\vdash \Box((A_1 \wedge A_2 \wedge \ldots \wedge A_n) \Rightarrow A)$ so $\vdash \Box(A_1 \wedge A_2 \wedge \ldots \wedge A_n) \Rightarrow \Box A$. Because $\Box A_1, \Box A_2 \ldots \Box A_n$ is valid in Γ, $\Gamma \vdash \Box A_1 \wedge \Box A_2 \wedge \ldots \wedge \Box A_n$, and $\Gamma \vdash \Box(A_1 \wedge A_2 \wedge \ldots \wedge A_n)$, so $\Gamma \vdash \Box A$, $\Box A \in \Gamma$.

(2) "⇒": Obviously.

"⇐": If $\Diamond A \in \Gamma$, then $\urcorner\Box\urcorner A \in \Gamma$, that is, to prove that a maximal consistent set Δ makes $\{A: \Box A \in \Gamma\} \subset \Delta$ and $A \in \Delta$. Otherwise, any maximal consistent set Δ of $\{A: \Box A \in \Gamma\} \subset \Delta$ make $A \notin \Delta$, by $\urcorner\Box\urcorner A \in \Gamma$, then $\Box\urcorner A \notin \Gamma$ and $\sim\Box\urcorner A \notin \Gamma$. The former and the conclusion (1) have a maximal consistent set Δ_0. And $\urcorner A \notin \Delta_0$, by the latter then $\Box\neg A \notin \Gamma$ or $\Diamond \sim A \notin \Gamma$. If $\Box\neg A \notin \Gamma$, then a maximal consistent set Δ_1 make $\{A: \Box A \in \Gamma\} \subset \Delta_1$ and $\neg A \notin \Delta_1, A \in \Delta_1$, this is contrary to the disproof assumption. If $\Diamond \sim A \notin \Gamma$, by $\{A: \Box A \in \Gamma\} \subset \Delta_0$, then $\{\Diamond A: A \in \Delta_0\} \subset \Gamma$, so $\sim A \notin \Delta_0$, and $A \in \Delta_0$, it is also contrary to the disproof assumption. So, a maximal consistent set Δ make $\{A: \Box A \in \Gamma\} \subset \Delta$ and $A \in \Delta$.

To prove the completeness of MKCOM, we construct the following normal structure $<W, R, V, T>$, and $W = \{T/\Gamma$ is the maximal consistent set in MK$\}$, R is the binary relation on W.

$<\Gamma, \Delta> \in R$ if and only if $\{A: \Box A \in \Gamma\} \subset \Delta$. For any propositional word P_i ($i = 1, 2, \ldots$) and $\Gamma \in W$:

$$V(P_i, \Gamma) = \begin{cases} 1, & \text{when } P_i \in \Gamma \\ 1/2, & \text{when } \sim P_i \in \Gamma \\ 0, & \text{when } \urcorner P_i \in \Gamma \end{cases}$$

Theorem 10 *For any formula A and $\Gamma \in W$,*

(1)$V(A, \Gamma) = 1$, if and only if $A \in \Gamma$.

(2)$V(A, \Gamma) = 1/2$, if and only if $\sim A \in \Gamma$.

(3)$V(A, \Gamma) = 0$, if and only if $\urcorner A \in \Gamma$.

Proof Induct on the structure of A and we only discuss the case of A shaped like $\Box B$, other cases can be classified into the case of $\Box B$ or similar to the completeness proof of FLCOM.

(1). $V(A, \Gamma) = 1$ if and only if $\forall\Box \in W, (\Gamma, \Delta) \in R_i \rightarrow V(B, \Delta) = 1$ if and only if $\forall\Delta \in W$, if $(\Gamma, \Delta) \in R_i$, then $B \in \Delta$ if and only if $\Box B \in \Gamma$ if and only if $A \in \Gamma$.

(3). $V(A, \Gamma) = 0$ if and only if $\exists\Delta \in W, (\Gamma, \Delta) \in R_i \wedge V(B, \Delta) = 0$ if and only if $\exists\, \Delta \in W, (\Gamma, \Delta) \in R_i$ and $\urcorner B \in \Delta$ if and only if $\Delta \in W$, $\{\Diamond A: A \in \Delta\} \subset \Gamma$ and $\urcorner B \in \Delta$ if and only if $\Diamond\urcorner B \in \Gamma$ if and only if $\urcorner\Box B \in \Gamma$.

We can obtain (2) by (1) and (3). □

Theorem 11 (Completeness of MKCOM).

(a) If $\Gamma \models A$, then $\Gamma \vdash A$ is provable in MKCOM.

(b) If $\models A$, then $\vdash A$ is provable in MKCOM.

Proof (a). If $\Gamma \models A$, then there is $A \in \Gamma$ by Lemma 10. According to Lemma 4, there is $\Gamma \vdash A$. (2) can be proved in the same way. □

Based on Fuzzy Modal Propositional Logic System with three kinds of negation (MKCOM), we can further obtain the following extension systems MTCOM, MS_4COM and MS_5COM of MKCOM.

Definition 9 Adding the inference rule $\Box A \vdash A$ in MKCOM, then the intermediate mode system MTCOM can be obtained.

Definition 10 Adding the inference rule $\Box A \vdash \Box\Box$ A in MKCOM, then the intermediate mode system MS_4COM can be obtained.

Definition 11 To T Adding the inference rule $\Diamond A \vdash \Box\Diamond A$, then the intermediate mode system MS_5COM can be obtained.

5 Conclusions

Base on fuzzy propositional logic system with three kinds of negation (FLCOM), this paper proposes fuzzy modal propositional logic system with three kinds of negations (MKCOM), and the extension systems MTCOM, MS_4COM and MS_5COM of MKCOM. These systems can be as foundation of the modal propositions and their different negations.

Acknowledgments This work is supported by the National Natural Science Foundation of China (61375004, 60973156).

References

1. Zhu, W.J., Xiao, X.: An extension of the propositional calculus system of medium logic (I) (II). J. Nanjing Univ. (Natural Sciences Edition) **26**(4), 564–578 (1990); **27**(2), 209–221 (1991)
2. Zhu, W.J., Xiao, X.A.: Propositional calculus system of medium logic (i) (ii) (iii). J. Math. Res. Expo. **8**(2), 327–332, (3), 457–466, (4), 617–631 (1998)
3. Zou, J., Qui, W. : Medium Modal logic formal system and semantics. In: Proceedings of 19th International Symposium Multiple-Valued Logic (1989)
4. Gong, N.S., Zhang, D.M., Zhu, W.J.: Theory and implementation of the medium automatic reasoning (IV)- a kind of modal logic system based on medium logic. Pattern Recognit. Artif. Intell. **8**(1), 1–8 (1995)
5. Zhang, D.M., Gong, N.S.: Theory and implementation of the medium automatic reasoning (V)- the table deduction system of the medium modal logic MK. Pattern Recognit. Artif. Intell. **8**(2), 114–120 (1995)
6. Pan, Z.H.: A logical description of different negative relation in knowledge. Prog. Natural Sci. **18**(11), 66–74 (2008)
7. Pan, Z.H.: Three kinds of negation of fuzzy knowledge and their base of set. Chin. J. Comput. **07**, 1421–1428 (2012)
8. Pan, Z.H.: Fuzzy propositional logic system of distinguish between three kinds of negation and its application. J. Softw. **25**(6), 1255–1272 (2014)

Robustness Analysis of Fuzzy Computation Tree Logic

Li Li, Hong-Juan Yuan and Hai-Yu Pan

Abstract Fuzzy computation tree logic is an extension of classical temporal logic computation tree logic, which is used to specify the properties of systems with uncertain information content. This paper investigates the robustness of fuzzy computation tree logic. Robustness results are proved based on complete Heyting algebra and standard Łukasiewicz algebra.

Keywords Temporal logic · Fuzzy computation tree logic · Model checking · Complete residuated lattices · Heyting algebra

1 Introduction

Temporal logics [1], such as linear temporal logic (LTL) and computation tree logic (CTL), are a useful formalism for specification of reactive systems such as discrete-event controllers. They have been successfully used in many situations, especially for model checking [1].

To specifying systems with fuzzy uncertainty, much efforts have been made to extend temporal logic to fuzzy setting in the past years. Chechik et al. [3, 4] defined CTL over a finite De Morgan algebra and investigated its model-checking problem. Li et al. gave LTL and CTL over possibility measures and their model-checking

L. Li · H.-J. Yuan · H.-Y. Pan (✉)
College of Computer Science and Technology, Taizhou University,
Taizhou 225300, China
e-mail: phyu76@126.com

L. Li
e-mail: lylitzxy@163.com

H.-J. Yuan
e-mail: yhj_blue@126.com

H.-Y. Pan
College of Computer Science, Shaanxi Normal University,
Xi'an 710062, China

T.-H. Fan et al. (eds.), *Quantitative Logic and Soft Computing 2016*,
Advances in Intelligent Systems and Computing 510,
DOI 10.1007/978-3-319-46206-6_12

problems are also discussed [8–10]. Pan et al. presented model-checking algorithms for semantics of CTL in the sense of fuzzy logic and finite lattices, respectively [13, 14].

The present work is a continuation of [13, 14]. We mainly focus on robustness analysis of semantics of fuzzy computation tree logic (FCTL) over complete residuated lattices [2]. In application the proposition values, together with fuzzy transition relations, of a model are somewhat imprecise and subjective, because they are often provided by experts in ad hoc (heuristic) manner from experience or intuition. Hence, it is necessary to require that two models close in some equivalence criterion should yield near truth values to formulas of FCTL in two models. In light of this, we will conduct robustness analysis of the semantics of FCTL based on a logically equivalence measure introduced in [2]. Moreover, we propose a new measure to calculate the perturbation degree of model based on normalized Minkowski distances [5]. Based on the measure, robustness of semantics of FCTL over standard Łukasiewicz algebra is carefully investigated.

2 Fuzzy Computation Tree Logic for Fuzzy Kripke Structures

We start with a short introduction to complete residuated lattices, and then present a fuzzy extension of Kripke structures and its specification language fuzzy computation tree logic. For more details on complete residuated lattices, the reader may refer to [2]. We write $\mathbb{N}$ for the set of natural numbers, I the index set, and $\mathcal{P}(S)$ the power set of S.

A complete residuated lattice [2] is an algebra $\mathcal{L} = (L, \vee, \wedge, \otimes, \to, 0, 1)$, where $(L, \vee, \wedge, 0, 1)$ is a complete lattice with the minimum element 0 and the maximum element 1, $(L, \otimes, 1)$ is a commutative monoid, and $\otimes$ and $\to$ satisfy the adjointness property, i.e. $x \otimes y \leq z$ iff $x \leq y \to z$, for any $x, y, z \in L$. A complete Heyting algebra can be defined as a complete residuated lattice with $\otimes = \wedge$. Standard Łukasiewicz algebra, where $L = [0, 1], x \otimes_L y = (x + y - 1) \vee 0, x \to_L y = (1 - x + y) \wedge 1$, is a complete residuated lattice.

Let X be a universal set. A (lattice-valued) fuzzy set A of X over $\mathcal{L}$ is defined by a function assigning to each element x of X a value $A(x)$ in L; $A(x)$ characterizes the degree of membership of x in A. We say that an $\mathcal{L}$-set A is crisp if $A(x) \in \{0, 1\}$ for all $x \in X$. We denote by $\mathcal{L}(X)$ the set of all fuzzy sets of X. We now turn to the concept of fuzzy Kripke structures (FKSs). To do so, let AP be a fixed finite set of atomic propositions.

Definition 1 ([11, 12]) A fuzzy Kripke structure over AP and $\mathcal{L}$ is a tuple $\mathcal{K} = (S, R, \mathcal{V})$, where

- S is a finite, non-empty set of states;
- R, which is the fuzzy transition function, is a mapping from $S \times S$ to L;
- $\mathcal{V}$ is a labeling function $\mathcal{V} : S \to \mathcal{L}(AP)$ that assigns a truth value in L to an atomic proposition in a state.

For states s and s', $R(s, s')$ is the possibility of making a transition to state s' given that the system is in state s. Kripke structures arise as a special case of FKSs if the transition function and labeling function are crisp.

A path π in an FKS $\mathcal{K}$ is a non-empty sequence of states $\pi = s_0 s_1 \cdots$, where $s_i \in S$ for all $i \geq 0$. A path can be either finite or infinite. We denote the $(i+1)$-st state of π by $\pi(i)$, and $|\pi|$ represents the length of π (i.e., the number of transitions). We write $\Pi(\mathcal{K}, s)$ for the set of all infinite paths starting from state s of $\mathcal{K}$. When the context is clear, we will drop the notation $\mathcal{K}$. Now, we define a logic, called fuzzy computation tree logic (FCTL), for expressing the behaviours of FKSs. For simplicity, we will use the same symbol for a binary logical connective and its interpretation in a model.

Definition 2 ([13, 14]) Let AP be a set of atomic propositions. All well-formed formulas of FCTL are inductively defined using the following grammar:

$$\varphi ::= \mathbf{true} \mid p \mid \varphi \vee \varphi \mid \varphi \wedge \varphi \mid \varphi \rightarrow \varphi \mid \exists \mathrm{X}\varphi \mid \forall \mathrm{X}\varphi \mid \exists \varphi \mathrm{U} \varphi \mid \forall \varphi \mathrm{U} \varphi,$$

where $p \in AP$.

We interpret FCTL formulas over the states of an FKS $\mathcal{K}$ that has the same propositions. Given an FKS $\mathcal{K}$, a state s of $\mathcal{K}$, we write $[\![\mathcal{K}, \varphi]\!](s)$ for the degree to which φ holds in the state s of $\mathcal{K}$.

Definition 3 ([11, 12]) Let $\mathcal{K}$ be an FKS. The valuation $[\![\mathcal{K}, \varphi]\!]$ is defined inductively as:

$$\begin{aligned}
[\![\mathcal{K}, \mathbf{true}]\!](s) &= 1,\\
[\![\mathcal{K}, p]\!](s) &= \mathcal{V}(s)(p),\\
[\![\mathcal{K}, \varphi_1 \wedge \varphi_2]\!](s) &= [\![\mathcal{K}, \varphi_1]\!](s) \wedge [\![\mathcal{K}, \varphi_2]\!](s),\\
[\![\mathcal{K}, \varphi_1 \vee \varphi_2]\!](s) &= [\![\mathcal{K}, \varphi_1]\!](s) \vee [\![\mathcal{K}, \varphi_2]\!](s),\\
[\![\mathcal{K}, \varphi_1 \rightarrow \varphi_2]\!](s) &= [\![\mathcal{K}, \varphi_1]\!](s) \rightarrow [\![\mathcal{K}, \varphi_2]\!](s),\\
[\![\mathcal{K}, \exists X\varphi]\!](s) &= \sup_{s' \in S}(R(s, s') \otimes [\![\mathcal{K}, \varphi]\!](s')),\\
[\![\mathcal{K}, \forall X\varphi]\!](s) &= \inf_{s' \in S}(R(s, s') \otimes [\![\mathcal{K}, \varphi]\!](s')),\\
[\![\mathcal{K}, \exists \varphi_1 \mathrm{U} \varphi_2]\!](s) &= \sup_{\pi \in \Pi(s)} \sup_{i \in \mathbb{N}} \Bigg(\bigotimes_{0 \leq j < i} ([\![\mathcal{K}, \varphi_1]\!](\pi(j)) \otimes R(\pi(j), \pi(j+1)))\\
&\qquad \otimes [\![\mathcal{K}, \varphi_2]\!](\pi(i)) \Bigg),\\
[\![\mathcal{K}, \forall \varphi_1 \mathrm{U} \varphi_2]\!](s) &= \inf_{\pi \in \Pi(s)} \sup_{i \in \mathbb{N}} \Bigg(\bigotimes_{0 \leq j < i} ([\![\mathcal{K}, \varphi_1]\!](\pi(j)) \otimes R(\pi(j), \pi(j+1)))\\
&\qquad \otimes [\![\mathcal{K}, \varphi_2]\!](\pi(i)) \Bigg).
\end{aligned}$$

The proof of the following theorem is similar to that of Lemma 3 in [14] and is omitted.

Theorem 1 *The following equalities hold.*
(i) $[\![\exists\varphi_1 U\varphi_2]\!] = \mu x.[\![\varphi_2]\!] \cup ([\![\varphi_1]\!] \cap [\![\exists Xx]\!]_\Pi)$.
(ii) $[\![\forall\varphi_1 U\varphi_2]\!] = \mu x.[\![\varphi_2]\!] \cup ([\![\varphi_1]\!] \cap [\![\forall Xx]\!])$.

3 Robustness Analysis

To formulate the robustness of semantics of FCTL, we need to measure the degree of equality of two FKSs. This measurement is similar to that of given in lattice-valued doubly labelled transition systems in [11].

Definition 4 Consider two FKSs $\mathcal{K}_1 = (S,\ R_1,\ \mathcal{V}_1)$ and $\mathcal{K}_2 = (S,\ R_2,\ \mathcal{V}_2)$ with the same state space S and the same set AP of propositions. The equality degree of $\mathcal{K}_1$ and $\mathcal{K}_2$, denoted $\mathcal{E}(\mathcal{K},\ \mathcal{K}')$, is defined by

$$\mathcal{E}(\mathcal{K},\ \mathcal{K}') = \inf_{s\in S}\Big(\inf_{p\in AP}(\mathcal{V}_1(s)(p) \leftrightarrow \mathcal{V}_2(s)(p)) \wedge \inf_{s'\in S}(R_1(s,\ s') \leftrightarrow R_2(s,\ s'))\Big).$$

Such a function first considers one state at a time. For each state, the equality of degree is given by the infimum of degrees of equality between the value of a proposition in said state in the two structures. The one-step degree of equality, considering the best way a transition from the first structure can be matched by a transition in the second, and vice versa. For each state, the infimum of the local degree of equality and the one-step degree of equality is taken. Finally, the infimum over all states is taken. The following observation follows immediately from Definition 4. The desirability of the definition is justified by the following fact, whose proof is easy and is thus omitted.

Lemma 1 *The equality degree $\mathcal{E}$ is a fuzzy equivalence relation on FKS, i.e., for any FKSs $\mathcal{K}_1,\ \mathcal{K}_2,\ \mathcal{K}_3$, the following conditions hold:*
(i) $\mathcal{E}(\mathcal{K}_1,\ \mathcal{K}_1) = 1$;
(ii) $\mathcal{E}(\mathcal{K}_1,\ \mathcal{K}_2) = \mathcal{E}(\mathcal{K}_2,\ \mathcal{K}_1)$;
(iii) $\mathcal{E}(\mathcal{K}_1,\ \mathcal{K}_2) \otimes \mathcal{E}(\mathcal{K}_2,\ \mathcal{K}_3) \leq \mathcal{E}(\mathcal{K}_1,\ \mathcal{K}_3)$.

One of the main results of this paper is the following theorem, which expresses the intuition "if two models are equivalent, then the truth values of every FCTL formula in the two models are also equivalent".

Theorem 2 *Let $\mathcal{L}$ be a complete Heyting algebra. Let $\mathcal{K}_1$ and $\mathcal{K}_2$ be two FKSs with the same state space S and the same set AP of propositions. Then $\mathcal{E}(\mathcal{K}_1,\ \mathcal{K}_2) \leq [\![\mathcal{K}_1,\ \varphi]\!](s) \leftrightarrow [\![\mathcal{K}_2,\ \varphi]\!](s)$ for all FCTL formulas φ and $s \in S$.*

Proof The proof can be given by structural induction on the structure of the formula φ.

Induction basis: The cases φ = true and $\varphi = p$ are trivial.

Induction step: Assume φ_1 and φ_2 are formulas for which the theorem holds.

- $\varphi = \varphi_1 \wedge \varphi_2$: We have from the induction hypothesis:

$$\begin{aligned}&\mathcal{E}(\mathcal{K}_1, \mathcal{K}_2)\\ &\leq ([\![\mathcal{K}_1, \varphi_1]\!](s) \rightarrow [\![\mathcal{K}_2, \varphi_1]\!](s)) \wedge ([\![\mathcal{K}_1, \varphi_2]\!](s) \rightarrow [\![\mathcal{K}_2, \varphi_2]\!](s))\\ &\leq ([\![\mathcal{K}_1, \varphi_1 \wedge \varphi_2]\!](s) \rightarrow [\![\mathcal{K}_2, \varphi_1]\!](s)) \wedge ([\![\mathcal{K}_1, \varphi_1 \wedge \varphi_2]\!](s) \rightarrow [\![\mathcal{K}_2, \varphi_2]\!](s))\\ &= [\![\mathcal{K}_1, \varphi_1 \wedge \varphi_2]\!](s) \rightarrow [\![\mathcal{K}_2, \varphi_1 \wedge \varphi_2]\!](s).\end{aligned}$$

  Similarly, we can also show that $\mathcal{E}(\mathcal{K}_1, \mathcal{K}_2) \leq [\![\mathcal{K}_2, \varphi_1 \wedge \varphi_2]\!](s) \rightarrow [\![\mathcal{K}_1, \varphi_1 \wedge \varphi_2]\!](s)$. Hence $\mathcal{E}(\mathcal{K}_1, \mathcal{K}_2) \leq [\![\mathcal{K}_1, \varphi_1 \wedge \varphi_2]\!](s) \leftrightarrow [\![\mathcal{K}_2, \varphi_1 \wedge \varphi_2]\!](s)$.
- $\varphi = \varphi_1 \vee \varphi_2$: We have from the induction hypothesis:

$$\begin{aligned}&\mathcal{E}(\mathcal{K}_1, \mathcal{K}_2)\\ &\leq ([\![\mathcal{K}_1, \varphi_1]\!](s) \rightarrow [\![\mathcal{K}_2, \varphi_1]\!](s)) \wedge ([\![\mathcal{K}_1, \varphi_2]\!](s) \rightarrow [\![\mathcal{K}_2, \varphi_2]\!](s))\\ &\leq ([\![\mathcal{K}_1, \varphi_1]\!](s) \rightarrow [\![\mathcal{K}_2, \varphi_1 \vee \varphi_2]\!](s)) \wedge ([\![\mathcal{K}_1, \varphi_2]\!](s) \rightarrow [\![\mathcal{K}_2, \varphi_1 \vee \varphi_2]\!](s))\\ &= [\![\mathcal{K}_1, \varphi_1 \vee \varphi_2]\!](s) \rightarrow [\![\mathcal{K}_2, \varphi_1 \vee \varphi_2]\!](s).\end{aligned}$$

  Similarly, we have $\mathcal{E}(\mathcal{K}_1, \mathcal{K}_2) \leq [\![\mathcal{K}_2, \varphi_1 \vee \varphi_2]\!](s) \rightarrow [\![\mathcal{K}_1, \varphi_1 \vee \varphi_2]\!](s)$. Hence $\mathcal{E}(\mathcal{K}_1, \mathcal{K}_2) \leq [\![\mathcal{K}_1, \varphi_1 \vee \varphi_2]\!](s) \leftrightarrow [\![\mathcal{K}_2, \varphi_1 \vee \varphi_2]\!](s)$.
- $\varphi = \varphi_1 \rightarrow \varphi_2$: We have from the induction hypothesis:

$$\begin{aligned}&\mathcal{E}(\mathcal{K}_1, \mathcal{K}_2)\\ &\leq [\![\mathcal{K}_2, \varphi_1]\!](s) \rightarrow [\![\mathcal{K}_1, \varphi_1]\!](s)\\ &\leq ([\![\mathcal{K}_1, \varphi_1]\!](s) \rightarrow [\![\mathcal{K}_1, \varphi_2]\!](s)) \rightarrow ([\![\mathcal{K}_2, \varphi_1 \rightarrow \varphi_2]\!](s)).\end{aligned}$$

  We also have

$$\mathcal{E}(\mathcal{K}_1, \mathcal{K}_2) \leq [\![\mathcal{K}_1, \varphi_1 \rightarrow \varphi_2]\!](s) \rightarrow ([\![\mathcal{K}_1, \varphi_1]\!](s) \rightarrow [\![\mathcal{K}_1, \varphi_2]\!](s)).$$

  Hence we have:

$$\mathcal{E}(\mathcal{K}_1, \mathcal{K}_2) \leq [\![\mathcal{K}_1, \varphi_1 \rightarrow \varphi_2]\!](s) \rightarrow [\![\mathcal{K}_2, \varphi_1 \rightarrow \varphi_2]\!](s).$$

  Similarly, we have $\mathcal{E}(\mathcal{K}_1, \mathcal{K}_2) \leq [\![\mathcal{K}_2, \varphi_1 \rightarrow \varphi_2]\!](s) \rightarrow [\![\mathcal{K}_1, \varphi_1 \rightarrow \varphi_2]\!](s)$. Whence we have $\mathcal{E}(\mathcal{K}_1, \mathcal{K}_2) \leq [\![\mathcal{K}_1, \varphi_1 \rightarrow \varphi_2]\!](s) \leftrightarrow [\![\mathcal{K}_2, \varphi_1 \rightarrow \varphi_2]\!](s)$.
- $\varphi = \exists X\varphi_1$: We have from the induction hypothesis that

$$\begin{aligned}&[\![\mathcal{K}_1, \exists X\varphi_1]\!](s) \rightarrow [\![\mathcal{K}_2, \exists X\varphi_1]\!](s)\\ &= \sup_{s' \in S}(R_1(s, s') \wedge [\![\mathcal{K}_1, \varphi_1]\!](s')) \rightarrow \sup_{s' \in S}(R_2(s, s') \wedge [\![\mathcal{K}_2, \varphi_1]\!](s'))\end{aligned}$$

$$
\begin{aligned}
&\geq \sup_{s' \in S}((R_1(s, s') \wedge [\![\mathcal{K}_1, \varphi_1]\!](s')) \rightarrow (R_2(s, s') \wedge [\![\mathcal{K}_2, \varphi_1]\!](s'))) \\
&\geq \sup_{s' \in S}((R_1(s, s') \rightarrow R_2(s, s')) \wedge ([\![\mathcal{K}_1, \varphi_1]\!](s') \rightarrow [\![\mathcal{K}_2, \varphi_1]\!](s'))) \\
&\geq \mathcal{E}(\mathcal{K}_1, \mathcal{K}_2).
\end{aligned}
$$

- $\varphi = \forall X \varphi_1$: We have from the induction hypothesis that

$$
\begin{aligned}
&[\![\mathcal{K}_1, \forall X \varphi_1]\!](s) \rightarrow [\![\mathcal{K}_2, \forall X \varphi_1]\!](s) \\
&= \inf_{s' \in S}(R_1(s, s') \wedge [\![\mathcal{K}_1, \varphi_1]\!](s')) \rightarrow \inf_{s' \in S}(R_2(s, s') \wedge [\![\mathcal{K}_2, \varphi_1]\!](s')) \\
&\geq \inf_{s' \in S}((R_1(s, s') \wedge [\![\mathcal{K}_1, \varphi_1]\!](s')) \rightarrow (R_2(s, s') \wedge [\![\mathcal{K}_2, \varphi_1]\!](s'))) \\
&\geq \inf_{s' \in S}((R_1(s, s') \rightarrow R_2(s, s')) \wedge ([\![\mathcal{K}_1, \varphi_1]\!](s') \rightarrow [\![\mathcal{K}_2, \varphi_1]\!](s'))) \\
&\geq \mathcal{E}(\mathcal{K}_1, \mathcal{K}_2).
\end{aligned}
$$

- $\varphi = \exists \varphi_1 U \varphi_2$: By Theorem 1, we know that the sequence

$$
\begin{aligned}
x_0(s) &= 0, \\
x_{n+1}(s) &= [\![\mathcal{K}_1, \varphi_2]\!](s) \vee ([\![\mathcal{K}_1, \varphi_1]\!](s) \wedge \sup_{s' \in S}(R(s, s') \wedge x_n(s'))),
\end{aligned}
$$

converges to $[\![\mathcal{K}_1, \exists \varphi_1 U \varphi_2]\!](s)$. Similarly, the sequence

$$
\begin{aligned}
y_0(s) &= 0, \\
y_{n+1}(s) &= [\![\mathcal{K}_1, \varphi_2]\!](s) \vee ([\![\mathcal{K}_1, \varphi_1]\!](s) \wedge \sup_{s' \in S}(R(s, s') \wedge y_n(s'))),
\end{aligned}
$$

converges to $[\![\mathcal{K}_2, \exists \varphi_1 U \varphi_2]\!](s)$. We shall use induction on n to show that $\mathcal{E}(\mathcal{K}_1, \mathcal{K}_2) \leq x_n(s) \leftrightarrow y_n(s)$. For the base case, we have $\mathcal{E}(\mathcal{K}_1, \mathcal{K}_2) \leq x_0(s) \leftrightarrow y_0(s) = 1$. Assume by induction on n that $\mathcal{E}(\mathcal{K}_1, \mathcal{K}_2) \leq x_n(s) \leftrightarrow y_n(s)$. Then $\mathcal{E}(\mathcal{K}_1, \mathcal{K}_2) \leq x_{n+1}(s) \leftrightarrow y_{n+1}(s)$. Hence

$$
\mathcal{E}(\mathcal{K}_1, \mathcal{K}_2) \leq [\![\mathcal{K}_1, \exists \varphi_1 U \varphi_2]\!](s) \leftrightarrow [\![\mathcal{K}_2, \exists \varphi_1 U \varphi_2]\!](s).
$$

- $\varphi = \forall \varphi_1 U \varphi_2$: The case for $\varphi = \forall \varphi_1 U \varphi_2$ is similar to the case for $\varphi = \exists \varphi_1 U \varphi_2$, hence the proof for the case is omitted. □

In the following, we introduce the concept of perturbation degree of fuzzy Kripke structures based on normalized Minkowski distances.

Definition 5 Let $L = [0, 1]$. Consider two FKSs $\mathcal{K}_1 = (S, R_1, \mathcal{V}_1)$ and $\mathcal{K}_2 = (S, R_2, \mathcal{V}_2)$ with the same state space S and the same set AP of propositions. The perturbation degree between $\mathcal{K}_1$ and $\mathcal{K}_2$, denoted $d_r(\mathcal{K}_1, \mathcal{K}_2)$, is defined by

$$d_r(\mathcal{K}_1, \mathcal{K}_2) = \sqrt[r]{\frac{1}{|AP|.|S|} \sum_{p \in AP, s \in S} |\mathcal{V}_1(s)(p) - \mathcal{V}_2(s)(p)|^r} \vee$$
$$\sqrt[r]{\frac{1}{|S|^2} \sum_{s,s' \in S} |R_1(s, s') - R_2(s, s')|^r},$$

where r is a parameter satisfying $1 \leq r \leq +\infty$.

Remark 1 If $r = 1$, i.e.,

$$d_1(\mathcal{K}_1, \mathcal{K}_2) = \frac{1}{|AP|.|S|} \sum_{p \in AP, s \in S} |\mathcal{V}_1(s)(p) - \mathcal{V}_2(s)(p)| \vee$$
$$\frac{1}{|S|^2} \sum_{s,s' \in S} |R_1(s, s') - R_2(s, s')|,$$

where d_1 is the Hamming metric, which has been used to measure the perturbation of Markov decision processes [6]. If $r = +\infty$, i.e.,

$$d_{+\infty}(\mathcal{K}_1, \mathcal{K}_2) = \sup_{p \in AP, s \in S} |\mathcal{V}_1(s)(p) - \mathcal{V}_2(s)(p)| \vee \sup_{s,s' \in S} |R_1(s, s') - R_2(s, s')|,$$

$d_{+\infty}$ is the uniform metric, which has been used to measure the perturbation of fuzzy automata [7].

The following observation follows immediately from Definition 5.

Remark 2 If $d_r(\mathcal{K}_1, \mathcal{K}_2) \leq \varepsilon$, $\varepsilon \in [0, 1]$, then for all $p \in AP$, $s, s' \in S$,

$$|\mathcal{V}_1(s)(p) - \mathcal{V}_2(s)(p)| \leq \sqrt[r]{|AP||S|}\varepsilon,$$
$$|R_1(s, s') - R_2(s, s')| \leq \sqrt[r]{|S|^2}\varepsilon.$$

We need the following lemma, which is a basic result of mathematical analysis.

Lemma 2 ([7]) *Let I be a set and $\{a_i\}_{i \in I}$, $\{b_i\}_{i \in I}$ be two sequences of numbers in* $[0, 1]$. *If $|a_i - b_i| \leq c$ for all $i \in I$, then*

$$|\sup_{i \in I} a_i - \sup_{i \in I} b_i| \leq c,$$
$$|\inf_{i \in I} a_i - \inf_{i \in I} b_i| \leq c.$$

The result below shows that if $\mathcal{L}$ is the Standard Łukasiewicz algebra and models are near enough, then the truth values of FCTL formulas defined over them are nearby.

Theorem 3 *Let $\mathcal{L}$ be the Standard Łukasiewicz algebra, φ an FCTL formula, and $\varepsilon > 0$. There exists a $\delta > 0$ such that for all states $s \in S$, if $d_r(\mathcal{K}_1, \mathcal{K}_2) < \delta$, then $|[\![\mathcal{K}_1, \varphi]\!](s) - [\![\mathcal{K}_2, \varphi]\!](s)| < \varepsilon$.*

Proof We show that the theorem holds by structural induction on φ.
Induction basis: The case for $\varphi = \textbf{true}$ is obvious. The case for $\varphi = p,\ p \in AP$ holds, because let $\delta = \frac{\varepsilon}{\sqrt[r]{|AP||S|}}$, then for any $\mathcal{K}_1$ and $\mathcal{K}_2$, $d_r(\mathcal{K}_1, \mathcal{K}_2) < \delta$, we have from Remark 2 that $|[\![\mathcal{K}_1, p]\!](s) - [\![\mathcal{K}_2, p]\!](s)| < \varepsilon$.
Induction step: Assume φ_1 and φ_2 are FCTL formulas for which the assertion holds. We treat the cases $\varphi = \exists X\varphi_2$ and $\varphi = \exists\varphi_1 U\varphi_2$ in detail. The other cases can be handled similarly.

- $\varphi = \exists X\varphi_1$: Notice that $\otimes_L$ is a continuous t-norm, moreover, it is uniformly continuous for two arguments. So for every $\varepsilon > 0$. there exists some $\delta_1 > 0$ such that
$$|R_1(s, s') \otimes_L [\![\mathcal{K}_1, \varphi_1]\!](s') - R_2(s, s') \otimes [\![\mathcal{K}_2, \varphi_1]\!](s')| < \varepsilon$$
where R_1 and R_2 are fuzzy transition functions of $\mathcal{K}_1$ and $\mathcal{K}_2$, respectively, $|R_1(s, s') - R_2(s, s')| < \delta_1$ and $|[\![\mathcal{K}_1, \varphi_1]\!](s') - [\![\mathcal{K}_2, \varphi_1]\!](s')| < \delta_1$ for all states $s' \in S$. Hence if $d_r(\mathcal{K}_1, \mathcal{K}_2) < \frac{\delta_1}{\sqrt[r]{|S|^2}}$, and $|[\![\mathcal{K}_1, \varphi_1]\!](s') - [\![\mathcal{K}_2, \varphi_1]\!](s')| < \delta_1$, then
$$|[\![\mathcal{K}_1, \exists X\varphi_1]\!](s) - [\![\mathcal{K}_2, \exists X\varphi_1]\!](s)| < \varepsilon.$$
Applying the induction hypothesis, we know that there exists some $\delta_2 > 0$ such that
$$|[\![\mathcal{K}_1, \varphi_1]\!](s') - [\![\mathcal{K}_2, \varphi_1]\!](s')| < \delta_1$$
where $d_r(\mathcal{K}_1, \mathcal{K}_2) < \delta_2$. Take $\delta = \frac{\delta_1}{\sqrt[r]{|S|^2}} \wedge \delta_2$. $d_r(\mathcal{K}_1, \mathcal{K}_2) < \delta$ implies that
$$|[\![\mathcal{K}_1, \exists X\varphi_1]\!](s) - [\![\mathcal{K}_2, \exists X\varphi_1]\!](s)| < \varepsilon.$$
- $\varphi = \exists\varphi_1 U\varphi_2$: According to Theorem 1 and the preceding case, the assertion clearly holds for the case $\varphi = \exists\varphi_1 U\varphi_2$. □

4 Conclusion

In this paper, we discussed robustness of semantics function of FCTL. One result showed that the semantics of FCTL is robust with respect with logically equivalence measure when the truth structure of FCTL is a Heyting algebra. Another result showed that the semantics of fuzzy computation tree logic is uniformly continuous with respect to the measure induced by normalized Minkowski distances when the truth structure of FCTL is Standard Łukasiewicz algebra.

Acknowledgments This work was supported by the Humanities and Social Sciences Research Project of Ministry of Education of China under Grant BR2015022, by the National Natural Science Foundation of China under Grants 61370053, and 11401361, and by China Postdoctoral Science Foundation under Grant 2014M552408.

References

1. Baier, C., Katoen, J.P.: Principles of Model Checking. MIT Press, Cambridge (2008)
2. Bělohlávek, R.: Fuzzy Relational Systems: Foundations and Principles. Kluwer Academic Publishers, New York (2002)
3. Chechik, M., Devereux, B., Easterbrook, S., et al.: Multi-valued symbolic model-checking. ACM Trans. Softw. Eng. Meth. **12**(4), 371–408 (2003)
4. Chechik, M., Gurfinkel, A., Devereaux, B., et al.: Data structures for symbolic multi-valued model-checking. Formal Meth. Syst. Des. **29**(3), 295–344 (2006)
5. Dai, S., Pei, D., Wang, S.: Perturbation of fuzzy sets and fuzzy reasoning based on normalized Minkowski distances. Fuzzy Sets Syst. **189**, 63–73 (2012)
6. De Alfaro, L., Faella, M., Henzinger, T.A., et al.: Model checking discounted temporal properties. Theor. Comput. Sci. **345**(1), 139–170 (2005)
7. Li, Y.: Approximation and robustness of fuzzy finite automata. Int. J. Approx. Reason. **47**, 247–257 (2008)
8. Li, Y., Li, Y., Ma, Z.: Computation tree logic model checking based on possibility measures. Fuzzy Sets Syst. **262**, 44–59 (2015)
9. Li, Y., Li, L.: Model checking of linear-time properties based on possibility measure. IEEE Trans. Fuzzy Syst. **21**(5), 842–854 (2013)
10. Li, Y., Ma, Z.: Quantitative computation tree logic model checking based on generalized possibility measures. IEEE Trans. Fuzzy Syst. **23**(6), 2034–2047 (2015)
11. Pan, H., Cao, Y., Zhang, M., Chen, Y.: Simulation for lattice-valued doubly labeled transition systems. Int. J. Approx. Reason. **55**, 797–811 (2014)
12. Pan, H., Zhang, M., Chen, Y., Wu, H.: Quantitative analysis of lattice-valued Kripke structures. Fundamenta Informaticae **135**, 269–293 (2014)
13. Pan, H., Li, Y., Cao, Y., Ma, Z.: Model checking fuzzy computation tree logic. Fuzzy Sets Syst. **262**, 60–77 (2015)
14. Pan, H., Li, Y., Cao, Y., Ma, Z.: Model checking computation tree logic over finite lattices. Theor. Comput. Sci. **612**, 45–62 (2016)

Localic Conuclei on Quantales

Fang-Fang Pan and Sheng-Wei Han

Abstract In this note, our main purpose is to investigate different kinds of quantic conuclei. Also, we shall consider a characterization for (strong) localic conuclei.

Keywords Quantale · Subquantale · Quantic conucleus · Localic conucleus

1 Introduction

As a generalization of frames, Quantales were introduced by Mulvey (see [12]) in order to provide a lattice theoretic setting for studying non-commutative C^*-algebras, as well as a constructive foundation for quantum mechanics. Quantales as the structures of membership truth values have been applied to enriched category, fuzzy set, fuzzy topology and fuzzy domain (see [3, 5, 7, 10, 11, 16–22]). At present, Quantales arise in a lot of structures like rings, topological spaces, von-Neumann algebras and inverse semigroups (see [2, 8, 9, 13, 14]). In quantale theory, Quantic nuclei and conuclei are two important concepts, and play a crucial role in studying the structure of quantales (see [4, 8, 9, 15]). Different kinds of quantic nuclei have been studied by Rosenthal in [15]. In order to describe a quotient of quantale which is a frame, Rosenthal introduced the concept of a localic nucleus, and, by means of commutative, right-sided and idempotent nuclei, gave a characterization for localic nuclei. In terms of the largest localic nucleus, Rosenthal proved that the category Frm of frames is a reflective subcategory of the category Quant of quantales. For quantic conuclei, there is not quite richness of theory and examples as exhibited by quantic nuclei, however it is important to analyze relationship between quantales and frames.

F.-F. Pan
Department of Mathematics, Xi'an University of Posts and Telecommunications, Xi'an 710121, People's Republic of China
e-mail: panfangfang@xupt.edu.cn

S.-W. Han (✉)
Department of Mathematics, Shaanxi Normal University, Xi'an 710062, People's Republic of China
e-mail: hansw@snnu.edu.cn

T.-H. Fan et al. (eds.), *Quantitative Logic and Soft Computing 2016*,
Advances in Intelligent Systems and Computing 510,
DOI 10.1007/978-3-319-46206-6_13

In order to investigate relationship between quantic nuclei and quantic conuclei, Han and Zhao introduced the concept of ideal conucleus, and proved that quantic nuclei and ideal conuclei are in one-to-one correspondence in Girard quantale (see [4]). In [15], Rosenthal introduced the concept of a localic conucleus, but he did not explore a deep study for the localic conuclei. In this note, our purpose is to investigate different kinds of localic conuclei. Furthermore, we shall give a characterization for (strong) localic conuclei.

Definition 1 A *quantale* is a complete lattice Q with an associative binary operation & satisfying:

$$a\&(\bigvee_i b_i) = \bigvee_i (a\&b_i) \text{ and } (\bigvee_i a_i)\&b = \bigvee_i (a_i\&b)$$

for all $a, b, a_i, b_i \in Q$.

Definition 2 Let Q be a quantale, $a \in Q$.
(i) a is right-sided iff $a\&1 \leq a$;
(ii) a is left-sided iff $1\&a \leq a$; (iii) a is two-sided iff a is both right-sided and left-sided;
(iv) a is idempotent iff $a\&a = a$;
(v) a is semiprime iff $b\&b \leq a \Longrightarrow b \leq a$ for all $b \in Q$.

A subset $S \subseteq Q$ is a subquantale of Q if it is closed under sups and &.

Definition 3 Let Q be a quantale. A quantic conucleus on Q is a coclosure operator g such that $g(a)\&g(b) \leq g(a\&b)$ for all $a, b \in Q$.

We denote by $CN(Q)$ the set of all quantic conuclei on Q. Then $CN(Q)$ is a complete lattice with the pointwise order.

Lemma 1 *Let Q be a quantale, $g, k \in CN(Q)$. Then $g \leq k \Longleftrightarrow Q_g \subseteq Q_k$.*

Theorem 1 ([15]) *Let Q be a quantale. If g is a quantaic conucleus (nucleus) on Q, then $Q_g = \{a \in Q : g(a) = a\}$ is a subquantale (quantic quotient) of Q. Moreover, if S is any subquantale (quantic quotient) of Q, then $S = Q_g$ for some quantic conucleus (nucleus) g.*

For the notions and concepts, which are not explained here, please refer to [1, 15].

2 Localic Conuclei

In [15], Roenthal introduced the concept of localic conucleus, but he did not explore a thorough study of localic conuclei. Although there is not quite richness of theory and examples as exhibited by localic nuclei, localic conuclei also play an important role in studying relationship between quantales and frames. In this section, we shall investigate properties of (strong) localic conuclei and give their characterization.

Let Q be a quantale, $S \subseteq Q$. $\langle S \rangle$ denotes the subquantale generated by S, that is, $\langle S \rangle = \bigcap \{K \subseteq Q \mid K \text{ is a subquantale}, \ S \subseteq K\}$, and (S) denote the semigroup generated by S. We denote by g_s the quantic conucleus determined by $\langle S \rangle$.

Lemma 2 ([6]) *Let Q be a quantale, $S \subseteq Q$. Then each element of $\langle S \rangle$ can be represented by elements of (S), that is, for each element $q \in \langle S \rangle$, there exists a subset $U \subseteq (S)$ such that $q = \bigvee U$.*

Corollary 1 *Let Q be a quantale, $S \subseteq Q$. Then $g_s(a) = \bigvee \{s \in (S) \mid s \leq a\}$ for all $a \in Q$.*

Let $\mathcal{P}(Q)$ denote the set of all subsets of Q. Then $\mathcal{P}(Q)$ is a complete lattice with the inclusion order. We have the following proposition.

Proposition 1 *Let Q be a quantale. Then $g_{()} : \mathcal{P}(Q) \to CN(Q)$ is left adjoint to $Q_{()} : CN(Q) \to \mathcal{P}(Q)$.*

Proof It suffices to prove that $\forall g \in CN(Q), \ S \in \mathcal{P}(Q), \ g_s \leq g \Longleftrightarrow S \subseteq Q_g$.

Let $g_s \leq g$. Then by Lemma 1 and Theorem 1 we have $Q_{g_S} \subseteq Q_g$, and $S \subseteq \langle S \rangle = Q_{g_S}$, which implies $S \subseteq Q_g$ Conversely, let $S \subseteq Q_g$, then $\langle S \rangle \subseteq Q_g$, and $g_s \leq g_{Q_g} = g$, that is, $g_s \leq g$. □

In what follows we shall consider different quantic conuclei.

Definition 4 Let Q be a quantale, g a quantic conucleus. Then g is called commutative (weak commutative) provided that $g(a\&b) = g(b\&a)$ $(g(a)\&g(b) = g(b)\&g(a))$ for all $a, b \in Q$.

Definition 5 Let Q be a quantale and $c \in Q$. Then c is called co-symmetric provided that $c \leq a\&b \Longleftrightarrow c \leq b\&a$ for all $a, b \in Q$.

We denote by $CS(Q)$ the set of all co-symmetric elements of Q.

Proposition 2 *Let Q be a quantale, g a quantic conucleus. Then*
(i) *If g is commutative, then g is weak commutative;*
(ii) *g is commutative if and only if $Q_g \subseteq CS(Q)$;*
(iii) *g is weak commutative if and only if Q_g is commutative.*

Proof (i) Let g be a commutative conucleus. Then for all $a, b \in Q$, we have $g(a)\&g(b) \in Q_g$ and $g(b)\&g(a) \in Q_g$, which implies $g(a)\&g(b) = g(g(a)\&g(b)) = g(g(b)\&g(a)) = g(b)\&g(a)$, that is, g is weak commutative.

(ii) Let g be commutative and $g(a) \leq b\&c$. Then $g(a) \leq g(b\&c) = g(c\&b) \leq c\&b$, which implies $g(a) \in CS(Q)$, that is, $Q_g \subseteq CS(Q)$. Conversely, let $Q_g \subseteq CS(Q)$. Then for all $a, b \in Q$, we have $g(a\&b) = \bigvee \{x \in Q_g \mid x \leq a\&b\} = \bigvee \{y \in Q_g \mid y \leq b\&a\} = g(b\&a)$, that is, g is commutative.

(iii) The proof is straightforward. □

Proposition 3 *Let Q be a quantale, g a quantic conucleus. Then g is commutative if and only if there exists a subset $K \subseteq CS(Q)$ such that $(K, \&)$ is a semigroup and $g = g_K$.*

Proof The proof directly follows from Theorem 1, Lemma 2 and Proposition 2. □

Definition 6 Let Q be a quantale, g a quantic conucleus. g is called right-sided (weak right-sided) provided that $g(a\&1) \leq g(a)$ $(g(a)\&g(1) \leq g(a))$ for all $a \in Q$.

Definition 7 Let Q be a quantale, $c \in Q$. c is called a co-right-sided prime element provided that $c \leq a\&1 \Longrightarrow c \leq a$ for all $a \in Q$.

Let $CRP(Q)$ denote the set of all co-right-sided prime elements of Q.

Proposition 4 *Let Q be a quantale and g be a quantic conucleus. Then*
(i) *If g is right-sided, then g is weak right-sided;*
(ii) *g is weak right-sided $\Longleftrightarrow Q_g$ is right-sided;*
(iii) *g is right-sided $\Longleftrightarrow Q_g \subseteq CRP(Q)$.*

Proof The proof is similar to that of Proposition 2. □

Definition 8 Let Q be a quantale and g a quantic conucleus. g is called idempotent (weak idempotent) provided that $g(a\&a) = g(a)$ $(g(a)\&g(a) = g(a))$ for all $a \in Q$.

Definition 9 Let Q be a quantale, $c \in Q$. c is called a co-semiprime element provided that $c \leq a\&a \Longrightarrow c \leq a$ for all $a \in Q$.

We denote by $CSP(Q)$ the set of all co-semiprime elements of Q.

Proposition 5 *Let Q be a quantale and g a quantic conucleus. Then*
(i) *If g is idempotent, then g is weak idempotent;*
(ii) *g is weak idempotent if and only if Q_g is idempotent;*
(iii) *g is idempotent if and only if $Q_g \subseteq CSP(Q)$ and Q_g is idempotent.*

Proof The proof is similar to that of Proposition 2. □

Lemma 3 *Let Q be a quantale and g a coclosure operator on Q. Then $g(a \wedge b) = g(a \wedge g(b)) = g(g(a) \wedge b) = g(g(a) \wedge g(b))$ for all $a, b \in Q$.*

Proof The proof is easy. □

Definition 10 Let Q be a quantale, g a coclosure operator on Q. g is called localic provided that $g(a \wedge b) = g(a)\&g(b)$ for all $a, b \in Q$.

Proposition 6 *Let g be a localic coclosure operator on a quantale Q. Then g is a quantic conucleus on Q.*

Proof It suffices to prove that $g(a)\&g(b) \leq g(a\&b)$ for all $a, b \in Q$.

For all $a, b \in Q$, by Lemma 3 we have $g(a)\&g(b) = g(a \wedge b) = g(g(a) \wedge g(b)) = g(g(g(a) \wedge g(b))) = g(g(g(a))\&g(g(b))) = g(g(a)\&g(b)) \leq g(a\&b)$. □

From Proposition 6, we can see that the concept of localic coclosure operator and the concept of localic conucleus introduced in [15] are equivalent.

Proposition 7 ([15]) *Let Q be a quantale and g a quantic conucleus. Then g is localic $\Longleftrightarrow Q_g$ is a frame with $\& = \wedge_g$.*

Proposition 8 *Let Q be a quantale and g be a quantic conucleus. Then the following statements are equivalent:*
(i) *g is localic;*
(ii) *Q_g is a frame with $\& = \wedge_g$;*
(iii) *g is weak commutative, weak right-sided and weak idempotent.*

Proof $(i) \Longleftrightarrow (ii)$ and $(i) \Longrightarrow (iii)$ are obvious.

$(iii) \Longrightarrow (i)$ For all $a, b \in Q$, we have that (1) $g(a \wedge b) \leq g(a), g(a \wedge b) \leq g(b) \Longrightarrow g(a \wedge b) = g(a \wedge b)\&g(a \wedge b) \leq g(a)\&g(b)$; (2) $g(a)\&g(b) \leq g(a) \leq a$ and $g(a)\&g(b) = g(b)\&g(a) \leq g(b) \leq b \Longrightarrow g(a)\&g(b) \leq a \wedge b \Longrightarrow g(a)\&g(b) \leq g(a \wedge b)$. Thus, $g(a \wedge b) = g(a)\&g(b)$, which indicates that g is localic. □

Definition 11 Let Q be a quantale and g a quantic conucleus. g is called *strong localic* provided that $g(a \wedge b) = g(a\&b)$ for all $a, b \in Q$.

Proposition 9 *Let Q be a quantale and g a strong localic conucleus. Then g is a semigroup homomorphism, that is, $g(a\&b) = g(a)\&g(b)$ for all $a, b \in Q$.*

Proof For all $a, b \in Q$, by Lemma 3 we have $g(a)\&g(b) \leq g(a\&b) = g(a \wedge b) = g(g(a) \wedge g(b)) = g(g(a)\&g(b)) \leq g(a)\&g(b)$, which implies $g(a\&b) = g(a)\&g(b)$. □

Proposition 10 *Let Q be a quantale and g a quantic conucleus. Then the following are equivalent*
(i) *g is strong localic;*
(ii) *g is commutative, right-sided and idempotent.*

Proof $(i) \Longrightarrow (ii)$ is obvious.

$(ii) \Longrightarrow (i)$ For all $a, b \in Q$, (1) Since g is idempotent, we have $(a \wedge b)\&(a \wedge b) \leq a\&b \Longrightarrow g(a \wedge b) = g((a \wedge b)\&(a \wedge b)) \leq g(a\&b)$; (2) Since g is right-sided and commutative, we see that $g(a\&b) \leq g(a\&1) \leq g(a) \leq a$ and $g(a\&b) = g(b\&a) \leq b \Longrightarrow g(a\&b) \leq a \wedge b \Longrightarrow g(a\&b) \leq g(a \wedge b)$. Thus, $g(a \wedge b) = g(a\&b)$, which indicates that g is strong localic. □

Proposition 11 *Let Q be a quantale and g a quantic conucleus. Then g is strong localic if and only if g is localic and g is also a semigroup homomorphism.*

Proof The proof directly follows from Proposition 10. □

Let $\varepsilon(Q)$ denote the set $\{a \in Q \mid a \text{ is idempotent and two-sided}\}$. In [15], Rosenthal gave the following proposition.

Proposition 12 ([15]) *The following are equivalent for a quantale Q.*
(i) *Q has the largest localic subquantale;*
(ii) *$\varepsilon(Q)$ is a frame with $\& = \wedge$;*
(iii) *$e\&f = f\&e$ for all $e, f \in \varepsilon(Q)$.*

In fact, the above proposition is not right. The following example indicates the case.

Example 1 Let $Q = \{0, a, 1\}$ be a quantale with a binary operation & defined by the table below

&	0	a	1
0	0	0	0
a	0	a	1
1	0	1	1

It is easy to verify that there only exists two localic subquantales $\varepsilon(Q) = \{0, 1\}$ and $S = \{0, a\}$ in Q, but there does not exist the largest localic subquantale.

If a quantale satisfies some condition, then the result of Proposition 12 holds.

Proposition 13 *If Q is a two-sided quantale, then Q has the largest localic subquantale if and only if $\varepsilon(Q)$ is a subquantale.*

Proposition 14 *If Q is a commutative two-sided quantale, then $\varepsilon(Q)$ is the largest localic subquantale.*

Acknowledgments This work was supported by the National Natural Science Foundation of China (Grant no. 11531009), the Fundamental Research Funds for the Central Universities (Grant no. GK201501001) and the Natural Science Program for Basic Research of Shaanxi Province, China (Grant nos. 2015JM1020, 15JK1667).

References

1. Birkhoff, G.: Lattice Theory. American Mathematical Society Colloquium Publications, Providence (1940)
2. Coniglio, M.E., Miraglia, F.: Non-commutative topology and quantales. Studia Logica **65**, 223–236 (2000)
3. Fan, L.: A new approach to quantitative domain theory. Electron. Notes Theor. Comput. Sci. **45**, 77–87 (2001)
4. Han, S.W., Zhao, B.: The quantic conuclei on quantales. Algebra Universalis **61**, 97–114 (2009)
5. Han, S.W., Zhao, B.: Q-fuzzy subsets on ordered semigroups. Fuzzy Sets Syst. **210**, 102–116 (2013)
6. Han, S.W., Pan, F.F.: The basis on quantales. Comput. Eng. Appl. **46**(9), 31–32,37 (2010)
7. Hofmann, D., Waszkiewicz, P.: Approximation in quantale-enriched categories. Topol. Appl. **158**(8), 963–977 (2011)
8. Kruml, D., Paseka, J.: Algebraic and categorical aspects of quantales. Handb. Algebra **5**, 323–362 (2008)
9. Kruml, D.: Points of quantales. Ph.D. Thesis, Masaryk University (2002)

10. Lai, H.L., Zhang, D.X.: Complete and directed complete Ω-categories. Theor. Comput. Sci. **388**, 1–25 (2007)
11. Liu, M., Zhao, B.: Two cartesian closed subcategories of fuzzy domains. Fuzzy Sets Syst. **238**, 102–112 (2014)
12. Mulvey, C.J.: Suppl. Rend. Circ. Mat. Palermo Ser. **II**(12), 99–104 (1986)
13. Pelletier, J.W.: Von Neumann algebras and Hilbert quantales. Apllied Categ. Struct. **5**, 249–264 (1997)
14. Resende, P.: A note on infinitely distributive inverse semigroups. Semigroup Forum **73**(1), 156–158 (2006)
15. Rosenthal, K.I.: Quantales and Their Applications. Longman Scientific and Technical, New York (1990)
16. Solovyov, S.A.: From quantale algebroids to topological spaces: fixed- and variable-basis approaches. Fuzzy Sets Syst. **161**, 1270–1287 (2010)
17. Stubbe, I.: Categorical structures enriched in a quantaloid: categories, distributors and functors. Theory Appl. Categ. **14**, 1–45 (2005)
18. Stubbe, I.: Categorical structures enriched in a quantaloid: tensored and cotensored categories. Theory Appl. Categ. **16**, 283–306 (2006)
19. Yao, W.: Quantitative domains via fuzzy sets: part I: continuity of fuzzy directed complete posets. Fuzzy Sets Syst. **161**, 973–987 (2010)
20. Yao, W.: A survey of fuzzifications of frames, the Papert–Papert–Isbell adjunction and sobriety. Fuzzy Sets Syst. **90**, 63–81 (2012)
21. Zhang, D.X.: An enriched category approach to many valued topology. Fuzzy Sets Syst. **158**, 349–366 (2007)
22. Zhang, Q.Y., Fan, L.: Continuity in quantitative domains. Fuzzy Sets Syst. **154**, 118–131 (2005)

An Equivalent Form of Uncertain Measure

Xing-Fang Zhang and Feng-Xia Zhang

Abstract A new measure, called uncertain measure, was presented by Liu in 2007 in order to deal with uncertainty, intelligently. The definition of uncertain measure contains Normality axiom, Duality axiom and Subadditivity axiom for any sequence of events. This paper gives an equivalent form of the definition by substituting Subadditivity axiom with a new Subadditivity axiom, Compared with the original axiom, the new axiom only requires that the subadditivity hold for mutually disjoint sets. The equivalent form illustrates that we have two methods to complete uncertain measure. One method is to preserve original axioms of uncertain measure, and to add the property which is equivalent to the conditions with Normality, Duality and Subadditivity for any sequence of mutually disjoint events. The other is to change the third axiom of uncertain measure into Subadditivity for any sequence of mutually disjoint events, and the front three axioms of uncertain measure is regarded as a property.

Keywords Uncertain measure · Axiom · Uncertainty theory · Equivalent condition

1 Introduction

In real life there are many kinds of uncertain parameters, for example, the price of a product at tomorrow, the life of a light bulb, the time of dealing with a task at tomorrow, etc. At the earliest, people gave it a real number directly by the most simple way. In 1933, A.N. Kolmogoroff put forward probabilistic measure according to the nature of the frequency that the random events happen. Probability theory was founded based on it. Naturally, people applied probability theory to make uncertain parameters [1, 2]. In 1965, Zadeh proposed fuzzy set theory. So fuzzy set theory is apply to such problems [4]. In 2007, Baoding Liu also put forward a new theory

X.-F. Zhang (✉) · F.-X. Zhang
School of Mathematical Sciences, Liaocheng University, Liaocheng 252059, China
e-mail: zhangxingfang2005@126.com

F.-X. Zhang
e-mail: fengxiazhang@163.com

T.-H. Fan et al. (eds.), *Quantitative Logic and Soft Computing 2016*,
Advances in Intelligent Systems and Computing 510,
DOI 10.1007/978-3-319-46206-6_14

called uncertainty theory to intelligently deal with uncertain parameters [4]. Later the theory was completed gradually [5–7]. At present, it have been recognized to be scientific. It has been widely applied to uncertain programming [3, 8–10], uncertain game [11, 12], uncertain process [13], Zhang, 2013 [14], uncertain logic [15–17] etc.

Probability theory origins from the statistical frequency. Naturally, for uncertain parameters with sufficient historical data, it is the most scientific to deal with uncertain parameters by using probability theory. This fact has been approved by many people. In the circumstances of no historical data or the lack of historical data, people have recognized to make decisions by experts' evaluation. However, it is still disputable that which method (subjective probability, fuzzy set theory and uncertainty theory) is the best. Fuzzy set theory origins from the fuzzy concepts, such as tall, young people, old people, bald, etc. While the characteristics of the uncertain parameters are different from fuzzy concepts. In objective recognition, uncertain parameter is unique, but not sure. So we should build this theory based on two-valued logic. However, the point of view of fuzzy set theory is: a fuzzy proposition is often neither true nor false, it has a third possibility. This shows that it firstly denies two-valued logic. Naturally, it does not satisfy the dual law and the law of contradiction. Therefore, fuzzy set theory is not appropriate to deal with uncertain parameters. Subjective probability theory and uncertainty theory are based on two-valued logic, and satisfy the dual law and the law of contradiction. So they are more scientific relatively to deal with uncertain parameters. The key of subjectively dealing with uncertain parameters is how to reasonably select: disjunction and conjunction. At present, people have recognized that triangle norm is an ideal operator to measure conjunction $\cap$. In uncertainty theory, the disjunction $\cup$ is restricted between triangle conorm-supremum and triangle conorm-sum, and it uses triangle norm-infimum to measure conjunction. Uncertainty theory is more scientific than subjective probability theory because it allows the hesitation of cognition.

Uncertain measure is a basic concept in uncertainty theory. It includes four axioms: Normality, Duality, Subadditivity and Product uncertainty measure. Where, Subadditivity axiom is not convenient for application. Therefore, the paper will give its an equivalent and convenient form.

2 Uncertain Measure and Its an Equivalent Form

Firstly, we introduce Liu's uncertain measure.

Definition 1 ([4–7]) Let L be a σ-algebra on nonempty set Γ. A set function M is called an uncertain measure on L if it satisfies the following axioms:

Axiom 1 (Normality) $M\{\Gamma\} = 1$, for the universal set Γ;

Axiom 2 (Duality) $M\{\Lambda\} + M\{\Lambda^c\} = 1$, for any event Λ;

Axiom 3 **(Subadditivity)** For every countable sequence of events $\{\Lambda_i\}$, we have

$$\bigvee_{i=1}^{\infty} M\{\Lambda_i\} \leq M\left\{\bigcup_{i=1}^{\infty} \Lambda_i\right\} \leq \sum_{i=1}^{\infty} M\{\Lambda_i\}.$$

In this case, the triple (Γ, L, M) is called an uncertainty space.

Liu (2009) further presented the following axiom.

Let (Γ_k, L_k, M_k) be uncertainty spaces for $k = 1, 2, \ldots$ Write

$$\Gamma = \Gamma_1 \times \Gamma_2 \times \cdots, L = L_1 \times L_2 \times \cdots$$

Then the product uncertain measure M on the product σ-algebra L is defined by the following product axiom.

Axiom 4 **(Product Axiom)** Let (Γ_k, L_k, M_k) be uncertainty spaces for $k = 1, 2, \ldots$. Then the product uncertain measure M is an uncertain measure satisfying

$$M\left\{\prod_{k=1}^{\infty} \Lambda_k\right\} = \bigwedge_{k=1}^{\infty} M_k\{\Lambda_k\},$$

where Λ_k are arbitrarily chosen events from L_k for $k = 1, 2, \ldots$. respectively.

We give an equivalent condition of uncertain measure in the following.

Theorem 1 *Let L be a σ-algebra on nonempty set Γ. A set function M is an uncertain measure if and only if:*

Condition 1 **(Normality)** $M\{\Gamma\} = 1$, *for the universal set* Γ;

Condition 2 **(Duality)** $M\{\Lambda\} + M\{\Lambda^c\} = 1$, *for each event* Λ;

Condition 3 **(Subadditivity)** *For every countable mutually disjoint sequence of events* $\{\Lambda_i\}$, *we have*

$$\bigvee_{i=1}^{\infty} M\{\Lambda_i\} \leq M\left\{\bigcup_{i=1}^{\infty} \Lambda_i\right\} \leq \sum_{i=1}^{\infty} M\{\Lambda_i\}.$$

Proof The necessity is obvious. We only prove the sufficiency.

For sufficiency, it is obvious that we only need to prove that for every countable sequence of events $\{\Lambda_i\}$,

$$\bigvee_{i=1}^{\infty} M\{\Lambda_i\} \leq M\left\{\bigcup_{i=1}^{\infty} \Lambda_i\right\} \leq \sum_{i=1}^{\infty} M\{\Lambda_i\}.$$

First monotony holds. In fact, for any event $A, B \in Ł, A \subseteq B$,
since

$$B = A \cup (B - A), A \cap (B - A) = \emptyset,$$

we have

$$M\{A\} \le M\{A\} \vee M\{B - A\} \le M\{A \cup (B - A)\} = M\{B\}$$

by conditions 3. Thus, $M\{A\} \le M\{B\}$.

We prove that condition 3 holds in the following.

Since $B \subseteq A \cup B$ and $A \subseteq A \cup B$, it follows that

$$M\{B\} \le M\{A \cup B\}, M\{A\} \le M\{A \cup B\}.$$

Therefore

$$M\{B\} \vee M\{A\} \le M\{A \cup B\}. \tag{1}$$

Since

$$B = A \cup B = A \cup (B - B \cap A), A \cap (B - B \cap A) = \emptyset,$$

we have

$$M\{A \cup B\} \le M\{A\} + M\{B - B \cap A\}$$

by conditions 3.

Note that $B - B \cap A \subseteq B$. Thus $M\{B \cap A \subseteq B\} \le M\{B\}$ by monotonicity. Therefore,

$$M\{A \cup B\} \le M\{A\} + M\{B\}. \tag{2}$$

Thus we have $M\{A\} \vee M\{B\} \le M\{A \cup B\} \le M\{A\} + M\{B\}$ by (1) and (2). □

In general, it is obvious that for every countable sequence of events $\{\Lambda_i\}$,

$$\bigvee_{i=1}^{\infty} M\{\Lambda_i\} \le M\left\{\bigcup_{i=1}^{\infty} \Lambda_i\right\} \le \sum_{i=1}^{\infty} M\{\Lambda_i\}.$$

The equivalent form is very necessary. Since, in practical application, it is possible that some scholars say that a set function with Normality, Duality and Subadditivity for any sequence of mutually disjoint events is an uncertain measure. It is obvious that this not verifies completely the first three axioms of uncertain measure. Now we may say that a set function with Normality, Duality and Subadditivity for any sequence of mutually disjoint events is an uncertain measure according to Theorem 1. It is obvious that it simplifies procedures to verify if a set function is an uncertain measure. Therefore the three conditions in the above theorem are more convenient than the first three axioms of uncertain measure in application.

3 Conclusions

The contribution of the paper is to give an equivalent form of the first three axioms of uncertain measure. This equivalent form illustrates that we have two methods to complete uncertain measure. One method is to preserve the original axioms of uncertain measure, and to add the property which is equivalent to the conditions with Normality, Duality and Subadditivity for any sequence of mutually disjoint events. The other is to change the third axiom of uncertain measure into Subadditivity for any sequence of mutually disjoint events, and the first three axioms of uncertain measure is regarded as a property.

Acknowledgments This work is supported by National Natural Science Foundation of China grant No.11471152 and No.61273044.

References

1. Lahdelma, R., Hokkanen, J., Salminen, P.: SMAA - stochastic multiobjective acceptability analysis. Eur. J. Oper. Res. **106**, 137–143 (1998)
2. Liu, Y., Fan, Z.P., Zhang, Y.: A method for stochastic multiple criteria decision making based on dominance degrees. Inf. Sci. **181**, 4139–4153 (2011)
3. Ahlatcioglu, M., Tiryaki, F.: Interactive fuzzy programming for decentralized two-level linear fractional programming (DTLLFP) problems. Omega **35**, 432–450 (2007)
4. Liu, B.: Uncertainty Theory, 2nd edn. Springer, Berlin (2007)
5. Liu, B.: Some research problems inuncertainty theory. J. Uncertain Syst. **3**(1), 3–10 (2009)
6. Liu, B.: Uncertainty Theory: A Branch of Mathematics for Modeling Human Uncertainty. Springer, Berlin (2010)
7. Liu, B.: Uncertainty Theory, 5 edn. (2015). http://orsc.edu.cn/liu
8. Gao, Y.: Uncertain models for single facility location problems on networks. Appl. Math. Model. **36**(6), 2592–2599 (2012)
9. Gao, Y., Qin, Z.: On computing the edge-connectivity of an uncertain graph. IEEE Trans. Fuzzy Syst., to be published (2016)
10. Gao,Y., Li, X., Yang, S., Kar, S.: On distribution function of the diameter in uncertain graph. Inf. Sci., to be published (2016)
11. Gao, J.: Uncertain bimatrix game with applications. Fuzzy Optim. Decision Mak. **12**, 65–78 (2013)
12. Yang, X., Gao, J.: Linear-quadratic uncertain differential game with application to resource extraction problem. IEEE Trans. Fuzzy Syst., to be published (2016)

13. Yao, K., Li, X.: Uncertain alternating renewal process and its application. IEEE Trans. Fuzzy Syst. **20**(6), 1154–1160 (2012)
14. Yang, X., Gao, J., Samarjit, K.: Uncertain calculus with Yao process. IEEE Trans. Fuzzy Syst., to be published (2016)
15. Li, X., Liu, B.: Hybrid logic and uncertain logic. J. Uncertain Syst. **3**, 83–94 (2009)
16. Zhang, X., Li, L., Meng, G.: A modified uncertain entailment model. J. Intell. Fuzzy Syst. **27**(1), 549–553 (2014)
17. Zhang, X., Li, X.: A semantic study of the first-order predicate logic with uncertainty involved. Fuzzy Optim. Decision Mak. **13**, 357–367 (2014)

A Propositional Logic System for Regular Double Stone Algebra

Hua-Li Liu and Yan-Hong She

Abstract Rough set theory is a mathematical theory dealing with uncertain and imprecise information. Since the inception of this theory, many approaches have come to the fore, and resulted in different "rough logic" as well. In this paper, a kind of propositional logic system with semantics based on regular double stone algebra is proposed, and its soundness and completeness theorems with respect to rough set semantics are also obtained.

Keywords Rough set · Rough logic · Regular double stone algebra · Soundness theorem · Completeness theorem

1 Introduction

Rough set theory [1, 2] is proposed by Pawlak to account for the definability of a concept in terms of some elementary ones in an approximation space. It captures and formalizes the basic phenomenon of information granulation. The finer the granulation is, the more concepts are definable in it. For those concepts not definable in an approximation space, the lower and upper approximations for them can be defined. Recent years have witnessed its wide application in intelligent data analysis, decision making, machine learning and other related fields [3–5].

Since the inception of rough set theory, many scholars have studied rough sets from the perspective of rough set pairs and seek to capture the abstract feature of rough set pairs, which resulted in a series of abstract structures for rough sets. For instance, Iwinski [6] suggested a lattice-theoretic approach. Iwinski's aim, which was later extended by Pomykala [7], was to endow the rough sets of (U, R) with a natural algebraic structure. Gehrke and Walker [8] extended the work of Pomykala

H.-L. Liu · Y.-H. She (✉)
College of Science, Xi'an Shiyou University, Xi'an 710065, China
e-mail: yanhongshe@yeah.net, yanhongshe@gmail.com

H.-L. Liu
e-mail: lhl1901@163.com

T.-H. Fan et al. (eds.), *Quantitative Logic and Soft Computing 2016*,
Advances in Intelligent Systems and Computing 510,
DOI 10.1007/978-3-319-46206-6_15

by proposing a precise structure theorem for the Stone algebra of rough sets, which is in a setting more general than that in [7]. The work of Pomykala was also improved by Comer [9] who noticed that the collection of rough sets of an approximation space is, in fact, a regular double Stone algebra by introducing another unary operator, i.e. the dual pseudo-complement operator. Comer [9] also showed the reverse result, i.e. every regular double Stone algebra is isomorphic to a subalgebra of the rough set algebra for some approximation space. Pagliani [10] investigated rough set systems within the framework of Nelson algebras, and he showed that for any approximation space, the corresponding rough set system can yield the structure of a semi-simple Nelson algebra. He also showed that the converse direction of the above results hold under the assumption of finite universe, i.e., any finite semi-simple Nelson algebra is isomorphic to the rough set system induced by an approximation space. What's more, Banerjee and Chakraborty [11] proposed two algebraic structures called pre-rough algebra and rough algebra, which are obtained by enriching the topological Boolean algebra [11] with additional axioms.

Algebra theory and logic systems are strongly coupled in the development of modern logic. Until now, diverse rough logics corresponding to rough set semantics have been proposed. The notion of rough logic was initially proposed by Pawlak in [12], in which five rough values, i.e., true, false, roughly true, roughly false and roughly inconsistency were also introduced. This work was subsequently followed by E. Ortowska and Vakarelov in a sequence of papers [13–15]. In [11], Banerjee proposed two logic systems $\mathcal{L}_1$ and $\mathcal{L}_2$ corresponding to pre-rough algebra and rough algebra, respectively. These two formal logics include axioms and inference rules, and in addition, the syntax and semantics are in perfect harmony, i.e., the corresponding soundness theorem and completeness theorem hold. Sen and Chakraborty [16] proposed a sequent calculus for topological quasi-Boolean algebras and pre-rough algebras. As to regular double Stone algebra, Dü ntsch presented a corresponding logic in [17], however, such a logic does not include axioms and inference rules, moreover, the soundness and completeness theorem corresponding to the rough semantics are not discussed. Dai [18] proposed a sequent calculus for rough sets with rough double Stone algebra semantics, and obtained the corresponding soundness and completeness theorems.

In this paper, we aim to present a formal logic system $\mathcal{L}_r$ for rough set with regular double Stone algebras. Our logic system contains axioms and inference rules, and is also sound and complete with respect to the class of the regular double Stone algebras. Moreover, the so-called standard completeness theorem for $\mathcal{L}_r$ is obtained, i.e., $\mathcal{L}_r$ is complete w.r.t. the class of rough regular double Stone algebras.

2 Regular Double Stone Algebra

Let us first review the basic notions of rough set theory initially proposed by Pawlak.

Let U be a non-empty set, and R a binary equivalence relation on U. Then we call (U, R) an approximation space. For any given subset A of U, we say that A

is a definable set, if it is the union of some equivalence blocks induced by R, and otherwise, a rough set. For any rough set $X \subseteq U$, two definable sets are employed to approximate it from below and above, respectively, i.e.,

$$\underline{R}(X) = \{x|[x] \subseteq X\}, \overline{R}(X) = \{x|[x] \cap X \neq \emptyset\}, \tag{1}$$

where $[x]$ is the equivalence block containing x. Then we call $\underline{R}(X)(\bar{R}(X))$ the lower approximation (upper approximation) of X. In what follows, we also identify a rough set X with such a pair $(\underline{R}(X), \overline{R}(X))$. An easy verification shows that X is definable if and only if $\underline{R}(X) = \overline{R}(X)$.

Definition 1 ([7]) A Stone algebra $(L, \vee, \wedge, *, 0, 1)$ is an algebra of type $(2, 2, 1, 0, 0)$ satisfying
(1) $(L, \vee, \wedge, 0, 1)$ is a bounded distributive lattice;
(2) x^* is the pseudo-complement of x, i.e., $y \leq x^*$ if and only $y \wedge x = 0$;
(3) $x^* \vee x^{**} = 1$.

Definition 2 ([9]) A double Stone algebra $(L, \vee, \wedge, *, +, 0, 1)$ is an algebra of the type $(2, 2, 1, 1, 0, 0)$ such that
(1) $(L, \vee, \wedge, 0, 1)$ a bounded distributive lattice with the least element 0 and the largest element 1;
(2) x^* is the pseudo-complement of x, i.e., $y \leq x^*$ if and only $y \wedge x = 0$;
(3) x^+ is the dual pseudo-complement of x, i.e., $y \geq x^+$ if and only $y \vee x = 1$;
(4) $x^* \vee x^{**} = 1, x^+ \wedge x^{++} = 0$.

$(L, \vee, \wedge, *, +, 0, 1)$ is regular, if it additionally satisfies

$$\forall x, y \in L, x \wedge x^+ \leq y \vee y^*. \tag{2}$$

This is also equivalent to the fact that $x^* = y^*$ and $x^+ = y^+$ imply $x = y$.

Let (U, R) be an approximation space, and $RS(U)$ the collection of all rough sets in (U, R), i.e., $RS(U) = \{(\underline{R}(X), \overline{R}(X))|X \subseteq U\}$. It was observed by Pomykala that for any given approximation space (U, R), $RS(U)$ can be made into a Stone algebra $\mathcal{RS} = (RS(U), \vee, \wedge, *, (\emptyset, \emptyset), (U, U))$, where

$(\underline{R}(X), \overline{R}(X)) \vee (\underline{R}(Y), \bar{R}(Y)) = (\underline{R}(X) \cup \underline{R}(Y), \bar{R}(X) \cap \bar{R}(Y))$,
$(\underline{R}(X), \overline{R}(X)) \wedge (\underline{R}(Y), \bar{R}(Y)) = (\underline{R}(X) \cap \underline{R}(Y), \bar{R}(X) \cap \bar{R}(Y))$,
$(\underline{R}(X), \bar{R}(X))^* = (U - \bar{R}(X), U - \bar{R}(X)) = (\bar{R}(X)^c, \bar{R}(X)^c)$.

It has been proved in the literature that the above operators are well defined. This work was subsequently improved by Comer [9], who noticed that $RS(U)$ is in fact a regular double Stone algebra, when one introduces the dual pseudo-complement $+$ by

$(\underline{R}(X), \bar{R}(X))^+ = (U - \underline{R}(X), U - \underline{R}(X)) = (\underline{R}(X)^c, \underline{R}(X)^c)$.

We will call $\mathcal{RS} = (RS(U), \vee, \wedge, *, +, (\emptyset, \emptyset), (U, U))$ rough double Stone algebra in what follows.

Regular double Stone algebra enjoys the following representation theorem.

Theorem 1 ([9]) *Each regular double Stone algebra is isomorphic to a subalgebra of* $\mathcal{RS}$.

3 A Propositional Logic System $\mathcal{L}_r$ for Regular Double Stone Algebra

The language of $\mathcal{L}_r$ consists of propositional variables $p_1, p_2, \ldots, p_n, \ldots$(also called atomic formulae) and logic symbols $\vee, \wedge, +, *$. The formation rules are as usual. In $\mathcal{L}_r$, one additional logic connective $\rightarrow$ is defined as follows:

$$A \rightarrow B = (A^+ \vee B^{++}) \wedge (A^* \vee B^{**}). \tag{3}$$

The set of logic formulae is denoted by $F(S)$.

Definition 3 The set of axioms in $\mathcal{L}_r$ consists of the formulae of the following forms:
(1) $A \wedge B \rightarrow A$;
(2) $A \wedge B \rightarrow B$;
(3) $A \rightarrow A$;
(4) $A \rightarrow A \vee B$;
(5) $A \vee B \rightarrow B \vee A$;
(6) $(A \vee B) \wedge (A \vee C) \rightarrow A \vee (B \wedge C)$;
(7) $A^* \vee A^{**}$;
(8) $A^+ \wedge A^{++} \rightarrow \bot$;
(9) $\bot \rightarrow A$.

The inference rules are as follows:

(1) MP rule: $\{A, A \rightarrow B\} \vdash B$;
(2) HS rule: $\{A \rightarrow B, B \rightarrow C\} \vdash A \rightarrow C$;
(3) $\{A \rightarrow B, A \rightarrow C\} \vdash A \rightarrow B \wedge C$,
(4) $\{A \rightarrow C, B \rightarrow C\} \vdash A \vee B \rightarrow C$;
(5) $\{B\} \vdash A \rightarrow B$;
(6) $\{A \wedge B \rightarrow \bot\} \vdash B \rightarrow A^*$;
(7) $\{B \rightarrow A^*\} \vdash A \wedge B \rightarrow \bot$;
(8) $\{A \vee B\} \vdash A^+ \rightarrow B$;
(9) $\{A^+ \rightarrow B\} \vdash A \vee B$;
(10) $\{A^* \rightarrow B^*, A^+ \rightarrow B^+\} \vdash B \rightarrow A$;
(11) $\{A \rightarrow B\} \vdash B^* \rightarrow A^*$;
(12) $\{A \rightarrow B\} \vdash B^+ \rightarrow A^+$.

In $\mathcal{L}_r$, syntactic notions such as theorems, Γ-consequence can be given in an usual manner, and hence is omitted here.

Presented below is the concept of valuation in $\mathcal{L}_r$.

Definition 4 Let $(L, \vee, \wedge, *, +, 0, 1)$ be a regular double Stone algebra. A mapping $v : F(S) \to L$ called a valuation in $\mathcal{L}_r$ if and only if it satisfies the following conditions, i.e., $\forall A, B \in F(S)$,

$$v(A \vee B) = v(A) \vee v(B), \tag{4}$$

$$v(A \wedge B) = v(A) \wedge v(B), \tag{5}$$

$$v(A^*) = v(A)^*, \tag{6}$$

$$v(A^+) = v(A)^+, \tag{7}$$

$$v(\bot) = 0. \tag{8}$$

In what follows, the set of all valuations $v : F(S) \to L$ is denoted by Ω.

Definition 5 Let $\Gamma \subseteq F(S)$, $B \in F(S)$, then B is called a semantic consequence of Γ(denoted by $\Gamma \models B$) if and only if $\forall A \in \Gamma$, $v(A) = 1$ implies $v(B) = 1$ for any valuation $v \in \Omega$. Particularly, if $\Gamma = \emptyset$, then we say that B is a valid formula.

Now, we are ready to present the most important results of this paper, i.e., the soundness theorem and completeness theorem of $\mathcal{L}_r$.

Theorem 2 $\forall A \in F(S), \Gamma \subseteq F(S)$, *$\Gamma \vdash A$ if and only if $\Gamma \models A$.*

To prove Theorem 2, we need the following lemma.

Lemma 1 *Let $(L, \vee, \wedge, *, +, 0, 1)$ be a regular double Stone algebra and $v : F(S) \longrightarrow L$ any valuation in Ω. Then for any two formulae $A, B \in F(S)$, $v(A \to B) = 1$ if and only if $v(A) \leq v(B)$.*

Proof We have from the definition of valuation that $v(A \to B) = v((A^+ \vee B^{++}) \wedge (A^* \vee B^{**})) = (v(A)^+ \vee v(B)^{++}) \wedge (v(A)^* \vee v(B)^{**})$. Moreover, it follows from the representation theorem of regular double Stone algebra that $(L, \vee, \wedge, *, +, 0, 1)$ is isomorphic to a subalgebra of some rough double Stone algebra $\mathcal{RS} = (RS(U), \vee, \wedge, *, (\emptyset, \emptyset), (U, U))$, i.e., $(L, \vee, \wedge, *, +, 0, 1)$ can be embedded into $\mathcal{RS}$. □

Assume that $f : L \longrightarrow RS(U)$ is the embedding mapping and $f(v(A)) = (\underline{X}, \bar{X})$, $f(v(B)) = (\underline{Y}, \bar{Y})$. Then

$$\begin{aligned} v(A \to B) = 1 &\Leftrightarrow f(v(A \to B)) = (U, U) \\ &\Leftrightarrow ((\underline{X}, \bar{X})^+ \vee (\underline{Y}, \bar{Y})^{++}) \wedge ((\underline{X}, \overline{X})^* \vee (\underline{Y}, \bar{Y})^{**}) = (U, U) \\ &\Leftrightarrow (\underline{X}^c \cup \underline{Y}, \underline{X}^c \cup \underline{Y}) \wedge (\bar{X}^c \cup \bar{Y}, \bar{X}^c \cup \bar{Y}) = (U, U) \\ &\Leftrightarrow \underline{X}^c \cup \underline{Y} = U, \bar{X}^c \cup \bar{Y} = U \\ &\Leftrightarrow \underline{X} \subseteq \underline{Y}, \bar{X} \subseteq \bar{Y} \\ &\Leftrightarrow f(v(A)) \leq f(v(B)) \\ &\Leftrightarrow v(A) \leq v(B). \end{aligned}$$

This completes the proof of Lemma 1.

Now let's turn to the proof of Theorem 2.

Proof "Necessity." We only need to show that each axiom is valid and each inference rule preserves validity.

It can be easily verified that each axiom is valid and hence is omitted here.

For inference rules, we will only show that both (1) and (10) preserve validity. The others can be proved similarly.

(1) Let v be any valuation in Ω, then we have from the validity of A and $A \to B$ that $v(A) = 1$ and $v(A \to B) = 1$. By Lemma 1, $v(A \to B) = 1$ immediately entails that $v(A) \leq v(B)$, which yields $v(B) = 1$, and hence, B is valid due to the arbitrariness of v.

(10) Let v be any valuation in Ω. To show that $B \to A$ is valid, it suffices to show that $v(B \to A) = 1$. Then by Lemma 1, it is also equivalent to prove that $v(B) \leq v(A)$.

We still denote by f the embedding mapping and $f(v(A)) = (\underline{X}, \bar{X})$, $f(v(B)) = (\underline{Y}, \bar{Y})$ as in the proof of Lemma 1. Then it follows immediately that $v(B) \leq v(A)$ if and only if $f(v(B)) \leq f(v(A))$.

We have from the validity of $A^* \to B^*$ and $A^+ \to B^+$ and Lemma 1 that $v(A^*) \leq v(B^*)$ and $v(A^+) \leq v(B^+)$, which implies that $f(v(A^*)) \leq f(v(B^*))$ and $f(v(A^+)) \leq f(v(B^+))$, i.e., $(\underline{X}^c, \underline{X}^c) \leq (\underline{Y}^c, \underline{Y}^c)$ and $(\bar{X}^c, \bar{X}^c) \leq (\bar{Y}^c, \bar{Y}^c)$, and hence, $f(v(B)) \leq f(v(A))$.

"Sufficiency." We will adopt the usual Lindenbaum algebra approach as follows.

To this end, we first define a binary relation $\approx$ on the set of logic formulae $F(S)$ in the following manner:

$$\forall A, B \in F(S), A \approx B \Leftrightarrow \Gamma \vdash A \leftrightarrow B.$$

It is routine to show that $\approx$ is a congruence relation, and an induced quotient algebra $(F(S)/\approx, \vee, \wedge, [\bot], [T])$($T$ is any theorem in $\mathcal{L}_r$) is therefore obtained.

We will further prove that $(F(S)/\approx, \vee, \wedge, [\bot], [T])$ is a regular double Stone algebra with the least element $[\bot]$ and the largest element $[T]$, respectively.

To do this, define a binary relation $\leq$ on $F(S)/\approx$ as follows:

$$\forall [A], [B] \in F(S)/\approx, [A] \leq [B] \Leftrightarrow \Gamma \vdash A \to B.$$

It can be easily verified that $\leq$ is a partial order on $F(S)/\approx$, and $[A \vee B]$, $[A \wedge B]$ are the least upper bound and the largest lower bound of $\{[A], [B]\}$, respectively. Moreover, axiom (9) follows and $\forall A \in F(S), \Gamma \vdash A \to T$ that $[\bot]$ and $[T]$ are the least element and the largest element of $F(S)/\approx$, respectively, which shows that $(F(S)/\approx, \vee, \wedge, [\bot], [T])$ is a bounded lattice.

We now will show that all the conditions in Definition 2 hold.

(1) It can be proved that in $\mathcal{L}_r$, $\vdash A \vee (B \wedge C) \to (A \vee B) \wedge (A \vee C)$, which together with axiom (6) immediately entail the distributivity of $(F(S)/\approx, \vee, \wedge, [\bot], [T])$.

(2) $[A] \wedge [B] = [\bot]$ if and only if $\Gamma \vdash (A \wedge B \to \bot) \wedge (\bot \to A \wedge B)$ if and only if $\Gamma \vdash A \wedge B \to \bot$ if and only if $\Gamma \vdash B \to A^*$ if and only if $[B] \leq [A]^*$, where the penultimate "if and only if" holds due to the inference rule (6) in Definition 3.

(3) It can be proved similarly as above.

(4) It follows from axiom (7) and (8) that $[A]^* \vee [A]^{**} = [A^* \vee A^{**}] = [T]$, and $[A]^+ \wedge [A]^{++} = [A^+ \wedge A^{++}] = [\bot]$.

(5) If $[A]^* = [B]^*, [A]^+ = [B]^+$, i.e., $[A^*] = [B^*], [A^+] = [B^+]$, then we have $\Gamma \vdash (A^* \to B^*) \wedge (B^* \to A^*), \Gamma \vdash (A^+ \to B^+) \wedge (B^+ \to A^+)$. It follows from the inference (10) in Definition 3 that $\Gamma \vdash A \to B$ and $\Gamma \vdash B \to A$, which immediately yield $[A] = [B]$.

The above argument shows that $(F(S)/\approx, \vee, \wedge, [\bot], [T])$ is indeed a regular double Stone algebra.

If $\Gamma \models A$, then by Definition 5, for any regular double Stone algebra $\mathcal{L} = (L, \vee, \wedge, *, +, 0, 1)$ and any valuation $v \in \Omega, \forall B \in \Gamma, v(B) = 1$ implies that $v(A) = 1$. Particularly, let $\mathcal{L} = (F(S)/\approx, \vee, \wedge, [\bot], [T])$, and $v_0 : S \to F(S)/\approx$ be a mapping satisfying $v_0(p) = [p]$ and $v_0(\bot) = [\bot]$. It can be extended to $F(S)$ in the following manner, and the obtained mapping is denoted by $v : F(S) \to F(S)/\approx$:

$$v(p) = v_0(p) = [p], v(\bot) = v_0(\bot) = [\bot],$$

$$v(A \wedge B) = v(A) \wedge v(B),$$

$$v(A \vee B) = v(A) \vee v(B),$$

$$v(A^*) = v(A)^*,$$

$$v(A^+) = v(A)^+.$$

Obviously, $v : F(S) \longrightarrow F(S)/\approx$ is a valuation. We will prove $\forall C \in F(S), v(C) = [C]$ by induction on the number n of logic connectives contained in C below.

In case $n = 0$, i.e., C is either an atomic proposition or $C = \bot$, then the conclusion holds obviously.

Suppose that the conclusion holds when $n \leq k$. Now, let's consider the case of $n = k + 1$.

There are still four subcases to be considered.

(1) If $C = C_1 \vee C_2$, where the number of logic connectives contained in both C_1 and C_2 are less or equal to k, then $v(C) = v(C_1 \vee C_2) = v(C_1) \vee v(C_2) = [C_1] \vee [C_2] = [C_1 \vee C_2] = [C]$.

(2) In case $C = C_1 \wedge C_2$, it can be proved similarly as above.

(3) If $C = D^*$, then $v(C) = v(D^*) = v(D)^* = [D]^* = [D^*] = [C]$.

(4) In case $C = [D^+]$, it can be proved similarly as above.

It can be easily verified that $\forall B \in \Gamma, v(B) = [B] = [T]$, i.e., $v(B)$ is the largest element of $F(S)/\approx$. Combining with $\Gamma \models A$, we have that $v(A) = [A] = [T]$, i.e., $\Gamma \vdash (A \to T) \wedge (T \to A)$, whence $\Gamma \vdash A$ immediately follows.

This completes the proof of Theorem 2. □

Rough double Stone algebra is a special kind of regular double Stone algebra. The following theorem will show that $\mathcal{L}_r$ enjoys the standard completeness theorem, i.e., it is complete with respect to the class of rough double Stone algebras. In what follows, $\Gamma \models_R A$ means that A is the consequence of Γ with respect to the class of rough double Stone algebras.

Theorem 3 *Let $A \in F(S)$, $\Gamma \subseteq F(S)$, then $\Gamma \vdash A$ if and only if $\Gamma \models_R A$.*

Proof If $\Gamma \vdash A$, then by Theorem 2, $\Gamma \models_R A$.

For the converse direction, we only need to show that for any regular double Stone algebra $(L, \vee, \wedge, *, +, 0, 1)$ and any valuation $v : F(S) \longrightarrow L$ satisfying $\forall B \in \Gamma$, $v(B) = 1$, $v(A) = 1$ holds. It follows from the representation theorem of regular double Stone algebra that $(L, \vee, \wedge, *, +, 0, 1)$ is isomorphic to a subalgebra of some rough double Stone algebra $\mathcal{RS}(U) = (RS(U), \vee, \wedge, *, +, (\emptyset, \emptyset), (U, U))$, i.e., $(L, \vee, \wedge, *, +, 0, 1)$ can be embedded into $\mathcal{RS}(U)$. Assume that the embedding mapping is i, then it can be easily verified that the compositional mapping $i \circ v : F(S) \longrightarrow \mathcal{RS}U$ is a valuation, and in addition, $\forall B \in \Gamma$, $(i \circ v)(B) = (U, U)$. Combining with $\Gamma \models_R A$, we have $(i \circ v)(A) = (U, U)$, which entails that $v(A) = 1$. Hence $\Gamma \models A$, then we have from Theorem 2 that $\Gamma \vdash A$. This completes the proof of Theorem 3. □

4 Concluding Remarks

In the present paper, a kind of propositional logic system with semantics based on regular double stone algebra is proposed, and its soundness theorem and completeness theorem with respect to rough set semantics are also obtained. As pointed by one reviewer of this paper, relationship between $\mathcal{L}_r$ and the pre-rough logic system $\mathcal{L}_1$ proposed by Banerjee and the rough set logic initiated in [17] is an attractive research topic, we will study them subsequently.

Acknowledgments Project supported by the National Nature Science Fund of China under Grant 61103133 and 61472471, the Natural Science Program for Basic Research of Shaanxi Province, China (No. 2014JQ1032).

References

1. Pawlak, Z.: Rough sets. Int. J. Comput. Inf. Sci. **11**(5), 341–356 (1982)
2. Pawlak, Z.: Rough Sets-Theoretical Aspects of Reasoning about Data. Knowledge academic Publishers, Dordrecht (1991)
3. Lin, T.Y.: Cercone N Rough Sets and Data Mining-Analysis of Imprecise Data. Kluwer Academic Publishers, Boston (1997)
4. Polkowski, L.: A Skowron, Rough Sets in Knowledge Discovery 1: Methodology and Applications. Studies of Fuzziness and Soft Computing, vol. 18. Physica-Verlag, Heidlberg (1998)
5. Polkowski, L.: A Skowron, Rough Sets in Knowledge Discovery 2: Applications, Case Studies and Software Systems. Studies of Fuzziness and Soft Computing, vol. 19. Physica-Verlag, Heidlberg (1998)

6. Iwiński, T.B.: Algebraic approach to rough sets. Bull. Pol. Acad. Sci. Math. **35**, 673–683 (1987)
7. Pomykala, J., Pomykala, J.A.: The stone algebra of rough sets. Bull. Pol. Acad. Sci. Math. **36**, 495–508 (1988)
8. Gehrke, M., Walker, E.: On the structure of rough sets. Bull. Pol. Acad. Sci. Math. **40**, 235–255 (1992)
9. Comer, S.: On connections between information systems. rough sets and algebraic logic. In: Algebraic Methods in Logic and Computer Science, pp. 117–124. Banach Center Publications (1993)
10. Pagliani, P.: Rough sets and Nelson algebras. Fundamenta Informaticae **27**, 205–219 (1996)
11. Banerjee, M., Chakraborty, M.K.: Rough sets through algebraic logic. Fundamenta Informaticae **28**, 211–221 (1996)
12. Pawlak, Z.: Rough logic. Bull. Pol. Acad. Sci. (Technical Sciences) **35**, 253–258 (1987)
13. Orlowska, E.: Kripke semantics for knowledge representation logics. Studia Logica XLIX **49**, 255–272 (1980)
14. Vakarelov, D.: A modal logic for similarity relations in Pawlak knowledge representaion systems. Fundamenta Informaticae **15**, 61–79 (1991)
15. Vakarelov, E.: Modal logics for knowledge representation systems. Theor. Comput. Sci. **90**, 433–456 (1991)
16. Sen, J., Chakraborty, M.K.: A study of interconnections between rough and 3-valued Łukasiewicz logics. Fundamenta Informaticae **51**, 311–324 (2002)
17. Düntsch, I.: A logic for rough sets. Theor. Comput. Sci. **179**, 427–436 (1997)
18. Dai, J.H.: Logic for rough sets with rough double Stone algebraic semantics. In: Slezak, D., Wang, G., Szczuka, M.S., Düntsch, I., Yao, Y. (eds.) Proceedings of RSFSDMGrC(1), Canada. LNCS, vol. 3641, pp. 141–148. Springer (2005)

The Comparison of Expressiveness Between LTL and IGPoLTL

Jia-Qi Dang and Yong-Ming Li

Abstract The expressiveness of linear temporal logic plays an important role in model checking. But the expressiveness of linear temporal logic based on generalized possibility measure has not been researched roundly. We compare the expressiveness of linear temporal logic (LTL) and interval generalized possibilistic linear temporal logic (IGPoLTL), and prove that LTL is a proper subclass of IGPoLTL. Besides, we define the α-equivalence between LTL formulae and IGPoLTL formulae and get some corresponding properties.

Keywords Interval generalized possibilistic linear temporal logic · Linear temporal logic · Expressiveness · Model checking

1 Introduction

Model checking [1–3] is an effective automated verification technique to analyze correctness of software and hardware design. The primary parts of model checking includes three steps: abstracting the mathematical model of the system, specifying the properties of the system, and verifying whether the system satisfies the properties using model-checking algorithms. Generally, a finite state model or Kripke structure is used to represent system. The properties of the system are specified using temporal logic, such as linear temporal logic (LTL) or computation tree logic (CTL). The verification phase gives a Boolean answer: the properties hold in the system or not hold in it with counterexample.

Classical models and temporal logic usually are qualitative and Boolean. However, in reality, the systems, such as computer hardware and software systems, are inevitably referred to a lot of uncertainty information, this will affect modeling and

J.-Q. Dang (✉) · Y.-M. Li
College of Computer Science, Shaanxi Normal University, Xi'an 710119, China
e-mail: dangjiaqi@snnu.edu.cn

Y.-M. Li
e-mail: liyongm@snnu.edu.cn

T.-H. Fan et al. (eds.), *Quantitative Logic and Soft Computing 2016*,
Advances in Intelligent Systems and Computing 510,
DOI 10.1007/978-3-319-46206-6_16

verifying of systems. In order to deal with the uncertain information, many quantitative extensions of the state-transition model have been proposed, such as models that embed state changes into times [1], models that assign probabilities [1], possibilities [4, 5], multi-valued [6–8], or fuzzy [9–11], etc., methods, the quantitative temporal logic such as probabilistic temporal logic [1], possibilistic temporal logic [4, 5], fuzzy temporal temporal logic [9–11] are also proposed to represent the quantitative properties of system.

Probabilistic computation tree logic (PCTL) and CTL are not comparable with each other [1]. This indicates that PCTL can be used for doing model checking of real-world problems, which cannot handle classical CTL model checking. Possibilistic CTL(PoCTL) is differ from PCTL, CTL is a proper subclass of PoCTL. However, the expressiveness of linear temporal logic based on generalized possibility measure has not been researched roundly, we have studied quantitative model checking of linear-time properties based on generalized possibility measures [12]. This paper is based on the above work to compare the expressiveness of LTL and interval generalized possibilistic linear temporal logic (IGPoLTL).

The content of this paper is organized as follows. Section 2 gives some preliminary knowledge. Section 3 gives the definition of equivalence between LTL formulae and IGPoLTL formulae, and proves that LTL is a proper subclass of IGPoLTL, and defines the α-equivalence between LTL formulae and IGPoLTL formulae.

2 Preliminaries

In this section, we introduce some notions of classical LTL and generalized possibilistic linear temporal logic (GPoLTL), including the notion of Kripke structure and the syntax and semantics of LTL as well as the notion of generalized possibilistic Kripke structure and the syntax and semantics of GPoLTL.

2.1 Linear Temporal Logic

Definition 1 ([2]) A Kripke structure is a tuple $M = (S, I, R, AP, L)$, where we have the following:
(1) S is a countable, nonempty set of states;
(2) $I \subseteq S$ is a set of initial states;
(3) $R \subseteq S \times S$ is a transition relation;
(4) AP is a set of atomic propositions;
(5) $L : S \longrightarrow 2^{AP}$ is a labeling function.

Definition 2 [1] (*Syntax of LTL*) LTL formulae over the set AP of atomic propositions are formed according to the following grammar:

$$\varphi ::= true \mid a \mid \varphi_1 \wedge \varphi_2 \mid \neg\varphi \mid \bigcirc\varphi \mid \varphi_1 \sqcup \varphi_2$$

where $a \in AP$.

Definition 3 [1] (*Semantics of LTL*) Assume that $\pi = s_0 s_1 s_2 \cdots$ is a path starting s_0 in a Kripke structure M, $\pi_i = s_i s_{i+1} s_{i+2} \cdots$ and $a \in AP$. The satisfaction relation ($\models$) is defined recursively as follows:

$\pi \models true$;
$\pi \models a$, iff $a \in L(s_0)$;
$\pi \models \varphi_1 \wedge \varphi_2$, iff $\pi \models \varphi_1$ and $\pi \models \varphi_2$;
$\pi \models \neg\varphi$, iff $\pi \not\models \varphi$;
$\pi \models \bigcirc\varphi$, iff $\pi_1 \models \varphi$;
$\pi \models \varphi_1 \sqcup \varphi_2$, iff $\exists\, j \geqslant 0$, $\pi_j \models \varphi_2$, and $\pi_i \models \varphi_1$ for all $0 \le i < k$.

2.2 Generalized Possibilistic Linear Temporal Logic

Definition 4 ([13]) A generalized possibilistic Kripke structure (GPKS, in short) is a tuple $M = (S, P, I, AP, L)$, where
(1) S is a countable, nonempty set of states;
(2) $P : S \times S \longrightarrow [0, 1]$ is a function, called possibilistic transition distribution function;
(3) $I : S \longrightarrow [0, 1]$ is a function, called possibilistic initial distribution function;
(4) AP is a set of atomic propositions;
(5) $L : S \times AP \longrightarrow [0, 1]$ is a possibilistic labeling function, which can be viewed as function mapping a state s to the fuzzy set of atomic propositions which are possible in the state s, i.e., $L(s, a)$ denotes the possibility or truth value of atomic proposition a that is supposed to hold in s.

Furthermore, if the set S and AP are finite sets, then $M = (S, P, I, AP, L)$ is called a finite generalized possibilistic Kripke structure.

Remark 1 (1) In Definition 4, if we require the transition possibility distribution and initial distribution to be normal, i.e., $\vee_{s' \in S} P(s, s') = 1$ and $\vee_{s \in S} I(s) = 1$, and the labeling function L is also crisp, i.e., $L : S \times AP \longrightarrow \{0, 1\}$. Then we obtain the notion of possibilistic Kripke structure (PKS) [4, 5]. In this case, we also say that M is normal.
(2) Paths in GPKS M are infinite paths. They are defined as infinite state sequences $\pi = s_0 s_1 s_2 \cdots \in S^\omega$ such that $P(s_i, s_{i+1}) > 0$ for all $i \ge 0$. Let $Paths(M)$ denotes the set of all paths in M, and $Paths_{fin}(M)$ denotes the set of finite path fragments $s_0 s_1 \cdots s_n$, where $n \ge 0$ and $P(s_i, s_{i+1}) > 0$ for $0 \le i \le n-1$. Let $Paths(s)$ denotes the set of all paths in M that start in state s. Similarly, $Paths_{fin}(s)$ denotes the set of finite fragments $s_0 s_1 \cdots s_n$ such that $s_0 = s$.

Definition 5 [12] (*Syntax of GPoLTL*) Generalized possibilistic linear temporal logic (GPoLTL, in short) formulae over the set AP of atomic propositions are the same as LTL formulae, which are formed according to the following grammar:

$$\varphi ::= true \mid a \mid \varphi_1 \wedge \varphi_2 \mid \neg\varphi \mid \bigcirc\varphi \mid \varphi_1 \sqcup \varphi_2$$

where $a \in AP$.

Other path formulae can be derived as follows:
$\varphi_1 \vee \varphi_2 = \neg(\neg\varphi_1 \wedge \neg\varphi_2)$;
$\varphi_1 \to \varphi_2 = \neg\varphi_1 \vee \varphi_2$;
$\lozenge\varphi = true \sqcup \varphi$;
$\square\varphi = \neg\lozenge\neg\varphi$;
$\varphi_1 R \varphi_2 = \neg(\neg\varphi_1 \sqcup \neg\varphi_2)$.

Definition 6 [12] (*Path semantics of GPoLTL*) Assume that $\pi = s_0 s_1 s_2 \cdots$ is a path starting s_0 in a GPKS M, $\pi_i = s_i s_{i+1} s_{i+2} \cdots$, $\pi[i] = s_i$, φ is a GPoLTL formula, its path semantics over M is a fuzzy set on $Paths(M)$, i.e., $||\varphi||_M : Paths(M) \longrightarrow [0, 1]$, which is defined recursively as follows:
$||true||_M(\pi) = 1$;
$||a||_M(\pi) = L(s_0, a)$;
$||\varphi_1 \wedge \varphi_2||_M(\pi) = ||\varphi_1||_M(\pi) \wedge ||\varphi_2||_M(\pi)$;
$||\neg\varphi||_M(\pi) = 1 - ||\varphi||_M(\pi)$;
$||\bigcirc\varphi||_M(\pi) = ||\varphi||_M(\pi_1)$;
$||\varphi_1 \sqcup \varphi_2||_M(\pi) = \bigvee_{j\geq 0}(||\varphi_2||_M(\pi_j) \wedge \bigwedge_{i<j} ||\varphi_1||_M(\pi_i))$.

Definition 7 [12] (*Language semantics of GPoLTL*) Let φ be a GPoLTL formula. The language semantics of φ over the alphabet $\Sigma = [0, 1]^{AP}$(or $\Sigma = l^{AP}$ for some finite subset $l \subseteq [0, 1]$) is a fuzzy ω-language, i.e., $||\varphi|| : \Sigma^\omega \longrightarrow [0, 1]$, which is defined iteratively as follows: for $\sigma = A_0 A_1 \cdots \in \Sigma^\omega$, write $\sigma_j = A_j A_{j+1} \cdots$,
$||true||(\sigma) = 1$;
$||a||(\sigma) = A_0(a)$;
$||\varphi_1 \wedge \varphi_2||(\sigma) = ||\varphi_1||(\sigma) \wedge ||\varphi_2||(\sigma)$;
$||\neg\varphi||(\sigma) = 1 - ||\varphi||(\sigma)$;
$||\bigcirc\varphi||(\sigma) = ||\varphi||(\sigma_1)$;
$||\varphi_1 \sqcup \varphi_2||(\sigma) = \bigvee_{j\geq 0}(||\varphi_2||(\sigma_j) \wedge \bigwedge_{i<j} ||\varphi_1||(\sigma_i))$.

3 The Comparison of Expressiveness Between LTL and IGPoLTL

In this section, we first define the equivalence between LTL formulae and IGPoLTL formulae and prove the relation of LTL and interval generalized possibilistic linear temporal logic (IGPoLTL). In addition, we define α-equivalence between LTL formulae and IGPoLTL formulae.

3.1 Definition of Equivalence Between LTL Formulae and IGPoLTL Formulae

Definition 8 (*Syntax of IGPoLTL*) IGPoLTL formulae over the set AP of atomic propositions are formed according to the following grammar:

$$\varphi ::= true \mid a_J \mid \varphi_1 \wedge \varphi_2 \mid \neg\varphi \mid \bigcirc\varphi \mid \varphi_1 \sqcup \varphi_2$$

where $a \in AP$, $J \subseteq [0, 1]$.

Definition 9 (*Semantics of IGPoLTL*) Assume that $\pi = s_0 s_1 s_2 \cdots$ is a path starting from s_0 in a GPKS M, $\pi_i = s_i s_{i+1} s_{i+2} \cdots$, $\pi[i] = s_i$, φ and ψ are IGPoLTL formulae, The satisfaction relation ($\models$) is defined recursively as follows:

$\pi \models true$;
$\pi \models a_J$, iff $L(s_0, a) \in J$;
$\pi \models \varphi \wedge \psi$, iff $\pi \models \varphi$ and $\pi \models \psi$;
$\pi \models \neg\varphi$, iff $\pi \not\models \varphi$;
$\pi \models \bigcirc\varphi$, iff $\pi_1 \models \varphi$;
$\pi \models \varphi \sqcup \psi$, iff $\exists\, j \geqslant 0$, $\pi_j \models \psi$, and $\pi_i \models \varphi$ for all $0 \leq i < k$.

Definition 10 For a generalized possibilistic Kripke structure M, if φ is an IGPoLTL formula over AP, let $Words_M(\varphi)$, or briefly $Words(\varphi)$, denote$\{\pi \in Paths(M) | \pi \models \varphi\}$.

Definition 11 IGPoLTL formulae φ and ψ are called equivalent, denoted $\varphi \equiv \psi$, if $Words(\varphi) = Words(\psi)$ for any GPKS M over AP.

Definition 12 An IGPoLTL formula φ is equivalent to a LTL formula ψ, denoted $\varphi \equiv \psi$, if $Words_M(\varphi) = Words_{TS(M)}(\psi)$ for any generalized possibilistic Kripke structure $M = (S, P, I, AP, L)$, where $TS(M) = (S, \rightarrow, I', AP, L')$ is defined by $s \rightarrow t$ iff $P(s, t) > 0$, $s \in I'$ iff $I(s) > 0$, $a \in L'(s)$ iff $L(s, a) > 0$. Obviously, $Paths_M(s) = Paths_{TS(M)}(s)$, so we use the same symbol $Paths(s)$ to denote $Paths_M(s)$ and $Paths_{TS(M)}(s)$ in the following.

Definition 13 The length of IGPoLTL formulae φ are denoted by $|\varphi|$, i.e., $|\varphi|$ denotes the number of subformulae of φ, which is defined as follows:

if $\varphi = true$, then $|\varphi| = 1$;
if $\varphi \in AP$, then $|\varphi| = 1$;
$|\neg\varphi| = |\varphi| + 1$;
$|\varphi_1 \wedge \varphi_2| = |\varphi_1| + |\varphi_2| + 1$;
$|\bigcirc\varphi| = |\varphi| + 1$;
$|\varphi_1 \sqcup \varphi_2| = |\varphi_1| + |\varphi_2| + 1$.

IGPoLTL is obviously a simple crispness of GPoLTL. Even so, IGPoLTL is powerful than LTL considering their expressiveness, we show this fact in the following subsection.

3.2 IGPoLTL is More Powerful than LTL Considering Their Expressiveness

Theorem 1 *For any LTL formula φ, there exists an IGPoLTL formula ψ such that $\varphi \equiv \psi$.*

Proof The proof is proceeded by induction on the length of formula φ.

For any GPKS M and any path $\pi = s_0 s_1 s_2 \cdots \in Paths(M)$, we have the following discussion.

There are six cases to be considered.

Case1: $\varphi = true$, then $\psi = true$;

Case2: $\varphi = a$.

Note that $\pi \models a$ iff $a \in L'(s_0)$ iff $L(s_0, a) > 0$ iff $a \models a_{>0}$. Therefore, $\varphi = a \equiv a_{>0} = \psi$;

Case3: $\varphi = \varphi_1 \wedge \varphi_2$.

By the induction hypothesis, there exist IGPoLTL formulae ψ_1 and ψ_2 such that $\varphi_1 \equiv \psi_1, \varphi_2 \equiv \psi_2$. Note that $\pi \models \varphi_1 \wedge \varphi_2$, iff $\pi \models \varphi_1$ and $\pi \models \varphi_2$, iff $\pi \models \psi_1$ and $\pi \models \psi_2$, iff $\pi \models \psi_1 \wedge \psi_2$. Therefore, $\varphi = \varphi_1 \wedge \varphi_2 \equiv \psi_1 \wedge \psi_2 = \psi$;

Case4: $\varphi = \bigcirc \varphi_1$.

By the induction hypothesis, there exists an IGPoLTL formula ψ_1 such that $\varphi_1 \equiv \psi_1$. Note that $\pi \models \bigcirc \varphi_1$ iff $\pi_1 \models \varphi_1$ iff $\pi_1 \models \psi_1$ iff $\pi \models \bigcirc \psi_1$. Therefore, $\varphi = \bigcirc \varphi_1 \equiv \bigcirc \psi_1 = \psi$;

Case5: $\varphi = \neg \varphi'$.

By the induction hypothesis, there exists an IGPoLTL formula ψ' such that $\varphi' \equiv \psi'$. Note that $\pi \models \neg \varphi'$ iff $\pi \not\models \varphi'$ iff $\pi \not\models \psi'$ iff $\pi \models \neg \psi'$. Therefore, $\varphi = \neg \varphi' \equiv \neg \psi' = \psi$;

Case6: $\varphi = \varphi_1 \sqcup \varphi_2$.

By the induction hypothesis, there exist IGPoLTL formulae ψ_1 and ψ_2 such that $\varphi_1 \equiv \psi_1, \varphi_2 \equiv \psi_2$. Note that $\pi \models \varphi_1 \sqcup \varphi_2$, iff there exists $j \geq 0$ such that $\pi_j \models \varphi_2$ and $\pi_i \models \varphi_1$ for all $0 \leq i < k$, iff there exists $j \geq 0$ such that $\pi_j \models \psi_2$ and $\pi_i \models \psi_1$ for all $0 \leq i < k$, iff $\pi \models \psi_1 \sqcup \psi_2$. Therefore, $\varphi = \varphi_1 \sqcup \varphi_2 \equiv \psi_1 \sqcup \psi_2 = \psi$. □

3.3 LTL is a Proper Subclass of IGPoLTL

Theorem 2 *There is no LTL formula that is equivalent to $a_{=1}$.*

Proof Assume that there is a LTL formula φ such that $\varphi \equiv a_{=1}$. Consider the following two finite generalized possibilistic Kripke structures M_1 and M_2, see Figs. 1 and 2. By the figures, we have $L(s_0, a) = 0.5$ in M_1. But $L(s_0, a) = 1$ in M_2. Path $s_0 s_1 s_3^\omega$ does not satisfy $a_{=1}$ in M_1, while $s_0 s_1 s_3^\omega$ satisfies $a_{=1}$ in M_2. Therefore, $s_0 s_1 s_3^\omega \notin Words_{M_1}(a_{=1})$, but $s_0 s_1 s_3^\omega \in Words_{M_2}(a_{=1})$. This implies that

$$Words_{M_1}(a_{=1}) \neq Words_{M_2}(a_{=1}). \tag{1}$$

Fig. 1 A finite GPKS M_1

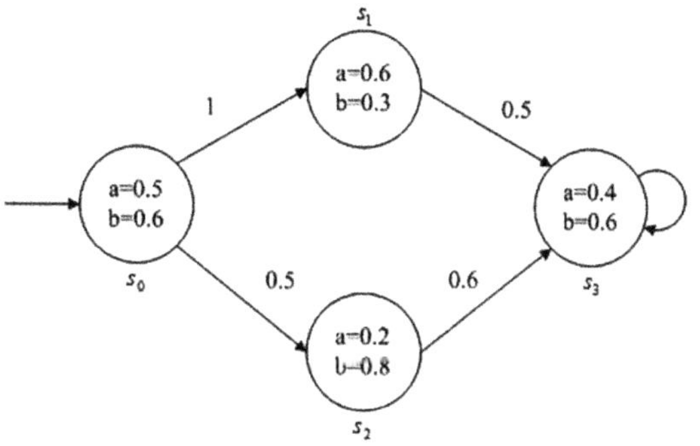

Fig. 2 A finite GPKS M_2

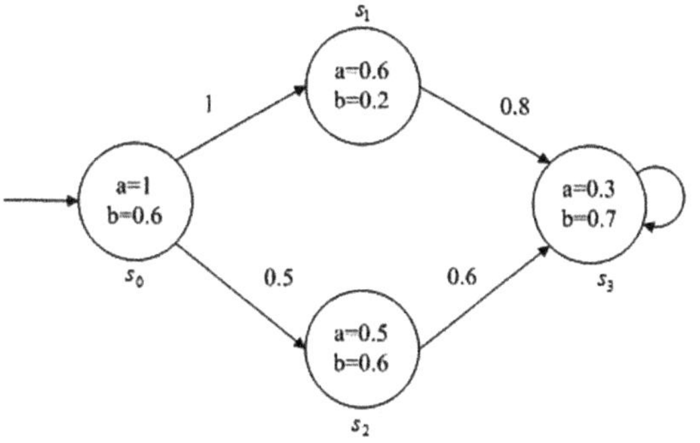

Since φ is an LTL formula, and $TS(M_1) = TS(M_2)$, we have

$$Words_{TS(M_1)}(\varphi) = Words_{TS(M_2)}(\varphi). \tag{2}$$

By the assumption $\varphi \equiv a_{=1}$, it follows that $Words_{TS(M)}(\varphi) = Words_M(a_{=1})$ for any generalized possibilistic Kripke structure M. Then we have

$$Words_{M_1}(a_{=1}) = Words_{M_2}(a_{=1}). \tag{3}$$

Equations (1) and (3) show a contradiction, which demonstrates that there is no LTL formula that is equivalent to $a_{=1}$. □

Combining Theorems 1 and 2, it follows that LTL is a proper subclass of IGPoLTL.

Using the same method, we can prove that following theorems also hold for any generalized possibilistic Kripke structure M.

Theorem 3 *There is no LTL formula that is equivalent to* $a_{=1} \wedge b_{=1}$.

Theorem 4 *There is no LTL formula that is equivalent to* $\bigcirc a_{=1}$.

Theorem 5 *There is no LTL formula that is equivalent to* $a_{=1} \sqcup b_{=1}$.

3.4 Alternative Way to Define Equivalence Between LTL Formulae and IGPoLTL Formulae

The definition of equivalence between LTL formulae and IGPoLTL formulae is not unique. In this subsection, We will give another way to define equivalence between LTL formulae and IGPoLTL formulae.

Definition 14 An IGPoLTL formula φ is α-equivalent to a LTL formula ψ, denoted $\varphi \equiv_\alpha \psi$, if $Words_M(\varphi) = Words_{TS_\alpha(M)}(\psi)$ for any GPKS $M = (S, P, I, AP, L)$ and $\alpha \in (0, 1]$, where $TS_\alpha(M) = (S, \rightarrow_\alpha, I_\alpha, AP, L_\alpha)$ is defined by $s \rightarrow_\alpha t$ iff $P(s, t) \geq \alpha$, $s \in I_\alpha$ iff $I(s) \geq \alpha$, $a \in L_\alpha(s)$ iff $L(s, a) \geq \alpha$. Obviously, $Paths_M(s) = Paths_{TS_\alpha(M)}(s)$, so we use the same symbol $Paths(s)$ to denote $Paths_M(s)$ and $Paths_{TS_\alpha(M)}(s)$ in the following.

We will give some properties of IGPoLTL using the definition of α-equivalence between IGPoLTL formulae and LTL formulae for $\alpha \in (0, 1]$. The proofs are similar to those in Sects. 3.2 and 3.3.

Proposition 1 *For all LTL formula φ and $\alpha \in (0, 1]$, there exists an IGPoLTL formula ψ such that $\varphi \equiv_\alpha \psi$.*

Proposition 2 *For all $\alpha \in (0, 1)$, there is no LTL formula that is α-equivalent to $a_{=1}$.*

4 Conclusion

In this paper, we mainly compare the expressiveness between linear temporal logic and interval generalized possibilistic linear temporal logic. We introduce the definition of equivalence between LTL formulae and IGPoLTL formulae, and obtain the conclusion that LTL is a proper subclass of IGPoLTL. In addition, we define α-equivalence between LTL formulae and IGPoLTL formulae, and get some corresponding properties.

Acknowledgments This work is supported by the NSFC (Grants No. 11271237, No. 61228305).

References

1. Baier, C., Katoen, J.P.: Principles of Model Checking. The MIT Press, Cambridge (2008)
2. Clark, E.M., Grumberg, O., Peled, D.: Model Checking. The MIT Press, Cambridge (1999)
3. Rozier, K.Y.: Linear temporal logic symbolic model checking. Comput. Sci. Rev. **5**(2), 163–203 (2011)
4. Li, Y.M., Li, L.J.: Model checking of linear-time properties based on possibility measure. IEEE Trans. Fuzzy Syst. **21**(5), 842–854 (2013)

5. Li, Y.M., Li, Y.L., Ma, Z.Y.: Computation tree logic model checking based on possibility measures. Fuzzy Sets Syst. **262**, 44–59 (2015)
6. Chechik, M., Devereux, B., Gurfinkel, A., Easterbrook, S.: Multi-valued symbolic model-checking. ACM Trans. Softw. Eng. Methodol. **12**(4), 371–408 (2003)
7. Chechik, M., Easterbrook, S., Petrovykh, V.: Model-checking over multi-valued logics. In: Proceeding of Formal Methods Europe (FME01). Lecture Notes in Computer Science, vol. 2021, pp. 72–98. Springer, Berlin (2001)
8. Chechik, M., Gurfinkel, A., Devereux, B., Lai, A., Easterbrook, S.: Data structures for symbolic multi-valued model-checking. Formal Methods Syst. Design **29**, 295–344 (2006)
9. Frigeri, A., Pasquale, L., Spoletini, P.: Fuzzy time in linear temporal logic. ACM Trans. Comput. Logic (TOCL) **15**(4), Article No. 30 (2014)
10. Mukherjee, S., Dasgupta, P.: A fuzzy real-time temporal logic. Int. J. Approx. Reason. **54**(9), 1452–1470 (2013)
11. Pan, H.Y., Li, Y.M., Cao, Y.Z., Ma, Z.Y.: Model checking fuzzy computation tree logic. Fuzzy Sets Syst. **262**, 60–77 (2015)
12. Li, Y.M.: Quantitative model checking of linear-time properties based on generalized possibility measures (2016). arXiv:1601.06504
13. Li, Y.M., Ma, Z.Y.: Quantitative computation tree logic model checking based on generalized possibility measures. IEEE Trans. Fuzzy Syst. **23**(6), 2034–2047 (2015)

Part III
Automata and Quantification of Software

Analyzing Quantitative Transition Systems

Guo-Wu Wang, Ying-Xiong Shen and Hai-Yu Pan

Abstract During the past years, substantial progress has been made towards developing quantitative formal verification methods. In this paper, we establish a lattice-valued relation between the states of a quantitative transition system(QTS), called lattice-valued language containment relation, to measure to what extent the language of one state is included by that of the other. We study the relationship between lattice-valued language containment relation and two lattice-valued versions of similarity defined previously, and we explore the properties of compositionality of the lattice-valued language containment relation. These properties suggest that our language containment relation provides an appropriate basis for a quantitative theory of concurrent and distributed systems.

Keywords Language containment · Fuzzy automata · Complete residuated lattices · Simulation

G.-W. Wang
College of Computer and Information, Anhui Polytechnical University,
Wuhu 241000, China
e-mail: wangguo5@126.com

G.-W. Wang
Key Laboratory of Computer Application Technology, Anhui Polytechnical University,
Wuhu 241000, China

Y.-X. Shen · H.-Y. Pan (✉)
College of Computer Science and Technology, Taizhou University,
Taizhou 225300, China
e-mail: phyu76@126.com

Y.-X. Shen
e-mail: syxtzxy2013@163.com

H.-Y. Pan
College of Computer Science, Shaanxi Normal University, Xi'an 710062, China

T.-H. Fan et al. (eds.), *Quantitative Logic and Soft Computing 2016*,
Advances in Intelligent Systems and Computing 510,
DOI 10.1007/978-3-319-46206-6_17

1 Introduction

Labelled transition systems (LTSs), a variant of automata, are a basic model for formal description, specification, and analysis of concurrent and distributed systems [19]. The most fundamental verification approach to comparing LTSs is by means of concepts of *preorders* and *equivalence relations*, among which *language containment* (equivalence) and (bi)similarity are one of the basic tools for verifying LTSs [19]. Classical formal verification techniques are largely qualitative; for example, given two LTSs, either the language of one of them is contained by the other or not.

In the past years, researchers have been devoted to developing a quantitative approach to formal model and verification. A great variety of LTSs whose transitions, states or actions contain *quantitative* data, have been proposed to model the quantitative properties of systems [2–4, 10, 20, 21]. Classical system relations such as language containment and equivalence have been adapted for these systems. Two general approaches can be recognized in the existing literature. One, from [7–9, 18], is based on *distance functions* over systems. More precisely, language containment and equivalence relations are replaced by the real-valued distances. For instance, the authors in [18] presented the weighted transition systems that are LTSs assigned weights from non-negative real numbers to transitions and actions, and introduced three types of distances on weighted transition systems as extensions of language containment relation of LTSs.

The other approach, from [12–16], is based on *lattice-valued (fuzzy) relations* on the state space of lattice-valued extensions of LTSs. For instance, the lattice automata of Kupferman and Lusting [12] assign to each word a value from a finite De Morgan lattice. To generalize the language containment to the lattice-valued setting, they defined the *implication value* of two lattice automata $\mathcal{A}$ and $\mathcal{B}$ as the truth value of the statement "for all words, the membership value in $\mathcal{A}$ implies the membership value in $\mathcal{B}$". The present work is closely related to this kind of approach.

In this paper, following the methodology in [14, 16], we concern ourselves with extending the notion of language containment relation to the complete residuated lattice-valued setting and discuss its properties. Complete residuated lattices are a very general algebraic structure with very important applications in different areas such as fuzzy automata [5, 6, 17].

In [14], Pan et al. considered the general quantitative model of finite-state transition systems in the framework of finite residuated lattices, and established the simulation semantics over the model. Some models have quantitative transition systems [16], where the actions of LTSs are equipped with a complete residuated lattice-valued equality relation, and lattice-valued Kripke structures [15] as special cases. Based on finite-state quantitative transition systems (QTS), in [16] we introduced a lattice-valued relation between states of a QTS, called approximate similarity, to quantify to what extent one state is simulated by the other.

To investigate the behaviours of QTSs from the view point of language, we introduce two lattice-valued language containment relations $\mathcal{TI}$ and $\mathcal{LI}$ (cf. Definition 2) over the state space of a QTS as extensions of language containment relation of an

LTS, to measure to what extent the language of one state is included by that of the other, where $\mathcal{TI}$ is motivated by the distance functions in [18], while $\mathcal{LI}$ is motivated by the implication value of lattice automata [12]. We show that the two lattice-valued relations coincide. It should be pointed out that although the distance functions and lattice-valued relations versions of language containment relation in the literature have counterparts of both $\mathcal{TI}$ and $\mathcal{LI}$, the relationships between them have not yet been studied.

We believe that our lattice-valued language containment relation is a natural extension of language containment relation in the complete residuated lattice-valued setting. As an evidence, we will provide some of its properties, from the perspectives of simulation, and compositionality. First, we relate lattice-valued language containment relation with two lattice-valued versions of similarity, lattice-valued similarity and approximate similarity previously introduced in [14, 16], showing that just as similarity implies language containment relation, so lattice-valued similarity is contained by lattice-valued language containment relation. However, while language containment relation coincides with similarity for deterministic LTSs, we show that lattice-valued language containment relation and lattice-valued similarity do not coincide for deterministic QTSs. Moreover, the usual relation between similarity and language containment relation cannot be transferred to the case for approximate similarity and lattice-valued language containment relation. We also show that the lattice-valued language containment relation is compositional for the approximate synchronous composition operator.

2 Preliminaries

In this section we first review basic concepts of complete residuated lattices and some properties of this kind of algebraic structures, then recall the basic elements of LTSs, which are used to give the notion of QTSs. We write $\mathbb{N}$ for the set of natural numbers, I the index set, and $\mathcal{P}(S)$ the power set of S.

A *complete residuated lattice* [1] is an algebra $\mathcal{L} = (L, \vee, \wedge, \otimes, \rightarrow, 0, 1)$ with four binary operations and two constants such that

- $(L, \vee, \wedge, 0, 1)$ is a complete lattice with the least element 0 and the largest element 1 with respect to the lattice ordering $\leq$;
- $(L, \otimes, 1)$ is a commutative monoid with the unit element 1, i.e. $\otimes$ is commutative, associative, and $1 \otimes x = x$ for all $x \in L$; and
- $\otimes$ and $\rightarrow$ form an adjoint pair, i.e. $x \otimes y \leq z$ iff $x \leq y \rightarrow z$, for all $x, y, z \in L$.

The complete residuated lattice $\mathcal{L}$ is *finite* if $(L, \vee, \wedge, 0, 1)$ is a finite lattice. A *complete Heyting algebra* can be defined as a complete residuated lattice with $\otimes = \wedge$. The notation $x \leftrightarrow y$ will be reserved as $(x \rightarrow y) \wedge (y \rightarrow x)$ (biimplication), which is used to model the equivalence of truth values. Given a finite sequence $x_0, x_1, \ldots, x_n \in L$, we abbreviate $x_0 \otimes \cdots \otimes x_n$ as $\bigotimes_{0 \leq i \leq n} x_i$. Moreover, for any $x \in$

L, we can inductively define the power of x as follows: $x^0 = 1$, $x^1 = x$, and $x^{n+1} = x^n \otimes x$ for any nonnegative integer n.

In the following let $\mathcal{L}$ be a complete residuated, let X a universal set. A *lattice-valued set* (for short, $\mathcal{L}$-set) A of X over $\mathcal{L}$, is defined by a function assigning to each element x of X a value $A(x)$ in L, $A(x)$ characterizes the degree of membership of A at x. We say that an $\mathcal{L}$-set A is *crisp* if $A(x) \in \{0, 1\}$ for all $x \in X$. We denote by $\mathcal{L}(X)$ the set of all lattice-valued subsets of X. For $A, B \in \mathcal{L}(X)$, we say that A is contained in B, denoted by $A \subseteq B$, if $A(x) \leq B(x)$ for all $x \in X$. We say that $A = B$ if $A \subseteq B$ and $B \subseteq A$.

Let X and Y be non-empty sets. A *lattice-valued relation* (for short, $\mathcal{L}$-relation) between sets X and Y is function from $X \times Y$ to L. A reflexive, symmetric, and transitive $\mathcal{L}$-relation on X is called an $\mathcal{L}$*-equivalence relation*. An $\mathcal{L}$-equivalence on X where $E(x, y) = 1$ implies $x = y$ will be called an $\mathcal{L}$*-equality*.

For every finite or countably infinite set Σ, the symbol Σ^* denotes the set of all finite words over Σ. Then length of a given word σ is denoted by $|\sigma|$, and the individual letters in σ are denoted by $\sigma(0), \ldots, \sigma(|\sigma| - 1)$. The empty word is denoted by ε, and we set $|\varepsilon| = 0$.

A *labelled transition system* (LTS) $\mathcal{A}$ is a quadruple (S, s_0, Σ, R) where S is a set of states with an initial state $s_0 \in S$, Σ is a set of actions, and $R \subseteq S \times \Sigma \times S$ is a transition relation. The transition relation R denotes possible state changes; if $(s, a, t) \in R$ we say that the system can move from state s to t by performing action a. As a more compact notation, we usually write $s \xrightarrow{a} t$ whenever, $(s, a, t) \in R$, $a \in \Sigma$, and $t \in S$. If for any state $s \in S$ and any action $a \in \Sigma$, there exists at most one transition $s \xrightarrow{a} s'$, then $\mathcal{A}$ is called *deterministic*. An LTS $\mathcal{A}$ is called *finite* if S and Σ are finite sets, and *infinite* otherwise.

The generalized transition relations $\xrightarrow{\sigma}$ for $\sigma \in \Sigma^*$ are defined recursively by:

- $s \xrightarrow{\varepsilon} s$ for any state s.
- $(s, a, s') \in R$ with $a \in \Sigma$ implies $s \xrightarrow{a} s'$ with $a \in \Sigma^*$.
- $s \xrightarrow{\sigma_1} s'$ and $s' \xrightarrow{\sigma_2} t$ imply $s \xrightarrow{\sigma_1\sigma_2} t$.

$\sigma \in \Sigma^*$ is a *trace* of a state s in $\mathcal{A}$ if there is a state s' such that $s \xrightarrow{\sigma} s'$. The set of traces of the state s in LTS $\mathcal{A}$ is called the *language* of state s, and is denoted by $L(\mathcal{A}, s)$. Where the context is clear, we will drop the notation $\mathcal{A}$. In particular, the set $L(\mathcal{A}, s_0)$ associated with the initial state s_0 of $\mathcal{A}$ is also called the language of $\mathcal{A}$. We use the notation $L(\mathcal{A})$ to denote the language of $\mathcal{A}$.

We define language containment relation for LTSs as usual. Let $\mathcal{A}_i = (S_i, s_{i,0}, \Sigma, R_i)$, $i = 1, 2$, be two LTSs and $s \in S_1$, $t \in S_2$. Then, s is contained by t with respect to language, written as $s \subseteq_L t$, if $L(\mathcal{A}_1, s) \subseteq L(\mathcal{A}_2, t)$. More precisely, if for every trace $\sigma_1 \in L(\mathcal{A}_1, s)$, there exists a trace $\sigma_2 \in L(\mathcal{A}_2, t)$ such that $\sigma_1 = \sigma_2$. This concept is common in the area of formal verification. An equivalent concept of language containment is usually illustrated from automata-theoretic view: s is contained by t with respect to language, if for every $\sigma \in \Sigma^*$ and $\sigma \in L(\mathcal{A}_1, s)$, $\sigma \in L(\mathcal{A}_2, t)$. We say that $\mathcal{A}_1$ is contained by $\mathcal{A}_2$, written $\mathcal{A}_1 \subseteq_L \mathcal{A}_2$, if $s_{1,0} \subseteq_L s_{2,0}$.

In this article, we sometime mention language equivalence. Two LTSs $\mathcal{A}_1$ and $\mathcal{A}_2$ are called language-equivalent if $\mathcal{A}_1 \subseteq_L \mathcal{A}_2$ and $\mathcal{A}_2 \subseteq_L \mathcal{A}_1$.

Labelled transition systems are generalized to quantitative transition systems by augmenting them with an additional structure in [16]. The labels are endowed with an $\mathcal{L}$-equality relation θ. An LTS may be viewed as a degenerate QTS, one in which θ is crisp.

Definition 1 A quantitative transition system (QTS) is $\mathcal{Q} = (\mathcal{A}, \theta)$, where $\mathcal{A} = (S, s_0, \Sigma, R)$, which is called the support set of QTS, is an LTS; θ is an $\mathcal{L}$-equality relation over Σ.

We say that $\mathcal{Q}$ is *finite* if the support set $\mathcal{A}$ of $\mathcal{Q}$ is finite and $\mathcal{L}$ is finite; $\mathcal{Q}$ is *deterministic* if the support set $\mathcal{A}$ of $\mathcal{Q}$ is deterministic.

3 Lattice-Valued Language Containment Relation

In this section we will generalize the language containment relation to the complete residuated lattice-valued setting, then relate it to two lattice-valued versions of similarity in [14, 16]. We proceed by giving a method to combine the similarity degree on actions to a similarity degree over the words, then between states. In what follows, unless specifically noted, we consider the fixed QTS $\mathcal{Q} = (\mathcal{A}, \theta)$ where $\mathcal{A} = (S, s_0, \Sigma, R)$ and θ is an $\mathcal{L}$-equality relation over Σ.

Definition 2 Let $\mathcal{Q}$ be a QTS. The word equality $\mathcal{E}_\theta$ on Σ^* is defined as, for all $\sigma_1, \sigma_2 \in \Sigma^*$, as follows:

$$\mathcal{E}_\theta(\sigma_1, \sigma_2) = \begin{cases} 1 & \text{if } \sigma_1 = \sigma_2 = \varepsilon; \\ \bigotimes\limits_{0 \le i < |\sigma_1|} \theta(\sigma_1(i), \sigma_2(i)) & \text{if } |\sigma_1| = |\sigma_2| \text{ and } |\sigma_1| > 0; \\ 0 & \text{otherwise.} \end{cases}$$

The truth value of σ_1 accepted by the state s in $\mathcal{Q}$, denoted by $Tv(s, \sigma_1)$, is defined as, $Tv(s, \sigma_1) = \bigvee\limits_{\sigma \in \mathrm{La}(s)} \mathcal{E}_\theta(\sigma, \sigma_1)$. We define two lattice-valued language containment relations among states as follows: for all $s, t \in S$,

$$\mathcal{TI}(s, t) = \bigwedge_{\sigma_1 \in L(s)} \bigvee_{\sigma_2 \in L(t)} \mathcal{E}_\theta(\sigma_1, \sigma_2), \ \mathcal{LI}(s, t) = \bigwedge_{\sigma_1 \in \Sigma^*} (Tv(s, \sigma_1) \to Tv(t, \sigma_1)).$$

Recall that in the complete residuated lattice-valued logics, supremum ($\bigvee$) and infimum ($\bigwedge$) are used to model the existential and general quantifiers, respectively, while the implication operator is used to model the implication of the corresponding logical calculus. Hence Definition 2 is the natural generalization of classical language containment relation in the framework of complete residuated lattices, where $\mathcal{TI}$ and

$\mathcal{LI}$ are the lattice-valued extensions of language containment relation from the views of formal verification and automata theory, respectively.

The following theorem shows that the two relations $\mathcal{TI}$ and $\mathcal{LI}$ coincide. Hence, we consider some properties of $\mathcal{TI}$ in the sequel.

Theorem 1 *Let $\mathcal{Q}$ be a QTS. Then, $\mathcal{TI} = \mathcal{LI}$.*

Proof First, we show that $\mathcal{LI}(s, t) \leq \mathcal{TI}(s, t)$ holds for all $s, t \in S$. By the properties of complete residuated lattices, we have that

$$\begin{aligned}
\mathcal{LI}(s, t) &\leq \bigwedge_{\sigma \in L(s_1)} \Big(\bigvee_{\sigma_1 \in L(s)} \mathcal{E}_\theta(\sigma_1, \sigma) \to \bigvee\nolimits_{\sigma_2 \in L(t)} \mathcal{E}_\theta(\sigma_2, \sigma)\Big) \\
&= \bigwedge_{\sigma \in L(s)} \bigwedge_{\sigma_1 \in L(s)} \Big(\mathcal{E}_\theta(\sigma_1, \sigma) \to \bigvee_{\sigma_2 \in L(t)} \mathcal{E}_\theta(\sigma_2, \sigma)\Big) \\
&\leq \bigwedge_{\sigma \in L(s)} \Big(\mathcal{E}_\theta(\sigma, \sigma) \to \bigvee_{\sigma_2 \in L(t)} \mathcal{E}_\theta(\sigma_2, \sigma)\Big) \\
&= \bigwedge_{\sigma \in L(s)} \bigvee_{\sigma_2 \in L(t)} \mathcal{E}_\theta(\sigma_2, \sigma) \\
&= \mathcal{TI}(s, t)
\end{aligned}$$

which implies that assertion (1) holds.

Now, we prove the other direction. We observe that for all $\sigma \in \Sigma^*$,

$$\begin{aligned}
\bigvee_{\sigma_1' \in L(s)} \bigwedge_{\sigma_1 \in L(s)} \bigvee_{\sigma_2 \in L(t)} (\mathcal{E}_\theta(\sigma_1, \sigma_2) \otimes \mathcal{E}_\theta(\sigma_1', \sigma)) &\leq \bigvee_{\sigma_1' \in L(s)} \bigvee_{\sigma_2 \in L(t)} (\mathcal{E}_\theta(\sigma_1', \sigma_2) \\
&\quad \otimes \mathcal{E}_\theta(\sigma_1', \sigma)) \\
&\leq \bigvee_{\sigma_2 \in L(t)} \mathcal{E}_\theta(\sigma, \sigma_2)
\end{aligned}$$

Using the properties of complete residuated lattices, we also have that for all $\sigma \in \Sigma^*$,

$$\begin{aligned}
\bigvee_{\sigma_1' \in L(s)} \bigwedge_{\sigma_1 \in L(s)} \bigvee_{\sigma_2 \in L(t)} (\mathcal{E}_\theta(\sigma_1, \sigma_2) \otimes \mathcal{E}_\theta(\sigma_1', \sigma)) &= \bigvee_{\sigma_1' \in L(s)} \bigwedge_{\sigma_1 \in L(s)} \Big(\bigvee_{\sigma_2 \in L(t)} \mathcal{E}_\theta(\sigma_1, \sigma_2) \\
&\quad \otimes \mathcal{E}_\theta(\sigma_1', \sigma)\Big) \\
&\geq \bigvee_{\sigma_1' \in L(s)} \Big(\bigwedge_{\sigma_1 \in L(s)} \bigvee_{\sigma_2 \in L(t)} \mathcal{E}_\theta(\sigma_1, \sigma_2) \\
&\quad \otimes \mathcal{E}_\theta(\sigma_1', \sigma)\Big) \\
&= \bigwedge_{\sigma_1 \in L(s)} \bigvee_{\sigma_2 \in L(t)} \mathcal{E}_\theta(\sigma_1, \sigma_2) \\
&\quad \otimes \bigvee_{\sigma_1' \in L(s)} \mathcal{E}_\theta(\sigma_1', \sigma)
\end{aligned}$$

Based on the above analysis, we can derive that for all $\sigma \in \Sigma^*$, $\bigwedge_{\sigma_1 \in L(s)} \bigvee_{\sigma_2 \in L(t)} \mathcal{E}_\theta(\sigma_1, \sigma_2) \otimes \bigvee_{\sigma_1' \in L(s)} \mathcal{E}_\theta(\sigma_1', \sigma) \leq \bigvee_{\sigma_2 \in L(t)} \mathcal{E}_\theta(\sigma, \sigma_2)$. According to the definition of complete residuated lattices, for all $\sigma \in \Sigma^*$, we have $\bigwedge_{\sigma_1 \in L(s)} \bigvee_{\sigma_2 \in L(t)} \mathcal{E}_\theta(\sigma_1, \sigma_2) \leq$

$\bigvee_{\sigma_1' \in L(s)} \mathcal{E}_\theta(\sigma_1', \sigma) \to \bigvee_{\sigma_2 \in L(t)} \mathcal{E}_\theta(\sigma, \sigma_2)$. Thus, $\mathcal{TI}(s, t) \leq \mathcal{LI}(s, t)$. Now the theorem follows. □

So far, $\mathcal{TI}$ has been defined as an $\mathcal{L}$-relation between states within a single QTS. An alternative perspective is to consider them as an $\mathcal{L}$-relations between QTSs. It can be done by defining the disjoint union of QTSs in the usual sense.

Definition 3 ([16]) Let $\mathcal{Q}_1 = ((S_1, s_{1,0}, \Sigma, R_1), \theta)$ and $\mathcal{Q}_2 = ((S_2, s_{2,0}, \Sigma, R_2), \theta)$ be two QTSs where $S_1 \cap S_2 = \emptyset$. The union of $\mathcal{Q}_1 \cup \mathcal{Q}_2$ is defined as a QTS $\mathcal{Q} = ((S, s_0, \Sigma, R), \theta)$ with $S = S_1 \cup S_2 \cup \{s_0\}$ where $s_0 \notin S_1 \cup S_2$; $R = R_1 \cup R_2 \cup \{(s_0, a, s_{1,0}) \mid a \in \Sigma\} \cup \{(s_0, a, s_{2,0}) \mid a \in \Sigma\}$.

With this definition, we can define the truth value of $\mathcal{Q}_1$ included by $\mathcal{Q}_2$ with respect to language as $\mathcal{TI}(\mathcal{Q}_1, \mathcal{Q}_2) = \mathcal{TI}(s_{1,0}, s_{2,0})$.

Now we are in a position to analyze the relationship between our lattice-valued language containment relation and classical language inclusion relation. We know from Definition 2 that if $s \subseteq_{La} t$, then $\mathcal{TI}(s, t) = 1$. The following counterexample shows that the converse does not always hold.

Example 1 Consider a QTS $\mathcal{Q} = ((\{s, t\}, s, \Sigma, R), \theta)$, where $\mathcal{L}$ is the standard Gödel algebra, $\Sigma = \{a\} \cup \{b_i : i \in \mathbb{N}\}$, the transition relation R consists of the transitions:
(s, a, s) and (t, b_i, t) for every $i \in \mathbb{N}$, θ is defined as, for all $c, d \in \Sigma$,

$$\theta(c, d) = \begin{cases} 1 & \text{if } c = d; \\ 1 - 0.5^i & \text{if } c = a,\ d = b_i, \text{ or } c = b_i,\ d = a,\ i \in \mathbb{N}; \\ 0 & \text{otherwise.} \end{cases}$$

Then, by Definition 2, we have that $\mathcal{TI}(s, t) = 1$, but $s \subseteq_{La} t$ does not hold.

In the following, we first recall the definitions of two lattice-valued versions of similarity over QTSs, which have originally appeared in [14, 16], then provide a comparison between lattice-valued language containment relation and lattice-valued versions of similarity. The motivations with respect to introducing the two lattice-valued versions of similarity, can be found in [14, 16].

Definition 4 ([14, 16]) Let $\mathcal{Q}$ be a QTS. An $\mathcal{L}$-relation $\mathcal{R} \in \mathcal{L}(S \times S)$ is called an $\mathcal{L}$-simulation, if for all $s, t \in S$, $\mathcal{R}(s, t) = \bigwedge_{s \xrightarrow{a} s'} \bigvee_{t \xrightarrow{b} t'} \big(\theta(a, b) \otimes \mathcal{R}(s', t')\big)$. The $\mathcal{L}$-similarity, written $\preceq_{\mathcal{L}}$, is the greatest fixed point of the above recursive equation.

Definition 5 ([16]) Let $\mathcal{Q}$ be a QTS, $s, t \in S$ and $\delta \in L$.
(1) A relation $\mathcal{R}_\delta \subseteq S \times S$ is called a δ-simulation if for any $(s, t) \in \mathcal{R}_\delta$ and for each $a \in \Sigma$, when $s \xrightarrow{a} s'$, there exist $b \in \Sigma$ and $t \xrightarrow{b} t'$ such that $\theta(a, b) \geq \delta$ and $(s', t') \in \mathcal{R}_\delta$. δ-similarity, written $\preceq_\delta$, is the union of all δ-simulations; we say that t δ-simulates s if $s \preceq_\delta t$.

(2) The degree that s is simulated by t with respect to $\preceq_\delta$ is defined as: $\mathcal{S}(s, t) = \vee\{\delta \in L \mid s \preceq_\delta t\}$. $\mathcal{S}$ is called an *approximate similarity*.

The relationship between the $\mathcal{L}$-similarity and lattice-valued language containment relation is captured by the following result which holds for all quantitative transition systems, not necessarily finite. The following theorem expresses the intuition: "Similarity implies language containment relation".

Theorem 2 *Let $\mathcal{Q}$ be a QTS. Then, $\preceq_L \subseteq \mathcal{TI}$.*

Proof Let $s, t \in S$. We have from the properties of complete residuated lattices and $\bigvee_{i\in I} \bigwedge_{j\in J} a_{ij} \leq \bigwedge_{j\in J} \bigvee_{i\in I} a_{ij}$ $(a_{ij} \in L)$ that

$$\begin{aligned}
\mathcal{TI}(s, t) &= \bigwedge_{\sigma_1\in L(s)} \bigvee_{\sigma_2\in L(t)} \mathcal{E}_\theta(\sigma_1, \sigma_2)\\
&= \bigwedge_{s\xrightarrow{a}s'} \bigwedge_{\sigma_1'\in L(s')} \bigvee_{t\xrightarrow{b}t'} \bigvee_{\sigma_2'\in L(t')} \left(\theta(a, b) \otimes \mathcal{E}_\theta(\sigma_1', \sigma_2')\right)\\
&\geq \bigwedge_{s\xrightarrow{a}s'} \bigvee_{t\xrightarrow{b}t'} \bigwedge_{\sigma_1'\in L(s')} \bigvee_{\sigma_2'\in L(t')} \left(\theta(a, b) \otimes \mathcal{E}_\theta(\sigma_1', \sigma_2')\right)\\
&= \bigwedge_{s\xrightarrow{a}s'} \bigvee_{t\xrightarrow{b}t'} \bigwedge_{\sigma_1'\in L(s')} \left(\theta(a, b) \otimes \bigvee_{\sigma_2'\in L(t')} \mathcal{E}_\theta(\sigma_1', \sigma_2')\right)\\
&\geq \bigwedge_{s\xrightarrow{a}s'} \bigvee_{t\xrightarrow{b}t'} \left(\theta(a, b) \otimes \bigwedge_{\sigma_1'\in L(s')} \bigvee_{\sigma_2'\in L(t')} \mathcal{E}_\theta(\sigma_1', \sigma_2')\right)\\
&= \bigwedge_{s\xrightarrow{a}s'} \bigvee_{t\xrightarrow{b}t} (\theta(a, b) \otimes \mathcal{TI}(s', t'))
\end{aligned}$$

Now, the result can be shown by using a structural-induction argument. □

As is well-known, in the boolean setting, language containment relation coincides with similarity on deterministic LTSs. The following example gives a counterexample to illustrate that the equality $\preceq_L = \mathcal{TI}$ does not hold for deterministic QTSs in general, moreover, unfortunately, the example also shows that Theorem 2 cannot hold for the case of approximate similarity $\mathcal{S}$, i.e., $\mathcal{S} \subseteq \mathcal{TI}$ does not hold necessarily.

Example 2 Consider two deterministic QTSs with the same $\mathcal{L}$-equality relation θ where $\mathcal{L}$ is the standard Łukasiewicz algebra, their support sets are shown in Fig. 1, and θ is defined by: $\theta(a, a) = \theta(b, b) = \theta(c, c) = 1$, $\theta(a, b) = \theta(b, a) = 0.6$, $\theta(a, c) = \theta(c, a) = 0.4$, $\theta(b, c) = \theta(c, b) = 0.7$. We have $\mathcal{TI}(s_0, t_0) = 0.4$, while $\preceq_L (s_0, t_0) = 0.3$ and $\mathcal{S}(s_0, t_0) = 0.6$.

Remark 1 We note that the distance function analogues of approximate similarity and $\mathcal{L}$-similarity coincide [9, 18, 20]. The fact that approximate similarity and $\mathcal{L}$-similarity do not coincide in the arbitrary residuated lattice setting in [16], together with Example 2, implies that the results obtained in the lattice-valued relations are more richer.

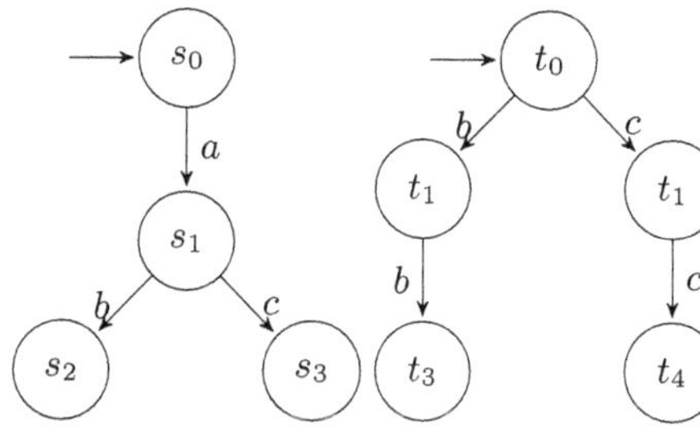

Fig. 1 Two deterministic LTSs

4 Compositionality

In the two-valued case, compositionality theorems is used to analyze large systems by decomposing them into smaller components. For example, if LTS $\mathcal{A}_1$ contains LTS $\mathcal{B}_1$ and $\mathcal{A}_2$ contains $\mathcal{B}_2$ with respect to language, we have that the composition of $\mathcal{A}_1$ and $\mathcal{A}_2$ contains the composition of $\mathcal{B}_1$ and $\mathcal{B}_2$. We show that our lattice-valued language containment relation has the corresponding results for approximate synchronous composition operator.

Recall that the synchronous composition operator requires that the systems are strictly synchronous, whereas the asynchronous composition operator requires that the systems act independently. Motivated by [11], for every $\delta \in L$, we can consider the δ-approximate synchronous operator, where Σ is equipped with an $\mathcal{L}$-equality relation.

Definition 6 Let $\mathcal{Q}_i = ((S_i, s_{i,0}, \Sigma, R_i), \theta)$, $i = 1, 2$, be two QTSs, and $\delta \in L$. The δ-approximate synchronization operator $\|_\delta$ acting on the two systems results in another transition system $\mathcal{Q} = \mathcal{Q}_1 \|_\delta \mathcal{Q}_2$, where $\mathcal{Q} = ((S_1 \times S_2, (s_{1,0}, s_{2,0}), \Sigma \times \Sigma, R), \theta')$; the transition relation R is such that $((s_1, s_2), (a, b), (s_1', s_2')) \in R$ iff $(s_1, a, s_1') \in R_1$, $(s_2, b, s_2') \in R_2$, $\theta(a, b) \geq \delta$, and $a, b \in \Sigma$; θ' is defined as, for each $(a_1, b_1), (a_2, b_2) \in \Sigma \times \Sigma$, $\theta'((a_1, b_1), (a_2, b_2)) = \theta(a_1, a_2) \otimes \theta(b_1, b_2)$.

The above definition is well-defined, since it is easy to prove that θ' is an $\mathcal{L}$-equality relation on $\Sigma \times \Sigma$. The idea of the approximate synchronization is to let two transition systems synchronize using actions that are close, but not necessarily the same. As the extreme cases, $\|_1$ and $\|_0$ mean synchronous and asynchronous composition operators, respectively. Approximate synchronization operator satisfies the following properties:

Proposition 1 *Suppose that $\mathcal{Q}_1 \|_\delta \mathcal{Q}_2$, $\mathcal{Q}_1' \|_\delta \mathcal{Q}_2'$ are well-defined. Let $\delta_1 \leq \delta_2$ and $\delta_3 \leq \delta_4$. Then*
(1) $\mathcal{TI}(\mathcal{Q}_1 \|_{\delta_1} \mathcal{Q}_2, \mathcal{Q}_1' \|_{\delta_4} \mathcal{Q}_2') \leq \mathcal{TI}(\mathcal{Q}_1 \|_{\delta_1} \mathcal{Q}_2, \mathcal{Q}_1' \|_{\delta_3} \mathcal{Q}_2')$;
(2) $\mathcal{TI}(\mathcal{Q}_1 \|_{\delta_1} \mathcal{Q}_2, \mathcal{Q}_1' \|_{\delta_3} \mathcal{Q}_2') \leq \mathcal{TI}(\mathcal{Q}_1 \|_{\delta_2} \mathcal{Q}_2, \mathcal{Q}_1' \|_{\delta_3} \mathcal{Q}_2')$.

To study further property of lattice-valued language containment relation with respect to approximate synchronization, we need to give some concepts on distributive lattices.

For a distributive lattice $\mathcal{L}$, an element $x \in L$ is called *join-irreducible* if for any $y, z \in L, x = y \vee z$ implies that either $x = y$ or $x = z$. In a distributive lattice, if x is join-irreducible and $x \leq y \vee z$, then it always holds that $x \leq y$ or $x \leq z$. Let $\mathcal{JI}$ denote the set of all join-irreducible elements of L. Then for any finite distributive lattice $\mathcal{L}$ and $x \in L$, $x = \vee\{y \in \mathcal{JI}(L) : y \leq x\}$. It should be pointed out that a *complete Heyting algebra* is a distributive lattice.

Proposition 2 *Let $\mathcal{L}$ be a finite Heyting algebra, $\delta \in L$, $l \in \mathcal{JI}(L)$ and $\mathcal{Q}_i$, $\mathcal{Q}'_i, i = 1, 2$, be QTSs. If $l \leq \mathcal{TI}(\mathcal{Q}_1, \mathcal{Q}_2) \wedge \mathcal{TI}(\mathcal{Q}'_1, \mathcal{Q}'_2)$. Then $l \leq \mathcal{TI}(\mathcal{Q}_1\|_\delta\mathcal{Q}'_1, \mathcal{Q}_2\|_{\delta\wedge l}\mathcal{Q}'_2)$.*

Proof By Definition 6, if $\sigma \in L(\mathcal{Q}_1\|_\delta\mathcal{Q}'_1)$, there exist $\sigma_1 \in L(\mathcal{Q}_1)$ and $\sigma_2 \in L(\mathcal{Q}_2)$ such that $\theta(\sigma_1(i), \sigma_2(i)) \geq \delta$ and $\sigma(i) = (\sigma_1(i), \sigma_2(i))$ for all $i < |\sigma|$. By the assumption that $l \leq \mathcal{TI}(\mathcal{Q}_1, \mathcal{Q}_2) \wedge \mathcal{TI}(\mathcal{Q}'_1, \mathcal{Q}'_2)$ and $l \in \mathcal{JI}(L)$, for every $\sigma_1 \in L(\mathcal{Q}_1)$ (resp. $\sigma_2 \in L(\mathcal{Q}_2)$), there exists $\sigma'_1 \in L(\mathcal{Q}_2)$ (resp. $\sigma'_2 \in L(\mathcal{Q}'_2)$) such that $l \leq \mathcal{E}_\theta(\sigma_1, \sigma'_1)$ (resp. $l \leq \mathcal{E}_\theta(\sigma_2, \sigma'_2)$). Let $\sigma' = (\sigma'_1, \sigma'_2)$ where $\sigma'(i) = (\sigma'_1(i), \sigma'_2(i))$ for all $i < |\sigma|$. Then $\sigma' \in L(\mathcal{Q}_2\|_{\delta\wedge l}\mathcal{Q}'_2)$. This assertion is proved as follows: since for every $i < |\sigma|$, $\delta \leq \theta(\sigma_1(i), \sigma_2(i))$, $l \leq \theta(\sigma_1(i), \sigma'_1(i))$, $l \leq \theta(\sigma_2(i), \sigma'_2(i))$, we have that for all $i < |\sigma|$, $\delta \wedge l \leq \theta(\sigma'_1(i), \sigma'_2(i))$. Therefore, $\sigma' \in L(\mathcal{Q}_2\|_{\delta\wedge l}\mathcal{Q}'_2)$, as required. Based on the above analysis, we know that for every $\sigma \in L(\mathcal{Q}_1\|_\delta\mathcal{Q}'_1)$, there exists a $\sigma' \in L(\mathcal{Q}_2\|_{\delta\wedge l}\mathcal{Q}'_2)$ such that $\mathcal{E}_\theta(\sigma, \sigma') = \mathcal{E}_\theta(\sigma_1, \sigma'_1) \wedge \mathcal{E}_\theta(\sigma_2, \sigma'_2) \geq l$. Hence the proposition holds. □

The following theorem shows that lattice-valued language containment relation has a compositional property with respect to approximate synchronization, which is a direct consequence of the above proposition.

Theorem 3 *Let δ, $\delta' \in L$ and $\mathcal{Q}_i$, $\mathcal{Q}'_i$, $i = 1, 2$, be QTSs and $\mathcal{L}$ be a finite Heyting algebra. Suppose $\delta' = \mathcal{TI}(\mathcal{Q}_1, \mathcal{Q}_2) \wedge \mathcal{TI}(\mathcal{Q}'_1, \mathcal{Q}'_2)$. Then $\mathcal{TI}(\mathcal{Q}_1, \mathcal{Q}_2) \wedge \mathcal{TI}(\mathcal{Q}'_1, \mathcal{Q}'_2) \leq \mathcal{TI}(\mathcal{Q}_1\|_\delta\mathcal{Q}'_1, \mathcal{Q}_2\|_{\delta\wedge\delta'}\mathcal{Q}'_2)$.*

5 Conclusion

Based on the notion of QTSs, we have introduced two lattice-valued language containment relations from different viewpoints to quantify a concept of "closeness" between the states of a QTS and showed that they coincide. The relation has compositionality under approximate synchronous operator. These properties suggest that our relation provides an appropriate basis for a quantitative theory of concurrent and distributed systems.

Acknowledgments This research is partly supported by National Natural Science Foundation of China under Grant 11401361, by China Postdoctoral Science Foundation under Grant 2014M552408, and by Natural Science promotion plan Foundation of Anhui Provincial Education Department under Grant TSKJ2016B02.

References

1. Bělohlávek, R.: Fuzzy Relational Systems: Foundations and Principles. Kluwer, New York (2002)
2. Cao, Y.: Reliability of mobile processes with noisy channels. IEEE Trans. Comput. **61**(9), 1217–1230 (2012)
3. Cao, Y., Sun, S., Wang, H., Chen, G.: A behavioral distance for fuzzy-transition systems. IEEE Trans. Fuzzy Syst. **21**(4), 735–747 (2013)
4. Cerný, P., Henzinger, T.A., Radhakrishna, A.: Simulation distances. Theor. Comput. Sci. **413**(1), 21–35 (2012)
5. Ćirić, M., Ignjatović, J., Damljanović, N., et al.: Bisimulations for fuzzy automata. Fuzzy Sets Syst. **186**, 100–139 (2012)
6. Ćirić, M., Ignjatović, J., Jancic, I., et al.: Computation of the greatest simulations and bisimulations between fuzzy automata. Fuzzy Sets Syst. **208**, 22–42 (2012)
7. De Alfaro, L., Faella, M., Stoelinga, M.: Linear and branching system metrics. IEEE Trans. Softw. Eng. **35**(2), 258–273 (2009)
8. Fahrenberg, U., Legay, A.: The quantitative linear-time-branching-time spectrum. Theor. Comput. Sci. **538**, 54–69 (2014)
9. Girard, A., Pappas, G.J.: Approximation metrics for discrete and continuous systems. IEEE Trans Autom. Control **52**(5), 782–798 (2007)
10. Henzinger, T.A.: Quantitative reactive modeling and verification. Comput. Sci. Res. Dev. **28**(4), 331–344 (2013)
11. Julius, A.A., D'Innocenzo, A., DiBenedetto, M.D., et al.: Approximate equivalence and synchronization of metric transition systems. Syst. Control Lett. **58**, 94–101 (2009)
12. Kupferman, O., Lustig, Y.: Lattice automata. In: Proceedings of VWCAI2007. LNCS, vol. 4349, pp. 199–213 (2007)
13. Kupferman, O., Lustig, Y.: Latticed simulation relations and games. Int. J. Found. Comput. Sci. **21**(2), 167–189 (2010)
14. Pan, H., Cao, Y., Zhang, M., Chen, Y.: Simulation for lattice-valued doubly labeled transition systems. Int. J. Appox. Reason. **55**, 797–811 (2014)
15. Pan, H., Zhang, M., Wu, H., Chen, Y.: Quantitative analysis of lattice-valued Kripke structures. Fundam. Inform. **135**(3), 269–293 (2014)
16. Pan, H., Li, Y., Cao, Y.: Lattice-valued simulations for quantitative transition systems. Int. J. Approx. Reason. **56**, 28–42 (2015)
17. Qiu, D.: Automata theory based on completed residuated lattice-valued logic (I) and (II). Sci. China Ser. F **44**, 419–429 (2001); **45**, 442–452 (2002)
18. Thrane, C.R., Fahrenberg, U., Larsen, K.G.: Quantitative analysis of weighted transition systems. J. Log. Algebr. Program. **79**(7), 689–703 (2010)
19. Van Glabbeek, R.J.: The linear time-branching time spectrum. I: The semantics of concrete, sequential processes. In: Bergstra, J.A., Ponse, A., Smolka, S.A. (eds.) Handbook of Process Algebra, pp. 3–99. Elsevier, Amsterdam (2001)
20. Zhang, J., Zhu, Z.: A behavioural pseudometric based on λ-bisimilarity. Electron. Notes Theor. Comput. Sci. **220**, 115–127 (2008)
21. Zhang, J., Zhu, Z.: Characterize branching distance in terms of $(\eta,\ \alpha)$-bisimilarity. Inf. Comput. **206**, 953–965 (2008)

An Approach of Matching Based on Performance Specification of Components

Bao-Hua Wang and Yi-Xiang Chen

Abstract Software component technology is important to construct software system. How to seek out required component efficiently and accurately is a challenge issue for component-based software development. The goal of this paper is to propose matching method, called specification-based performance matching. At first, performance specification is formally defined. And then, based on performance specification, we give two kinds of matchings: the Boolean matching and quantitative matching. Finally, properties of specification-based performance matching are presented and some examples are given to illustrate effectiveness of the machining method we propose.

Keywords Software components · Performance specification · Quantitative · Matching

1 Introduction

Component-based software development (CBD) [1–3] is the mainstream technology of software development. Software components are assembled into a composite component or combined into a complex software system so that one can reduces the development cost. The concept of software component has had several definitions up to date. CMU/SEI defines software component as an opaque implementation of functionality, subject to third-party composition and conformation with a component model in [4]. Szyperski defines in [2] software component as a unit of composition

B.-H. Wang
School of Computer Science and Software Engineering, East China Normal University, Shanghai, China
e-mail: wbh1028@163.com

Y.-X. Chen (✉)
MOE Engineering Research Center for Software Hardware, Co-Design Engineering and Its Application, Shanghai, China
e-mail: yxchen@sei.ecnu.edu.cn

T.-H. Fan et al. (eds.), *Quantitative Logic and Soft Computing 2016*,
Advances in Intelligent Systems and Computing 510,
DOI 10.1007/978-3-319-46206-6_18

with contractually specified interfaces and explicit context dependencies only. Reference [5] defines a software component as unit which can be deployed independently subject to composition by third parties. References [6, 7] adopt Szyperski's definition. In this paper, we also adopt Szyperski's definition.

It is well known that how to seek out required components efficiently and accurately for component is a basic issue. Specifying software component is one method of exploring this issue. The specification of software component can help users to understand and use software component. Specification-based component matching includes syntax and semantic matching [8]. Syntax matching is based on information of the functions and interfaces [9, 10]. Semantics refers to constraints and behaviors information of software component interaction [11].

In this paper, we introduce an approach of matching of software components based on performance specification of software components. Performance is made of facets of performance. The specification of a performance (called performance specification) is the set of specifications of performance's facets. Specifications of performance's facet consists of four parts: name, checkpoint, flag and value. Then, we introduce the refinement relationship between performance's facet based specification of performance's facet. After then, we introduce the matching between performances based on the refinement relationship of performances and further set up two kinds of matchings between requisite components and candidate components: Boolean matching and quantitative matching.

The remainder of this paper is organized as follows. Section 2 presents the software performance and the formal definition of performance specification. Section 3 introduces the Boolean matching of performance. Based on the Boolean matching of performance, we define the Boolean matching between requisite components and candidate components. Section 4 introduces the quantitative matching between requisite component and candidate component. Section 5 is the conclusion section.

2 Performance Specification and Refinement

Matching is a way to retrieve software component or compare two software components. Specification is a foundation of matching technology of software components and plays a crucial role in the method and efficiency of software component matching. Performance matching is very key to the success of software development. The matching of performance specification helps us to select the appropriate software component in performance. In this paragraph, we define the performance specification as follows.

Definition 1 (*Specification of performance's facet*) The specification of a performance's facet consists of four parts: the name of performance's facet, the checkpoint of performance's facet, the flag of performance's facet, the value of performance's facet. It is denoted as a quadruple facet = (name, checkpoint, flag, value).

The flag of a performance's facet takes the value 1 and -1. The flag value of a performance's facet represents the relation between this performance's facet and its value. The value of a performance's facet takes real numbers from the interval $(0, \infty)$. When facet.flag $= 1$, then the bigger the value of a performance's facet is, the better this performance's facet is. But, when facet.flag $= -1$, then the smaller the value of performance's facet is, the better this performance's facet is.

Definition 2 (*Specification of performance*) We use per $= \{facet_1, \ldots, facet_n\}$ to represent specification of performance, where $facet_i$ is a facet of performance for each i.

Example 1 Suppose that a performance per_1 is about the throughout and the response time. Its description is that the throughout of the payment function is no less than 20 users per seconds when 1000 users use the payment function and the response time of the payment function is no more than 6 seconds when 1000 users use the payment function. We use $facet_1$ = ("pay-throughout", "1000users", 1, 20) to represent the throughout and $facet_2$ = ("pay-response", "1000users", −1, 6) to represent the response time. Then, the specification of the performance is per_1 = $\{facet_1, facet_2\}$ = {("pay-throughout", "1000users", 1, 20), ("pay-response", "1000users", −1, 6)}.

2.1 Refinement Relation

Definition 3 (*Refinement relation*) For two given performance's facets $facet_1$ and $facet_2$, we say that $facet_1$ refines $facet_2$, denoted by $facet_1 \succ facet_2$, if $(facet_1.name = facet_2.name) \wedge (facet_1.checkpoint = facet_2.checkpoint) \wedge (facet_1.flag = facet_2.flag) \wedge (((facet_1.flag = 1) \wedge (facet_1.value \geq facet_2.value)) \vee ((facet_1.flag = -1) \wedge (facet_1.value \leq facet_2.value)))$.

Example 2 If the performance's facet $facet_1$ is ("pay-throughout", "1000users", 1, 20) and another performance's facet $facet_2$ is ("pay-throughout", "1000users", 1, 18), then we have $facet_1 \succ facet_2$. If a performance's facet $facet_3$ is ("pay-response", "1000users", −1, 5)}) and another performance's facet $facet_4$ is ("pay-response", "1000users", −1, 6)}), then we have $facet_3 \succ facet_4$.

The refinement relation has the following desirable properties.

Proposition 1 (Reflexivity) *The refinement relation $\succ$ satisfies reflexivity.*

Proof Obvious. □

Proposition 2 (Antisymmetry) *If $facet_i \succ facet_j$ and $facet_j \succ facet_i$, then $facet_i = facet_j$.*

Proof Since $(facet_i \succ facet_j)$, we have that $facet_i.name = facet_j.name$, $facet_i.checkpoint = facet_j.checkpoint$ and $facet_i.flag = facet_j.flag$.

We discuss the two cases of flag.

Case 1: ($facet_i.flag = 1$)

Since $facet_i \succ facet_j$, we have $facet_i.value \geq facet_j.value$. Since $facet_j \succ facet_i$, we have $facet_j.value \geq facet_i.value$.

As a result, we get $facet_i.value = facet_j.value$.

Case 2: ($facet_i.flag = -1$)

Similarly, we have $facet_i.value = facet_j.value$.

Therefore, we have shown that ($facet_i.name = facet_j.name$)$\wedge$ ($facet_i.checkpoint = facet_j.checkpoint$) $\wedge$ ($facet_i.flag = facet_j.flag$)$\wedge$ ($facet_j.value = facet_i.value$), then we get ($facet_i = facet_j$). □

Proposition 3 (Transitivity) *If* $facet_i \succ facet_j$ *and* $facet_j \succ facet_k$, *then* $facet_i \succ facet_k$.

Proof Since ($facet_i \succ facet_j$), we have that $facet_i.name = facet_j.name$, $facet_i.checkpoint = facet_j.checkpoint$ and $facet_i.flag = facet_j.flag$. Since ($facet_j \succ facet_k$), we have $facet_j.name = facet_k.name$, $facet_j.checkpoint = facet_k.checkpoint$ and $facet_j.flag = facet_k.flag$. As a result, we have $facet_i.name = facet_k.name$, $facet_i.checkpoint = facet_k.checkpoint$ and $facet_i.flag = facet_k.flag$.

We discuss the cases of different flags: Case 1: ($facet_i.flag = 1$) and Case 2: ($facet_i.flag = -1$). Their proofs are the same as the previous proposition.

So, we have ($facet_i.name = facet_k.name$) $\wedge$ ($facet_i.checkpoint = facet_k.checkpoint$) $\wedge$ ($facet_i.flag = facet_k.flag$) $\wedge$ ((($facet_i.flag = 1$)$\wedge$ ($facet_i.value \geq facet_k.value$)) $\vee$ (($facet_i.flag = -1$) $\wedge$ ($facet_i.value \leq facet_k.value$))). Therefore ($facet_i \succ facet_k$). □

From the propositions, we get the fact that the refinement relation is a partial order relation.

3 Boolean Matching of Software Components

Matching is to determine the relationship between the candidate component and the requisite component. Based on performance specification of software component, we introduce the Boolean matching of software component.

3.1 The Matching Framework of the Performance Specification

The matching framework of the performance specification is shown in Fig. 1.

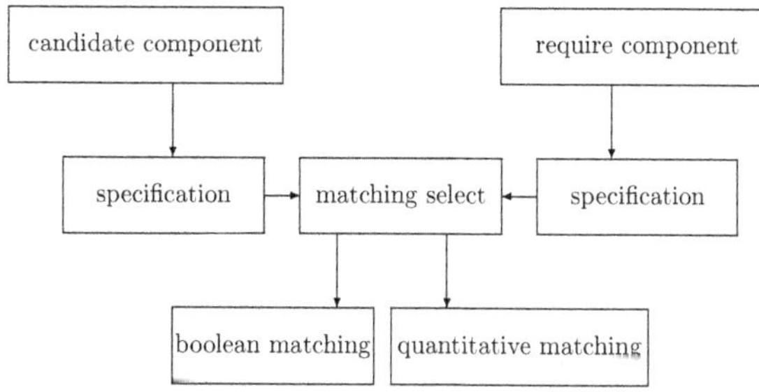

Fig. 1 The matching framework of the performance specification

3.2 *Boolean Matching of Software Components*

Definition 4 (*Boolean matching*) For a given requisite component R and a candidate component C, we say that the candidate component C matches the requisite component R in a Boolean way, denoted by $C \models R$, if $\forall x \exists y((x \in per_r \bigwedge y \in per_c) \rightarrow (y \succ x))$, where per_r is the performance specification of R and per_c is the performance specification of C.

Example 3 For given a requisite component R_1 and a candidate component C_1, if the performance specification of R_1 is $\{(\text{"}pay - response\text{"}, \text{"}1000users\text{"}, -1, 6)\}$ and the performance specification of C_1 is $\{(\text{"}pay - response\text{"}, \text{"}1000users\text{"}, -1, 5)\}$, then $C_1 \models R_1$. For another example, for a given requisite component R_2 and a candidate component C_2, the performance specification of R_2 is represented as {("pay-response", "1000users", $-1, 6$), ("$pay - throughout$", "1000users", 1, 20)} and the performance specification of C_2 is $\{(\text{"}pay - response\text{"}, \text{"}1000users\text{"}, -1, 5), (\text{"}pay - throughout\text{"}, \text{"}1000users\text{"}, 1, 15)\}$, then $C_2 \models R_2$ is false.

3.3 *Examples of the Boolean Matching*

Examples of the Boolean matching are shown in Table 1.

4 Quantitative Matching of Components

When we select software component by considering performance specification, it is a rare case that a candidate component matches completely a requisite component under all performance's facets. In this paragraph, we introduce the quantitative matching method which helps us to select the optimal software component.

Table 1 Boolean matching of components

Performance specification of requisite component	Performance specification of candidate component	Boolean matching
{("pay-response", "1000users", −1, 6)}	{("pay-response", "1000users", −1, 5)}	*TRUE*
	{("pay-response", "1000users", −1, 8)}	*FLASE*
	{("order-response","1000users", −1, 4)}	*FLASE*
{("pay-response", "1000users", −1, 6), ("pay-throughout", "1000users", 1, 20)}	{("pay-response", "1000users', −1, 7)}	*FLASE*
	{("pay-response", "1000users", −1, 6)} ("pay-throughout","1000users", 1, 30)}	*TRUE*
	{ ("pay-response", "1000users", −1, 6)} {("pay-throughout", "1000users", 1, 15)}	*FLASE*

4.1 Quantitative Refinement of Performance's Facet

Definition 5 (*Quantitative matching of performance's facet*) For two given performance's facets $facet_1$ and $facet_2$, we say that $facet_1$ refines $facet_2$ quantitatively, donated by $facet_1 \succ_\delta facet_2$), if $facet_1$ and $facet_2$ has the same name, the same checkpoint and the same flag, and δ is calculated by $\delta = m(facet_1, facet_2)$ as follows:

When $facet_1.flag = 1$,

$$m(facet_1, facet_2) = \begin{cases} \left(\frac{facet_1.value}{facet_2.value}\right)^{\frac{1}{2}}, & facet_1.value < facet_2.value \\ 1, & facet_1.value \geq facet_2.value \end{cases}$$

When $facet_1.flag = -1$, $m(facet_1, facet_2)$

$$= \begin{cases} 1, & facet_1.value \leq facet_2.value \\ \left(\frac{2*facet_2.value - facet_1.value}{facet_2.value}\right)^{\frac{1}{2}}, & facet_2.value < facet_1.value \\ & < 2*facet_2.value \\ 0, & facet_1.value \geq 2*facet_2.value \end{cases}$$

Inspired by the research results of multi-dimensional software trustworthiness [12], we get the following properties.

Proposition 4 (Boundness) *It is used to describe that* $m(facet_1, facet_2)$ *is in a certain range.*

Proof Since $m(facet_1, facet_2)$ is different for the flag, we discuss alternative cases as follows:

Case 1: $(facet_1.flag = 1)$

When $facet_1.value < facet_2.value$, we have $0 < (\frac{facet_1.value}{facet_2.value})^{\frac{1}{2}} < 1$. When $facet_1.value \geq facet_2.value$ we have $m(facet_1, facet_2) = 1$. As a result, we get $0 < m(facet_1, facet_2) \leq 1$.

Case 2: $(facet_1.flag = -1)$

When $facet_1.value \leq facet_2.value$, we have $m(facet_1, facet_2) = 1$. When $facet_2.value \leq facet_1.value < 2 * facet_2.value \Rightarrow 0 < \frac{2*facet_2.value - facet_1.value}{facet_2.value} \leq 1$, we have $0 < (\frac{2*facet_2.value - facet_1.value}{facet_2.value})^{\frac{1}{2}} \leq 1$. When $facet_1.value \geq 2 * facet_2.value$, we have $m(facet_1, facet_2) = 0$. As a result, we get $0 \leq m(facet_1, facet_2) \leq 1$.

In a word, $m(facet_1, facet_2)$ is a bounded function. □

Proposition 5 (Consistency) *Consistency means that if* $facet_1 \succ facet_3$ *then* $m(facet_1, facet_2) \geq m(facet_3, facet_2)$.

Proof We prove this result in the flags' cases.

Case 1: $(facet_1.flag = 1)$

Since $\frac{\partial(m(facet_1, facet_2))}{\partial(facet_1.value)} = \frac{1}{2} * (\frac{facet_1.value}{facet_2.value})^{-\frac{1}{2}} \geq 0$, $m(facet_1 facet_2)$ is an increasing function for $facet_1$. Since $facet_1 \succ facet_3$ implies $facet_1.value \geq facet_3.value$, we have $m(facet_1, facet_2) \geq m(facet_3, facet_2)$. As a result, if $facet_1 \succ facet_3$ then $m(facet_1, facet_2) \geq m(facet_3, facet_2)$.

Case 2: $(facet_1.flag = -1)$

Since $\frac{\partial(m(facet_1, facet_2))}{\partial(facet_1.value)} \leq 0$, $m(facet_1, facet_2)$ is a decreasing function for $facet_1$. Since $facet_1 \succ facet_3$ implies $facet_1.value \leq facet_3.value$, we have $m(facet_1, facet_2) \geq m(facet_3, facet_2)$. Then, $facet_1 \succ facet_3$ implies $m(facet_1, facet_2) > m(facet_3, facet_2)$. □

Proposition 6 (Approximation) *Approximation means that if* $facet_1$ *is closing in a refinement way to* $facet_2$*, then* $m(facet_1, facet_2)$ *is increasing.*

Proof We shall prove it in the cases of the flag of $facet_1$.

Case 1: $(facet_1.flag = 1)$

In this case, we have $0 \leq facet_1.value \leq facet_2.value$. Set $x = facet_2.value - facet_1.value$, i.e., x denotes the distance from $facet_1.value$ to $facet_2.value$, then

$$m(facet_1, facet_2) = \left(\frac{facet_2.value - x}{facet_2.value}\right)^{\frac{1}{2}}.$$

Since $0 \leq facet_1.value \leq facet_2.value$, we have

$$\frac{\partial(m(facet_1, facet_2))}{\partial(x)} = \left(-\frac{1}{2 * facet_2.value}\right) * \left(\frac{facet_2.value - x}{facet_2.value}\right)^{-\frac{1}{2}} \leq 0.$$

Case 2: $(facet_1.flag = -1)$

This case implies that $facet_2.value \le facet_1.value < 2 * facet_2.value$. We set $x = facet_1.value - facet_2.value$, i.e., x denotes the distance from $facet_2.value$ to $facet_1.value$. Then $m(facet_1, facet_2) = (\frac{facet_2.value - x}{facet_2.value})^{\frac{1}{2}}$.

Since $facet_2.value \le facet_1.value < 2 * facet_2.value$, we have

$$\frac{\partial(m(facet_1, facet_2))}{\partial(x)} = \left(-\frac{1}{2 * facet_2.value}\right) * \left(\frac{facet_2.value - x}{facet_2.value}\right)^{-\frac{1}{2}} \le 0.$$

Both cases show that $m(facet_1, facet_2)$ is a decreasing function of x. Thus, we have that the degree of the matching increases when x decreases of x. Therefore, when $facet_1.value$ is closing to $facet_2.value$ in a refinement way, the degree of matching increases. □

Proposition 7 (Acceleration) *Acceleration means that if* $facet_1$ *is closing to* $facet_2$ *in a refinement way, the increasing rate of* $m(facet_1, facet_2)$ *decreases.*

Proof For different flags, separately discuss all cases separately.

Case 1: $(facet_1.flag = 1)$

Similar to Claim 4, we have

$$\frac{\partial^2(m(facet_1, facet_2))}{\partial(x)^2} = \frac{1}{4 * facet_2.value} * \left(\frac{facet_2.value - x}{facet_2.value}\right)^{-\frac{3}{2}} > 0$$

where $x = facet_2.value - facet_1.value$.

As a result, $m(facet_1, facet_2)$ is a decreasing function in the range $(facet_2.value, 2 * facet_2.value)$, and the rate of increasing is decreasing when $facet_1.value$ changes from $2 * facet_2.value$ to $facet_2.value$. The decreasing of x is the fact that $facet_1.value$ is closing to $facet_2.value$. In other word, $m(facet_1, facet_2)$ is increasing in the approximation of $facet_2.value$.

Case 2: $(facet_1.flag = -1)$

Similarly, we have that

$$\frac{\partial^2(m(facet_1, facet_2))}{\partial(x)^2} = \frac{1}{4 * facet_2.value} * \left(\frac{facet_2.value - x}{facet_2.value}\right)^{-\frac{3}{2}} > 0.$$

where $x = facet_1.value - facet_2.value$.

We can get that the increasing rate of $m(facet_1, facet_2)$ decreases, when $facet_1.value$ is closing to $facet_2.value$ in a refinement way. □

4.2 Quantitative Matching of Components

Definition 6 (*Quantitative matching of components*) Suppose that a requisite component R and a candidate component C have the same number of performance's facets. We say that the candidate component C matches the requisite component R in a quantitative matching way, denoted by $C \models_{\delta} R$, where $\delta = \delta_1 * \delta_2 * \cdots * \delta_n$ and for each i, $per_c.facet_i \succ_{\delta_i} per_r.facet_i$.

Example 4 We have a requisite component R and a candidate component C. The performance specification of R is $\{(\text{“}pay - response\text{”}, \text{“}1000users\text{”}, -1, 6)\}$ and the performance specification of C is $\{(\text{“}pay - response\text{”}, \text{“}1000users\text{”}, -1, 5)\}$. We have that $C \models_{\delta} R$ with $\delta = 1$.

For another example,we are given a requisite component R_1 and a candidate component C_1. The performance specification of R_1 is $\{(\text{“}pay - response\text{”}, \text{“}1000users\text{”}, -1, 6), (\text{“}pay - throughout\text{”}, \text{“}1000users\text{”}, 1, 20)\}$ and the performance specification of C_1 is $\{(\text{“}pay - response\text{”}, \text{“}1000users\text{”}, -1, 5), (\text{“}pay - throughout\text{”}, \text{“}1000users\text{”}, 1, 15)\}$. Therefore, after computing δ, we get that $C_1 \models_{\delta} R_1$ with $\delta = \delta_1 \times \delta_2 = 1 \times (\frac{15}{20})^{\frac{1}{2}} = 0.87$.

Example 5 Here are examples of quantitative matching of software components (Table 2).

Table 2 Ranks of Matching of Components

Requisite Components	Performance Specification of Candidate Components	Quantitative Matching
{(“pay-response”, “1000users”, −1, 6)}	{(“pay-response”, “1000users”, −1, 5)}	1
	{(“pay-response”, “1000users”, −1, 9)}	0.71
	{(“order-response”, “1000users”, −1, 4)}	0
{(“pay-response”,“1000users”,−1, 6), (“pay-throughout”, “1000users”, 1, 20)}	{(“pay-response”, “1000users’, −1, 7)}	0
	{(“pay-response”, “1000users”, −1, 6)} (“pay-throughout”, “1000users”, 1, 30)}	1
	{(“pay-response”, “1000users”, −1, 6)} { (“pay-throughout”, “1000users”, 1, 15)}	0.87

5 Conclusions

Software component matching and assembly are an important activity in component-based software development. This paper presents a kind of matching between requisite components and candidate components based on performance specification of software components. Two kinds of matchings: the Boolean matching and quantitative matching are introduced in this paper. Some simple examples show the effectiveness of our matching method. Practice use of the method will be our future work.

Acknowledgments This work is supported by the Innovation Group Project of the National Natural Science Foundation (Grant No. 61321064), the National Natural Science Foundation of China (Grant No. 61370100), the National Natural Science Foundation of Anhui Province (Grant No. KJ2012B160), and Shanghai Knowledge Service Platform for Trustworthy Internet of Things (Grant No. ZF1213) as well as Shanghai Municipal Science and Technology Commission Project (No.14511100400).

References

1. McIlroy, M.D.: Mass-produced software components. Software Engineering Concepts and Techniques, pp. 88–98 (1968)
2. Szyperski, C.: Component technology what, where and how. In: Proceedings of the 25th International Conference on Software Engineering, pp. 684–693. IEEE Computer Society Press, Washington (2003)
3. Sametinger, J.: Software Engineering with Reusable Components. Springer, Berlin (1997)
4. Bachman, F., et al.: Technical concepts of component-based software engineering. In: CMU/SEI-2000-TR-008, pp. 1–36 (2000)
5. Szyperski, C.: Component Softwarebeyond Object-oriented Programming, 2nd edn, pp. 1–22. Addison Wesley Professional, New York (2003)
6. Bertrand, M.: The grand challenge of trusted components. In: The International Conference on Software Engineering (ICSE2003), pp. 660–667. IEEE Computer Press, Portland (2003)
7. Abdellatief, M., Sulta, A.B.M., Ghani, A.A.A., et al.: Component-based software system dependency metrics based on component information flow measurements. In: The Sixth International Conference on Software Engineering Advances ICSEA2011, Barcelona, Spain, pp. 76–83 (2011)
8. Jeng, J.J., Cheng, B.H.C.: Specification matching for software reuse: a foundation. In: SSR Proceedings of the ACM SIGSOFT Symposium on Software Re-usability, pp. 28–30, 131–138 (1995)
9. Zaremski, A.M., Wing, J.M.: Signature matching: a tool for using software libraries. ACM Trans. Softw. Eng. Methodol. **4**(2), 146–170 (1995)
10. Zaremski, A.M., Wing, J.M.: Specification matching of software components. ACM Trans. Softw. Eng. Methodol. **6**(4), 333–369 (1997)
11. Hameurlain, N.: On compatibility and behavioral substitutability of component protocols. Softw. Eng. Form. Methods **30**, 394–403 (2005)
12. Tao, H.W., Chen, Y.X.: A new metric model for trustworthiness of software. Telecommun. Syst. **51**(2–3), 95–105 (2011)

Bisimulation Relations for Weighted Automata over Valuation Monoids

Chao Yang and Yong-Ming Li

Abstract In this paper, we firstly give a definition of bisimulation relations for weighted automata over valuation monoids and prove that weighted automata A and B are equivalent with behavior under a bisimulation relation. Then we put forward the notion of surjective functional bisimulation relations and construct the related equivalence relations on A. We also give the properties of these related equivalence relations and define bisimulations relations for A with these properties. Finally, we prove the existence of the greatest bisimulation relation and give the method of constructing the greatest bisimulation relation.

Keywords Valuation monoid · Weighted automata · Bisimulation relation · The greatest bisimulation relation

1 Introduction

Bisimulation relations are very powerful tools that have been used in computer science to solve practical problems. Park and Milner [1, 2] introduced bisimulation relations and pointed out that they can be used in discrete event systems like process algebras, Petri nets or automata models. In automata theory, bisimulation relations can be used to reduce the states of an automaton by combining equivalent states to generate an aggregated automaton with an equivalent behavior but with fewer states [3–6].

People have investigated bisimulation relations in the context of deterministic, nondeterministic, fuzzy, weighted and other kinds of automata. One type of bisimulation relations for weighted automata has been introduced by Ésik and Kuich [7].

C. Yang (✉) · Y.-M. Li
College of Mathematics and Information Science, Shaanxi Normal University, Xi'an 710119, China
e-mail: yangch12@snnu.edu.cn

Y.-M. Li
e-mail: liyongm@snnu.edu.cn

T.-H. Fan et al. (eds.), *Quantitative Logic and Soft Computing 2016*, Advances in Intelligent Systems and Computing 510,
DOI 10.1007/978-3-319-46206-6_19

A universal definition of bisimulation relations introduced by Buchholz [8] can be applied to a wide class of automata models. A new approach to bisimulation relations has been proposed in [9–11] in the framework of fuzzy automata and it has been applied to ordinary nondeterministic automata in [12]. In 2011, M. Droste and I. Meinecke [13] firstly put forward the notion of valuation monoid and weighted automata with weights in a valuation monoid. In semirings, we define bisimulation relations by relational matrixs which is constructed with the identity element. Since there is no identity element with multiplication in valuation monoids, the definition of bisimulation relations by relational matrixs for weighted automata over semirings is not suitable for that over valuation monoids. We also find that the method of constructing the greatest surjective functional bisimulation relation with matrixs for weighted automata over semirings is also not suitable for weighted automata over valuation monoids.

The structure of the paper is as follows. In Sect. 2 we introduce basic notions related to valuation monoids and weighted autonata over them. Then in Sect. 3 we give the definition of bisimulation relations for weighted automata over valuation monoids and prove that weighted automata A and B are equivalent with behavior under a bisimulation relation. Then we put forward the notion of surjective functional bisimulation relations and construct the related equivalence relations for A. We also give properties of these related equivalence relations and define bisimulations relations for A with these properties. Section 4 contains proofs of existence of the greatest bisimulation relation and construction of the greatest bisimulation relation.

2 Preliminaries

Throughout this paper, S denotes the non-empty set, $\sum^+$ and $\sum^*$ denote respectively the free semigroup and the free monoid over an alphabet $\sum$, and ε denotes the identity element in $\sum^*$.

A valuation monoid $(S, +, val, 0)$ consists of a commutative monoid $(S, +, 0)$ and a valuation function $val : S^+ \to S$ with $val(s) = s$ and $val(s_1, \ldots, s_n) = 0$ whenever $s_i = 0$ for some $i \in \{1, 2, \ldots, n\}$.

Note that $(S, +, val, (\cdot_{m,n}|m, n \in N), 0)$ is a Cauchy valuation monoid if $(S, +, val, 0)$ is a valuation monoid and $\cdot_{m,n} : S \times S \to S$ with $m, n \in N$ is a family of products such that for all $s, s_i, s'_j \in S$ and all finite subsets $A, B \subseteq_{fin} S$:

$$0 \cdot_{m,n} s = s \cdot_{m,n} 0 = 0, \tag{1}$$

$$val(s_1, \ldots, s_m, s_1{}', \ldots, s_n{}') = val(s_1, \ldots, s_m) \cdot_{m,n} val(s_1{}', \ldots, s_n{}'), \tag{2}$$

$$\left(\sum\{s|s \in A\}\right) \cdot_{m,n} \left(\sum\{s'|s' \in B\}\right) = \sum\{s \cdot_{m,n} s'|s \in A,\ s' \in B\}. \tag{3}$$

A weighted automaton $A = (Q, \sum, \delta, I, F)$ over a Cauchy valuation monoid $(S, +, val, (\cdot_{m,n}|m, n \in N), 0)$ consists of a finite state set Q, an alphabet $\sum$, a set $I \subseteq Q$ of initial states, a set $F \subseteq Q$ of final states and a weight function $\delta : Q \times \sum \times Q \to S$. For all $w = a_1a_2 \cdots a_n \in \sum^+$, $T = (t_i)_{1 \le i \le n}$ are defined as finite sequences of matching transitions $t_i = (q_{i-1}, a_i, q_i)$ where $i = 1, 2, \ldots, n$. Moreover, $\delta(T) = (\delta(t_i))_{1 \le i \le n}$ is the sequence of the transition weights of T and $wgt(T) = val(\delta(T))$ is the weight of T. A run is successful if it starts in an initial state from I and ends in a final state from F. We denote the set of successful runs of A by $succ(A)$. The behavior of A is the function $||A|| : \sum^+ \to S$, defined by $(||A||, w) = \sum\{val(\delta(T))|T \in succ(A)\}$, for all $w \in \sum^+$, if there is no successful run on w, $(||A||, w) = 0$.

The above definition of weighted automata over a Cauchy valuation monoid $(S, +, val, (\cdot_{m,n}|m, n \in N), 0)$ does not consider the initial or final weights. In this paper, we define weighted automata over a Cauchy valuation monoid with initial and final weights.

Definition 1 Let $A = (Q, \sum, \delta, I, F)$ be a weighted automaton over a Cauchy valuation monoid $(S, +, val, (\cdot_{m,n}|m, n \in N), 0)$ where Q is a finite state set, $\sum$ is an alphabet, $\delta : Q \times \sum \times Q \to S$ is a weight function, $I : Q \to S$ and $F : Q \to S$ are initial weight function and final weight function respectively.

The behavior of A is the function $||A|| : \sum^* \to S$, defined by $(||A||, \varepsilon) = \sum_{q_0 \in Q} val(I(q_0), F(q_0))$, for all $w = a_1a_2 \cdots a_n \in \sum^+$, $(||A||, w) = \sum_{q_0,\ldots,q_n \in Q} val(I(q_0), \delta(q_0, a_1, q_1), \ldots, \delta(q_{n-1}, a_n, q_n), F(q_n))$.

Definition 2 Let $A = (Q, \sum, \delta, I, F)$ and $B = (P, \sum, \eta, J, G)$ be weighted automata over a Cauchy valuation monoid $(S, +, val, (\cdot_{m,n}|m, n \in N), 0)$. There is a function $\varphi : Q \to P$ between A and B if it is bijective and for all $q, q' \in Q$, $a \in \sum$, the following holds:

$$J(\varphi(q)) = I(q), \tag{4}$$

$$\eta(\varphi(q), a, \varphi(q')) = \delta(q, a, q'), \tag{5}$$

$$G(\varphi(q)) = F(q). \tag{6}$$

Then we say that A and B are isomorphic.

3 Bisimulation Relations for Weighted Automata

Definition 3 Let $A = (Q, \sum, \delta, I, F)$ and $B = (P, \sum, \eta, J, G)$ be weighted automata over a Cauchy valuation monoid $(S, +, val, (\cdot_{m,n}|m, n \in N), 0)$. $R \subseteq Q \times P$ is a relation from the state Q of A to the state P of B. For all $q \in Q$, $p \in P$, $a \in \sum$, if

$$J(p) = \sum \{I(q')|(q', p) \in R\}, \tag{7}$$

$$\sum \{\delta(q, a, q')|(q', p) \in R\} = \sum \{\eta(p', a, p)|(q, p') \in R\}, \tag{8}$$

$$F(q) = \sum \{G(p')|(q, p') \in R\}. \tag{9}$$

Then R is called a bisimulation relation between A and B.

Lemma 1 *Let $(S, +, val, (\cdot_{m,n}|m, n \in N), 0)$ be a Cauchy valuation monoid. For all $s_1, \ldots, s_{i-1}, s_{ij}, s_{i+1}, \ldots, s_n \in S$, $j \in J$, then*

$$val\left(s_1, \ldots, \sum_{j \in J} s_{ij}, \ldots, s_n\right) = \sum_{j \in J} val(s_1, \ldots, s_{i-1}, s_{ij}, s_{i+1}, \ldots, s_n). \tag{10}$$

Proof

$$val\left(s_1, \ldots, \sum_{j \in J} s_{ij}, \ldots, s_n\right)$$

$$= val\left(s_1, \ldots, \sum_{j \in J} s_{ij}\right) \cdot_{i,n-i} val(s_{i+1}, \ldots, s_n) \left(val(s_1, \ldots, s_{i-1}) \cdot_{i-1,1} \sum_{j \in J} s_{ij}\right) \cdot_{i,n-i} \cdot$$
$$val(s_{i+1}, \ldots, s_n)$$

$$= \left(\sum_{j \in J} val(s_1, \ldots, s_{i-1}) \cdot_{i-1,1} s_{ij}\right) \cdot_{i,n-i} val(s_{i+1}, \ldots, s_n)$$

$$= \left(\sum_{j \in J} val(s_1, \ldots, s_{i-1}, s_{ij})\right) \cdot_{i,n-i} val(s_{i+1}, \ldots, s_n)$$

$$= \sum_{j \in J} val(s_1, \ldots, s_{i-1}, s_{ij}) \cdot_{i,n-i} val(s_{i+1}, \ldots, s_n)$$

$$= \sum_{j \in J} val(s_1, \ldots, s_{i-1}, s_{ij}, s_{i+1}, \ldots, s_n)$$

□

Theorem 1 *Let $A = (Q, \sum, \delta, I, F)$ and $B = (P, \sum, \eta, J, G)$ be weighted automata over a Cauchy valuation monoid $(S, +, val, (\cdot_{m,n}|m, n \in N), 0)$. $|Q| = n$, $|P| = m$. If R is a bisimulation relation between A and B, then $||A|| = ||B||$.*

Proof If $w = \varepsilon$

$$
\begin{aligned}
(||A||, \varepsilon) &= \sum_{q_0 \in Q} val(I(q_0), F(q_0)) = \sum_{q_0 \in Q} val(I(q_0), \sum\{G(p_0) | (q_0, p_0) \in R\}) \\
&= \sum_{q_0 \in Q} \sum_{(q_0, p_0) \in R} val(I(q_0), G(p_0)) = \sum_{p_0 \in P} \sum_{(q_0, p_0) \in R} val(I(q_0), G(p_0)) \\
&= \sum_{p_0 \in P} val(\sum_{(q_0, p_0) \in R} I(q_0), G(p_0)) = \sum_{p_0 \in P} val(J(p_0), G(p_0)) = (||B||, \varepsilon).
\end{aligned}
$$

$$
\begin{aligned}
&(||A||, w) \\
&= \sum_{q_0, \ldots, q_k \in Q} val(I(q_0), \delta(q_0, a_1, q_1), \ldots, \delta(q_{k-1} a_k, q_k), F(q_k)) \\
&= \sum_{q_k \in Q} \sum_{q_0, \ldots, q_{k-1} \in Q} val(I(q_0), \delta(q_0, a_1, q_1), \ldots, \delta(q_{k-1}, a_k, q_k), F(q_k)) \\
&= \sum_{q_k \in Q} \sum_{q_0, \ldots, q_{k-1} \in Q} val(I(q_0), \delta(q_0, a_1, q_1), \ldots, \delta(q_{k-1}, a_k, q_k) \sum_{(q_k, p_k) \in R} G(p_k)) \\
&= \sum_{q_k \in Q} \sum_{q_0, \ldots, q_{k-1} \in Q} \sum_{(q_k, p_k) \in R} val(I(q_0), \delta(q_0, a_1, q_1), \ldots, \delta(q_{k-1}, a_k, q_k), G(p_k)) \\
&= \sum_{q_k \in Q} \sum_{(q_k, p_k) \in R} \sum_{q_0, \ldots, q_{k-1} \in Q} val(I(q_0), \delta(q_0, a_1, q_1), \ldots, \delta(q_{k-1}, a_k, q_k), G(p_k))
\end{aligned}
$$

$$
\begin{aligned}
&= \sum_{p_k \in P} \sum_{(q_k, p_k) \in R} \sum_{q_0, \ldots \in Q} val(I(q_0), \delta(q_0, a_1, q_1), \ldots, \delta(q_{k-1}, a_k, q_k), G(p_k)) \\
&= \sum_{p_k \in P} \sum_{q_0, \ldots, q_{k-1} \in Q} \sum_{(q_k, p_k) \in R} val(I(q_0), \delta(q_0, a_1, q_1), \ldots, \delta(q_{k-1}, a_k, q_k), G(p_k)) \\
&= \sum_{p_k \in P} \sum_{q_0, \ldots, q_{k-1} \in Q} val(I(q_0), \delta(q_0, a_1, q_1), \ldots, \sum_{(q_k, p_k) \in R} \delta(q_{k-1}, a_k, q_k), G(p_k)) \\
&= \sum_{p_k \in P} \sum_{q_0, \ldots, q_{k-1} \in Q} val(I(q_0), \delta(q_0, a_1, q_1), \ldots, \sum_{(q_k, p_k) \in R} \delta(q_{k-1}, a_k, q_k)) \cdot_{k+1,1} \cdot \\
&\quad val(G(p_k)) \\
&= \sum_{p_k \in P} \left(\sum_{q_0, \ldots, q_{k-1} \in Q} val(I(q_0) \delta(q_0, a_1, q_1), \ldots, \sum_{(q_k, p_k) \in R} \delta(q_{k-1}, a_k, q_k)) \right) \cdot_{k+1,1} \cdot \\
&\quad G(p_k).
\end{aligned}
$$

Let $d(w) = \sum_{q_0, \ldots, q_{k-1} \in Q} val(I(q_0), \delta(q_0, a_1, q_1), \ldots, \sum_{(q_k, p_k) \in R} \delta(q_{k-1}, a_k, q_k))$. By induction on the length of the input string w, it follows that for all $w = a_1 a_2 \cdots a_k \in \sum^+$, $d(w) = \sum_{p_0, \ldots, p_{k-1} \in P} val(J(p_0), \eta(p_0, a_1, p_1), \ldots, \eta (p_{k-1}, a_k, p_k))$.

Hence, for all $w = a_1 a_2 \cdots a_k \in \sum^+$

$$(||A||, w) = \sum_{p_k \in P} d(w) \cdot_{k+1,1} G(p_k)$$

$$= \sum_{p_k \in P} \left(\sum_{p_0,\ldots,p_{k-1} \in P} val(J(p_0), \eta(p_0, a_1, p_1), \ldots, \eta(p_{k-1}, a_k, p_k)) \right) \cdot_{k+1,1} G(p_k)$$

$$= \sum_{p_k \in P} \sum_{p_0,\ldots,p_{k-1} \in p} val(J(p_0), \eta(p_0, a_1, p_1), \ldots, \eta(p_{k-1}, a_k, p_k)) \cdot_{k+1,1} G(p_k)$$

$$= \sum_{p_0,\ldots,p_k \in P} val(J(p_0), \eta(p_0, a_1, p_1), \ldots, \eta(p_{k-1}, a_k, p_k), G(p_k)) = (||B||, w).$$

Thus, $||A|| = ||B||$. □

Definition 4 Let $A = (Q, \sum, \delta, I, F)$ and $B = (P, \sum, \eta, J, G)$ be weighted automata over a Cauchy valuation monoid $(S, +, val, (\cdot_{m,n}|m, n \in N), 0)$. R is a bisimulation relation between A and B. If $R : Q \rightarrow P$ is a surjective function, we say that R is a surjective functional bisimulation relation between A and B and B is an aggregated automata of A.

Corollary 1 *Let* $A = (Q, \sum, \delta, I, F)$ *and* $B = (P, \sum, \eta, J, G)$ *be weighted automata over a Cauchy valuation monoid* $(S, +, val, (\cdot_{m,n}|m, n \in N), 0)$. R *is a surjective functional bisimulation relation between A and B if and only if for all* $q \in Q, p \in P, a \in \sum$, *the following holds:*

$$R : Q \rightarrow P \; is \; a \; surjective \; function, \tag{11}$$

$$J(p) = \sum \{I(q')|R(q') = p\}, \tag{12}$$

$$\eta(R(q), a, p) = \sum \{\delta(q, a, q')|R(q') = p\}, \tag{13}$$

$$F(q) = G(R(q)). \tag{14}$$

Proof Combining Definitions 1 and 4, we can see that the conclusion holds. □

If R is a surjective functional bisimulation relation between A and B, we can construct an equivalence relation $\widetilde{R}$ on Q by R where $\widetilde{R} = \{(q, q')|R(q) = R(q')\}$.

Theorem 2 *Let* $A = (Q, \sum, \delta, I, F)$ *and* $B = (P, \sum, \eta, J, G)$ *be weighted automata over a Cauchy valuation monoid* $(S, +, val, (\cdot_{m,n}|m, n \in N), 0)$. $\widetilde{R}$ *is an equivalence relation on* Q. *Then* $\widetilde{R}$ *is constructed by a surjective functional bisimulation relation between A and B if and only if the following holds:*
(a) *If* $(q, q') \in \widetilde{R}$, *for all* $q'' \in Q, a \in \sum$,

$$\sum \{\delta(q, a, q''')|(q'', q''') \in \widetilde{R}\} = \sum \{\delta(q', a, q''')|(q'', q''') \in \widetilde{R}\}, \tag{15}$$

(b) *If* $(q, q') \in \widetilde{R}$, $F(q) = F(q')$;

(c) *Constructing a weighted automaton* $A_{\tilde{R}} = (Q_{\tilde{R}}, \sum, \delta_{\tilde{R}}, I_{\tilde{R}}, F_{\tilde{R}})$ *where for all* $[q], [q''] \in Q_{\tilde{R}}, a \in \sum$,

$$Q_{\tilde{R}} = \{[q] | q \in Q\}, \tag{16}$$

$$I_{\tilde{R}}([q]) = \sum \{I(q') | (q, q') \in \tilde{R}\}, \tag{17}$$

$$\delta_{\tilde{R}}([q], a, [q'']) = \sum \{\delta(q, a, q''') | (q'', q''') \in \tilde{R}\}, \tag{18}$$

$$F_{\tilde{R}}([q]) = F(q). \tag{19}$$

Then $A_{\tilde{R}}$ *and* B *are isomorphic.*

Proof ($\Rightarrow$) (a) If $\tilde{R}$ is an equivalence relation on Q which is constructed by a surjective functional bisimulation relation R between A and B, for all $q \in Q$, there exists unique $p \in P$ such that $R(q) = p$. Then if $(q, q') \in \tilde{R}$, there exists unique $p \in P$ such that $R(q) = R(q') = p$. Therefore, for all $q'' \in Q$, $a \in \sum$, $\sum \{\delta(q, a, q''') | (q'', q''') \in \tilde{R}\} = \sum \{\delta(q, a, q''') | R(q'') = R(q''') = p''\} = \sum\{\delta(q, a, q''') | R(q''') = p''\} = \eta(R(q), a, p'') = \eta(p, a, p'')$,
$\sum \{\delta(q', a, q''') | (q'', q''') \in \tilde{R}\} = \sum \{\delta(q', a, q''') | R(q'') = R(q''') = p''\} = \sum \{\delta(q', a, q''') | R(q''') = p''\} = \eta(R(q'), a, p'') = \eta(p, a, p'')$.

Thus, if $(q, q') \in \tilde{R}$, for all $q'' \in Q, a \in \sum$,

$$\sum \{\delta(q, a, q''') | (q'', q''') \in \tilde{R}\} = \sum \{\delta(q', a, q''') | (q'', q''') \in \tilde{R}\}. \tag{20}$$

(b) Since $\tilde{R}$ is an equivalence relation on Q which is constructed by a surjective functional bisimulation relation R between A and B, for all $q \in Q$, there exists unique $p \in P$ such that $R(q) = p$. Then if $(q, q') \in \tilde{R}$, there exists unique $p \in P$ such that $R(q) = R(q') = p$. Therefore, $F(q) = G(R(q)) = G(p)$, $F(q') = G(R(q')) = G(p)$. Thus, if $(q, q') \in \tilde{R}$, $F(q) = F(q')$.

(c) Let $\varphi : Q_{\tilde{R}} \to P$(i.e.$\varphi : [q] \mapsto p$) where $R(q) = p$. Since R is a surjective functional bisimulation relation between A and B, for all $q \in Q$, there exists unique $p \in P$ such that $R(q) = p$. Then for all $[q] \in Q_{\tilde{R}}$, there exists unique $p \in P$ such that $\varphi([q]) = p$. So, φ is a map. Since R is a surjective functional bisimulation relation between A and B, for all $p \in P$, there exists $q \in Q$ such that $R(q) = p$. Then there exists $[q] \in Q_{\tilde{R}}$ such that $\varphi([q]) = p$. So, φ is a surjective map. If $\varphi([q]) = \varphi([q']) = p$, $R(q) = R(q') = p$. Hence, $(q, q') \in \tilde{R}$, and evidently, it is equivalent to $[q] = [q']$. Thus, φ is an injective map. $\varphi : Q_{\tilde{R}} \to P$ is a bijective map.

Furthermore, for all $[q], [q''] \in Q_{\tilde{R}}, a \in \sum$, $J(\varphi([q])) = J(p) = \sum\{I(q') | R(q') = p\} = \sum \{I(q') | R(q) = R(q') = p\} = \sum \{I(q') | \ (q, q') \in \tilde{R}\} = I_{\tilde{R}}([q])$, $\eta(\varphi([q]), a, \varphi([q''])) = \eta(p, a, p'') = \eta(R(q), a, p'') = \sum\{\delta(q, a, q''') | R(q''') = p''\} = \sum\{\delta(q, a, q''') | R(q'') = R(q''') = p''\} = \sum\{\delta(q, a, q''') | (q'', q''') \in \tilde{R}\} = \delta_{\tilde{R}}([q], a, [q''])$, $G(\varphi([q])) = G(p) = G(R(q)) = F(q) = F_{\tilde{R}}([q])$.

Thus, $A_{\tilde{R}}$ and$B$ are isomorphic.

($\Leftarrow$) Let $\tilde{R}$ be an equivalence relation on Q and suppose that it satisfies conditions (a) and (b). We can construct a weighted automaton $A_{\tilde{R}} = (Q_{\tilde{R}}, \sum, \delta_{\tilde{R}}, I_{\tilde{R}}, F_{\tilde{R}})$.

Since $A_{\widetilde{R}}$ and B are isomorphic, for all $[q], [q''] \in Q_{\widetilde{R}}, a \in \sum$, there exists a bijective map $\varphi : A_{\widetilde{R}} \to B$ (i.e.$\varphi : [q] \mapsto p$) which satisfies

$$J(\varphi([q])) = I_{\widetilde{R}}([q]) \tag{21}$$

$$\eta(\varphi([q]), a, \varphi([q''])) = \delta_{\widetilde{R}}([q], a, [q'']) \tag{22}$$

$$G(\varphi([q]) = F_{\widetilde{R}}([q]) \tag{23}$$

Let $R_0 = \{(q, p)|\varphi([q]) = p\}$. Now, we prove that R_0 is a surjective functional bisimulation relation between A and B.

Since $\widetilde{R}$ is an equivalence relation on Q, for all $q \in Q$, there exists unique $[q] \in Q_{\widetilde{R}}$. Since φ is a bijective map, for all $[q] \in Q_{\widetilde{R}}$, there exists unique $p \in P$ such that $\varphi([q]) = p$. Then for all $q \in Q$, there exists unique $p \in P$ such that $(q, p) \in R_0$. Since φ is a bijective map, for all $p \in P$, there exists $[q] \in Q_{\widetilde{R}}$ such that $\varphi([q]) = p$. Then there exists $q \in Q$ such that $(q, p) \in R_0$. Thus, $R_0 : Q \to P$ is a surjective map.

Furthermore, for all $p \in P$, $J(p) = J(\varphi([q])) = I_{\widetilde{R}}([q]) = \sum\{I(q')|(q, q') \in \widetilde{R}\} = \sum\{I(q')|[q'] = [q]\} = \sum\{I(q')|\varphi([q']) = \varphi([q])\} = \sum\{I(q')|\varphi([q']) = p\} = \sum\{I(q')|(q', p) \in R_0\} = \sum\{I(q')|R_0(q') = p\}$.

For all $q \in Q, p'' \in P, a \in \sum$, $\eta(R_0(q), a, p'') = \eta(\varphi([q]), a, \varphi([q''])) = \delta_{\widetilde{R}}([q], a, [q'']) = \sum\{\delta(q, a, q''')|(q'', q''') \in \widetilde{R}\} = \sum\{\delta(q, a, q''')|[q''] = [q''']\} = \sum\{\delta(q, a, q''')|\varphi([q'']) = \varphi([q'''])\} = \sum\{\delta(q, a, q''')|\varphi([q''']) = p''\} = \sum\{\delta(q, a, q''')|(q''' p'') \in R_0\} = \sum\{\delta(q, a, q''')|R_0(q''') = p''\}$.

For all $q \in Q$, $F(q) = F_{\widetilde{R}}([q]) = G(\varphi([q])) = G(p) = G(R_0(q))$.

Therefore, R_0 is a surjective functional bisimulation relation between A and B.

Now, we prove $\widetilde{R_0} = \widetilde{R}$.

Since $R_0 = \{(q, p)|\varphi([q]) = p\}$ is a surjective functional bisimulation relation between A and B, $\widetilde{R_0} = \{(q, q')|R_0(q) = R_0(q')\} = \{(q, q')|\varphi([q]) = \varphi([q'])\} = \{(q, q')|[q] = [q']\} = \{(q, q')|(q, q') \in \widetilde{R}\} = \widetilde{R}$.

Thus, $\widetilde{R}$ is an equivalence relation on Q which is constructed by a surjective functional bisimulation relation R_0 between A and B. □

According to above the theorem, we can get $||A|| = ||A_{\widetilde{R}}||$ and $|Q| \geq |Q_{\widetilde{R}}|$.

Definition 5 Let $A = (Q, \sum, \delta, I, F)$ be a weighted automaton over a Cauchy valuation monoid $(S, +, val, (\cdot_{m,n}|m, n \in N), 0)$. $\widetilde{R}$ an equivalence relation on Q. If $\widetilde{R}$ satisfies the following conditions:
(a) If $(q, q') \in \widetilde{R}$, for all $q'' \in Q, a \in \sum$,

$$\sum\{\delta(q, a, q''')|(q'', q''') \in \widetilde{R}\} = \sum\{\delta(q', a, q''')|(q'', q''') \in \widetilde{R}\}, \tag{24}$$

(b) If $(q, q') \in \widetilde{R}$, $F(q) = F(q')$.

Then we say that $\widetilde{R}$ is a bisimulation relation for A and $A_{\widetilde{R}}$ is an aggregated automaton of A.

4 Construction of the Greatest Bisimulation Relation

Before constructing the greatest bisimulation relation for A, we firstly prove the existence of the greatest bisimulation relation for A.

Theorem 3 *Let $A = (Q, \sum, \delta, I, F)$ be a weighted automaton over a Cauchy valuation monoid $(S, +, val, (\cdot_{m,n}|m, n \in N), 0)$. Then the set of the bisimulation relations for A has the greatest element.*

Proof It is clear that the equality relation is a bisimulation relation. So, the set of bisimulation relations for A is not empty. Since the set of states of A is finite, the set of bisimulation relations for A is finite. Let $\widetilde{R}_1$, $\widetilde{R}_2$ be two bisimulation relation for A. We can construct the minimum equivalence relation $\widetilde{R}$ which contains $\widetilde{R}_1$ and $\widetilde{R}_2$. Notice that $(q, q') \in \widetilde{R}$ if and only if there exists finite states $q_{i1}, q_{i2}, \ldots, q_{ik-1} \in Q$ such that $(q, q_{i1}) \in \widetilde{R}_{i1}, (q_{i1}, q_{i2}) \in \widetilde{R}_{i2}, \ldots, (q_{ik-1}, q') \in \widetilde{R}_{ik}$, where $\widetilde{R}_{i1}, \widetilde{R}_{i2}, \ldots, \widetilde{R}_{ik} \in \{\widetilde{R}_1, \widetilde{R}_2\}$.

Now, we prove that $\widetilde{R}$ is also a bisimulation relation for A.

Since $\widetilde{R}$ is the minimum equivalence relation which contains $\widetilde{R}_1$ and $\widetilde{R}_2$, for all $q \in Q$, we use $[q], [q]_1, [q]_2$ to represent the equivalence class in $\widetilde{R}$, $\widetilde{R}_1$ and $\widetilde{R}_2$ respectively. Then there exists $q_{j1}, q_{j2}, \ldots, q_{jk_1}, q_{l1}, q_{l2}, \ldots, q_{lk_2} \in Q$ such that

$$[q] = [q_{j1}]_1 \cup [q_{j2}]_1 \cup \cdots \cup [q_{jk_1}]_1 = [q_{l1}]_2 \cup [q_{l2}]_2 \cup \cdots \cup [q_{lk_2}]_2 \tag{25}$$

Hence, $(q'', q''') \in \widetilde{R}$ if and only if there exists $j \in \{j1, j2, \ldots, jk_1\}$ such that $(q''_j, q''') \in \widetilde{R}_1$, if and only if there exists $l \in \{l1, l2, \ldots, lk_2\}$ such that $(q''_l, q''') \in \widetilde{R}_2$.

If $(q, q') \in \widetilde{R}_1$, for all $q'' \in Q, a \in \sum$, $\sum\{\delta(q, a, q''')|(q'', q''') \in \widetilde{R}\} = \sum\limits_{i=1}^{k_1} \sum\{\delta(q, a, q''')|(q''_{ji}, q''') \in \widetilde{R}_1\} = \sum\limits_{i=1}^{k_1} \sum\{\delta(q', a, q''')|(q''_{ji}, q''') \in \widetilde{R}_1\} = \sum\{\delta(q', a, q''')|(q'', q''') \in \widetilde{R}\}$. If $(q, q') \in \widetilde{R}_2$, for all $q'' \in Q, a \in \sum$, $\sum\{\delta(q, a, q''')|(q'', q''') \in \widetilde{R}\} = \sum\limits_{i=1}^{k_2} \sum\{\delta(q, a, q''')|(q''_{li}, q''') \in \widetilde{R}_2\} = \sum\limits_{i=1}^{k_2} \sum\{\delta(q', a, q''')|(q''_{li}, q''') \in \widetilde{R}_2\} = \sum\{\delta(q', a, q''')|(q'', q''') \in \widetilde{R}\}$. If $(q, q') \in \widetilde{R}$, there exists finite states $q_{i1}, q_{i2}, \ldots, q_{ik-1} \in Q$ such that $(q, q_{i1}) \in \widetilde{R}_{i1}, (q_{i1}, q_{i2}) \in \widetilde{R}_{i2}, \ldots, (q_{ik-1}, q') \in \widetilde{R}_{ik}$ where $\widetilde{R}_{i1}, \widetilde{R}_{i2}, \ldots, \widetilde{R}_{ik} \in \{\widetilde{R}_1, \widetilde{R}_2\}$. Then for all $q'' \in Q, a \in \sum$, $\sum\{\delta(q, a, q''')|(q'', q''') \in \widetilde{R}\} = \sum\{\delta(q_{i1}, a, q''')|(q'', q''') \in \widetilde{R}\} = \cdots = \sum\{\delta(q_{ik-1}, a, q''')|(q'', q''') \in \widetilde{R}\} = \sum\{\delta(q', a, q''')|(q'', q''') \in \widetilde{R}\}$. Hence, if $(q, q') \in \widetilde{R}$, for all $q'' \in Q, a \in \sum$, $\sum\{\{\delta(q, a, q''')|(q'', q''') \in \widetilde{R}\} = \sum\{\delta(q', a, q''')|(q'', q''') \in \widetilde{R}\}$.

If $(q, q') \in \widetilde{R}$, there exists finite states $q_{i1}, q_{i2}, \ldots, q_{ik-1} \in Q$ such that $(q, q_{i1}) \in \widetilde{R}_{i1}, (q_{i1}, q_{i2}) \in \widetilde{R}_{i2}, \ldots, (q_{ik-1}, q') \in \widetilde{R}_{ik}$ where $\widetilde{R}_{i1}, \widetilde{R}_{i2}, \ldots, \widetilde{R}_{ik} \in \{\widetilde{R}_1, \widetilde{R}_2\}$. Then $F(q) = F(q_{i1}) = F(q_{i2}) = \cdots = F(q_{ik-1}) = F(q')$. Hence, if $(q, q') \in \widetilde{R}$, $F(q) = F(q')$.

$\widetilde{R}$ is a bisimulation relation for A. Thus, we can draw the conclusion that the set of bisimulation relations for A has the greatest element. □

Theorem 4 *Let* $A = (Q, \sum, \delta, I, F)$ *be a weighted automaton over a Cauchy valuation monoid* $(S, +, val, (\cdot_{m,n}|m, n \in N), 0)$. *We define equivalence relations on* Q *by induction as follows:*
$\widetilde{R}_0 = \{(q, q') \in Q \times Q | F(q) = F(q')\}$;
$\widetilde{R}_{i+1} = \{(q, q') \in \widetilde{R}_i | \sum\{\delta(q, a, q''')|(q'', q''') \in \widetilde{R}_i\} = \sum\{\delta(q', a, q''')|(q'', q''') \in \widetilde{R}_i\},\ for\ all\ q'' \in Q, a \in \sum\}$.
Then
(a) $\widetilde{R}_0 \supseteq \widetilde{R}_1 \supseteq \cdots \supseteq \widetilde{R}_i \supseteq \widetilde{R}_{i+1} \supseteq \cdots$;
(b) *If* $\widetilde{R}_k = \widetilde{R}_{k+1}$, *for all* $n \in N$, $\widetilde{R}_k = \widetilde{R}_{k+n}$;
(c) *There exists* $k \in N$ *such that* $\widetilde{R}_k = \widetilde{R}_{k+1}$. *Then* $\widetilde{R}_k$ *is the greatest bisimulation relation for* A.

Proof (a) According to the definition of $\widetilde{R}_i$, the conclusion holds.

(b) We prove by induction on n. For $n = 1$, it is clear that $\widetilde{R}_k = \widetilde{R}_{k+1}$. Now, we assume that if $n = m(m \geq 1)$, $\widetilde{R}_k = \widetilde{R}_{k+m}$ holds. If $n = m + 1$, $\widetilde{R}_{k+m+1} = \{(q, q') \in \widetilde{R}_{k+m} | \sum\{\delta(q, a, q''')|(q'', q''') \in \widetilde{R}_{k+m}\} = \sum\{\delta(q', a, q''')|(q'', q''') \in \widetilde{R}_{k+m}\}, for\ all\ q'' \in Q, a \in \sum\} = \{(q, q') \in \widetilde{R}_k | \sum\{\delta(q, a, q''')|(q'', q''') \in \widetilde{R}_k\} = \sum\{\delta(q', a, q''')|(q'', q''') \in \widetilde{R}_k\},\ for\ all\ q'' \in Q, a \in \sum\} = \widetilde{R}_{k+1} = \widetilde{R}_k$. Thus, if $\widetilde{R}_k = \widetilde{R}_{k+1}$, for all $n \in N$, $\widetilde{R}_k = \widetilde{R}_{k+n}$.

(c) Since the set of states of A is finite, there are finite equivalence relations on Q. Since $\widetilde{R}_0 \supseteq \widetilde{R}_1 \supseteq \widetilde{R}_2 \supseteq \cdots \supseteq \widetilde{R}_i \ldots$, there exists $k \in N$ such that $\widetilde{R}_k = \widetilde{R}_{k+1}$. If $(q, q') \in \widetilde{R}_k$, $(q, q') \in \widetilde{R}_0$. Then $F(q) = F(q')$. Since $\widetilde{R}_k = \widetilde{R}_{k+1}$, if $(q, q') \in \widetilde{R}_k$, $(q, q') \in \widetilde{R}_{k+1}$. Then for all $q'' \in Q, a \in \sum$, $\sum\{\delta(q, a, q''')|(q'', q''') \in \widetilde{R}_k\} = \sum\{\delta(q', a, q''')|(q'', q''') \in \widetilde{R}_k\}$. Thus, $\widetilde{R}_k$ is a bisimulation relation for A.

We can prove by induction on i that $\widetilde{R}_k$ is the greatest bisimulation relation for A. Assume that $\widetilde{R}$ is also a bisimulation relation for A. for $i = 0$, it is clear that $\widetilde{R} \subseteq \widetilde{R}_0$. Now, we assume that if $i = m(m \geq 0)$, $\widetilde{R} \subseteq \widetilde{R}_m$. If $i = m + 1$, for all $q \in Q$, let $[q]_m, [q]$ be the equivalence class in $\widetilde{R}_m, \widetilde{R}$ respectively, there exists $q_{j1}, q_{j2}, \ldots, q_{jk_1} \in Q$ such that $[q]_m = [q_{j1}] \cup [q_{j2}] \cup \cdots \cup [q_{jk_1}]$. Then $(q'', q''') \in \widetilde{R}_m$ if and only if there exists $j \in \{j1, j2, \ldots, jk_1\}$ such that $(q''_j, q''') \in \widetilde{R}$. Thus, if $(q, q') \in \widetilde{R}$, for all $q'' \in Q, a \in \sum$, $\sum\{\delta(q, a, q''')|(q'', q''') \in \widetilde{R}_m\} = \sum_{i=1}^{k_1} \sum\{\delta(q, a, q''')|(q''_{ji}, q''') \in \widetilde{R}\} = \sum_{i=1}^{k_1} \sum\{\delta(q', a, q''')|(q''_{ji}, q''') \in \widetilde{R}\} = \sum\{\delta(q', a, q''')|(q'', q''') \in \widetilde{R}_m\}$. Since $\widetilde{R} \subseteq \widetilde{R}_m$, if $(q, q') \in \widetilde{R}$, $(q, q') \in \widetilde{R}_m$. Then $(q, q') \in \widetilde{R}_{m+1}$. Hence, for all $i \in N$, $\widetilde{R} \subseteq \widetilde{R}_i$ especially $\widetilde{R} \subseteq \widetilde{R}_k$. Thus, $\widetilde{R}_k$ is the greatest bisimulation relation for A. □

Algorithm 1 Let $A = (Q, \sum, \delta, I, F)$ be a weighted automaton over a Cauchy valuation monoid $(S, +, val, (\cdot_{m,n}|m, n \in N), 0)$ and $Q = \{q_1, q_2, \ldots, q_n\}$.

Step 1. Let $X = \{(q_i, q_j)|F(q_i) = F(q_j)\}$.

Step 2. For all $q_j \in Q, a \in \sum$, let $Max_a(q_i, q_j) = \sum\{\delta(q_i, a, q_k)|(q_j, q_k) \in X\}$.

Step 3. For all $(q_i, q_m) \in X$, if there exists $a \in \sum$ such that $Max_a(q_i, q_j) \neq Max_a(q_m, q_j)$, then we delete (q_i, q_m) from X and get a new X.

Step 4. We execute Step 2 and Step 3 until X becomes constant.

Step 5. If X becomes constant, then $\widetilde{R}_k = X$.

5 Conclusions

In this paper, we mainly study bisimulation relations for weighted automata over valuation monoids. Bisimulation relations for weighted automaton A over valuation monoids are equivalence relations with two additional properties as shown in Definition 5 (a) and (b). If we get the greatest bisimulation relation, we can construct an aggregated automaton of A with an equivalent behavior but with fewer states.

References

1. Park, D.: Concurrency and automata on infinite sequence. In: Deussen P. (ed.) Proceedings of the 5th GI Conference, Karlsruhe, Germany. Lecture Notes in Computer Science, vol. 104, pp. 167–183. Springer, Berlin (1981)
2. Milner, R.: A calculus of communicating systems. In: Goos, G., Hartmanis J. (eds.) Lecture Notes in Computer Science. vol. 92, Springer, Berlin (1980)
3. Li, Y.M., Pedrycz, W.: Fuzzy finite automata and fuzzy regular expressions with membership values in lattice-ordered monoids. Fuzzy Sets Syst. **156**, 68–92 (2005)
4. Sun, D.D., Li, Y.M., Yang, W.W.: Bisimulation relations for fuzzy finite automata. Fuzzy Syst. Math. **23**, 92–99 (2009)
5. Li, Y.M.: Finite automata theory with membership values in lattices. Inf. Sci. **181**, 1003–1017 (2011)
6. Li, L., Li, Y.M.: On reversible fuzzy automaton. Fuzzy Syst. Math. **27**, 35–41 (2013)
7. Ésik, Z., Kuich, W.: A generalization of Kozen's axiomatization of the equational theory of the regular sets. In: Words, Semigroups and Transductions, World Scientific, River Edge, pp. 99–114 (2011)
8. Buchholz, P.: Bisimulation relations for weighted automata. Theor. Comput. Sci. **393**, 109–123 (2008)
9. Ćirić, M., Ignjatović, J., Damljanović, N., Bašić, M.: Bisimulations for fuzzy automata. Fuzzy Sets Syst. **186**, 100–139 (2012)
10. Ćirić, M., Ignjatović, J., Jančić, I., Damljanović, N.: Computation of the greatest simulations and bisimulations between fuzzy automata. Fuzzy Sets Syst. **208**, 22–42 (2012)
11. Jančić, I.: Weak bisimulations for fuzzy automata. Fuzzy Sets Syst. **249**, 49–72 (2014)
12. Ćirić, M., Ignjatović, J., Bašić, M., Jančić, I.: Nondeterministic automata: equivalence, bisimulations and uniform relations. Inf. Sci. **261**, 185–218 (2014)
13. Droste, M., Meinecke, I.: Weighted automata and regular expressions over valuation monoid. Int. J. Found. Comput. Sci. **22**, 1829–1844 (2011)

Conclusions

References

Weighted Automata Over Valuation Monoids with Input and Multi-output Characteristics

Jian-Hua Jin, Dong-Xue Li and Chun-Quan Li

Abstract Weighted automata are significant modelling notions of discrete dynamic systems. This paper aims to study weighted automata over valuation monoids with input and multi-output characteristics, whose truth values involve a wide range of algebraic structures such as semirings and strong bimonoids. In particular, if these domains are Cauchy double unital valuation monoids, it is pointed out that weighted sequential-like automata and weighted generalized Moore automata are equivalent in the sense of the same input and multi-output behaviors.

Keywords Weighted automata · Valuation monoids · Weighted sequential-like automata · Weighted generalized moore automata · Equivalence

1 Introduction

Fuzzy automata have been increasingly important models in computer science and artificial intelligence, which initiated by Wee and Santos [16–18, 21] as early as in the late 1960s. It is noticeable that fuzzy automata explore new frontiers and help to provide sound formal theory for discrete dynamic systems [11, 19]. Not only these automata are constantly concerned about theories like mathematical models [14, 20], its have also been widely applied to many fields in machine intelligence [22], digital image compression [5], natural language processing [8, 13] and model checking [12, 15].

J.-H. Jin (✉) · D.-X. Li
School of Computer Science, Shaanxi Normal University, Xi'an 710119, China
e-mail: jjh2006ok@aliyun.com
URL: http://www.snnu.edu.cn

D.-X. Li
e-mail: lidongxueever@163.com

C.-Q. Li
School of Sciences, Southwest Petroleum University, Chengdu 610500, China
e-mail: lichunquan@swpu.edu.cn

T.-H. Fan et al. (eds.), *Quantitative Logic and Soft Computing 2016*,
Advances in Intelligent Systems and Computing 510,
DOI 10.1007/978-3-319-46206-6_20

Weighted automata are developed based on general algebraic systems recently. Chatterjee et al. [1–4] established probablistic weighted automata and study quantitative languages which compute objectives such as the long-run average cost and maximal reward. Li [10] studies finite automata theory with membership values in lattices, where the role of the distributive law for the underlying lattice was analyzed. Droste et al. [6] initiated weighted finite automata theory over strong bimonoids based on run semantics, initial algebra semantics and free monoid semantics. Droste and Meinecke [7] investigated weighted automata and regular expressions over valuation monoids, where the weight of a successful run is computed by a valuation function.

It is interesting to study the behaviors of weighted automata over valuation monoids associated with input-output or input-multioutput characteristics. As is pointed out by Li and Pedrycz [9], a complex system may be modeled by two different fuzzy machines. Li and Pedrycz proved the equivalence relationship between lattice-valued sequential-like machines and lattice-valued finite Moore machines over lattice-ordered monoids, which means that they exhibit the same input-output characteristics. However, little research is done on the relationships among weighted automata over valuation monoids in the sense of input-multioutput features. The purpose of this paper is to investigate these automata, in which the input feature means a single-input.

The rest of the paper is arranged as follows. Section 2 introduces three kinds of weighted automata and their response functions, including weighted sequential-like automata, weighted generalized Mealy automata and weighted generalized Moore automata over valuation monoids. Section 3 studies the relationship between weighted sequential-like automata and weighted generalized Moore automata with the same input and multi-output sets, whose codomains are Cauchy double unital valuation monoids. Moreover, some illuminated examples are presented. Section 4 gives conclusions finally.

2 Weighted Automata Over Valuation Monoids with Input and Multi-output Characteristics

Let N be the set of positive integers and Σ an alphabet. By Σ^* we denote the set of all finite words over Σ, containing the empty word ε. Then Σ^* is the free monoid generated by Σ with the concatenation operation. For $\theta \in \Sigma^*$, $|\theta|$ is the length of θ. By Σ^+ we denote the set of non-empty finite words in Σ^*. Let A^Q denote the set of all functions from set Q to set A. For every $F \in A^{Q\times Q}$ and $\sigma \in A^Q$, we write F_{q_1,q_2} instead of $F(q_1, q_2)$ and σ_{q_1} instead of $\sigma(q_1)$ for all $q_1, q_2 \in Q$.

A monoid $(A, +, 0)$ is complete [7] if it has infinitary sum operations $\sum_I : A^I \to A$ for any index set I such that $\sum_{i\in\emptyset} d_i = 0$, $\sum_{i\in\{k\}} d_i = d_k$, $\sum_{i\in\{j,k\}} d_i = d_j + d_k$, for $j \neq k$. $\sum_{j\in J}(\sum_{i\in I_j} d_i) = \sum_{i\in I} d_i$ if $\bigcup_{j\in J} I_j = I$ and $I_j \bigcap I_k = \emptyset$ for $j \neq k$.

Definition 1 A valuation monoid $(A, +, val, 0)$ consists of a commutative monoid $(A, +, 0)$ and a valuation function val: $A^+ \to A$ with $val(d) = d$ for all $d \in A$ and

$$val(d_1, \ldots, d_n) = 0$$

whenever $d_1, \ldots, d_n \in A$ and $d_i = 0$ for some $i \in \{1, \ldots, n\}$, $A^+ = \{(d_1, \ldots, d_n) \in A^n \mid d_1, \ldots, d_n \in A, n \in N^*\}$. Moreover, a valuation monoid $(A, +, val, 0)$ is called a double unital valuation monoid if there exists an element $e \in A$ such that for any $d_1, \ldots, d_{i-1}, d_{i+1}, \ldots, d_n \in A$,

$$val(d_1, \ldots, d_{i-1}, e, d_{i+1}, \ldots, d_n) = val(d_1, \ldots, d_{i-1}, d_{i+1}, \ldots, d_n).$$

The double unital valuation monoid will be denoted by $(A, +, val, 0, e)$.

A valuation monoid includes a variety of algebraic systems such as bounded lattices, strong bimonoids and semirings, examples can be found in [7, 10].

Definition 2 The algebraic system $(A, +, val, \{\cdot_{m,n} | m, n \in N\}, 0, e)$ is a Cauchy double unital valuation monoid if $(A, +, val, 0, e)$ is a double unital valuation monoid and $\cdot_{m,n} : A \times A \to A$ with $m, n \in N$ is a family of products such that for all $d, d_i, d'_j \in A$ and all finite subsets $B, C \subseteq A$:

$$0 \cdot_{m,n} d = d \cdot_{m,n} 0 = 0$$

$$val(d_1, \ldots, d_m, d'_1, \ldots, d'_n) = val(d_1, \ldots, d_m) \cdot_{m,n} val(d'_1, \ldots, d'_n),$$

and

$$\sum(d | d \in B) \cdot_{m,n} \sum(d' | d' \in C) = \sum(d \cdot_{m,n} d' | d \in B, d' \in C).$$

Definition 3 A weighted sequential-like automaton $M = (Q, \sigma, \tau)$ over a finite input alphabet Σ, a finite multi-output set $Y = Y_1 \times \cdots \times Y_k$ and a valuation monoid $(A, +, val, 0, e)$ is a triple (for short WSLAM), consisting of a finite nonempty state set Q, an initial weight vector $\sigma \in A^Q$ and a transition mapping $\tau : \Sigma \times Y \to A^{Q \times Q}$ such that:

$$\sum_{y \in Y} \sum_{p \in Q} \tau(u, y)_{q,p} \neq 0, \forall q \in Q, \ \forall u \in \Sigma.$$

To compute with words, τ is extended to a function from $\Sigma^* \times Y^*$ to $A^{Q \times Q}$, denoted by $\tau^* : \Sigma^* \times Y^* \to A^{Q \times Q}$,

$$\tau^*(\theta, \omega)_{q,p} = \begin{cases} e, & \text{if } l = m_i = 0, \ \forall i \in \{1, \ldots, k\} \text{ and } q = p \\ 0, & \text{if } l = m_i = 0, \ \forall i \in \{1, \ldots, k\} \text{ and } q \neq p \\ a, & \text{if } l = m_i \neq 0 \text{ and } \forall i \in \{1, \ldots, k\} \\ 0, & \text{otherwise} \end{cases}$$

where $\theta = u_1u_2\ldots u_l$, $|\theta| = l$, $u_i \in \Sigma$, $\omega = (\omega_1, \ldots, \omega_k) \in Y^* = Y_1^* \times Y_2^* \times \cdots Y_k^*$,
$\omega_i = y_{i1}y_{i2}\ldots y_{im_i}$, $|\omega_i| = m_i$, $i = \{1, \ldots, k\}$, and
$a = \sum_{q_1,\ldots,q_{l-1}\in Q} val(\tau(u_1, (y_{1l}, \ldots, y_{kl}))_{q,q_1}, \tau(u_2, (y_{12}, \ldots, y_{k2}))_{q_1,q_2}, \ldots, \tau(u_l, (y_{1l}, \ldots, y_{kl}))_{q_{l-1},p})$.

Noting that if the above $\omega = (\omega_1, \ldots, \omega_k)$ satisfies the condition $|\omega_i| = m$, $\forall i \in \{1, \ldots, k\}$, then we denote ω by
$\omega = (\omega_1, \ldots, \omega_k) = (y_{11}y_{12}\ldots y_{1m}, y_{21}y_{22}\ldots y_{2m}, \ldots, y_{k1}y_{k2}\ldots y_{km}) = (y_{11}, y_{21}, \ldots, y_{k1})(y_{12}, y_{22}, \ldots, y_{k2})\ldots(y_{1m}, y_{2m}, \ldots, y_{km})$.

The input and multi-output (I/MO) behavior of M, denoted by $\varphi = \varphi_M : \Sigma^* \times Y^* \to A$, is defined for all $\theta \in \Sigma^*$ and $\omega = (\omega_1, \ldots, \omega_k) \in Y^* = Y_1^* \times Y_2^* \times \cdots \times Y_k^*$,

$$\varphi(\theta, \omega) = \sum_{p\in Q}\sum_{q\in Q} val(\sigma_q, (\tau^*(\theta, \omega))_{q,p}).$$

Definition 4 A weighted generalized Mealy automaton $N = (Q, \sigma, \delta, h)$ over Σ, Y and a double unital valuation monoid $(A, +, val, 0, e)$ is a quadruple, where Q and Σ are finite nonempty sets, a finite set $Y = Y_1 \times \cdots \times Y_k$, $\delta : \Sigma \to A^{Q\times Q}$ and $h : \Sigma \times Y \to A^Q$ are mappings, and $\sigma \in A^Q$ is an initial weight vector, which satisfy the following condition:

$$\sum_{y\in Y}\sum_{p\in Q} val((\delta(u))_{q,p}, (h(u, y))_q) \neq 0, \forall q \in Q, \forall u \in \Sigma.$$

For a weighted generalized Mealy automaton N, we could define a corresponding input-transition-output function $f : \Sigma \times Y \to A^{Q\times Q}$ as follows: for all $q, p \in Q$, $u \in \Sigma$ and $y \in Y$,

$$(f(u, y))_{q,p} = val((\delta(u))_{q,p}, (h(u, y))_q).$$

Then the I/MO behavior of N can be given as follows:

$$\varphi = \varphi_N : \Sigma^* \times Y^* \to A,$$

for any $\theta \in \Sigma^*$ and all $\omega = (\omega_1, \ldots, \omega_k) \in Y^* = Y_1^* \times Y_2^* \times \cdots \times Y_k^*$,

$$\varphi(\theta, \omega) = \sum_{p\in Q}\sum_{q\in Q} val(\sigma_q, (f^*(\theta, \omega))_{q,p}).$$

Definition 5 A weighted Moore automaton $M_2 = (Q, \sigma, \delta, g)$ over a finite input alphabet Σ, a finite multi-output set $Y = Y_1 \times \cdots \times Y_k$ and a valuation monoid $(A, +, val, 0, e)$ is a quadruple (for short WMAM), where $\sigma \in A^Q$ is an initial weight vector, the transition mapping $\delta : \Sigma \to A^{Q\times Q}$ and output function $g : Y \to A^Q$ satisfy the following condition:

$$\sum_{y \in Y} \sum_{p \in Q} val((\delta(u))_{q,p}, (g(y))_q) \neq 0, \forall q \in Q, \forall u \in \Sigma.$$

The response function $\varphi : \Sigma^* \times Y^* \to A$ of M_2 is defined in the following form:

$$\varphi(\theta, \omega) = \begin{cases} a, \text{ if } m_i = l+1, \ \forall i \in \{1, \ldots, k\} \\ 0, \qquad \text{otherwise} \end{cases}$$

where $\theta = u_1 u_2 \ldots u_l$, $|\theta| = l$, $u_i \in \Sigma$, $\omega = (\omega_1, \ldots, \omega_k) \in Y^{*} = Y_1^{*} \times Y_2^{*} \times \cdots Y_k^{*}$,
$\omega_i = y_{i1} y_{i2} \ldots y_{im_i}$, $|\omega_i| = m_i$, $i = \{1, \ldots, k\}$, and
$a = \varphi(\theta, \omega) = \sum_{q, q_1, \ldots, q_k \in Q} val((\sigma)_q, g(y_0)_q, (\delta(u_1))_{q,q_1}, (g(y_1))_{q_1}, (\delta(u_2))_{q_1,q_2}, (g(y_2))_{q_2}, \ldots, (\delta(u_k))_{q_{k-1},q_k}, (g(y_k))_{q_k})$.

3 The Relationship Between WSLAM and WMAM

In this section, consider WSLAM M_1 and WMAM M_2 over valuation monoids with the same input set Σ and the same multi-output set Y. If their response functions $\varphi_1 = \varphi_{M_1}$ and $\varphi_2 = \varphi_{M_2}$ satisfy the following equation:

$$\varphi_1(\theta, (y_1, y_2, \ldots, y_k)) = \sum_{y_{i0} \in Y_i, i \in \{1, \ldots, k\}} \varphi_2(\theta, (y_{10} y_1, y_{20} y_2, \ldots, y_{k0} y_k)),$$

for any input sequence $\theta = u_1 u_2 \ldots u_l \in \Sigma^*$ and output sequence $(y_1, y_2, \ldots, y_k) \in Y_1^* \times Y_2^* \times \cdots \times Y_k^*$ with $y_i = y_{i1} y_{i2} \ldots y_{il}$ ($i \in \{1, \ldots, k\}$), then M_1 and M_2 are said to be equivalent.

Proposition 1 *If $(A, +, val, 0, e)$ is a Cauchy double unital valuation monoid $(A, +, val, \{\cdot_{m,n} | m, n \in N\}, 0, e)$, then for every WMAM M, there exists an equivalent WSLAM M_1.*

Proof Suppose a WMAM $M = (Q, \sigma, \delta, g)$ over Σ, Y and $(A, +, val, 0, e)$. Then a WSLAM $M_1 = (Q, \sigma_1, \tau)$ over Σ, Y and $(A, +, val, 0, e)$ is constructed as follows: $\sigma_1 \in A^Q$ is given by

$$(\sigma_1)_q = \sum_{y_0 \in Y} val((\sigma)_q, (g(y_0))_q), \forall q \in Q,$$

$\tau : \Sigma \times Y \to A^{Q \times Q}$ is given by

$$(\tau(u, y))_{q,p} = val((\delta(u))_{q,p}, (g(y))_p), \forall (q, p) \in Q \times Q.$$

Owing to the Cauchy double unital valuation monoid $(A, +, val, \{\cdot_{m,n} | m, n \in N\}, 0, e)$, the response function for M_1 is computed by Definitions 3 and 5 as follows:

for each $\theta = u_1u_2 \ldots u_l \in \Sigma^*$, $\omega = (\omega_1, \ldots, \omega_k) \in Y^*$, $\omega_i = y_{i1}y_{i2} \ldots y_{il}$, $i \in \{1, \ldots, k\}$. Denote $\omega = Z_1Z_2 \ldots Z_l$, where $Z_j = (y_{1j}, y_{2j}, \ldots, y_{kj}), j \in \{1, 2, \ldots, l\}$.
$\varphi_1(\theta, \omega) = \sum_{q,q_1,\ldots,q_k \in Q} val((\sigma_1)_q, (\tau(u_1, Z_1))_{q,q_1}, (\tau(u_2, Z_2))_{q_1,q_2}, \ldots, (\tau(u_k, Z_k))_{q_{k-1},q_k})$
$= \sum_{y_0 \in Y} \sum_{q,\ldots,q_k \in Q} val((\sigma)_q, (g(y_0))_q, (\delta(u_1))_{q,q_1}, (g(Z_1))_{q_1}, \ldots, (\delta(u_k))_{q_{k-1},q_k}, (g(Z_k))_{q_k})$
$= \sum_{y_0 \in Y} \varphi(\theta, y_0Z_1 \ldots Z_k) = \sum_{y_0 \in Y} \varphi(\theta, y_0\omega)$,

otherwise, $\varphi_1(\theta, \omega) = 0 = \sum_{y_0 \in Y} \varphi(\theta, y_0\omega)$. Therefore, M_1 is equivalent to the given automaton M. □

The following example illuminates the construction above.

Example 1 Let $L = [0, 1]$, $a + b = sup\{a, b\}$, $val(a, b) = a \wedge b, \forall a, b \in [0, 1]$. Then $(L, +, val, 0, 1)$ is a Cauchy double unital valuation monoid. A WMAM $M = (Q, \sigma, \delta, g)$ over Σ, Y and $(A, +, val, 0, 1)$ is given by, $Q = \{q_1, q_2, q_3\}$, $\Sigma = \{u, v\}, Y = Y_1 \times Y_2, Y_1 = Y_2 = \{a, b\}$. $\sigma = \frac{0.8}{q_1} + \frac{0.6}{q_2} + \frac{0.3}{q_3}$, $g(a, a) = \frac{0.5}{q_1} + \frac{0.1}{q_2} + \frac{0.2}{q_3}, g(a, b) = \frac{0.7}{q_1} + \frac{0.2}{q_2} + \frac{0.2}{q_3}, g(b, a) = \frac{0.1}{q_1} + \frac{0.3}{q_2} + \frac{0.4}{q_3}, g(b, b) = \frac{0.2}{q_1} + \frac{0.3}{q_2} + \frac{0.7}{q_3}$,

$$\delta(u) = ((\delta(u))_{i,j})_{3\times3} = \begin{pmatrix} 1 & 0.8 & 0.7 \\ 0.7 & 0.5 & 0.4 \\ 0.2 & 0.3 & 1 \end{pmatrix}, \ \delta(v) = ((\delta(v))_{i,j})_{3\times3} = \begin{pmatrix} 0.7 & 0.8 & 0.9 \\ 0.2 & 0.3 & 0.4 \\ 0.3 & 0.2 & 0.5 \end{pmatrix}.$$

The corresponding WSLAM $M_1 = (Q, \sigma_1, \tau)$ is constructed as follows: $\sigma_1 = \frac{(\sigma_1)_{q_1}}{q_1} + \frac{(\sigma_1)_{q_2}}{q_2} + \frac{(\sigma_1)_{q_3}}{q_3}$, where $(\sigma_1)_{q_1} = \sum_{y \in Y} val((\sigma)_{q_1}, (g(y))_{q_1}) = (0.8 \wedge 0.5) \vee (0.8 \wedge 0.7) \vee (0.8 \wedge 0.1) \vee (0.8 \wedge 0.2) = 0.7$;
$(\sigma_1)_{q_2} = (0.6 \wedge 0.1) \vee (0.6 \wedge 0.2) \vee (0.6 \wedge 0.3) \vee (0.6 \wedge 0.3) = 0.3$;
$(\sigma_1)_{q_3} = (0.3 \wedge 0.2) \vee (0.3 \wedge 0.2) \vee (0.3 \wedge 0.4) \vee (0.3 \wedge 0.7) = 0.3$.

$$\tau(u, (a, a)) = \begin{pmatrix} 0.5 & 0.1 & 0.2 \\ 0.5 & 0.1 & 0.2 \\ 0.2 & 0.1 & 0.2 \end{pmatrix}, \ \tau(u, (a, b)) = \begin{pmatrix} 0.7 & 0.2 & 0.2 \\ 0.7 & 0.2 & 0.2 \\ 0.2 & 0.2 & 0.2 \end{pmatrix},$$
$$\tau(u, (b, a)) = \begin{pmatrix} 0.1 & 0.3 & 0.4 \\ 0.1 & 0.3 & 0.4 \\ 0.1 & 0.3 & 0.4 \end{pmatrix}, \ \tau(u, (b, b)) = \begin{pmatrix} 0.2 & 0.3 & 0.7 \\ 0.2 & 0.3 & 0.4 \\ 0.2 & 0.3 & 0.7 \end{pmatrix},$$
$$\tau(v, (a, a)) = \begin{pmatrix} 0.5 & 0.1 & 0.2 \\ 0.2 & 0.1 & 0.2 \\ 0.3 & 0.1 & 0.2 \end{pmatrix}, \ \tau(v, (a, b)) = \begin{pmatrix} 0.7 & 0.2 & 0.2 \\ 0.2 & 0.2 & 0.2 \\ 0.3 & 0.2 & 0.2 \end{pmatrix},$$
$$\tau(v, (b, a)) = \begin{pmatrix} 0.1 & 0.3 & 0.4 \\ 0.1 & 0.3 & 0.4 \\ 0.1 & 0.2 & 0.4 \end{pmatrix}, \ \tau(v, (b, b)) = \begin{pmatrix} 0.2 & 0.3 & 0.7 \\ 0.2 & 0.3 & 0.4 \\ 0.2 & 0.2 & 0.5 \end{pmatrix},$$

Take $\theta = uv, \omega = (ba, aa) \in Y_1^* \times Y_2^*$, Then
$\varphi_1(\theta, \omega) = \sum_{p_1,p_2,p_3 \in Q} val((\sigma_1)_{p_1}, \tau(u, (b, a))_{p_1,p_2}, \tau(v, (a, a))_{p_2,p_3}) = (0.7 \wedge 0.1 \wedge 0.5) \vee (0.7 \wedge 0.3 \wedge 0.1) \vee (0.7 \wedge 0.4 \wedge 0.2) = 0.3$. $\varphi(\theta, (aba, aaa)) = 0.3, \varphi(\theta, (aba, baa)) = 0.3, \varphi(\theta, (bba, baa)) = 0.2, \varphi(\theta, (bba, aaa)) = 0.3$.

Thus $\varphi_1(\theta, \omega) = \sum_{x_1, x_2 \in Y_1} \varphi(\theta, (x_1 ba, x_2 aa))$.

Proposition 2 *Let* $(A, +, val, \{\cdot_{m,n} | m, n \in N\}, 0, e)$ *be a Cauchy double unital valuation monoid. Then for every WSLAM* M*, there exists an equivalent WMAM* M_1.

Proof Suppose WSLAM $M = (Q, \sigma, \tau)$ over Σ, Y and $(A, +, val, \{\cdot_{m,n} | m, n \in N\}, 0, e)$. Then a WMAM $M_1 = (Q_1, \sigma_1, \delta, g)$ over Σ, Y and $(A, +, val, \{\cdot_{m,n} | m, n \in N\}, 0, e)$ is constructed as follows: $Q_1 = Q \times Y$, $\sigma_1 \in A^{Q_1}$ is given by

$$\sigma_1(q, y) = \begin{cases} \sigma(q), & \textit{if } y = y_0 \\ 0, & \textit{otherwise} \end{cases}$$

where $y_0 = (y_{01}, y_{02}, \ldots, y_{0k})$ is a particular output symbol in Y. $\delta : \Sigma \to A^{Q_1 \times Q_1}$ is designed as follows:

$$\delta(u_1)_{(q,y),(q_1,y_1)} = (\tau(u_1, y_1))_{q,q_1}, \ \forall u_1 \in \Sigma, \forall (q, y), (q_1, y_1) \in Q_1 \times Q_1.$$

$g : Y \to A^{Q_1}$ is given by for any $y \in Y$,

$$(g(y))_{(q,y_1)} = \begin{cases} 1, & \textit{if } y = y_1 \\ 0, & \textit{if} y \neq y_1 \end{cases}$$

Then for any $\theta = u_1 u_2 \ldots u_l \in \Sigma^*$, $\omega = (\omega_1, \ldots, \omega_k) \in Y^*$, $\omega_i = y_{i1} y_{i2} \ldots y_{il}$, $i \in \{1, \ldots, k\}$, denote $\omega = z_1 z_2 \ldots z_l$, where $z_j = (y_{1j}, y_{2j}, \ldots, y_{kj})$, $j \in \{1, 2, \ldots, l\}$. $\varphi_1(\theta, y_0\omega) = \sum_{q,q_1,\ldots,q_k \in Q} val((\sigma_1)_{(q,y_0)}, (g(y_0))_{(q,y_0)}, (\delta(u_1))_{(q,y_0),(q_1,z_1)}, (g(z_1))_{(q_1,z_1)}, (\delta(u_2))_{(q_1,z_1),(q_2,z_2)}, (g(z_2))_{(q_2,z_2)}, \ldots, (\delta(u_k))_{(q_{k-1},z_{k-1}),(q_k,z_k)}, (g(z_k))_{(q_k,z_k)}) = \sum_{q,q_1,\ldots,q_k \in Q} val((\sigma_1)_{(q,y_0)}, (g(y_0))_{(q,y_0)}, (\delta(u_1))_{(q,y_0),(q_1,z_1)}, \ldots, (g(z_k))_{(q_k,z_k)}) = \sum_{q,q_1,\ldots,q_k \in Q} val((\sigma)_q, \tau(u_1, z_1)_{q,q_1}, \tau(u_2, z_2)_{q_1,q_2}, \ldots, \tau(u_k, z_k)_{q_{k-1},q_k}) = \varphi(\theta, \omega)$.

Obviously, for all $\theta \in \Sigma^*, \omega \in Y^*, y \in Y$,

$$\sum_{y \in Y} \varphi_1(\theta, y\omega) = \varphi(\theta, \omega).$$

This completes the proof. □

Example 2 Given a WSLAM $M = (Q, \sigma, \tau)$ over Σ, Y and $([0, 1], +, val, 0, 1)$, where $([0, 1], +, val, 0, 1)$ is the same as in Example 1, $Q = \{q_1, q_2, q_3\}$, $\Sigma = \{u, v\}$, $Y = Y_1 \times Y_2$, $Y_1 = Y_2 = \{a, b\}$, $A = [0, 1]$, $\sigma \in A^Q$ is given as,
$\sigma = \frac{1}{q_1} + \frac{0}{q_2} + \frac{0.4}{q_3}$; $\tau : \Sigma \times Y \to A^{Q \times Q}$ is given by,

$$\tau(u, (a, b)) = \begin{pmatrix} 1 & 0.8 & 0.6 \\ 0.3 & 0.3 & 0.5 \\ 0.4 & 0.5 & 0.7 \end{pmatrix}, \ \tau(u, (a, a)) = \begin{pmatrix} 0.5 & 0.7 & 0.5 \\ 0.2 & 0.4 & 0.3 \\ 0.3 & 0.6 & 0.8 \end{pmatrix},$$

$$\tau(u, (b, a)) = \begin{pmatrix} 0.5 & 0.3 & 0.7 \\ 0.2 & 0.4 & 0.6 \\ 0.8 & 0.5 & 0.7 \end{pmatrix}, \ \tau(u, (b, b)) = \begin{pmatrix} 0.5 & 0.3 & 0.4 \\ 0.1 & 0.2 & 0.5 \\ 0.6 & 0.5 & 0.7 \end{pmatrix},$$

$$\tau(v,(a,b)) = \begin{pmatrix} 0.5\ 0.7\ 0.9 \\ 0.3\ 0.4\ 0.5 \\ 0.7\ 0.6\ 0.8 \end{pmatrix}, \tau(v,(a,a)) = \begin{pmatrix} 0.3\ 0.2\ 0.5 \\ 0.4\ 0.6\ 0.7 \\ 0.3\ 0.4\ 0.5 \end{pmatrix},$$

$$\tau(v,(b,a)) = \begin{pmatrix} 0.7\ 0.6\ 0.5 \\ 0.3\ 0.5\ 0.7 \\ 0.5\ 0.6\ 0.7 \end{pmatrix}, \tau(v,(b,b)) = \begin{pmatrix} 0.3\ 0.5\ 0.7 \\ 0.4\ 0.4\ 0.6 \\ 0.5\ 0.7\ 0.8 \end{pmatrix}.$$

Let $\theta = uv, y = (ba, ab)$, Then
$\varphi(\theta, y) = \sum val((\sigma)_{p_1}, \tau(u,(b,a))_{p_1,p_2}, \tau(v,(a,b))_{p_2,p_3}) = (1 \wedge 0.5 \wedge 0.9) \vee (1 \wedge 0.3 \wedge 0.5) \vee (1 \wedge 0.7 \wedge 0.8) \vee (0.4 \wedge 0.8 \wedge 0.9) \vee (0.4 \wedge 0.5 \wedge 0.5) \vee (0.4 \wedge 0.7 \wedge 0.8) = 0.7$.

A WMAM $M_2 = (Q_1, \sigma_1, \delta, g)$ over Σ, Y and $([0, 1], +, val, 0, 1)$ is constructed as follows: $Q_1 = Q \times Y, \sigma_1 \in A^{Q_1}$, $\sigma_1 = \frac{1}{(q_1,(b,b))} + \frac{0.4}{(q_3,(b,b))}$, $\delta : \Sigma \to A^{Q_1 \times Q_1}$ is given by

$$\delta(u_1)_{(q,y),(q_1,y_1)} = (\tau(u_1, y_1))_{q,q_1}, \ \forall u_1 \in \Sigma, \forall (q, y), (q_1, y_1) \in Q_1 \times Q_1.$$

$g : Y \to A^{Q_1}$ is constructed as follows:
$g(a,b) = \frac{1}{(q_1,(a,b))} + \frac{1}{(q_2,(a,b))} + \frac{1}{(q_3,(a,b))}$, $g(a,a) = \frac{1}{(q_1,(a,a))} + \frac{1}{(q_2,(a,a))} + \frac{1}{(q_3,(a,a))}$,
$g(b,a) = \frac{1}{(q_1,(b,a))} + \frac{1}{(q_2,(b,a))} + \frac{1}{(q_3,(b,a))}$, $g(b,b) = \frac{1}{(q_1,(b,b))} + \frac{1}{(q_2,(b,b))} + \frac{1}{(q_3,(b,b))}$.

Then we can derive the results that $\varphi_2(u, v, (aba, aab)) = 0$, $\varphi_2(u, v, (aba, bab)) = 0$,$\varphi_2(uv, (bba, aab)) = 0$ and
$\varphi_2(uv, (bba, bab)) = \sum_{p_1,p_2,p_3 \in Q} val((\sigma_1)_{(p_1,(b,b))}, g(b,b)_{(p_1,(b,b))}, \delta(u)_{(p_1,(b,b)),(p_2,(b,a))}, g(b,a)_{(p_2,(b,a))}, \delta(v)_{(p_2,(b,a)),(p_3,(a,b))}, g(b,a)_{(p_3,(a,b))}) = 0.7$.

Hence, $\varphi(\theta, y) = \sum_{x_1,x_2 \in Y_1} \varphi_2(uv, (x_1ba, x_2ab))$.

4 Conclusions

The paper introduces the notions of weighted sequential-like automata and weighted generalized automata over valuation monoids. Considering these automata's behaviors accompanied by input and multioutput characteristics, we have demonstrated the equivalence between weighted sequential-like automata and weighted generalized automata over Cauchy double unital valuation monoids. It is worthwhile for further research the modeling applications of complex dynamic systems described by the proposed automata.

Acknowledgments This work is supported by National Natural Science Foundation of China (Grant No. 11401495).

References

1. Chatterjee, K., Doyen, L., Henzinger, T.A.: Quantitative Languages, in CSL 2008. LNCS. **5213**, 385–400 (2008)
2. Chatterjee, K., Doyen, L., Henzinger, T.A.: Alternating weighted automata, in FCT 2009. LNCS. **5699**, 3–13 (2009)
3. Chatterjee, K., Doyen, L., Henzinger, T.A.: Expressiveness and closure properties for quantitative languages pp. 199–208 (2009)
4. Chatterjee, K. , Doyen, L., Henzinger, T.A.: Probabilistic weighted automata. in CONCUR 2009, LNCS. **5710**, 244–258 (2009)
5. Culik, K., Kari, J.: Image compression using weighted finite automata. Comput. Graph. **17**, 305–313 (1993)
6. Droste, M., Stber, T., Vogler, H.: Weighted Finite Automata over Strong Bimonids. Inf. Sci. **180**, 156–166 (2010)
7. Droste, M., Meinecke, I.: Weighted automata and regular expressions over valuation monoids. Int. J. Found. Comput. Sci. **22**, 1829–1844 (2011)
8. Knight, K., May, J.: Applications of Weighted Automata in Natural Language Processing. In: Droste, M., Kuich, W., Vogler, H. (eds.) Handbook of Weighted Automata, pp. 571–591. Springer, Berlin (2009)
9. Li, Y.M., Pedrycz, W.: The Equivalence between Fuzzy Mealy and Fuzzy Moore Machines. Soft Comput. **10**, 953–959 (2006)
10. Li, Y.M.: Finite automata theory with membership values in lattices. Inf. Sci. **181**, 1003–1017 (2011)
11. Liu, F.: Diagnosability of fuzzy discrete-event systems: a fuzzy approach. IEEE Trans. Fuzzy Syst. **17**(2), 372–384 (2009)
12. Meinecke, I., Quaas, K.: Parameterized Model Checking of Weighted Networks. Theor. Comput. Sci. **534**, 69–85 (2014)
13. Mohri, M.: Finite-State Transducers in Language and Speech Processing. Comput Linguist. **23**, 269–311 (1997)
14. Mordeson, J.N., Malik, D.S.: Fuzzy Automata and Languages: Theory and Applications. Chapman & Hall/CRC, Boca Raton (2002)
15. Pan, H.Y., Li, Y.M., Cao, Y.Z., Ma, Z.Y.: Model Checking Computation Tree Logic over Finite Lattices. Theor. Comput. Sci. **612**, 45–62 (2016)
16. Santos, E.S.: Maximin automata. Inf. Control. **12**, 367–377 (1968)
17. Santos, E.S.: Max-product machines. J. Math. Anal. Appl. **37**, 677–686 (1972)
18. Santos, E.S., Wee, W.G.: General formulation of sequential machines. Inf. Control. **12**(1), 5–10 (1968)
19. Ushio, T., Takai, S.: Supervisory Control of Discrete Event Systems Modeled by Mealy Automata with Nondeterministic Output Functions. In: Proceedings of the American Control Conference Hyatt Regency Riverfront, St. Louis, MO, USA June. 10–12(2009)
20. Ying, M.S.: A formal model of computing with words. IEEE Trans Fuzzy Syst. **10**(5), 640–652 (2002)
21. Wee, W.G.: On generalizations of adaptive algorithm and application of the fuzzy sets concept to pattern classification, PhD Thesis, Purdue University (1967)
22. Zadeh, L.A.: Fuzzy languages and their relation to human and machine intelligence. Electronic Research Laboratory, University of California, Berkeley. Tech. Rep. ERL-M302 (1971)

References

Splitting Algorithm of Valuation Algebra and Its Application in Automaton Representation of Semiring Valued Constraint Problems

Bang-He Han, Yong-Ming Li and Qiong-Na Chen

Abstract A splitting algorithm is developed for solving single-query projection problems in valuation algebras. This method is based on a generalized combination theorem. It is shown that by using this new kind of combination property, a given single-query projection problem can be broken into pieces of subprojection problems which might be solved simultaneously by different computational resources. At last, as an application of splitting algorithms, we develop an optimized procedure for automaton representation of semiring valued constraint problems.

Keywords Valuation algebra · Automaton representation · Semiring valued constraint

1 Introduction

Two main factors were proposed [15] in formalisms for representing information. These two main factors are combination and projection operations. A valuation algebra [17] is one of this kind of models, where the projection problems also called local computation problem [15, 16, 23, 28–30]. Constraint systems [22], relations, probabilistic networks and logic all have these unifying structure. A kind of valuation algebra named semiring induced valuation algebra such as the soft constraint satisfaction problems (SCSPs) has been studied in [1–7, 11, 17, 20, 31]. The algebraic

B.-H. Han (✉)
School of Mathematics and Statistics, Xidian University, Xi'an 710071, China
e-mail: bhhan@mail.xidian.edu.cn

Y.-M. Li
College of Mathematics and Information Science, Shaanxi Normal University, Xi'an 710119, China
e-mail: liyongm@snnu.edu.cn

Q.-N. Chen
School of Telecommunications Engineering, Xidian University, Xi'an 710071, China
e-mail: 447045259@qq.com

T.-H. Fan et al. (eds.), *Quantitative Logic and Soft Computing 2016*, Advances in Intelligent Systems and Computing 510,
DOI 10.1007/978-3-319-46206-6_21

structure semiring is used to define combination and focusing operations by its two operations $+$, $\times$. Valuations in SCSPs mean preferences for different assignments [4, 25, 26].

There are four major local computation architectures called Shenoy-Shafer,[29] Lauritzen-Spiegelhalter [19], HUGIN [13] and Idempotent Architecture [15]. All these methods are based on join tree [8, 15, 24, 27]. Frameworks for computing approximately with upper or lower bounds have been discussed in [9, 10, 12, 14, 17].

Based on the initial combination axiom, we are going to present the Strong Combination Property in valuation algebras. Then this paper is mainly concerned with what we can make use of the Strong Combination Property in solving projection problems. After we give a method for breaking the initial projection problem, an algorithm named splitting algorithm is proposed.

A method of representing the soft constraint problems as fuzzy finite automaton was proposed in [21]. There exists one disadvantage: the representation method mainly depends on the initial projection problem. As one application of our splitting algorithm, we develop an optimized procedure for automaton representation of semiring valued constraint problems, by which we can reduce the complexity to some extent.

For more details of valuation algebras we refer to [15]. For preliminaries of semrings and soft constraint systems we refer to [2, 4, 6]. For semiring-valued Finite automaton, we refer to [18].

2 Main Result 1: Splitting Algorithm of Valuation Algebra

In this section, we will give a splitting algorithm for single marginalization problem in valuation algebras.

2.1 A Generalized Combination Property

In this subsection, we present a generalized combination property in labeled valuation algebras.

Lemma 1 *[15] Suppose that (Φ, D) is a valuation algebra. If $\phi, \psi \in \Phi$ such that $d(\phi) = x, d(\psi) = y, x \subseteq z \subseteq x \cup y$, then*

$$(\phi \otimes \psi)^{\downarrow z} = \phi \otimes \psi^{\downarrow y \cap z}.$$

Theorem 1 (Generalized Combination Property) *In labeled valuation algebras, $\forall \phi, \psi \in \Phi, t \in r$, if*

$$d(\phi) \cap d(\psi) \subseteq t \subseteq d(\phi) \cup d(\psi), \tag{1}$$

then

$$(\phi \otimes \psi)^{\downarrow t} = \phi^{\downarrow t} \otimes \psi^{\downarrow t}. \tag{2}$$

Proof Denote $d(\phi) = x, d(\psi) = y$, note that $x \cap y \subseteq t \subseteq x \cup y$. It suffices to show that

$$(\phi \otimes \psi^{\downarrow (x \cup t) \cap y})^{\downarrow t} = \phi^{\downarrow x \cap t} \otimes \psi^{\downarrow y \cap t}.$$

According to the Transitivity Axiom, we have

$$(\phi \otimes \psi)^{\downarrow t} = ((\phi \otimes \psi)^{\downarrow x \cup t})^{\downarrow t}.$$

Since $x \subseteq x \cup t \subseteq x \cup y$, by Lemma 1,

$$((\phi \otimes \psi)^{\downarrow x \cup t})^{\downarrow t} = (\phi \otimes \psi^{\downarrow (x \cup t) \cap y})^{\downarrow t}.$$

By expression (1), it is easy to show that $(x \cup t) \cap y = y \cap t \subseteq t \subseteq x \cup (y \cap t)$. Thus by using Lemma 1 again, we have

$$(\phi \otimes \psi^{\downarrow (x \cup t) \cap y})^{\downarrow t} = (\phi \otimes \psi^{\downarrow y \cap t})^{\downarrow t} = \phi^{\downarrow x \cap t} \otimes \psi^{\downarrow y \cap t}.$$

This completes the proof. □

The generalized combination property requires that the set of variables to which the projection operation marginalizes contains the intersection of the domains of the two factors. It shows that in order to marginalize to t which satisfies expression (1), it is not necessary to compute first the combination but that we can as well first marginalize each factor to t and then combine the two results.

Suppose that (Φ, D) is a semiring induced valuation algebra. Consider the projection problem: $\phi, \psi \in \Phi, d(\phi) \cap d(\psi) \subseteq t \subseteq d(\phi) \cup d(\psi)$, compute $(\phi \otimes \psi)^{\downarrow t}$.

$\forall \mathbf{x} \in \Omega_t$, we can compute $(\phi \otimes \psi)^{\downarrow t}(\mathbf{x})$ by either the expressions (1), (2) or the generalized combination property. Denote the numbers of the operations needed in computing $(\phi \otimes \psi)^{\downarrow t}(\mathbf{x})$ by $\mathcal{N}_{def}, \mathcal{N}_{gcp}$ correspondingly. We make a quantitative comparison as follows:

Theorem 2 (1) $\mathcal{N}_{def} = 2 \prod_{X \in d(\phi) \cup d(\psi) - t} \mid \Omega_X \mid -1;$
(2) $\mathcal{N}_{scp} = \prod_{X \in d(\phi) - t} \mid \Omega_X \mid + \prod_{Y \in d(\psi) - t} \mid \Omega_Y \mid -1.$

Example 1 Suppose that (Φ, D) is a semiring induced valuation algebra. $\phi, \psi \in \Phi$, $d(\phi) = \{X_1, \ldots, X_5\}, d(\psi) = \{X_2, X_4, X_6, X_7, X_8\}$ and $t = \{X_2, X_4, X_7\}$, clearly we have $d(\phi) \cap d(\psi) = \{X_2, X_4\} \subseteq t$, then by the above discussion, we have

$$\mathcal{N}_{def} = 2 \prod_{i \in \{1,3,5,6,8\}} \mid \Omega_{X_i} \mid -1,$$

$$\mathcal{N}_{scp} = \prod_{i \in \{1,3,5\}} \mid \Omega_{X_i} \mid + \prod_{j \in \{6,8\}} \mid \Omega_{X_j} \mid -1.$$

For instance, assume that $\Omega_{X_i} = 5, \forall i = i, \ldots, 8$, then $\mathcal{N}_{def} = 6249$, while $\mathcal{N}_{scp} = 149$.

2.2 t-Factorization, t-Partition, t-Splitting

In this subsection we are going to bring in some basic concepts which are based on the generalized combination property.

Denote $I = \{1, 2, \ldots, n\}$. Suppose $I_i \subseteq I$, $I_i \neq \emptyset$, $i = 1, 2, \ldots, m$. Recall that $\{I_1, I_2, \ldots, I_m\}$ is called a partition of I if for $i \neq j$, $I_i \cap I_j = \emptyset$, $\bigcup_{i=1}^{m} I_i = I$.

Definition 1 Suppose that (Φ, D) is a labeled valuation algebra. $\phi_i \in \Phi$, $i = 1, 2, \ldots, n$. $\phi = \phi_1 \otimes \phi_2 \cdots \otimes \phi_n$, $t \subseteq d(\phi) = \bigcup_{i=1}^{n} d(\phi_i)$. $I = \{1, 2, \ldots, n\}$. $\{I_1, I_2, \ldots, I_m\}$ is a partition of I, $m \geq 2$. Then $\{I_1, I_2, \ldots, I_m\}$ is said to be a t-partition of I with respect to ϕ if $\forall h, k \in \{1, 2, \ldots, m\}$, $h \neq k$,

$$\bigcup_{i \in I_h} d(\phi_i) \cap \bigcup_{j \in I_k} d(\phi_j) \subseteq t. \tag{3}$$

Particularly, if $m = 2$, then we call $\{I_1, I_2\}$ a binary t-partition of I with respect to ϕ.

Suppose $\{I_1, I_2, \ldots, I_m\}$ is a t-partition of I with respect to ϕ, $\phi = \phi_1 \otimes \phi_2 \cdots \otimes \phi_n$. Denote $\psi_{I_j} = \bigotimes_{i \in I_j} \phi_i$, $j = 1, 2, \ldots, m$, obviously by the Semigroup Axiom we have

$$\phi = \psi_{I_1} \otimes \psi_{I_2} \otimes \cdots \otimes \psi_{I_m}. \tag{4}$$

We call the right side of expression (4) a t-factorization of ϕ with respect to the t-partition $\{I_1, I_2, \ldots, I_m\}$ of I. When $m = 2$, then expression (4) is called a binary t-factorization of ϕ.

According to the generalized combination property we have the following theorem:

Theorem 3 *Suppose that (Φ, D) is a labeled valuation algebra. $\phi_i \in \Phi$, $i = 1, 2, \ldots, n$. $\phi = \phi_1 \otimes \phi_2 \cdots \otimes \phi_n$, $t \subseteq d(\phi)$. $I = \{1, 2, \ldots, n\}$. If $\{I_1, I_2\}$ is a binary t-partition of I with respect to ϕ, then*

$$\phi^{\downarrow t} = \psi_{I_1}^{\downarrow t \cap d(\psi_{I_1})} \otimes \psi_{I_2}^{\downarrow t \cap d(\psi_{I_2})}. \tag{5}$$

We see that $\phi^{\downarrow t}$ splits into the combination of two valuations, thus we call the right sides of the equality (5) binary t-splitting of $\phi^{\downarrow t}$ with respect to the binary t-partition $\{I_1, I_2\}$ of I. And the performance of changing $\phi^{\downarrow t}$ to $\psi_{I_1}^{\downarrow t \cap d(\psi_{I_1})} \otimes \psi_{I_2}^{\downarrow t \cap d(\psi_{I_2})}$ is said to be a binary t-splitting transformation.

Lemma 2 *Suppose that* (Φ, D) *is a labeled valuation algebra.* $\phi_i \in \Phi$, $i = 1, 2, \ldots, n$. $\phi = \phi_1 \otimes \phi_2 \cdots \otimes \phi_n$, $t \subseteq d(\phi)$. $I = \{1, 2, \ldots, n\}$. $\{I_1, I_2, \ldots, I_m\}$ *is a* t-*partition of* I *with respect to* ϕ. *Then* $\forall u, v \subseteq \{I_1, I_2, \ldots, I_m\}$, $u, v \neq \emptyset$ *and* $u \cap v = \emptyset$,

$$\bigcup_{i \in \bigcup_{I_h \in u} I_h} d(\phi_i) \quad \cap \bigcup_{j \in \bigcup_{I_k \in v} I_k} d(\phi_j) \subseteq t. \tag{6}$$

Theorem 4 *Suppose that* (Φ, D) *is a labeled valuation algebra.* $\phi_i \in \Phi$, $i = 1, 2, \ldots, n$. $\phi = \phi_1 \otimes \phi_2 \otimes \cdots \otimes \phi_n$, $t \subseteq d(\phi)$. $I = \{1, 2, \ldots, n\}$. $\{I_1, I_2, \ldots, I_m\}$ *is a* t-*partition of* I *with respect to* ϕ. *Then*

$$\phi^{\downarrow t} = \psi_{I_1}^{\downarrow t \cap d(\psi_{I_1})} \otimes \psi_{I_2}^{\downarrow t \cap d(\psi_{I_2})} \otimes \cdots \otimes \psi_{I_m}^{\downarrow t \cap d(\psi_{I_m})}. \tag{7}$$

Proof By Theorem 3 and Lemma 2, we have $\psi_{I_1}^{\downarrow t} \otimes \psi_{I_2}^{\downarrow t} \otimes \cdots \otimes \psi_{I_m}^{\downarrow t} = (\psi_{I_1} \otimes \psi_{I_2})^{\downarrow t} \otimes \psi_{I_3}^{\downarrow t} \otimes \cdots \otimes \psi_{I_m}^{\downarrow t} = \cdots = (\psi_{I_1} \otimes \psi_{I_2} \otimes \cdots \otimes \psi_{I_m})^{\downarrow t}$. □

We call the right sides of the expressions (7) t-splitting of $\phi^{\downarrow t}$ with respect to the t-partition $\{I_1, I_2, \ldots, I_m\}$ of I. And the performance of changing $\phi^{\downarrow t}$ to $\psi_{I_1}^{\downarrow t \cap d(\psi_{I_1})} \otimes \cdots \otimes \psi_{I_m}^{\downarrow t \cap d(\psi_{I_m})}$ is said to be t-splitting transformation. $\{I_1, I_2, \ldots, I_m\}$ is called the corresponding t-partition with this t-splitting transformation of $\phi^{\downarrow t}$.

Definition 2 Suppose that (Φ, D) is a labeled valuation algebra. $\phi_i \in \Phi$, $i = 1, 2, \ldots, n$. $\phi = \phi_1 \otimes \phi_2 \otimes \cdots \otimes \phi_n$, $t \subseteq d(\phi)$. $I = \{1, 2, \ldots, n\}$. $\{I_1, I_2, \cdots, I_m\}$ is a t-partition of I with respect to ϕ. Then the t-splitting $\psi_{I_1}^{\downarrow t} \otimes \psi_{I_2}^{\downarrow t} \otimes \cdots \otimes \psi_{I_m}^{\downarrow t}$ is said to be a final t-splitting of $\phi^{\downarrow t}$ if $\forall I_j \in \{I_1, I_2, \ldots, I_m\}$, there exists no t-partition of I_j. And the corresponding t-partition $\{I_1, I_2, \ldots, I_m\}$ of I is called a final t-partition of I.

We say that two t-splitting of $\phi^{\downarrow t}$ are equal iff the corresponding t-partitions of I are equal. Given a projection problem $\phi^{\downarrow t} = (\phi_1 \otimes \phi_2 \cdots \otimes \phi_n)^{\downarrow t}, I = \{1, 2, \ldots, n\}$. A natural question is: does there exist different final t-splitting of $\phi^{\downarrow t}$ (i.e., final t-partition of I)? The next theorem tells us that if there exists one final t-splitting of $\phi^{\downarrow t}$, then it is unique.

Definition 3 $I = \{1, 2, \ldots, n\}$, suppose that $Part1 = \{I_1, I_2, \ldots, I_m\}$, $Part2 = \{J_1, J_2, \cdots, J_k\}$ are two different partitions of I. Then the intersection of them, denoted by $Par1 \cap Par2$ is defined as follows: $\forall i, j \in I$, i, j belong to the same part of $Par1 \cap Par2$ iff i and j belong to the same part in either partition.

Lemma 3 *Suppose that* (Φ, D) *is a labeled valuation algebra.* $\phi_i \in \Phi$, $i = 1, 2, \ldots, n$. $\phi = \phi_1 \otimes \phi_2 \cdots \otimes \phi_n$, $t \subseteq d(\phi)$. $I = \{1, 2, \ldots, n\}$. *If there exist two different* t-*partitions* $Part1$ *and* $Part2$ *of* I *with respect to* ϕ, *then* $Par1 \cap Par2$ *is also a* t-*partition of* I *with respect to* ϕ *thinner than at least one of them.*

By Definition 3 and Lemma 3, we have

Theorem 5 (Uniqueness of final t-splitting) *Suppose that* (Φ, D) *is a labeled valuation algebra.* $\phi_i \in \Phi$, $i = 1, 2, \ldots, n$. $\phi = \phi_1 \otimes \phi_2 \otimes \cdots \otimes \phi_n$, $t \subseteq d(\phi)$. $I = \{1, 2, \ldots, n\}$. *If there exists one final* t*-splitting of* $\phi^{\downarrow t}$, *then there exists exact only one* t*-splitting of* $\phi^{\downarrow t}$.

Now let us have a look at how the final t-splitting of $\phi^{\downarrow t}$ helps in reducing the computation complexity. Similar with what we do in Sect. 2.1, suppose that (Φ, D) is a semiring induced valuation algebra. $\phi_i \in \Phi, i = 1, 2, \ldots, n$. $\phi = \phi_1 \otimes \phi_2 \cdots \otimes \phi_n$, $t \subseteq d(\phi)$. $I = \{1, 2, \ldots, n\}$. $\{I_1, I_2, \ldots, I_m\}$ is the final t-splitting of $\phi^{\downarrow t}$. $\forall \mathbf{x} \in \Omega_t$, we can compute $(\phi_1 \otimes \phi_2 \cdots \otimes \phi_n)^{\downarrow t}(\mathbf{x})$ by either definition or expression (7). Denote the numbers of operations needed in computing $(\phi_1 \otimes \phi_2 \cdots \otimes \phi_n)^{\downarrow t}(\mathbf{x})$ by $\mathcal{N}_{def}$, $\mathcal{N}_{scp}$ correspondingly. We make a quantitative comparison as follows:

Theorem 6 (1) $\mathcal{N}_{def} = n \prod_{X \in d(\phi)-t} \mid \Omega_X \mid -1$;
(2) $\mathcal{N}_{scp} = \sum_{i=1}^{m} \mid I_i \mid (\prod_{X \in d(\psi_{I_i})-t} \mid \Omega_X \mid) - 1$.

2.3 Extended t-Splitting and Extended t-Splitting Transformation

Given an initial factorization of ϕ, $\phi = \phi_1 \otimes \phi_2 \otimes \cdots \otimes \phi_n$, $\phi_i \in \Phi, i = 1, 2, \ldots, n$. $t \subseteq d(\phi)$. $I = \{1, 2, \ldots, n\}$. Bad situation happens if there exists no final t-partition of I. This means that we can't perform any t-splitting transformation to $\phi^{\downarrow t}$. So we can't enjoy the efficiency in reducing the computation complexity by using the final t-splitting transformation. Therefore, we have to turn to other methods that maybe less efficient in reducing the computation complexity.

Theorem 7 *Suppose that* (Φ, D) *is a labeled valuation algebra.* $\phi_i \in \Phi$, $i = 1, 2, \ldots, n$. $\phi = \phi_1 \otimes \phi_2 \cdots \otimes \phi_n$, $t \subseteq d(\phi)$. $I = \{1, 2, \ldots, n\}$. *If there exists a set* s *of variables contained in* $d(\phi)$ *which is disjoint with* t *such that the final* $t \cup s$*-partition of* I *with respect to* ϕ *exists, denoted by* $\{I_1, I_2, \ldots, I_m\}$, *then*

$$\phi^{\downarrow t} = (\psi_{I_1}^{\downarrow (t \cup s) \cap d(\psi_{I_1})} \otimes \cdots \otimes \psi_{I_m}^{\downarrow (t \cup s) \cap d(\psi_{I_m})})^{\downarrow t}. \tag{8}$$

Definition 4 We call the right sides of expressions (8) extended t-splitting of $\phi^{\downarrow t}$ with respect to the partition $\{I_1, I_2\}$ of I. This process of changing $\phi^{\downarrow t}$ to expressions (8) is called extended t-splitting transformation. The set of variables s is said to be the adjoint set of variables in this extended t-splitting transformation.

Can we reduce the computation complexity by using Theorem 7? The answer is positive. For instance, suppose that (Φ, D) is a semiring induced valuation algebra. Consider the marginalization problem $(\phi_1 \otimes \phi_2)^{\downarrow t}$, where $\phi_1, \phi_2, \in \Phi, t \subseteq d(\phi_1) \cup$

$d(\phi_2)$. $I = \{1, 2\}$. Let $s = d(\phi_1) \cap d(\phi_2) - t$. Then it is easy to see that $d(\phi_1) \cap d(\phi_2) \subseteq s \cup t$, thus $\{\{1\}, \{2\}\}$ is a $t \cup s$-partition of I.

$\forall$ $\mathbf{x} \in \Omega_t$, we can compute $(\phi \otimes \psi)^{\downarrow t}(\mathbf{x})$ by either definition or expression (8). Denote the numbers of operations needed in computing $(\phi \otimes \psi)^{\downarrow t}(\mathbf{x})$ by $\mathcal{N}_{def}, \mathcal{N}_{scp}$ correspondingly. Similar with the discussion in Sect. 3, we can see that:

$$\mathcal{N}_{def} = 2 \mid \Omega_{d(\phi_1) \cup d(\phi_2) - t} \mid -1.$$

Since $s = d(\phi_1) \cap d(\phi_2) - t$, it can be shown that

$$\mid \Omega_{d(\phi_1) \cup d(\phi_2) - t} \mid = \mid \Omega_s \mid \times \mid \Omega_{d(\phi_1) - t \cup s} \mid \times \mid \Omega_{d(\phi_2) - t \cup s} \mid .$$

Therefore

$$\mathcal{N}_{def} = \mid \Omega_s \mid \times (2 \times \mid \Omega_{d(\phi_1) - t \cup s} \mid \times \mid \Omega_{d(\phi_2) - t \cup s} \mid) - 1,$$

thus we have

$$\mathcal{N}_{scp} = \mid \Omega_s \mid \times (\mid \Omega_{d(\phi_1) - t \cup s} \mid + \mid \Omega_{d(\phi_2) - t \cup s} \mid) - 1.$$

Remark 1 The key point in extended t-splitting transformation is to find an adjoint set s of variables such that $t \cup s \neq d(\phi)$ and the final $t \cup s$-partition of I exists with respect to a given initial factorization of ϕ, $\phi = \phi_1 \otimes \phi_2 \cdots \otimes \phi_n$. At the end of this subsection, we are going to introduce two types of algorithms for solving this problem.

2.4 *Splitting Algorithm of Valuation Algebras*

Based on the above discussion, in this subsection we are going to give a splitting algorithm of labeled valuation algebras.

Definition 5 Suppose that $\phi = \phi_1 \otimes \phi_2 \otimes \cdots \otimes \phi_n$, $t \subseteq d(\phi)$, then a sequence of valuations $\psi_1, \psi_2, \ldots, \psi_m$ is called a splitting sequence of $\phi^{\downarrow t}$ if
(1) $\psi_1 = \phi^{\downarrow t}$;
(2) $\forall i = 2$ to m, ψ_i is the result of ψ_{i-1} by using the (extended) t-splitting transformation.

Definition 6 (*Splitting Algorithm*) Suppose that (Φ, D) is an labeled valuation algebra. $\phi_i \in \Phi$, $i = 1, 2, \ldots, n$. $\phi = \phi_1 \otimes \phi_2 \otimes \cdots \otimes \phi_n$, denote $d(\phi) = V$, $t \subseteq V$. The *splitting algorithm* means constructing a splitting consequences of $\phi^{\downarrow t}$ by (extended) t-splitting transformations.

3 Main Result 2: Automaton Representation of Semiring Valued Constraint Problem

Given a constraint system $CS =< S, D, V >$, where S is a positively ordered semiring, $V = \{X_1, X_2, \cdots, X_N\}$, $D(X_i) = \{0_i, 1_i\}$. $c = (con, def)$, $con = \{X_{i_1}, \ldots, X_{i_d}\}$, $d = |con|$. $def : \prod_{j=i_1}^{i_d} D(X_j) \to S$.

By combing the method in [21] and our splitting algorithm, we give the following procedures for automaton representation of semiring valued constraint problem:

First of all, we can construct an S-DFA $\mathcal{A}_c = (Q, \delta, q_0, F)$ on $\Sigma = \bigcup\{D(X_i)|i = 1, \cdots, N\}$ which accept def as follows:
(1) $Q = \{p_0, p_1\} \cup \{q_\alpha | \alpha$ is a 0-1 string, $1 \leq |\alpha| \leq d\}$;
(2) $\delta(p_0, 0_1) = q_0$, $\delta(p_0, 1_1) = q_1$; $\forall \alpha$ which is a 0-1 string and $1 \leq |\alpha| \leq d$, $\delta(q_\alpha, 0_{|\alpha|+1}) = q_{\alpha 0}$, $\delta(q_\alpha, 1_{|\alpha|+1}) = q_{\alpha 1}$.
(3) p_0 is the initial state.
(4) If $|\alpha| = d$, $F(q_\alpha) = def(\alpha)$. Here for instance, if $con = \{X_1, X_2, X_3\}$, then $F(q_{010}) = def(0_1, 1_2, 0_3)$.

For those we have not specified, $\delta(s, \sigma) = p_1$; $F(s) = 0$, $s \in Q$.

Secondly, denote the corresponding S-DFA with c_1 and c_2 by $\mathcal{A}_1$ and $\mathcal{A}_2$, respectively. Then we can show that $\mathcal{A}_1 \times \mathcal{A}_2$ is an automaton representation of $c_1 \otimes c_2$.

Thirdly, given a constraint $c = (con, def)$ with S-DFA representation $\mathcal{A}_c = (Q, \delta, q_0, F)$. Assume $con = \{X_1, X_2, \cdots, X_N\}$. $con' = \{X_{i_1}, X_{i_2}, \cdots, X_{i_k}\}$. $i_1 < \cdots < i_k$. We try to get an S-FFA $\pi(\mathcal{A}) = (\pi(Q), \pi(\delta), \pi_{0(Q)}, \pi(F))$ which satisfies $\mathcal{R}ec_{\pi(\mathcal{A})} = c \Downarrow con'$ as follows.

(1) $\forall i = 1, \cdots, n+1$, let $Q_i = \{\delta^*(q_0, \theta) | \theta \in \prod_{j=1}^{i-1} D(X_j)$. Here, for instance, $(0_1, 0_2, 1_3)$ is treated as the string $0_1 0_2 1_3$. In particular, $Q_1 = \{q_0\}$. Denote $\pi(Q) = \bigcup_{j=1}^{k} Q_{i_j} \cup Q_{i_k+1}$.

(2) $\forall \sigma \in \Sigma, \forall q \in Q_{i_j}$,

• When $j = 1, \cdots, k-1$, if $i_{j+1} - 2 \geq i_j$, then define $\pi(\delta)(q, \sigma) = \{\delta^*(q, \sigma\theta) | \theta \in \prod_{h=i_j}^{i_{j+1}-2} D(X_h)\} \cap Q_{i_{j+1}}$; otherwise $\pi(\delta)(q, \sigma) = \{\delta^*(q, \sigma)\} \cap Q_{i_{j+1}}$.

• When $j = i_k$, define $\pi(\delta)(q, \sigma) = \{\delta^*(q, \sigma)\} \cap Q_{i_k+1}$.

(3) $\pi_0(Q) = Q_{i_1}$.

(4) $\forall q \in Q_{i_k+1}$, define $\pi(F)(q) = \sum\{F(\delta^*(q, \theta) | \theta \in \prod_{h=i_k+1}^{n} D(X_h)\}$. For those we have not specified, $\pi(F)(q) = 0$.

Remark 2 $\pi(\mathcal{A})$ is an S-FFA, and we can translate it as an DFA $\pi_D(\mathcal{A})$ without using the powerset construction method.

Given a constraint system $CS = (S, D, V)$, $c_1 =< def_1, con_1 >$, $c_2 =< def_2, con_2 >$. We want to get an S-DFA representation of $(c_1 \bigotimes c_2) \Downarrow_{con}$. According to what we have discussed above, we can do this in the following way:

Step 1 construct $\mathcal{A}_{c_1}$ and $\mathcal{A}_{c_2}$.
Step 2 construct $\mathcal{A}_{c_1} \times \mathcal{A}_{c_2}$.
Step 3 construct $\pi(\mathcal{A}_{c_1} \times \mathcal{A}_{c_2})$ with respect to con.
Step 4 construct $\pi_D(\mathcal{A}_{c_1} \times \mathcal{A}_{c_2})$.

According to expression (3) in Definition 1, we have

Theorem 8 *Given a constraint system* $CS = (S, D, V)$, $c_1 =< def_1, con_1 >$, $c_2 =< def_2, con_2 >$. *We can get an S-DFA representation of* $(c_1 \bigotimes c_2) \Downarrow_{con}$ *in the following steps:*

Step 1 *construct* $\mathcal{A}_{c_1}$ *and* $\mathcal{A}_{c_2}$.
Step 2^* *construct* $\pi(\mathcal{A}_{c_1})$ *and* $\pi(\mathcal{A}_{c_2})$ *with respect to con.*
Step 3^* *construct* $\pi_D(\mathcal{A}_{c_1})$ *and* $\pi_D(\mathcal{A}_{c_2})$.
Step 4^* *construct* $\pi_D(\mathcal{A}_{c_1}) \times \pi_D(\mathcal{A}_{c_2})$.

Definition 7 (*Automaton representation for single query of semiring-valued constraint satisfaction problems*) Given a constraint system $CS = (S, D, V)$, $C = \{c_1, \cdots, c_n\}$, $c_i =< def_i, con_i >$. $con \subseteq V$, $V = \bigcup\{con_i\}$. Give an S-DFA representation of $(c_1 \otimes \cdots \otimes c_n) \Downarrow con$.

Splitting algorithm for Automaton representation for single query of semiring-valued constraint satisfaction problems

Step 1 First get a splitting sequence of $(c_1 \otimes \cdots \otimes c_n) \Downarrow con$ by using splitting algorithm.

Step 2 Get the order for combination and projection operations according to the last valuation in the splitting sequence we get in *Step 1*.

Step 3 By the order we have in *Step 2*, we get the order for constructing product automatons and projection automatons.

Acknowledgments This work is supported by the National Natural Science Foundation of China (Grant No. 61403290, 61261047) and the horizontal subject (Grant No. HX0112071615) of Xidian University. And this work is also funded by the Fundamental Research Funds for the Central Universities (JB150704).

References

1. Biso, A., Rossi, F., Sperduti, A.: Experimental results on Learning Soft Constraints. In: Proceedings of the 7th International Conference on Principles of Knowledge Representation and Reasoning (KR 2000)
2. Bistarelli, S.: Semirings for soft constraint solving and programming. Lecture Notes in Computer Science, vol. 2962. Springer, Heidelberg (2004)
3. Bistarelli, S., Montanari, U., Rossi, F.: Constraint solving over semirings. In: Proceedings of IJCAI 1995, Morgan Kaufman, San Francisco (1995)
4. Bistarelli, S., Montanari, U., Rossi, F.: Semiring-based constraint solving and optimization. J. ACM **44**, 201–236 (1997)
5. Bistarelli, S., Montanari, U., Rossi, F.: Semiring-based constraint logic programming: syntax and semantics. ACM Trans. Program. Lang. Syst. (TOPLAS) **23**, 1–29 (2001)
6. Bistarelli, S., Codognet, P., Rossi, F.: Abstracting soft constraints: framework, properties, examples. Artif. Intell. **139**, 175–211 (2002)
7. Cooper, M.C., Schiex, T.: Arc consistency for soft constraints: reduction opertions in fuzzy or valued constraint satisfaction. Artif. Intell. **154**, 199–227 (2004)
8. Dechter, R.: Bucket elimination: a unifying framework for reasoning. Artif. Intell. **113**, 41–85 (1999)

9. Dechter, R., Kask, K., Larrosa, J.: A general scheme for multiple lower bound computation in constraint optimization. In: Proceedings of CP2001, 346–360 (2001)
10. Dechter, R., Rish, I.: Mini-buckets: a general scheme for bounded inference. J. ACM **50**, 107–153 (2003)
11. Fargier, H., Lang, J., Schiex, T.: Selecting prefered solutions in fuzzy constraint satisfaction problems. In: Proceedings of the 1st European Congress on Fuzzy and Intelligent Technologies (EUFIT) (1993)
12. Haenni, R.: Ordered valuation algebras: A generic framework for approximation inference. Internat. J. Approx. Reason. **37**, 1–41 (2004)
13. Jensen, F.V., Lauritzen, S.L., Olesen, K.G.: Bayesian updating in causal probabilistic networks by local computation. Comput. Stat. Quart. **4**, 269–282 (1990)
14. Kask, K., Dechter, R.: Branch and bound with mini-bucket heuristics. In: Proceedings of the International Joint Conference on Artificial Intelligence(IJCAI99), 426–433 (1999)
15. Kohlas, J.: Information algebras: Generic structures for inference. Springer, Heidelberg (2003)
16. Kohlas, J., Shenoy, P.P.: Computation in valuation algebras. In: Gabbay, D.M., Smets, P. (eds.) Handbook of Defeasible Reasoning and Uncertainty Management Systems. Algorithms for Uncertainty and Defeasible Reasoning, vol. 5, pp. 5–39. Kluwer Academic Publishers, Dordrecht (2000)
17. Kohlas, J., Wilson, N.: Semiring induced valuation algebras: exact and approximate local computation algorithms. Artif. Intell. **172**, 1360–1398 (2008)
18. Kolda, T.G.: Partitioning sparse rectangular matrices for parallel processing, in: Solving Irregularly Structured Problems in Parallel, Irregualr'98, number 1457 in Lecture Notes in Computer Science. Springer, Berlin, 68–79 (1998)
19. Lauritzen, S.L., Spiegelhalter, D.J.: Local computations with probabilities on graphical structures and their apllication to expert systems. J. Royal Statis. Soc. B **50**, 157–224 (1988)
20. Leenen, L., Meyer, T., Ghose, A.K.: Relaxations of semiring constraint satisfaction problems. In: Beek, P.V. (ed.) CP 2005 . LNCS, vol. 3709, Springer, Heidelberg (2005)
21. Li, Y. M.: The representation of semiring-based constraint satisfaction problems using fuzzy finite automata. In: Proceedings of the Quantitative Logic and Soft Computing-the QlSC'2012, Springer, Xi'an (2012)
22. Montanari, U.: Networks of constraints: fundamental properties and applications to Picture Processing, Information Science, 95–132 (1974)
23. Pouly, M., Kohlas, J.: Local computation dynamic programming. Technical Report 07-02, University of Fribourg (2007)
24. Pouly, M., Haenni, R., Wachter, M.: Compiling solution configurations in semiring valuation systems. In: Gelbukh, A., Kuri Morales, A.F., (eds.). MICAI 2007, LNAI 4827, pp. 248–259 (2007)
25. Rossi, F., Sperduti, A.: Learning solution preferences in constraint problems, Journal of Theoretical and Experimental Artificial Intelligence (JETAI)10 (1998)
26. Rossi, F., Sperduti, A.: Acquiring both constraint and solution preferences in interactive constraint systems. Constraints 9, (2004)
27. Schneuwly, C., Pouly, M., Kohlas, J.: Local computation in covering join trees. Technical Report 04-16, University of Fribourg (2004)
28. Shafer, G.: An axiomatic study of computation in hypertrees. Working Paper 232, School of Business. University of Kansas (1991)
29. Shenoy, P.P., Shafer, G.: Axioms for probability and belief-function propagation. In: Shafer, R.D., Levitt, T.S., Kanal, L.N., Lemmer, J.F. (eds.). Uncertainty in Artificial Intelligence 4. Machine intelligience and pattern recognition 9169–198 (1990)
30. Shenoy, P.: Valuation-based systems: a framework for managing uncertainty in expert systems. In L.A. Zadeh, Kacprzyk (eds), Fuzzy Logic for the Management of Uncertainty, 83–104 (1992)
31. Wachter, M., Haenni, R., Pouly, M.: Optimizing inference in Bayesian netwoks and semiring valuation algebras. In: Gelbukh, A., Kuri Morales, A.F. (eds.), MICAI 2007, LNAI 4827, 236–247 (2007)

Topological Constructions of Epsilon-Bisimulation

Yan-Fang Ma and Liang Chen

Abstract ϵ-bisimulation provides a kind of abstraction description for the correctness of software with probabilistic information. ϵ-limit bisimulation had been proposed, which entails that specification is the limit of implementations based on ϵ-bisimulation. In this paper, we only focus on the topological properties of ϵ-limit bisimulation. According to the definition of ϵ-limit bisimulation, several closure constructions are established, such as subnet closure, tail closure, natural extension and iteration. These closure constructions are useful to characterize properties of software.

Keywords Topology · Bisimulation · Correctness of software · Probabilistic process calculus

1 Introduction

According to the theory of process algebra, correctness can be described by relation between implementation and its specification. There exists many relations which can be used to describe correctness, such as bisimulation equivalence, trace equivalence, failure equivalence, and so on. In real world situations, some softwares are often approximate correctness. This is mainly caused by some probabilistic phenomenons of system. For example, in network system, unreliability of hardware may make a site to crash with probability 0.0034 [1, 2]. The correctness of these softwares with probabilistic phenomenons may be abstracted by probabilistic bisimulation in probabilistic process algebra [3–5].

Y.-F. Ma (✉)
School of Computer and Science, Huaibei Normal University, Huaibei 235000, China
e-mail: clmyf@163.com

L. Chen
School of Mathematical Science, Huaibei Normal University, Huaibei 235000, China
e-mail: clmyf2@163.com

T.-H. Fan et al. (eds.), *Quantitative Logic and Soft Computing 2016*,
Advances in Intelligent Systems and Computing 510,
DOI 10.1007/978-3-319-46206-6_22

ϵ-bisimulation is an important relation in the deterministic probabilistic processes, which extends the classical probabilistic bisimulation [6, 7] to approximate case. If a deterministic probabilistic process P simulate Q, but the absolute difference between the probability of executing the same action is less than or equal to ϵ, then P simulates Q with a bound ϵ, and vice versa. Furthermore, a quantitative model of deterministic probabilistic processes is presented. In software development and design, if the real specification is described as a probabilistic process, then one may use probabilistic bisimulation to prove the correctness of software. However, in the real world, implementations often approximate the specification, so ϵ-bisimulation of probabilistic processes set may be used to verify the correctness of software.

In the course of developing and designing software, many reasons can lead to the fact that the first implementation does not satisfy the specification completely, such as the technology of design, hardware equipment, and so on. Therefore, the developer or designer should modify the implementation step by step. Thus, a lots of implementation versions are obtained. These implementations have the same characterization, that is close to the specification more and more. Since the procedure of modifying implementation might be concurrent, steps of modification can be treated as a partial order.

For describing the relation that implementations are close to its specification step by step, Ying proposed strong limit bisimulations and weak limit bisimulation. Some important topological characteristics are proved in [8, 9]. The first author established the limit and topology of two-thirds bisimulation and parameterized bisimulation [10, 11].

ϵ-bisimulation can be used to verify the correctness of software with probabilistic information. In [12], the first author presented ϵ-limit bisimulation and ϵ-bisimulation limit to describe the situation that implementation approximates specification dynamically. In topology, closure is an important definition, the closure of a set A is the least closed set that including A. A set and its closure have the same limit points. Since the ϵ-bisimulation limit characterizes the fact that the limit of implementation is its specification under ϵ-bisimulation, in this paper, we will show whether or not the set of probabilistic process and its closure have the same ϵ-bisimulation limit.

In Sect. 2, the definition of ϵ- bisimulation is introduced. In Sect. 3, some topological constructions of ϵ-limit bisimulation are presented, such as subnet closure of ϵ-limit bisimulation, tail closure of ϵ-limit bisimulation, natural extension of ϵ-limit bisimulation and iteration structure of ϵ-limit bisimulation. In Sect. 4, we state our future work.

2 Preliminaries

Let $\mathcal{A}$ be the *names* set, $\bar{\mathcal{A}}$ the *co-names* set and $\Gamma = \mathcal{A} \cup \bar{\mathcal{A}}$ the *labels* set. 1 is the "idle action". Define $Act = \Gamma \cup \{1\}$ to be the set of all *actions*; $\alpha, \beta \ldots \in Act$.

The syntax of probabilistic process algebra and semantics can be found in [7]. Define DPr to be the set of all deterministic probabilistic processes. That means that for each $W \in DPr$, $W \overset{\alpha[p]}{\rightarrow} Q$, where the probabilistic derivation of type α has at most one. Next, we review the definition of ϵ-bisimulation.

Definition 1 (*ϵ-bisimulation*) [7] Let $\epsilon \in [0, 1)$. $\mathcal{R}_\epsilon \subseteq DPr \times DPr$ is a binary relation. If for any $\alpha \in Act$, $(W, Q) \in \mathcal{R}_\epsilon$ implies:

(1) when $W \overset{\alpha[p]}{\rightarrow} W'$, $Q' \in DPr$ can be found, such that $Q \overset{\alpha[q]}{\rightarrow} Q'$, the absolute difference $|p - q|$ less than or equal to ϵ and the next states satisfy $(W', Q') \in \mathcal{R}_\epsilon$;

(2) when $Q \overset{\alpha[q]}{\rightarrow} Q'$, $W' \in DPr$ can be found, such that $W \overset{\alpha[p]}{\rightarrow} W'$, the absolute difference $|p - q|$ less than or equal to ϵ and the next states satify $(W', Q') \in \mathcal{R}_\epsilon$, then $\mathcal{R}$ is called an ϵ-bisimulation.

If an ϵ-bisimulation $\mathcal{R}_\epsilon$ can be found such that $(W, Q) \in \mathcal{R}_\epsilon$, then the processes W and Q are called ϵ-bisimilar, written $W \overset{\epsilon}{\sim} Q$. Define $\overset{\epsilon}{\sim} = \bigcup\{R_\epsilon : R_\epsilon$ is an ϵ-bisimulation $\}$.

3 The Topological Construction of ϵ-Limit Bisimulation

When ϵ-bisimulation is chosen as the criteria of verifying the correctness between specification and implementation, the first implementation may not satisfy the specification. So, the implementation will be modified. Thus, a series of implementations are produced. For a simple system, these implementations can form a sequence $\{P_n : n \in N\}$, where N is natural number set. In $\{P_n : n \in N\}$, $P_1, P_2, \ldots$ can not satisfy the specification, but $n_0 \in N$ can be found, such that for each $n \geq n_0$, $P_n \overset{\epsilon}{\sim} Q$, where Q is the specification.

From the view of topology, specification can be treated as limit of implementations. Generally, for a complex system, the design pattern may be net, for example, the system includes several module. These module can be revised at the same time. In this case, sequence is not sufficient to describe this kind of pattern. So, we can appeal to net in topology. In [12], the author presented ϵ- limit bisimulation and ϵ-bisimulation limit to describe dynamic correctness. In this work, some important topological constructions of ϵ-limit bisimulation will be established in order to find some mathematical tools to explain the developing and designing of software.

Suppose that $(B, \leq)$ be a directed set, U be a nonempty set. A net V in U over D can be expressed as $\{V_n : n \in B\}$, where for every $n \in B$, $V_n = V(n) \in U$. For the detailing information of directed set and net can be found in the paper [13].

Let DPr_N be the set of all nets on DPr. For any $\{V_n : n \in B\} \in DPr_N$, where B is directed set, $V_n \in DPr$ for any $n \in B$. In the next discuss, we will use the definition of cofinality, cofinal subset and subnet, that are important to establish the closure construction of ϵ- limit bisimulation. For the detailing information, please see the paper [14].

Definition 2 (*ϵ-limit bisimulation*) [12] For $\epsilon \in [0, 1)$, a relation $S_\epsilon \subseteq DPr \times DPr_N$ is a relation between processes set and processes net set. If for each $\alpha \in Act$, $(W, \{V_n : n \in B\}) \in S_\epsilon$ implies:

(1) when $W \overset{\alpha[p]}{\rightarrow} W', \{V_n' : n \in B\} \in DPr_N$ and $n_0 \in B$ can be found, such that $V_n \overset{\alpha[q_n]}{\rightarrow} V_n'$, for each $n \geq n_0$, the absolute difference $|p - q_n|$ less than or equal to ϵ and the next stats satisfy $(W', \{V_n' : n \in B\}) \in S_\epsilon$;

(2) when H is a cofinal subset of B and $V_m \overset{\alpha[q_n]}{\rightarrow} V_m'$ for any $m \in H$, $W' \in DPr$ and a cofinal subset K of H can be found, such that $W \overset{\alpha[p]}{\rightarrow} W'$, for each $k \in K$, the absolute difference $| p - q_k |$ less than or equal to ϵ and the next stats satisfy $(W', \{V_k : k \in K\}) \in S_\epsilon$.

From the definition of ϵ- limit bisimulation, we know that ϵ-limit bisimulation is the limit version of ϵ-bisimulation. When the specification is described as deterministic probabilistic process P, the implementations are characterized by the deterministic probabilistic process net $\{Q_n : n \in D\}$, and P and $\{Q_n : n \in D\}$ have the relation of ϵ-bisimulation, then the ultimate aim of modifying implementations $\{Q_n : n \in D\}$ is to satisfy the specification P, therefore P is treated as the limit behavior of $\{Q_n : n \in D\}$.

Definition 3 (*ϵ-bisimulation limit*) [12]
(1) Suppose that $\epsilon \in [0, 1)$, $W \in DPr$ and $\{V_n : n \in B\} \in DPr_N$. If an ϵ-limit bisimulation $R_\epsilon \subseteq DPr \times DPr_N$ can be found such that $(W, \{V_n : n \in B\}) \in R_\epsilon$, then W is an ϵ-bisimulation limit of $\{V_n : n \in B\}$, written $W \overset{\epsilon}{\sim} \lim_{n \in B} V_n$.
(2)Suppose that $\{V_n : n \in B\} \in DPr_N$. If $W \in DPr$ and $\epsilon \in [0, 1)$ can be found such that $W \overset{\epsilon}{\sim} \lim_{n \in D} V_n$, then $\{V_n : n \in B\}$ is said to be ϵ- bisimulation convergent.

From the paper [7], we know that the identical relation between deterministic probabilistic process is ϵ-bisimulation. Next, we will try to establish a relation between deterministic probabilistic process and deterministic probabilistic process net, which is the extension of identical binary relation between deterministic probabilistic process. Suppose that $\epsilon \in [0, 1)$,

$$Ilim_{S_\epsilon} = \{W, \{V_n : n \in B\}) : W \in DPr, \{V_n : n \in B\} \in DPr_N, \text{and there is } n_0 \in D \text{ satisfies } V_n = W \text{ for each } n \geq n_0\}.$$

Proposition 1 $\epsilon \in [0, 1)$, $S_\epsilon \subseteq DPr \times DPr_N$. *Then $Ilim_{S_\epsilon}$ is an ϵ-limit bisimulation.*

This proposition states that, in the real development and designing of software, if the implementations obtained are the same with specification, then from the view of topology, specification is the limit behavior of these implementations. We don't need to modify the implementation. More generally, we can obtain the following property.

$$S_\epsilon = \{(W, \{V_n : n \in B\}) : W \in DPr, \{V_n : n \in DPr_N\} \text{and there exists } n_0 \in B \text{ such that } V_n \overset{\epsilon}{\sim} W \text{ for any } n \geq n_0\}.$$

Proposition 2 *S_ϵ is an ϵ-limit bisimulation.*

Next, we will show some topological constructions of ϵ-limit bisimulations, which are useful to describe some designing module of software from mathematical view.

3.1 Subnet Closure

Let $\epsilon \in [0, 1)$, $S_\epsilon \subseteq DPr \times DPr_N$. The subnet closure of ϵ- limit bisimulation is defined as following.

$$sub(S_\epsilon) = \{(W, \{V_n : n \in B\}) : \ (W, \{U_m : m \in H\}) \in S_\epsilon \text{ can be found such that } \{V_n : n \in B\} \text{ is the subnet of } \{U_m : m \in H\}\}.$$

Theorem 1 *Suppose that $\epsilon \in [0, 1)$ and S_ϵ be a relation between deterministic probabilistic process and deterministic probabilistic process net, i.e. $S_\epsilon \subseteq DPr \times DPr_N$. If for any $\alpha \in Act$, $(W, \{V_n : n \in B\}) \in S_\epsilon$ satisfies:*
(1) when $W \stackrel{\alpha[p]}{\rightarrow} W'$, $\{V_n' : n \in B\} \in DPr_N$ and $n_0 \in B$ can be found, such that $V_n \stackrel{\alpha[q_n]}{\rightarrow} V_n'$, the absolute difference $|p - q_n|$ less than or equal to ϵ for every $n \geq n_0$ and the next states satisfy $(P, \{Q_n : n \in D\}) \in sub(S)$;
(2) when H is a cofinal subset of B and $V_m \stackrel{\alpha[q_m]}{\rightarrow} V_m'$ for every $m \in H$, then $W' \in DPr$ and a cofinal subset M of H can be obtained, such that $W \stackrel{\alpha[p]}{\rightarrow} W'$, the absolute difference $|p - q_k|$ less than or equal to ϵ for $k \in M$ and $(W', \{V_k : k \in M\}) \in sub(S_\epsilon)$ holds, then $sub(S_\epsilon)$ becomes an ϵ-limit bisimulation.

Proof "$\Rightarrow$" is obvious.

"$\Leftarrow$" Let $(W, \{U_m : m \in H\}) \in sub(S_\epsilon)$. Then $(W, \{V_n : n \in B\}) \in S_\epsilon$ can be found, that satisfies $\{U_m : m \in H\}$ is a subnet of $\{V_n : n \in B\}$. According to the definition of subnet, there exists a function $N : H \rightarrow B$, which makes (H, N) is a cofinality of B and for each $m \in H$, $U_m = V_{N_m}$. By the properties of subnet, we can assume the function N is increasing.

When $W \stackrel{\alpha[p]}{\rightarrow} W'$, then $\{V_n' : n \in B\}$ and $n_0 \in B$ can be found that makes for all $n \geq n_0$, $V_n \stackrel{\alpha[q_n]}{\rightarrow} V_n'$, the absolute difference $|p - q_n| \leq \epsilon$ and the next states satisfy $(W', \{V_n' : n \in B\}) \in sub(S_\epsilon)$. (H, N) is a cofinality of B leads to there exists $m_0 \in H$ such that for each $m \geq m_0$, $N_{m_0} \geq n_0$. For each $m \in H$, let $U_m' = V_{N_m}'$. So, $U_m = V_{N_m} \stackrel{\alpha[q_{N_m}]}{\rightarrow} V_{N_m}' = U_m'$ for each $m \geq m_0$, and $\{U_m' : m \in H\}$ is a subnet of $\{V_n' : n \in B\}$ and $(W', \{U_m' : m \in H\}) \in sub(S_\epsilon)$. That means that when $W \stackrel{\alpha[p]}{\rightarrow} W'$, $\{U_m' : m \in H\}$ and $m_0 \in H$ can be obtained such that $U_m \stackrel{\alpha[q_{N_m}]}{\rightarrow} U_m'$ holds, and for all $m \geq m_0$, the absolute difference $| p - q_{N_m} | \leq \epsilon$, $(W', \{U_m' : m \in H\}) \in sub(S_\epsilon)$.

When M is a cofinal subset of H, and for each $f \in M$, $U_f \stackrel{\alpha[q_f]}{\rightarrow} U_f'$, then according to the definition of subset, we can obtain $N(M)$ is a cofinal subset of B. For every

$f \in M$, suppose that $V'_{N_f} = U'_f$. Then for each $f \in M$, $V_{N_f} = U_f \stackrel{\alpha[q_f]}{\rightarrow} U'_f = V'_{N_f}$. $W' \in DPr$ and a cofinal subset G of $N(M)$ can be obtained, that satisfy $W \stackrel{\alpha[p]}{\rightarrow} W'$ and $(W', \{V'_r : r \in G\}) \in sub(S_\epsilon)$. Furthermore, N is increasing that leads to $N^{-1}(G)$ is a cofinal subset of M. By the definition of ϵ- limit bisimulation, it holds that $\{v'_r : r \in G\} = \{U'_f : f \in N^{-1}(G)\}$ and $(W', \{U'_f : f \in N^{-1}(G)\}) \in sub(S_\epsilon)$. The theorem is proved. □

Proposition 3 *Let $\epsilon \in [0, 1)$. If S_ϵ is an ϵ-limit bisimulation, then $sub(S_\epsilon)$ is also an ϵ-limit bisimulation.*

3.2 Tail Closure

Let B be a directed set, $n \in B$ and $\epsilon \in [0, 1)$. Then $B[n) = \{m \in B : n \leq m\}$ and $B[n)$ is a cofinal subsets of B. Suppose that $S_\epsilon \subseteq DPr \times DPr_N$. Then

$$tail(S_\epsilon) = \{(W, \{V_n : n \in B\}) : (W, \{V_n : n \in B[n_0)\}) \in S_\epsilon \text{ for some } n_0 \in B\}.$$

Theorem 2 *Let $\epsilon \in [0, 1)$, $S_\epsilon \subseteq DPr \times DPr_N$. If for any $\alpha \in Act$, $(W, \{V_n \mid n \in B\}) \in S$ satisfies:*
(1) when $W \stackrel{\alpha[p]}{\rightarrow} W'$, $\{V'_n \mid n \in B\} \in DPr_N$ and $n_0 \in B$ can be obtained, that satisfies for any $n \geq n_0$, $V_n \stackrel{\alpha[q_n]}{\rightarrow} V'_n$, the absolute difference $\mid p - q_n \mid$ less than or equal to ϵ and the next states satisfy $(W', \{V'_n \mid n \in B\}) \in tail(S_\epsilon)$.
(2) when H is a cofinal subset of B and for every $m \in H$, $V_m \stackrel{\alpha[q_m]}{\rightarrow} V'_m$, $W' \in DPr$ and a cofinal subset M of H can be found, that make $W \stackrel{\alpha[p]}{\rightarrow} W'$, the absolute difference $\mid p - q_k \mid$ less than or equal to ϵ, where $k \in M$ and the next states satisfy $(W', \{V'_k : k \in M\}) \in tail(S_\epsilon)$, then $tail(S_\epsilon)$ is an ϵ- limit bisimulation.

Proof "⇒" is easy to prove. Now we prove "⇐".

Let $(W, \{V_n : n \in B\}) \in tail(S_\epsilon)$. Then, some $n_0 \in B$ can be found such that $(W, \{V_n : n \in B[n_0)\}) \in S_\epsilon$. When $W \stackrel{\alpha[p]}{\rightarrow} W'$, $\{V'_n : n \in B[n_0)\} \in DPr_N$ and $n_1 \in B[n_0)$ can be obtained such that $V_n \stackrel{\alpha[q_n]}{\rightarrow} V'_n$, and for any $n \in B[n_0)$, the absolute difference $\mid p - q_n \mid \leq \epsilon$ with $n \geq n_1$, and the next states satisfy $(W', \{V'_n : n \in B[n_0)\}) \in tail(S_\epsilon)$. That is to say, there is $n_2 \in B[n_0)$ leads to $(W', \{V'_n : n \in B[n_2)\}) \in S_\epsilon$. For any $n \in B$ with $n \ngeq n_0$, we can choose an arbitrary elements of DPr as V'_n. Thus, we can obtain $\{V'_n : n \in B\} \in DPr_N$, and $V_n \stackrel{\alpha[q_n]}{\rightarrow} V'_n$, for each $n \geq n_1$, the absolute difference $\mid p - q_n \mid \leq \epsilon$ and $(W', \{V'_n : n \in B\}) \in tail(S_\epsilon)$ holds.

On the other hand, when H is a cofinal subset of B and for each $m \in H$, $V_m \stackrel{\alpha[q_m]}{\rightarrow} V'_m$, then $m_0 \in C$ can be obtained such that $m_0 \geq n_0$, $H[m_0)$ is a cofinal subset of

$B[n_0)$ and for each $m \in H[m_0)$, $V_m \overset{\alpha[q_m]}{\rightarrow} V'_m$. Thus, there are $V' \in DPr$ and a cofinal subset of $H[m_0)$, M, such that $W \overset{\alpha[p]}{\rightarrow} W'$, the absolute difference $| p - q_k | \leq \epsilon$ and $(W', \{V'_k : k \in M\}) \in tail(S_\epsilon)$ holds. Furthermore, M is also a cofinal sunset of H that leads to the proof of the theorem. □

Proposition 4 *Let* $\epsilon \in [0, 1)$. *When* S_ϵ *is an* ϵ*-limit bisimulation,* $tail(S_\epsilon)$ *is also an* ϵ*-limit bisimulation.*

3.3 Nature Extension

Natural extension is a kind of closure property on the processes and nets of processes. Next, this definition is reviewed.

Definition 4 ([8]) Let $\{U_m : m \in H\}, \{V_n : n \in B\} \in DPr_N$. If (H, N) is a cofinality of B and for every $n \in B$, $V_n = U_{m_n}$ for some $m_n \in H$ with $N_{m_n} \geq n$, then $\{V_n : n \in B\}$ is a natural extension of $\{U_m : m \in H\}$.

For any $S_\epsilon \subseteq DPr \times DPr_N$,

$$
\begin{aligned}
ext(S_\epsilon) = \{(W, \{V_n : n \in B\}) :\ & \text{there exists}\{U_m : m \in H\} \in DPr_N \text{ such that} \\
& (W, \{U_m : m \in\}) \in S_\epsilon \text{ and}\{V_n : n \in B\} \text{ is natural extensions of} \\
& \{U_m : m \in H\}\}.
\end{aligned}
$$

Theorem 3 *Let* $\epsilon \in [0, 1)$. *When* S_ϵ *is an* ϵ*-limit bisimulation,* $ext(S_\epsilon)$ *is also an* ϵ*-limit bisimulation.*

Proof Let $(W, \{V_n : n \in B\}) \in ext(S_\epsilon)$. Then $\{U_m : m \in H\} \in DPr_N$ can be obtained such that $(W, \{U_m : m \in H\}) \in S_\epsilon$ and $\{V_n : n \in B\}$ is a natural extension of $\{U_m : m \in H\}$. By the definition of natural extension, there is a function N such that (H, N) is a cofinality of B. Generally, assume that N is increasing. At same time, for any $n \in B$, $V_n = U_{m_n}$ holds. And, there is some $m_n \in H$, such that $N_{m_n} \geq n$, $n_1 \leq n_2 \Rightarrow m_{n_1} \leq m_{n_2}$.

(1) When $W \overset{\alpha[p]}{\rightarrow} W'$, $\{U'_m : m \in H\} \in DPr_N$ and $m_0 \in H$ can be obtained such that $U_m \overset{\alpha[p_m]}{\rightarrow} U'_m$, for any $m \geq m_0$, the absolute difference$| p - p_m | \leq \epsilon$ and the next states satify $(W', \{U'_m \mid m \in H\}) \in S$. For any $n \in B$, let $V'_n = U'_{m_n}$, $N_{m_n} \geq n$. For any $n \geq N_{m_0}$, $N_{m_n} \geq n \geq N_{m_0}$ holds and $m_n \geq m_0$. Then for all $n \geq N_{m_0}$, $V_n = U_{m_n} \overset{\alpha[p_{m_n}]}{\rightarrow} U'_{m_n} = V'_n$, the absolute difference $| p - p_{m_n} | \leq \epsilon$, $V'_n \mid n \in B\}$ is a natural extension of $\{U'_m \mid m \in H\}$ and $(W', \{V'_n \mid n \in B\}) \in ext(S_\epsilon)$.

(2) When M is a cofinal subset of B and for all $l \in M$, $V_l \overset{\alpha[q_l]}{\rightarrow} V'_l$, then there exists $M' = \{m_l : l \in M\}$ that is a cofinal subset of H. For each $l \in M$, let $U'_{m_l} = V'_l$. Thus, for all $l \in M$, $U_{m_l} = V_l \overset{\alpha[q_l]}{\rightarrow} V'_l = U'_{m_l}$. Therefore, there are $W' \in DPr$ and a

cofinal subset of M,K, satisfying $W \overset{\alpha[p]}{\rightarrow} W'$, for each $k \in K$, the absolute difference $| p - q_k | \leq \epsilon$ and $(W', \{V'_k : k \in K\}) \in S_\epsilon$ holds. Assume that $F = \{l \in M : m_l \in K\}$. Then F is a cofinal subset of M, and $\{V'_l : l \in F\} = \{U'_k : k \in K\}$. So, we finish the proof. □

3.4 Iteration Structure

In this subsection, we will discuss the iteration structure of ϵ-limit bisimulation. This construction can express the dynamic counterpart of the composition of binary relation on processes set.

Suppose that for every $j \in J$, $(B_j, \leq_j)$ be a directed set. $\{(B_j, \leq_j) \mid j \in J\}$ is defined as $\times_{j\in J}(B_j, \leq_j) = (\times_{j\in J} B_j, \leq)$, where $\leq$ means that for all $d, e \in \times_{j\in J} B_j$, $d \leq e$ if and only if $d(j) \leq e(j)$ for every $j \in J$. From the definition of directed set, we can get $(\times_{j\in J} B_j, \leq)$ is a directed set. Assume that B be a directed set. For each $m \in B$, D_m be a directed set and $C = B \times \times_{m\in B} D_m$. If for any $m \in B$, $\{R(m, n) : n \in D_m\}$ is a net over D_m, then the iteration $\prod_{m\in B} \{R(m, n) : n \in D_m\}$ of $\{R(m, n) : n \in D_m\}(m \in B)$ is the net $\{R(m, f(m)) : (m, f) \in C\}$ over C.

Now let $\epsilon \in [0, 1)$, $S_\epsilon \subseteq DPr \times DPr_N$ and $(T_m)_\epsilon \subseteq DPr \times DPr_N$ for every $m \in B$. Then the composition of $S_\epsilon \circ \{(T_m)_\epsilon : m \in B\}$ of S_ϵ and $\{(T_m)_\epsilon : m \in B\}$ is defined as

$$S_\epsilon \circ \{(T_m)_\epsilon : m \in B\} = \{(W, \prod_{m\in B} \{R(m, n) : n \in D_m\}) : \text{there exists } V_m \in DPr \ (m \in B) \text{ such that} (W, \{V_m : m \in B\}) \in S_\epsilon \text{ and for each } m \in B, (V_m, \{R(m, n) : m \in D_m\}) \in (T_m)_\epsilon\}.$$

Theorem 4 *Suppose that $\epsilon \in [0, 1)$. Let S_ϵ and $(T_m)_\epsilon$ $(m \in B)$ be both $\frac{1}{2}\epsilon$-limit bisimulations. Then $S_\epsilon \circ \{(T_m)_\epsilon : m \in D\}$ is also an ϵ-limit bisimulation.*

Proof Suppose that $(W, \prod_{m\in D} \{R(m, n) : n \in D_m\}) \in S_\epsilon \circ \{(T_m)_\epsilon : m \in B\}$. Then there exists $V_m \in DPr$ $(m \in B)$ such that $(W, \{V_m : m \in B\}) \in S_\epsilon$, and for each $m \in B$, $(V_m, \{R(m, n) : m \in D_m\}) \in (T_m)_\epsilon$.

Suppose that $W \overset{\alpha[p]}{\rightarrow} W'$. Then $\{V'_m : m \in B\} \in DPr_N$ and $m_0 \in B$ can be acquired such that $V_m \overset{\alpha[q_m]}{\rightarrow} V'_m$, and for each $m \geq m_0$, the absolute difference $| p - q_m | \leq \frac{1}{2}\epsilon$. At the same time, $(W', \{V'_m : m \in B\}) \in S_\epsilon$. For every $m \geq m_0$, $\{R'(m, n) : n \in D_m\}$ and $f_0(m) \in D_m$ can be obtained that make $R(m, n) \overset{\alpha[r_{(m,n)}]}{\rightarrow} R'(m, n)$, for $n \geq f_0(m)$, the absolute difference $| q_m - r_{(m,n)} | \leq \frac{1}{2}\epsilon$ and the next states hold $(V'_m, \{R'(m, n) \mid n \in D_m\} \in (T_m)_\epsilon$. The arbitrary element of D_m can be choose to be $f_0(m)$ for $m \ngeq m_0$. Therefore, $f_0 \in \times_{m\in B} D_m$ is well defined and $(m, f) \geq (m_0, f_0)$. So

$m \geq m_0, f(m) \geq f_0(m)$, and $R(m, f(m)) \overset{\alpha[r_{(m,f(m))}]}{\rightarrow} R'(m, f(m))$, and $\mid p - r_{(m,f(m))} \mid \leq \mid p - q_m \mid + \mid q_m - r_{(m,f(m))} \mid \leq \epsilon$ for any $(m, f) \geq (m_0, f_0)$. Furthermore, $(W', \{R'(m, f(m)) \mid (m, f) \in C\}) \in S_\epsilon \circ \{(T_m)_\epsilon \mid m \in B\}$.

(2) Let U be a cofinal subset of C. Define

$$Proj_B U = \{m \in B : (m, f) \in U \text{ for some} f \in X_{m \in B} D_m\}.$$
$$Proj_m U = \{f(m) : (m, f) \in U\}, \text{for any } m \in B.$$

By the definition of cofinal subset, we can get $Proj_B U$ is a cofinal subset of B. At the same time, for every $m \in Proj_B U$, $Proj_m U$ is also a cofinal subset of D_m. For all $(m, f) \in U$, let $R(m, f(m)) \overset{\alpha[r(m,f(m))]}{\rightarrow} R'(m, f(m))$. Then $(T_m)_\epsilon$ is a $\frac{1}{2}\epsilon$-limit bisimulation leads to for every $m \in Proj_B U$, $R(m, n) \overset{\alpha[r(m,n)]}{\rightarrow} R'(m, n)$ for all $n \in Proj_m U$ and $V'_m \in DPr$, a cofinal subset K_m of $Proj_m U$ and $q_m \in [0, 1]$ can be found to make $V_m \overset{\alpha[q_m]}{\rightarrow} V'_m$, the absolute difference $\mid r(m, n) - q_m \mid \leq \frac{1}{2}\epsilon$ and $(V'_m, \{R(m, k)' : n \in K_m\}) \in (T_m)_\epsilon$ holds. Furthermore, S_ϵ is a $\frac{1}{2}\epsilon$-limit bisimulation and $Proj_B U$ is a cofinal subnet of B make there are $W' \in DPr$ and a cofinal subnet H of $Proj_B U$ and $p \in [0, 1]$ such that $W \overset{\alpha[p]}{\rightarrow} W'$ and $\mid q_h - p \mid \leq \frac{1}{2}\epsilon$, $h \in H$ and $(W', \{V'_h : h \in H\}) \in S_\epsilon$. So, $(W', \Pi_{h \in H}\{R(h, k)' : k \in K_h\}) \in S_\epsilon \circ \{(T_m)_\epsilon : m \in B\}$, and $\mid r(m, n) - p \mid \leq \mid r(m, n) - q_h \mid + \mid q_h - p \mid \leq \epsilon$. By the definition of cofinal subset of U, we can get $H \times \times_{h \in H} K_h$ is a cofinal subset of U. Therefore, $S_\epsilon \circ \{(T_m)_\epsilon : m \in B\}$ is also an ϵ-limit bisimulation. □

Theorem 4 states the iteration property between modulars of system. If given the specification P, and the first designment of implementation $\{Q_n \mid n \in D\}$ may not satisfy the specification, then the more concrete designment of implementation will be developed. Thus, the iteration between the first implementation and the more concrete implementation can satisfy the specification.

Proposition 5 *For each $j \in J$, let S_j is the ϵ-limit bisimulation. Then $\bigcup_{j \in J} S_j$ is also the ϵ-limit bisimulation.*

Theorem 5 (1) *Let $W \in DPr$. If there is $n_0 \in B$ that makes $V_n \overset{\epsilon}{\sim} W$ for each $n \geq n_0$, then $W \overset{\epsilon}{\sim} \lim_{n \in B} V_n$.*

(2) *When $\{V_n : n \in B\}$ is a subnet of $\{U_m : m \in H\}$ and $W \overset{\epsilon}{\sim} \lim_{m \in H} U_m$, then $W \overset{\epsilon}{\sim} \lim_{n \in B} V_n$.*

(3) *Suppose that B be a directed set. For each $m \in B$, let D_m be a directed set. Assume that $C = B \times \times_{m \in B} D_m$ and $R(m, f) = (m, f(m))$ for each $(m, f) \in C$. When for each $m \in B$, $V_m \overset{\frac{1}{2}\epsilon}{\sim} \lim_{n \in D_m} W(m, n)$, and $V \overset{\frac{1}{2}\epsilon}{\sim} \lim_{m \in B} V_m$, then $V \overset{\epsilon}{\sim} \lim_{(m,f) \in C} (W \circ R)(m, f)$.*

Proof (1), (2) and (3) can be obtained by Propositions 2, 3 and 4. □

This theorem states that the set of probabilistic processes and its closure have the same ϵ-bisimulation limit.

4 Conclusion and Future Work

In this work, the topological constructions of ϵ-limit bisimulation are mainly discussed. These topological constructions obtained in this paper can help the developer and designer of software to verify the correctness of software.

Notice that our results are based on deterministic probabilistic processes. However, there are many nondeterministic phenomena during the development and designing of software. So, in the future, we will focus on the other methods to establish the dynamic correctness for the nondeterministic processes.

Acknowledgments The work is supported by the NSFC grant (61300048), the Anhui Provincial Natural Science Foundation grant (1508085MA14), the Key Natural Science Foundation grant of Universities of Anhui Province (KJ2014A223), the Excellent Young Talents in Universities of Anhui Province, the major teaching reform project of Anhui higher education revitalization plan (2014ZDJY058), the provincial teaching research project of Anhui province (2015JYXM157), the teaching research project of Huaibei Normal Unversity (JY15118) and the excellent teacher project of Huaibei Normal University (2015ZYJS185). The author would like to thank Professor S.A. Smolka for his invaluable suggestions about the topological properties of ϵ- limit bisimulation.

References

1. Larsen, K.G., Skou, A.: Compositional verification of probabilistic processes. In: Cleaveland, W. R. (ed.) Proceedings of the CONCUR'92, vol. 630, pp. 46–471. Springer, Berlin (1992)
2. Smolka, S.A., Steffen, B.U.: Priority as extremal probability. In: Baeten, J.C.M., Klop, J. W. (eds.) Proceedings of the CONCUR'90. Lecture Notes in Computer Science, vol. 458, pp. 456–466. Springer, Berlin (1990)
3. Segala, R., Lynch, N.: Probabilistic simulations for probabilistic processes. In Jonsson, B., Parrow, J. (eds.) Proceedings of the Concur' 94. Lecture Notes in Computer Science, vol. 836, pp. 481–496. Springer, Berlin (1994)
4. Deng, Y.X., Glabbeek, R., Hennessy, M., Morgan, C.: Testing finitary probabilistic processes. Lecture Notes in Computer Science, vol. 5710, pp. 274–288. Springer, Berlin (2009)
5. Song, L., Deng, Y.X., Cai, X.J.: Towards automatic measurement of probabilistic processes. In: Proceedings of the 7th International Conference on Quality Software, pp. 50–59 (2007)
6. van Glabbeek, R.J., Smolka, S.A., Steffen, B.: Reactive, generative, and stratified models of probabilistic processes. Inf. Comput. **121**(1), 59–80 (1995)
7. Giacalone, A., Jou, C.C., Smolka, S.A.: Algebraic Reasoning for Probabilistic Concurrenct Systems. In: Proceedings of the IFIP TC2 Working Conference on Programming Concepts and Methods, Tiberias, Israel, pp. 443–458 (1990)
8. Ying, M.S.: Topology in Process Calculus: Approximate Correctness and Infinite Evolution of Concurrency Programs. Springer, Berlin (2001)

9. Ying, M.S.: Bisimulation indexes and their applications. Theor. Comput. Sci. **275**(1–2), 1–68 (2002)
10. Ma, Y.F., Zhang, M.: Topological construction of parameterized bisimulation limit. Electr. Notes Theor. Comput. Sci. **257**, 55–70 (2009)
11. Ma, Y.F., Zhang, M., Chen, Y.X.: Formal description of software dynamic correctness. J. Comput. Res. Dev. **50**(3), 626–635 (2013)
12. Ma, Y.F., Zhang, M.: The infinite evolution mechnism of ϵ- bisimilarity. J. Comput. Sci. Technol. **28**(6), 1097–1105 (2013)
13. Engelking, R.: General Topology. Polish Scientific Publisher, Warszawa (1977)
14. Kelley, J.L.: General Topology. Springer, New York (1975)

An Outline of 4-Valued Transition Statement Calculus

Long Hong

Abstract This paper aims to establish a framework of 4-valued transition statement calculus so that we can characterize transition states and the changing process among them from logic. We briefly introduce interval adjacency and transition as preliminaries; create transition connectives that can reflect the direction and the multi-state of transition, and interpret the intuition meaning of truth-value of them. Emphasis here is on an analysis to the intension of a transition. We establish a formal system of 4-valued transition statement calculus L^{T4}, and focus on the discussions of the valuation and the characteristics of L^{T4}, in which soundness theorem and adequacy theorem are given.

Keywords Transition phenomenon · Semantic · Syntax · Soundness · Adequacy

1 Introduction

Transition means a gradually changing process from one state to another [1]. For example, dawn is the transition from night to day; middle age is the transition from youth to old age. These examples show an unidirectional type of transition. Another type of transition is bidirectional such as gray from black to white or from white to black, and the amplifying region of the transistor from cut-off to saturation and vice versa. Moreover, transition has also a multi state-ness, i.e., many states are in the transfer process between two states. For example, in the changing process from spring to winter, it goes through summer between spring and autumn, and autumn between summer and winter. The examples mentioned above can be perceived by human senses, so transition is a universal phenomenon in nature or society. Since the 1900 s transition has been widely concerned and millions of published papers using

L. Hong (✉)
College of Computer, Nanjing University of Posts and Telecommunications, Nanjing, China
e-mail: hongl@njupt.edu.cn

L. Hong
Institute of Modern Logic and Application, Nanjing University, Nanjing, China

T.-H. Fan et al. (eds.), *Quantitative Logic and Soft Computing 2016*,
Advances in Intelligent Systems and Computing 510,
DOI 10.1007/978-3-319-46206-6_23

transition as a subject covered almost all disciplines [2, 3]. Though there are quite a number of contributions for dealing with transition, most of these papers studied the transition in specific. Transition is of state, and studying state from logic point of view is a basic method. There have been many kinds of non-classic logic that describe state, such as modal logic [4], tense logic [5], medium logic [6] and so on. We shall introduce a novel multi-valued logic for characterizing transition and the changing process among states in order to set a logical foundation on transition.

The remainder of this paper is organized as follows: in Sect. 2, we briefly introduce interval adjacency and transition as preliminaries. Incorporating the intension of a transition, we create transition connectives in Sect. 3, and intuitively describe the functions of them. In Sect. 4, we establish formal system of 4-valued transition statement calculus (L^{T4}). We introduce the valuation of L^{T4}, and give soundness theorem and adequacy theorem in Sect. 5.

2 Preliminaries

2.1 *Interval Adjacency*

The left hand symbol and the right hand symbol of an interval are respectively denoted by ⟦ and ⟧, i.e. ⟦ ∈ {(, [} and ⟧ ∈ {),]}. The group of three symbols ⟧, ⟦ in the middle of ⟦a, b⟧, ⟦b, c⟧ is called interval adjacency; the two intervals are called neighboring intervals, and b is called the adjacent element. The interval adjacency is called a **Type I** adjacency if both ⟧ and ⟦ in the adjacency are closed or open, otherwise that is called a **Type II** adjacency. Therefore, a sequence of neighboring intervals is as follows:

$$\ldots, ⟦r_i, r_{i+1}⟧, ⟦r_{i+1}, r_{i+2}⟧, \ldots, ⟦r_{i+n-1}, r_{i+n}⟧, \ldots$$

where r is a real number. Clearly, there are n-1 interval adjacencies in the sequence consisted of n neighboring intervals.

2.2 *Transition*

We describe transition in the real number field.

Definition 1 Given $f : X \to R_i \cup R_{i+1} \cup R_{i+2} \subset \boldsymbol{R}$, where X is a non empty set. Let $a \in R_i$, $b \in R_{i+1}$ and $c \in R_{i+2}$, then a< b<c. If f is monotonic and continuous, the variable process of its value in R_i is called the transition of f at somewhere between R_{i-1} and R_{i+1}; f is called a transition function, and R_{i+1} is called a shift area.

Denote the ordered triple T(y, μ, β) as a transition. Where, $y = f(x)$ and $x \in X$, and y is a transition variable; the μ is a shift area mentioned above; the β that is the

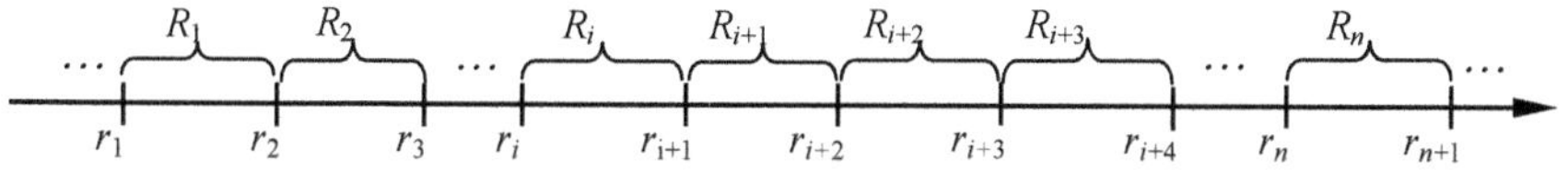

Fig. 1 The expression of transition in number line. From increasing transition, R_{i+1} is the shift area from R_i to R_{i+2}; R_i is the initial area and R_{i+2} is the arrival area. However, R_i and $R_{i=2}$ are arrival area and initial area in decreasing transition, respectively. For multi-step transition, R_i is the initial area and R_{i+3} is the arrival area; R_{i+1} and R_{i+2} are shift area.

first point of a shift area reached by f is called the beginning point of the transition. Moreover, let $x \in \mu$ and $x \neq \beta$, then T(y, μ, β) is an increasing transition if $\beta < x$, or a decreasing transition if $\beta > x$. T(y, μ, β) is called a single-step transition if $\mu = R_i$, or multi-step transition if $\mu = \mathrm{R}_i \cup R_{i+1} \cup \ldots \cup R_{i+n}$.

Theorem 1 *The interval adjacencies among the codomain of a transition function are **Type II** adjacencies.*

By Theorem 1, the form of the neighbor intervals of an increasing transition is as follows:

$$\ldots, [r_i, r_{i+1}), [r_{i+1}, r_{i+2}), \ldots, [r_{i+n-1}, r_{i+n}), \ldots$$

and that of a decreasing transition is

$$\ldots, (r_i, r_{i+1}], (r_{i+1}, r_{i+2}], \ldots, (r_{i+n-1}, r_{i+n}], \ldots$$

Transition occurs one after the other and continuously; the description of it in real line is shown in Fig. 1. Increasing and decreasing transition are for showing direction of the transition. Consider increasing transition, in Fig. 1, R_{i+1} is the shift area (μ) from R_i to R_{i+2}; R_i is the initial area of the transition and R_{i+2} is the arrival area. For decreasing transition, on the contrary, R_i and R_{i+2} are arrival area and initial area, respectively. These three areas form a complete transition.

3 Connectives of Transition Logic

3.1 *Transition Connectives*

To characterize transition, firstly, we create four transition connectives besides classic connectives, which are shown in Table 1.

Table 1 Names and symbols of transition connectives

Name	Symbol
Single-step increasing connective	$\uparrow$
Single-step decreasing connective	$\downarrow$
Multi-step increasing connective	$\nearrow$
Multi-step decreasing connective	$\swarrow$

Then, we assign truth-value to initial area so that ensure whether a transition is complement. Let p be a statement and it stands for 'R_i is the initial area of the transition', then there are five statements: $\uparrow$, $\downarrow$, p, $\nearrow p$ and $\swarrow p$. Where $\uparrow$ stands for 'R_i is the initial area of the single-step increasing transition'; $\swarrow p$ stands for 'R_i is the initial area of the multi-step decreasing transition', etc. If assignment of the compound statements above is true, it means that there are three areas related to R_i, initial, shift and arrival, which form a complete transition.

3.2 Intuitive Interpretation of Some Compound Statements

(1) $\nearrow\alpha \vee \swarrow\alpha$, $\uparrow\alpha \vee \downarrow\alpha$.

In multi-step or single-step, transition is either increasing or decreasing.

(2) $\nearrow\alpha \rightarrow \uparrow\alpha$, $\swarrow\alpha \rightarrow \downarrow\alpha$.

If there exist a multi-step increasing (decreasing) transition, then there is single-step increasing (decreasing) transition.

(3) $\nearrow \alpha \rightarrow \neg \swarrow \alpha$, $\swarrow \alpha \rightarrow \neg \nearrow \alpha$.

If there exist multi-step increasing (decreasing) transition, then no multi-step decreasing (increasing) transition.

(4) $\neg(\swarrow \alpha \wedge \nearrow \alpha)$.

There is neither multi-step decreasing nor increasing transition.

(5) $\neg \nearrow \alpha \leftrightarrow \downarrow \alpha$.

Being single-step decreasing transition if and only if not multi-step increasing transition. Knowable, the negation directly to transition connectives is dual. As above, single-step and decreasing are obtained by negating both multi-step and increasing. The other compound statements like this explanation are as follows:

(6) $\uparrow\alpha \leftrightarrow \neg \swarrow\alpha$.

(7) $\nearrow\alpha \leftrightarrow \neg \downarrow\alpha$.

(8) $\swarrow\alpha \leftrightarrow \neg \uparrow\alpha$.

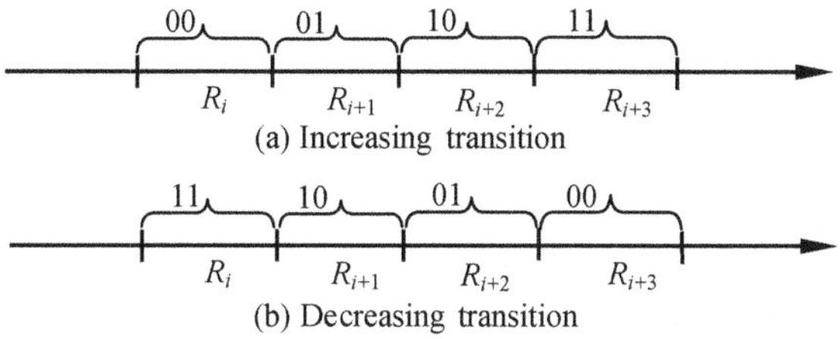

Fig. 2 Relation between truth-value and interval

3.3 *Truth-Value and Their Meaning*

Denote 00, 10, 01 and 00 as the truth-value of 4-valued transition logic. '11' is the same meaning as true in 2-value logic, and '00' as false. '10' and '01' are neither true nor false. Moreover, the scale of true of '10' is higher than '01'.

We firstly discuss where truth-value is in number line. As shown in Fig. 1, transition occurs one after the other and continuously, in which we may take four neighboring intervals to associating with four truths, so that can reflect direction and multi-state. The relation between truth-values and intervals is intuitively shown in Fig. 2. According to Theorem 1 and the caption of Fig. 1, '00' in Fig. 2 relates to initial area, and '11' to arrival area. From multi-step transition, '10' and '01' associate with transition area, but in single-step '10' relates to initial area and '01' to arrival area. For the same interval, furthermore, truth-value on increasing transition is different from decreasing area, e.g. the R_i in Fig. 2 relates to '00' on increasing transition, but to '11' on decreasing.

We then explain truths of transition connectives. If truth-value of transition connective is 11, then the transition is at initial area in a complete transition. If truth is 00, then the transition is at arrival area. If truth is 10, then the transition is at shift area on multi-step transition or at initial area on single-step. If the truth is 01, then it is at shift area, or at arrival area on single-step.

3.4 *Relation Between Increasing and Decreasing Transition*

The relation between truths and intervals in increasing and decreasing has been illustrated by Fig. 2(a) and Fig. 2(b). However, we must make relation between increasing and decreasing clear to use one of them as basic model for transition logic.

Definition 2 Let truth-value set of 4-valued transition logic $s_4 = \{00, 01, 10, 11\}$ and interval set interval$=\{R_i, R_{i+1}, R_{i+2}, R_{i+3}\}$, then

$$f_I : s_4 \to interval$$

is called increasing mapping, and $f_I(00) = R_i$, $f_I(01) = R_{i+1}$, $f_I(10) = R_{i+2}$ and $f_I(11) = R_{i+3}$;

$$f_D : s_4 \rightarrow interval$$

is called decreasing mapping, and $f_D(00) = R_{i+3}$, $f_D(01) = R_{i+2}$, $f_D(10) = R_{i+1}$ and $f_D(11) = R_i$.

Theorem 2 $\neg f_I^{-1}(R_j) = f_D^{-1}(R_j)$. *Where* $i \leq j \leq i+3$ *and symbol* $\neg$ *is bit negation connective.*

Theorem 2 gives a guarantee of the ability to describe transition direction by selecting either increasing or decreasing. And we use increasing transition as basic model.

4 Formal 4-Valued Transition Statement Calculus L^{T4}

4.1 Syntax of L^{T4}

(1) Alphabet of symbols

a. Statement variables: p; p_1, p_2, ...

b. Connectives: $\neg$, $\rightarrow$.

c. Transition connectives: $\nearrow$, $\swarrow$.

d. Others:), (.

(2) Set of well-formed formulas (*wff*)

a. p is a *wff* and p_i is a *wff*, where $i \geq 1$.

b. If α and β are *wffs*, then $\neg\alpha$, $\nearrow\alpha$, $\swarrow\alpha$ and $\alpha\rightarrow\beta$ are *wffs*.

c. The set of all *wffs* is merely generated by *a* and *b*.

(3) Axioms

Let Γ be the set of *wffs*, and α, β, $\varphi \in \Gamma$, then following *wffs* are axioms of L^{T4}:

(Ax1) $\alpha\rightarrow(\beta\rightarrow\alpha)$.

(Ax2) $(\neg\alpha \rightarrow \neg\beta) \rightarrow (\beta \rightarrow \alpha)$.

(Ax3) $(\alpha\rightarrow(\beta \rightarrow\varphi))\rightarrow((\alpha\rightarrow\beta)\rightarrow(\alpha\rightarrow\varphi))$.

(Ax4) $\swarrow\alpha\rightarrow\alpha$.

(Ax5) $\nearrow\alpha\rightarrow \neg\alpha$.

(Ax6) $\nearrow(\alpha\rightarrow\beta)\rightarrow (\nearrow\alpha\rightarrow\nearrow\beta)$.

(Ax7) $\swarrow(\alpha\rightarrow\beta)\rightarrow (\swarrow\alpha\rightarrow\swarrow\beta)$.

(4) Rule of deduction

(MF) α, $\alpha\rightarrow\beta \vdash \beta$.

Where '$\vdash$ 'is a symbol, but not a symbol of L^{T4}. If $\Gamma\vdash\alpha$, we say Γ yields α.

4.2 Basic Definitions

For simplicity of presentation, some connectives are defined as follows.

Definition 3

(1) $\alpha \wedge \beta$ is the abbreviation of $\neg(\alpha \rightarrow \neg\beta)$.
(2) $\alpha \vee \beta$ is the abbreviation of $\neg\alpha \rightarrow \beta$.
(3) $\alpha \leftrightarrow \beta$ is the abbreviation of $(\alpha \rightarrow \beta) \wedge (\beta \rightarrow \alpha)$.
(4) $\uparrow\alpha$ is the abbreviation of $\neg \swarrow \alpha$.
(5) $\downarrow\alpha$ is the abbreviation of $\neg \nearrow \alpha$.

4.3 Some Theorems

Theorem 3

(1) $\neg\neg\alpha \leftrightarrow \alpha$
(2) $\neg(\alpha \vee \beta) \leftrightarrow (\neg\alpha \wedge \neg\beta)$
(3) $\neg(\alpha \wedge \beta) \leftrightarrow (\neg\alpha \vee \neg\beta)$
(4) $(\alpha \rightarrow \beta) \leftrightarrow (\neg\alpha \vee \beta)$
(5) $(\alpha \vee \beta) \wedge \varphi) \leftrightarrow ((\alpha \wedge \varphi) \vee (\beta \wedge \varphi))$
(6) $(\alpha \wedge \beta) \vee \varphi) \leftrightarrow ((\alpha \vee \varphi) \wedge (\beta \vee \varphi))$

Theorem 4

(1) $\alpha \vdash \beta,\ \beta \vdash \varphi \Rightarrow \alpha \vdash \varphi$
(2) $\Gamma,\ \neg\alpha \vdash \beta,\ \neg\beta \Rightarrow \Gamma \vdash \alpha$
(3) $\Gamma,\ \alpha \vdash \beta \Rightarrow \Gamma \vdash \alpha \rightarrow \beta$
(4) $\alpha \wedge \beta \vdash \alpha,\ \beta$
(5) $\alpha,\ \beta \vdash \alpha \wedge \beta$
(6) $\neg\beta,\ \alpha \rightarrow \beta \vdash \neg\alpha$
Where '⇒' is a symbol of nature language. If $\alpha \Rightarrow \beta$, *we say 'if* α *then* β*'.*

Theorem 5

(1) $(\neg\alpha \rightarrow \alpha) \rightarrow \alpha$
(2) $\alpha \rightarrow (\neg\alpha \rightarrow \beta)$
(3) $\neg\alpha \rightarrow (\alpha \rightarrow \beta)$
(4) $\beta \rightarrow (\neg\alpha \rightarrow \beta)$

Theorem 6

(1) $\nearrow(\alpha \rightarrow \beta) \rightarrow \nearrow(\neg\beta \rightarrow \neg\alpha)$
(2) $\nearrow(\alpha \rightarrow \beta) \rightarrow (\nearrow \neg\beta \rightarrow \nearrow \neg\alpha)$
(3) $\nearrow\alpha \rightarrow (\alpha \rightarrow \beta)$
(4) $\alpha \rightarrow (\nearrow\alpha \rightarrow \beta)$

(5) $\uparrow(\alpha\rightarrow\beta)\rightarrow\ \uparrow(\neg\beta\rightarrow\neg\alpha)$
(6) $\nearrow(\alpha\rightarrow\beta)\rightarrow(\uparrow\alpha\rightarrow\uparrow\beta)$
(7) $\swarrow(\alpha\rightarrow\beta)\rightarrow\swarrow(\neg\beta\rightarrow\neg\alpha)$
(8) $\neg\alpha\rightarrow(\swarrow\alpha\rightarrow\beta)$
(9) $\swarrow\alpha\rightarrow(\neg\alpha\rightarrow\beta)$
(10) $\downarrow(\alpha\rightarrow\beta)\rightarrow\ \uparrow(\neg\beta\rightarrow\neg\alpha)$
(11) $\nearrow(\nearrow\alpha\rightarrow\nearrow\beta)\rightarrow(\swarrow\alpha\rightarrow\swarrow\beta)$
(12) $\uparrow(\nearrow\alpha\rightarrow\nearrow\beta)\rightarrow(\swarrow\alpha\rightarrow\swarrow\beta)$
(13) $\nearrow(\nearrow\alpha\rightarrow\nearrow\beta)\rightarrow\neg(\nearrow\alpha\rightarrow\nearrow\beta)$
(14) $\nearrow\alpha\rightarrow\nearrow(\alpha\wedge\beta)$
(15) $\nearrow\beta\rightarrow\nearrow(\alpha\wedge\beta)$
(16) $\nearrow(\alpha\vee\beta)\rightarrow\nearrow\alpha$
(17) $\nearrow(\alpha\vee\beta)\rightarrow\nearrow\beta$
(18) $\swarrow(\alpha\wedge\beta)\rightarrow\swarrow\alpha$
(19) $\swarrow(\alpha\wedge\beta)\rightarrow\swarrow\beta$
(20) $\swarrow\alpha\rightarrow\swarrow(\alpha\vee\beta)$
(21) $\swarrow\beta\rightarrow\swarrow(\alpha\vee\beta)$
(22) $\nearrow(\alpha\vee\beta)\rightarrow(\nearrow\alpha\vee\nearrow\beta)$
(23) $(\swarrow\alpha\vee\swarrow\beta)\rightarrow\swarrow(\alpha\vee\beta)$
(24) $(\nearrow\alpha\wedge\nearrow\beta)\rightarrow\nearrow(\alpha\wedge\beta)$

Proof For (3).

(1) $\nearrow\alpha$	assumption
(2) $\nearrow\alpha\rightarrow\neg\alpha$	Ax5
(3) $\neg\alpha$	(1), (2), MP
(4) $\neg\alpha\rightarrow(\alpha\rightarrow\beta)$	Theorem 5 (3)
(5) $\alpha\rightarrow\beta$	(3), (4), MP
(6) $\nearrow\alpha\rightarrow(\alpha\rightarrow\beta)$	(1), (5), Theorem 4 (3) □

For (9).

(1) $\swarrow\alpha$	assumption
(2) $\swarrow\alpha\rightarrow\alpha$	Ax4
(3) α	(1), (2), MP
(4) $\alpha\rightarrow(\neg\beta\rightarrow\alpha)$	Ax1
(5) $\neg\beta\rightarrow\alpha$	(3), (4), MP
(6) $(\neg\beta\rightarrow\alpha)\rightarrow(\neg\alpha\rightarrow\beta)$	Ax2, Theorem 3 (1)
(7) $\neg\alpha\rightarrow\beta$	(5), (6), MP
(8) $\swarrow\alpha\rightarrow(\neg\alpha\rightarrow\beta)$	(1), (7), Theorem 4 (3) □

Theorem 7

(1) $\alpha\vdash\ \downarrow\alpha$
(2) $\neg\alpha\vdash\ \uparrow\alpha$
(3) $\Gamma,\nearrow\alpha\vdash\beta,\neg\beta\Rightarrow\Gamma\vdash\alpha$

Theorem 8

(1) $\nearrow\alpha\rightarrow\uparrow\alpha$
(2) $\swarrow\alpha\rightarrow\downarrow\alpha$
(3) $\nearrow\alpha\rightarrow\neg\swarrow\alpha$
(4) $\swarrow\alpha\rightarrow\neg\nearrow\alpha$
(5) $\neg(\nearrow\alpha\wedge\swarrow\alpha)$
(6) $\neg\nearrow\alpha\vdash\swarrow\alpha$
(7) $\neg\swarrow\alpha\vdash\nearrow\alpha$

Proof For (1).

(1) $\neg(\nearrow\alpha\rightarrow\uparrow\alpha)$	assumption
(2) $\nearrow\alpha\wedge\neg\uparrow\alpha$	Definition 3 (2), Theorem 3 (2)
(3) $\nearrow\alpha$	(2), Theorem 4 (4)
(4) $\nearrow\alpha\rightarrow\neg\alpha$	Ax5
(5) $\neg\alpha$	(3), (4), MP
(6) $\neg\uparrow\alpha$	(2), Theorem 4 (4)
(7) $\swarrow\alpha$	Definition 3 (4), Theorem 3 (1)
(8) $\swarrow\alpha\rightarrow\alpha$	Ax4
(9) α	(3), (7), MP
(10) $\nearrow\alpha\rightarrow\uparrow\alpha$	(5), (9), (1), Theorem 4 (2) □

5 Discussion

5.1 Valuation of L^{T4}

Denote x_1x_0 as a truth-value form of L^{T4}, where $x_1, x_0\in\{0, 1\}$.

Definition 4 Let Γ be the set of wffs of L^{T4}, and s_4={11, 10, 01, 00} be the truth-value set of L^{T4}. The mapping

$$v:\Gamma\rightarrow s_4$$

is called a valuation of L^{T4} such that, for all $\alpha,\beta\in\Gamma$,

(1) $v(\alpha)=\alpha_1\alpha_0$, $v(\beta)=\beta_1\beta_0$;

(2) $v(\neg\alpha)={}^{\neg}v(\alpha)={}^{\neg}\alpha_1{}^{\neg}\alpha_0$, where ${}^{\neg}\alpha_i=\begin{cases}1 & \alpha_i=0\\ 0 & \alpha_i=1\end{cases}$, $0\leq i\leq 1$.

(3) $v(\nearrow\alpha)=\begin{cases}00 & \alpha_1=1\\ 1^{\neg}\alpha_0 & \alpha_1=0\end{cases}$.

(4) $v(\swarrow\alpha)=\begin{cases}00 & v(\alpha)=01\\ v(\alpha) & otherwise\end{cases}$.

(5) $v(\alpha\to\beta) =^{\to}(v(\alpha)v(\beta))=^{\to}\alpha_1\beta_1^{\to}\alpha_0\beta_0$, where $\to\alpha_i\beta_i = \begin{cases} 0 & \alpha_i = 1,\ \beta_i = 0 \\ 1 & otherwise \end{cases}$, $0 \le i \le 1$.

Theorem 9

(1) $v(\alpha\wedge\beta)= {}^{\wedge}(v(\alpha)v(\beta))= {}^{\wedge}\alpha_1\beta_1^{\wedge}\alpha_0\beta_0$, *where* ${}^{\wedge}\alpha_i\beta_i = \begin{cases} 1 & \alpha_i = \beta_i = 1 \\ 0 & otherwise \end{cases}$, $0 \le i \le 1$.

(2) $v(\alpha\vee\beta)={}^{\vee}(v(\alpha)v(\beta))= {}^{\vee}\alpha_1\beta_1^{\vee}\alpha_0\beta_0$, *where* ${}^{\vee}\alpha_i\beta_i = \begin{cases} 0 & \alpha_i = \beta_i = 0 \\ 1 & otherwise \end{cases}$, $0 \le i \le 1$.

(3) $v(\alpha\leftrightarrow\beta) ={}^{\leftrightarrow}(v(\alpha)v(\beta))={}^{\leftrightarrow}\alpha_1\beta_1^{\leftrightarrow}\alpha_0\beta_0$, *where* ${}^{\leftrightarrow}\alpha_i\beta_i = \begin{cases} 0 & \alpha_i \ne \beta_i \\ 1 & \alpha_i = \beta_i \end{cases}$, $0 \le i \le 1$.

(4) $v(\uparrow\alpha)= \begin{cases} 0^{\neg}\alpha_0 & \alpha_1 = 1 \\ 11 & \alpha_1 = 0 \end{cases}$.

(5) $v(\downarrow\alpha)= \begin{cases} 11 & v(\alpha) = 10 \\ v(\alpha) & otherwise \end{cases}$.

(6) ${}^{\to}\alpha_i\beta_i ={}^{\vee\neg}\alpha_i\beta_i$.

Theorem 10

(1) $v(\alpha \vee\neg\alpha)= 11$
(2) $v(\alpha \vee 00)= v(\alpha)$
(3) $v(\alpha \vee 11)= 11$
(4) $v(\alpha \wedge\neg\alpha)= 00$
(5) $v(\alpha \wedge 00)= 00$
(6) $v(\alpha \wedge 11)= v(\alpha)$
(7) $v(\alpha\to\beta) =v(\neg\alpha\vee\beta)$.

Theorem 11 *$v(\alpha\to\beta) =11$ if $v(\alpha)=v(\beta)$.*

Theorem 12 *$v(\alpha\to\beta) =v(\beta)$ if $v(\alpha)= 11$.*

5.2 Soundness Theorem for L^{T4}

Definition 5 Let $\alpha\in\Gamma$. If for every valuation v, v (α)=11, then α is called a tautology and denoted by $\models$ A; if α is a last member of some deductions from Γ, then α is yielded by Γ, denoted by $\Gamma \vdash \alpha$.

Theorem 13 *If $v(\alpha) =11$ and $v(\alpha\to\beta) =11$, then $v(\beta) =11$.*

Theorem 14 *$v(\alpha \to (\beta\to \alpha)) =11$.*

Theorem 15 *$v((\alpha\to(\beta\to\varphi))\to((\alpha\to\varphi)\to(\alpha\to\varphi))) =11$.*

Theorem 16 $v((\neg\alpha\to\neg\beta)\to(\beta\to\alpha))=11$.

Theorem 17 $v(\swarrow\alpha\to\alpha)=11$.

Theorem 18 $v(\nearrow\alpha\to\neg\alpha)=11$.

Theorem 19 $v(\nearrow(\alpha\to\beta)\to(\nearrow\alpha\to\nearrow\beta))=11$.

Proof $v(\nearrow(\alpha\to\beta)\to(\nearrow\alpha\to\nearrow\beta))$
$=v(\neg\nearrow(\alpha\to\beta)\vee\neg\nearrow\alpha\vee\nearrow\beta)$
$={}^{\vee}({}^{\vee}({}^{\neg}(v(\nearrow(\alpha\to\beta)))v(\nearrow\beta))^{\neg}(v(\nearrow\alpha)))$
Clearly, if $v(\nearrow(\alpha\to\beta))=00$, or $v(\nearrow\alpha)=00$, or $v(\nearrow\beta)=11$, then this theorem is valid. Consider $v(\nearrow(\alpha\to\beta))\neq00$. By Definition 4 (3), $v(\nearrow(\alpha\to\beta))=1^{\rightarrow}\alpha_0\beta_0$. There are two cases.
Case 1: When $v(\nearrow(\alpha\to\beta))=11$, $v(\alpha\to\beta)=00$, i.e. $v(\beta)=00$, so $v(\nearrow\beta)=11$.
Case 2: When $v(\nearrow(\alpha\to\beta))=10$, $v(\alpha\to\beta)=01$, i.e. $v(\beta)=00$ or $v(\beta)=01$. Thanks to $v(\beta)=01$, $v(\nearrow\beta)=10$. Hence
${}^{\vee}({}^{\neg}(v(\nearrow(\alpha\to\beta)))v(\nearrow\beta))$
$={}^{\vee}({}^{\neg}(10)\,(1^{\neg}\beta_0))$
$={}^{\vee}((01)\,(1^{\neg}\beta_0))$
$=11$. □

Theorem 20 $v(\swarrow(\alpha\to\beta)\to(\swarrow\alpha\to\swarrow\beta))=11$.

Theorem 21 $\Gamma\vdash\alpha\Rightarrow\models\alpha$. *(Soundness theorem)*

5.3 Adequacy Theorem for L^{T4}

Theorem 22 *L^{T4} is consistent.*

Definition 6 An extension L^{T4*} of L^{T4} is consistent if for $\alpha\in\Gamma$, in the theorems of L^{T4*} no α are both α and $\neg\alpha$, or both α and $\nearrow\alpha$, or both $\neg\alpha$ and $\swarrow\alpha$.

Theorem 23 *An extension L^{T4*} of L^{T4} is consistent if and only if there exist a wff that is not a theorem of L^{T4*}.*

Moreover, changing truth-value into decimal number from binary number, the valuation of α, for any $\alpha\in\Gamma$, is as follows:

$$V(\alpha)=\begin{cases}0\\ m & 0<m<3\\ 3\end{cases}.$$

Clearly, when $V(\alpha)=3$, α is a tautology; when $V(\alpha)=m$, non of α, $\neg\alpha$, $\nearrow\alpha$ and $\swarrow\alpha$ is a tautology. To describe briefly below, let $\Gamma^m\subset\Gamma$, and for any $\alpha\in\Gamma^m$, $V(\alpha)=m$.

Theorem 24 *Let* $\alpha \in \Gamma - \Gamma^m$*, and* α *is not a theorem of* L^{T4*}*. Then, an extension* L^{T4**} *of* L^{T4} *is consistent if it is obtained by adding* $\neg\, \alpha$ *or* $\nearrow \alpha$ *as an axiom.*

Theorem 25 $\models \alpha \Rightarrow \Gamma \vdash \alpha$. *(Adequacy theorem)*

Theorem 26 L^{T4} *is decidable.*

Acknowledgments This work was supported by the National Natural Science Foundation of China under Grant No. 61170322.

References

1. Simpson, J., Weiner, E.: The Oxford English Dictionary., 2nd edn. Clarendon Press, Oxford (1989)
2. http://www.engineeringvillage.com/controller/servlet
3. http://thomsonreuters.com/web-of-science
4. Kripke, S.A.: Semantical considerations on modal logic. Acta Philos. Fenn. **16**, 83–94 (1963)
5. Prior, A.: Time and Modality. Oxford University Press, Oxford (1957)
6. Xiao, Xi-An, Zhu, Wu-Jia: Propositional calculus of medium logic (I). Nat. Mag. **8**(4), 315–316 (1985)

Part IV
Fuzzy Connectives and Fuzzy Reasoning

Generalized G-Generated Implications

Yue Zhu and Dao-Wu Pei

Abstract A new class of fuzzy implications, which are called the generalized g-generated implications, generated from decreasing functions and g-generators, are proposed. This class of fuzzy implications are generalizations of g-generated implications proposed by Professor Yager in 2004. Naturally, some basic properties of these new fuzzy implications are investigated. The law of importation and the distributive equations for these fuzzy implications are studied in detail. Relations of this class of implications with other known fuzzy implications are discussed.

Keywords Fuzzy logic · Fuzzy implication · g-generated implication · Law of importation · Distributive equation

1 Introduction

Fuzzy implications are important operations in many fields. In particular, in fuzzy logic, as propositional connectives, fuzzy implications can be used to model fuzzy conditionals "If p, then q" where p and q are fuzzy statements [4]. As we know, fuzzy implications have a significant role in approximate reasoning and fuzzy control [2, 8, 13]. In addition, fuzzy implications in many other fields are also very important. These fields including many-valued logic, fuzzy decision making, image processing, expert system, data mining, fuzzy relational equation, fuzzy mathematical morphology, fuzzy DI-subsethood measures, and so on [2]. All these applications have led to generate more fuzzy implications.

The main research topics of fuzzy implications are the characterizations and construction methods.

Characterizations of fuzzy implications in terms of algebraic properties are important so that we can better understand the behavior of fuzzy implications.

Y. Zhu · D.-W. Pei (✉)
School of Science, Zhejiang Sci-Tech University, Hangzhou 310018, China
e-mail: peidw@163.com

T.-H. Fan et al. (eds.), *Quantitative Logic and Soft Computing 2016*,
Advances in Intelligent Systems and Computing 510,
DOI 10.1007/978-3-319-46206-6_24

Many kinds of fuzzy implications have been proposed. Among them, the most popular ones are R-implications, (S, N)-implications, QL-implications and D-implications, which are obtained from t-norms, s-norms and negations [2].

Unlike the above methods derived from binary operators, fuzzy implications can be constructed based on generating functions such as *f*- and *g*-generated implications and *h*-implications. *f*- and *g*-generated implications were proposed by Professor Yager [17] in 2004.

Baczyński et al. [1] studied some properties and relations between Yager's implications and the other kinds of fuzzy implications. And characterizations of Yager's implications based on distributive equations have been shown. Massanet et al. [11] obtained some characterizations of Yager's implications based on the law of importation. Recently, Xie and Liu [16] proposed a generalization of Yager's *f*-generated implications, and discussed the basic algebraic properties of the class of implications and studied some classical logic tautologies for them.

Hilnena et al. [5] obtained a class of new implications in terms of two fuzzy negations and a uninorm. This is the other way to generalize *f*-generated implications. Massanet et al. [12] introduced *h*-implications by means of additive generators of representable uninorms. Liu [7] proposed a new class of fuzzy implications in terms of the so-called generalized *h*-generators, and also discussed their properties.

In this paper, we propose a new class of fuzzy implications as generalization of Yager's *g*-generated implications. We investigate properties of these new implications, study the law of importation and distributive equations for them and discuss the relations between this new class of implications with other known fuzzy implications.

2 Preliminaries

This section recalls some necessary concepts and known results used in the rest of the paper (see [2] or [15]). We denote $U = [0, 1]$ in this paper.

A function $N : U \to U$ is called a fuzzy negation (shortly negation), if $N(0) = 1$, $N(1) = 0$ and N is decreasing.

A fuzzy negation is strict if it is strictly decreasing and continuous.

A fuzzy negation is strong if it is an involution.

There are three important fuzzy negations: the standard fuzzy negation, the least (or Gödel) and the greatest fuzzy negations are respectively defined as follows:

$$N_C(x) = 1 - x, \quad N_1(x) = \begin{cases} 1, & \text{if } x = 0, \\ 0, & \text{if } x \in (0, 1]. \end{cases} \quad N_2(x) = \begin{cases} 0, & \text{if } x = 1, \\ 1, & \text{if } x \in [0, 1). \end{cases}$$

A function $I : U^2 \to U$ is a fuzzy implication, (shortly implication), if it is decreasing about the first variable, increasing about the second variable, and $I(0, 0) = I(1, 1) = 1$, $I(1, 0) = 0$.

The set of all fuzzy implications will be denoted by $\mathcal{FI}$.

A fuzzy implication I is said to satisfy
(NP) if $I(1, y) = y, \;\; y \in U$;
(EP) if $I(x, I(y, z)) = I(y, I(x, z)), \;\; x, \; y, \; z \in U$;
(IP) if $I(x, x) = 1, \;\; x \in U$;
(OP) if $I(x, y) = 1 \iff x \leq y, \;\; x, \; y \in U$;
(CP(N)) if $I(x, y) = I(N(y), N(x))$ where N is a given negation, $x, \; y \in U$.

A binary operation T on U is a triangular norm (shortly t-norm), if it is commutative, associative, increasing and has a neutral element 1. Dually, a binary operation S on U is a triangular conorm (shortly t conorm), if it is commutative, associative, increasing and has a neutral element 0.

A pair of important t-norm and s-norm are as follows:

$$T_M(x, y) = \min(x, y), \;\; S_M(x, y) = \max(x, y), \;\; x, y \in U.$$

Two interesting fuzzy implications will be used to describe our main results: Weber implication and the largest (S,N)-implication ([2]):

$$I_{WB}(x, y) = \begin{cases} 1, \text{ if } x < 1, \\ y, \text{ if } x = 1; \end{cases} \quad I_D(x, y) = \begin{cases} 1, \text{ if } x = 0, \\ y, \text{ if } x > 0. \end{cases}$$

In 2004, Professor Yager [17] introduced two new kinds of implications as follows.

Definition 1 Let $f : \; U \to [0, \infty]$ be a strictly decreasing and continuous function with $f(1) = 0$. The function $I : \; U^2 \to U$ defined by,

$$I(x, y) = f^{-1}(x \cdot f(y)), \quad x, \; y \in U,$$

with convention $0 \cdot \infty = \infty$, is called an f- generated implication. The function f itself is called an f-generator of I. In such a case, to emphasize the apparent relation we will write I_f instead of I.

Definition 2 Let $g : \; U \to [0, \infty]$ be a strictly increasing and continuous function with $g(0) = 0$. The function $I : \; U^2 \to U$ defined by,

$$I(x, y) = g^{(-1)}(\frac{1}{x} \cdot g(y)), \quad x, \; y \in U,$$

with convention $\infty \cdot 0 = \infty$, $\frac{1}{0} = \infty$, is called a g-generated implication, where $g^{(-1)}$ is the pseudo-inverse of g given by

$$g^{(-1)}(x) = \begin{cases} g^{-1}(x), & \text{if } x \in [0, g(1)), \\ 1, & \text{if } x \in [g(1), \infty]. \end{cases}$$

g itself is called a g-generator of I. In such a case, we will similarly write I^g instead of I.

In this paper, we call a function $g: U \to [0, \infty]$ a g-generator if it is strictly increasing and continuous with $g(0) = 0$.

By Φ we denote the set of all increasing bijections on U.

Definition 3 For two n-ary functions $f, g : U^n \to U$, g is called a Φ-conjugate of f, denoted $g = f_\varphi$, if there exists a $\varphi \in \Phi$ such that

$$g(x_1, x_2, ..., x_n) = \varphi^{-1}(f(\varphi(x_1), \varphi(x_2), ..., \varphi(x_n))), \quad x_1, x_2, ..., x_n \in U.$$

Let $I \in \mathcal{FI}$. The function $N_I : U \to U$ defined by $N_I = I(x, 0)$ is called the natural negation of I or the negation induced by I.

By [2] we know that if $I_{S,N}$ is an (S,N)-implication and $I_{T,S,N}$ is a QL-operation, then

$$N_{I_{S,N}} = N, \quad N_{I_{T,S,N}} = N.$$

Proposition 1 ([2]) *If $I \in \mathcal{FI}$ satisfies (EP) and (NP), then I satisfies the law of contraposition CP(N) if and only if $N = N_I$ and N_I is strong.*

3 Definition and Properties of the New Implications

Yager's definition of g-generated implication can be naturally generalized as follows.

Definition 4 A function $I : U^2 \to U$ defined by

$$I(x, y) = g^{(-1)}(f(x) \cdot g(y)), \quad x, y \in U,$$

with the conventions $\infty \cdot 0 = \infty$, $\frac{1}{0} = \infty$, is called a generalized g-generated operation, where $f : U \to [1, \infty]$ is a decreasing and continuous function satisfying $f(0) = \infty, f(1) = 1$, g is a g-generator, and $g^{(-1)}$ is the pseudo-inverse of the function g in the sense of Proposition 2.

In such case, to emphasize the apparent relation we will write $I^{f,g}$ instead of I.

If a generalized g-generated operation is an implication, then we call it generalized g-generated implication. The set of all generalized g-generated implications will be denoted by $\mathcal{GGI}$.

Remark 1 (i) If $f(x) = \frac{1}{x}$, then $I^{f,g}$ is the same as g-generated implication I^g. This fact shows that the new implication $I^{f,g}$ is a generalization of Yager's g-generated implication.
(ii) It should be noted that in [16] Xie et al. introduced *generalized f-generated implication* $I_{f,g}(x, y) = f^{-1}(g(x) \cdot f(y))$, where f is an f-generator, g is an increasing function satisfy $g(0) = 0$ and $g(1) = 1$. This class of implications are very similar to our implication $I^{f,g}$ defined by Definition 4. However, they indeed are completely

different. The readers can clearly see this fact from Theorem 12 in Sect. 5.
(iii) According to the definition of pseudo-inverse, $I^{f,g}$ can be rewritten as

$$I^{f,g}(x, y) = g^{-1}(\min(f(x) \cdot g(y), g(1))), \quad x, \ y \in U.$$

(iv) Similar to the corresponding results and proofs of g-generated implications (see [17] or [2]), we can prove some results hold for generalized g-generated implications. For the sake of simplicity, we will omit proofs of these new results in this and next sections.

Example 1 (i) Let $f(x) = \frac{1}{x^2}$ and $g(x) = -\ln(1-x)$. Then for $x, \ y \in U$ we have

$$I^{f,g}(x, y) = \begin{cases} 1, & \text{if } x = y = 0, \\ 1 - (1-y)^{\frac{1}{x^2}}, & \text{otherwise.} \end{cases}$$

(ii) Let $f(x) = 1 - \ln x$ and $g(x) = x$. Then for $x, \ y \in U$ we have

$$I^{f,g}(x, y) = \begin{cases} (1 - \ln x) \cdot y, & \text{if } y \leq \frac{1}{1-\ln x}, \\ 1, & \text{otherwise.} \end{cases}$$

The next proposition shows that generalized g-generated implications are indeed fuzzy implications.

Proposition 2 *If $I^{f,g}$ is a generalized g-generated operation, then it is a fuzzy implication, i.e., $I^{f,g} \in \mathcal{FI}$.*

The following theorem shows the fact that when function f is fixed, a necessary and sufficient condition under which two generalized g-generated implications are equal is: the g-generators are unique up to a positive multiplicative constant.

Theorem 1 *Let $f : U \to [1, \infty]$ be a decreasing and continuous function satisfying $f(0) = \infty, f(1) = 1$, and $g_1, \ g_2$ two g-generators. Then the following statements are equivalent:*
(i) $I^{f,g_1} = I^{f,g_2}$;
(ii) *There exists a constant $c \in (0, \infty)$ such that $g_2(x) = c \cdot g_1(x), \ x \in U$.*

Remark 2 From the above result it follows that, if g is a g-generator such that $g(1) < \infty$, then function $g_1 : U \to U$, defined by $g_1(x) = \frac{g(x)}{g(1)}$, is a well defined g-generator such that $I^{f,g} = I^{f,g_1}$ and $g_1(1) = 1$. In other words, it is enough to consider only increasing generators with $g(1) = \infty$ or $g(1) = 1$.

The next proposition shows some properties of generalized g-generated implications.

Proposition 3 *If $I = I^{f,g} \in \mathcal{GGI}$, then*
(i) *I satisfies (NP) and (EP);*
(ii) *I satisfies (IP) if and only if*

$$f(x) \begin{cases} \geq \frac{g(1)}{g(x)}, & \text{if } g(1) < \infty \\ = 1, & \text{if } x = 1,\ g(1) = \infty \\ = \infty, & \text{if } x \in [0, 1),\ g(1) = \infty \end{cases} ;$$

(iii) *if $g(1) = \infty$, then I does not satisfy the ordering property (OP);*
rm(iv) $I(x, y) \geq y,\ \ x, y \in U$.

The following proposition shows some properties of natural negations of generalized g-generated implications.

Proposition 4 *If $I = I^{f,g} \in \mathcal{GGI}$, then for allx $\in U$ we have*

$$N_I(x) = \begin{cases} 1, & \text{if } f(x) = \infty, \\ 0, & \text{if } f(x) < \infty. \end{cases}$$

Thus, N_I is not continuous. In particular,
(i) *if $f(x)$ is strictly decreasing, then $N_I = N_1$.*
(ii) *if*

$$f(x) = \begin{cases} 1, & \text{if } x = 1, \\ \infty, & \text{if } x \in [0, 1), \end{cases}$$

then $N_I = N_2$.

Corollary 1 *If $I \in \mathcal{GGI}$, then I does not satisfy the law of contraposition CP(N) for any fuzzy negation N.*

Following this, we explore the continuity of generalized g-generated implications.

Proposition 5 *If $I = I^{f,g} \in \mathcal{GGI}$ where f is strictly decreasing, then I is continuous except at the point (0,0). Moreover, $I(x, y)$ is right-continuous at 0 with respect to both arguments.*

Proposition 2 shows that when $g(1) = \infty$, $I^{f,g}$ does not satisfy ordering property (OP). When $f(x)$ is strictly decreasing, however, the following theorem provides a necessary and sufficient condition under which $I^{f,g}$ satisfies ordering property (OP).

Theorem 2 *If $I = I^{f,g} \in \mathcal{GGI}$ where f is strictly decreasing, then the following statements are equivalent:*
(i) *I satisfies the ordering property (OP).*
(ii) *$g(1) < \infty$ and $g(x) = \frac{g(1)}{f(x)}$.*

For the Φ-conjugates of a generalized g-generated implication, we have:

Proposition 6 *If $I = I^{f,g} \in \mathcal{GGI}$, then each Φ-conjugate of I is also a generalized g-generated implication and $I_\varphi = I^{f\circ\varphi, g\circ\varphi}$.*

4 The Law of Importation and the Distributive Equations for New Implications

In classical logic, $(p \wedge q) \to r \equiv (p \to (q \to r))$ is a tautology which is called the *law of importation* (shortly LI). The general form of the above equivalence is given by

$$I(T(x, y), z) = I(x, I(y, z)), \quad x,\ y,\ z \in U,$$

where $I \in \mathcal{FI}$, T is a t-norm. In this case, we say that implication I satisfies the *law of importation with respect to* T (see [2, 9, 10]).

The following conclusion shows the relationship between generalized g-generated implications and the law of importation (LI).

Proposition 7 *Suppose that $I = I^{f,g} \in \mathcal{GGI}$ where f is strictly decreasing and $g(1) = \infty$. Define*

$$F: U^2 \to U, \quad (x, y) \mapsto f^{-1}(f(x) \cdot f(y)), \quad x,\ y \in U,$$

and denote $Ran(f) = \{f(x) \mid x \in U\}$ to represent the range of f. Then I satisfy $I(F(x, y), z) = I(x, I^{f,g}(y, z))$ if and only if $f(x) \cdot f(y) \in Ran(f)$, and F is a commutative and increasing function with a neutral element 1.

From Corollary 3.35 in [6], we easily know that the function F in Proposition 7 is a t-norm and then we can obtain the following proposition.

Proposition 8 *Suppose that T is a t-norm, $I^{f,g} \in \mathcal{GGI}$ where f is strictly decreasing and $g(1) = \infty$. If $f(x) \cdot f(y) \in Ran(f)$, then $I^{f,g}$ satisfies the law of importation (LI) with respect to T if and only if $T(x, y) = f^{-1}(f(x) \cdot f(y))$.*

When $g(1) < \infty$, we get the following conclusion.

Proposition 9 *Suppose that T is a t-norm, $I = I^{f,g} \in \mathcal{GGI}$ where f is strictly decreasing and $g(1) < \infty$, then I satisfies the law of importation (LI) with respect to T if and only if*

$$T(x, y) = f^{-1}(f(x) \cdot f(y)), \quad x, y \in U.$$

From Propositions 8 and 9, the following theorem can be deduced.

Theorem 3 *Suppose that T is a t-norm and $I = I^{f,g} \in \mathcal{GGI}$ where f is strictly decreasing. If $f(x) \cdot f(y) \in Ran(f)$, then I satisfies the law of importation (LI) with respect to T if and only if $T(x, y) = f^{-1}(f(x) \cdot f(y))$.*

In fuzzy logic, many authors discussed the distributive equations. If I is a fuzzy implication, T, T_1 and T_2 are t-norms, S, S_1 and S_2 are s-norms, then four kinds of distributivity for fuzzy implications over t-norms and s-norms are defined as follows (see [2, 3, 14, 15]),

(D1) $I(S(x, y), z) = T(I(x, z), I(y, z))$ for $x,\ y,\ z \in U$;
(D2) $I(T(x, y), z) = S(I(x, z), I(y, z))$ for $x,\ y,\ z \in U$;
(D3) $I(x, T_1(y, z)) = T_2(I(x, y), I(x, z))$ for $x,\ y,\ z \in U$;
(D4) $I(x, S_1(y, z)) = S_2(I(x, y), I(x, z))$ for $x,\ y,\ z \in U$.

The following two propositions are important for our discussion.

Proposition 10 ([2]) *Let a function $I : U^2 \to U$ satisfy the left neutrality property (NP), T be a t-norm and S a s-norm.*

(i) If the triple (I, T, S) satisfies (D1), then $T = T_M = \min(x, y)$.
(ii) If the triple (I, T, S) satisfies (D2), then $S = S_M = \max(x, y)$.

Proposition 11 ([2]) *For a function $I : U^2 \to U$ the following statements are equivalent:*

(i) I is decreasing in the first variable, i.e., I satisfies (I1).
(ii) I satisfies $I(\max(x, y), z) = \min(I(x, z), (y, z))$ for $x,\ y,\ z \in U$.
(iii) I satisfies $I(\min(x, y), z) = \max(I(x, z), (y, z))$ for $x,\ y,\ z \in U$.

In the following, we deal with the above distributive equations based on our implications.

Theorem 4 *If $I = I^{f,g} \in \mathcal{GGI}$ where f is strictly decreasing, then the triple (I, T, S) satisfies (D1) if and only if $S = S_M$ and $T = T_M$.*

Similarly to Theorem 4, we have the following conclusion about (D2).

Theorem 5 *If $I = I^{f,g} \in \mathcal{GGI}$ where f is strictly decreasing, then the triple (I, T, S) satisfy (D2) if and only if $S = S_M$ and $T = T_M$.*

Now, let us discuss (D3) and (D4) for $I^{f,g}$ with $g(1) < \infty$.

Proposition 12 ([2]) *Let T_1 and T_2 be t-norms, S_1 and S_2 s-norms and function $I : U^2 \to U$ satisfy the left neutrality property (NP).*

(i) If the triple (I, T_1, T_2) satisfies (D3) for all $x,\ y,\ z \in U$, then $T_1 = T_2$.
(ii) If the triple (I, S_1, S_2) satisfies (D4) for all $x,\ y,\ z \in U$, then $S_1 = S_2$.

Proposition 13 ([2]) *For a function $I : U^2 \to U$ the following statements are equivalent:*
(i) *I is increasing in the second variable, i.e., I satisfies (I2).*
(ii) *I satisfies $I(x, \min(y, z)) = \min(I(x, y), (x, z))$ for $x,\ y,\ z \in U$.*
(iii) *I satisfies $I(x, \max(y, z)) = \max(I(x, y), (x, z))$ for $x,\ y,\ z \in U$.*

Theorem 6 *If $I = I^{f,g} \in \mathcal{GGI}$ where f is strictly decreasing and $g(1) < \infty$, then the triple (I, T_1, T_2) satisfies (D3) if and only if $T_1 = T_2 = T_M$.*

Similarly to Theorem 6, we have following conclusion about (D4).

Theorem 7 *If $I = I^{f,g} \in \mathcal{GGI}$ where f is strictly decreasing and $g(1) < \infty$, then the triple (I, S_1, S_2) satisfies (D4) if and only if $S_1 = S_2 = S_M$.*

5 Intersection of $\mathcal{GGI}$ with Other Known Classes of Implications

Next theorem shows fact that generalized g-generated implications are not (S,N)-implications.

Theorem 8 *If* $I = I^{f,g} \in \mathcal{GGI}$ *where* f *is strictly decreasing, then* I *is not an (S,N)-implication.*

Proof Assume that I is a (S,N)-implication obtained form a s-norm S and a fuzzy negation N. Thus, we have $N_I = N$. However, by Proposition 4, when f is strictly decreasing, $N_I = N_1$, i.e., $N = N_I = N_1$. From [2], we know that the (S,N)-implication obtained from N_1 is the largest (S,N)-implication I_D, i.e.,

$$I(x, y) = I_D(x, y) = \begin{cases} 1, & \text{if } x = 0, \\ y, & \text{if } x > 0. \end{cases}$$

Then when $x,\ y \in (0, 1)$, we have

$$I(x, y) = g^{(-1)}(f(x) \cdot g(y)) = g^{(-1)}(\min(f(x) \cdot g(y), g(1))) = y.$$

We get $f(x) \cdot g(y) = g(y)$, a contradiction, i.e., I is not a (S,N)-implication. □

Similarly, for QL-implications we can obtain the following theorem.

Theorem 9 *If* $I = I^{f,g} \in \mathcal{GGI}$ *where* f *is strictly decreasing, then* I *is not an QL-implication.*

Proof If I is a QL-implication obtained from a s-norm S, a t-norm T and a fuzzy negation N. From Remark 1 and Proposition 3, we know that $N = N_{I^{f,g}} = N_1$. By [2], the QL-operation $I^{f,g}$ generated from N_1 is not a implication, a contradiction, i.e., $I^{f,g}$ is not a QL-implication. □

Proposition 14 ([2]) *For a function* $I:\ U^2 \to U$*, the following statements are equivalent:*

(i) I is an R-implication generated from a left-continuous t-norm.
(ii) I satisfies (I2), the exchange principle (EP), the ordering property (OP) and it is right-continuous with respect to the second variable.

Theorem 10 *If* $I = I^{f,g} \in \mathcal{GGI}$ *where* f *is strictly decreasing, then the following statements are equivalent:*

(i) I is an R-implication obtained from a left-continuous t-norm.
(ii) $g(1) < \infty$ *and* $g(x) = \frac{g(1)}{f(x)}$.

Proof "(i) $\Longrightarrow$ (ii)" Assume that I is an R-implication obtained from a left-continuous t-norm. From Proposition 14, we know that I satisfies the ordering property (OP). Hence, by Theorem 2, we have $g(1) < \infty$, so $g(x) = \frac{g(1)}{f(x)}$.

"(ii) $\Longrightarrow$ (i)" Assume $g(1) < \infty$ such that $g(x) = \frac{g(1)}{f(x)}$. From Theorem 2, Propositions 1, 2 and 4, $I^{f,g}$ satisfies (I2), the exchange principle (EP), the ordering property (OP) and it is right-continuous with respect to the second variable. Then, by Proposition 14, I is an R-implication obtained from a left-continuous t-norm. □

The following we will investigate the intersections between generalized g-generated implications and f-generated implications.

Proposition 15 ([2]) *Let f be an f-generator, I_f an f-generated implication. Then*

(i) the natural negation N_{I_f} is a strict negation if and only if $f(0) < \infty$;
(ii) $I_f(x, y) = 1$ if and only if $x = 0$ or $y = 1$, i.e., I_f does not satisfy the identity principle (IP) and the ordering property (OP).

Theorem 11 *If $I = I^{f,g} \in \mathcal{GGI}$ where f is strictly decreasing and $g(1) < \infty$, then I is not an f-generated implication.*

Proof Let I be an f-generated implication. From Proposition 15, we know that $I(x, y) = 1$ if and only if $x = 0$ or $y = 1$. On the other hand, I is a generalized g-generated implication with $g(1) < \infty$. Meanwhile, fix arbitrarily $y \in (0, 1)$. Then there exists $x_0 \in (0, 1)$ such that $f(x_0) \cdot g(y) \geq g(1)$ we get

$$I(x_0, y) = g^{(-1)}(f(x_0) \cdot g(y)) \geq g^{(-1)}(g(1)) = 1.$$

Then $I(x_0, y) = 1$, therefore there exists $x,\ y \in (0, 1)$ such that $I(x, y) = 1$. Thus I is not an f-generated implication. □

Theorem 12 *If $I = I^{f,g} \in \mathcal{GGI}$, then I is not an f-generated implication with $f(0) < \infty$.*

Proof Let I be an f-generated implication with $f(0) < \infty$. Then from Proposition 15, we get that the natural negation N_I is a strict negation. However, by Proposition 3, N_I is not continuous, a contradiction. Then, we have I is not an f-generated implication with $f(0) < \infty$. □

6 Conclusions

In this paper, we proposed a class of new fuzzy implications, called generalized Yager's g-generated implications, which is generated from a decreasing function f and a g-generator. We investigated some basic properties of these new implications. We also studied the law of importation and the distributive equations for them, discussed the relations between this new class and other known fuzzy implications such as (S,N)-implications, R-implications and QL-implications.

As future work, we will investigate other ways to generalize Yager's implications, apply the proposed new implications to approximate reasoning and fuzzy control.

Acknowledgments This work is supported by the National Natural Science Foundation of China (Grant Nos. 11171308, 51305400, 61379018 and 61472471).

References

1. Baczyński, M., Jayaram, B.: Yager's classes of fuzzy implications: some properties and intersections. Kybern. **43**, 157–182 (2007)
2. Baczyński, M., Jayaram, B.: Fuzzy Implications. Springer, Berlin (2008)
3. Baczyński, M., Jayaram, B.: On the distributivity of fuzzy implications over nilpotent or strict triangular conorms. IEEE Trans. Fuzzy Syst. **17**, 590–603 (2009)
4. Gottwald, S.: A Treatise on Many-Valued Logics. Research Studies Press LTD, Baldock (2001)
5. Hilnena, D., Kalina, M., Kral, P.: A class of implications related to Yager's f-implications. Inform. Sci. **260**, 171–184 (2014)
6. Klement, E.P., Mesiar, R., Pap, E.: Triangular Norms. Kluwer, Dordrecht (2000)
7. Liu, H.W.: A new class of fuzzy implications derived from generalized h-generators. Fuzzy Sets Syst. **24**, 63–92 (2013)
8. Mas, M., Monserrat, M., Torrens, J.: The law of importation for discrete implications. Inform. Sci. **179**, 4208–4218 (2009)
9. Mas, M., Monserrat, M., Torrens, J., Trillas, E.: A survey on fuzzy implications functions. IEEE Trans. Fuzzy Syst. **15**, 1107–1121 (2007)
10. Massanet, S., Torrens, J.: The law of importation versus the exchange principle on fuzzy implications. Fuzzy Sets Syst. **168**, 47–69 (2011)
11. Massanet, S., Torrens, J.: On the characterization of Yager's implications. Inform. Sci. **201**, 1–18 (2012)
12. Massanet, S., Torrens, J.: On a new class of fuzzy implications: h-implications and generalizations. Inform. Sci. **181**, 2111–2127 (2014)
13. Pei, D.W.: Theory and Applications of Triangular Norm Based Fuzzy Logics. Science Press, Beijing (2013). (in Chinese)
14. Qin, F., Baczynski, M.: On distributivity equations of implications and contrapositive symmetry equations of implications. Fuzzy Sets Syst. **247**, 81–91 (2014)
15. Trillas, E., Alsina, C.: On the law $(x \wedge y) \to z \equiv (x \to (y \to z))$ in fuzzy logic. IEEE Trans. Fuzzy Syst. **10**, 84–88 (2002)
16. Xie, A.F., Liu, H.W.: A generalization of Yager's f-generated implications. Int. J. Approx. Reason. **54**, 35–46 (2013)
17. Yager, R.R.: On some new classes of implication operators and their role in approximate Reasoning. Inform. Sci. **167**, 193–216 (2004)

On Relations Between Several Classes of Uninorms

Gang Li and Hua-Wen Liu

Abstract Uninorms are an important class of aggregation functions in information aggregation. It is well known that there exist many different classes of uninorms in references. In this paper, the relationships among several classes of uninorms are discussed. Moreover, a complete characterization of the class of almost equitable uninorms is presented. As a byproduct, a characterization of the class of representable uninorms is obtained.

Keywords Uninorms · Continuous underlying operators · Almost equitable uninorms · Boundary

1 Introduction

Uninorms constitute an important class of aggregation functions in information aggregation. Since their introduction by Yager and Rybalov [23], they have attracted lots of research activities, ranging from theoretical study to practical applications. The first deep study by Fodor et al. revealed the structure of uninorms in [10]. Later on it is justified that uninorms are useful in many fields like expert systems [4], fuzzy logic [11], fuzzy mathematical morphology [6] and bipolar aggregation [24]. On the other hand, the theoretical study of uninorms is even more extensive [14, 18, 20, 21].

Nowadays, several classes of uninorms are available. For example, the four usual classes of uninorms: $\mathcal{U}_{\min}$(or $\mathcal{U}_{\max}$), idempotent uninorms, representable uninorms

G. Li (✉)
Faculty of Science, Qilu University of Technology,
Ji'nan 250353, Shandong, People's Republic of China
e-mail: sduligang@163.com

H.-W. Liu
School of Mathematics, Shandong University,
Ji'nan 250100, Shandong, People's Republic of China
e-mail: hw.liu@sdu.edu.cn

T.-H. Fan et al. (eds.), *Quantitative Logic and Soft Computing 2016*,
Advances in Intelligent Systems and Computing 510,
DOI 10.1007/978-3-319-46206-6_25

and uninorms continuous in $]0, 1[^2$. Moreover, in order to discuss the migrativity of uninorms, a class of uninorms which are locally internal on the boundary appeared in [18], and the class of almost equitable uninorms [19] was introduced to describe the equitable behavior of uninorms when receiving contradictory information. So, it is interesting to discuss the relations among different classes of uninorms.

2 Preliminaries

We assume that the reader is familiar with some basic notions concerning t-norms and t-conorms which can be found for instance in [1, 13]. Also, some results on uninorms can be found in [10].

Definition 1 ([13]) A function $N : [0, 1] \to [0, 1]$ is said to be a *negation* if it is decreasing and satisfies $N(0) = 1, N(1) = 0$. Moreover, if N is continuous and strictly decreasing, then it is called a *strict negation*. If a strict negation N is involutive, i.e., $N(N(x)) = x$ for all $x \in [0, 1]$, then it is called a *strong negation*.

Definition 2 ([23]) A *uninorm* is a two-place function: $U : [0, 1]^2 \to [0, 1]$ which is associative, commutative, increasing in each variable and there exists some element $e \in [0, 1]$, called *neutral element*, such that $U(e, x) = x$ for all $x \in [0, 1]$.

We summarize some fundamental results from [10].

It is clear that the function U becomes a t-norm when $e = 1$ and a t-conorm when $e = 0$. For any uninorm we have $U(0, 1) \in \{0, 1\}$. A uninorm U such that $U(0, 1) = 0$ is called *conjunctive* and if $U(0, 1) = 1$ then it is called *disjunctive*.

Throughout this paper, we exclusively consider uninorms with a neutral element e strictly between 0 and 1.

With any uninorm U with neutral element $e \in]0, 1[$, we can associate two binary operations $T_U, S_U : [0, 1]^2 \to [0, 1]$ defined by

$$T_U(x, y) = \frac{1}{e} \cdot U(ex, ey)$$

and

$$S_U(x, y) = \frac{1}{1-e}(U(e + (1 - e)x, e + (1 - e)y) - e).$$

It is easy to see that T_U is a t-norm and that S_U is a t-conorm. In other words, on $[0, e]^2$ any uninorm U is determined by a t-norm T_U, and on $[e, 1]^2$ any uninorm U is determined by a t-conorm S_U; T_U is called *the underlying t-norm*, and S_U is called *the underlying t-conorm*. Let us denote the remaining part of the unit square by $A(e)$, i.e., $A(e) = [0, 1]^2 \backslash ([0, e]^2 \cup [e, 1]^2)$. On the set $A(e)$, any uninorm U is bounded by the minimum and maximum of its arguments, i.e. for any $(x, y) \in A(e)$ it holds that

$$\min(x, y) \le U(x, y) \le \max(x, y). \tag{1}$$

The most studied classes of uninorms are:

- Uninorms in $\mathcal{U}_{\min}$ (or $\mathcal{U}_{\max}$) [10], those given by minimum (or maximum) in $A(e)$.
- Uninorms in $\mathcal{U}_{loc}$ [7], those local internal in the area $A(e)$, i.e., $U(x, y) \in \{x, y\}$ for all $(x, y) \in A(e)$.
- Idempotent uninorms in $\mathcal{U}_{id}$ [3, 22], those that satisfy $U(x, x) = x$ for all $x \in [0, 1]$.
- Representable uninorms in $\mathcal{U}_{rep}$ [9, 10], those that have an additive generator (or multiplicative generator).
- Uninorms in $\mathcal{CU}$ [12], those that are continuous in the open square $]0, 1[^2$.
- Uninorms in $\mathcal{COU}$ [9, 14, 16], those that are with continuous underlying operators.
- Uninorm in $\mathcal{U}_{bli}$ [2, 15, 18], those that are locally internal on the boundary of $[0, 1]^2$, i.e., $U(0, x) = U(x, 0) \in \{0, x\}$, $U(1, x) = U(x, 1) \in \{1, x\}$ for all $x \in [0, 1]$.
- Uninorms in $\mathcal{U}_{aeq}$ [17, 19], those satisfy $U(x, N(x)) = e$ for all $x \in [0, 1]$ and a strong negation $N : [0, 1] \to [0, 1]$.

Note that a uninorm U with neutral element $e \in]0, 1[$ is often called *almost equitable with respect to* N if $U(x, N(x)) = e$ for all $x \in]0, 1[$ and a strong negation N.

The relationships among the classes of uninorms above will be studied in detail in the following section.

3 Main Results

Our first result is about the class of uninorms with continuous underlying operators.

Proposition 1 *The following statements hold:*

(i) $\mathcal{U}_{rep} \subseteq \mathcal{CU} \subseteq \mathcal{COU}$;
(ii) $\mathcal{U}_{\min} \cap \mathcal{COU} \neq \emptyset, \mathcal{U}_{\max} \cap \mathcal{COU} \neq \emptyset, \mathcal{U}_{loc} \cap \mathcal{COU} \neq \emptyset$;
(iii) $\mathcal{U}_{id} \subseteq \mathcal{COU}$;
(iv) $\mathcal{U}_{\min} \subseteq \mathcal{U}_{loc}, \mathcal{U}_{\max} \subseteq \mathcal{U}_{loc}, \mathcal{U}_{id} \subseteq \mathcal{U}_{loc}$.

Remark 1 (i) It is obvious that the uninorm $U \in \mathcal{U}_{id} \cap \mathcal{U}_{\min}$ has the following form

$$U(x, y) = \begin{cases} \max(x, y) & (x, y) \in [e, 1]^2, \\ \min(x, y) & \text{otherwise.} \end{cases}$$

where $e \in]0, 1[$ is the neutral element of U. Uninorm $U \in \mathcal{U}_{id} \cap \mathcal{U}_{\max}$ has form

$$U(x, y) = \begin{cases} \min(x, y) & (x, y) \in [0, e]^2, \\ \max(x, y) & \text{otherwise.} \end{cases}$$

where $e \in]0, 1[$ is the neutral element of U. Uninorm $U \in \mathcal{U}_{\min} \cap \mathcal{COU}$ has form

$$U(x, y) = \begin{cases} eT(\frac{x}{e}, \frac{y}{e}) & (x, y) \in [0, e]^2, \\ e + (1 - e)S(\frac{x-e}{1-e}, \frac{y-e}{1-e}) & (x, y) \in [e, 1]^2, \\ \min(x, y) & \text{otherwise,} \end{cases}$$

where $e \in]0, 1[$ is the neutral element of U and T, S are continuous t-norm, t-conorm, respectively. Uninorm $U \in \mathcal{U}_{\max} \cap \mathcal{COU}$ has form

$$U(x, y) = \begin{cases} eT(\frac{x}{e}, \frac{y}{e}) & (x, y) \in [0, e]^2, \\ e + (1-e)S(\frac{x-e}{1-e}, \frac{y-e}{1-e}) & (x, y) \in [e, 1]^2, \\ \max(x, y) & \text{otherwise}, \end{cases}$$

where $e \in]0, 1[$ is the neutral element of U and T, S are continuous t-norm, t-conorm, respectively.

(ii) A complete characterization of the uninorms in $\mathcal{U}_{loc} \cap \mathcal{COU}$ was given in [8].

Now, we discuss the class of uninorms which are locally internal on the boundary.

Lemma 1 ([17, 18]) *Let U be a uninorm with neutral element $e \in]0, 1[$. Then the following two statements hold:*

(i) If U is a conjunctive uninorm with continuous underlying t-norm, then $U(0, y) = U(y, 0) = 0$, $U(1, y) = U(1, y) \in \{1, y\}$ for all $y \in [0, 1]$.
(ii) If U is a disjunctive uninorm with continuous underlying t-conorm, then $U(y, 0) = U(0, y) \in \{0, y\}$, $U(1, y) = U(y, 1) = 1$ for all $y \in [0, 1]$.

Based on Lemma 1, we can obtain the second result which is about the class of uninorms which are locally internal on the boundary.

Proposition 2 *The following three statements hold:*

(i) $\mathcal{U}_{rep} \subseteq \mathcal{CU} \subseteq \mathcal{COU} \subseteq \mathcal{U}_{bli}$;
(ii) $\mathcal{U}_{id} \subseteq \mathcal{U}_{loc} \subseteq \mathcal{U}_{bli}$;
(iii) $\mathcal{U}_{\min} \subseteq \mathcal{U}_{bli}, \mathcal{U}_{\max} \subseteq \mathcal{U}_{bli}$.

Remark 2 (i) There exists uninorm $U \in \mathcal{U}_{bli}, U \notin \mathcal{U}_{loc}$, for example, the representable uninorm.
(ii) There exists uninorm $U \in \mathcal{U}_{bli}$, but $U \notin \mathcal{COU}, U \notin \mathcal{U}_{\min}, U \notin \mathcal{U}_{\max}$. Furthermore, there exists uninorm $U \notin \mathcal{U}_{bli}$. Two uninorms are given in the following examples.

Example 1 Suppose that $U : [0, 1]^2 \to [0, 1]$ is defined as follows

$$U(x, y) = \begin{cases} 2xy & (x, y) \in [0, \frac{1}{2}[^2, \\ 1 & (x, y) \in]\frac{1}{2}, 1]^2, \\ 0 & x = 0 \text{ or } y = 0, \\ y & x = \frac{1}{2}, \\ x & y = \frac{1}{2}, \\ \max(x, y) & \text{otherwise}. \end{cases}$$

By Theorem 4 in [16], U is a uninorm of which the underlying t-norm T_U and the underlying t-conorm S_U are defined as follows:

$$T_U(x,y) = \begin{cases} \frac{1}{2}xy & (x,y) \in [0,1[^2, \\ \min(x,y) & \text{otherwise}, \end{cases}$$

$$S_U(x,y) = \begin{cases} 1 & (x,y) \in]0,1]^2, \\ \max(x,y) & \text{otherwise}, \end{cases}$$

It is obvious that $U \in \mathcal{U}_{bli}$, $U \notin \mathcal{U}_{\max}$. But the underlying operators T_U, S_U are not continuous. Hence, $U \notin \mathcal{COU}$.

Example 2 Suppose that $U : [0,1]^2 \to [0,1]$ is defined by

$$U(x,y) = \begin{cases} eT_D(\frac{x}{e}, \frac{y}{e}) & (x,y) \in [0,e]^2, \\ e + (1-e)S_D(\frac{x-e}{1-e}, \frac{y-e}{1-e}) & (x,y) \in [e,1]^2, \\ 1 & x = 1 \text{ or } y = 1, \\ y_0 & (x,y) \in [0,e[\times]y_0,1) \cup]y_0,1) \times [0,e[, \\ \max(x,y) & \text{otherwise} \end{cases} \tag{2}$$

where $e \in]0,1[$, $y_0 \in]e,1[$ and

$$T_D(x,y) = \begin{cases} 0 & (x,y) \in [0,1[^2, \\ \min(x,y) & \text{otherwise} \end{cases}, \quad S_D(x,y) = \begin{cases} 1 & (x,y) \in]0,1]^2, \\ \max(x,y) & \text{otherwise.} \end{cases}$$

U is a uninorm. In fact, it is obvious that U is commutative, increasing in each variable and with neutral element e. Only the associativity of U need to be verified, i.e., for all $(x,y,z) \in [0,1]^3$

$$U(x, U(y,z)) = U(U(x,y), z).$$

Without loss of generality, we assume that $x \le y \le z$.

- $0 < x \le b < e < z \le y_0$. Then $U(y,z) = z$, $U(x, U(y,z)) = U(x,z) = z$ and $U(U(x,y),z) = U(0,z) = z$.
- $0 < x \le y < e < y_0 < z < 1$. Then $U(y,z) = y_0$. So, $U(x, U(y,z)) = U(x, y_0) = y_0$ and $U(U(x,y),z) = U(0,z) = y_0$.
- $0 < x < e < y \le y_0$. Then $U(y,z) = 1$. So, $U(x, U(y,z)) = U(x,1) = 1$ and $U(U(x,y),z) = U(y,z) = 1$.
- $0 < x < e < y_0 < y \le z < 1$. Then $U(y,z) = 1$. So, $U(x, U(y,z)) = U(x,1) = 1$ and $U(U(x,y),z) = U(y_0,z) = 1$.

For the remaining cases. $U(x, U(y, z)) = U(U(x, y), z)$ holds obviously.

It is obvious that $U \notin \mathcal{U}_{bli}$, $U \notin \mathcal{U}_{\max}$. Furthermore, the underlying operators T_U, S_U are not continuous. Hence, $U \notin \mathcal{COU}$.

Now, we discuss the class of almost equitable uninorms.

Lemma 2 ([17]) *Let $U : [0, 1]^2 \to [0, 1]$ be a uninorm with neutral element $e \in]0, 1[$ and N be a strong negation. If U is almost equitable with respect to N, then e is the only fixed point of N, i.e., $N(e) = e$.*

Lemma 3 ([17]) *Let $U : [0, 1]^2 \to [0, 1]$ be a uninorm with neutral element $e \in]0, 1[$. If U is locally internal on $[0, e[\times]e, 1]\cup]e, 1] \times [0, e[$ (i.e., $U(x, y) \in \{x, y\}$ for any (x, y) in this region), then there does not exist strong negation N such that U is almost equitable with respect to N.*

Lemma 4 ([17]) *Let $U : [0, 1]^2 \to [0, 1]$ be a uninorm with neutral element $e \in]0, 1[$ and N be a strong negation. If U is continuous in $]0, 1[^2$, then U is almost equitable with respect to N if and only if U is a representable uninorm with additive generator $h : [0, 1] \to [-\infty, +\infty]$ and $N = N_U$ is a strong negation, where $N_U(x) = h^{-1}(-h(x))$ for all $x \in [0, 1]$.*

Based on above Lemmas, we give a complete characterization theorem of almost equitable uninorms as follows:

Theorem 1 *Let $U : [0, 1]^2 \to [0, 1]$ be a uninorm with neutral element $e \in]0, 1[$ and N be a strong negation. U is almost equitable with respect to N if and only if U is a representable uninorm with additive generator $h : [0, 1] \to [-\infty, +\infty]$ and $N = N_U$ is a strong negation, where $N_U(x) = h^{-1}(-h(x))$ for all $x \in [0, 1]$.*

Proof If U is a representable uninorm and $N = N_U$ then $U(x, N(x)) = e$ for all $x \in]0, 1[$ by Proposition 6 in [5].

Conversely, let U be a uninorm which is almost equitable with respect to N. Then $U(x, N(x)) = e$ for all $x \in]0, 1[$. We can prove the result in the following steps.

Step 1: $U(x, y) \neq e$ for all $x \in]0, 1[$, $y \in [0, 1]$ such that $y \neq N(x)$. On the contrary, suppose that there exists $x_0 \in]0, 1[$, $y_0 \in [0, 1]$ such that $y_0 \neq N(x_0)$ and $U(x_0, y_0) = e$. Then, $U(U(x_0, y_0), N(x_0)) = U(e, N(x_0)) = N(x_0)$ and $U(U(x_0, N(x_0)), y_0) = U(e, y_0) = y_0$, a contradiction with the associativity of U. Hence, by the monotonicity of U, we have $U(x, y) < e$ for all $x \in]0, 1[$, $y \in [0, 1]$ such that $y < N(x)$, and $U(x, y) > e$ for all $x \in]0, 1[$, $y \in [0, 1]$ such that $y > N(x)$.

Step 2: There does not exist $(x, y) \in]0, e[^2$ such that $U(x, y) = 0$. On the contrary, suppose that there exist $x_1, x_2 \in]0, e[$ such that $x_1 < x_2$, $U(x_1, x_2) = 0$. Since U is almost equitable with respect to N, we have $U(x_1, N(x_1)) = e$, $U(x_2, N(x_2)) = e$ and

$$e \leq N(x_2) < N(x_1) < 1, U(N(x_1), N(x_2)) \geq e$$

by Lemma 2. Furthermore, we have

$$U(U(x_1, N(x_1)), U(x_2, N(x_2))) = U(e, e) = e$$

and

$$U(U(x_1, x_2), U(N(x_1), N(x_2))) = U(0, U(N(x_1), N(x_2))).$$

If $U(N(x_1), N(x_2)) = 1$ then

$$U(U(x_1, x_2), U(N(x_1), N(x_2))) = U(0, 1) \in \{0, 1\}.$$

On the other hand, if $U(N(x_1), N(x_2)) < 1$, then

$$U(U(x_1, x_2), U(N(x_1), N(x_2))) = U(0, U(N(x_1), N(x_2))) < e$$

by the result in (Step 1) and the monotonicity of U.

So, in both cases, we get a contradiction with the associativity and the commutativity of U.

Step 3: There does not exist $(x, y) \in]e, 1[^2$ such that $U(x, y) = 1$. On the contrary, suppose that there exist $y_1, y_2 \in]e, 1[$ such that $y_1 < y_2$, $U(y_1, y_2) = 1$. Since U is almost equitable with respect to N, we have $U(y_1, N(y_1)) = e$, $U(y_2, N(y_2)) = e$. By Lemma 2, $0 < N(y_2) < N(y_1) < e$, $U(N(y_1), N(y_2)) \leq e$. Hence, $U(N(y_1), N(y_2)) > 0$ by the result in (Step 2). Furthermore, we have

$$U(U(y_1, N(y_1)), U(y_2, N(y_2))) = U(e, e) = e,$$

$$U(U(y_1, y_2), U(N(y_1), N(y_2))) = U(1, U(N(y_1), N(y_2))),$$

and

$$U(1, U(N(y_1), N(y_2))) > U(N(U(N(y_1), N(y_2))), U(N(y_1), N(y_2))) = e$$

by the result in (Step 1) and the monotonicity of U. So, we obtain a contradiction with the associativity and commutativity of U.

Step 4: U is continuous in $]0, 1[^2$. By Lemma 2.1.2 in [1], we only need to prove that for all $x_0, y_0 \in]0, 1[$ both the vertical section $U(x_0, \cdot) :]0, 1[\to [0, 1]$ and the horizontal section $U(\cdot, y_0) :]0, 1[\to [0, 1]$ are continuous functions of one variable. Due to the commutativity of U, the continuity of horizontal section $U(\cdot, y_0)$ is proved here. For all $y_0 \in]0, 1[$, we know that $U(x, y_0) \in [0, U(1, y_0)]$ for all $x \in [0, 1]$. By Eq. (1), $U(1, y_0) \in [y_0, 1]$. For every $x_0 \in]0, U(1, y_0)[$, we have

$$U(U(x_0, N(y_0)), y_0) = U(x_0, U(N(y_0), y_0)) = U(x_0, e) = x_0.$$

Since $y_0 \in]0, 1[$, $N(y_0) \in]0, 1[$. Now, we prove that $U(x_0, N(y_0)) \in]0, 1[$. We divide the proof in four cases.

- $(x_0, N(y_0)) \in]0, e]^2$. Then $U(x_0, N(y_0)) \in]0, e]$ by the result in (Step 2).
- $(x_0, N(y_0)) \in]e, 1[^2$. Then $U(x_0, N(y_0)) \in]e, 1[$ by the result in (Step 2).

- $x_0 \in]0, e], N(y_0) \in [e, 1[$. Then $U(x_0, N(y_0)) \in [x_0, N(y_0)]$ by Eq. (1). Hence, $U(x_0, N(y_0)) \in]0, 1[$.
- $x_0 \in [e, 1[, N(y_0) \in]0, e]$. Then $U(x_0, N(y_0)) \in [N(y_0), x_0]$ by Eq. (1). Hence, $U(x_0, N(y_0)) \in]0, 1[$.

Hence, the result holds by Lemma 4. □

Example 3 ([13]) Suppose that conjunctive uninorm $U : [0, 1]^2 \to [0, 1]$ is defined by

$$U(x, y) = \begin{cases} \frac{xy}{xy+(1-x)(1-y)} & (x, y) \neq \{0, 1\}, \\ 0 & \text{otherwise} \end{cases} \tag{3}$$

It is obvious that U is a representable uninorm with additive generator $h(x) = \log(\frac{x}{1-x})$ and neutral element $e = \frac{1}{2}$. It is easy to verify that U is almost equitable with respect to the strong negation N, where $N(x) = N_U(x) = h^{-1}(-h(x)) = 1 - x$ for all $x \in [0, 1]$.

By Theorem 1, a new characterization for the class of representable uninorms can be obtained as follows:

Corollary 1 *Let* $U : [0, 1]^2 \to [0, 1]$ *be uninorm with neutral element* $e \in]0, 1[$. *Then* U *is a representable uninorm if and only if there exists a strong negation* $N : [0, 1] \to [0, 1]$ *such that* $U(x, N(x)) = e$ *for all* $x \in [0, 1]$.

Finally, relations between the class of almost equitable uninorms and the other classes of uninorms are summarized in the following proposition.

Proposition 3 *The following two statements hold:*

(i) $\mathcal{U}_{\min} \cap \mathcal{U}_{aeq} = \emptyset, \mathcal{U}_{\max} \cap \mathcal{U}_{aeq} = \emptyset, \mathcal{U}_{id} \cap \mathcal{U}_{aeq} = \emptyset, \mathcal{U}_{loc} \cap \mathcal{U}_{aeq} = \emptyset$;
(ii) $\mathcal{U}_{rep} \cap \mathcal{U}_{aeq} = \mathcal{CU} \cap \mathcal{U}_{aeq} = \mathcal{COU} \cap \mathcal{U}_{aeq} = \mathcal{U}_{bli} \cap \mathcal{U}_{aeq} = \mathcal{U}_{aeq} = \mathcal{U}_{rep}$.

Acknowledgments This work is supported by the National Natural Science Foundation of China (Grant Nos. 61403220, 61573211).

References

1. Alsina, C., Frank, M.J., Schweizer, B.: Associative Functions. Triangular Norms and Copulas. World Scientific, New Jersey (2006)
2. Csiszár, O., Fodor, J.: On uninorms with fixed values along their border. Ann. Univ. Sci. Bp. Sect. Comput. **42**, 93–108 (2014)
3. De Baets, B.: Idempotent uninorms. Eur. J. Oper. Res. **118**, 631–642 (1998)
4. De Baets, B., Fodor, J.: Van Melle's combining function in MYCIN is a representable uninorm: an alternative proof. Fuzzy Sets Syst. **104**, 133–136 (1999)
5. De Baets, B., Fodor, J.: Residual operators of uninorms. Soft Comput. **3**, 89–100 (1999)
6. De Baets, B., Kwasnikowska, N., Kerre, E.: Fuzzy morphology based on uninorms. In: Seventh IFSA World Congress, Prague, vol. 220, pp. 215–220 (1997)
7. Drygaś, P.: Discussion of the structure of uninorms. Kybernetika **41**, 213–226 (2005)

8. Drygaś, P., Ruiz-Aguilera, D., Torrens, J.: A characterization of a class of uninorms with continuous underlying operators. Fuzzy Sets Syst. **287**, 137–153 (2016)
9. Fodor, J., De Baets, B.: A single-point characterization of representable uninorms. Fuzzy Sets Syst. **202**, 89–99 (2012)
10. Fodor, J., Yager, R., Rybalov, A.: Structure of uninorms. Int. J. Uncertain. Fuzziness Knowl.-Based Syst. **5**, 411–427 (1997)
11. Gabbay, D.M., Metcalfe, G.: Fuzzy logics based on [0, 1)-continuous uninorms. Arch. Math. Log. **46**, 425–449 (2007)
12. Hu, S., Li, Z.: The structure of continuous uninorms. Fuzzy Sets Syst. **124**, 43–52 (2001)
13. Klement, E.P., Mesiar, R., Pap, E.: Triangular Norms. Kluwer Academic Publishers, Dordrecht (2000)
14. Li, G., Liu, H.W.: Distributivity and conditional distributivity of a uninorm with continuous underlying operators over a continuous t-conorm. Fuzzy Sets Syst. **287**, 154–171 (2016)
15. Li, G., Liu, H.W.: On properties of uninorms locally internal on the boundary. Fuzzy Sets Syst. Submitted
16. Li, G., Liu, H.W., Fodor, J.: Single-point characterization of uninorms with nilpotent underlying t-norm and t-conorm. Int. J. Uncertain. Fuzziness Knowl.-Based Syst. **22**, 591–604 (2014)
17. Li, G., Liu, H.W., Fodor, J.: On almost equitable uninorms. Kybernetika **51**, 699–711 (2015)
18. Mas, M., Monserrat, M., Ruiz-Aguilera, D., Torrens, J.: Migrative uninorms and nullnorms over t-norms and t-conorms. Fuzzy Sets Syst. **261**, 20–32 (2015)
19. Pradera, A., Beliakov, G., Bustince, H.: Aggregation functions and contradictory information. Fuzzy Sets Syst. **191**, 41–61 (2012)
20. Qin, F., Zhao, B.: The distributive equations for idempotent uninorms and nullnorms. Fuzzy Sets Syst. **155**, 446–458 (2005)
21. Ruiz, D., Torrens, J.: Distributivity and conditional distributivity of a uninorm and a continuous t-conorm. IEEE Trans. Fuzzy Syst. **14**, 180–190 (2006)
22. Ruiz-Aguilera, D., Torrens, J., De Baets, B., Fodor, J.C.: Some remarks on the characterization of idempotent uninorms. In: Hüllermeier, E., Kruse, R., Hoffmann, F. (eds.) IPMU 2010, LNAI 6178, pp. 425–434. Springer, Berlin (2010)
23. Yager, R., Rybalov, A.: Uninorm aggregation operators. Fuzzy Sets Syst. **80**, 111–120 (1996)
24. Yager, R., Rybalov, A.: Bipolar aggregation using the uninorms. Fuzzy Optim. Decis. Mak. **10**, 59–70 (2011)

A p-R$_0$ Type Triple I Method for Interval Valued Fuzzy Reasoning

Li-Na Ma and Shuo Liu

Abstract In this paper, we introduce the concept of p-relative degrees of activation, and a p-R_0 type triple I method for interval valued fuzzy reasoning model is proposed which is proved to be continuous. Furthermore, we showed that this method has a good transmissible performance for approximate errors.

Keywords Interval valued fuzzy reasoning · p-R_0 type triple I method · Continuity

1 Introduction

Fuzzy reasoning is a kind of approximate reasoning models for simulating human reasoning. As core content of the fuzzy control technology, it has gained broad attention after being proposed. However, its theoretical basis is not perfect. Full implication triple I method for fuzzy inference is proposed by Wang in [1], which made it possible to provide logical basis for fuzzy reasoning.

The general form of fuzzy reasoning can be expressed as follows [1]

$$\begin{array}{lll} \text{Rule1:} & A_{11} \text{ and } A_{12} \text{ and } \cdots \text{ and } A_{1n} & \text{imply } B_1 \\ \text{Rule2:} & A_{21} \text{ and } A_{22} \text{ and } \cdots \text{ and } A_{2n} & \text{imply } B_2 \\ & \cdots \quad \cdots \quad \cdots & \\ \text{Rule}l: & A_{l1} \text{ and } A_{l2} \text{ and } \cdots \text{ and } A_{ln} & \text{imply } B_l \\ \text{and} & A_1^* \text{ and } A_2^* \text{ and } \cdots \text{ and } A_n^* & \\ \hline \text{Obtain} & & B^* \end{array} \tag{1}$$

L.-N. Ma (✉)
School of Mathematics and Information Science, Shaanxi Normal University, Xi'an 710062, China
e-mail: malina@snnu.edu.cn

S. Liu
Faculty of Biomedical Engineering, The Fourth Military Medical University, Xi'an 710032, China

T.-H. Fan et al. (eds.), *Quantitative Logic and Soft Computing 2016*, Advances in Intelligent Systems and Computing 510,
DOI 10.1007/978-3-319-46206-6_26

where $A_{ij}, A_j^* \in \mathcal{F}(X_j)$, $B_i, B^* \in \mathcal{F}(Y)$ ($\mathcal{F}(X_j), \mathcal{F}(Y)$ respectively denote the set of all fuzzy subsets of domain X_j and Y), $i = 1, 2, \ldots, l$; $j = 1, 2, \ldots, n$.

As to the fuzzy reasoning model (1), a commonly used approach is to transform the multiple premises of each rule into a fuzzy set by use product. A new method to solve the model (2) is given in [1], which can be converted in to the following interval-valued fuzzy reasoning model (short as IVFS reasoning model):

$$\begin{array}{lll} \text{Rule1:}\ \mathcal{U}_1 & \text{imply} & B_1 \\ \text{Rule2:}\ \mathcal{U}_2 & \text{imply} & B_2 \\ & \cdots & \\ \text{Rule}l:\ \mathcal{U}_l & \text{imply} & B_l \\ \text{and}\quad A^* & & \\ \hline \text{Obtain} & & B^* \end{array} \tag{2}$$

where $\mathcal{U}_i = [a_{i1}, c_{i1}] \times \cdots \times [a_{in}, c_{in}]$ are interval-valued fuzzy sets of a domain $X = \{1, 2, \ldots, n\}$, $A^* \in \mathcal{F}(X)$, $B_1, \ldots, B_l, B^* \in \mathcal{F}(Y)$.

The principle of "Fire One Or Leave (in brief, FOOL)" was proposed by Wang in [1], and on the basis of this principle triple I method was used to solve the IVFS reasoning model. The concept of p-sensitive parameter was defined after an analysis of sensitivity of the consequents with respect to the antecedents of rules, and the (p-θ) method for (2) was given in [2]. In this paper we introduce the concept of p-relative degrees of activation, hence the (p-θ) method for the IVFS reasoning model has been refined to be the p-R_0 type triple I method. Furthermore, we prove that this method is continuous [4–7] and has a good transmissible performance for approximate errors.

2 The Full Implication Triple I Inference Method

In this section, we briefly overview the full implication triple I inference method of fuzzy inference. For more details, please see [3].

Triple I Method of FMP Suppose that a fuzzy rule "if x *is* A, then y *is* B" and an input "x *is* A^*" are given, where $A, A^* \in \mathcal{F}(X)$, $B \in \mathcal{F}(Y)$. Then the triple I solution B^* of FMP is the smallest fuzzy subset of Y such that $(A(x) \to B(y)) \to (A^*(x) \to B^*(y))$ gains the greatest value for all $x \in X$ and $y \in Y$.

In particular, this method is called R_0 type triple I method when "$\to$" is taken as the implication operator R_0, and the R_0 type triple I solution B^* can be calculated as follows:

$$B^*(y) = \sup_{x \in X}\{A^*(x) \otimes_0 R_0(A(x), B(y))\},\ y \in Y, \tag{3}$$

where $\otimes_0$ is the residuum of the operator R_0.

3 p-R$_0$ Type Triple I Method for Interval Valued Fuzzy Reasoning

In this section, the domain X is always refers to $X = \{1, 2, \ldots, n\}$, and we use the notation $\mathcal{IF}(X)$ to denote the set of all interval valued fuzzy set of X.

In the following, we use the Hamming distance to estimate the approximate degree of fuzzy sets, that is, $\|A_1 - A_2\| = \frac{1}{n}\sum_{i=1}^{n} |A_1(x_i) - A_2(x_i)|$, in which $A_1, A_2 \in \mathcal{F}(X)$.

Definition 1 ([1]) Let $\mathcal{U} \in \mathcal{IF}(X)$, $\mathcal{U} = [a_1, c_1] \times \cdots \times [a_n, c_n]$, $x^* = (x_1^*, \ldots, x_n^*) \in [0, 1]^n$. Define

$$d^p(x^*, \mathcal{U}) = \frac{1}{n}\sum_{j=1}^{n}\left(\frac{|x_j^* - e_j|}{w_j}\right)^p, p > 0. \tag{4}$$

$d^p(x^*, \mathcal{U})$ is called the p-sensitive distance from x^* to $\mathcal{U}$, where $e_j = \frac{c_j + a_j}{2}$, $w_j = c_j - a_j$, $j = 1, 2, \ldots, n$.

Definition 2 ([1]) Let $\mathcal{U}, \mathcal{V} \in \mathcal{IF}(X)$, $\mathcal{U} = [a_1, c_1] \times \cdots \times [a_n, c_n]$, $\mathcal{V} = [s_1, t_1] \times \cdots \times [s_n, t_n]$. Then the p-sensitive distance between $\mathcal{U}$ and $\mathcal{V}$ is defined as follows:

$$d^p(\mathcal{U}, \mathcal{V}) = \inf\{d^p(x, \mathcal{V}) | x \in \mathcal{U}\}, \tag{5}$$

where $x \in \mathcal{U}$ means $x_j \in [a_j, c_j]$, $j = 1, 2, \ldots, n$.

Definition 3 ([1]) The IVFS reasoning model is said to be in p-good condition, if for all $x \in \mathcal{U}_i$, $d^p(x, \mathcal{U}_i) < d^p(\mathcal{U}_j, \mathcal{U}_i)$, $i, j = 1, 2, \ldots, l$, $j \neq i$.

Theorem 1 ([1]) *The IVFS reasoning model is in p-good condition if and only if* $d^p(\mathcal{U}_i, \mathcal{U}_j) > (\frac{1}{2})^p$, $i, j = 1, 2, \ldots, l$, $i \neq j$.

Definition 4 ([2]) Let a domain $Y = \{y_1, \ldots, y_s\}$ be a nonempty set, $B, C \in \mathcal{F}(Y)$. Define

$$M^p(B, C) = \frac{1}{s}\sum_{r=1}^{s} |B(y_r) - C(y_r)|^p. \tag{6}$$

Then $M^p(B, C)$ is called the p-distance between B and C.

Definition 5 ([2]) In the IVFS reasoning model, a p-sensitive parameter of the kth rule is defined by

$$s_k^p = \max\left\{\frac{M^p(B_i, B_k)}{d^p(\mathcal{U}_i, \mathcal{U}_k)} \vee \frac{M^p(B_k, B_i)}{d^p(\mathcal{U}_k, \mathcal{U}_i)} | i = 1, \ldots, l, i \neq k\right\}, k = 1, 2, \ldots, l. \tag{7}$$

Let $A^* = (x_1, \ldots, x_n) \in [0, 1]^n$. Define the p-synthetical distance from A^* to $\mathcal{U}_k$ as follows:

$$\rho^p(A^*, \mathcal{U}_k) = s_k^p d^p(A^*, \mathcal{U}_k), k = 1, 2, \ldots, l. \tag{8}$$

Remark 1 According to the actual situation, we often assume that there does not exist two rules which have the same antecedents but different consequents in the IVFS reasoning model, otherwise we can combine them into a single rule. So although $d^p(\mathcal{U}_i, \mathcal{U}_k)$ and $d^p(\mathcal{U}_k, \mathcal{U}_i)$ all appear in the denominator, both are not zero. Therefore, the s_k^p definition is reasonable.

Definition 6 ([2]) Let $p > 0$, and $\theta(> 0)$ be a threshold value. The IVFS reasoning model is said to be $(p\text{-}\theta)$ solvable, if it is in p-good condition and $K(\theta) = \{k \leqslant l | \rho^p(A^*, \mathcal{U}_k) < \theta\} \neq \emptyset$.

Definition 7 ([2]) In the IVFS reasoning model, the $(p\text{-}\theta)$ relative activation degree of the input A^* which is about the k_ith rule is defined by

$$r(p\text{-}\theta, k_i) = \frac{(\rho^p(A^*, \mathcal{U}_{k_i}))^{-1}}{\sum_{j=1}^{t} (\rho^p(A^*, \mathcal{U}_{k_j}))^{-1}}. \tag{9}$$

Remark 2 In Definition 7, when the input A^* is exactly the central data fuzzy set of the antecedents of the k_ith rule, $\rho^p(A^*, \mathcal{U}_{k_i}) = 0$, then the result of the expression (9) is meaningless. Furthermore, in Definition 6, the specific method of setting threshold value is not given. In fact, the threshold settings is random and with a strong subjectivity in some cases, which would result in the loss of data. So we refine Definition 7 and propose the concept of p-relative activation degree, which can overcome the shortcomings.

Definition 8 In the IVFS reasoning model, the p-relative activation degree of the input A^* about the ith$(i = 1, 2, \ldots, l)$ rule is defined by

$$r(p, i) = \begin{cases} 1, & A^* \text{ is the central data fuzzy set of } \mathcal{U}_i, \\ \frac{(\rho^p(A^*, \mathcal{U}_i))^{-1}}{\sum_{j=1}^{n} (\rho^p(A^*, \mathcal{U}_j))^{-1}}, & \text{otherwise.} \end{cases} \tag{10}$$

Next, we give the p-R_0 type triple I method for the IVFS reasoning model.

Definition 9 Let $p > 0$ and the IVFS reasoning model is in p-good condition. Then the p-R_0 type triple I method for the IVFS reasoning model is defined as follows:
(i) To solve the central data fuzzy set A_k of the antecedents of the kth$(k = 1, 2, \ldots, l)$, that is, $A_k = (A_{k1}, A_{k2}, \ldots, A_{kn})$ where $A_{kt} = \frac{1}{2}(a_{kt} + c_{kt}), t = 1, 2, \ldots, n$. Then, to calculate the R_0 type triple I solution B_k^* of the following FMP model:

$$\text{from } A_k \longrightarrow B_k \text{ and } A^*, \text{ calculate } B_k^* \ (k = 1, 2, \ldots, l). \tag{11}$$

(ii) To compute the weighted average by multiplying each B_k^* by a corresponding p-relative activation degree, then the p-R_0 type triple I solution B^* for the IVFS reasoning model can be calculated by $B^* = \sum_{j=1}^{l} r(p, j)B_j^*$.

Example 1 In the IVFS reasoning model, let $X = \{1, 2, 3\}$, $Y = \{1, 2\}$. The rules and the input A^* are given as follows:

$$
\begin{array}{lll}
\text{Rule1} & \mathcal{U}_1 = [0.10, 0.20] \times [0.00, 0.20] \times [0.50, 0.60] \longrightarrow & B_1 = (0.30, 0.20) \\
\text{Rule2} & \mathcal{U}_2 = [0.30, 0.40] \times [0.50, 0.70] \times [0.50, 0.60] \longrightarrow & B_2 = (0.50, 0.40) \\
\text{Rule3} & \mathcal{U}_3 = [0.50, 0.60] \times [0.40, 0.60] \times [0.60, 0.70] \longrightarrow & B_3 = (0.60, 0.80) \\
\text{Rule4} & \mathcal{U}_4 = [0.00, 0.20] \times [0.20, 0.60] \times [0.10, 0.20] \longrightarrow & B_4 = (0.40, 0.70) \\
\text{Rule5} & \mathcal{U}_5 = [0.50, 0.60] \times [0.40, 0.45] \times [0.30, 0.40] \longrightarrow & B_5 = (0.30, 0.40) \\
\text{and} & A^* = (0.20, 0.40, 0.60) & \\
\hline
\text{Calculate} & & B^* = ?
\end{array} \tag{12}
$$

Let $p = 1$. Calculate the 1-R_0 type triple I solution B^* for the above IVFS reasoning model (12).

Solution. By (4), (5), we have

$$
d^p(\mathcal{U}_i, \mathcal{U}_j) = \begin{pmatrix} 0 & 1.17 & 1.83 & 1.33 & 3.17 \\ 1.17 & 0 & 0.67 & 1.58 & 1.5 \\ 1.83 & 0.67 & 0 & 2.17 & 0.83 \\ 1.33 & 1.67 & 2.67 & 0 & 1.67 \\ 2.17 & 1.25 & 0.92 & 1.17 & 0 \end{pmatrix}
$$

The smallest element is $0.67 > \frac{1}{2}$ in the above matrix whenever $i \neq j$. So the above IVFS reasoning model (4) is in 1-good condition. By (6), we have

$$
(M^p(B_i, B_j)) = \begin{pmatrix} 0 & 0.2 & 0.45 & 0.3 & 0.1 \\ 0.2 & 0 & 0.25 & 0.2 & 0.1 \\ 0.45 & 0.25 & 0 & 0.15 & 0.35 \\ 0.3 & 0.2 & 0.15 & 0 & 0.2 \\ 0.1 & 0.1 & 0.35 & 0.2 & 0 \end{pmatrix} \tag{13}
$$

And by (7), (13), we obtain the 1-sensitive parameter of each rule one by one: $s_1^p = 0.25$, $s_2^p = 0.37$, $s_3^p = 0.42$, $s_4^p = 0.23$, $s_5^p = 0.42$. Then we obtain the 1-sensitive distance between A^* and the antecedents of every rule, respectively, $d^p(A^*, \mathcal{U}_1) = 0.67, d^p(A^*, \mathcal{U}_2) = 0.67, d^p(A^*, \mathcal{U}_3) = 1.5, d^p(A^*, \mathcal{U}_4) = 1.67, d^p(A^*, \mathcal{U}_5) = 2.17$. Thus, we have the 1-synthetical distance from A^* to $\mathcal{U}_k)(k = 1, 2, 3, 4, 5.)$: $\rho^p(A^*, \mathcal{U}_1) = 0.17$, $\rho^p(A^*, \mathcal{U}_2) = 0.25$, $\rho^p(A^*, \mathcal{U}_3) = 0.63$, $\rho^p(A^*, \mathcal{U}_4) = 0.38$, $\rho^p(A^*, \mathcal{U}_5) = 0.91$. By (9), we obtain the 1-relative activation degree of the input A^* which is about each rule, in order, $r(p, 1) = 0.39$, $r(p, 2) = 0.26$, $r(p, 3) = 0.1$, $r(p, 4) = 0.17$, $r(p, 5) = 0.07$.

Furthermore, it is easy to obtain the central data fuzzy set of each rule: $A_1 = (0.15, 0.1, 0.55)$, $A_2 = (0.35, 0.6, 0.55)$, $A_3 = (0.55, 0.5, 0.65)$, $A_4 = (0.1, 0.4, 0.15)$, A_5=$(0.55, 0.425, 0.35)$. By (3), we have $B_1^*(1)$=$(0.2 \otimes_0 R_0(0.15, 0.3)) \vee (0.4 \otimes_0 R_0(0.1, 0.3)) \vee (0.6 \otimes_0 R_0(0.55, 0.3)) = 0.2 \vee 0.4 \vee 0.45 = 0.45$, and $B_1^*(2) = (0.2 \otimes_0 R_0(0.15, 0.2)) \vee (0.4 \otimes_0 R_0(0.1, 0.2)) \vee (0.6 \otimes_0 R_0(0.55, 0.2))$= $0.2 \vee 0.4 \vee 0.45 = 0.45$, i.e., $B_1^* = (0.45, 0.45)$. By a similar argument we have $B_2^* = (0.5, 0.45)$, $B_3^* = (0.6, 0.6)$, $B_4^* = (0.6, 0.6)$, $B_5^* = (0.6, 0.6)$. Thus, we obtain the 1-R_0 type triple I solution $B^* = 0.39 \times B_1^* + 0.26 \times B_2^* + 0.1 \times B_3 *$ $+0.17 \times B_4^* + 0.07 \times B_5^* = (0.51, 0.49)$.

Definition 10 Let X, Y be nonempty finite domains, $\mathcal{U}_i \to B_i (i = 1, 2, \ldots, l)$ be the known fuzzy rules, where $\mathcal{U}_i \in \mathcal{IF}(X)$, $B_i \in \mathcal{F}(Y)$, and A_i denote the central data fuzzy set of $\mathcal{U}_i$. If we have the p-R_0 type triple I method f is continuous at A_i, then f is said to be continuous at $\mathcal{U}_i$.

Lemma 1 ([4]) *Let $a, b_1, b_2 \in [0, 1]$. Then we have the following:*

(i) $|a \wedge b_1 - a \wedge b_2| \leqslant |b_1 - b_2|$;
(ii) $|a \vee b_1 - a \vee b_2| \leqslant |b_1 - b_2|$.

Lemma 2 *Let $a_i, b_i \in [0, 1], i \in \{1, 2, \ldots, n\}$. Then*

(i) $|\bigwedge_{i=1}^{n} a_i - \bigwedge_{i=1}^{n} b_i| \leqslant \bigvee_{i=1}^{n} |a_i - b_i|$;
(ii) $|\bigvee_{i=1}^{n} a_i - \bigvee_{i=1}^{n} b_i| \leqslant \bigvee_{i=1}^{n} |a_i - b_i|$.

Proof We only prove (i), (ii) similarly follows. Let $\bigwedge_{i=1}^{n} a_i = a_k$, $\bigwedge_{i=1}^{n} b_i = b_j$. Then for all $i, i \in \{1, 2, \ldots, n\}$, we have $a_k \leqslant a_i$, $b_j \leqslant b_i$. In particular, we have $a_k \leqslant a_j, b_j \leqslant b_k$. If $a_k \leqslant b_j \leqslant b_k$, then $|\bigwedge_{i=1}^{n} a_i - \bigwedge_{i=1}^{n} b_i|$=$|a_k - b_j| \leqslant |a_k - b_k| \leqslant \bigvee_{i=1}^{n} |a_i - b_i|$. And if $b_j < a_k \leqslant a_j$, then $|\bigwedge_{i=1}^{n} a_i - \bigwedge_{i=1}^{n} b_i| = |a_k - b_j| \leqslant |a_j - b_j| \leqslant \bigvee_{i=1}^{n} |a_i - b_i|$. This proves (i). □

Lemma 3 ([4]) *Let X, Y be nonempty finite domains, $A^*, A \in \mathcal{F}(X)$, $B \in \mathcal{F}(Y)$. Then there exists $\delta_0 > 0$ such that for all $y \in Y$, $EA_y^* = EA_y (\subset X)$ whenever $\|A^* - A\| < \delta_0$, where $EA_y^* = \{x \in X | (A^*(x))' < R_0(A(x), B(y))\}$, $EA_y = \{x \in X | (A(x))' < R_0(A(x), B(y))\}$.*

Lemma 4 *Let X, Y be nonempty finite domains, A^*, $A_k \in \mathcal{F}(X)$, $B_k \in \mathcal{F}(Y)$, $k = 1, 2, \ldots, l$, and B_k^* is the R_0 type triple I solution for FMP model (11). Then for each $\varepsilon > 0$, there is a $\delta > 0$ such that $|B_k^*(y) - B_k(y)| < \varepsilon$ whenever $\|A^* - A_k\| < \delta$.*

Proof Obviously, for an arbitrary $x \in X$, we have $A_k(x) \otimes_0 R_0(A_k(x), B_k(y)) \leqslant B_k(y)$ and $\bigvee_{x \in X} \{A_k(x) \otimes_0 R_0(A_k(x), B_k(y))\} \leqslant B_k(y)$. Then

$$\begin{aligned} |B_k^*(y) - B_k(y)| &= |\bigvee_{x\in X}\{A^*(x) \otimes_0 R_0(A_k(x), B_k(y))\} - B_k(y)| \\ &\leqslant |\bigvee_{x\in X}\{A^*(x) \otimes_0 R_0(A_k(x), B_k(y))\} \\ &\quad - \bigvee_{x\in X}\{A_k(x) \otimes_0 R_0(A_k(x), B_k(y))\}| \\ &= |\bigvee_{x\in EA_y^*}\{A^*(x) \wedge R_0(A_k(x), B_k(y))\} \end{aligned} \tag{14}$$

$$- \bigvee_{x \in EA_{ky}} \{A_k(x) \wedge R_0(A_k(x), B_k(y))\}|,$$

where $EA_y^* = \{x \in X | (A^*(x))' < R_0(A_k(x), B_k(y))\}$, $EA_{ky} = \{x \in X | (A_k(x))' < R_0(A_k(x), B_k(y))\}$. By Lemma 3, there exists $\delta_0 > 0$ such that $EA_y^* = EA_{ky}$ whenever $\|A^* - A\| < \delta_0$. So for each $\varepsilon > 0$, putting $\delta = \min\{\frac{\delta_0}{n+1}, \frac{\varepsilon}{n+1}\}$. Then for all $y \in Y$, we have $EA_y^* = EA_{ky}$ whenever $\|A^* - A_k\| < \delta$. Obviously, for all $x \in X$, $|A^*(x) - A_k(x)| < n\delta$, then by (14), we get

$$\begin{aligned} |B_k^*(y) - B_k(y)| &\leqslant |\bigvee_{x\in EA_{ky}}\{A^*(x) \wedge R_0(A_k(x), B_k(y))\} \\ &\quad - \bigvee_{x\in EA_{ky}}\{A_k(x) \wedge R_0(A_k(x), B_k(y))\}| \\ &\leqslant \bigvee_{x\in EA_{ky}} |A^*(x) \wedge R_0(A_k(x), B_k(y)) \\ &\quad - A_k(x) \wedge R_0(A_k(x), B_k(y))| \\ &\qquad\qquad \text{(by Lemma 2(i))} \\ &\leqslant \bigvee_{x\in EA_{ky}} |A^*(x) - A_k(x)| \qquad \text{(by Lemma 1(i))} \\ &\leqslant \bigvee_{x\in X} |A^*(x) - A_k(x)| < n\delta < \varepsilon. \end{aligned}$$

□

In the following, we use the notation A_k to denote the central data fuzzy set of $\mathcal{U}_k \in \mathcal{IF}(X)$.

Lemma 5 *Let $\mathcal{U}_1, \mathcal{U}_2, \ldots, \mathcal{U}_l \in \mathcal{IF}(X)$, $\mathcal{U}_k = [a_1, c_1] \times \cdots \times [a_n, c_n]$, $A^* = (x_1^*, \ldots, x_n^*) \in [0, 1]^n$, $A^* \neq A_k$, $A_i = (x_{i1}, \ldots, x_{in}) \in [0, 1]^n (i \neq k, i = 1, 2, \ldots, l)$, $p \geqslant 1$, and putting $w_{jk} = c_j - a_j (j = 1, 2, \ldots, n)$, $b_k = min\{\rho^p(A^*, \mathcal{U}_k), \rho^p(A_i, \mathcal{U}_k)\}$, w_{jk}. Then we have $|d^p(A^*, \mathcal{U}_k) - d^p(A_i, \mathcal{U}_k)| \leqslant \frac{1}{n} \sum_{j=1}^{n} \frac{p \cdot |x_j^* - x_{ij}|}{w_{jk}^p}$.*

Proof Putting $e_{jk} = \frac{c_j + a_j}{2}$, $j = 1, 2, \ldots, n$. Then we have

$$
\begin{aligned}
|d^p(A^*,\mathcal{U}_k)-d^p(A_i,\mathcal{U}_k)| &= \frac{1}{n}\sum_{j=1}^{n}\left(\frac{|x_j^*-e_{jk}|^p-|x_{ij}-e_{jk}|^p}{w_{jk}^p}\right)\\
&\leqslant \frac{1}{n}\sum_{j=1}^{n}\left(\frac{|(x_j^*-e_{jk})^p-(x_{ij}-e_{jk})^p|}{w_{jk}^p}\right)\\
&= \frac{1}{n}\sum_{j=1}^{n}\left(\frac{|p(\xi_{jk}-e_{jk})^{p-1}\cdot(x_j^*-x_{ij})|}{w_{jk}^p}\right)\\
&\leqslant \frac{1}{n}\sum_{j=1}^{n}\left(\frac{p\cdot|x_j^*-x_{ij}|}{w_{jk}^p}\right). \qquad (\text{by } (\xi_{jk}-e_{jk})^{p-1}\leqslant 1)
\end{aligned}
$$

(where point ξ_{jk} is between x_j^* and x_{ij}).

Hence, by (8) we get $|\rho^p(A^*,\mathcal{U}_k)^{-1}-\rho^p(A_i,\mathcal{U}_k)^{-1}| = |\frac{\rho^p(A_i,\mathcal{U}_k)-\rho^p(A^*,\mathcal{U}_k)}{\rho^p(A_i,\mathcal{U}_k)\cdot\rho^p(A^*,\mathcal{U}_k)}| \leqslant \frac{1}{b_k^2}\cdot|\rho^p(A^*,\mathcal{U}_k)-\rho^p(A_i,\mathcal{U}_k)| = \frac{1}{b_k^2}\cdot|s_k^p\cdot d^p(A^*,\mathcal{U}_k)-s_k^p\cdot d^p(A_i,\mathcal{U}_k)| = \frac{s_k^p}{b_k^2}|d^p(A^*,\mathcal{U}_k)-d^p(A_i,\mathcal{U}_k)| \leqslant \frac{1}{n}\sum_{j=1}^{n}(\frac{p\cdot|x_j^*-x_{ij}|}{w_{jk}^p})$. □

Lemma 6 *Let* $\mathcal{U}_1,\mathcal{U}_2,\ldots,\mathcal{U}_l\in\mathcal{IF}(X)$, $A^*=(x_1^*,\ldots,x_n^*)\in[0,1]^n$, *and* $A^*\neq A_k$, $A_i=(x_{i1},\ldots,x_{in})\in[0,1]^n (i\neq k, i=1,2,\ldots,l)$, $p\geqslant 1$, $b_k=min\{\rho^p(A^*,\mathcal{U}_k),\rho^p(A_i,\mathcal{U}_k)\}$, $w_k^p{=}min\{w_{1k}^p,\ldots,w_{nk}^p\}$. *Putting* $w_{jk}{=}c_j-a_j (j=1,2,\ldots,n)$, $s^p=max\{s_1^p,\ldots,s_l^p\}$, $b=min\{b_1,\ldots,b_l\}$, $w^p=min\{w_1^p,\ldots,w_l^p\}$, $a=min\{\sum_{k=1}^{l}(\rho^p(A^*,\mathcal{U}_k))^{-1},\sum_{k=1}^{l}(\rho^p(A_i,\mathcal{U}_k))^{-1}\}$, $c=\sum_{k=1}^{l}(\rho^p(A_i,\mathcal{U}_k))^{-1}$. *Then we have*

$$
\left|\frac{1}{\sum_{k=1}^{l}(\rho^p(A^*,\mathcal{U}_k))^{-1}}-\frac{1}{\sum_{k=1}^{l}(\rho^p(A_i,\mathcal{U}_k))^{-1}}\right|\leqslant\frac{l\cdot s^p}{a^2\cdot b^2\cdot n}\cdot\sum_{j=1}^{n}\frac{p\cdot|x_j^*-x_{ij}|}{w^p}.
$$

Proof By Lemma 5, we get

$$
\begin{aligned}
&\left|\frac{1}{\sum_{k=1}^{l}(\rho^p(A^*,\mathcal{U}_k))^{-1}}-\frac{1}{\sum_{k=1}^{l}(\rho^p(A_i,\mathcal{U}_k))^{-1}}\right|\\
&=\left|\frac{\sum_{k=1}^{l}(\rho^p(A_i,\mathcal{U}_k))^{-1}-\sum_{k=1}^{l}(\rho^p(A^*,\mathcal{U}_k))^{-1}}{\sum_{k=1}^{l}(\rho^p(A^*,\mathcal{U}_k))^{-1}\cdot\sum_{k=1}^{l}(\rho^p(A_i,\mathcal{U}_k))^{-1}}\right|\\
&\leqslant\frac{1}{a^2}\cdot\left|\sum_{k=1}^{l}(\rho^p(A^*,\mathcal{U}_k))^{-1}-\sum_{k=1}^{l}(\rho^p(A_i,\mathcal{U}_k))^{-1}\right|\\
&\leqslant\frac{1}{a^2}\sum_{k=1}^{l}|(\rho^p(A^*,\mathcal{U}_k))^{-1}-(\rho^p(A_i,\mathcal{U}_k))^{-1}|\\
&\leqslant\frac{1}{a^2}\sum_{k=1}^{l}\left(\frac{s_k^p}{b_k^2}\cdot\frac{1}{n}\cdot\sum_{j=1}^{n}\frac{p\cdot|x_j^*-x_{ij}|}{w_{jk}^p}\right)\\
&\leqslant\frac{1}{a^2}\cdot l\cdot\frac{s^p}{b^2}\cdot\frac{1}{n}\cdot\sum_{j=1}^{n}\frac{p\cdot|x_j^*-x_{ij}|}{w^p}=\frac{l\cdot s^p}{a^2\cdot b^2\cdot n}\cdot\sum_{j=1}^{n}\frac{p\cdot|x_j^*-x_{ij}|}{w^p}.
\end{aligned}
$$

Thus, by (10) we have

$$
\begin{aligned}
|r(p,j)-r_i(p,j)| &= \Big| \frac{(\rho^p(A^*,\mathcal{U}_j))^{-1}}{\sum\limits_{k=1}^{l}(\rho^p(A^*,\mathcal{U}_k))^{-1}} - \frac{(\rho^p(A_i,\mathcal{U}_j))^{-1}}{\sum\limits_{k=1}^{l}(\rho^p(A_i,\mathcal{U}_k))^{-1}} \Big| \\
&\leqslant \Big| \frac{(\rho^p(A^*,\mathcal{U}_j))^{-1}\cdot\sum\limits_{k=1}^{m}(\rho^p(A_i,\mathcal{U}_k))^{-1}-(\rho^p(A_i,\mathcal{U}_j))^{-1}\cdot\sum\limits_{k=1}^{m}(\rho^p(A^*,\mathcal{U}_k))^{-1}}{\sum\limits_{k=1}^{l}(\rho^p(A^*,\mathcal{U}_k))^{-1}\cdot\sum\limits_{k=1}^{l}(\rho^p(A_i,\mathcal{U}_k))^{-1}} \Big| \\
&\leqslant \tfrac{1}{a^2}\Big(\Big|\sum_{k=1}^{l}(\rho^p(A_i,\mathcal{U}_k))^{-1}\Big|\cdot|(\rho^p(A^*,\mathcal{U}_j))^{-1}-(\rho^p(A_i,\mathcal{U}_j)^{-1}| \\
&+ |(\rho^p(A_i,\mathcal{U}_j))^{-1}|\cdot\Big|\sum_{k=1}^{l}(\rho^p(A_i,\mathcal{U}_k))^{-1}-\sum_{k=1}^{l}(\rho^p(A^*,\mathcal{U}_k))^{-1}\Big|\Big) \\
&\leqslant \tfrac{1}{a^2}\Big|\sum_{k=1}^{l}(\rho^p(A_i,\mathcal{U}_k))^{-1}\Big(|(\rho^p(A^*,\mathcal{U}_j))^{-1}-(\rho^p(A_i,\mathcal{U}_j)^{-1}| \\
&+ \Big|\sum_{k=1}^{l}(\rho^p(A_i,\mathcal{U}_k))^{-1}-\sum_{k=1}^{l}(\rho^p(A^*,\mathcal{U}_k))^{-1}\Big|\Big) \\
&\leqslant \tfrac{c}{a^2}\Big(\tfrac{s^p}{b^2}\cdot\tfrac{1}{n}\cdot\sum_{j=1}^{n}\frac{p\cdot|x_j^*-x_{ij}|}{w^p}+l\cdot\Big(\tfrac{s^p}{b^2}\cdot\tfrac{1}{n}\cdot\sum_{j=1}^{n}\frac{p\cdot|x_j^*-x_{ij}|}{w_{jk}^p}\Big)\Big) \\
&= \frac{c\cdot(l+1)\cdot s^p}{a^2\cdot b^2\cdot n}\cdot\sum_{j=1}^{n}\frac{p\cdot|x_j^*-x_{ij}|}{w^p}.
\end{aligned}
$$

□

Theorem 2 *The p-R_0 type triple I method for the IVFS reasoning model is continuous at $\mathcal{U}_i (i = 1, 2, \ldots, l)$.*

Proof We use the notation f to denote the p-R_0 type triple I method for the IVFS reasoning model. Let $A^* = (x_1^*, \ldots, x_n^*)$, $A_i = (x_{i1}, \ldots, x_{in})$. Then we have

$$
\begin{aligned}
\|f(A^*)-f(A_i)\| &= \tfrac{1}{m}\sum_{y\in Y}|f(A^*)(y)-f(A_i)(y)| \\
&= \tfrac{1}{m}\sum_{y\in Y}\Big|\sum_{j=1}^{l}r(p,j)\cdot B_j^*(y)-\sum_{j=1}^{l}r_i(p,j)\cdot B_j(y)\Big| \\
&\leqslant \tfrac{1}{m}\sum_{y\in Y}\sum_{j=1}^{l}|r(p,j)\cdot B_j^*(y)-r_i(p,j)\cdot B_j(y)| \\
&\leqslant \tfrac{1}{m}\sum_{y\in Y}\sum_{j=1}^{l}(|r(p,j)||B_j^*(y)-B_j(y)| \\
&+ |B_j(y)||r(p,j)-r_i(p,j)|) \\
&\leqslant \tfrac{1}{m}\sum_{y\in Y}\sum_{j=1}^{l}(|B_j^*(y)-B_j(y)|+|r(p,j)-r_i(p,j)|).
\end{aligned}
\tag{15}
$$

By Lemma 4, we get that for each $\varepsilon > 0$ there is a $\delta > 0$ such that $|B_j^*(y) - B_j(y)| < \frac{\varepsilon}{2l}$ whenever$\| A^* - A_i\| < \delta$. Putting $\delta_1 = \frac{a^2b^2w^p\varepsilon}{2lpc(l+1)s^p}$, then by Lemma 5 we have $|r(p,j) - r_i(p,j)| < \frac{\varepsilon}{2l}$ whenever $\|A^* - A_i\| < \delta_1$. For each $\varepsilon > 0$, we take $\delta = \min\{\delta_0, \delta_1\}$, then by (15) we get $\|f(A^*) - f(A_i)\| < \frac{1}{m}\sum_{y\in Y}\sum_{j=1}^{l}(\frac{\varepsilon}{2l} + \frac{\varepsilon}{2l}) = \varepsilon$. This proves Theorem 2. □

In the following, we will investigate the transmissible performance for approximate errors.

Definition 11 Let $\mathcal{U} \in \mathcal{IF}(X)$, $A^* = (x_1^*, \ldots, x_n^*) \in [0,1]^n$. Define $w(A^*, A) = \bigvee_{i=1}^{n} |x_i^* - x_i|$, where A denotes the central data fuzzy set of $\mathcal{U}$. $w(A^*, A)$ is called the maximum number of point by point error between A^* and $\mathcal{U}$.

Proposition 1 *Let f be the p-R_0 type triple I method for the IVFS reasoning model. If the maximum number of point by point error between the input A^* and $\mathcal{U}(i = 1, 2, \ldots, l)$ is less than ε, then the maximum number of point by point error between the output $f(A^*)$ and $f(A_i)$ is also less than ε, where A_i denotes the central data fuzzy set of $\mathcal{U}_i$.*

Proof By Lemmas 4 and 6, we obtain that for each $\varepsilon > 0$, there exist $\delta_1 > 0$, $\delta_2 > 0$ such that $|B_j^*(y) - B_j(y)| < \frac{\varepsilon}{2l}$ whenever $\|A^* - A_i\| < \delta_1$, and $|r(p,j) - r_i(p,j)| < \frac{\varepsilon}{2l}$ whenever $\|A^* - A_i\| < \delta_2$. Putting $\delta = \min\{\delta_1, \delta_2\}$. Then we get $|B_j^*(y) - B_j(y)| < \frac{\varepsilon}{2l}$ and $|r(p,j) - r_i(p,j)| < \frac{\varepsilon}{2l}$ whenever $\|A^* - A_i\| < \delta$. For each $\varepsilon > 0$, if $w(A^*, A_i) < \varepsilon < \delta$ then $\|A^* - A_i\| < \delta$. Hence we have

$$
\begin{aligned}
w(f(A^*), f(A_i)) &= \bigvee_{y\in Y} |f(A^*)(y) - f(A_i)(y)| \\
&= \bigvee_{y\in Y} \Big|\sum_{j=1}^{l} r(p,j)\cdot B_j^*(y) - \sum_{j=1}^{l} r_i(p,j)\cdot B_j(y)\Big| \\
&\leqslant \bigvee_{y\in Y} \sum_{j=1}^{l} |r(p,j)\cdot B_j^*(y) - r_i(p,j)\cdot B_j(y)| \\
&\leqslant \bigvee_{y\in Y} \sum_{j=1}^{l} (|r(p,j)||B_j^*(y) - B_j(y)| + |B_j(y)||r(p,j) - r_i(p,j)|) \\
&\leqslant \bigvee_{y\in Y} \sum_{j=1}^{l} (|B_j^*(y) - B_j(y)| + |r(p,j) - r_i(p,j)|) \\
&< \bigvee_{y\in Y} \sum_{j=1}^{l} (\tfrac{\varepsilon}{2l} + \tfrac{\varepsilon}{2l}) = \varepsilon.
\end{aligned}
$$

□

This result states that the p-R_0 type triple I method for the IVFS reasoning model has a good transmissible performance for approximate errors.

Acknowledgments This paper was supported by the Fundamental Research Funds for the Central Universities (GK201503013)

References

1. Wang, G.J.: Triple I method and interval valued fuzzy reasoning. Sci. China Ser. E. **30**(4), 331–340 (2000) (in chinese)
2. Wang, G.J., Wang, X.Y., Song, J.S.: Refinement of FOOL method for interval valued fuzzy reasonging. Shaanxi Normal Univ. **29**(3), 6–19 (2001) (Natural Science Edition)
3. Wang, G.J.: The full implicational triple I method of fuzzy reasoning. Sci. China Ser. E. **29**(1), 44–53 (1999) (in chinese)
4. Xu, W.H., Xie, Z.K., Yang, J.Y., Ye, Y.P.: Continuity and approximation properties of two classes of algorithms for fuzzy inference. J. Soft. **15**(10), 1485–1492 (2004) (in chinese)
5. Pan, H.Y., Pei, D.W., Huang, A.M.: Continuity and approximation property of triple I method. Sci. J. Comput. Sci. **35**(12), 158–162 (2008) (in Chinese)
6. Jenei, S.: Continuity in Zadeh's compositional rule of inference. Fuzzy Sets Syst. **104**(2), 333–339 (1999)
7. Liu, H.W., Wang, G.J.: Continuity of triple I methods based on several implications. Comput. Math. Appl. **56**(8), 2079–2087 (2008)

Multiple Fuzzy Implications and Their Generating Methods

Fang Li and Dao-Wu Pei

Abstract By means of multiple iteration about the first variable of a known fuzzy implication, a new fuzzy implication is introduced. It is proved that the new implication preserves some important properties of the original implication. These properties include the identity principle (IP), the left neutrality property (NP), the exchange principle (EP), and so on. Based on this idea, by considering multiple iteration about the second variable of the known fuzzy implications, another new kind of fuzzy implications are similarly introduced, main properties of this kind of new implications are analyzed. This work is beneficial to enhance applications of fuzzy implications in other fields.

Keywords Fuzzy logic · Fuzzy implication · Multiple fuzzy implication · Mixed fuzzy implication

1 Introduction

Fuzzy implications, as a kind of most important logical connectives, play a kernel role in many fields (see [2, 4, 5, 10, 13]).

In the literature, there are three main directions on the study of fuzzy implication (see [2] or [4]): properties and characterizations of the known fuzzy implications (see [1, 2]), construction methods of new fuzzy implications (see [2, 3, 6, 8, 11, 12]), and practical applications of fuzzy implications (see [2, 9, 10, 13]).

In [8], the authors proposed a new construction method of fuzzy implications by multiple iterations of the known fuzzy implications about the first variable. A kind of new fuzzy implications are obtained, and various properties of these new implications are discussed in detail.

F. Li · D.-W. Pei (✉)
School of Science, Zhejiang Sci-Tech University, Hangzhou 310018, China
e-mail: peidw@163.com

F. Li
e-mail: 1670777133@163.com

T.-H. Fan et al. (eds.), *Quantitative Logic and Soft Computing 2016*, Advances in Intelligent Systems and Computing 510,
DOI 10.1007/978-3-319-46206-6_27

In this paper, by multiple iteration about the second variable of the known fuzzy implications, we can generate new classes of fuzzy implications. In addition, we will study properties of the new implications, and study which properties they have when the known implications are (S, N)-implication and R-implication respectively.

Finally, we will integrate these two types of implications together, get a new kind of fuzzy implications when the time of iteration is a special number, the above two types of implications are all special cases of these implications.

The content of this paper is organized as follows: the second part recalls some useful definitions and properties for the paper. In the third part, we propose a new class of functions, which are fuzzy implications under certain conditions, analyze their properties, and study which properties the new implications posses when the known implications are (S, N)-implications or R-implications, respectively. The fourth part, by multiple iteration about two variables of the known implications, we proposed another new kind of fuzzy implications. Lastly, the fifth part concludes the paper.

2 Preliminaries

In this section, we recall some basic notations and facts used later in the paper (see [2, 5, 7, 10, 13]). Let $U = [0, 1]$.

Definition 1 A binary operation T on U is a t-norm if it fulfills commutativity, associativity, monotonicity and boundary condition $T(1, x) = x$ for all $x \in U$.

Definition 2 A binary operation S on U is a t-conorm if it fulfills commutativity, associativity, monotonicity and boundary condition $S(0, x) = x$ for all $x \in U$.

Definition 3 A unary operation N on U is a fuzzy negation if $N(0) = 1, N(1) = 0$ and N is decreasing.

Definition 4 A binary operation I on U is an (fuzzy) implication if it is decreasing in the first variable, increasing in the second variable, and

$$I(0, 0) = I(1, 1) = 1, \quad I(1, 0) = 0.$$

Several common used fuzzy implications are given by Table 1.

In this paper, we use T, S, N and I to denote t-norm, t-conorm, fuzzy negation and fuzzy implication, respectively.

For some special types of implications, they also have some of the following properties.

Let I be an implication, and $x,\ y,\ y_1,\ y_2 \in U$.

(i) **The left neutrality principle, (NP)**: $I(1, y) = y$;
(ii) **The identity principle, (IP)**: $I(x, x) = 1$;
(iii) **The ordering principle, (OP)**: $I(x, y) = 1 \Longleftrightarrow x \le y$;

Table 1 Examples of fuzzy implications

Name	Notation	Formula
Lukasiewicz	I_{LK}	$I_{LK}(x, y) = \min(1, 1 - x + y)$
Kleene-Dienes	I_{KD}	$I_{KD}(x, y) = \max(1 - x, y)$
Rescher	I_{RS}	$I_{RS}(x, y) = \begin{cases} 0, & x > y \\ 1, & x \le y \end{cases}$
Godel	I_{GD}	$I_{GD}(x, y) = \begin{cases} y, & x > y \\ 1, & x \le y \end{cases}$
Weber	I_{WB}	$I_{WB}(x, y) = \begin{cases} 1, & x < 1 \\ y, & x = 1 \end{cases}$
R_0	I_0	$I_0(x, y) = \begin{cases} \max(1 - x, y), & x > y \\ 1, & x \le y \end{cases}$

(iv) **The consequent boundary, (CB)**: $I(x, y) = y$;
(v) **The law of excluded middle, (LEM)**: $S(N(x), x) = 1$;
(vi) **1-Lipschitz property, (1-Li)**: $|I(x, y_2) - I(x, y_1)| \le |y_2 - y_1|$;
(vii) **Special property, (SP)**: $I(x, y) \le I(x + \varepsilon, y + \varepsilon), \quad \forall \varepsilon \ge 0$.

In the literature [8], a new class of fuzzy implications, namely multiple fuzzy implications, are constructed by means of multiple iteration about the first variable of the known fuzzy implications as follows.

Definition 5 ([8]) Let $I_1, I_2, \ldots, I_n$ be fuzzy implications. Then $J_{I_1,I_2,\ldots,I_n}$, called the n-dimension function generated by $(I_1, I_2, \ldots, I_n)$, is defined as

$$J_{I_1,I_2,\ldots,I_n}(x, y) = I_1(x, I_2(x, I_3(x, \ldots, I_n(x, y) \ldots))), \quad x, \ y \in U.$$

It has been proved that this function is a fuzzy implication [8]. Therefore, we call it a multiple fuzzy implication about x.

3 Multiple Fuzzy Implications About the Second Variable

Similar to the multiple fuzzy implications about x, in this section we define a class of multiple fuzzy implications about y, and discuss their properties.

Definition 6 Let $I_1, I_2, \ldots, I_n$ be fuzzy implications. Then $G_{I_1,I_2,\ldots,I_n}$ defined below is called a n-dimension function generated by $(I_1, I_2, \ldots, I_n)$,

$$G_{I_1,I_2,\ldots,I_n}(x, y) = I_1(x, I_2(I_3(I_4(\ldots I_n(x, y), \ldots, y), y), y)), \quad x, \ y \in U.$$

The following proposition shows that this function is also a fuzzy implication. We call it multiple fuzzy implication about y.

Proposition 1 *Let $I_1, \ldots, I_n$ be fuzzy implications satisfied (1-Li) and (SP), n a positive even number. Then $G_{I_1,\ldots,I_n}$ is also a fuzzy implication.*

Proof The conclusion holds obviously for the case $n = 2$ (see [8] or [11]). Without loss of generality, we take $n = 4$.

(i) Take $x_1,\ x_2,\ y \in U$ with $x_1 \le x_2$. Since I_1, I_2, I_3, I_4 are fuzzy implications, we have

$I_4(x_1, y) \ge I_4(x_2, y), I_3(I_4(x_1, y), y) \le I_3(I_4(x_2, y), y)$;

$I_2(I_3(I_4(x_1, y), y), y) \ge I_2(I_3(I_4(x_1, y), y), y)$;

$I_1(x_1, I_2(I_3(I_4(x_1, y), y), y)) \ge I_1(x, I_2(I_3(I_4(x_1, y), y), y))$.

Moreover, we have $G_{I_1,I_2,I_3,I_4}(x_1, y) \ge G_{I_1,I_2,I_3,I_4}(x_2, y)$.

(ii) Take $x,\ y_1,\ y_2 \in U$ with $y_1 \le y_2$. Since I_1, I_2, I_3, I_4 satisfy (SP), we have for any $\varepsilon > 0$ with $x + \varepsilon,\ y + \varepsilon \in U$,

$$I_i(x + \varepsilon, y) \le I_i(x, y) \le I_i(x + \varepsilon, y + \varepsilon) \le I_i(x, y + \varepsilon).$$

Since I_1, I_2, I_3, I_4 all satisfy (1-Li), we have $I_i(x, y_2) - I_i(x, y_1) \le y_2 - y_1,\ i = 1, 2, 3, 4$. Let $y_2 - y_1 \doteq \varepsilon$. Thus

$$I_3(I_4(x, y_1), y_1) \le I_3(I_4(x, y_1) + \varepsilon, y_1 + \varepsilon) \le I_3(I_4(x, y_2), y_2).$$

Similarly, we have $I_2(I_3(I_4(x, y_1), y_1), y_1) \le I_2(I_3(I_4(x, y_2), y_2), y_2)$. Thus we obtain

$$G_{I_1,I_2,I_3,I_4}(x, y_1) \le G_{I_1,I_2,I_3,I_4}(x, y_2).$$

(iii) By simple calculation we have,

$$G_{I_1,I_2,I_3,I_4}(0, 0) = 1 = G_{I_1,I_2,I_3,I_4}(1, 1); \quad G_{I_1,I_2,I_3,I_4}(1, 0) = 0.$$

By the definition of fuzzy implication, G_{I_1,I_2,I_3,I_4} is a fuzzy implication.

Obviously, the above proof is valid for all even numbers, thus when n is a positive even number, $G_{I_1,\ldots,I_n}$ is a fuzzy implication. □

Example 1 Let $n = 4$, $I_1 = I_{KD}$, $I_2 = I_{GD}$, $I_3 = I_{WB}$ and $I_4 = I_{RS}$. We have

$$G_{I_1,I_2,I_3,I_4}(x, y) = \begin{cases} 1, & \text{if } x \le y, \\ \max(1 - x, y), & \text{if } x > y. \end{cases}$$

Thus $G_{I_1,I_2,I_3,I_4}(x, y) = R_0(x, y)$.

Remark 1 When n is a positive odd number, e.g., $n = 3$, we have

$$I_1(1, I_2(I_3(1, 0), 0)) = 1.$$

Thus, G_{I_1,I_2,I_3} is not an implication.

Proposition 2 *Let $I_1, \ldots, I_n$ be fuzzy implications, $G_{I_1,\ldots,I_n}$ a fuzzy implication generated by the above implications.*

(i) *If $I_1, \ldots, I_n$ all satisfy (IP) and (OP), then $G_{I_1,\ldots,I_n}$ also satisfies (IP).*
(ii) *If $I_1, \ldots, I_n$ all satisfy (NP) and (IP), then $G_{I_1,\ldots,I_n}$ also satisfies (NP).*
(iii) *If $I_1, \ldots, I_n$ all satisfy (OP), then G_{I_1,I_2,I_3,I_4} also satisfies (OP).*

Proof (i) Let $n = 4$ and $x \in U$. Since $I_1, \ldots, I_4$ satisfy (IP) and (OP), we have

$$I_1(x, I_2(I_3(I_4(x, x), x), x)) = I_1(x, I_2(I_3(1, x), x)) = I_1(x, 1) = 1.$$

(ii) Let $n = 4$ and $x \in U$. Since $I_1, \ldots, I_4$ satisfy (NP) and (IP), we have

$$I_1(1, I_2(I_3(I_4(1, x), x), x)) = I_1(1, I_2(I_3(x, x), x)) = I_1(1, x) = x.$$

(iii) Let $n = 4$ and $x,\ y \in U$ with $x \leq y$. Since $I_1, \ldots, I_4$ satisfy (OP), we have

$$I_1(x, I_2(I_3(I_4(x, y), y), y)) = 1.$$

When $I_1(x, I_2(I_3(I_4(x, y), y), y)) = 1$ and $x > y$, we have

$$I_1(x, I_2(I_3(I_4(x, y), y), y)) = 0.$$

This is a contradiction. So $x \leq y$. Thus $I_1(x, I_2(I_3(I_4(x, y), y), y))$ satisfies (OP).

Similar to the case $n = 4$, we can easily see that when n is any even number, this proposition still holds. □

Proposition 3 *Let $I_1, \ldots, I_n$ be (S, N)-implications with $I_i(x, y) = S_i(N_i(x), y)$ ($i = 1, \ldots, n$), $G_{I_1,\ldots,I_n}$ a fuzzy implication generated by above implications. Then for $x,\ y \in U$, we have*

$$G_{I_1,\ldots,I_n}(x, y) = S_1(N_1(x), S_2(N_2(S_3(N_3(\ldots S_n(N_n(x), y)\ldots, y), y), y), y), y),$$

where S_i and N_i are t-conorm and fuzzy negation, respectively, n is a positive even number.
(i) *When $I_1, \ldots, I_n$ satisfy (CB), $G_{I_1,\ldots,I_n}$ also satisfies (CB);*
(ii) *When all I_i satisfies (LEM), $G_{I_1,\ldots,I_n}$ satisfies (NP).*

Proof (i) Let $x, y \in U$, and I_i satisfy (CB) for each i. Then we have $S_i(N_i(x), y) \geq y$. So

$$S_2(N_2(S_3(N_3(\ldots S_n(N_n(x), y)\ldots, y), y), y), y) \geq y,$$

$$S_1(N_1(x), S_2(N_2(S_3(N_3(\ldots S_n(N_n(x), y)\ldots, y), y), y), y)) \geq y.$$

Thus, $G_{I_1,\ldots,I_n}$ satisfies (CB).

(ii) We know that every (S, N)-implication satisfies (NP) (see [2]). Now by the known condition, every I_i satisfies (LEM), for (S, N)-implications, equivalently, every I_i satisfies (IP), we have that $G_{I_1,\ldots,I_n}$ satisfies (NP) by Proposition 2 (ii). □

Proposition 4 *Let $I_1, \ldots, I_n$ be R-implications with*

$$I_i(x, y) = \sup\{t \in U \mid T_i(x, t) \leq y\}, \quad i = 1, \ldots, n.$$

Then $G_{I_1,\ldots,I_n}$ satisfies (NP).

Proof Since all of I_i are R-implications, we have that I_i satisfies (IP) and (NP). By Proposition 2(ii), $G_{I_1,\ldots,I_n}$ also satisfies (NP). □

4 The Mixed Fuzzy Implication

Combining the multiple fuzzy implications about both x and y, in this section we define a class of multiple fuzzy implications about both variables x and y, and study their properties.

Definition 7 Let $I_1, \ldots, I_n$ be fuzzy implications. Then $H_{I_1,\ldots,I_n}$ defined below is called an n-dimension mixed function generated by the above implications,

$$H_{I_1,\ldots,I_n}(x, y) = J_{I_1,\ldots,I_i}(x, G_{I_{i+1},\ldots,I_n}(x, y)), \quad x,\ y \in U.$$

Remark 2 Obviously, both the multiple fuzzy implications about x and y are special cases of the new functions defined by Definition 7. In fact, in Definition 7, when $i = n$, we have $H_{I_1,\ldots,I_n}(x, y) = J_{I_1,I_2,\ldots,I_n}(x, y)$; when $i = 1$, we have $H_{I_1,\ldots,I_n}(x, y) = G_{I_1,I_2,\ldots,I_n}(x, y)$.

Proposition 5 *Let $I_1, \ldots, I_n$ be fuzzy implications satisfying (1-Li) and (SP), $n - i$ be a positive even number. Then $H_{I_1,\ldots,I_n}$ is also a fuzzy implication.*

Proof From Proposition 1, when $I_{i+1}, \ldots, I_n$ are fuzzy implications satisfying (1-Li) and (SP), and $n - i$ is a positive even number, then $G_{I_{i+1},\ldots,I_n}$ is a fuzzy implication. Also, by Proposition 5.1 of [8], when $I_1, \ldots, I_i$ are fuzzy implications, G is a fuzzy implications. Thus, $J_{I_1,\ldots,I_i,G}$ is also a fuzzy implication. Therefore, $H_{I_1,\ldots,I_n}$ is a fuzzy implication. □

In this paper we call $H_{I_1,\ldots,I_n}$ the mixed multiple fuzzy implication.

Example 2 Let $n = 5, I_1 = I_{LK}, I_2 = I_{KD}, I_3 = I_{GD}, I_4 = I_{WB}, I_5 = I_{RS}$.
By Example 1, we have

$$I_2(x, I_3(I_4(I_5(x, y), y), y)) = \begin{cases} 1, & x \leq y, \\ \max(1 - x, y), & x > y. \end{cases}$$

Since $I_1(x, y) = \min(1, 1 - x + y)$, we have

$$H_{I_1,I_2,I_3,I_4,I_5}(x, y) = \begin{cases} 1, & x \le y, \\ \min(1, 2 - 2x), & 1 - y \ge x > y, \\ \min(1, 1 - x + y), & x \ge \max(y, 1 - y). \end{cases}$$

Proposition 6 *Let $I_1, \ldots, I_n$ be fuzzy implications, $H_{I_1,\ldots,I_n}$ the fuzzy implication generated.*

(i) *If $I_1, \ldots, I_n$ satisfy (IP) and (OP), then $H_{I_1,\ldots,I_n}$ also satisfies (IP).*
(ii) *If $I_1, \ldots, I_n$ satisfy (NP) and (IP), then $H_{I_1,\ldots,I_n}$ also satisfies (NP).*
(iii) *If $I_1, \ldots, I_n$ satisfy (OP), then $H_{I_1,\ldots,I_n}$ also satisfies (OP).*

Proof From Propositions 2 and 5.2 of [8], this proposition can be proven. □

Proposition 7 *Let $I_1, \ldots, I_n$ be (S, N)-implications, $H_{I_1,\ldots,I_n}$ the corresponding fuzzy implication generated. If $I_1, \ldots, I_n$ satisfy (CB), then $H_{I_1,\ldots,I_n}$ also satisfies (CB).*

Proof From Proposition 3, we have $G_{I_{i+1},\ldots,I_n}(x, y) \ge y$. Since $I_1, \ldots, I_i$ satisfy (CB), we have

$$H_{I_1,\ldots,I_n}(x, y) = S_1(N_1(x), S_2(N_2(x), \ldots, S_i(N_i(x), G_{I_{i+1},I_{i+2},\ldots,I_n}(x, y)))) \ge y.$$

□

Proposition 8 *Let $I_1, \ldots, I_n$ be R-implications, $H_{I_1,\ldots,I_n}$ the corresponding fuzzy implication generated. Then $H_{I_1,\ldots,I_n}$ satisfies (NP).*

Proof Since $I_1, \ldots, I_n$ are R-implications, so all of them satisfy (NP). From Proposition 4, $G_{I_{i+1},\ldots,I_n}(x, y)$ satisfies (NP). Also, by Proposition 5.2 of [8], $H_{I_1,\ldots,I_n}$ also satisfies (NP). □

5 Conclusions

Fuzzy implications have important applications in many fields. So it is very necessary to study construction methods and properties of fuzzy implications.

In this paper, we mainly continue the study of multiple fuzzy implications studied in [8]. Two kinds of new construction methods of fuzzy implications are proposed. The first method is multiple iteration about the second variable of the fuzzy implication. The second method is multiple iteration about both variables of the fuzzy implication, main properties of the second kind of the new implications are investigated. Our results show that the new implications have good behaviors.

In the future, we will discuss relations between the two kinds of new implications and the existing implications such as (S, N)-implications and R-implications. Naturally, applications of the new implications in fuzzy control, data mining, fuzzy decision and other fields should be important topics for further study.

Acknowledgments This work is supported by the National Natural Science Foundation of China (No. 11171308; 61379018; 61472471; 51305400).

References

1. Baczynski, M., Jayaram, B.: On the characterizations of (S, N)-implications. Fuzzy Sets Syst. **158**, 1713–1727 (2007)
2. Baczynski, M., Jayaram, B.: Fuzzy Implications. Springer, Heidelberg (2008)
3. Benjamin, C.B.: On interval fuzzy negations. Fuzzy Sets Syst. **161**, 2290–2313 (2010)
4. Fodor, J., Torrens, J.: An overview of fuzzy logic connectives on the unit interval. Fuzzy Sets Syst. **281**, 183–187 (2015)
5. Gottwald, S.: A Treatise on Many-Valued Logic. Research Studies Press, Baldock (2001)
6. Jayaram, B., Mesiar, R.: On special fuzzy implications. Fuzzy Sets Syst. **160**, 2063–2084 (2009)
7. Klement, E.P., Mesiar, R., Pap, E.: Triangular Norms. Kluwer, Dordrecht (2000)
8. Li, F., Pei, D.W.: Multiple fuzzy implications: a novel method to generate fuzzy implications. Submitted for publication (2016)
9. Pei, D.W.: R_0 implication: characteristics and applications. Fuzzy Sets Syst. **131**, 297–302 (2002)
10. Pei, D.W.: T-Norm Based Fuzzy Logics: Theory and Applications. Science Press, Beijing (2013) (in Chinese)
11. Vemuri, N.R., Jayaram, B.: The $\otimes$-composition of fuzzy implications: closures with respect to properties, powers and families. Fuzzy Sets Syst. **275**, 58–87 (2015)
12. Vemuri, N.R., Jayaram, B.: Homomorphisms on the monoid of fuzzy implications and the iterative functional equation $I(x, I(x, y)) = I(x, y)$. Inf. Sci. **298**, 1–21 (2015)
13. Wang, G.J.: Non-classical Mathematical Logic and Approximate Reasoning, 2nd edn. Science Press, Beijing (2008) (in Chinese)

Extended Threshold Generation of a New Class of Fuzzy Implications

Zhi-Hong Yi and Feng Qin

Abstract The threshold generation of a new implication from two given ones is introduced by Massanet and Torrens. Along the lines of the ordinal sum method in the construction of fuzzy connectives, the paper deals with generalization of the e-threshold generation method by applying the scaling method in the second variable, which can generate some implications different from usual implications, such as f-, g-implications, R-implications, S-implications and the ones derived from the ordinal sum method. And generated implications are characterized.

Keywords Fuzzy connectives · Fuzzy implications · Threshold generation

1 Introduction

Fuzzy implications [5], with which fuzzy conditional statements of the type "if-then" can be well modeled, play an important role in both approximating reasoning and fuzzy control. There exists several different models of fuzzy implications that can be more or less adequate in any case, depending on the behavior of the conditional rule they have to model, or depending on the inference rule that is going to be applied. Among these different models, the most established and well-studied classes of fuzzy implications are the so-called $R-$, $(S, N)-$, $QL-$, $D-$ implications, usually obtained from fuzzy connectives, such as t-norms, t-conorms and negations. Moreover, fuzzy implications can also be obtained from other binary functions, such as copulas, quasi-copulas, conjunctors in general [11], representable aggrega-

Z.-H. Yi (✉)
College of Modern Economics and Management, Jiangxi University of Finance and Economics, Nanchang 330013, People's Republic of China
e-mail: yizhihong1206@163.com

F. Qin
College of Mathematics and Information Science, Jiangxi Normal University, Nanchang 330022, People's Republic of China
e-mail: qinfeng923@163.com

T.-H. Fan et al. (eds.), *Quantitative Logic and Soft Computing 2016*, Advances in Intelligent Systems and Computing 510,
DOI 10.1007/978-3-319-46206-6_28

tion functions [9] and uninorms [1, 7, 10, 20]. The typical ones are $f-$, $g-$ generated implications [4, 14, 22, 23] by Yager [33], $h-$implications by Massanet and Torrens [21], $(g, \min)-$implications, $(h, \min)$-implications and $h^{-1}-$implications by Liu [17–19], (f, g)-implications by Xie and Liu [31], $(g, u)-$implications by Zhang and Liu [34], fuzzy implications $I_{U f,g}$ by Hliněná et al. [13].

Moreover, there is another method of generating new fuzzy implications from already existing ones. The remarkable ones are φ-conjugation, min and max operations, convex combinations and composition over given one or two fuzzy implication(s). In the recent years, in [24, 25], Massanet and Torrens provide a new construction method to get fuzzy implication from two given ones, called threshold generation method, similar to the threshold generation method, they [26] introduced the vertical threshold generation method of fuzzy implication through an adequate scaling on the first variable of the given fuzzy implications. Su et al. [30] introduced a new class of fuzzy implications, called ordinal sum implications, from a family of given implications, which is similar to ordinal sum of t-norms (or t-conorms) [16].

Intrigued by those referred studies, we generalize the e-threshold generation method from a single value to multi-value by applying the scaling method in the second variable of the initial implications, and some implications different from the existing implications can be obtained.

The paper is organized as follows. In Sect. 2, some known concepts and results to be used are listed. In Sect. 3, we give the extended threshold generation of a fuzzy implication from a family of fuzzy implications and show that the $E-$generated fuzzy implications are really a new class of implications different from the usually referred ones. Finally, some concluding remarks and comments on future work are given.

2 Preliminary

For some necessary results and notations about fuzzy operators, we recommend [2, 12, 16].

Definition 1 ([5, 12]) A function $I : [0, 1]^2 \to [0, 1]$ is called a *fuzzy implication* if it satisfies the following conditions:

$$I \text{ is decreasing in the first variable,} \tag{I1}$$

$$I \text{ is increasing in the second variable, and} \tag{I2}$$

$$I(0, 0) = 1, \tag{I3}$$

$$I(1, 1) = 1, \tag{I4}$$

$$I(1, 0) = 0. \tag{I5}$$

It can be easily verified that $I(0,0) = I(0,1) = I(1,1) = 1$. Hence a fuzzy implication is usually considered as an extension of the corresponding classical one. Throughout the paper, we use the same notations for the most referred implications and some related properties as those appeared in [24].

In the following, we list the threshold generation method.

Theorem 1 ([24]) (Threshold generation method) *Let I_1, I_2 be two implications and $e \in]0,1[$, then the binary function $I_{I_1-I_2} : [0,1]^2 \to [0,1]$, given by*

$$I_{I_1-I_2}(x,y) = \begin{cases} 1, & x = 0; \\ eI_1(x, \frac{y}{e}), & x > 0, 0 \le y \le e; \\ e + (1-e)I_2(x, \frac{y-e}{1-e}), & x > 0, y > e \end{cases} \tag{1}$$

is a fuzzy implication, called the e-generated implication from I_1 to I_2.

3 Extended Threshold Generation Method

Now we consider a special sequence $\{e_i\}_{i\in\mathbb{N}}$ in $]0,1[$, which is strictly increasing, i.e., $e_i < e_j$ if $i, j \in \mathbb{N}$ with $i < j$; or strictly increasing if $i \le n$ and constant if $i > n$ for some $n \in \mathbb{N}$, i.e., $0 < e_1 < e_2 < \ldots < e_n = e < 1$ and $e_i = e$ if $i > n$. If $\{e_i\}_{i\in\mathbb{N}}$ is strictly increasing, then there is some $e \in]0,1]$ such that $\lim_{i\to\infty} e_i = e \in]0,1]$. For a sequence $\{e_i\}_{i\in\mathbb{N}}$ with limit, define

$$E = \begin{cases} \{e_i \mid i \in \mathbb{N}\}, \{e_i\}_{i\in\mathbb{N}} \text{ is strictly increasing if } i \le n \text{ and constant if } i > n \\ \{e_i \mid i \in \mathbb{N}\} \bigcup \{e\}, \quad \{e_i\}_{i\in\mathbb{N}} \text{ is strictly increasing.} \end{cases} \tag{2}$$

the set E will be used throughout the paper. Thus the set E is either finitely or infinitely countable. For the generating method, we will use a family of implications $\{I_0\} \bigcup \{I_i\}_{i\in\mathbb{N}} \bigcup \{I^0\}$ denoted by $\mathcal{I}$ throughout the paper. For consistency, if $|E| = n$, we assume, $I_i = I_{n-1}$ for all $i \ge n$, and somewhere $e_0 = 0$. Using the denotation, we give the extended threshold generation method in the following.

Theorem 2 *Let E be the set defined by Eq. 2 where $\{e_i\}_{i\in\mathbb{N}}$ is the aforementioned sequence, and $\mathcal{I}$ the referred family of fuzzy implications, then $I_E : [0,1]^2 \to [0,1]$ defined by*

$$I_E(x,y) = \begin{cases} 1, & x = 0; \\ e_1 I_0(x, \frac{y}{e_1}), & x > 0, 0 \le y \le e_1; \\ e_i + (e_{i+1} - e_i)I_i(x, \frac{y-e_i}{e_{i+1}-e_i}), & x > 0, e_i < y \le e_{i+1} \text{ for } i \in \mathbb{N}, \\ e, & x > 0, y = e; \\ e + (1-e)I^0(x, \frac{y-e}{1-e}), & x > 0, y > e; \end{cases} \tag{3}$$

is a fuzzy implication.

Proof We only provide the proof for the case when $\{e_i\}_{i\in\mathbb{N}}$ in $]0,1[$ is strictly increasing with $\lim_{i\to\infty} e_i = e < 1$, the others can be checked similarly. We proceed the proof in the following steps:

(i) (I1) It can be checked easily.

(ii) (I2) Let $y_1 \le y_2$, if $x = 0$, then, $I_E(x, y_1) = 1 = I_E(x, y_2)$; if $x > 0$, there are several cases to be considered.

  (a) If $y_1, y_2 \in]e_i, e_{i+1}]$ for some $i \in \mathbb{N}$, then $I_E(x, y_1) = e_i + (e_{i+1} - e_i)I_i(x, \frac{y_1-e_i}{e_{i+1}-e_i}) \le e_i + (e_{i+1} - e_i)I_i(x, \frac{y_2-e_i}{e_{i+1}-e_i}) = I_E(x, y_2)$ by the increasingness of I_i. Similarly if $y_1, y_2 \in [0, e_1]$ or $y_1, y_2 \in]e, 1]$ the increasingness of I_E can be checked by using the increasingness of I_0 and I^0, respectively. In addition, $I_E(x, y_1) = e \ge I_E(x, y_2) = e$ if $y_1 = y_2 = e$.

  (b) For the case $y_1 \in [0, e_1]$, if $y_2 \in]e_i, e_{i+1}]$ for some $i \in \mathbb{N}$, then $I_E(x, y_1) \le I_E(x, y_2)$ since $I_E(x, y_1) = e_1 I_1(x, \frac{y_1}{e_1}) \in [0, e_1]$ and $I_E(x, y_2) = e_i + (e_{i+1} - e_i)I_i(x, \frac{y_2-e_j}{e_{i+1}-e_i}) \in [e_i, e_{i+1}]$; Similarly we can check (I2) for the case $y_2 \in]e, 1]$. Additionally, $I_E(x, y_1) \le e_1 \le e = I_E(x, y_2)$ if $y_2 = e$.

  (c) For the case $y_1 \in]e_i, e_{i+1}]$ with some $i \in \mathbb{N}$, if $y_2 \in]e_j, e_{j+1}]$ with $i < j$ and $j \in \mathbb{N}$, then $I_E(x, y_1) \le I_E(x, y_2)$ since $I_E(x, y_1) = e_i + (e_{i+1} - e_i)I_i(x, \frac{y_1-e_i}{e_{i+1}-e_i}) \in [e_i, e_{i+1}]$ and $I_E(x, y_2) = e_j + (e_{j+1} - e_j)I_j(x, \frac{y_2-e_j}{e_{j+1}-e_j}) \in [e_j, e_{j+1}]$. Similarly we can check (I2) for the case $y_2 \in]e, 1]$. Additionally, $I_E(x, y_1) \le e_{i+1} \le e = I_E(x, y_2)$ if $y_2 = e$.

  (d) If $y_1 = e$ and $y_2 \in]e, 1]$, then $I_E(x, y_1) \le I_E(x, y_2)$ since $I_E(x, y_1) = e$ and $I_E(x, y_2) = e + (1 - e)I^0(x, \frac{y_2-e}{1-e}) \in [e, 1]$;

(iii) (I3, I4, I5)

$$I_E(0, 0) = 1 \text{ by Eq. 3;}$$

$$I_E(1, 1) = e + (1 - e)I^0(1, 1) = 1 \text{ since } I^0 \text{ is a fuzzy implication;}$$

$$I_E(1, 0) = e_1 I_0(1, 0) = 0 \text{ since } I_0 \text{ is a fuzzy implication.}$$

Therefore, I_E is a fuzzy implication. □

Remark 1 For the case when E is finite, if $E = \{e\}$, i.e., $e_i = e \in]0, 1[$ for all $i \in \mathbb{N}$, then the generation method reduces to the e-threshold generation method with $I_0 = I_1$ and $I^0 = I_2$ in Eq. 1; if $E = \{e_1, e_2, \ldots, e_n = e\}$ with $n > 1$, then Eq. 3 reduces to

$$I_E(x, y) = \begin{cases} 1, & x = 0; \\ e_1 I_0(x, \frac{y}{e_1}), & x > 0, 0 \le y \le e_1; \\ e_i + (e_{i+1} - e_i)I_i(x, \frac{y-e_i}{e_{i+1}-e_i}), & x > 0, e_i < y \le e_{i+1} \text{ for } 1 \le i \le n-1, \\ e + (1 - e)I^0(x, \frac{y-e}{1-e}), & x > 0, e < y \le 1; \end{cases} \quad (4)$$

To sum up, the aforementioned method of generating fuzzy implications is a generalization of $e-$generation method. In the sequel, the fuzzy implication I_E is called E-generated implication or E-generation of a family of fuzzy implications $\mathcal{I}$.

Example 1 We give an $E-$generated implications when E is infinite, where the classes of $f-$implications can be found in [14, 33]. The Yager's class of $f-$implications I_f^λ is given by $I_f^\lambda(x, y) = 1 - x^{\frac{1}{\lambda}}(1 - y)$ with $f-$generator $f(x) = (1 - x)^\lambda$, $(\lambda > 0)$. Let $E = \{e_i = 1 - \frac{1}{2^i} \mid i \in \mathbb{N}\}$ with $\lim\limits_{i\to\infty} e_i = e = 1$ and $\lambda = i$ for $i \in \mathbb{N}$. Then the E-generated fuzzy implication I_E from the family of fuzzy implications $\{I_{YG}\} \bigcup \{I_f^{\lambda=i}\}_{i\in\mathbb{N}}$ is given by

$$I_E(x, y) = \begin{cases} 1, & x = 0; \\ \frac{1}{2}(2y)^x, & x > 0, 0 \le y \le \frac{1}{2}; \\ 1 - \frac{1}{2^i} + (\frac{1}{2^i} - \frac{1}{2^{i+1}})x^{\frac{1}{i}}(1 - \frac{y-1+\frac{1}{2^i}}{\frac{1}{2^i} - \frac{1}{2^{i+1}}}), & x > 0, 1 - \frac{1}{2^i} < y \le \\ & 1 - \frac{1}{2^{i+1}} \text{for each } \ i \in \mathbb{N}; \\ 1, & x > 0, y = 1; \end{cases}$$

Example 2 We give an $E-$generated implications when E is finite. With the implications I_{YG}, I_{GG}, I_{GD} and $E = \{\frac{1}{3}, \frac{2}{3}\}$, the generated implication I_E by Eq. 4 is given by

$$I_E(x, y) = \begin{cases} \frac{1}{3}(3y)^x, & x > 0, 0 \le y \le \frac{1}{3}; \\ \frac{1}{3} + \frac{1}{3}\frac{3y-1}{x}, & x > 0, \frac{1}{3} < y \le \frac{2}{3}, x > 3y - 1, \\ \frac{2}{3}, & x > 0, \frac{1}{3} < y \le \frac{2}{3}, x \le 3y - 1, \\ y, & x > 0, \frac{2}{3} < y \le 1, x > 3y - 2, \\ 1, & else, \end{cases} \tag{5}$$

Remark 2 (i) The E-threshold generation method is different from the ordinal sum methods in [30], which can be illustrated by Fig. 1.

(ii) Throughout this paper, we only focus on the aforementioned class of sequences in $]0, 1[$, specifically. In fact, the strictly decreasing sequence $\{e_i\}_{i\in\mathbb{N}}$ with $\lim\limits_{i\to\infty} e_i = e$ can also be applied to the generation method. For the case $e > 0$, with the same family of implications in Theorem 2, the generation operation given by

$$I_E(x, y) = \begin{cases} 1, & x = 0; \\ e_1 + (1 - e_1)I^0(x, \frac{y-e_1}{1-e_1}), & x > 0, e_1 < y \le 1; \\ e_{i+1} + (e_i - e_{i+1})I_i(x, \frac{y-e_{i+1}}{e_i-e_{i+1}}), & x > 0, e_{i+1} < y \le e_i \text{ for } i \in \mathbb{N}, \\ eI_0(x, \frac{y}{e}), & x > 0, 0 \le y \le e; \end{cases}$$

is also a fuzzy implication.

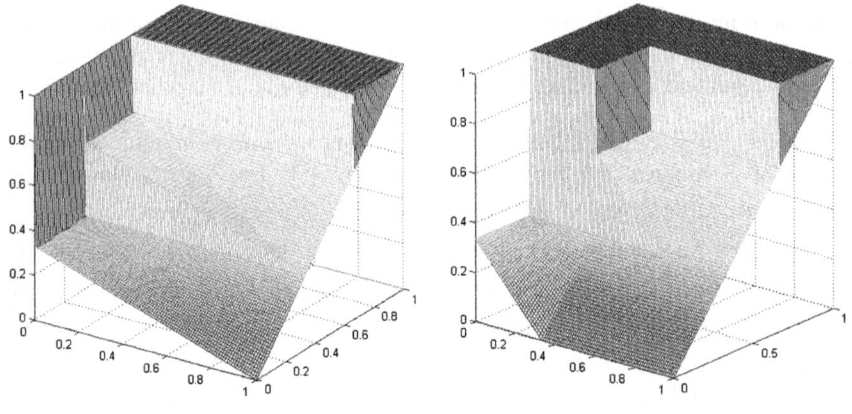

Fig. 1 The *left figure* and the *right one* are the surface plot of the generated implications whose original implications are I_{RC}, I_{KD}, I_{WB} with $E = \{\frac{1}{3}, \frac{2}{3}\}$ in the E-threshold generation and the disjoint intervals be $\{[0, \frac{1}{3}], [\frac{1}{3}, \frac{2}{3}], [\frac{2}{3}, 1]\}$ in the ordinal sum construction, respectively

Proposition 1 *Let I_E be the E-generation of a family of fuzzy implications $\mathcal{I}$, then*

(i) $I_E(x, y) \in [0, e_1]$ *if* $x > 0, y \in [0, e_1]$*; for each* $e_i \in E, I_E(x, y) \in [e_i, e_{k+1}]$ *if* $x > 0, y \in]e_i, e_{k+1}]$ *and* $I_E(x, y) \in [e, 1]$ *if* $x > 0, y \in]e, 1]$.
(ii) *For each* $e_i \in E$, $I_E(x, e_i) = e_i$ *if* $x > 0$; $I_E(x, e) = e$ *if* $x > 0$.
(iii) *For each* $e_i \in E$, $I_E(x, y) < e_{i+1}$ *for* $x > 0, y \in]e_i, e_{i+1}[\Leftrightarrow I_i(a, b) < 1$ *for* $a > 0, b < 1$;
(iv) *For each* $e_i \in E$, $I_E(x, y) > e_i$ *for* $x > 0, y \in]e_i, e_{i+1}[\Leftrightarrow I_i(a, b) > 0$ *for* $a > 0, b > 0$.

Proof The results can be got by direct calculation by using Eq. 3. □

Proposition 2 *Let I_E be the E-generation of a family of fuzzy implications $\mathcal{I}$, then*
(i) *I_E is neither an f-generated implication nor a g-generated implication.*
(ii) *If $|E| \geq 2$ then I_E is not an h-implication. If $E = \{e\}$, then I_E is an h-implication if and only if I_0 is an f-generated implication and I^0 is a g-generated implication with generators satisfying $f(0) = g(1) = \infty$.*

Proof (i) Assume I_E is an f-generated implication, then there is a strictly decreasing, continuous function $f : [0, 1] \to [0, +\infty]$ with $f(1) = 0$ such that $I(x, y) = f^{-1}(xf(y))$. Hence, for arbitrary $x > 0$ and each $e_i \in E$, $I_E(x, e_i) = f^{-1}(xf(e_i)) = e_i$ by Proposition 1(ii). Then $xf(e_i) = f(e_i)$, thus we can get $e_i = 1$ or $e_i = 0$, which contradicts the fact that $e_i \in]0, 1[$. For the case when I_E is a g-generated implication, we can get a similar contradiction.
(ii) If $|E| \geq 2$, assume that I_E is an h-implication, then there is a strictly increasing, continuous function $h : [0, 1] \to [-\infty, +\infty]$ such that $I(x, y) = h^{-1}(xh(y))$ with $h(0) = -\infty$, $h(e) = 0$ and $h(1) = +\infty$ and $e \in]0, 1[$. Similar as above, we have $xh(e_i) = h(e_i)$ for arbitrary $x > 0$ and each $e_i \in E$. Thus for each $e_i \in E$,

$e_i = 1$ or $e_i = 0$ or $e_i = e$, which implies $E = \{e\}$, which contradicts the fact that $|E| \geq 2$. For the case $E = \{e\}$, the result is that of Theorem 2 in [24].

□

By direct calculation, we can get the following results.

Proposition 3 *Let I_E be the E-generation of a family of fuzzy implications $\mathcal{I}$, then the natural negation of I_E is* $N_{I_E}(x) = I_E(x, 0) = \begin{cases} 1, & x = 0; \\ e_1 N_{I_0}(x), & x > 0; \end{cases}$ *and is always non-continuous.*

Remark 3 As is well known, an (S, N)-implication $I_{S,N}$ is given by $I_{S,N}(x, y) = S(N(x), y)$ with S being a t-conorm and N a negation, while its natural negation $N_{I_{S,N}} = N$. Since the natural negation of I_E is always non-continuous, then I_E can never be an (S, N)-implication derived from a t-conorm S and a continuous negation N. Whence I_E can never be an S-implication.

Proposition 4 *Let I_E be the E-generation of a family of fuzzy implications $\mathcal{I}$, then*
(i) $I_E(x, y) = 1$ *if and only if either* $x = 0$ *or* ($x > 0$ *and* $e < y \leq 1, I^0(x, \frac{y-e}{1-e}) = 1$) *or* ($e = 1, x > 0, y = 1$)*, i.e., I_E does not satisfy (OP);*
(ii) $I_E(x, x) = 1$ *if and only if either* $x = 0$ *or* ($e < x \leq 1$ *with* $I^0(x, \frac{x-e}{1-e}) = 1$) *or* ($e = 1, x = 1$)*, i.e., I_E does not satisfy (IP);*

Remark 4 If I is the residual implication [12] derived from an arbitrary t-norm(not necessarily left-continuous), then I satisfies (IP). Since I_E does not satisfy both (OP) and (IP), then I_E is not an R-implication derived from any t-norms.

Much attention are paid to the characterization of generated implications [1, 3, 7, 23]. Here we can characterize all the implications generated by the extended threshold generation method.

Theorem 3 (Characterization of the E-generated implications) *Let I be a fuzzy implication and E be the aforementioned set.*
(i) *For the case when E is finite, i.e., $E = \{e_i \mid i = 1, 2, \ldots n\}$ with $0 < e_1 < e_2 < \ldots < e_n = e < 1$ for some $n \in \mathbb{N}$, then I is an E-generated implication if and only if for each $e_i \in E$, $I(x, e_i) = e_i$ whenever $x > 0$. In this case, the operations $I_i : [0, 1]^2 \to [0, 1]$ defined by*

$$I_i(x, y) = \begin{cases} 1, & x = 0; \\ \frac{I(x, e_i + (e_{i+1} - e_i)y) - e_i}{e_{i+1} - e_i}, & x > 0; \end{cases} \tag{6}$$

for $0 \leq i \leq n - 1$ with $e_0 = 0$ are fuzzy implications and $I^0(x, y) = \frac{I(x, e + (1-e)y) - e}{1-e}$ is also a fuzzy implication. By these fuzzy implications, I can be generated.

(ii) *For the case that E is infinite, i.e.,* $\{e_i\}_{i\in\mathbb{N}}$ *is a strictly increasing sequence in* $]0, 1[$ *with* $\lim_{i\to\infty} e_i = e \in]0, 1]$. *then I is an E-generated implication if and only if for each* $e_i \in E$, $I(x, e_i) = e_i$ *and* $I(x, e) = e$ *whenever* $x > 0$. *In this case, the operations* $I_i : [0, 1]^2 \to [0, 1]$ *by*

$$I_i(x, y) = \begin{cases} 1, & x = 0; \\ \frac{I(x, e_i + (e_{i+1} - e_i)y) - e_i}{e_{i+1} - e_i}, & x > 0; \end{cases} \tag{7}$$

are fuzzy implications for $i \in \mathbb{N} \bigcup \{0\}$ *with* $e_0 = 0$ *and* $I^0(x, y) = \frac{I(x, e+(1-e)y)-e}{1-e}$ *is also a fuzzy implication for* $e < 1$. *By these fuzzy implications, I can be generated.*

Proof We only focus on the first case.

($\Rightarrow$): It is a direct result of Proposition 1;

($\Leftarrow$): Firstly, we prove that the operations I_i defined by Eq. 6 are fuzzy implications. By Proposition 1, I_i is well-defined. For each $0 \le i \le n - 1$, we have the following results.

1. (I1) Let $x_1 \le x_2$, if $x_1 = 0$, then, $I_i(x_1, y) = 1 \ge I_i(x_2, y)$ for all $y \in [0, 1]$ by Eq. 3; if $x_1 > 0$, then for all $y \in [0, 1]$, $I_i(x_1, y) = \frac{I(x_1, e_i + (e_{i+1} - e_i)y) - e_i}{e_{i+1} - e_i} \ge \frac{I(x_2, e_i + (e_{i+1} - e_i)y) - e_i}{e_{i+1} - e_i}$ $= I_i(x_2, y)$ by the decreasingness of I on the first variable; Hence, J_i is decreasing in the first variable.
2. (I2) Let $y_1 \le y_2$. by Eq. 6, it is enough to check the case when $x > 0$, then $I_i(x, y_1) = \frac{I(x, e_i + (e_{i+1} - e_i)y_1) - e_i}{e_{i+1} - e_i} \le \frac{I(x, e_i + (e_{i+1} - e_i)y_2) - e_i}{e_{i+1} - e_i} = I_i(x, y_2)$, since I is increasing on the second variable. Therefore, I_i is increasing in the second variable.
3. (I3, I4, I5) With Eq. 6
$I_i(0, 0) = 1$;
$I_i(1, 1) = \frac{I(1, e_{i+1}) - e_i}{e_{i+1} - e_i} = 1$;
$I_i(1, 0) = \frac{I(1, e_k) - e_i}{e_{i+1} - e_i} = \frac{e_i - e_i}{e_{i+1} - e_i} = 0$;

Thus, I_i is a fuzzy implication. Similarly, it can be checked that I^0 is also a fuzzy implication.

Next, we check that the E-generation of the fuzzy implications $\{I_i \mid 0 \le k \le n - 1\} \bigcup \{I^0\}$ reduces to I, i.e., $I_E = I$.

$$I_E(x, y) = \begin{cases} 1, & x = 0; \\ e_1 I_0(x, \frac{y}{e_1}), & x > 0, 0 \le y \le e_1; \\ e_i + (e_{i+1} - e_i) I_i(x, \frac{y - e_i}{e_{i+1} - e_i}), & x > 0, e_i < y \le e_{i+1} \\ & \text{for each } 0 \le i \le n - 1; \\ e + (1 - e) I^0(x, \frac{y - e}{1 - e}), & x > 0, e < y \le 1. \end{cases}$$

$$= \begin{cases} 1, & x = 0; \\ e_1 \frac{I(x, e_1 \frac{y}{e_1})}{e_1}, & x > 0, 0 \le y \le e_1; \\ e_i + (e_{i+1} - e_i) \frac{I(x, e_i + (e_{i+1} - e_i)\frac{y - e_i}{e_{i+1} - e_i}) - e_i}{e_{i+1} - e_i}, & x > 0, e_i < y \le e_{i+1} \\ & \text{for each } 0 \le i \le n-1; \\ e + (1 - e) \frac{I(x, (1-e)\frac{y-e}{1-e} + e) - e}{1 - e}, & x > 0, e < y \le 1. \end{cases}$$

$$= I(x, y).$$

□

Remark 5 The initial implications I_i and I^0 except I_0 can not be unique since these operations on $\{(x, 0) \mid 0 < x \le 1\}$ are not involved in the generation method. For instance, in Example 2(ii) if T_{LK} is replaced by the fuzzy implication I_4, where $I_4(x, y) = \begin{cases} 0, & 0 < x \le 1, y = 0; \\ T_{LK}(x, y), & \text{else}; \end{cases}$, then the same implication can be generated.

4 Concluding Remarks and Future Work

In this paper, threshold generated implication by a single point is extended to multi-valued(countable) case. We present the construction theorem for the generation method and show that the generated implications, which are different from the ones derived from the ordinal sum method, are neither $f-$, $g-$implications, $R-$implications nor $S-$implications. Moreover, a characterization theorem for the generated implications is provided. In the future work, we will concentrate on some related properties, such as the exchange law and distributivity [7, 8, 15, 27–29, 32] over fuzzy connectives.

Acknowledgments The work is supported by National Natural Science Foundation of China (Grants 61165014 and 61563020), Jiangxi Natural Science Foundation (Grant 20151BAB201019) and Project of Science and Technology of Educational Commission of Jiangxi Province of China(Grants GJJ151601 and GJJ151602). The authors are grateful to the anonymous referees for their comments to improve the paper.

References

1. Aguiló, I., Suñer, J., Torrens, J.: A characterization of residual implications derived from left-continuous uninorms. Inf. Sci. **180**, 3992–4005 (2010)
2. Alsina, C., Frank, M., Schweizer, B.: Associative Functions: Triangular norms and Copulas. World Scientific Publishing Co, Singapore (2006)
3. Baczyński, M., Jayaram, B.: On the characterizations of (S, N)-implications. Fuzzy Sets Syst. **158**(15), 1713–1727 (2007)

4. Baczyński, M., Jayaram, B.: Yager's classes of fuzzy implications: some properties and intersections. Kybernetika **43**(2), 157–182 (2007)
5. Baczyński, M., Jayaram, B.: Fuzzy Implications. Springer, Berlin, Heidelberg (2008)
6. Baczyński, M., Jayaram, B.: On the distributivity of fuzzy implications over nilpotent or strict triangular conorms. IEEE Trans. Fuzzy Syst. **17**, 590–603 (2009)
7. Baczyński, M., Jayaram, B.: (U, N)-implications and their characterizations. Fuzzy Sets Syst. **160**, 2049–2062 (2009)
8. Baczyński, M., Qin, F.: Some remarks on the distributive equation of fuzzy implication and the contrapositive symmetry for continuous, archimedean t-norms. Int. J. Approx. Reason. **54**, 290–296 (2013)
9. Carbonell, M., Torrens, J.: Continuous R-implications generated from representable aggregation functions. Fuzzy Sets Syst. **161**, 2276–2289 (2010)
10. De Baets, B., Fodor, J.: Residual operators of uninorms. Soft Comput. **3**, 89–100 (1999)
11. Durante, F., Klement, E., Mesiar, R., Sempi, C.: Conjunctors and their residual implicators: characterizations and construction methods. Med. J. Math. **4**, 343–356 (2007)
12. Fodor, J., Roubens, M.: Fuzzy Preference Modelling and Multicriteria Decision Support. Springer, Berlin (1994)
13. Hliněná, D., Kalina, M., Král, P.: A class of implications related to yagers $f-$implications. Inf. Sci. **260**, 171–184 (2014)
14. Jayaram, B.: Yager's new class of implications J_f and some classical tautologies. Inf. Sci. **177**, 930–946 (2007)
15. Jayaram, B., Rao, C.: On the distributivity of implication operators over t-and s-norms. IEEE Trans. Fuzzy Syst. **12**, 194–198 (2004)
16. Klement, E.P., Mesiar, R., Pap, E.: Triangular Norms. Kluwer Academic Publishers, Dordrecht, Netherlands (2000)
17. Liu, H.: Fuzzy implications derived from generalized additive generators of representable uninorms. IEEE Trans. Fuzzy Syst. **21**, 555–566 (2013)
18. Liu, H.: On a new class of implications:$(g, \min)$-implications and several classical tautologies. Int. J. Uncertain. Fuzz. Knowl-B. **20**, 1–20 (2012)
19. Liu, H.: A new class of fuzzy implications derived from generalized h -generators. Fuzzy Sets Syst. **224**, 63–92 (2013)
20. Mas, M., Monserrat, M., Torrens, J.: Two types of implications derived from uninorms. Fuzzy Sets Syst. **158**, 2612–2626 (2007)
21. Massanet, S., Torrens, J.: On a new class of fuzzy implications: $h-$implications and generalizations. Inf. Sci. **181**, 2111–2127 (2011)
22. Massanet, S., Torrens, J.: Intersection of Yager's implications with $QL-$ and $D-$implications. Int. J. Approx. Reason. **53**, 467–479 (2012)
23. Massanet, S., Torrens, J.: On the characterization of Yager's implications. Inf. Sci. **201**, 1–18 (2012)
24. Massanet, S., Torrens, J.: Threshold generation method of construction of a new implication from two given ones. Fuzzy Sets Syst. **205**, 50–75 (2012)
25. Massanet, S., Torrens, J.: On some properties of threshold generated implications. Fuzzy Sets Syst. **205**, 30–49 (2012)
26. Massanet, S., Torrens, J.: On the vertical threshold generation method of fuzzy implication and its properties. Fuzzy Sets Syst. **226**, 32–52 (2013)
27. Qin, F., Baczyński, M.: Distributivity equations of implications based on continuous triangular conorms (II). Fuzzy Sets Syst. **240**, 86–102 (2014)
28. Qin, F., Baczyński, M., Xie, A.: Distributive equations of implications based on continuous triangular norms (I). IEEE Trans. Fuzzy Syst. **20**, 153–167 (2012)
29. Qin, F., Yang, L.: Distributive equations of implications based on nilpotent triangular norms. Int. J Approx. Reason. **51**, 984–992 (2010)
30. Su, Y., Xie, A., Liu, H.: On ordinal sum implications. Inf. Sci. **293**, 251–262 (2015)
31. Xie, A., Liu, H.: A generalization of Yager's f-generated implications. Int. J Approx. Reason. **54**, 35–46 (2013)

32. Xie, A., Liu, H., Zhang, F., Li, C.: On the distributivity of fuzzy implications over continuous archimedean t-conorms and continuous t-conorms given as ordinal sums. Fuzzy Sets Syst. **205**, 76–100 (2012)
33. Yager, R.R.: On some new classes of implication operators and their role in approximate reasoning. Inf. Sci. **167**, 193–216 (2004)
34. Zhang, F., Liu, H.: On a new class of implications: $(g, u)-$implications and the distributive equations. Int. J. Approx. Reason. **54**, 1049–1065 (2013)

Mechanisms of Mixed Fuzzy Reasoning for Asymmetric Types

Yan Liu and Mu-Cong Zheng

Abstract In the basic models of fuzzy reasoning, the fuzzy propositions are the same type of fuzzy sets. In the paper, we intend to investigate the inference mechanisms of mixed fuzzy reasoning for asymmetric types such that the fuzzy propositions are the different type of fuzzy sets. We establish the two new models for the asymmetric type approximate reasoning problems and present the corresponding methods to solve the new models. Furthermore, we analyze the characterizations of the solutions and give their reductivity.

Keywords Fuzzy reasoning · Intuitionistic fuzzy sets · Asymmetric type · Triple I method · Reductivity

1 Introduction

The theory of fuzzy sets introduced by Zadeh [1] has been found to be useful to deal with uncertainty, imprecision and vagueness of information. It is well known that fuzzy reasoning is a significant part of the theory of fuzzy sets. Two fundamental inference models of fuzzy reasoning are fuzzy modus ponens (FMP) and fuzzy modus tollens (FMT). The most widespread reasoning principle for FMP and FMT is Zadeh's method Composition Rule of Inference (CRI) [2–4]. The Triple I method given by Wang [5] is a very important method to solve the problems of FMP and FMT [4–8]. Tang et al. [9, 10] generalize the Triple I method by selecting different implications in the solutions of Triple I method for FMP and FMT.

Y. Liu
College of Science, Xi'an University of Science and Technology,
Xi'an 710054, China
e-mail: dearly@sina.com

M.-C. Zheng (✉)
School of Mathematics and Information Science, Shaanxi Normal University,
Xi'an 710119, China
e-mail: zhengmucong@aliyun.com

T.-H. Fan et al. (eds.), *Quantitative Logic and Soft Computing 2016*,
Advances in Intelligent Systems and Computing 510,
DOI 10.1007/978-3-319-46206-6_29

Intuitionistic fuzzy sets introduced by Atanassov [11], as a generation of fuzzy set, is a pair of fuzzy sets, namely a membership and a non-membership function which represent positive and negative aspects of the given information. FMP and FMT are extended to models of deductive processes with intuitionistic fuzzy sets which are IFMP (intuitionistic fuzzy modus ponens) and IFMT (intuitionistic fuzzy modus tollens) [12–15]. The CRI method of intuitionistic fuzzy reasoning was discussed in [12]. Zheng et al. [1] presented the Triple I method of intuitionistic fuzzy reasoning, proved the reductivity of the Triple I method for IFMP and showed that the Triple I method of IFMT satisfied the local reductivity instead of reductivity. In order to improve the quality of the Triple I method for lack of reductivity, Liu and Zheng [14] proposed the dual Triple I approximate reasoning method for IFMT. The decomposition methods for solving the IFMP problem and the IFMT problem were presented in [15] and [14] respectively.

In FMP and IFMP, the inference modalities are the same as follows:

$$\begin{array}{lll} \text{Given} & \text{If } x\ is\ A & \text{then } y\ is\ B \\ \text{input} & x\ is\ A^* & \\ \hline \text{output} & & y\ is\ B^* \end{array} \tag{1.1}$$

the inference former A, A^* and rear B, B^* are symmetric, i.e., they all are the fuzzy sets and the intuitionistic fuzzy sets in FMP and IFMP respectively. In the process of approximate reasoning, "if-then" rule is looked on as implication relations between the two fuzzy propositions and it would be represented as fuzzy implication operator and intuitionistic fuzzy implication operator in solving FMP problem and IFMP problem respectively.

In fact, the information is varied and the inference rules are hybrid in our everyday reasoning. Taking account of the variety of information and inference rules in the real world, we intend to investigate the inference mechanisms of mixed fuzzy reasoning for asymmetric types such that formers and rears are the different type of fuzzy sets in this paper. We first discuss the type that the formers are the intuitionistic fuzzy sets and the rears are the fuzzy sets (FIRF type for short), then discuss the type that the formers are the fuzzy sets and the rears are the intuitionistic fuzzy sets (FFRI type for short).

2 Two Methods for Solving the FIRF Type Problem

We consider the inference method of the model with intuitionistic fuzzy formers and fuzzy rears. The inference model is as follows:

$$\begin{array}{lll} \text{Given} & \text{If } x\ is\ A & \text{then } y\ is\ B \\ \text{input} & x\ is\ A^* & \\ \hline \text{output} & & y\ is\ B^* \end{array} \tag{2.1}$$

where A, A^* are intuitionistic fuzzy sets on $X, x \in X, A = (A_t, A_f)$, $A^* = (A_t^*, A_f^*)$, B, B^* are fuzzy sets instead of the intuitionistic fuzzy sets on Y, $y \in Y$.

2.1 First Decomposition Then Aggregation Method (FDTA Method)

Based on the decomposition method of intuitionistic fuzzy reasoning (see [15]), we first decompose the model (2.1) into two FMP models, then solve the two FMP problems, finally aggregate the two solutions.

Step1: The model (2.1) is decomposed into the following two FMP models:

$$\begin{array}{ll} \text{Given} & A_t(x) \to B(y) \\ \text{input} & A_t^*(x) \\ \hline \text{output} & \qquad B_1^*(y) \end{array} \tag{2.2}$$

$$\begin{array}{ll} \text{Given} & A_{-f}(x) \to B(y) \\ \text{input} & A_{-f}^*(x) \\ \hline \text{output} & \qquad B_2^*(y) \end{array} \tag{2.3}$$

where $\to$ is usually selected as a residual fuzzy implication, $x \in X$, $y \in Y$. $A_{-f} = 1 - A_f$, $A_{-f}^* = 1 - A_f^*$.

Step2: Solving the problems (2.2) and (2.3) respectively.

$$B_1^*(y) = \vee_{x \in X}\{A_t^*(x) \otimes (A_t(x) \to B(y))\}$$

$$B_2^*(y) = \vee_{x \in X}\{A_{-f}^*(x) \otimes (A_{-f}(x) \to B(y))\}$$

Step3: Aggregating solutions $B_1^*(y)$ and $B_2^*(y)$. The output of model (2.1) is

$$B^*(y) = B_1^*(y) \wedge B_2^*(y) \tag{2.4}$$

2.2 First Expansion Then Restriction Method (FETR Method)

We first take the fuzzy set B as a degenerate intuitionistic fuzzy set, and convert B into $B' = (B_t', B_f')$ where $B_t' = B$, $B_f' = 1 - B$, so model (2.1) is transformed into an intuitionistic fuzzy reasoning model. By the method of intuitionistic fuzzy reasoning, then we get the solution B'^*. Since B'^* is a intuitionistic fuzzy set, we need to convert B'^* into a fuzzy set also.

Step1: Model (2.1) is transformed into a intuitionistic fuzzy reasoning model as follows:

$$\begin{array}{ll} \text{Given} & A(x) \rightarrow_{L^*} B'(y) \\ \text{input} & A^*(x) \\ \hline \text{output} & \qquad B'^*(y) \end{array} \tag{2.5}$$

where A, A^*, B', B'^* are intuitionistic fuzzy sets and $\rightarrow_{L^*}$ is usually selected as a residual intuitionistic fuzzy implication induced by t-norm $\otimes$, $x \in X$, $y \in Y$.

Step2: We solve the output B'^* of problem (2.5) by the Triple I method given by [13].

$$B'^*(y) = (\vee_{x \in X}\{A_t^*(x) \otimes (A_{-f}(x) \rightarrow B(y))\}, \wedge_{x \in X}\{A_f^*(x) \oplus (1 - A_{-f}(x) \rightarrow B(y))\})$$

Step3: By the two degeneration operations $B^* = B_t'^*$ and $B^* = 1 - B_f'^*$, we can finally get two solutions

$$B^*(y) = \vee_{x \in X}\{A_t^*(x) \otimes (A_{-f}(x) \rightarrow B(y))\} \tag{2.6}$$

and

$$\begin{aligned} B^*(y) &= 1 - \wedge_{x \in X}\{A_f^*(x) \oplus (1 - A_{-f}(x) \rightarrow B(y))\} \\ &= \vee_{x \in X}\{A_{-f}^*(x) \otimes (A_{-f}(x) \rightarrow B(y))\} \end{aligned} \tag{2.7}$$

respectively.

2.3 Character Analysis of Solutions

In the above two subsections, we can obtain a solution by FDTA method and obtain two solutions by FETR method. What is the relationship between these solutions? Now we denote $B_3^*(y) = \vee_{x \in X}\{A_t^*(x) \otimes (A_{-f}(x) \rightarrow B(y))\}$, and denote the solutions given by (2.4), (2.6) and (2.7) as $B^1(y)$, $B^2(y)$ and $B^3(y)$ respectively. Then $B^1 = B_1^* \wedge B_2^*$, $B^2 = B_3^*$, $B^3 = B_2^*$.

Theorem 1 *If* $\rightarrow_{L^*}$ *in model (2.5) is generated by residual fuzzy implication* $\rightarrow$ *in the models (2.2) and (2.3), then* $B^2 \leq B^1 \leq B^3$.

Proof Obviously, $B^1 \leq B^3$. As $A_t^*(x) \leq A_{-f}^*(x)$, $\forall x \in X$, we can get that

$$\begin{aligned} A_t^*(x) \otimes (A_{-f}(x) \rightarrow B(y)) &\leq A_{-f}^*(x) \otimes (A_{-f}(x) \rightarrow B(y)) \\ A_t^*(x) \otimes (A_{-f}(x) \rightarrow B(y)) &\leq A_t^*(x) \otimes (A_t(x) \rightarrow B(y)) \end{aligned}$$

It is clear that $B_3^* \leq B_2^*$. Thus $B^2 \leq B^3$.

$$
\begin{aligned}
&B_1^* \wedge B_2^* \\
&= (\vee_{x\in X}\{A_t^*(x) \otimes (A_t(x) \to B(y))\}) \wedge (\vee_{x\in X}\{A_{-f}^*(x) \otimes (A_{-f}(x) \to B(y))\}) \\
&= \vee_{x\in X}\{(A_t^*(x) \otimes (A_t(x) \to B(y))) \wedge (A_{-f}^*(x) \otimes (A_{-f}(x) \to B(y)))\} \\
&\geq \vee_{x\in X}\{A_t^*(x) \otimes (A_{-f}(x) \to B(y))\} \\
&= B_3^*
\end{aligned}
$$

i.e., $B_3^* \leq B_1^* \wedge B_2^*$. So $B^2 \leq B^1$. The proof is completed. □

In fuzzy reasoning, the reductivity of inference methods is an important topic (see [6]). The following theorem tell us the reductivity of solutions by FDTA and FETR methods.

Theorem 2 *If $\to_{L^*}$ is a residual intuitionistic implication induced by a left-continuous t-norm, then the solutions given by (2.4) and (2.7) for the model (2.1) are reductive, i.e., $B^* = B$ whenever $A^* = A$ satisfies the condition $\exists x_0 \in X$ such that $A(x_0) = 1^* = (1, 0)$.*

Proof $\exists x_0 \in X$ such that $A(x_0) = 1^* = (1, 0)$, i.e., $A_t^*(x_0) = 1$, $A_{-f}^*(x_0) = 1$. It follows from Theorem 6 in [6] that B_1^* and B_2^* are reductive, therefore B^1 and B^3 are reductive. The proof is completed.

3 Two Methods for Solving the FFRI Type Problem

Now we consider the inference method of the model with fuzzy formers and fuzzy intuitionistic rears. The inference model is as follows:

$$
\begin{array}{lll}
\text{Given} & \text{If } x\ is\ A & \text{then } y\ is\ B \\
\text{input} & x\ is\ A^* & \\
\hline
\text{output} & & y\ is\ B^*
\end{array}
\tag{3.1}
$$

where A, A^* are fuzzy sets on X, $x \in X$, $B = (B_t, B_f)$, $B^* = (B_t^*, B_f^*)$ are intuitionistic fuzzy sets on Y, $y \in Y$.

3.1 First Decomposition Then Aggregation Method (FDTA Method)

Similar to Sect. 2.1 we first decompose model (3.1) into two FMP models, then solve the two FMP problems, finally aggregate the two solutions into a intuitionistic fuzzy solution.

Step1: The model (3.1) is decomposed into the following two FMP models:

$$\begin{array}{ll} \text{Given} & A(x) \to B_t(y) \\ \text{input} & A(x) \\ \hline \text{output} & \qquad B_t^*(y) \end{array} \tag{3.2}$$

$$\begin{array}{ll} \text{Given} & A(x) \to B_{-f}(y) \\ \text{input} & A(x) \\ \hline \text{output} & \qquad B_{-f}^*(y) \end{array} \tag{3.3}$$

where $\to$ is usually selected as a residual fuzzy implication.

Step2: Solving problems (3.2) and (3.3) respectively.

$$B_t^*(y) = \vee_{x \in X}\{A^*(x) \otimes (A(x) \to B_t(y))\}$$

$$B_{-f}^*(y) = \vee_{x \in X}\{A^*(x) \otimes (A(x) \to B_{-f}(y))\}$$

Step3: Aggregating solutions $B_1^*(y)$ and $B_2^*(y)$ into a intuitionistic fuzzy solution. The output of model (3.1) is:

$$\begin{aligned} B^* &= (B_t^*, B_f^*) = (B_t^*, 1 - B_{-f}^*) \\ &= (\vee_{x \in X}\{A^*(x) \otimes (A(x) \to B_t(y))\}, 1 - \vee_{x \in X}\{A^*(x) \otimes (A(x) \to B_{-f}(y))\}) \end{aligned} \tag{3.4}$$

3.2 *Direct Expansion Method (DE Method)*

Its similar to Sect. 2.2 we convert fuzzy sets A, A^* into intuitionistic fuzzy sets $A' = (A, 1 - A)$, $A'^* = (A^*, 1 - A^*)$ respectively, so model (3.1) is transformed into a intuitionistic fuzzy reasoning model. By the method of intuitionistic fuzzy reasoning, then we get the solution B^*.

Step1: The model (3.1) is transformed into a intuitionistic fuzzy reasoning model as follows:

$$\begin{array}{ll} \text{Given} & A'(x) \to_{L^*} B(y) \\ \text{input} & A'^*(x) \\ \hline \text{output} & \qquad B^*(y) \end{array} \tag{3.5}$$

where A', A'^*, B, B^* are the intuitionistic fuzzy sets and $\to_{L^*}$ is usually selected as a residual intuitionistic fuzzy implication induced by t-norm $\otimes$.

Step2: We solve the output B^* of the problem (3.5) by the Triple I method given by [13].

$$\begin{aligned} B^*(y) &= (\vee_{x \in X}\{A^*(x) \otimes ((A(x) \to B_t(y)) \wedge (A(x) \to B_{-f}(y)))\}, \\ &\qquad \wedge_{x \in X}\{(1 - A^*(x)) \oplus (1 - A(x) \to B_{-f}(y))\}) \\ &= (\vee_{x \in X}\{A^*(x) \otimes (A(x) \to B_t(y))\}, \\ &\qquad 1 - \vee_{x \in X}\{A^*(x) \otimes (A(x) \to B_{-f}(y))\}) \end{aligned} \tag{3.6}$$

3.3 Character Analysis of Solutions

We denote the solutions given by (3.4) and (3.6) as $B^4(y)$, $B^5(y)$ respectively. Obviously, $B^5(y) = B^4(y)$. So we get the following theorem.

Theorem 3 *The FDTA method is equivalent to the DE method for model (3.1).*

Theorem 4 *If* $\rightarrow_{L^*}$ *is a residual intuitionistic implication induced by a left-continuous t-norm, then the solutions of the FDTA method and the DE method for model (3.1) are reductive, i.e.,* $B^* = B$ *whenever* $A^* = A$ *satisfies the condition* $\exists x_0 \in X$ *such that* $A(x_0) = 1^* = (1, 0)$.

4 Conclusions

In current research, the basic models of fuzzy reasoning are symmetric type, i.e., the formers and rears are the same type sets such as FMP and IFMP. In this paper, we consider new models where the formers and rears are different types of sets. We first present two models of mixed fuzzy reasoning for different types, then we propose the corresponding methods to solve the new models. Moreover, we obtain the reductivity of the solutions. We provide alternative approximate reasoning mechanisms for mixed fuzzy reasoning under the multiple information representations. The methods may be useful tools to be applied in a wide variety of fields related to fuzzy reasoning in future.

Acknowledgments The authors acknowledge their supports from the National Natural Science Foundation of China (Nos. 11401361, 61473336, 61572016).

References

1. Zadeh, L.A.: Fuzzy sets. Inf. Control **8**, 338–353 (1965)
2. Zadeh, L.A.: Outline of a new approach to the analysis of complex systems and decision processes. IEEE Trans. Syst. Man Cybern. **3**, 28–44 (1973)
3. Wang, L.X.: A Course in Fuzzy Systems and Control. Prentice Hall PTR, Upper Saddle River (1997)
4. Wang, G.J.: Non-classical Mathematical Logic and Approximate Reasoning, 2nd edn. Science Press, Beijing (2008). (in Chinese)
5. Wang, G.J.: Full implicational triple I method for fuzzy reasoning. Sci. China Ser. E **29**(1), 43–53 (1999)
6. Pei, D.W.: Unified full implication algorithms of fuzzy reasoning. Inf. Sci. **178**(2), 520–530 (2008)
7. Liu, H.W., Wang, G.J.: Unified forms of fully implicational restriction methods for fuzzy reasoning. Inf. Sci. **177**, 956–966 (2007)
8. Wang, G.J., Duan, J.Y.: On robustness of the full implication triple I inference method with respect to finer measurements. Int. J. Approx. Reason. **55**, 787–796 (2014)

9. Tang, Y.M., Liu, X.P.: Differently implicational universal triple I method of (1,2,2) type. Comput. Math. Appl. **59**, 1965–1984 (2010)
10. Tang, Y.M., Yang, X.Z.: Symmetric implicational method of fuzzy reasoning. Int. J. Approx. Reason. **54**, 1034–1048 (2013)
11. Atanassov, K.: Intuitionistic fuzzy sets. Fuzzy Sets Syst. **20**, 87–96 (1986)
12. Cornelis, C., Deschrijver, G., Kerre, E.E.: Implication inintuitionistic fuzzy and interval-valued fuzzy set theory: construction. classification, application. Int. J. Approx. Reason. **35**, 55–95 (2004)
13. Zheng, M.C., Shi, Z.K., Liu, Y.: Triple I method of approximate reasoning on Atanassov's intuitionistic fuzzy sets. Int. J. Approx. Reason. **55**, 1369–1382 (2014)
14. Liu, Y., Zheng, M.C.: The Dual Triple I Methods of FMT and IFMT. Math. Prob. Eng. **2014**, Article ID 507401, 8 (2014). doi:10.1155/2014/507401
15. Zheng, M.C., Shi, Z.K., Liu, Y.: Triple I method of intuitionistic fuzzy reasoning based on residual implicator. Sci. China Inf. Sci. **43**, 810–820 (2013). (in Chinese)

Two Kinds of Modifications of Implications

Wen-Wen Zhang and Dao-Wu Pei

Abstract The identity principle (IP) and ordering property (OP) are two important properties of fuzzy implications. They have important role in the applications of fuzzy implications. However, many fuzzy implications do not satisfy these two properties. In this paper, two kinds of new modifications of fuzzy implications are proposed such that every new implication satisfies one of the two properties, respectively. Then properties of the modified fuzzy implications are explored.

Keywords Fuzzy logic · Fuzzy implication · Identity principle · Ordering property

1 Introduction

Fuzzy implications are widely studied as one kind of the most important operators in fuzzy logic and approximate reasoning. The five kinds of basic fuzzy implications are (S, N)-implications, R-implications, QL-implications, and Yager's f-implications and g-implications [1–5].

Fuzzy implications have played an important role in many applied fields, such as intelligent control, image processing, data mining, fuzzy mathematical morphology, and so on. Therefore, in recent years many scholars have constructed different types of fuzzy implications to meet the needs of applications. On the other hand, in order to better understand these different types of fuzzy implications, it is very necessary to characterize them by their algebraic properties [1–5].

It is needless to say, we hope that the fuzzy implications have very good properties, such as identity principle, ordering property, left neutrality property, law of contraposition and exchange principle etc. However, many implications do not satisfy some of these properties. So Fodor [6] proposed some methods to modify fuzzy

W.-W. Zhang · D.-W. Pei (✉)
School of Sciences, Zhejiang Sci-Tech University, Hangzhou 310018, China
e-mail: peidw@163.com

W.-W. Zhang
e-mail: 18705580706@163.com

T.-H. Fan et al. (eds.), *Quantitative Logic and Soft Computing 2016*,
Advances in Intelligent Systems and Computing 510,
DOI 10.1007/978-3-319-46206-6_30

implications so that the modified fuzzy implications satisfy the law of contraposition with respect to a strong fuzzy negation. Based on this work, Aguilo et al. [7] also put forward some new methods to modify fuzzy implications, the modified fuzzy implications satisfy the law of contraposition with respect to a strong fuzzy negation and even a family of not strong fuzzy negations.

So, how to modify the fuzzy implications so that they satisfy the other properties? This is what we are interested in. In this paper we first introduce two new methods to modify fuzzy implications, and then we explored the properties of the modified fuzzy implications.

2 Preliminaries

This section reviews some necessary concepts and examples (see [8–11]). In the whole, we denote $U = [0, 1]$.

Definition 1 ([8, 11]) A binary operation T (or S) on U is said to be a t-norm (resp. t-conorm), if it is commutative, associative, non-decreasing in both variables and 1 (resp. 0) is its neutral element.

Definition 2 ([8, 11]) A unary operation N on U is a (fuzzy) negation if it is decreasing, $N(0) = 1$ and $N(1) = 0$.

Moreover, a fuzzy negation is strict if it is strictly decreasing and continuous, and strong if it is an involution, i.e., $N(N(x)) = x$ for all $x \in U$.

Example 1 The standard fuzzy negation N_0 and the least fuzzy negation N_1 are as follows.

$$N_0(x) = 1 - x, \quad x \in U; \quad N_1(x) = \begin{cases} 1, \ x = 0 \\ 0, \ x \in (0, 1]. \end{cases}$$

Definition 3 ([8]) A binary operation I on U is a fuzzy implication, or implication, if the following conditions hold:
(I1) I is decreasing in the first variable;
(I2) I is increasing in the second variable;
(I3) $I(0, 0) = I(1, 1) = 1, I(1, 0) = 0$.

We will denote by $\mathcal{FI}$ the set of all implications.

Some basic properties of fuzzy implications are given below (see [8]).
The left neutrality principle **(NP)**: $I(1, y) = y, y \in U$;
The identity property **(IP)**: $I(x, x) = 1, x \in U$;
The ordering property **(OP)**: $I(x, y) = 1 \Longleftrightarrow x \le y, x, \ y \in U$;
The contrapositive symmetry with respect to a fuzzy negation N **(CP(N))**: $I(x, y) = I(N(y), N(x)), x, \ y \in U$;
The exchange principle **(EP)**: $I(x, I(y, z)) = I(y, I(x, z)), x, \ y, \ z \in U$.

Table 1 Several basic fuzzy implications

Name	Symbol	Formula
Lukasiewicz	I_L	$I_L(x, y) = \min(1,\ 1 - x + y)$
Reichenbach	I_{RC}	$I_{RC}(x, y) = 1 - x + xy$
Yager	I_{YG}	$I_{YG}(x, y) = \begin{cases} y^x, & x > y \\ 1, & x \le y \end{cases}$
Godel	I_{GD}	$I_{GD}(x, y) = \begin{cases} y, & x > y \\ 1, & x \le y \end{cases}$

Table 1 give some basic fuzzy implications.

Definition 4 ([8]) Let I be a fuzzy implication. The function N_I, defined by $N_I(x) = I(x, 0)$, $x \in U$, is called the natural negation of I.

3 The IP-Modifications of Fuzzy Implications

This section presents the first kind of modifications of fuzzy implications, so that the modified fuzzy implications satisfy (IP). Some properties of the modified implications are also studied.

First, we can imitate Professor Wang's method to construct a good implication (Revised Kleene implication, or R_0 implication) from a familiar implication (Kleene implication) (see [9], or [10, 11]), and give the following definition.

Definition 5 Let $I \in \mathcal{FI}$. Then the binary operation $I^{(1)}$ on U, called the first kind of modification, or IP-modification of I, is defined as

$$I^{(1)}(x, y) = \begin{cases} 1, & x \le y \\ I(x, y), & x > y. \end{cases} \tag{1}$$

Example 2 (i) If $I = I_{RC}$, then

$$I^{(1)}_{RC}(x, y) = \begin{cases} 1, & x \le y \\ 1 - x + xy, & x > y. \end{cases}$$

(ii) If $I = I_{YG}$, then

$$I^{(1)}_{YG}(x, y) = \begin{cases} 1, & x \le y \\ y^x, & x > y > 0. \end{cases}$$

(iii) If $I_0(x, y) = \min\{1,\ 1 - x + \sqrt{y}\}$, then

$$I_0^{(1)}(x,y) = \begin{cases} 1, & x \le y \\ \min\{1, 1-x+\sqrt{y}\}, & x > y. \end{cases}$$

Theorem 1 *If $I \in \mathcal{FI}$, then $I^{(1)} \in \mathcal{FI}$.*

Proof Suppose that $x_1,\ x_2 \in U$ with $x_1 \le x_2$.
If $x_1 \le x_2 \le y$, then $I^{(1)}(x_1, y) = I^{(1)}(x_2, y) = 1$.
If $x_1 \le y < x_2$ then $1 = I^{(1)}(x_1, y) \ge I^{(1)}(x_2, y)$.
If $y < x_1 \le x_2$, then $I^{(1)}(x_1, y) = I(x_1, y) \ge I(x_2, y) = I^{(1)}(x_2, y)$.
This shows that $I^{(1)}$ satisfies (I1).
In the same way we can prove that $I^{(1)}$ satisfies (I2). Also

$$I^{(1)}(1,1) = I^{(1)}(0,0) = 1,\quad I^{(1)}(1,0) = I(1,0) = 0.$$

I.e., $I^{(1)}$ satisfies (I3). Therefore, $I^{(1)} \in \mathcal{FI}$. □

Remark 1 By Theorem 1 and the definition of $I^{(1)}$, it is not difficult to see that $I^{(1)}$ satisfies (IP) and $I^{(1)} \ge I$.

Proposition 1 *If $I \in \mathcal{FI}$ satisfies (IP) or (OP), then $I^{(1)} = I$.*

Proof (i) Suppose that I satisfy (IP). Then $I(x, x) = 1$. Thus, to prove $I^{(1)} = I$, it is only need to consider $x,\ y \in U$ with $x \le y$. In this case, $1 = I(x, x) \le I(x, y) = 1$. So we have $I(x, y) = 1 = I^{(1)}(x, y)$.
(ii) Suppose that I satisfy (OP). Then obviously, we have $I = I^{(1)}$. □

Remark 2 Based on Proposition 1, we know that if I satisfies (OP), then $I^{(1)}$ satisfies (OP). However, the converse is not true. In Example 2(ii), $I_{YG}^{(1)}$ satisfies (OP), but I_{YG} does not satisfy (OP).

Below we will explore the conditions for I under which $I^{(1)}$ satisfy (OP).

In Example 2, $I_{RC}^{(1)}$ and $I_{YG}^{(1)}$ satisfy (OP) but $I_0^{(1)}$ does not satisfy (OP). We observe that when $x,\ y \in U$ with $x > y$, $I_{RC}(x, y) < 1$ and $I_{YG}(x, y) < 1$. However, when $x > y \ge x^2$, $I_0^{(1)}(x, y) = 1$. This is the reason why $I_0^{(1)}$ does not satisfies (OP). So if $I^{(1)}$ satisfies (OP), then I always less than 1 with $x > y$.

From the above analysis, we can obtain the following proposition.

Proposition 2 *Let $I \in \mathcal{FI}$. Then $I^{(1)}$ satisfies (OP) if and only if for $x,\ y \in U$, $I(x, y) < 1$ whenever $x > y$.*

Proof "$\Longrightarrow$" Suppose that $I^{(1)}$ satisfy (OP). Consider $x,\ y \in U$ with $x > y$. Then $I(x, y) = I^{(1)}(x, y) < 1$.
"$\Longleftarrow$" Suppose that $I(x, y) < 1$ whenever $x > y$. If $I^{(1)}(x_0, y_0) = 1$ for some $x_0 > y_0$, then $I(x_0, y_0) = I^{(1)}(x_0, y_0) < 1$. This is a contradiction. □

By Proposition 2 we can get the following corollary.

Corollary 1 *Let $I \in \mathcal{FI}$. Then the following statements hold:*
(i) *If I is strictly monotonic in the first variable with $x > 0$ and $y < 1$, then $I^{(1)}$ satisfies (OP).*
(ii) *If I is strictly monotonic in the second variable with $x > 0$ and $y < 1$, then $I^{(1)}$ satisfies (OP).*
(iii) *If I satisfies (IP) and is strictly monotonic when $0 < x \leq 1$, $0 \leq y \leq x$, then $I^{(1)}$ satisfies (OP).*

Proof (i) Suppose that I be strictly monotonic in the first variable with $x > 0$ and $y < 1$. Then $I(x, y) < I(0, y) = 1$. Therefore, for all $x,\ y \in U$ with $x > y$, we have $I(x, y) < 1$. Thus by Proposition 2, $I^{(1)}$ satisfies (OP).

(ii) Suppose that I be strictly monotonic in the second variable with $x > 0$ and $y < 1$. Then $I(x, y) < I(x, 1) = 1$. Therefore, for all $x,\ y \in U$ with $x > y$, we have $I(x, y) < 1$. Thus by Proposition 2, $I^{(1)}$ satisfies (OP).

(iii) Suppose that I satisfy (IP), that is $I(x, x) = 1, x \in U$, and when $0 < x \leq 1$ and $0 \leq y \leq x$, I be strictly monotonic. Then, $I(x, y) < I(x, x) = 1$ whenever $x > y$. Thus by Proposition 2, $I^{(1)}$ satisfies (OP). □

Remark 3 (i) Since f-implications and g-implications are all strictly monotonic in two variables with $x > 0$ and $y < 1$, therefore, the IP-modifications of f-implications and g-implications are all satisfy (OP);
(ii) If I is a residual implication induced by a left continuous t-norm, then I satisfy (OP). By Proposition 1, $I^{(1)} = I$. Naturally, $I^{(1)}$ satisfies (OP).

Now we discuss properties of I which are preserved by $I^{(1)}$.

Proposition 3 *Let $I \in \mathcal{FI}$ and N be a strict negation.*
(i) *If I satisfies (NP) [or (CB), (CP(N))], then $I^{(1)}$ satisfies (NP) [resp. (CB), (CP(N))].*
(ii) *The natural negation of $I^{(1)}$ is the same as the natural negation of I. That is, $N_{I^{(1)}} = N_I$.*

Proof (i) Suppose that I satisfy (NP). If $y = 1$, $I^{(1)}(1, 1) = 1$; If $y \in [0, 1)$, $I^{(1)}(1, y) = I(1, y) = y$. Therefore, $\forall y \in U,\ I^{(1)}(1, y) = y$.

Suppose that I satisfy (CB). For $x,\ y \in U$, if $x \leq y$ we have $I^{(1)}(x, y) = 1 \geq y$. If $x > y$ then $I^{(1)}(x, y) = I(x, y) \geq y$. Therefore, $\forall x,\ y \in U$, we always have $I^{(1)}(x, y) \geq y$. That is, $I^{(1)}$ satisfies (CB).

Suppose that I satisfies (CP(N)) with respect to a strict negation N. Because of the negation N is strict, we have

$$x \leq y \Longleftrightarrow N(y) \leq N(x), \quad x > y \Longleftrightarrow N(y) > N(x).$$

Table 2 Comparison of properties of I and $I^{(1)}$

Property	IP	OP	NP	CB	CP(N)	EP
I	U	Y	Y	Y	Y	Y
$I^{(1)}$	Y	Y	Y	Y	Y	U

Thus,

$$\begin{aligned} I^{(1)}(N(y), N(x)) &= \begin{cases} 1, & N(y) \leq N(x) \\ I(N(y), N(x)), & N(y) > N(x) \end{cases} \\ &= \begin{cases} 1, & x \leq y \\ I(x, y), & x > y \end{cases} \\ &= I^{(1)}(x, y), \quad x, \ y \in [0, 1]. \end{aligned}$$

Therefore, $I^{(1)}$ satisfies (CP(N)).
(ii) The natural negation of $I^{(1)}$ is defined as $N_{I^{(1)}}(x) = I^{(1)}(x, 0)$.
If $0 < x \leq 1$, then $N_{I^{(1)}}(x) = I^{(1)}(x, 0) = I(x, 0) = N_I(x)$.
If $x = 0$, then $N_{I^{(1)}}(0) = I^{(1)}(0, 0) = 1 = N_I(0)$.
Therefore, we have $N_{I^{(1)}} = N_I$. □

Remark 4 If I satisfies (IP) or (OP), then $I^{(1)} = I$. Furthermore, if I also satisfies (EP) then $I^{(1)}$ satisfies (EP). However, if I does not satisfy (IP) or (OP), it is uncertain that whether $I^{(1)}$ satisfies (EP) or not. In Example 2, I_{RC} satisfies (EP) but $I_{RC}^{(1)}$ does not satisfy (EP). In fact, taking $x = 0.5$, $y = 0.8$ and $z = 0.4$, we have

$$I^{(1)}(0.5, I^{(1)}(0.8, 0.4)) = 1, \quad I^{(1)}(0.8, I^{(1)}(0.5, 0.4)) = 0.76.$$

That is, $I_{RC}^{(1)}$ does not satisfy (EP).

Table 2 shows the properties preserved by $I^{(1)}$, where "Y" stands for "Yes", "U" for "Uncertain", N is the strict negation.

4 The Second Kind of Modification of Fuzzy Implications

In Sect. 3 we present the first kind of modifications of fuzzy implications. Although they satisfy (OP) under certain conditions, but there are some implications which do not satisfy (OP). For example, there are x_0 and y_0 in U with $x_0 > y_0$, such that $I(x_0, y_0) = 1$. In this case, the first kind of modifications do not satisfy (OP).

In this section we modify these implications and more general implications, so that the modified implications always satisfy (OP). The basic properties of the modified implications will also be discussed.

First, we propose the following concept.

Definition 6 Let $I \in \mathcal{FI}$, and N a fuzzy negation with $N(x) < 1$ for $x > 0$. Then the binary operation $I^{(2)}$ on U, called the second kind of modification, or OP-modification of I, is defined as

$$I^{(2)}(x, y) = \begin{cases} 1, & x \le y \\ I(x, y) \wedge (N(x) \vee y), & x > y. \end{cases} \tag{2}$$

Example 3 (i) If we take $I = I_{YG}$ and $N = N_1$, then

$$I_{YG}^{(2)}(x, y) = \begin{cases} 1, & x \le y \\ y^x, & x > y > 0. \end{cases}$$

(ii) If $I = I_L$ and $N = N_1$, then

$$I_L^{(2)}(x, y) = \begin{cases} 1, & x \le y \\ y, & x > y \end{cases} = I_G(x, y).$$

(iii) If $I = I_0$ and $N = N_0$, then

$$I_0^{(2)}(x, y) = \begin{cases} 1, & x \le y \\ 1 - x, & y < x \le 1 - y \\ y, & x > y, x > 1 - y. \end{cases}$$

Theorem 2 *If $I \in \mathcal{FI}$, then $I^{(2)} \in \mathcal{FI}$.*

Proof Let $x_1,\ x_2 \in U$ with $x_1 \le x_2$.
If $x_1 \le x_2 \le y$ then $I^{(2)}(x_1, y) = 1 = I^{(2)}(x_2, y)$.
If $x_1 \le y < x_2$ then $I^{(2)}(x_1, y) = 1 \ge I^{(2)}(x_2, y)$.
If $y < x_1 \le x_2$ then

$$I^{(2)}(x_1, y) = I(x_1, y) \wedge (N(x_1) \vee y) \ge I(x_2, y) \wedge (N(x_2) \vee y) = I^{(2)}(x_2, y).$$

Therefore, $I^{(2)}$ satisfies (I1). In the same way we can prove that $I^{(2)}$ satisfies (I2). Also, it is easy to verify the condition (I3):

$$I^{(2)}(1, 1) = I^{(2)}(0, 0) = 1, \quad I^{(2)}(1, 0) = 0.$$

Therefore, $I^{(2)} \in \mathcal{FI}$. □

Remark 5 By Theorem 2 and the definition of $I^{(2)}$, we see that $I^{(2)}$ satisfies (OP), (IP) and $I^{(2)} \le I^{(1)}$. So we call $I^{(2)}$ the (OP-IP)-modification of I.

Next we discuss properties preserved by $I^{(2)}$.

Proposition 4 *Suppose that $I \in \mathcal{FI}$ and N a fuzzy negation with $N(x) < 1$ for $x > 0$, $I^{(2)}$ is defined by (2). We have the following conclusions.*
(i) If I satisfies (NP) or (CB), then $I^{(2)}$ satisfies (NP) or (CB). Moreover, if N be a strong negation and I satisfies (CP(N)), then $I^{(2)}$ satisfies (CP(N)).
(ii) $N_{I^{(2)}} = N_I \wedge N$.

Proof (i) First, suppose that I satisfy (NP).
If $y = 1$, then $I^{(2)}(1, 1) = 1$.
If $y \in [0, 1)$, then $I^{(2)}(1, y) = I(1, y) \wedge (N(1) \vee y) = y \wedge y = y$.
Therefore, $\forall y \in U$, $I^{(2)}(1, y) = y$. That is, $I^{(2)}$ satisfies (NP).
Next, suppose that I satisfy (CB).
If $\forall x,\ y \in U$ with $x \leq y$, $I^{(2)}(x, y) = 1 \geq y$.
If $x > y$, $I^{(2)}(x, y) = I(x, y) \wedge (N(x) \vee y) \geq y \wedge y = y$.
Therefore, $\forall x,\ y \in U$, $I^{(2)}(x, y) \geq y$. That is, $I^{(2)}$ satisfies (CB).
Now Suppose that N is a strong negation and I satisfies (CP(N)). Since the negation N is strong, we have

$$x \leq y \Longleftrightarrow N(y) \leq N(x),\quad x > y \Longleftrightarrow N(y) > N(x).$$

Thus we obtain

$$\begin{aligned} I^{(2)}(N(y), N(x)) &= \begin{cases} 1, & N(y) \leq N(x) \\ I(N(y), N(x)) \bigwedge (N(N(y)) \bigvee N(x)), & N(y) > N(x) \end{cases} \\ &= \begin{cases} 1, & x \leq y \\ I(x, y) \bigwedge (N(x) \bigvee y), & x > y \end{cases} \\ &= I^{(2)}(x, y),\quad x,\ y \in U. \end{aligned}$$

Therefore, $I^{(2)}$ also satisfies (CP(N)).
(ii) If $x = 0$, then $N_{I^{(2)}}(0) = 1 = N_I(0) \bigwedge N(0)$.
If $x > 0$, we have

$$N_{I^{(2)}}(x) = I^{(2)}(x, 0) = I(x, 0) \bigwedge (N(x) \bigvee 0) = N_I(x) \bigwedge N(x).$$

Therefore, $\forall x \in U,\ N_{I^{(2)}} = N_I \wedge N$. □

Remark 6 Generally speaking, $I^{(2)}$ does not preserve (EP). However, if I satisfies (CB) and $N = N_1$, then $I^{(2)}$ satisfies (EP) even if I does not satisfy (EP). In fact,

$$\begin{aligned} I^{(2)}(x, I^{(2)}(y, z)) &= \begin{cases} 1, & y \leq z \\ I^{(2)}(x, I(y, z) \bigwedge ((N(y) \bigvee z))), & y > z \end{cases} \\ &= \begin{cases} 1,\ y \leq z,\ \text{or}\ x \leq z,\ y > z \\ z,\ x > z,\ y > z. \end{cases} \end{aligned}$$

Table 3 Comparison of properties of I and $I^{(2)}$

Property	IP	OP	NP	CB	CP(N)	EP
I	U	U	Y	Y	Y	Y
$I^{(2)}$	Y	Y	Y	Y	Y	U

Thus, we have

$$I^{(2)}(y, I^{(2)}(x, z)) = \begin{cases} 1, \ x \leq z, \ \text{or } y \leq z, \ x > z \\ z, \ x > z, \ y > z. \end{cases}$$

Therefore, $\forall x, \ y, \ z \in U, I^{(2)}(x, I^{(2)}(y, z)) = I^{(2)}(y, I^{(2)}(x, z))$. This shows that $I^{(2)}$ satisfies (EP).

Table 3 shows the properties preserved by $I^{(2)}$, where N is a strong negation.

5 Conclusions

In this paper we introduced two kinds of modified fuzzy implications. The first kinds of modified fuzzy implications satisfy (IP) and (OP) under certain conditions. The second kinds of modified fuzzy implications satisfy (OP), certainly, also satisfy (IP). It was found that the two kinds of modified fuzzy implications preserve most properties of the original fuzzy implications.

It is worth noting that, in practical applications when we want that the implications have property (IP) or (OP), we can suitably modify the given implications which do not have these properties. If the given implication satisfies the conditions of Proposition 2, the first kind of modification is a better choice, otherwise, we should choose the second kind.

In the future work, we shall consider other modifications, and study further properties satisfied by the modifications.

Acknowledgments This work is supported by the National Natural Science Foundation of China (No. 11171308; 61379018; 61472471; 51305400).

References

1. Baczynski, M., Jayaram, B.: On the characterizations of (S, N)-implications. Fuzzy Sets Syst. **158**, 1713–1727 (2007)
2. Baczynski, M., Jayaram, B.: (S, N)-implications and R-implications: a state-of-the-art survey. Fuzzy Sets Syst. **159**, 1836–1859 (2008)
3. Shi, Y., Van Gasse, B., Ruan, D., et al.: On the first place antitonicity in QL-implications. Fuzzy Sets Syst. **159**, 2988–3013 (2008)

4. Massanet, S., Torrens, J.: On the characterization of Yager's implications. Inf. Sci. **201**, 1–18 (2012)
5. Mas, M., Monserrat, M., Torrens, J.: A characterization of (U, N), RU, QL and D-implications derived from uninorms satisfying the law of importation. Fuzzy Sets Syst. **161**, 1369–1387 (2010)
6. Fodor, J.C.: Contrapositive symmetry of fuzzy implications. Fuzzy Sets Syst. **69**, 1836–1859 (1995)
7. Aguilo, I., Suner, J., Torrens, J.: New types of contrapositivisation of fuzzy implications with respect to fuzzy negations. Inf. Sci. **322**, 158–188 (2015)
8. Baczynski, M., Jayaram, B.: Fuzzy Implications. Springer, Berlin (2008)
9. Wang, G.J.: Non-classical Mathematical Logic and Approximate Reasoning. Science Press, Beijing (2000). (In Chinese)
10. Pei, D.W.: R_0 implication: characteristics and applications. Fuzzy Sets Syst. **131**, 297–302 (2002)
11. Pei, D.W.: T-norm Based Fuzzy Logics: Theory and Applications. Science Press, Beijing (2013). (In Chinese)

Distributivity for 2-Uninorms over Semi-uninorms

Ya-Ming Wang and Feng Qin

Abstract This paper is devoted to solving the distributivity equations for 2-uninorms over semi-uninorms. Our investigations are motivated by the couple of distributive logical connectives and their generalizations, such as t-norms, t-conorms, uninorms, nullnorms, and fuzzy implications, which are often used in fuzzy set theory. There are two generalizations of them. One is a 2-uninorm covering both a uninorm and a null-norm, which forms a class of commutative, associative and increasing operators on the unit interval with an absorbing element that separates two subintervals with neutral elements. Another is a semi-uninorm, which generalizes a uninorm by omitting commutativity and associativity. In this work, all possible solutions of the distributivity equation for the three defined subclasses of 2-uninorms over semi-uninorms are characterized.

Keywords Distributivity equations · Semi-t-norms · Semi-t-conorms · Semi-uninorms · 2-uninorms

1 Introduction

The functional equations involving aggregation operators [4, 5, 9, 19–22] play an important role in theories of fuzzy sets and fuzzy logic. As Ref. [12] pointed out, a new direction of investigations is concerned with distributivity equation and inequalities for uninorms and nullnorms [3, 7, 8, 11–13, 22–25, 30]. Uninorms, introduced by Yager and Rybalov [30], and studied by Fodor et al. [15], are special aggregation operators that have proven to be useful in many fields like fuzzy logic, expert systems, neural networks, utility theory and fuzzy system modeling [14, 16, 18, 26–29]. Uninorm is interesting because its structure is a special combination of a t-norm and a t-conorm having a neutral element lying somewhere in the unit interval.

Y.-M. Wang · F. Qin (✉)
College of Mathematics and Information Science, Jiangxi Normal University,
Nanchang 330022, People's Republic of China
e-mail: qinfeng923@163.com

T.-H. Fan et al. (eds.), *Quantitative Logic and Soft Computing 2016*,
Advances in Intelligent Systems and Computing 510,
DOI 10.1007/978-3-319-46206-6_31

This paper is mainly devoted to solving the distributivity equations for 2-uninorms over semi-uninorms. Our investigations are motivated by the couple of distributive logical connectives and their generalizations, which are used in fuzzy set theory. Recently, there appeared two kind of their generalizations. One is a 2-uninorm, which generalizes both a nullnorm and a uninorm. Such generalization, further extending to the n-uninorm, was introduced by P. Akella in [2]. A 2-uninorm belongs to the class of increasing, associative and commutative binary operators on the unit interval with an absorbing element separating two subintervals having their own neutral elements. The other is a semi-uninorm—a generalization of a uninorm by omitting commutativity and associativity, which was introduced by Drewniak et al. to study distributivity between a uninorm and a nullnorm [9].

This paper is organized as follows. In Sect. 2, we recall the structures of uninorms, semi-uninorms and 2-uninorms. Over here, we also recall the characterization of three subclasses of 2-uninorms and the functional equation of distributivity. In Sect. 3, the main section, we investigate distributivity for a 2-uninorm over a semi-uninorm and give full characterization for the above-mentioned three subclasses. Section 4 is conclusion and further work.

2 Preliminaries

In this section, we recall basic definitions and facts to be used later in the paper.

Definition 1 ([30]) Let $s \in [0, 1]$. A binary operator $U : [0, 1]^2 \to [0, 1]$ is called a *uninorm* if it is commutative, associative, non-decreasing in each variable, and there exists an element $s \in [0, 1]$ called neutral element such that $U(s, x) = x$ for all $x \in [0, 1]$.

It is clear that U becomes a *t-norm* when $s = 1$, while U becomes a *t-conorm* when $s = 0$ (see [17]). For any uninorm U, we have $U(0, 1) \in \{0, 1\}$, and U is called *conjunctive* when $U(0, 1) = 0$ and *disjunctive* when $U(0, 1) = 1$. By $\mathbf{U}_s$ we denote the family of all uninorms with the neutral element $s \in [0, 1]$. Now we recall the general structure of a uninorm (for more details see [6, 10, 15]) by using the notation $D_s = [0, s) \times (s, 1] \cup (s, 1] \times [0, s)$ for $s \in [0, 1]$.

Theorem 1 ([15]) *Let* $s \in [0, 1]$. *Then,* $U \in \mathbf{U}_s$ *if and only if*

$$U = \begin{cases} T_U & \text{if } (x, y) \in [0, s]^2, \\ S_U & \text{if } (x, y) \in [s, 1]^2, \\ C & \text{if } (x, y) \in D_s, \end{cases} \tag{1}$$

where T_U *and* S_U *are respectively isomorphic with a t-norm and a t-conorm, the increasing operation* $C : D_s \to [0, 1]$ *fulfills* $\min(x, y) \leqslant C(x, y) \leqslant \max(x, y)$ *for* $(x, y) \in D_s$.

Theorem 2 ([15]) *Let* $U : [0, 1]^2 \to [0, 1]$ *be a uninorm with the neutral element* $s \in (0, 1)$.

(i) *If* $U(0, 1) = 0$ *and the function* $U(x, 1)$ *is continuous except for the point* $x = s$, *then* $C = \min$ *in Eq.* (1) *and the class of such uninorms is denoted by* $\mathbf{U}_s^{\min}$.
(ii) *If* $U(0, 1) = 1$ *and the function* $U(x, 0)$ *is continuous except for the point* $x = s$, *then* $C = \max$ *in Eq.* (1) *and the class of such uninorms is denoted by* $\mathbf{U}_s^{\max}$.

Definition 2 ([6]) An element $a \in [0, 1]$ is called an *idempotent element* of $F : [0, 1]^2 \to [0, 1]$ if $F(a, a) = a$. The operation F is called *idempotent* if all elements from $[0, 1]$ are idempotent.

Theorem 3 ([6]) *Let* $s \in [0, 1]$. *Then, the operations*

$$U_s^{\min} = \begin{cases} \max & \text{if } (x, y) \in [s, 1]^2, \\ \min & \text{elsewhere}, \end{cases} \tag{2}$$

and

$$U_s^{\max} = \begin{cases} \min & \text{if } (x, y) \in [0, s]^2, \\ \max & \text{elsewhere}, \end{cases} \tag{3}$$

are unique idempotent uninorms in $\mathbf{U}_s^{\min}$ *and* $\mathbf{U}_s^{\max}$, *respectively.*

Definition 3 ([9]) A binary operator $V : [0, 1]^2 \to [0, 1]$ is called a *semi-uninorm* if it is non-decreasing in each variable, and there exists an element $s \in [0, 1]$ called neutral element such that $V(s, x) = x$ for all $x \in [0, 1]$.

It is clear that a commutative and associative semi-uninorm is a uninorm. By $\mathbf{V}_s$ we denote the family of all semi-uninorms with the neutral element $s \in [0, 1]$. A semi-uninorm V is called a *semi-t-norm* if $e = 1$, while V is called a *semi-t-conorm* if $e = 0$. For any $V \in \mathbf{V}_s$, we have $V(0, 0) = 0$ and $V(1, 1) = 1$. Therefore, V is a binary aggregation operator with the neutral element $s \in [0, 1]$. Moreover, a semi-uninorm V is said to be *conjunctive* if $V(0, 1) = V(1, 0) = 0$ and *disjunctive* if $V(0, 1) = V(1, 0) = 1$. In fact, for any conjunctive semi-uninorm V, it holds that $V(0, x) = V(x, 0) = 0$ for all $x \in [0, 1]$, while for any disjunctive semi-uninorm V, it follows that $V(1, x) = V(x, 1) = 1$ for all $x \in [0, 1]$.

Now we recall the structure of a semi-uninorm, which has also been given by Drewniak et al. [9] in another form.

Theorem 4 ([9]) *Let* $s \in [0, 1]$. *Then,* $V \in \mathbf{V}_s$ *if and only if*

$$V = \begin{cases} T_V & \text{if } (x, y) \in [0, s]^2, \\ S_V & \text{if } (x, y) \in [s, 1]^2, \\ C & \text{if } (x, y) \in D_s, \end{cases} \tag{4}$$

where T_V *and* S_V *are respectively isomorphic with a semi-t-norm and a semi-t-conorm, the increasing operation* $C\colon D_s \to [0,1]$ *fulfills* $\min(x,y) \leqslant C(x,y) \leqslant \max(x,y)$ *for* $(x,y) \in D_s$*. Moreover,* T_V *and* S_V *is called the underlying semi-t-norm and semi-t-conorm of* V*, respectively.*

By $\mathbf{V}_s^{\max}$ ($\mathbf{V}_s^{\min}$) we denote the family of all semi-uninorms with the same neutral element $s \in (0,1)$ fulfilling the additional condition: $V(0,x) = V(x,0) = 0$ for all $x \in (s,1]$ ($V(1,x) = V(x,1) = 1$ for all $x \in [0,s)$) [9].

From Theorems 4 and 7 in [9], we can obtain structures of elements in $\mathbf{V}_s^{\max}$ and $\mathbf{V}_s^{\min}$.

Theorem 5 ([9]) *Let* $V \in \mathbf{V}_s$ *with the neutral element* $s \in (0,1)$*. Then,*

(i) $V \in \mathbf{V}_s^{\min}$ *if and only if*

$$V = \begin{cases} T_V & \text{if } (x,y) \in [0,s]^2, \\ S_V & \text{if } (x,y) \in [s,1]^2, \\ \min & \text{if } (x,y) \in D_s, \end{cases} \tag{5}$$

(ii) $V \in \mathbf{V}_s^{\max}$ *if and only if*

$$V = \begin{cases} T_V & \text{if } (x,y) \in [0,s]^2, \\ S_V & \text{if } (x,y) \in [s,1]^2, \\ \max & \text{if } (x,y) \in D_s, \end{cases} \tag{6}$$

where T_V *and* S_V *are isomorphic with a semi-t-norm and a semi-t-conorm, respectively.*

Theorem 6 ([9]) *The operators* $U_s^{\min}$ *and* $U_s^{\max}$ *in Theorem 3 are unique idempotent semi-uninorms in* $\mathbf{V}_s^{\min}$ *and* $\mathbf{V}_s^{\max}$*, respectively.*

Now we recall the definitions and some results of 2-uninorms.

Definition 4 ([2]) Let $0 \leqslant e \leqslant k \leqslant f \leqslant 1$. An operator $G : [0,1]^2 \to [0,1]$ is called a 2-uninorm, if it is commutative, associative, non-decreasing with respect to both variables, and fulfilling

$$G(e,x) = x \text{ for all } x \in [0,k] \text{ and } G(f,x) = x \text{ for all } x \in [k,1]. \tag{7}$$

By $\mathbf{U}_{k(e,f)}$ we denote the class of all 2-uninorms.

Remark 1 ([12]) Any operator $G \in \mathbf{U}_{k(e,f)}$ fulfills the condition:

$$G(k,x) = k \text{ for all } x \in [e,f]. \tag{8}$$

Lemma 1 ([2]) *Let $G \in \mathbf{U}_{k(e,f)}$ be a 2-uninorm with $0 \leqslant e \leqslant k \leqslant f \leqslant 1$ and $k \in (0, 1)$. Then, two mappings U_1 and U_2 defined by, for $x, y \in [0, 1]$,*

$$U_1(x, y) = \frac{G(kx, ky)}{k} \tag{9}$$

and

$$U_2(x, y) = \frac{G(k + (1 - k)x, k + (1 - k)y) - k}{1 - k} \tag{10}$$

are uninorms with the neutral elements $\frac{e}{k}$ and $\frac{f-k}{1-k}$, respectively.

Lemma 2 ([12]) *Let $G \in \mathbf{U}_{k(e,f)}$ be a 2-uninorm with $0 \leqslant e \leqslant k \leqslant f \leqslant 1$. Then,*

(i) *$G(x, 0)$ is continuous except for the point e if and only if $U_1(x, 0)$ is continuous except for the point $\frac{e}{k}$.*
(ii) *$G(x, 1)$ is continuous except for the point f if and only if $U_2(x, 1)$ is continuous except for the point $\frac{f-k}{1-k}$.*

Lemma 3 ([2]) *Let $G \in \mathbf{U}_{k(e,f)}$ be a 2-uninorm with $0 \leqslant e \leqslant k \leqslant f \leqslant 1$. Then, $G(0, 1) \in \{0, k, 1\}$.*

Depending on the values of $G(0, 1)$, which plays a zero element role for the operator G, we obtain from the above lemmas that three subclass of operators in $\mathbf{U}_{k(e,f)}$ are respectively denoted by $\mathbf{C}^0_{k(e,f)}$, $\mathbf{C}^k_{k(e,f)}$, $\mathbf{C}^1_{k(e,f)}$ (or simplifying them $\mathbf{C}^0$, $\mathbf{C}^k$, $\mathbf{C}^1$).

Theorem 7 ([2]) *Let $G \in \mathbf{U}_{k(e,f)}$ be a 2-uninorm such that $G(x, 1)$ is discontinuous only at the points e and f. Then, $G \in \mathbf{C}^0$ and $G(1, k) = k$ if and only if $0 < e \leqslant k < f \leqslant 1$, and G has the following form*

$$G(x, y) = \begin{cases} T^{c_1}(x, y) & \text{if } (x, y) \in [0, e]^2, \\ S^{c_1}(x, y) & \text{if } (x, y) \in [e, k]^2, \\ T^{c_2}(x, y) & \text{if } (x, y) \in [k, f]^2, \\ S^{c_2}(x, y) & \text{if } (x, y) \in [f, 1]^2, \\ \min(x, y) & \text{if } (x, y) \in [0, e) \times (e, 1] \cup (e, 1] \times [0, e) \\ & \quad \cup [k, f) \times (f, 1] \cup (f, 1] \times [k, f), \\ k & \text{if } (x, y) \in [e, k) \times (k, 1] \cup (k, 1] \times [e, k), \end{cases} \tag{11}$$

where T^{c_1} and T^{c_2} are isomorphic with t-norms, S^{c_1} and S^{c_2} are isomorphic with t-conorms. We denote the set of all such 2-uninorms by $\mathbf{C}^0_k$.

Theorem 8 ([2]) *Let $G \in \mathbf{U}_{k(e,f)}$ be a 2-uninorm such that $G(x, 1)$ is discontinuous only at the point e, and $G(x, e)$ is discontinuous only at the point f. Then, $G \in \mathbf{C}^0$ and $G(1, k) = 1$ if and only if $0 < e \leqslant k \leqslant f < 1$, and G has the following form*

$$G(x,y)=\begin{cases} T^c(x,y) & \textit{if } (x,y)\in[0,e]^2,\\ S^c(x,y) & \textit{if } (x,y)\in[e,k]^2,\\ T^d(x,y) & \textit{if } (x,y)\in[k,f]^2,\\ S^d(x,y) & \textit{if } (x,y)\in[f,1]^2,\\ \min(x,y) & \textit{if } (x,y)\in[0,e)\times(e,1]\cup(e,1]\times[0,e),\\ \max(x,y) & \textit{if } (x,y)\in[e,f)\times(f,1]\cup(f,1]\times[e,f),\\ k & \textit{if } (x,y)\in[e,k)\times(k,f]\cup(k,f]\times[e,k), \end{cases} \tag{12}$$

where T^c and T^d are isomorphic with t-norms, S^c and S^d are isomorphic with t-conorms. We denote the set of all such 2-uninorms by $\mathbf{C}_1^0$.

Theorem 9 ([2]) *Let $G \in \mathbf{U}_{k(e,f)}$ be a 2-uninorm such that $G(x,0)$ is discontinuous only at the points e and f. Then, $G \in \mathbf{C}^1$ and $G(0,k) = k$ if and only if $0 \leqslant e < k \leqslant f < 1$, and G has the following form*

$$G(x,y)=\begin{cases} T^{d_1}(x,y) & \textit{if } (x,y)\in[0,e]^2,\\ S^{d_1}(x,y) & \textit{if } (x,y)\in[e,k]^2,\\ T^{d_2}(x,y) & \textit{if } (x,y)\in[k,f]^2,\\ S^{d_2}(x,y) & \textit{if } (x,y)\in[f,1]^2,\\ \max(x,y) & \textit{if } (x,y)\in[0,f)\times(f,1]\cup(f,1]\times[0,f)\\ & \quad\cup[0,e)\times(e,k]\cup(e,k]\times[0,e),\\ k & \textit{if } (x,y)\in[0,k)\times(k,f]\cup(k,f]\times[0,k), \end{cases} \tag{13}$$

where T^{d_1} and T^{d_2} are isomorphic with t-norms, S^{d_1} and S^{d_2} are isomorphic with t-conorms. We denote the set of all such 2-uninorms by $\mathbf{C}_k^1$.

Theorem 10 ([2]) *Let $G \in \mathbf{U}_{k(e,f)}$ be a 2-uninorm such that $G(x,0)$ is discontinuous only at the point f, and $G(x,f)$ is discontinuous only at the point e. Then, $G \in \mathbf{C}^1$ and $G(0,k) = 0$ if and only if $0 < e \leqslant k \leqslant f < 1$, and G has the following form*

$$G(x,y)=\begin{cases} T^c(x,y) & \textit{if } (x,y)\in[0,e]^2,\\ S^c(x,y) & \textit{if } (x,y)\in[e,k]^2,\\ T^d(x,y) & \textit{if } (x,y)\in[k,f]^2,\\ S^d(x,y) & \textit{if } (x,y)\in[f,1]^2,\\ \min(x,y) & \textit{if } (x,y)\in[0,e)\times(e,f]\cup(e,f]\times[0,e),\\ \max(x,y) & \textit{if } (x,y)\in[0,f)\times(f,1]\cup(f,1]\times[0,f),\\ k & \textit{if } (x,y)\in[e,k)\times(k,f]\cup(k,f]\times[e,k), \end{cases} \tag{14}$$

where T^c and T^d are isomorphic with t-norms, S^c and S^d are isomorphic with t-conorms. We denote the set of all such 2-uninorms by $\mathbf{C}_0^1$.

Theorem 11 ([2]) *Let $G \in \mathbf{U}_{k(e,f)}$ be a 2-uninorm such that $G(x, 0)$ is discontinuous only at the point e, and $G(x, 1)$ is discontinuous only at the point f. $G \in \mathbf{C}^k$ if and only if $0 \leqslant e < k < f \leqslant 1$, and G has the following form*

$$G(x, y)) = \begin{cases} T^d(x, y) & \text{if } (x, y) \in [0, e]^2, \\ S^d(x, y) & \text{if } (x, y) \in [e, k]^2, \\ T^c(x, y) & \text{if } (x, y) \in [k, f]^2, \\ S^c(x, y) & \text{if } (x, y) \in [f, 1]^2, \\ \max(x, y) & \text{if } (x, y) \in [0, e) \times (e, k] \cup (e, k] \times [0, e), \\ \min(x, y) & \text{if } (x, y) \in [k, f) \times (f, 1] \cup (f, 1] \times [k, f), \\ k & \text{if } (x, y) \in [0, k) \times (k, 1] \cup (k, 1] \times [0, k), \end{cases} \tag{15}$$

where T^c and T^d are isomorphic with t-norms, S^c and S^d are isomorphic with t-conorms.

For convenience, we assume that all of the underlying operators T^c, T^d, T^{c_1}, T^{c_2}, T^{d_1}, T^{d_2} and S^c, S^d, S^{c_1}, S^{c_2}, S^{d_1}, S^{d_2} of 2-uninorms in this paper are continuous. Next, we consider the distributivity equation.

Definition 5 ([1]) Let $F,\ G : [0, 1]^2 \to [0, 1]$. We say that G is *distributive* over F, if for all $x,\ y, z \in [0, 1]$,

$$G(x, F(y, z)) = F(G(x, y), G(x, z)). \tag{16}$$

Lemma 4 ([25]) *Let $F : X^2 \to X$ have the right (left) neutral element e in a subset $\emptyset \neq Y \subset X$ (i.e., $\forall_{x \in Y}, F(x, e) = x$ ($F(e, x) = x$)). If the operation F is distributive over another operation $G : X^2 \to X$ fulfilling $G(e, e) = e$, then G is idempotent in Y.*

Lemma 5 ([25]) *If an operation $F : [0, 1]^2 \to [0, 1]$ with the neutral element $e \in [0, 1]$ is distributive over another operation $G : [0, 1]^2 \to [0, 1]$ fulfilling $G(e, e) = e$, then G is idempotent.*

Lemma 6 ([25]) *Every increasing operation $G : [0, 1]^2 \to [0, 1]$ is distributive over* max *and* min.

3 Distributivity for a 2-Uninorm over a Semi-uninorm

Lemma 7 *Let $0 \leqslant e \leqslant k \leqslant f \leqslant 1$. If a 2-uninorm G is distributive over a semi-uninorm F with the neutral element $s \in [0, 1]$, then $G(s, s) = s$.*

Remark 2 In this paper, we always assume that the semi-uninorm F has neutral element $s \in (0, 1)$ because Ref. [12] has discussed the cases $s = 0$ and $s = 1$. That is, F is a semi-t-norm or a semi-t-conorm.

To completely characterize distributivity for a 2-uninorm G over a semi-uninorm F, according to Theorems from 7 to 11, there are five cases to be consider: (1) $G \in \mathbf{C}_k^0$; (2) $G \in \mathbf{C}_1^0$; (3) $G \in \mathbf{C}_k^1$; (4) $G \in \mathbf{C}_0^1$; (5) $G \in \mathbf{C}^k$. First, let us consider Case (1): $G \in \mathbf{C}_k^0$.

3.1 $G \in \mathbf{C}_k^0$

Lemma 8 *Let $G \in \mathbf{C}_k^0$ be a 2-uninorm with $0 < e \leqslant k < f \leqslant 1$, and let $F \in \mathbf{V}_s$ be a semi-uninorm with the neutral element $s \in (0, 1)$ such that the underlying semi-t-norm T_F and semi-t-conorm S_F are continuous. If G is distributive over F, then F is idempotent.*

So far, we know from Lemma 8 that the structure of F is completely determined. Therefore, the rest of this investigation requires that we characterize the operator G. Furthermore, we can only consider the case $e < k$ because the case $e = k$ is fully similar and much easier. Note that the assumption $0 < e < k < f \leqslant 1$ and the order relationship between s and e, k, f, then there are four cases: (1) $s \in (0, e]$; (2) $s \in (e, k]$; (3) $s \in (k, f]$; (4) $s \in (f, 1)$. The following lemma shows that two cases (2) and (4) are impossible.

Lemma 9 *Let $G \in \mathbf{C}_k^0$ be a 2-uninorm with $0 < e < k < f \leqslant 1$, and let $F \in \mathbf{V}_s$ be a semi-uninorm with the neutral element $s \in (0, 1)$ and $F(1, x) = F(x, 1) = x$ for $x \in [0, s)$ such that the underlying semi-t-norm T_F and semi-t-conorm S_F are continuous. If G is distributive over F, then $s \in (0, e]$ or $s \in (k, f]$.*

Theorem 12 *Let $G \in \mathbf{C}_k^0$ be a 2-uninorm with $0 < e < k < f \leqslant 1$, and let $F \in \mathbf{V}_s$ be a semi-uninorm with the neutral element $s \in (0, 1)$ and $F(1, x) = F(x, 1) = x$ for $x \in [0, s)$ such that the underlying semi-t-norm T_F and semi-t-conorm S_F are continuous. Then G is distributive over F if and only if $F = U_s^{\min}$ and one of the following two cases holds.*

(i) *$s \leqslant e$ and the structure of G is*

$$G(x, y) = \begin{cases} T_1^{c_1}(x, y) & \text{if } (x, y) \in [0, s]^2, \\ T_2^{c_1}(x, y) & \text{if } (x, y) \in [s, e]^2, \\ S^{c_1}(x, y) & \text{if } (x, y) \in [e, k]^2, \\ U^{c_2}(x, y) & \text{if } (x, y) \in [k, 1]^2, \\ k & \text{if } (x, y) \in [e, k) \times (k, 1] \cup (k, 1] \times [e, k), \\ \min(x, y) & \text{otherwise}, \end{cases} \tag{17}$$

where $T_1^{c_1}$ and $T_2^{c_1}$ are isomorphic with t-norms, S^{c_1} is isomorphic with a t-conorm.

(ii) $k < s \leqslant f$ *and the structure of* G *is*

$$G(x,y)=\begin{cases} U^{c_1}(x,y) & \text{if } (x,y)\in[0,k]^2,\\ T_1^{c_2}(x,y) & \text{if } (x,y)\in[k,s]^2,\\ T_2^{c_2}(x,y) & \text{if } (x,y)\in[s,f]^2,\\ S^{c_2}(x,y) & \text{if } (x,y)\in[f,1]^2,\\ k & \text{if } (x,y)\in[e,k)\times(k,1]\cup(k,1]\times[e,k),\\ \min(x,y) & \text{otherwise}, \end{cases} \tag{18}$$

where $T_1^{c_2}$ *and* $T_2^{c_2}$ *are isomorphic with t-norms,* S^{c_2} *is isomorphic with a t-conorm.*

Theorem 13 *Let* $G \in \mathbf{C}_k^0$ *be a 2-uninorm with* $0 < e \leqslant k < f \leqslant 1$*, and let* $F \in \mathbf{V}_s$ *be a semi-uninorm with the neutral element* $s \in (0,1)$ *and* $F(0,x) = F(x,0) = x$ *for* $x \in (s,1]$ *such that the underlying semi-t-norm* T_F *and semi-t-conorm* S_F *are continuous. Then* G *is not distributive over* F.

3.2 $G \in \mathbf{C}_k^1$

Lemma 10 *Let* $G \in \mathbf{C}_k^1$ *be a 2-uninorm with* $0 \leqslant e < k \leqslant f < 1$*, and let* $F \in \mathbf{V}_s$ *be a semi-uninorm with the neutral element* $s \in (0,1)$ *such that the underlying semi-t-norm* T_F *and semi-t-conorm* S_F *are continuous. If* G *is distributive over* F*, then* F *is idempotent.*

Next, we only consider the case $e < k$ because the case $e = k$ is similar and much easier.

Theorem 14 *Let* $G \in \mathbf{C}_k^1$ *be a 2-uninorm with* $0 \leqslant e < k < f < 1$*, and let* $F \in \mathbf{V}_s$ *be a semi-uninorm with the neutral element* $s \in (0,1)$ *and* $F(1,x) = F(x,1) = x$ *for* $x \in [0,s)$ *such that the underlying semi-t-norm* T_F *and semi-t-conorm* S_F *are continuous. Then* G *is not distributive over* F.

Lemma 11 *Let* $G \in \mathbf{C}_k^1$ *be a 2-uninorm with* $0 \leqslant e < k < f < 1$*, and let* $F \in \mathbf{V}_s$ *be a semi-uninorm with the neutral element* $s \in (0,1)$ *and* $F(0,x) = F(x,0) = x$ *for* $x \in (s,1]$ *such that the underlying semi-t-norm* T_F *and semi-t-conorm* S_F *are continuous. If* G *is distributive over* F*, then* $s \in [e,k)$ *or* $s \in [f,1)$.

Theorem 15 *Let* $G \in \mathbf{C}_k^1$ *be a 2-uninorm with* $0 \leqslant e < k < f < 1$*, and let* $F \in \mathbf{V}_s$ *be a semi-uninorm with the neutral element* $s \in (0,1)$ *and* $F(0,x) = F(x,0) = x$ *for* $x \in (s,1]$ *such that the underlying semi-t-norm* T_F *and semi-t-conorm* S_F *are continuous. Then* G *is distributive over* F *if and only if* $F = U_s^{\max}$ *and one of the following two cases holds.*

(i) $e \leqslant s < k$ *and the structure of* G *is*

$$G(x,y)=\begin{cases} T^{d_1}(x,y) & \textit{if } (x,y)\in[0,e]^2,\\ S_1^{d_1}(x,y) & \textit{if } (x,y)\in[e,s]^2,\\ S_2^{d_1}(x,y) & \textit{if } (x,y)\in[s,k]^2,\\ U^{d_2}(x,y) & \textit{if } (x,y)\in[k,1]^2,\\ k & \textit{if } (x,y)\in[0,k)\times(k,f]\cup(k,f]\times[0,k),\\ \max(x,y) & \textit{otherwise}, \end{cases} \tag{19}$$

where T^{d_1} *is isomorphic with a t-norm,* $S_1^{d_1}$ *and* $S_2^{d_1}$ *are isomorphic with t-conorms.*

(ii) $s \geqslant f$ *and the structure of* G *is*

$$G(x,y)=\begin{cases} U^{d_1}(x,y) & \textit{if } (x,y)\in[0,k]^2,\\ T^{d_2}(x,y) & \textit{if } (x,y)\in[k,f]^2,\\ S_1^{d_2}(x,y) & \textit{if } (x,y)\in[f,s]^2,\\ S_2^{d_2}(x,y) & \textit{if } (x,y)\in[s,1]^2,\\ k & \textit{if } (x,y)\in[0,k)\times(k,f]\cup(k,f]\times[0,k),\\ \max(x,y) & \textit{otherwise}, \end{cases} \tag{20}$$

where T^{d_2} *is isomorphic with a t-norm,* $S_1^{d_2}$ *and* $S_2^{d_2}$ *are isomorphic with t-conorms.*

3.3 $G \in \mathbf{C}_1^0$

Lemma 12 *Let* $G \in \mathbf{C}_1^0$ *be a 2-uninorm with* $0 < e \leqslant k \leqslant f < 1$, *and let* $F \in \mathbf{V}_s$ *be a semi-uninorm with the neutral element* $s \in (0,1)$ *such that the underlying semi-t-norm* T_F *and semi-t-conorm* S_F *are continuous. If* G *is distributive over* F, *then* F *is idempotent.*

Lemma 13 *Let* $G \in \mathbf{C}_1^0$ *be a 2-uninorm with* $0 < e < k < f < 1$, *and let* $F \in \mathbf{V}_s$ *be a semi-uninorm with the neutral element* $s \in (0,1)$ *and* $F(1,x) = F(x,1) = x$ *for* $x \in [0,s)$ *such that the underlying semi-t-norm* T_F *and semi-t-conorm* S_F *are continuous. If* G *is distributive over* F, *then* $s \in (0,e]$.

Theorem 16 *Let* $G \in \mathbf{C}_1^0$ *be a 2-uninorm with* $0 < e < k < f < 1$, *and let* $F \in V_s$ *be a semi-uninorm with the neutral element* $s \in (0,1)$ *and* $F(1,x) = F(x,1) = x$ *for* $x \in [0,s)$ *such that the underlying semi-t-norm* T_F *and semi-t-conorm* S_F *are* *continuous. Then G is distributive over F if and only if $F = U^{\min}$ and the structure of G is*

$$G(x, y) = \begin{cases} T_1^c(x, y) & \text{if } (x, y) \in [0, s]^2, \\ T_2^c(x, y) & \text{if } (x, y) \in [s, e]^2, \\ S^c(x, y) & \text{if } (x, y) \in [e, k]^2, \\ U^d(x, y) & \text{if } (x, y) \in [k, 1]^2, \\ k & \text{if } (x, y) \in [e, k) \times (k, f] \cup (k, f] \times [e, k), \\ \max(x, y) & \text{if } (x, y) \in [e, k) \times (f, 1] \cup (f, 1] \times [e, k), \\ \min(x, y) & \text{otherwise}, \end{cases} \tag{21}$$

where $s \leqslant e$ and T_1^c and T_2^c are isomorphic with t-norms, S^c is isomorphic with a t-conorm.

Theorem 17 *Let $G \in \mathbf{C}_1^0$ be a 2-uninorm with $0 < e < k < f < 1$, and let $F \in \mathbf{V}_s$ be a semi-uninorm with the neutral element $s \in (0, 1)$ and $F(0, x) = F(x, 0) = x$ for $x \in (s, 1]$ such that the underlying semi-t-norm T_F and semi-t-conorm S_F are continuous. Then G is not distributive over F.*

3.4 $G \in \mathbf{C}_0^1$

Lemma 14 *Let $G \in \mathbf{C}_0^1$ be a 2-uninorm with $0 < e \leqslant k \leqslant f < 1$, and let $F \in \mathbf{V}_s$ be a semi-uninorm with the neutral element $s \in (0, 1)$ such that the underlying semi-t-norm T_F and semi-t-conorm S_F are continuous. If G is distributive over F, then F is idempotent.*

Next, we only consider the case $e < k < f$, the other cases $e = k$ ae $k = f$ are similar.

Theorem 18 *Let $G \in \mathbf{C}_0^1$ be a 2-uninorm with $0 < e < k < f < 1$, and let $F \in \mathbf{V}_s$ be a semi-uninorm with the neutral element $s \in (0, 1)$ and $F(1, x) = F(x, 1) = x$ for $x \in [0, s)$ such that the underlying semi-t-norm T_F and semi-t-conorm S_F are continuous. Then G is not distributive over F.*

Lemma 15 *Let $G \in \mathbf{C}_0^1$ be a 2-uninorm with $0 < e < k < f < 1$, and let $F \in \mathbf{V}_s$ be a semi-uninorm with the neutral element $s \in (0, 1)$ and $F(0, x) = F(x, 0) = x$ for $x \in (s, 1]$ such that the underlying semi-t-norm T_F and semi-t-conorm S_F are continuous. If G is distributive over F, then $s \in [f, 1)$.*

Theorem 19 *Let $G \in \mathbf{C}_0^1$ be a 2-uninorm with $0 < e < k < f < 1$, and let $F \in \mathbf{V}_s$ be a semi-uninorm with the neutral element $s \in (0, 1)$ and $F(0, x) = F(x, 0) = x$ for $x \in (s, 1]$ such that the underlying semi-t-norm T_F and semi-t-conorm S_F are* *continuous. Then G is distributive over F if and only if* $F = U_s^{\max}$ *and the structure of G is*

$$G(x,y)=\begin{cases} U^c(x,y) & \text{if } (x,y)\in[0,k]^2,\\ T^d(x,y) & \text{if } (x,y)\in[k,f]^2,\\ S_1^d(x,y) & \text{if } (x,y)\in[f,s]^2,\\ S_2^d(x,y) & \text{if } (x,y)\in[s,1]^2,\\ k & \text{if } (x,y)\in[e,k)\times(k,f]\cup(k,f]\times[e,k),\\ \min(x,y) & \text{if } (x,y)\in[0,e)\times(k,f]\cup(k,f]\times[0,e),\\ \max(x,y) & \text{otherwise}, \end{cases} \tag{22}$$

where $s \geqslant f$ *and* T^d *is isomorphic with a t-norm,* S_1^d *and* S_2^d *are isomorphic with t-conorms.*

3.5 $G \in \mathbf{C}^k$

Lemma 16 *Let* $G \in \mathbf{C}^k$ *be a 2-uninorm with* $0 < e \leqslant k < f \leqslant 1$, *and let* $F \in \mathbf{V}_s$ *be a semi-uninorm with the neutral element* $s \in (0, 1)$ *such that the underlying semi-t-norm* T_F *and semi-t-conorm* S_F *are continuous. If G is distributive over F, then F is idempotent.*

Next, without loss of generality, we only consider the case $e < k$.

Lemma 17 *Let* $G \in \mathbf{C}^k$ *be a 2-uninorm with* $0 < e < k < f \leqslant 1$, *and let* $F \in \mathbf{V}_s$ *be a semi-uninorm with the neutral element* $s \in (0, 1)$ *and* $F(1, x) = F(x, 1) = x$ *for* $x \in [0, s)$ *such that the underlying semi-t-norm* T_F *and semi-t-conorm* S_F *are continuous. If G is distributive over F, then* $s \in (k, f]$.

Theorem 20 *Let* $G \in \mathbf{C}^k$ *be a 2-uninorm with* $0 < e < k < f \leqslant 1$, *and let* $F \in \mathbf{V}_s$ *be a semi-uninorm with the neutral element* $s \in (0, 1)$ *and* $F(1, x) = F(x, 1) = x$ *for* $x \in [0, s)$ *such that the underlying semi-t-norm* T_F *and semi-t-conorm* S_F *are continuous. Then G is distributive over F if and only if* $F = U_s^{\min}$ *and the structure of G is*

$$G(x,y)=\begin{cases} U^d(x,y) & \text{if } (x,y)\in[0,k]^2,\\ T_1^c(x,y) & \text{if } (x,y)\in[k,s]^2,\\ T_2^c(x,y) & \text{if } (x,y)\in[s,f]^2,\\ S^c(x,y) & \text{if } (x,y)\in[f,1]^2,\\ k & \text{if } (x,y)\in[0,k)\times(k,1]\cup(k,1]\times[0,k),\\ \min(x,y) & \text{otherwise}, \end{cases} \tag{23}$$

where $k < s \leqslant f$ *and* T_1^c *and* T_2^c *are isomorphic with t-norms,* S^c *is isomorphic with a t-conorm.*

Lemma 18 *Let* $G \in \mathbf{C}^k$ *be a 2-uninorm with* $0 < e < k < f \leqslant 1$*, and let* $F \in \mathbf{V}_s$ *be a semi-uninorm with the neutral element* $s \in (0, 1)$ *and* $F(0, x) = F(x, 0) = x$ *for* $x \in (s, 1]$ *such that the underlying semi-t-norm* T_F *and semi-t-conorm* S_F *are continuous. If G is distributive over F, then* $s \in [e, k)$.

Theorem 21 *Let* $G \in \mathbf{C}^k$ *be a 2-uninorm with* $0 < e < k < f \leqslant 1$*, and let* $F \in \mathbf{V}_s$ *be a semi-uninorm with the neutral element* $s \in (0, 1)$ *and* $F(0, x) = F(x, 0) = x$ *for* $x \in (s, 1]$ *such that the underlying semi-t-norm* T_F *and semi-t-conorm* S_F *are continuous. Then G is distributive over F if and only if* $F = U_s^{\max}$ *and one of the structure of G is*

$$G(x, y) = \begin{cases} T^d(x, y) & \textit{if } (x, y) \in [0, e]^2, \\ S_1^d(x, y) & \textit{if } (x, y) \in [e, s]^2, \\ S_2^d(x, y) & \textit{if } (x, y) \in [s, k]^2, \\ U^c(x, y) & \textit{if } (x, y) \in [k, 1]^2, \\ k & \textit{if } (x, y) \in [0, k) \times (k, 1] \cup (k, 1] \times [0, k), \\ \max(x, y) & \textit{otherwise}, \end{cases} \tag{24}$$

where $e \leqslant s < k$ *and* T^d *is isomorphic with a t-norm,* S_1^d *and* S_2^d *are isomorphic with t-conorms.*

4 Conclusions and Further Work

In this paper, we have investigated distributivity for 2-uninorms over semi-uninorms. In fact, 2-uninorms cover both uninorms and nullnorms, which form a class of commutative, associative and increasing operators on the unit interval with an absorbing element separating two subintervals having their own neutral elements. While semi-uninorms are generalizations of uninorms by omitting commutativity and associativity. Moreover, all possible solutions of the distributivity equation for the three defined subclasses of 2-uninorms over semi-uninorms are characterized. In future work, we will concentrate on the converse, that is, distributivity for semi-uninorms over 2-uninorms. Indeed, this problem is very difficult because we are not sure that the second operator, namely, the 2-uninorm, is idempotent.

Acknowledgments This research is supported by the National Natural Science Foundation of China (No. 61563020) and Jiangxi Natural Science Foundation (No. 20151BAB201019).

References

1. Aczél, J.: Lectures on Functional Equations and Their Applications. Academic Press, New York (1966)
2. Akella, P.: Structure of n-uninorms. Fuzzy Sets Syst. **159**, 1631–1651 (2007)
3. Calvo, T.: On some solutions of the distributivity equation. Fuzzy Sets Syst. **104**, 85–96 (1999)
4. Calvo, T., De Baets, B.: On the generalization of the absorption equation. Fuzzy Math. **8**(1), 141–149 (2000)
5. Calvo, T., De Baets, B., Fodor, J.C.: The functional equations of Frank and Alsina for uninorms and nullnorms. Fuzzy Sets Syst. **120**, 385–394 (2001)
6. De Baets, B.: Idempotent uninorms. Eur. J. Op. Res. **118**, 631–642 (1999)
7. Drewniak, J., Rak, E.: Subdistributivity and superdistributivity of binary operations. Fuzzy Sets Syst. **161**, 189–210 (2010)
8. Drewniak, J., Rak, E.: Distributivity inequalities of monotonic operations. Fuzzy Sets Syst. **191**, 62–71 (2012)
9. Drewniak, J., Drygaś, P., Rak, E.: Distributivity equations for uninorms and nullnorms. Fuzzy Sets Syst. **159**, 1646–1657 (2008)
10. Drygaś, P.: On properties of uninorms with underlying t-norm and t-conorm given as ordinal sums. Fuzzy Sets Syst. **161**, 149–157 (2010)
11. Drygaś, P.: Distributivity between semi-t-operators and semi-nullnorms. Fuzzy Sets Syst. **264**, 100–109 (2015)
12. Drygaś, P., Rak, E.: Distributivity equation in the class of 2-uninorms. Fuzzy Sets Syst. **291**, 82–97 (2016)
13. Drygaś, P., Rak, E., Zedam, L.: Distributivity of aggregation operators with 2-neutral elements. In: Baczynski M., De Baets, B., Mesiar, R. (eds.) Proceedings of the 8th International Summer School on Aggregation Operators (AGOP 2015), University of Silesia, Katowice, Poland (2015)
14. Dubois, D., Prade, H.: Fundamentals of Fuzzy Sets. Kluwer Academic Publishers, Boston (2000)
15. Fodor, J.C., Yager, R., Rybalov, A.: Structure of uninorms. Int. J. Uncertain. Fuzziness Knowl.-Based Syst. **5**, 411–427 (1997)
16. Gabbay, D., Metcalfe, G.: Fuzzy logics based on [0, 1)-continuous uninorms. Arch. Math. Log. **46**, 425–449 (2007)
17. Klement, E.P., Mesiar, R., Pap, E.: Triangular Norms. Kluwer, Dordrecht (2000)
18. Klir, G.J., Yuan, B.: Fuzzy Sets and Fuzzy Logic. Theory and Application. Prentice Hall PTR, Upper Saddle River (1995)
19. Liu, H.W.: Semi-uninorms and implications on a complete lattice. Fuzzy Sets Syst. **191**, 72–82 (2012)
20. Liu, H.W.: Distributivity and conditional distributivity of semi-uninorms over continuous t-conorms and t-norms. Fuzzy Sets Syst. **268**, 27–43 (2015)
21. Mas, M., Mayor, G., Torrens, J.: The distributivity condition for uninorms and t-operators. Fuzzy Sets Syst. **128**, 209–225 (2002)
22. Mas, M., Mayor, G., Torrens, J.: The modularity condition for uninorms and t-operators. Fuzzy Sets Syst. **126**, 207–218 (2002)
23. Qin, F.: Uninorm solutions and (or)nullnorm solutions to the modularity condition equations. Fuzzy Sets Syst. **148**, 231–242 (2004)
24. Qin, F., Zhao, B.: The distributive equations for idempotent uninorms and nullnorms. Fuzzy Sets Syst. **155**, 446–458 (2005)
25. Rak, E.: Distributivity equation for nullnorms. J. Electron. Eng. **56**(12/s), 53–55 (2005)
26. Wang, S.M.: Uninorm logic with the n-potency axiom. Fuzzy Sets Syst. **205**, 116–126 (2012)
27. Wang, S.M.: Involutive uninorm logic with the n-potency axiom. Fuzzy Sets Syst. **218**, 1–23 (2013)

28. Wang, S.M.: Logics for residuated pseudo-uninorms and their residua. Fuzzy Sets Syst. **218**, 24–31 (2013)
29. Yager, R.R.: Uninorms in fuzzy system modeling. Fuzzy Sets Syst. **122**, 167–175 (2001)
30. Yager, R.R., Rybalov, A.: Uninorm aggregation operators. Fuzzy Sets Syst. **80**, 111–120 (1996)

Further Studies of Left (Right) Semi-uninorms on a Complete Lattice

Yuan Wang, Ke-Ming Tang and Zhu-Deng Wang

Abstract In this paper, we lay bare some new formulas for calculating the upper and lower approximation left (right) semi-uninorms of a binary operation and further discuss the relations between the upper and lower approximation left (right) semi-uninorms of a given binary operation and the lower and upper approximation left (right) semi-uninorms of its dual operation, respectively.

Keywords Fuzzy connective · Uninorm · Left (right) semi-uninorm · Upper (lower) approximation

1 Introduction

Uninorms, introduced by Yager and Rybalov [14], and studied by Fodor et al. [3], are special aggregation operators that have been proven useful in many fields like fuzzy logic, expert systems, neural networks, aggregation, and fuzzy system modeling [12, 13]. Uninorms are interesting because their structure is a special combination of t-norms and t-conorms [3]. It is well known that a uninorm U can be conjunctive or disjunctive whenever $U(0, 1) = 0$ or 1, respectively. This fact allows us to use uninorms in defining fuzzy implications and coimplications [7, 8].

There are real-life situations when truth functions can not be associative or commutative. By throwing away the commutativity from the axioms of uninorms,

Y. Wang · K.-M. Tang
College of Information Engineering, Yancheng Teachers University, Jiangsu 224002, People's Republic of China
e-mail: yctuwangyuan@163.com

K.-M. Tang
e-mail: tkmchina@126.com

Z.-D. Wang (✉)
School of Mathematics and Statistics, Yancheng Teachers University, Jiangsu 224002, People's Republic of China
e-mail: zhudengwang2004@163.com

T.-H. Fan et al. (eds.), *Quantitative Logic and Soft Computing 2016*,
Advances in Intelligent Systems and Computing 510,
DOI 10.1007/978-3-319-46206-6_32

Mas et al. [5, 6] introduced the concepts of left and right uninorms, Wang and Fang [10, 11] studied the residual operators and the residual coimplicators of left (right) uninorms on a complete lattice. By removing the associativity and commutativity from the axioms of uninorms, Liu [4] introduced the concept of semi-uninorms on a complete lattice.

Recently, Su et al. [9] generalized the concepts of both left (right) uninorms and semi-uninorms, introduced the left (right) semi-uninorm, and laid bare the formulas for calculating the upper and lower approximation left (right) semi-uninorms of a given binary operation on a complete lattice. In this paper, we further study the concepts of left (right) semi-uninorms on a complete lattice. We will improve Theorems 3.6, 3.7, 4.6 and 4.7 in [9], correct Example 3.8 in [9], give out some new formulas for calculating the upper and lower approximation left (right) semi-uninorms of a binary operation, and further discuss the relations between the upper and lower approximation left (right) semi-uninorms of a given binary operation and the lower and upper approximation left (right) semi-uninorms of its dual operation, respectively.

The knowledge about lattices required in this paper can be found in [1].

Throughout this paper, unless otherwise stated, L always represents any given complete lattice with maximal element 1 and minimal element 0; J stands for any index set.

2 Left (Right) Semi-uninorms on a Complete Lattice

In this section, we briefly recall some necessary concepts about the left (right) semi-uninorms and illustrate these notions by means of an example and two theorems.

Definition 1 (*Su et al.* [9]) A binary operation U on L is called a left (right) semi-uninorm if it satisfies the following two conditions:
(U1) there exists a left (right) neutral element, i.e., an element $e_L \in L$ $(e_R \in L)$ satisfying $U(e_L, x) = x$ $(U(x, e_R) = x)$ for all $x \in L$,
(U2) U is non-decreasing in each variable.

If a left (right) semi-uninorm U on L is associative, then U is a left (right) uninorm [10]. If a left (right) semi-uninorm U with left (right) neutral element e_L (e_R) has a right (left) neutral element e_R (e_L), then $e_L = U(e_L, e_R) = e_R$. Let $e = e_L = e_R$. Then U is a semi-uninorm [4].

Definition 2 (*Wang and Fang* [10, 11]) Let J be any index set. A binary operation U on L is called left (right) infinitely $\vee$-distributive if

$$U\Big(\bigvee_{j\in J} x_j, y\Big) = \bigvee_{j\in J} U(x_j, y) \ \Big(U\Big(x, \bigvee_{j\in J} y_j\Big) = \bigvee_{j\in J} U(x, y_j)\Big) \quad \forall x, y, x_j, y_j \in L;$$

left (right) infinitely $\wedge$-distributive if

$$U\Big(\bigwedge_{j\in J} x_j, y\Big) = \bigwedge_{j\in J} U(x_j, y) \ \Big(U\Big(x, \bigwedge_{j\in J} y_j\Big) = \bigwedge_{j\in J} U(x, y_j)\Big) \ \forall x, y, x_j, y_j \in L.$$

If a binary operation U is left infinitely $\vee$-distributive ($\wedge$-distributive) and also right infinitely $\vee$-distributive ($\wedge$-distributive), then U is said to be infinitely $\vee$-distributive ($\wedge$-distributive).

Noting that the least upper bound of the empty set is 0 and the greatest lower bound of the empty set is 1, we have that

$$U(0, y) = U\Big(\bigvee_{j\in\emptyset} x_j, y\Big) = \bigvee_{j\in\emptyset} U(x_j, y) = 0$$
$$\Big(U(x, 0) = U\Big(x, \bigvee_{j\in\emptyset} y_j\Big) = \bigvee_{j\in\emptyset} U(x, y_j) = 0\Big)$$

for any $x, y \in L$ when U is left (right) infinitely $\vee$-distributive and

$$U(1, y) = U\Big(\bigwedge_{j\in\emptyset} x_j, y\Big) = \bigwedge_{j\in\emptyset} U(x_j, y) = 1$$
$$\Big(U(x, 1) = U\Big(x, \bigwedge_{j\in\emptyset} y_j\Big) = \bigwedge_{j\in\emptyset} U(x, y_j) = 1\Big)$$

for any $x, y \in L$ when U is left (right) infinitely $\wedge$-distributive.

Example 1 (*Su et al.* [9]) Let $e_L, e_R \in L$,

$$U_{sW}^{e_L}(x, y) = \begin{cases} y \text{ if } x \geq e_L, \\ 0 \text{ otherwise}, \end{cases} \qquad U_{sM}^{e_L}(x, y) = \begin{cases} y \text{ if } x \leq e_L, \\ 1 \text{ otherwise}, \end{cases}$$

$$U_{sW}^{e_R}(x, y) = \begin{cases} x \text{ if } y \geq e_R, \\ 0 \text{ otherwise}, \end{cases} \qquad U_{sM}^{e_R}(x, y) = \begin{cases} x \text{ if } y \leq e_R, \\ 1 \text{ otherwise}, \end{cases}$$

where x and y are elements of L. Then $U_{sW}^{e_L}$ ($U_{sW}^{e_R}$) and $U_{sM}^{e_L}$ ($U_{sM}^{e_R}$) are, respectively, the smallest and greatest left (right) semi-uninorms; $U_{sW}^{e_L}$ and $U_{sW}^{e_R}$ are, respectively, the smallest right infinitely $\vee$-distributive left semi-uninorm and left infinitely $\vee$-distributive right semi-uninorm; and $U_{sM}^{e_L}$ and $U_{sM}^{e_R}$ are, respectively, the greatest right infinitely $\wedge$-distributive left semi-uninorm and left infinitely $\wedge$-distributive right semi-uninorm.

Definition 3 (*Su et al.* [9]) Let $A \in L^{L\times L}$. Define the upper approximation A_u and the lower approximation A_l of A as follows:

$$A_u(x, y) = \bigvee\{A(u, v) \mid u \leq x, v \leq y\} \ \forall x, y \in L,$$
$$A_l(x, y) = \bigwedge\{A(u, v) \mid u \geq x, v \geq y\} \ \forall x, y \in L.$$

By Definition 3, if A is non-decreasing in its first variable, then

$$A_u(x, y) = \bigvee\{A(x, v) \mid v \leq y\},\ A_l(x, y) = \bigwedge\{A(x, v) \mid v \geq y\}\ \forall x, y \in L;$$

if A is non-decreasing in its second variable, then

$$A_u(x, y) = \bigvee\{A(u, y) \mid u \leq x\},\ A_l(x, y) = \bigwedge\{A(u, y) \mid u \geq x\}\ \forall x, y \in L.$$

Theorem 1 *Let* $A \in L^{L\times L}$.
(1) *If* A *is left (right) infinitely* $\vee$*-distributive, then* A_u *is left (right) infinitely* $\vee$*-distributive.*
(2) *If* A *is left (right) infinitely* $\wedge$*-distributive, then* A_l *is left (right) infinitely* $\wedge$*-distributive.*

Proof We only prove that statement (1) holds.
If A is left infinitely $\vee$-distributive, then A is non-decreasing in its first variable and for any index set J,

$$A_u(x, y) = \bigvee\{A(u, v) \mid u \leq x, v \leq y\} = \bigvee\{A(x, v) \mid v \leq y\}\ \forall x, y \in L,$$

$$A_u\Big(\bigvee_{j\in J} x_j, y\Big) = \bigvee\{A(\bigvee_{j\in J} x_j, v) \mid v \leq y\} = \bigvee\{\bigvee_{j\in J} A(x_j, v) \mid v \leq y\}$$

$$= \bigvee_{j\in J}\Big(\bigvee\{A(x_j, v) \mid v \leq y\}\Big) = \bigvee_{j\in J} A_u(x_j, y)\ \forall x_j, y \in L,$$

i.e., A_u is left infinitely $\vee$-distributive.
Similarly, we can show that A_u is right infinitely $\vee$-distributive when A is right infinitely $\vee$-distributive. □

Definition 4 (*De Baets* [2]) Consider a strong negation N on L. The N-dual operation of a binary operation A on L is the binary operation A_N on L defined by

$$A_N(x, y) = N^{-1}\Big(A(N(x), N(y))\Big)\ \forall x, y \in L.$$

Note that $(A_N)_{N^{-1}} = (A_N)_N = A$ for any binary operation A on L.
The following theorem collects some properties of the N-dual operation.

Theorem 2 (Su et al. [9]) *Let* A, B *be two binary operations and* N *a strong negation on* L*. Then the following statements hold:*
(1) $(A \wedge B)_N = A_N \vee B_N$ *and* $(A \vee B)_N = A_N \wedge B_N$.
(2) *If* A *is left (right) infinitely* $\vee$*-distributive, then* A_N *is left (right) infinitely* $\wedge$*-distributive.*
(3) *If* A *is left (right) infinitely* $\wedge$*-distributive, then* A_N *is left (right) infinitely* $\vee$*-distributive.*

(4) *If A is increasing (decreasing) in its ith variable, then A_N is increasing (decreasing) in its ith variable (i = 1, 2).*
(5) *The N-dual operation of a left (right) semi-uninorm with a left (right) neutral element e_L (e_R) is a left (right) semi-uninorm with a left (right) neutral element $N(e_L)$ ($N(e_R)$).*
(6) $(U_{sW}^{e_L})_N = U_{sM}^{N(e_L)}$, $(U_{sM}^{e_L})_N = U_{sW}^{N(e_L)}$, $(U_{sW}^{e_R})_N = U_{sM}^{N(e_R)}$ *and* $(U_{sM}^{e_R})_N = U_{sW}^{N(e_R)}$.
(7) $(A_N)_u = (A_l)_N$ *and* $(A_N)_l = (A_u)_N$.

3 Main Results

In this section, we give out some new formulas for calculating the upper and lower approximation left (right) semi-uninorms of a binary operation and further discuss the relations between the upper and lower approximation left (right) semi-uninorms of a given binary operation and the lower and upper approximation left (right) semi-uninorms of its dual operation, respectively.

For a binary operation A on L, if there exists a left semi-uninorm U with the left neutral element e_L such that $A \leq U$, then it follows from Theorem 3.1 in [9] that

$$\bigwedge\{U \mid A \leq U, U \text{ is a left semi} - \text{uninorm with the left neutral element } e_L\}$$

is the smallest left semi-uninorm that is stronger than A on L, we call it the upper approximation left semi-uninorm of A and written as $[A)_s^{e_L}$; if there exists a left semi-uninorm U with the left neutral element e_L such that $U \leq A$, then

$$\bigvee\{U \mid U \leq A, U \text{ is a left semi} - \text{uninorm with the left neutral element } e_L\}$$

is the largest left semi-uninorm that is weaker than A on L, we call it the lower approximation left semi-uninorm of A and written as $(A]_s^{e_L}$.

Similarly, we introduce the following symbols:
$[A)_s^{e_R}$: the upper approximation right semi-uninorm of A;
$(A]_s^{e_R}$: the lower approximation right semi-uninorm of A;
$(A]_{s\wedge}^{e_L}$: the right infinitely $\wedge$-distributive lower approximation left semi-uninorm of A;
$(A]_{\wedge s}^{e_R}$: the left infinitely $\wedge$-distributive lower approximation right semi-uninorm of A;
$[A)_{s\vee}^{e_L}$: the right infinitely $\vee$-distributive upper approximation left semi-uninorm of A;
$[A)_{\vee s}^{e_R}$: the left infinitely $\vee$-distributive upper approximation right semi-uninorm of A.

By removing the condition A is non-decreasing in its first variable in Theorem 3.6 in [9], we have the following theorem.

Theorem 3 *Let $A \in L^{L\times L}$ and $e_L \in L$.*
(1) If $A \leq U_{sM}^{e_L}$, then $[A)_s^{e_L} = U_{sW}^{e_L} \vee A_u$.
(2) If $U_{sW}^{e_L} \leq A$, then $(A]_s^{e_L} = U_{sM}^{e_L} \wedge A_l$.
(3) If $A \leq U_{sM}^{e_L}$ and A is right infinitely $\vee$-distributive, then $[A)_{s\vee}^{e_L} = U_{sW}^{e_L} \vee A_u$.
(4) If $U_{sW}^{e_L} \leq A$ and A is right infinitely $\wedge$-distributive, then $(A]_{s\wedge}^{e_L} = U_{sM}^{e_L} \wedge A_l$.

Proof The proofs of the statements (1) and (2) refer to the proofs of Theorem 3.6 (1) and (2) in [9].

(3) Let $U_3 = U_{sW}^{e_L} \vee A_u$. If $A \leq U_{sM}^{e_L}$, then $U_3 \in \mathcal{U}_s^{e_L}(L)$ by statement (1). Noting that A is right infinitely $\vee$-distributive, we can see that A_u is also right infinitely $\vee$-distributive by Theorem 1. Thus, U_3 is right infinitely $\vee$-distributive and $U_3 \in \mathcal{U}_{s\vee}^{e_L}(L)$. By the proof of Theorem 3.6 (1) in [9], we have that $[A)_{s\vee}^{e_L} = U_{sW}^{e_L} \vee A_u$.

(4) Let $U_4 = U_{sM}^{e_L} \wedge A_l$. If $U_{sW}^{e_L} \leq A$, then $U_3 \in \mathcal{U}_s^{e_L}(L)$ by statement (2). Noting that $U_{sM}^{e_L}$ and A are all right infinitely $\wedge$-distributive, we can see that U_3 is also right infinitely $\wedge$-distributive, i.e., $U_3 \in \mathcal{U}_{s\wedge}^{e_L}(L)$. Moreover, by the proof of Theorem 3.6 (2) in [9], we have $(A]_{s\wedge}^{e_L} = U_{sM}^{e_L} \wedge A_l$. □

Analogous to Theorem 3, by throwing away the condition that A is non-decreasing in its second variable in Theorem 3.7 in [9], we get the following theorem.

Theorem 4 *Let $A \in L^{L\times L}$ and $e_R \in L$.*
(1) If $A \leq U_{sM}^{e_R}$, then $[A)_s^{e_R} = U_{sW}^{e_R} \vee A_u$.
(2) If $U_{sW}^{e_R} \leq A$, then $(A]_s^{e_R} = U_{sM}^{e_R} \wedge A_l$.
(3) If $A \leq U_{sM}^{e_R}$ and A is left infinitely $\vee$-distributive, then $[A)_{\vee s}^{e_R} = U_{sW}^{e_R} \vee A_u$.
(4) If $U_{sW}^{e_R} \leq A$ and A is left infinitely $\wedge$-distributive, then $(A]_{\wedge s}^{e_R} = U_{sM}^{e_R} \wedge A_l$.

We see that $A(b, 1) = 0 \neq 1$ and $B(a, 0) = 1 \neq 0$ from Example 3.8 in [9]. Thus, A isn't right infinitely $\wedge$-distributive and B isn't right infinitely $\vee$-distributive.

Below, we correct Example 3.8 in [9].

Example 2 Let $L = \{0, a, b, 1\}$ be a lattice, where $0 < a < 1, 0 < b < 1, a \vee b = 1$ and $a \wedge b = 0$. Define two binary operations A and B on L as follows:

A	0	a	b	1
0	0	0	0	1
a	a	1	a	1
b	0	0	a	1
1	a	1	a	1

B	0	a	b	1
0	0	0	a	a
a	0	0	a	a
b	0	1	a	1
1	0	1	1	1

Clearly, $A \leq U_{sM}^{0_L}$, $U_{sW}^{1_L} \leq B$, A is non-decreasing in its first variable and right infinitely $\wedge$-distributive, and B is non-decreasing in its first variable and right infinitely $\vee$-distributive. Thus, $A_u = A$ and $B_l = B$. Let $U_1 = U_{sW}^{0_L} \vee A$ and $U_2 = U_{sM}^{1_L} \wedge B$. Then

U_1	0	a	b	1
0	0	a	b	1
a	a	1	1	1
b	0	a	1	1
1	a	1	1	1

U_2	0	a	b	1
0	0	0	0	a
a	0	0	0	a
b	0	a	0	1
1	0	a	b	1

Noting that $U_1(a, a \wedge b) = U_1(a, 0) = a \neq 1 = U_1(a, a) \wedge U_1(a, b)$ and $U_2(b, a \vee b) = U_2(b, 1) = 1 \neq a = U_2(b, a) \vee U_2(b, b)$, we see that U_1 isn't right infinitely $\wedge$-distributive and U_2 isn't right infinitely $\vee$-distributive. This shows that U_1 isn't the upper approximation right infinitely $\wedge$-distributive left semi-uninorm of A and U_2 isn't the lower approximation right infinitely $\vee$-distributive left semi-uninorm of B.

This example illustrates that analogous to Theorems 3 and 4 may not hold for calculating the upper approximation right (left) infinitely $\wedge$-distributive left (right) semi-uninorm and the lower approximation right (left) infinitely $\vee$-distributive left (right) semi-uninorm of a binary operation.

Now, we further investigate the relations between the upper and lower approximation left (right) semi-uninorms of a given binary operation and the lower and upper approximation left (right) semi-uninorms of its dual operation, respectively.

By virtue of Theorem 3, we can improve Theorem 4.6 in [9] by removing the condition that A is non-decreasing in its first variable and have the following theorem.

Theorem 5 *Let A, N and e_L be a binary operation, strong negation and fixed element on L, respectively. Then the following statements hold:*
(1) If $A \leq U_{sM}^{e_L}$, then $[A)_s^{e_L} = ((A_N]_s^{N(e_L)})_N$.
(2) If $U_{sW}^{e_L} \leq A$, then $(A]_s^{e_L} = ([A_N)_s^{N(e_L)})_N$.
(3) If $A \leq U_{sM}^{e_L}$ and A is right infinitely $\vee$-distributive, then

$$[A)_{s\vee}^{e_L} = ((A_N]_{s\wedge}^{N(e_L)})_N.$$

(4) If $U_{sW}^{e_L} \leq A$ and A is right infinitely $\wedge$-distributive, then

$$(A]_{s\wedge}^{e_L} = ([A_N)_{s\vee}^{N(e_L)})_N.$$

Proof We only prove that statements (1) and (3) hold.

(1) If $A \leq U_{sM}^{e_L}$, then $[A)_s^{e_L} = U_{sW}^{e_L} \vee A_u$ by Theorem 3 and $A_N \geq (U_{sM}^{e_L})_N = U_{sW}^{N(e_L)}$ by Theorem 2. Thus, $(A_N]_s^{N(e_L)} = U_{sM}^{N(e_L)} \wedge (A_N)_l$ by Theorem 3. Moreover, by virtue of Theorems 2 and 3, we see that

$$\begin{aligned}\big((A_N]_s^{N(e_L)}\big)_N &= \big(U_{sM}^{N(e_L)} \wedge (A_N)_l\big)_N = \big(U_{sM}^{N(e_L)} \wedge (A_u)_N\big)_N \\ &= (U_{sM}^{N(e_L)})_N \vee \big((A_u)_N\big)_N = U_{sW}^{e_L} \vee A_u = [A)_s^{e_L}.\end{aligned}$$

(3) If $A \leq U_{sM}^{e_L}$ and A is right infinitely $\vee$-distributive, then $[A)_{s\vee}^{e_L} = U_{sW}^{e_L} \vee A_u$ by Theorem 3, $A_N \geq (U_{sM}^{e_L})_N = U_{sW}^{N(e_L)}$ and A_N is right infinitely $\wedge$-distributive by Theorem 2. Thus, $(A_N]_{s\wedge}^{N(e_L)} = U_{sM}^{N(e_L)} \wedge (A_N)_l$ by Theorem 3. Moreover, we see that $[A)_{s\vee}^{e_L} = ((A_N]_{s\wedge}^{N(e_L)})_N$ by the proof of statement (1). □

Analogous to Theorem 5, by throwing away the condition that A is non-decreasing in its second variable in Theorem 4.7 in [9], we have the following theorem.

Theorem 6 *Let A, N and e_R be a binary operation, strong negation and fixed element on L, respectively. Then the following statements hold:*
(1) *If $A \leq U^{e_R}_{sM}$, then $[A)^{e_R}_s = ((A_N]^{N(e_R)}_s)_N$.*
(2) *If $U^{e_R}_{sW} \leq A$, then $(A]^{e_R}_s = ([A_N)^{N(e_R)}_s)_N$.*
(3) *If $A \leq U^{e_R}_{sM}$ and A is left infinitely $\vee$-distributive, then*

$$[A)^{e_R}_{\vee s} = ((A_N]^{N(e_R)}_{\wedge s})_N.$$

(4) *If $U^{e_R}_{sW} \leq A$ and A is left infinitely $\wedge$-distributive, then*

$$(A]^{e_R}_{\wedge s} = ([A_N)^{N(e_R)}_{\vee s})_N.$$

Acknowledgments The authors wish to thank Professor Daowu Pei from Zhejiang Sci-Tech University and the anonymous referees for their valuable comments and suggestions. This work is supported by the Jiangsu Provincial Natural Science Foundation of China (BK20161313) and National Natural Science Foundation of China (61379064).

References

1. Birkhoff, G.: Lattice Theory. American Mathematical Society Colloquium Publishers, Providence (1967)
2. De Baets, B.: Coimplicators, the forgotten connectives. Tatra Mt. Math. Publ. **12**, 229–240 (1997)
3. Fodor, J., Yager, R.R., Rybalov, A.: Structure of uninorms. Int. J. Uncertain., Fuzziness Knowl.-Based Syst. **5**, 411–427 (1997)
4. Liu, H.W.: Semi-uninorm and implications on a complete lattice. Fuzzy Sets Syst. **191**, 72–82 (2012)
5. Mas, M., Monserrat, M., Torrens, J.: On left and right uninorms. Int. J. Uncertain., Fuzziness Knowl.-Based Syst. **9**, 491–507 (2001)
6. Mas, M., Monserrat, M., Torrens, J.: On left and right uninorms on a finite chain. Fuzzy Sets Syst. **146**, 3–17 (2004)
7. Mas, M., Monserrat, M., Torrens, J.: Two types of implications derived from uninorms. Fuzzy Sets Syst. **158**, 2612–2626 (2007)
8. Ruiz, D., Torrens, J.: Residual implications and co-implications from idempotent uninorms. Kybernetika. **40**, 21–38 (2004)
9. Su, Y., Wang, Z.D., Tang, K.M.: Left and right semi-uninorms on a complete lattice. Kybernetika. **49**, 948–961 (2013)
10. Wang, Z.D., Fang, J.X.: Residual operators of left and right uninorms on a complete lattice. Fuzzy Sets Syst. **160**, 22–31 (2009)
11. Wang, Z.D., Fang, J.X.: Residual coimplicators of left and right uninorms on a complete lattice. Fuzzy Sets Syst. **160**, 2086–2096 (2009)
12. Yager, R.R.: Uninorms in fuzzy system modeling. Fuzzy Sets Syst. **122**, 167–175 (2001)
13. Yager, R.R.: Defending against strategic manipulation in uninorm-based multi-agent decision making. Eur. J. Oper. Res. **141**, 217–232 (2002)
14. Yager, R.R., Rybalov, A.: Uninorm aggregation operators. Fuzzy Sets Syst. **80**, 111–120 (1996)

Part V
Fuzzy Logical Algebras

Interval-Valued Intuitionistic (T, S)-Fuzzy LI-Ideals in Lattice Implication Algebras

Chun-Hui Liu

Abstract Combining interval-valued intuitionistic fuzzy sets, t-norm T and s-norm S on $D[0, 1]$ with the notion of ideal in lattice implication algebras, the concepts of interval-valued intuitionistic (T, S)-fuzzy LI-ideal and interval-valued intuitionistic (T, S)-fuzzy lattice ideal are introduced, some their properties are investigated. Some characterization theorems of interval-valued intuitionistic (T, S)-fuzzy LI-ideals are obtained. It is proved that the notion of interval-valued intuitionistic (T, S)-fuzzy LI-ideal is equivalent to the notion of interval-valued intuitionistic (T, S)-fuzzy lattice ideal in a lattice H implication algebra.

Keywords Many-valued logic · Lattice implication algebra · Interval-valued intuitionistic (T, S)-fuzzy LI-ideal

1 Introduction

With the developments of mathematics and computer science, non-classical mathematical logic has been actively studied. At present, many-valued logic has always been a kind of important non-classical logic. In order to research the many-valued logical system whose propositional value is given in a lattice, Xu proposed the concept of lattice implication algebras [1]. This structure was studied subsequently by many others [1–9]. Among them, Jun etc. introduced the notion of LI-ideals in lattice implication algebras and investigated their properties in [6]. The present author [9] studied the lattice structural feature of the set containing all of LI-ideals in a given lattice implication algebra.

Since the emergence of fuzzy set by Zadeh [10] in 1965, fuzzy set was studied from many viewpoint algebraically [11–13]. Later in 1986, Atanassov generalized Zadeh' fuzzy set and introduced the concept of intuitionistic fuzzy sets, he also introduced the concept of interval-valued intuitionistic fuzzy sets shortly after [22].

C.-H. Liu (✉)
Department of Mathematics and Statistics, Chifeng University, Chifeng 024001, China
e-mail: chunhuiliu1982@163.com

T.-H. Fan et al. (eds.), *Quantitative Logic and Soft Computing 2016*,
Advances in Intelligent Systems and Computing 510,
DOI 10.1007/978-3-319-46206-6_33

The algebraic aspect of this generalization was studied extensively afterward(Peng [14], Jun [15], Xue [16]) and [17]). Liu etc. studied the theories of interval-valued intuitionistic (T, S)-fuzzy filters on residuated lattices by applying this new notion in [18].

In this paper, We will continue to study the problem of ideal in lattice implication algebras by applying the notions of interval valued intuitionistic fuzzy sets, t-norm T and s-norm S on $D[0, 1]$, introduce the notions of interval-valued intuitionistic (T, S)-fuzzy LI-ideals and interval-valued intuitionistic (T, S)-fuzzy lattice ideals and discuss some of their properties. We believe that our work would serve as a foundation for enriching corresponding lattice-valued logical system based on lattice implication algebras.

2 Preliminaries

Definition 1 ([2]) Let $(L, \vee, \wedge, ', \rightarrow, O, I)$ be a bounded lattice with an order-reversing involution $'$, where I and O are the greatest and the smallest elements of L respectively, $\rightarrow: L \times L \rightarrow L$ is a mapping. Then $(L, \vee, \wedge, ', \rightarrow, O, I)$ is called a lattice implication algebra if the following conditions hold for all $x, y, z \in L$:
(I_1) $x \rightarrow (y \rightarrow z) = y \rightarrow (x \rightarrow z)$;
(I_2) $x \rightarrow x = I$;
(I_3) $x \rightarrow y = y' \rightarrow x'$;
(I_4) $x \rightarrow y = y \rightarrow x = I$ implies $x = y$;
(I_5) $(x \rightarrow y) \rightarrow y = (y \rightarrow x) \rightarrow x$;
(I_6) $(x \vee y) \rightarrow z = (x \rightarrow z) \wedge (y \rightarrow z)$;
(I_7) $(x \wedge y) \rightarrow z = (x \rightarrow z) \vee (y \rightarrow z)$.
In the sequel, for the sake of simplicity, a lattice implication algebra $(L, \vee, \wedge, ', \rightarrow, O, I)$ will be denoted by L.

Lemma 1 ([2]) *Let L be a lattice implication algebra. Then for all $x, y, z \in L$*
(p1) $x \leqslant y$ *if and only if* $x \rightarrow y = I$;
(p2) $O \rightarrow x = I, I \rightarrow x = x$ *and* $x \rightarrow I = I$;
(p3) $x \rightarrow y \leqslant (y \rightarrow z) \rightarrow (x \rightarrow z)$ *and* $x \rightarrow y \leqslant (z \rightarrow x) \rightarrow (z \rightarrow y)$;
(p4) $y \leqslant z$ *implies* $x \rightarrow y \leqslant x \rightarrow z$, *and* $x \leqslant y$ *implies* $y \rightarrow z \leqslant x \rightarrow z$;
(p5) $x \vee y = (x \rightarrow y) \rightarrow y$;
(p6) $x \rightarrow (y \wedge z) = (x \rightarrow y) \wedge (x \rightarrow z)$ *and* $x \rightarrow (y \vee z) = (x \rightarrow y) \vee (x \rightarrow z)$;
(p7) $x \oplus y = y \oplus x$ *and* $(x \oplus y) \oplus z = x \oplus (y \oplus z)$;
(p8) $O \oplus x = x, I \oplus x = I$ *and* $x \oplus x' = I$;
(p9) $x \vee y \leqslant x \oplus y$ *and* $x \leqslant (x \rightarrow y)' \oplus y$;
(p10) $x \leqslant y$ *implies* $x \oplus z \leqslant y \oplus z$;
where $x \oplus y = x' \rightarrow y$ *and* $x' = x \rightarrow O$.

Definition 2 ([1]) A lattice implication algebra L is said to be a lattice H implication algebra, if the following condition holds for all $x, y, z \in L$,

$$x \vee y \vee ((x \wedge y) \rightarrow z) = I. \qquad \text{(H)}$$

Remark 1 In a lattice H implication algebra L, it is easy to check that $x \vee y = x' \rightarrow y$ for all $x, y \in L$.

Definition 3 ([6]) Let L be a lattice implication algebra. An LI-ideal K is a non-empty subset of L such that for all $x, y \in L$
(LI1) $O \in K$;
(LI2) $(x \rightarrow y)' \in K$ and $y \in K$ imply $x \in K$.

Now we fix some notations for interval-valued fuzzy set. $\bar{a} = [a^-, a^+]$ is a closed interval of $[0, 1]$, where $0 \leqslant a^- \leqslant a^+ \leqslant 1$, $D[0, 1]$ is the set containing all such closed intervals of $[0, 1]$. The interval $[a, a]$ is simply identified as a. Define on $D[0, 1]$ an order relation $\preccurlyeq$ as follows
(1) $\bar{a}_1 \preccurlyeq \bar{a}_2 \Longleftrightarrow a_1^- \leqslant a_2^-$ and $a_1^+ \leqslant a_2^+$;
(2) $\bar{a}_1 = \bar{a}_2 \Longleftrightarrow a_1^- = a_2^-$ and $a_1^+ = a_2^+$;
(3) $\bar{a}_1 \prec \bar{a}_2 \Longleftrightarrow \bar{a}_1 \preccurlyeq \bar{a}_2$ and $\bar{a}_1 \neq \bar{a}_2$;
(4) $k\bar{a} = [ka^-, ka^+]$, whenever $0 \leqslant k \leqslant 1$;
(5) $\max\{\bar{a}_1, \bar{a}_2\} = [\max\{a_1^-, a_2^-\}, \max\{a_1^+, a_2^+\}]$;
(6) $\min\{\bar{a}_1, \bar{a}_2\} = [\min\{a_1^-, a_2^-\}, \min\{a_1^+, a_2^+\}]$;
(7) $\sup\{\bar{a}_\lambda\}_{\lambda \in \Lambda} = [\sup\{a_\lambda^-\}_{\lambda \in \Lambda}, \sup\{a_\lambda^+\}_{\lambda \in \Lambda}]$;
(8) $\inf\{\bar{a}_\lambda\}_{\lambda \in \Lambda} = [\inf\{a_\lambda^-\}_{\lambda \in \Lambda}, \inf\{a_\lambda^+\}_{\lambda \in \Lambda}]$;
where $\bar{a}_1 = [a_1^-, a_1^+], \bar{a}_2 = [a_2^-, a_2^+] \in D[0, 1]$ and $\{\bar{a}_\lambda\}_{\lambda \in \Lambda} \subseteq D[0, 1]$. Then, $D[0, 1]$ with $\preccurlyeq$ forms a complete lattice, with $\vee = \max$, $\wedge = \min$, $\bar{0} = [0, 0]$ and $\bar{1} = [1, 1]$ being its the smallest element and the greatest element, respectively.

Definition 4 ([19]) Let X be a non-empty set. An interval-valued fuzzy set on X is a mapping $\bar{\Upsilon} : X \rightarrow D[0, 1]$ such that for all $x \in X$, $\bar{\Upsilon}(x) = [\Upsilon^-(x), \Upsilon^+(x)]$, where Υ^- and Υ^+ are two fuzzy sets on X with $\Upsilon^-(x) \leqslant \Upsilon^+(x)$. Let $\bar{\Upsilon}$ be an interval-valued fuzzy set on X. For all $[0, 0] \prec \bar{\alpha} \preccurlyeq [1, 1]$, the crisp set $\bar{\Upsilon}_{\bar{\alpha}} = \{x \in X | \bar{\Upsilon}(x) \succcurlyeq \bar{\alpha}\}$ is called a level subset of $\bar{\Upsilon}$.

Definition 5 ([19]) Let X be a non-empty set. An interval-valued intuitionistic fuzzy set $\mathscr{A}$ on X is defined as an object of the form

$$\mathscr{A} = \{(x, \bar{\Phi}_{\mathscr{A}}, \bar{\Psi}_{\mathscr{A}}) | x \in X\},$$

where $\bar{\Phi}_{\mathscr{A}}$ and $\bar{\Psi}_{\mathscr{A}}$ are two interval-valued fuzzy sets on X with $[0, 0] \preccurlyeq \bar{\Phi}_{\mathscr{A}}(x) + \bar{\Psi}_{\mathscr{A}}(x) \preccurlyeq [1, 1]$, for all $x \in X$.

In the sequel, for the sake of simplicity, an interval-valued intuitionistic fuzzy set on X will be denoted by the symbol $\mathscr{A} = (\bar{\Phi}_{\mathscr{A}}, \bar{\Psi}_{\mathscr{A}})$.

Definition 6 ([20, 21]) Let T (resp. S) be a mapping form $D[0, 1] \times D[0, 1]$ to $D[0, 1]$. T (resp. S) is called a t-norm (resp. s-norm) on $D[0, 1]$, if it satisfies the following conditions: for all $\bar{a}, \bar{b}, \bar{c} \in D[0, 1]$,
(1) $T(\bar{a}, \bar{1}) = \bar{a}$ (resp. $S(\bar{a}, \bar{0}) = \bar{a}$);
(2) $T(\bar{a}, \bar{b}) = T(\bar{b}, \bar{a})$ (resp. $S(\bar{a}, \bar{b}) = S(\bar{b}, \bar{a})$);
(3) $T(T(\bar{a}, \bar{b}), \bar{c}) = T(\bar{a}, T(\bar{b}, \bar{c}))$ (resp. $S(S(\bar{a}, \bar{b}), \bar{c}) = S(\bar{a}, S(\bar{b}, \bar{c}))$);

(4) $\bar{a} \preccurlyeq \bar{b}$ implies $T(\bar{a}, \bar{c}) \preccurlyeq T(\bar{b}, \bar{c})$ (resp. $S(\bar{a}, \bar{c}) \preccurlyeq S(\bar{b}, \bar{c})$).
The set of all T-idempotent elements (resp. S-idempotent elements) denoted by $D_T = \{\bar{a} \in D[0,1] | T(\bar{a}, \bar{a}) = \bar{a}\}$ (resp. $D_S = \{\bar{a} \in D[0,1] | S(\bar{a}, \bar{a}) = \bar{a}\}$). An interval-valued fuzzy set $\bar{\Upsilon}$ is said to satisfy the imaginable property under T (resp. S), if $\mathrm{Im}\bar{\Upsilon} \subseteq D_T$ (resp. $\mathrm{Im}\bar{\Upsilon} \subseteq D_S$).

3 Interval-Valued Intuitionistic (S, T)-Fuzzy LI-Ideals

In the sequel, We will use the symbols T and S to denote a t-norm and a s-norm on $D[0, 1]$ unless otherwise specified, and assume that all the t-norm and s-norm are idempotent.

Definition 7 Let L be a lattice implication algebra. An interval-valued intuitionistic fuzzy set $\mathscr{A} = (\bar{\Phi}_{\mathscr{A}}, \bar{\Psi}_{\mathscr{A}})$ on L is said to be an interval-valued intuitionistic (T, S)-fuzzy LI-ideal (IVI-(T, S)-fuzzy LI-ideal for short) of L, if it satisfies the following conditions: for all $x, y \in L$,
(a1) $\bar{\Phi}_{\mathscr{A}}(O) \succcurlyeq \bar{\Phi}_{\mathscr{A}}(x)$ and $\bar{\Psi}_{\mathscr{A}}(O) \preccurlyeq \bar{\Psi}_{\mathscr{A}}(x)$;
(a2) $\bar{\Phi}_{\mathscr{A}}(x) \succcurlyeq T(\bar{\Phi}_{\mathscr{A}}((x \to y)'), \bar{\Phi}_{\mathscr{A}}(y))$ and $\bar{\Psi}_{\mathscr{A}}(x) \preccurlyeq S(\bar{\Psi}_{\mathscr{A}}((x \to y)'), \bar{\Psi}_{\mathscr{A}}(y))$.
denote by **IVILI**(L) is the set containing all IVI-(T, S)-fuzzy LI-ideals of L.

Example 1 Let $L = \{O, a, b, c, d, I\}$, $O' = I, a' = c, b' = d, c' = a, d' = b, I' = O$, the Hasse diagram of L is defined as in Fig. 1 and its implication operator $\to$ is defined as in Table 1.

Then $(L, \vee, \wedge, ', \to)$ is a lattice implication algebra. Define an interval-valued intuitionistic fuzzy set $\mathscr{A} = (\bar{\Phi}_{\mathscr{A}}, \bar{\Psi}_{\mathscr{A}})$ on L by $\bar{\Phi}_{\mathscr{A}}(O) = \bar{\Phi}_{\mathscr{A}}(c) = [0.7, 0.8]$, $\bar{\Phi}_{\mathscr{A}}(x) = [0.3, 0.4]$, $x \in L \backslash \{O, c\}$, and $\bar{\Psi}_{\mathscr{A}}(O) = \bar{\Psi}_{\mathscr{A}}(c) = [0.1, 0.2]$, $\bar{\Phi}_{\mathscr{A}}(x) = [0.5, 0.6]$, $x \in L \backslash \{O, c\}$. It is easily to verify that $\mathscr{A}$ is an interval-valued intuitionistic (T, S)-fuzzy LI-ideal of L. Where, for all $\bar{a} = [a^-, a^+], \bar{b} = [b^-, b^+] \in D[0, 1]$, $T(\bar{a}, \bar{b}) = [a^- b^-, \max\{a^- b^+, a^+ b^-\}]$ and $S(\bar{a}, \bar{b}) = \max\{\bar{a}, \bar{b}\}$.

Proposition 1 *Let L be a lattice implication algebra and $\mathscr{A} = (\bar{\Phi}_{\mathscr{A}}, \bar{\Psi}_{\mathscr{A}}) \in$ **IVILI**(L). Then for all $x, y \in L$,*
(a3) $x \leqslant y \Longrightarrow \bar{\Phi}_{\mathscr{A}}(x) \succcurlyeq \bar{\Phi}_{\mathscr{A}}(y)$ and $\bar{\Psi}_{\mathscr{A}}(x) \preccurlyeq \bar{\Psi}_{\mathscr{A}}(y)$.

Fig. 1 Hasse diagram of L

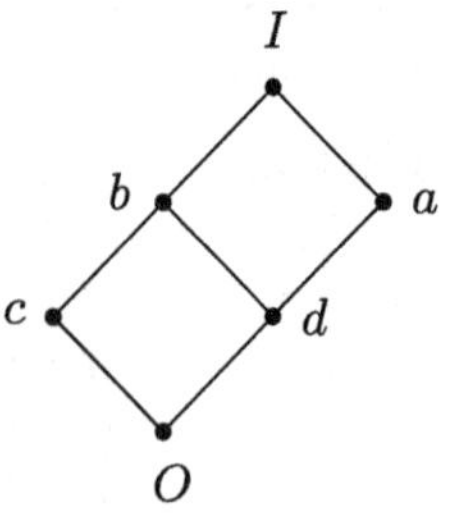

Table 1 Definition of "→"

→	O	a	b	c	d	I
O	I	I	I	I	I	I
a	c	I	b	c	b	I
b	d	a	I	b	a	I
c	a	a	I	I	a	I
d	b	I	I	b	I	I
I	O	a	b	c	d	I

Proof Assume that $x, y \in L$ and $x \leqslant y$, then $(x \to y)' = I' = O$. Since $\mathscr{A} \in$ **IVILI**(L), by using (a2) and (a1), we have that $\bar{\Phi}_{\mathscr{A}}(x) \succcurlyeq T(\bar{\Phi}_{\mathscr{A}}((x \to y)'), \bar{\Phi}_{\mathscr{A}}(y)) = T(\bar{\Phi}_{\mathscr{A}}(O), \bar{\Phi}_{\mathscr{A}}(y)) \succcurlyeq T(\bar{\Phi}_{\mathscr{A}}(y), \bar{\Phi}_{\mathscr{A}}(y)) = \bar{\Phi}_{\mathscr{A}}(y)$ and $\bar{\Psi}_{\mathscr{A}}(x) \preccurlyeq S(\bar{\Psi}_{\mathscr{A}}((x \to y)'), \bar{\Psi}_{\mathscr{A}}(y)) = S(\bar{\Psi}_{\mathscr{A}}(O), \bar{\Psi}_{\mathscr{A}}(y)) \preccurlyeq S(\bar{\Psi}_{\mathscr{A}}(y), \bar{\Psi}_{\mathscr{A}}(y)) = \bar{\Psi}_{\mathscr{A}}(y)$. Hence (a3) is valid. □

Theorem 1 *Let L be a lattice implication algebra and* $\mathscr{A} = (\bar{\Phi}_{\mathscr{A}}, \bar{\Psi}_{\mathscr{A}})$ *an interval-valued intuitionistic fuzzy set on L. Then* $\mathscr{A} \in$ **IVILI**(L) *if and only if, for all* $x, y, z \in L$, *it satisfies the conditions (IVI1) and*
(a4) $\bar{\Phi}_{\mathscr{A}}((x \to z)') \succcurlyeq T(\bar{\Phi}_{\mathscr{A}}((x \to y)'), \bar{\Phi}_{\mathscr{A}}((y \to z)'))$ *and* $\bar{\Psi}_{\mathscr{A}}((x \to z)') \preccurlyeq S(\bar{\Psi}_{\mathscr{A}}((x \to y)'), \bar{\Psi}_{\mathscr{A}}((y \to z)'))$.

Proof Assume that $\mathscr{A} \in$ **IVILI**(L) and $x, y, z \in L$. Then $\mathscr{A}$ satisfies (a1) by Definition 7. Since by (I_3) and Lemma 1,

$$((x \to z)' \to (y \to z)')' \to (x \to y)' = (x \to y) \to ((y \to z) \to (x \to z)) = I,$$

we have $(x \to z)' \to (y \to z)')' \leqslant (x \to y)'$. Thus, by (a2) and (a3) we have $\bar{\Phi}_{\mathscr{A}}((x \to z)') \succcurlyeq T(\bar{\Phi}_{\mathscr{A}}(((x \to z)' \to (y \to z)')'), \bar{\Phi}_{\mathscr{A}}((y \to z)')) \succcurlyeq T(\bar{\Phi}_{\mathscr{A}}((x \to y)'), \bar{\Phi}_{\mathscr{A}}((y \to z)'))$ and $\bar{\Psi}_{\mathscr{A}}((x \to z)') \preccurlyeq S(\bar{\Psi}_{\mathscr{A}}(((x \to z)' \to (y \to z)')'), \bar{\Psi}_{\mathscr{A}}((y \to z)')) \preccurlyeq S(\bar{\Psi}_{\mathscr{A}}((x \to y)'), \bar{\Psi}_{\mathscr{A}}((y \to z)'))$, Hence $\mathscr{A}$ also satisfies (a4).

Conversely, assume that $\mathscr{A}$ satisfies (a1) and (a4). In order to show that $\mathscr{A} \in$ **IVILI**(L), it is sufficient to show that $\mathscr{A}$ satisfies (a2) by Definition 7. In fact, since $(x \to O)' = x$ for any $x \in L$, by (a4) we have $\bar{\Phi}_{\mathscr{A}}(x) = \bar{\Phi}_{\mathscr{A}}((x \to O)') \succcurlyeq T(\bar{\Phi}_{\mathscr{A}}((x \to y)'), \bar{\Phi}_{\mathscr{A}}((y \to O)')) = T(\bar{\Phi}_{\mathscr{A}}((x \to y)'), \bar{\Phi}_{\mathscr{A}}(y))$ and $\bar{\Psi}_{\mathscr{A}}(x) = \bar{\Psi}_{\mathscr{A}}((x \to O)') \preccurlyeq S(\bar{\Psi}_{\mathscr{A}}((x \to y)'), \bar{\Psi}_{\mathscr{A}}((y \to O)')) = S(\bar{\Psi}_{\mathscr{A}}((x \to y)'), \bar{\Psi}_{\mathscr{A}}(y))$. Hence $\mathscr{A}$ satisfies (a2) and the proof is completed. □

Theorem 2 *Let L be a lattice implication algebra and* $\mathscr{A} = (\bar{\Phi}_{\mathscr{A}}, \bar{\Psi}_{\mathscr{A}})$ *an interval-valued intuitionistic fuzzy set on L. Then* $\mathscr{A} \in$ **IVILI**(L) *if and only if it satisfies the condition: for all* $x, y, z \in L$,
(a5) $z \leqslant x \oplus y \Longrightarrow \bar{\Phi}_{\mathscr{A}}(z) \succcurlyeq T(\bar{\Phi}_{\mathscr{A}}(x), \bar{\Phi}_{\mathscr{A}}(y))$ *and* $\bar{\Psi}_{\mathscr{A}}(z) \preccurlyeq S(\bar{\Psi}_{\mathscr{A}}(x), \bar{\Psi}_{\mathscr{A}}(y))$.

Proof Assume that $\mathscr{A} \in \mathbf{IVILI}(L)$ and $x, y, z \in L$. If $z \leqslant x \oplus y$, then by Lemma 1 and (I_5) we can get that

$$I = z \to (x \oplus y) = z \to (x' \to y) = x' \to (z \to y) = (z \to y)' \to x,$$

and so $((z \to y)' \to x)' = O$. Thus, by using (a2),we have that

$$\begin{aligned}\bar{\Phi}_{\mathscr{A}}(z) \succcurlyeq &T(\bar{\Phi}_{\mathscr{A}}((z \to y)'), \bar{\Phi}_{\mathscr{A}}(y))\\ \succcurlyeq &T(T(\bar{\Phi}_{\mathscr{A}}(((z \to y)' \to x)'), \bar{\Phi}_{\mathscr{A}}(x)), \bar{\Phi}_{\mathscr{A}}(y))\\ = &T(T(\bar{\Phi}_{\mathscr{A}}(O), \bar{\Phi}_{\mathscr{A}}(x)), \bar{\Phi}_{\mathscr{A}}(y))\\ \succcurlyeq &T(T(\bar{\Phi}_{\mathscr{A}}(x), \bar{\Phi}_{\mathscr{A}}(x)), \bar{\Phi}_{\mathscr{A}}(y))\\ = &T(\bar{\Phi}_{\mathscr{A}}(x), \bar{\Phi}_{\mathscr{A}}(y)),\end{aligned}$$

and

$$\begin{aligned}\bar{\Psi}_{\mathscr{A}}(z) \preccurlyeq &S(\bar{\Psi}_{\mathscr{A}}((z \to y)'), \bar{\Psi}_{\mathscr{A}}(y))\\ \preccurlyeq &S(S(\bar{\Psi}_{\mathscr{A}}(((z \to y)' \to x)'), \bar{\Psi}_{\mathscr{A}}(x)), \bar{\Psi}_{\mathscr{A}}(y))\\ = &S(S(\bar{\Psi}_{\mathscr{A}}(O), \bar{\Psi}_{\mathscr{A}}(x)), \bar{\Psi}_{\mathscr{A}}(y))\\ \preccurlyeq &S(S(\bar{\Psi}_{\mathscr{A}}(x), \bar{\Psi}_{\mathscr{A}}(x)), \bar{\Psi}_{\mathscr{A}}(y))\\ = &S(\bar{\Psi}_{\mathscr{A}}(x), \bar{\Psi}_{\mathscr{A}}(y)),\end{aligned}$$

that is, $\mathscr{A}$ satisfies condition (a5).

Conversely, assume that $\mathscr{A}$ satisfies the condition (a5). On the one hand, since $O \leqslant x \oplus x$ for any $x \in L$, we have $\bar{\Phi}_{\mathscr{A}}(O) \succcurlyeq T(\bar{\Phi}_{\mathscr{A}}(x), \bar{\Phi}_{\mathscr{A}}(x)) = \bar{\Phi}_{\mathscr{A}}(x)$ and $\bar{\Psi}_{\mathscr{A}}(O) \preccurlyeq S(\bar{\Psi}_{\mathscr{A}}(x), \bar{\Psi}_{\mathscr{A}}(x)) = \bar{\Psi}_{\mathscr{A}}(x)$, that is, $\mathscr{A}$ satisfies condition (a1). On the other hand, for all $x, y \in L$, it follows from $x \leqslant (x \to y)' \oplus y$ that $\bar{\Phi}_{\mathscr{A}}(x) \succcurlyeq T(\bar{\Phi}_{\mathscr{A}}((x \to y)'), \bar{\Phi}_{\mathscr{A}}(y))$ and $\bar{\Psi}_{\mathscr{A}}(x) \preccurlyeq S(\bar{\Psi}_{\mathscr{A}}((x \to y)'), \bar{\Psi}_{\mathscr{A}}(y))$, that is, $\mathscr{A}$ also satisfies condition (a2). By Definition 7, we have $\mathscr{A} \in \mathbf{IVILI}(L)$. □

Remark 2 Let L be a lattice implication algebra. for all $x, y, z \in L$, since $z \leqslant x \oplus y$ if and only if $(z \to x)' \leqslant y$ if and only if $x' \to (y' \to z') = I$, from Theorem 2, we have the following:

Corollary 1 *Let L be a lattice implication algebra and $\mathscr{A} = (\bar{\Phi}_{\mathscr{A}}, \bar{\Psi}_{\mathscr{A}})$ an interval-valued intuitionistic fuzzy set on L. Then $\mathscr{A} \in$ **IVILI**(L) if and only if it satisfies one of the following conditions: for all $x, y, z \in L$,*
(a6) $(z \to x)' \leqslant y$ implies $\bar{\Phi}_{\mathscr{A}}(z) \succcurlyeq T(\bar{\Phi}_{\mathscr{A}}(x), \bar{\Phi}_{\mathscr{A}}(y))$ and $\bar{\Psi}_{\mathscr{A}}(z) \preccurlyeq S(\bar{\Psi}_{\mathscr{A}}(x), \bar{\Psi}_{\mathscr{A}}(y))$;
(a7) $x' \to (y' \to z') = I$ implies $\bar{\Phi}_{\mathscr{A}}(z) \succcurlyeq T(\bar{\Phi}_{\mathscr{A}}(x), \bar{\Phi}_{\mathscr{A}}(y))$ and $\bar{\Psi}_{\mathscr{A}}(z) \preccurlyeq S(\bar{\Psi}_{\mathscr{A}}(x), \bar{\Psi}_{\mathscr{A}}(y))$.

Theorem 3 *Let L be a lattice implication algebra and $\mathscr{A} = (\bar{\Phi}_{\mathscr{A}}, \bar{\Psi}_{\mathscr{A}})$ an interval-valued intuitionistic fuzzy set on L. Then $\mathscr{A} \in$ **IVILI**(L) if and only if, for all*

$x, y \in L$, it satisfies conditions (IVI3) and
(a8) $\bar{\Phi}_{\mathscr{A}}(x \oplus y) \succcurlyeq T(\bar{\Phi}_{\mathscr{A}}(x), \bar{\Phi}_{\mathscr{A}}(y))$ *and* $\bar{\Psi}_{\mathscr{A}}(x \oplus y) \preccurlyeq S(\bar{\Psi}_{\mathscr{A}}(x), \bar{\Psi}_{\mathscr{A}}(y))$.

Proof Assume that $\mathscr{A} \in \mathbf{IVILI}(L)$, then $\mathscr{A}$ satisfies the condition (a5) by Theorem 2. Let $x, y \in L$, if $x \leqslant y$, then $x \leqslant y \leqslant y \oplus y$ by (p9). From (a5), it follows that $\bar{\Phi}_{\mathscr{A}}(x) \succcurlyeq T(\bar{\Phi}_{\mathscr{A}}(y), \bar{\Phi}_{\mathscr{A}}(y)) = \bar{\Phi}_{\mathscr{A}}(y)$ and $\bar{\Psi}_{\mathscr{A}}(x) \preccurlyeq S(\bar{\Psi}_{\mathscr{A}}(y), \bar{\Psi}_{\mathscr{A}}(y)) = \bar{\Psi}_{\mathscr{A}}(y)$, that is, $\mathscr{A}$ satisfies condition (a3). Since $x \oplus y \leqslant x \oplus y$, by (a5) we have $\bar{\Phi}_{\mathscr{A}}(x \oplus y) \succcurlyeq T(\bar{\Phi}_{\mathscr{A}}(x), \bar{\Phi}_{\mathscr{A}}(y))$ and $\bar{\Psi}_{\mathscr{A}}(x \oplus y) \preccurlyeq S(\bar{\Psi}_{\mathscr{A}}(x), \bar{\Psi}_{\mathscr{A}}(y))$, that is, $\mathscr{A}$ also satisfies condition (a8).

Conversely, assume that $\mathscr{A}$ satisfies the conditions (a3) and (a8). Obviously, $\bar{\Phi}_{\mathscr{A}}(O) \succcurlyeq \bar{\Phi}_{\mathscr{A}}(x)$ and $\bar{\Psi}_{\mathscr{A}}(O) \preccurlyeq \bar{\Psi}_{\mathscr{A}}(x)$ by $O \leqslant x$ and (a3), that is, $\mathscr{A}$ satisfies condition (a1). Let $x, y \in L$, since $x \leqslant (x \to y)' \oplus y$, by using (a3) and (a8), we have $\bar{\Phi}_{\mathscr{A}}(x) \succcurlyeq \bar{\Phi}_{\mathscr{A}}((x \to y)' \oplus y) \succcurlyeq T(\bar{\Phi}_{\mathscr{A}}((x \to y)'), \bar{\Phi}_{\mathscr{A}}(y))$ and $\bar{\Psi}_{\mathscr{A}}(x) \preccurlyeq \bar{\Psi}_{\mathscr{A}}((x \to y)' \oplus y) \preccurlyeq S(\bar{\Psi}_{\mathscr{A}}((x \to y)'), \bar{\Psi}_{\mathscr{A}}(y))$, that is, $\mathscr{A}$ satisfies condition (a2). Hence $\mathscr{A} \in \mathbf{IVILI}(L)$ by Definition 7. □

Theorem 4 *Let L be a lattice implication algebra and $\mathscr{A} = (\bar{\Phi}_{\mathscr{A}}, \bar{\Psi}_{\mathscr{A}})$ an interval-valued intuitionistic fuzzy set on L. Then $\mathscr{A} \in \mathbf{IVILI}(L)$ if and only if, for all $\bar{\alpha}, \bar{\beta} \in D[0, 1]$ with $\bar{\alpha} + \bar{\beta} \preccurlyeq [1, 1]$, the sets $U(\bar{\Phi}_{\mathscr{A}}; \bar{\alpha})(\neq \emptyset)$ and $V(\bar{\Psi}_{\mathscr{A}}; \bar{\beta})(\neq \emptyset)$ are* LI*-ideals of L. Where $U(\bar{\Phi}_{\mathscr{A}}; \bar{\alpha}) = \{x \in L \mid \bar{\Phi}_{\mathscr{A}}(x) \succcurlyeq \bar{\alpha}\}$ and $V(\bar{\Psi}_{\mathscr{A}}; \bar{\beta}) = \{x \in L \mid \bar{\Psi}_{\mathscr{A}}(x) \preccurlyeq \bar{\beta}\}$.*

Proof Assume that $\mathscr{A} \in \mathbf{IVILI}(L)$, then for all $x \in L$, by (a1) we have $\bar{\Phi}_{\mathscr{A}}(O) \succcurlyeq \bar{\Phi}_{\mathscr{A}}(x)$ and $\bar{\Psi}_{\mathscr{A}}(O) \preccurlyeq \bar{\Psi}_{\mathscr{A}}(x)$. From conditions $U(\bar{\Phi}_{\mathscr{A}}; \bar{\alpha}) \neq \emptyset$ and $V(\bar{\Psi}_{\mathscr{A}}; \bar{\beta}) \neq \emptyset$, it follows that there exists $a, b \in L$ such that $\bar{\Phi}_{\mathscr{A}}(O) \succcurlyeq \bar{\Phi}_{\mathscr{A}}(a) \succcurlyeq \bar{\alpha}$ and $\bar{\Psi}_{\mathscr{A}}(O) \preccurlyeq \bar{\Psi}_{\mathscr{A}}(b) \preccurlyeq \bar{\beta}$. Hence $O \in U(\bar{\Phi}_{\mathscr{A}}; \bar{\alpha})$ and $O \in V(\bar{\Psi}_{\mathscr{A}}; \bar{\beta})$.

For all $x, y \in L$, let $(x \to y)' \in U(\bar{\Phi}_{\mathscr{A}}; \bar{\alpha})$ and $y \in U(\bar{\Phi}_{\mathscr{A}}; \bar{\alpha})$, then $\bar{\Phi}_{\mathscr{A}}((x \to y)') \succcurlyeq \bar{\alpha}$ and $\bar{\Phi}_{\mathscr{A}}(y) \succcurlyeq \bar{\alpha}$. Since $\mathscr{A} \in \mathbf{IVILI}(L)$, by using (a2) we have $\bar{\Phi}_{\mathscr{A}}(x) \succcurlyeq T(\bar{\Phi}_{\mathscr{A}}((x \to y)'), \bar{\Phi}_{\mathscr{A}}(y)) \succcurlyeq T(\bar{\alpha}, \bar{\alpha}) = \bar{\alpha}$, thus $x \in U(\bar{\Phi}_{\mathscr{A}}; \bar{\alpha})$. Therefore $U(\bar{\Phi}_{\mathscr{A}}; \bar{\alpha})$ is an *LI*-ideal of L.

For all $x, y \in L$, let $(x \to y)' \in V(\bar{\Psi}_{\mathscr{A}}; \bar{\beta})$ and $y \in V(\bar{\Psi}_{\mathscr{A}}; \bar{\beta})$, then $\bar{\Psi}_{\mathscr{A}}((x \to y)') \preccurlyeq \bar{\beta}$ and $\bar{\Psi}_{\mathscr{A}}(y) \preccurlyeq \bar{\beta}$. Since $\mathscr{A} \in \mathbf{IVILI}(L)$, by using (a2) we have $\bar{\Psi}_{\mathscr{A}}(x) \preccurlyeq S(\bar{\Psi}_{\mathscr{A}}((x \to y)'), \bar{\Psi}_{\mathscr{A}}(y)) \preccurlyeq T(\bar{\beta}, \bar{\beta}) = \bar{\beta}$, thus $x \in V(\bar{\Psi}_{\mathscr{A}}; \bar{\beta})$. Therefore $V(\bar{\Psi}_{\mathscr{A}}; \bar{\beta})$ is an *LI*-ideal of L too.

Conversely, assume that $U(\bar{\Phi}_{\mathscr{A}}; \bar{\alpha})(\neq \emptyset)$ and $V(\bar{\Psi}_{\mathscr{A}}; \bar{\beta})(\neq \emptyset)$ are *LI*-ideals of L. Since for any $x \in L$, $x \in U(\bar{\Phi}_{\mathscr{A}}; \bar{\Phi}_{\mathscr{A}}(x))$ and $x \in V(\bar{\Psi}_{\mathscr{A}}; \bar{\Psi}_{\mathscr{A}}(x))$, we have $U(\bar{\Phi}_{\mathscr{A}}; \bar{\Phi}_{\mathscr{A}}(x)) \neq \emptyset$ and $V(\bar{\Psi}_{\mathscr{A}}; \bar{\Psi}_{\mathscr{A}}(x)) \neq \emptyset$, thus $U(\bar{\Phi}_{\mathscr{A}}; \bar{\Phi}_{\mathscr{A}}(x))$ and $V(\bar{\Psi}_{\mathscr{A}}; \bar{\Psi}_{\mathscr{A}}(x))$ are *LI*-ideals of L, and thus $O \in U(\bar{\Phi}_{\mathscr{A}}; \bar{\Phi}_{\mathscr{A}}(x))$ and $O \in V(\bar{\Psi}_{\mathscr{A}}; \bar{\Psi}_{\mathscr{A}}(x))$, this shows that $\bar{\Phi}_{\mathscr{A}}(O) \succcurlyeq \bar{\Phi}_{\mathscr{A}}(x)$ and $\bar{\Psi}_{\mathscr{A}}(O) \preccurlyeq \bar{\Psi}_{\mathscr{A}}(x)$, that is, $\mathscr{A}$ satisfies condition (a1).

For all $x, y \in L$, let $\bar{\alpha} = T(\bar{\Phi}_{\mathscr{A}}((x \to y)'), \bar{\Phi}_{\mathscr{A}}(y))$ and $\bar{\beta} = S(\bar{\Psi}_{\mathscr{A}}((x \to y)'), \bar{\Psi}_{\mathscr{A}}(y))$, then $(x \to y)', y \in U(\bar{\Phi}_{\mathscr{A}}; \bar{\alpha})$ and $(x \to y)', y \in V(\bar{\Psi}_{\mathscr{A}}; \bar{\beta})$. Therefore $\bar{\Phi}_{\mathscr{A}}(x) \succcurlyeq T(\bar{\Phi}_{\mathscr{A}}((x \to y)'), \bar{\Phi}_{\mathscr{A}}(y))$ and $\bar{\Psi}_{\mathscr{A}}(x) \preccurlyeq S(\bar{\Psi}_{\mathscr{A}}((x \to y)'), \bar{\Psi}_{\mathscr{A}}(y))$, that is, $\mathscr{A}$ also satisfies condition (a2). Hence $\mathscr{A} \in \mathbf{IVILI}(L)$ by Definition 7. □

Suppose L be a lattice implication algebra and $\mathscr{A}$ an interval-valued intuitionistic fuzzy set on L. For all $\bar{\alpha}, \bar{\beta} \in D[0,1]$, we define $\mathscr{A}_{(\bar{\alpha},\bar{\beta})} = \{x \in L | \bar{\Phi}_{\mathscr{A}}(x) \succcurlyeq \bar{\alpha} \text{ and } \bar{\Psi}_{\mathscr{A}}(x) \preccurlyeq \bar{\beta}\}$.

Theorem 5 *Let L be a lattice implication algebra, and* $\mathscr{A} = (\bar{\Phi}_{\mathscr{A}}, \bar{\Psi}_{\mathscr{A}}) \in$ **IVILI** (L). *Then For all* $\bar{\alpha}, \bar{\beta} \in D[0,1]$, $\mathscr{A}_{(\bar{\alpha},\bar{\beta})} (\neq \emptyset)$ *is an LI-ideal of L.*

Proof Assume that $\mathscr{A}_{(\bar{\alpha},\bar{\beta})} \neq \emptyset$, then there exists $x \in L$ such that $\bar{\Phi}_{\mathscr{A}}(x) \succcurlyeq \bar{\alpha}$ and $\bar{\Psi}_{\mathscr{A}}(x) \preccurlyeq \bar{\beta}$. Since $\mathscr{A} \in$ **IVILI**(L), by (a1) we have $\bar{\Phi}_{\mathscr{A}}(O) \succcurlyeq \bar{\Phi}_{\mathscr{A}}(x) \succcurlyeq \bar{\alpha}$ and $\bar{\Psi}_{\mathscr{A}}(O) \preccurlyeq \bar{\Psi}_{\mathscr{A}}(x) \preccurlyeq \bar{\beta}$, therefore $O \in \mathscr{A}_{(\bar{\alpha},\bar{\beta})}$.

Let $x, y \in L$ and $(x \to y)' \in \mathscr{A}_{(\bar{\alpha},\bar{\beta})}$, $y \in \mathscr{A}_{(\bar{\alpha},\bar{\beta})}$, then $\bar{\Phi}_{\mathscr{A}}((x \to y)') \succcurlyeq \bar{\alpha}$, $\bar{\Psi}_{\mathscr{A}}((x \to y)') \preccurlyeq \bar{\beta}$ and $\bar{\Phi}_{\mathscr{A}}(y) \succcurlyeq \bar{\alpha}$, $\bar{\Psi}_{\mathscr{A}}(y) \preccurlyeq \bar{\beta}$. Since $\mathscr{A} \in$ **IVILI**(L), by (a2) we have $\bar{\Phi}_{\mathscr{A}}(x) \succcurlyeq T(\bar{\Phi}_{\mathscr{A}}((x \to y)'), \bar{\Phi}_{\mathscr{A}}(y)) \succcurlyeq T(\bar{\alpha}, \bar{\alpha}) = \bar{\alpha}$ and $\bar{\Psi}_{\mathscr{A}}(x) \preccurlyeq S(\bar{\Psi}_{\mathscr{A}}((x \to y)'), \bar{\Psi}_{\mathscr{A}}(y)) \preccurlyeq S(\bar{\beta}, \bar{\beta}) = \bar{\beta}$, it follows that $x \in \mathscr{A}_{(\bar{\alpha},\bar{\beta})}$.

Hence $\mathscr{A}_{(\bar{\alpha},\bar{\beta})}$ is an *LI*-ideal of L by Definition 3. □

Remark 3 Let L be a lattice implication algebra, the *LI*-ideal $\mathscr{A}_{(\bar{\alpha},\bar{\beta})}$ in Theorem 5 is called an IVI-(T, S)-cut *LI*-ideal of IVI-(T, S)-fuzzy *LI*-ideal $\mathscr{A}$ of L.

Theorem 6 *Let L be a lattice implication algebra. Then any LI-ideal K of L is an IVI-*(T, S)*-cut LI-ideal of some IVI-*(T, S)*-fuzzy LI-ideal of L.*

Proof Define an interval-valued intuitionistic fuzzy set $\mathscr{A} = (\bar{\Phi}_{\mathscr{A}}, \bar{\Psi}_{\mathscr{A}})$ on L by

$$\bar{\Phi}_{\mathscr{A}}(x) = \begin{cases} \bar{\alpha}, & x \in K \\ [0,0], & x \notin K \end{cases} \quad \text{and} \quad \bar{\Psi}_{\mathscr{A}}(x) = \begin{cases} [1,1] - \bar{\alpha}, & x \in K \\ [1,1], & x \notin K \end{cases}$$

where $\bar{\alpha} \in D[0,1]$. Since K is an *LI*-ideal of L, we have $O \in K$. Therefore $\bar{\Phi}_{\mathscr{A}}(O) = \bar{\alpha} \succcurlyeq \bar{\Phi}_{\mathscr{A}}(x)$ and $\bar{\Psi}_{\mathscr{A}}(O) = [1,1] - \bar{\alpha} \preccurlyeq \bar{\Psi}_{\mathscr{A}}(x)$, for all $x \in L$. This shows that $\mathscr{A}$ satisfies (a1).

For all $x, y \in L$, if $x \in K$, then we have $\bar{\Phi}_{\mathscr{A}}(x) = \bar{\alpha} = T(\bar{\alpha}, \bar{\alpha}) \succcurlyeq T(\bar{\Phi}_{\mathscr{A}}((x \to y)'), \bar{\Phi}_{\mathscr{A}}(y))$ and $\bar{\Psi}_{\mathscr{A}}(x) = [1,1] - \bar{\alpha} = S([1,1] - \bar{\alpha}, [1,1] - \bar{\alpha}) \preccurlyeq T(\bar{\Psi}_{\mathscr{A}}((x \to y')), \bar{\Psi}_{\mathscr{A}}(y))$. If $x \notin K$, then $(x \to y)' \notin K$ or $y \notin K$ by K is an *LI*-ideal of L. Hence $\bar{\Phi}_{\mathscr{A}}(x) = [0,0] = T(\bar{\Phi}_{\mathscr{A}}((x \to y)'), \bar{\Phi}_{\mathscr{A}}(y))$ and $\bar{\Psi}_{\mathscr{A}}(x) = [1,1] = S(\bar{\Psi}_{\mathscr{A}}((x \to y')), \bar{\Psi}_{\mathscr{A}}(y))$. These show that $\mathscr{A}$ satisfies (a2).

Therefore $\mathscr{A}$ is an IVI-(T, S)-fuzzy *LI*-ideal of L. □

Theorem 7 *Let L be a lattice implication algebra, and* $\mathscr{A} = (\bar{\Phi}_{\mathscr{A}}, \bar{\Psi}_{\mathscr{A}}) \in$ **IVILI** (L). *Then* $K(\mathscr{A}) = \{x \in L | \bar{\Phi}_{\mathscr{A}}(x) = \bar{\Phi}_{\mathscr{A}}(O) \text{ and } \bar{\Psi}_{\mathscr{A}}(x) = \bar{\Psi}_{\mathscr{A}}(O)\}$ *is an LI-ideal of L.*

Proof Obviously, $O \in K(\mathscr{A})$ by the definition of $K(\mathscr{A})$. For any $x, y \in L$, let $(x \to y)' \in K(\mathscr{A})$ and $y \in K(\mathscr{A})$, then $\bar{\Phi}_{\mathscr{A}}((x \to y)') = \bar{\Phi}_{\mathscr{A}}(y) = \bar{\Phi}_{\mathscr{A}}(O)$ and $\bar{\Psi}_{\mathscr{A}}((x \to y)') = \bar{\Psi}_{\mathscr{A}}(y) = \bar{\Psi}_{\mathscr{A}}(O)$. It follows from $\mathscr{A} \in$ **IVILI**(L) that $\bar{\Phi}_{\mathscr{A}}(x) \succcurlyeq T(\bar{\Phi}_{\mathscr{A}}((x \to y)'), \bar{\Phi}_{\mathscr{A}}(y)) = T(\bar{\Phi}_{\mathscr{A}}(O), \bar{\Phi}_{\mathscr{A}}(O)) = \bar{\Phi}_{\mathscr{A}}(O)$ and $\bar{\Psi}_{\mathscr{A}}(x) \preccurlyeq S$

$(\bar{\Psi}_{\mathscr{A}}((x \to y)'), \bar{\Psi}_{\mathscr{A}}(y)) = S(\bar{\Psi}_{\mathscr{A}}(O), \bar{\Psi}_{\mathscr{A}}(O)) = \bar{\Psi}_{\mathscr{A}}(O)$, these together with $\bar{\Phi}_{\mathscr{A}}(O) \succcurlyeq \bar{\Phi}_{\mathscr{A}}(x)$ and $\bar{\Psi}_{\mathscr{A}}(O) \preccurlyeq \bar{\Psi}_{\mathscr{A}}(x)$, we have that $\bar{\Phi}_{\mathscr{A}}(x) = \bar{\Phi}_{\mathscr{A}}(O)$ and $\bar{\Psi}_{\mathscr{A}}(x) = \bar{\Psi}_{\mathscr{A}}(O)$, thus $x \in K(\mathscr{A})$. It follows that $K(\mathscr{A})$ is an *LI*-ideal of L. □

4 Interval-Valued Intuitionistic (S, T)-Fuzzy Lattice Ideals

Definition 8 Let L be a lattice implication algebra. An interval-valued intuitionistic fuzzy set $\mathscr{A} = (\bar{\Phi}_{\mathscr{A}}, \bar{\Psi}_{\mathscr{A}})$ on L is said to be an interval-valued intuitionistic (T, S)-fuzzy lattice ideal (IVI-(T, S)-fuzzy lattice ideal for short) of L, if for all $x, y \in L$, it satisfies the conditions (IVI3) and
(a9) $\bar{\Phi}_{\mathscr{A}}(x \vee y) \succcurlyeq T(\bar{\Phi}_{\mathscr{A}}(x), \bar{\Phi}_{\mathscr{A}}(y))$ and $\bar{\Psi}_{\mathscr{A}}(x \vee y) \preccurlyeq S(\bar{\Psi}_{\mathscr{A}}(x), \bar{\Psi}_{\mathscr{A}}(y))$.

Example 2 Let L be the lattice implication algebra given in Example 1. We define an interval-valued intuitionistic fuzzy set $\mathscr{A} = (\bar{\Phi}_{\mathscr{A}}, \bar{\Psi}_{\mathscr{A}})$ on L by $\bar{\Phi}_{\mathscr{A}}(O) = \bar{\Phi}_{\mathscr{A}}(d) = [0.7, 0.8]$, $\bar{\Phi}_{\mathscr{A}}(x) = [0.3, 0.4]$, $x \in L\backslash\{O, d\}$, and $\bar{\Psi}_{\mathscr{A}}(O) = \bar{\Psi}_{\mathscr{A}}(d) = [0.1, 0.2]$, $\bar{\Phi}_{\mathscr{A}}(x) = [0.5, 0.6]$, $x \in L\backslash\{O, d\}$. It is easy to verify that $\mathscr{A}$ is an interval-valued intuitionistic (T, S)-fuzzy lattice ideal of L. Where, for all $\bar{a} = [a^-, a^+], \bar{b} = [b^-, b^+] \in D[0, 1]$, $T(\bar{a}, \bar{b}) = [a^-b^-, \max\{a^-b^+, a^+b^-\}]$ and $S(\bar{a}, \bar{b}) = \max\{\bar{a}, \bar{b}\}$.

Theorem 8 *Let L be a lattice implication algebra, and* $\mathscr{A} = (\bar{\Phi}_{\mathscr{A}}, \bar{\Psi}_{\mathscr{A}}) \in$ **IVILI**(L). *Then* $\mathscr{A}$ *is an IVI-*(T, S)*-fuzzy lattice ideal of L.*

Proof Assume that $\mathscr{A} \in$ **IVILI**(L). Proposition 1 shows that $\mathscr{A}$ satisfies the condition (a3). Let $x, y \in L$, since $((x \vee y) \to y)' = ((x \to y) \wedge (y \to y))' = (x \to y)' \leqslant x$, by (a2) and (a3) we have $\bar{\Phi}_{\mathscr{A}}(x \vee y) \succcurlyeq T(\bar{\Phi}_{\mathscr{A}}(((x \vee y) \to y)'), \bar{\Phi}_{\mathscr{A}}(y)) \succcurlyeq T(\bar{\Phi}_{\mathscr{A}}(x), \bar{\Phi}_{\mathscr{A}}(y))$ and $\bar{\Psi}_{\mathscr{A}}(x \vee y) \preccurlyeq S(\bar{\Psi}_{\mathscr{A}}(((x \vee y) \to y)'), \bar{\Psi}_{\mathscr{A}}(y)) \preccurlyeq S(\bar{\Psi}_{\mathscr{A}}(x), \bar{\Psi}_{\mathscr{A}}(y))$.

Hence $\mathscr{A}$ satisfies the condition (a9). Therefore $\mathscr{A}$ is IVI-(T, S)-fuzzy lattice ideal of L by Definition 8. □

Remark 4 In general, the converse of Theorem 8 is not true. For example, the IVI-(T, S)-fuzzy lattice ideal $\mathscr{A}$ of L given in Example 2 is not an IVI-(T, S)-fuzzy *LI*-ideal of L. Since $\bar{\Phi}_{\mathscr{A}}(a) = [0.3, 0.4] \prec [0.7, 0.8] = T(\bar{\Phi}_{\mathscr{A}}((a \to d)'), \bar{\Phi}_{\mathscr{A}}(d))$.

Theorem 9 *Let L be a lattice H implication algebra and* $\mathscr{A} = (\bar{\Phi}_{\mathscr{A}}, \bar{\Psi}_{\mathscr{A}})$ *an IVI-*(T, S)*-fuzzy lattice ideal of L. Then* $\mathscr{A} \in$ **IVILI**(L).

Proof Assume that $\mathscr{A}$ is an IVI-(T, S)-fuzzy lattice ideal of lattice H implication algebra L. Since $O \leqslant x$ for any $x \in L$, by (a3) we have $\bar{\Phi}_{\mathscr{A}}(O) \succcurlyeq \bar{\Phi}_{\mathscr{A}}(x)$ and $\bar{\Psi}_{\mathscr{A}}(O) \preccurlyeq \bar{\Psi}_{\mathscr{A}}(x)$. Thus $\mathscr{A}$ satisfies the condition (a1). Let $x, y \in L$, we have

$$\begin{aligned}\bar{\Phi}_{\mathscr{A}}(x) \succcurlyeq \bar{\Phi}_{\mathscr{A}}(x \vee y) &= \bar{\Phi}_{\mathscr{A}}((x \to y) \to y) && \text{[by (a3) and Lemma 1]}\\ &= \bar{\Phi}_{\mathscr{A}}(y \vee (x \to y)') && \text{[by property H]}\\ &\succcurlyeq T(\bar{\Phi}_{\mathscr{A}}((x \to y)'), \bar{\Phi}_{\mathscr{A}}(y)) && \text{[by (a9)]}\end{aligned}$$

$$\begin{aligned}\bar{\Psi}_{\mathscr{A}}(x) \preccurlyeq \bar{\Psi}_{\mathscr{A}}(x \vee y) &= \bar{\Psi}_{\mathscr{A}}((x \to y) \to y) && \text{[by (a3) and Lemma 1]}\\ &= \bar{\Psi}_{\mathscr{A}}(y \vee (x \to y)') && \text{[by property H]}\\ &\preccurlyeq S(\bar{\Psi}_{\mathscr{A}}((x \to y)'), \bar{\Psi}_{\mathscr{A}}(y)) && \text{[by (a9)]}\end{aligned}$$

thus $\mathscr{A}$ satisfies condition (a2).

Hence we have $\mathscr{A} \in \mathbf{IVILI}(L)$ by Definition 7. □

5 Concluding Remarks

In this paper, we combined the notions of interval valued intuitionistic fuzzy sets, t-norm T and s-norm S on $D[0, 1]$ with the notion of ideals in lattice implication algebras, introduced the concepts interval-valued intuitionistic (T, S)-fuzzy LI-ideals and interval-valued intuitionistic (T, S)-fuzzy lattice ideals. Some their properties, characterizations and the relations between them were discussed. Several interesting results were obtained. This work gives interactions between Lattice-Valued Logic and the theory of interval-valued intuitionistic fuzzy sets. It should be noticed that other type interval-valued intuitionistic (T, S)-fuzzy ideals can also be considered in lattice implication algebras and other logical algebras by using the similar methods. We hope that more research topics of Lattice-Valued Logic will arise along this line.

Acknowledgments The work is supported by the Higher School Research Foundation of Inner Mongolia (NJSY14283).

References

1. Xu, Y.: Lattice implication algebras. J. Southwest Jiaotong Univ. **1**, 20–27 (1993)
2. Xu, Y., Ruan, D., Qin, K.Y.: Lattice-Valued Logics. Springer-Verlag, Berlin (2003)
3. Liu, J., Xu, Y.: On the representation of lattice implication algebras. J. Fuzzy Math. **7**, 251–258 (1999)
4. Qin, K.Y., Pei, Z., Jun, Y.B.: On normed lattice implication algebras. J. Fuzzy Math. **14**, 673–681 (2006)
5. Liu, C.H., Xu, L.S.: The representative theorems of lattice implication algebras by implication operator. Fuzzy Syst. and Math. **4**, 26–32 (2010)
6. Jun, Y.B., Roh, E.H., Xu, Y.: LI-ideals in lattice implication algebras. Bull. Korean Math. Soc. **35**, 13–24 (1998)
7. Jun, Y.B.: On LI-ideals and prime LI-ideals of lattice implication algebras. Bull. Korean Math. **36**, 369–380 (1999)
8. Liu, Y.L., Liu, S.Y., Xu, Y.: ILI-ideals and prime LI-ideals in lattice implication algebras. Inf. Sci. **155**, 157–175 (2003)

9. Liu, C.H.: LI-ideals lattice and its prime elements characterizations in a lattice implication algebra. Appl. Math. J. Chinese Univ. **29**, 475–482 (2014)
10. Zadeh, L.A.: Fuzzy sets. Inf. Control. **8**, 338–353 (1965)
11. Liu, L.Z., Li, K.T.: Fuzzy filters of BL-algbras. Inf. Sci. **173**, 141–154 (2005)
12. Zhan, J.M., Dudek, W.A.: Fuzzy h-ideals of hemirings. Inf. Sci. **177**, 876–886 (2007)
13. Liu, C.H.: $(\in, \in \vee q)$-fuzzy prime filters in BL-algebras. J. Zhejiang Univ. **41**, 489–493 (2014)
14. Peng, J.Y.: Intuitionistic fuzzy filters of effect algebras. Fuzzy Syst. and Math. **4**, 35–41 (2011)
15. Jun, Y.B., Kim, H.K.: Intuitionistic fuzzy approach to topological BCK-algebras. J. Mult-Valued Log. Soft Comput. **12**, 509–516 (2006)
16. Xue, Z.A., Xiao, Y.H., Li, W.H.: Intuitionistic fuzzy filters theory of BL-algebras. Int. J. Math. Learn. Cyber. **4**, 659–669 (2013)
17. Liu, C.H.: On intuitionistic fuzzy li -ideals in lattice implication algebras. J. Math. Res. Appl. **35**, 355–367 (2015)
18. Liu, Y., Qin, X.Y., Xu, Y.: Interval-valued intuitionistic (T, S)-fuzzy filters theory on residuated lattices. Int. J. Mach. Learn. Cybern. **5**, 683–696 (2014)
19. Atanassov, K.T., Gargov, G.: Interval valued intuitionistic fuzzy sets. Fuzzy Sets and Syst. **31**, 343–349 (1989)
20. Deschrijver, G.: A representation of t-norm in interval valued L-fuzzy set theory. Fuzzy Sets and Syst. **159**, 1597–1618 (2008)
21. Deschrijver, G.: Arithmetic operators in interval-valued fuzzy set theory. Inf. Sci. **177**, 2906–2924 (2007)
22. Atanassov, K.T.: Intuitionistic fuzzy sets. Fuzzy Sets and Syst. **20**, 87–96 (1986)

Tense Operators on Pseudo-MV Algebras

Wen-Juan Chen

Abstract In this paper the concept of tense operators on a pseudo-MV algebra is introduced. Since a pseudo-MV algebra can be regarded as an axiomatization of non-commutative infinite-valued Łukasiewicz logic, these tense operators are considered to quantify the dimension, i.e. one expresses "it is always going to be the case that" and the other expresses "it has always been the case that". We investigate basic properties of tense operators on pseudo-MV algebras and characterize the homomorphism of tense pseudo-MV algebras. Finally, we define a stronger version of tense pseudo-MV algebras and discuss the properties of filters under tense operators.

Keywords Tense operators · Pseudo-MV algebras · Tense pseudo-MV algebras · Filters

1 Introduction

It is well known that propositional logics do not incorporate dimension of time. To obtain a tense logic, the propositional calculus is enriched by adding new unary operations G, H, F and P in [3] which are called tense operators. The tense operator G usually expresses "it is always going to the case that" and H expresses "it has always been the case that", while the operators F and P can be defined by means of G and H, they usually express "it will at some time be the case that" and "it has at some time been the case that".

Study of tense operators was originated in the 1980s. In [3], tense operators were firstly introduced for the classical propositional logic as operators on the corresponding Boolean algebra satisfying the axioms:

W.-J. Chen (✉)
School of Mathematical Sciences, University of Jinan, No. 336, West Road of Nan Xinzhuang, Jinan 250022, Shandong, China
e-mail: wjchenmath@gmail.com

T.-H. Fan et al. (eds.), *Quantitative Logic and Soft Computing 2016*, Advances in Intelligent Systems and Computing 510, DOI 10.1007/978-3-319-46206-6_34

(B1) $G(1) = 1$, $H(1) = 1$;
(B2) $G(x \wedge y) = Gx \wedge Gy$, $H(x \wedge y) = Hx \wedge Hy$;
(B3) $x \leq GPx$, $x \leq HFx$.

Subsequently, to introduce tense operators in non-classical logics, the list of axioms for tense operators has been enlarged. For example, for Heyting algebras it was done in [4], for effect algebras see [5, 6], for basic algebras it was done in [2], for other interesting algebras the reader is referred to [7–10]. Among them, let us mention that tense MV-algebras were introduced in [8] which offered the algebraic framework in order to develop some tense many-valued logics. Many authors have investigated the representation of tense MV-algebras in [1, 8, 16].

On the other hand, the concepts of pseudo-MV algebras were introduced by Georgescu and Iorgulescu in [11, 12] and Rachunek in [17], respectively, as a non-commutative generalization of MV-algebras. Meanwhile, the pseudo-MV algebra can be regarded as an algebraic structure for the non-commutative infinite-valved Łukasiewicz propositional calculus [15]. Hence to introduce non-commutative tense many-valued logic, we investigate tense operators on pseudo-MV algebras in this paper.

The paper is organized as follows. In Sect. 2, we recall some definitions and results which will be used in what follows. In Sect. 3, we define tense pseudo-MV algebras and show the main results of our paper.

2 Preliminaries

In this section, we recall some definitions and results which will be used in what follows. Tense MV-algebras were introduced by Diagonescu and Georgescu in [8].

Definition 1 [8] Let $\mathbf{A} = \langle A; \oplus, ', 1\rangle$ be an MV-algebra. We say that $(\mathbf{A}, G, H)$ is a *tense MV-algebra*, if G and H are unary operations on $\mathbf{A}$ which are called *tense operators* and the following axioms are satisfied for all $x, y \in A$:
(TM1) $G(1) = H(1) = 1$;
(TM2) $G(x \to y) \leq G(x) \to G(y)$, $H(x \to y) \leq H(x) \to H(y)$, where $x \to y = x' \oplus y$;
(TM3) $Gx \oplus Gy \leq G(x \oplus y)$, $Hx \oplus Hy \leq H(x \oplus y)$;
(TM4) $G(x \oplus x) \leq Gx \oplus Gx$, $H(x \oplus x) \leq Hx \oplus Hx$;
(TM5) $Fx \oplus Fx \leq F(x \oplus x)$, $Px \oplus Px \leq P(x \oplus x)$, where F and P are the unary operations on A defined by $Fx = (Gx')'$, $Px = (Hx')'$;
(TM6) $x \leq GP(x)$, $x \leq HF(x)$.

In [8], the authors showed that if $(\mathbf{A}, G, H)$ is a tense MV-algebra, then axiom (TM2) is equivalent to the following conditions for all $x, y \in A$, (TM2$'$) $Gx \odot Gy \leq G(x \odot y)$ and $Hx \odot Hy \leq H(x \odot y)$; (TM2$''$) G and H are increasing, i.e., $x \leq y$ implies $Gx \leq Gy$ and $Hx \leq Hy$. Hence based on axiom (TM2$''$), for all $x \in A$, we have $G(0) \leq Gx$ and $H(0) \leq Hx$.

Let $\mathbf{A} = \langle A; \oplus, ', 1\rangle$ be an MV-algebra. Then we can define some operations as follows: $x \odot y = (x' \oplus y')'$, $x \ominus y = y' \odot x$ and $0 = 1'$. Below we give a new characterization of tense MV-algebra.

Proposition 1 *Let $\mathbf{A} = \langle A; \oplus, ', 1\rangle$ be an MV-algebra. If F and P are unary operations on $\mathbf{A}$ and the following axioms are satisfied for all $x, y \in A$:*
(DTM1) $F(0) = P(0) = 0$;
(DTM2) $F(x \ominus y) \geq F(x) \ominus F(y)$, $P(x \ominus y) \geq P(x) \ominus P(y)$;
(DTM3) $Fx \odot Fy \geq F(x \odot y)$, $Px \odot Py \geq P(x \odot y)$;
(DTM4) $F(x \odot x) \geq Fx \odot Fx$, $P(x \odot x) \geq Px \odot Px$;
(DTM5) $Gx \odot Gx \geq G(x \odot x)$, $Hx \odot Hx \geq H(x \odot x)$, *where G and H are the unary operations on A defined by* $Gx = (Fx')'$, $Hx = (Px')'$;
(DTM6) $x \geq FH(x)$, $x \geq PG(x)$.
Then $(\mathbf{A}, G, H)$ is a tense MV-algebra.

Proof It is easy to verify that $(\mathbf{A}, G, H)$ satisfies axioms (TM1)-(TM6) in Definition 1. We omit the details for brevity. □

Let $(\mathbf{A}, G, H)$ be a tense MV-algebra. Define a unary operation ρ on A by $\rho x = x \odot Gx \odot Hx$ for $x \in A$. Then $\rho x \leq x$ for $x \in A$. For $n \in \mathbb{N}$, we can define $\rho^n x$ by induction: $\rho^0 x = x$ and $\rho^{n+1} x = \rho(\rho^n x)$.

Proposition 2 ([8]) *Let $(\mathbf{A}, G, H)$ be a tense MV-algebra. Then the following two conditions are equivalent:*
(1) *$(\mathbf{A}, G, H)$ is a simple tense MV-algebra;*
(2) *For each $a \in A \setminus \{1\}$ there exists $n \in \mathbb{N}$ such that $\rho^n(a^n) = 0$.*

Proposition 3 ([8]) *Let $(\mathbf{A}, G, H)$ be a tense MV-algebra. Then the following two conditions are equivalent:*
(1) *$(\mathbf{A}, G, H)$ is a subdirectly irreducible tense MV-algebra;*
(2) *There exists $b \in A \setminus \{1\}$ such that for all $a \in A \setminus \{1\}$ there exists $n \in \mathbb{N}$ such that $\rho^n(a^n) \leq b$.*

3 Tense Pseudo-MV Algebras

In this section we define tense pseudo-MV algebras and discuss some basic properties of these structures. Then we give a strong version of tense pseudo-MV algebras, called strong tense pseudo-MV algebras and characterize filters under tense operators.

Recall that a pseudo-MV algebra is an algebra $\mathbf{A} = \langle A; \oplus, ^-, ^\sim, 1\rangle$ satisfying the following axioms for all $x, y, z \in A$:
(P1) $1^- = 1^\sim$ (is denoted by 0);
(P2) $x \oplus (y \oplus z) = (x \oplus y) \oplus z$;
(P3) $x \oplus 1^- = 1^- \oplus x = x$;
(P4) $x \oplus 1 = 1 \oplus x = 1$;

(P5) $(x^- \oplus y^-)^\sim = (x^\sim \oplus y^\sim)^-$;
(P6) $x \oplus (x^\sim \odot y) = y \oplus (y^\sim \odot x) = (x \odot y^-) \oplus y = (y \odot x^-) \oplus x$;
(P7) $x \odot (x^- \oplus y) = (x \oplus y^\sim) \odot y$;
(P8) $x^{-\sim} = x$, where $y \odot x = (x^- \oplus y^-)^\sim$.

A pseudo-MV algebra **A** is an MV-algebra if the operation $\oplus$ is commutative, i.e., $x \oplus y = y \oplus x$.

On pseudo-MV algebra **A**, we can define the following operations $x \to^L y = x^- \oplus y$, $x \to^R y = y \oplus x^\sim$, $x \ominus^L y = y^\sim \odot x$, $x \ominus^R y = x \odot y^-$, $x \vee y = x \oplus (x^\sim \odot y)$ and $x \wedge y = (x^- \vee y^-)^\sim$. We also define a natural partial order relation: $x \leq y$ if and only if $x \vee y = y$. With respect to this relation, $\langle A; \vee, \wedge, \leq \rangle$ is a bounded lattice.

Now, we give the definition of tense pseudo-MV algebras.

Definition 2 Let $\mathbf{A} = \langle A; \oplus, ^-, ^\sim, 1 \rangle$ be a pseudo-MV algebra and G, H unary operations on **A** satisfying the following axioms for all $x, y \in A$:
(TP1) $G(1) = H(1) = 1$;
(TP2) $Gx \oplus Gy \leq G(x \oplus y)$, $Hx \oplus Hy \leq H(x \oplus y)$;
(TP3) $G(x \to^L y) \leq Gx \to^L Gy$, $G(x \to^R y) \leq Gx \to^R Gy$,
$H(x \to^L y) \leq Hx \to^L Hy$, $H(x \to^R y) \leq Hx \to^R Hy$;
(TP4) $G(x \oplus x) \leq Gx \oplus Gx$, $H(x \oplus x) \leq Hx \oplus Hx$;
(TP5) $Fx \oplus Fx \leq F(x \oplus x)$, $Px \oplus Px \leq P(x \oplus x)$, where F and P are the unary operations on A defined by $Fx = (Gx^-)^\sim = (Gx^\sim)^-$ and $Px = (Hx^\sim)^- = (Hx^-)^\sim$;
(TP6) $x \leq GPx$, $x \leq HFx$.

Then $(\mathbf{A}, G, H)$ is called a *tense pseudo-MV algebra* and G, H are called *tense operators*.

Remark 1 Let **A** be a pseudo-MV algebra. If the operation $\oplus$ is commutative, then the unary operations $^-$ and $^\sim$ coincide, we have $x \to^L y = x \to^R y$, so $(\mathbf{A}, G, H)$ is a tense MV-algebra.

Example 1 Let **A** be a pseudo-MV algebra. Define unary operations $G = H$ such that $G(1) = 1$ and $G(x) = 0$ for $1 \neq x \in A$. Then $(\mathbf{A}, G, H)$ is a tense pseudo-MV algebra.

Example 2 Let **A** be a pseudo-MV algebra. Define the unary operations $G(x) = H(x) = x$ for each $x \in A$. Then $(\mathbf{A}, G, H)$ is a tense pseudo-MV algebra.

A *frame* is a pair (X, R), where X is a non-empty set and R is a binary relation on X. The notion of frame allows us to construct another example of tense pseudo-MV algebra.

Example 3 Let **L** be a complete pseudo-MV chain, (X, R) a frame and G^*, H^* the unary operations on the pseudo-MV algebra $\mathbf{L}^X$ defined by $G^*(p)(x) = \bigwedge_{y \in X}\{p(y) | xRy\}$, $H^*(p)(x) = \bigwedge_{y \in X}\{p(y) | yRx\}$ for $p \in L^X$ and $x \in X$. Then $(\mathbf{L}^X, G^*, H^*)$ is a tense pseudo-MV algebra.

Below we list elementary properties of tense pseudo-MV algebras.

Proposition 4 *Let $(\mathbf{A}, G, H)$ be a tense pseudo-MV algebra. Then we have*

(1) *if $x \leq y$, then $Gx \leq Gy$, $Hx \leq Hy$, $Fx \leq Fy$, $Px \leq Py$;*
(2) $G(x \to^L y) \leq Fx \to^L Fy$, $G(x \to^R y) \leq Fx \to^R Fy$, $H(x \to^L y) \leq Px \to^L Py$, $H(x \to^R y) \leq Px \to^R Py$;
(3) $Fx \ominus^L Fy \leq F(x \ominus^L y)$, $Fx \ominus^R Fy \leq F(x \ominus^R y)$, $Px \ominus^L Py \leq P(x \ominus^L y)$, $Px \ominus^R Py \leq P(x \ominus^R y)$;
(4) $Gx \odot Gy \leq G(x \odot y)$, $Hx \odot Hy \leq H(x \odot y)$, $F(x \odot y) \leq Fx \odot Fy$, $P(x \odot y) \leq Px \odot Py$;
(5) $F(x \oplus y) \leq Fx \oplus Fy$, $P(x \oplus y) \leq Px \oplus Py$;
(6) $G(x \vee y) \leq Fx \vee Gy$, $H(x \vee y) \leq Px \vee Hy$; $Gx \wedge Fy \leq F(x \wedge y)$, $Hx \wedge Py \leq P(x \wedge y)$;
(7) $G(x \oplus x) = Gx \oplus Gx$, $G(x \odot x) = Gx \odot Gx$, $H(x \oplus x) = Hx \oplus Hx$, $H(x \odot x) = Hx \odot Hx$, $F(x \oplus x) = Fx \oplus Fx$, $F(x \odot x) = Fx \odot Fx$, $P(x \oplus x) = Px \oplus Px$, $P(x \odot x) = Px \odot Px$;
(8) $Fx \odot y \leq F(x \odot Py)$, $Px \odot y \leq P(x \odot Fy)$, $G(Hx \oplus y) \leq x \oplus Gy$, $H(Gx \oplus y) \leq x \oplus Hy$;
(9) $PG(x) \leq x$, $FH(x) \leq x$;
(10) $PGP = P, GPG = G, HFH = H, FHF = F$;
(11) *G and H preserve arbitrary existing infima; While F and P preserve arbitrary existing suprema.*

Proof (1) Since $x \leq y \Leftrightarrow x \to^L y = 1 \Leftrightarrow x \to^R y = 1$, it follows that $1 = G(1) = G(x \to^L y) \leq Gx \to^L Gy$ by (TP1) and (TP3). Hence $Gx \to^L Gy \leq 1$, we have $Gx \to^L Gy = 1$, so $Gx \leq Gy$. Similarly, we thus have $Hx \leq Hy$. If $x \leq y$, then $y^- \leq x^-$, we have $G(y^-) \leq G(x^-)$ and $(Fx)^- = G(x^-)$, it turns out that $(Fy)^- \leq (Fx)^-$, thus $Fx \leq Fy$. Dually, we have $Px \leq Py$.

(2) Since $x \to^L y = y^- \to^R x^-$, we have $G(x \to^L y) = G(y^- \to^R x^-) \leq G(y^-) \to^R G(x^-) = (Fy)^- \to^R (Fx)^- = Fx \to^L Fy$. The rest can be proved similarly.

(3) We have $Fx \ominus^L Fy = (Fy)^\sim \odot Fx = ((Fx)^- \oplus Fy)^\sim = (Fx \to^L Fy)^\sim \leq (G(x \to^L y))^\sim = F((x \to^L y)^\sim) = F(y^\sim \odot x) = F(x \ominus^L y)$ using (2). The rest can be proved similarly.

(4) We have $Hx \odot Hy \leq Hx \odot H(x^\sim \vee y) = Hx \odot H(x \to^R (x \odot y)) \leq Hx \odot (Hx \to^R H(x \odot y)) = Hx \wedge H(x \odot y) \leq H(x \odot y)$. Similarly, $Gx \odot Gy \leq G(x \odot y)$. On the other hand, $F(x \odot y) = (G(x \odot y)^-)^\sim = (G(y^- \oplus x^-))^\sim \leq (G(y^-) \oplus G(x^-))^\sim = ((Fy)^- \oplus (Fx)^-)^\sim = Fx \odot Fy$. Similarly, $P(x \odot y) \leq Px \odot Py$.

(5) We have $(F(x \oplus y))^- = G((x \oplus y)^-) = G(y^- \odot x^-) \geq G(y^-) \odot G(x^-) = (Fy)^- \odot (Fx)^- = (Fx \oplus Fy)^-$, thus $F(x \oplus y) \leq Fx \oplus Fy$. Similarly, $P(x \oplus y) \leq Px \oplus Py$.

(6) We have $Fx \vee Gy = Gy \oplus ((Fx)^- \oplus Gy)^\sim = Gy \oplus (G(x^-) \oplus Gy)^\sim \geq Gy \oplus (G(x^- \oplus y))^\sim = Gy \oplus (G(x \to^L y))^\sim = G(x \to^L y) \to^R Gy \geq G((x \to^L y) \to^R y) = G(x \vee y)$. Similarly, $H(x \vee y) \leq Px \vee Hy$. On the other hand, we

have $F(x \wedge y) = (G(x \wedge y)^-)^\sim = (G(x^- \vee y^-))^\sim \geq (Fx^- \vee Gy^-)^\sim = (Fx^-)^\sim \wedge (Gy^-)^\sim = Gx \wedge Fy$. Similarly, $Hx \wedge Py \leq P(x \wedge y)$.

(7) By (TP2) and (TP4), we have $G(x \oplus x) = Gx \oplus Gx$. On the other hand, $G(x \odot x) = G((x^\sim \oplus x^\sim)^-) = (F(x^\sim \oplus x^\sim))^- \leq (F(x^\sim) \oplus F(x^\sim))^- = ((Gx)^\sim \oplus (Gx)^\sim)^- = Gx \odot Gx$. And by (4), we have $Gx \odot Gx \leq G(x \odot x)$. Thus $G(x \odot x) = Gx \odot Gx$. The rest can be proved similarly.

(8) Since $y \leq x^- \vee y$, we have $Gy \leq G(x^- \vee y) = G(x \to^L x \odot y) \leq Fx \to^L F(x \odot y) = (Fx)^- \oplus F(x \odot y)$ by (2), thus $1 = (Gy)^- \oplus Gy \leq (Gy)^- \oplus ((Fx)^- \oplus F(x \odot y)) = ((Gy)^- \oplus (Fx)^-) \oplus F(x \odot y) = (Fx \odot Gy)^- \oplus F(x \odot y)$, it turns out that $(Fx \odot Gy)^- \oplus F(x \odot y) = 1$, which implies that $Fx \odot Gy \leq F(x \odot y)$. Hence $Fx \odot y \leq Fx \odot GPy \leq F(x \odot Py)$. Similarly, $Px \odot y \leq P(x \odot Fy)$. On the other hand, $G(Hx \oplus y) = (F(Hx \oplus y)^\sim)^- = (F(y^\sim \odot (Hx)^\sim))^- = (F(y^\sim \odot Px^\sim))^- \leq (Fy^\sim \odot x^\sim)^- = x \oplus (Fy^\sim)^- = x \oplus Gy$. Similarly, $H(Gx \oplus y) \leq x \oplus Hy$.

(9) Since $Fx = (Gx^-)^\sim$, we have $Fx^\sim = (Gx)^\sim$. By (TP6), it follows that $x^\sim \leq HF(x^\sim) = H(Gx)^\sim$, then $PGx = (H(Gx)^\sim)^- \leq x^{\sim -} = x$. Similarly, $FHx \leq x$.

(10) Since $x \leq GPx$, we have $Px \leq PGPx$ by (1). Using (9), it follows that $PGPx \leq Px$, thus $PGPx = Px$. The rest can be proved similarly.

(11) By (1), $G(\wedge_{i \in I} x_i) \leq G(x_i)$ for all $i \in I$, we have $G(\wedge_{i \in I} x_i) \leq \wedge_{i \in I} Gx_i$. Suppose that $y = \wedge_{i \in I} Gx_i$. Then $y \leq G(x_i)$ for all $i \in I$. Using (1) and (9), $Py \leq PG(x_i) \leq x_i$ for all $i \in I$, which implies $Py \leq \wedge_{i \in I} x_i$, it follows that $y \leq GPy \leq G(\wedge_{i \in I} x_i)$. Hence $G(\wedge_{i \in I} x_i) = \wedge_{i \in I} Gx_i$. Similarly, $H(\wedge_{i \in I} x_i) = \wedge_{i \in I} Hx_i$. On the other hand, $F(\vee_{i \in I} x_i) = (G(\vee_{i \in I} x_i)^-)^\sim = (G(\wedge_{i \in I} x_i^-))^\sim = (\wedge_{i \in I}(Gx_{i \in I}^-))^\sim = \vee_{i \in I} Fx_i$. Similarly, $P(\vee_{i \in I} x_i) = \vee_{i \in I} Px_i$. □

Following from Proposition 4, we will see that the tense operators F and P satisfy conditions (DTP2)-(DTP5) in the next proposition. Moreover, it is easy to show that $F(0) = P(0) = 0$ and $FHx \leq x$, $PGx \leq x$. Conversely, if the operations F and P satisfy conditions (DTP1)-(DTP6), then $(\mathbf{A}, G, H)$ is a tense pseudo-MV algebra. The proof is straightforward.

Proposition 5 *Let* $\mathbf{A} = \langle A; \oplus, ^-, ^\sim, 1\rangle$ *be a pseudo-MV algebra and* F, P *unary operations on* $\mathbf{A}$ *satisfying the following conditions for all* $x, y \in A$:
(DTP1) $F(0) = P(0) = 0$;
(DTP2) $Fx \odot Fy \leq F(x \odot y)$, $Px \odot Py \leq P(x \odot y)$;
(DTP3) $F(x \ominus^L y) \leq Fx \ominus^L Fy$, $F(x \ominus^R y) \leq Fx \ominus^R Fy$,
$P(x \ominus^L y) \leq Px \ominus^L Py$, $P(x \ominus^R y) \leq Px \ominus^R Py$;
(DTP4) $F(x \odot x) \leq Fx \odot Fx$, $P(x \odot x) \leq Px \odot Px$;
(DTP5) $Gx \odot Gx \leq G(x \odot x)$, $Hx \odot Hx \leq H(x \odot x)$, *where* G *and* H *are the unary operations of* $\mathbf{A}$ *defined by* $Gx = (F(x^-))^\sim$ *and* $Hx = (P(x^\sim))^-$;
(DTP6) $FHx \leq x$, $PGx \leq x$.

Then $(\mathbf{A}, G, H)$ *is a tense pseudo-MV algebra.*

Let $(\mathbf{A}, G, H)$ be a tense pseudo-MV algebra. We define a unary operation d on $\mathbf{A}$ by $dx = x \wedge Gx \wedge Hx$ for any $x \in A$. Then $dx \leq x$ for any $x \in A$. For $n \in \mathbb{N}$, we define $d^n x$ by induction $d^0 x = x$ and $d^{n+1} x = d(d^n x)$.

Lemma 1 *Let $(\mathbf{A}, G, H)$ be a tense pseudo-MV algebra. Then for $x, y \in A$ and $n \in \mathbb{N}$, the following statements hold:*
(1) $d^n 0 = 0, d^n 1 = 1, d^{n+1}x \leq d^n x$;
(2) *if* $x \leq y$, *then* $d^n x \leq d^n y$;
(3) $x \leq d^n(d^n x^-)^\sim$ *and* $x \leq d^n(d^n x^\sim)^-$;
(4) *if* $dx = x$, *then* $dx^- = x^-$ *and* $dx^\sim = x^\sim$;
(5) $x = dx$ *if and only if* $d^n x = x$.

Proof We only prove (3). The rest are obvious. Since $x \leq x \wedge GPx \wedge HFx \leq (x \vee Fx \vee Px) \wedge G(x \vee Fx \vee Px) \wedge H(x \vee Fx \vee Px) = d(x \vee Fx \vee Px) = d(x \vee (Gx^\sim)^- \vee (Hx^\sim)^-) = d((x^\sim \wedge Gx^\sim \wedge Hx^\sim)^-) = d(dx^\sim)^-$, we have $(d^n x^\sim)^- \leq d(d(d^n x^\sim))^- = d(d^{n+1}x^\sim)^-$, it follows that $d^n(d^n x^\sim)^- \leq d^n(d(d^{n+1}x^\sim)^-) = d^{n+1}(d^{n+1}x^\sim)^-$. Thus (3) is true by induction. □

Proposition 6 *Let $(\mathbf{A}, G, H)$ be a tense pseudo-MV algebra. Then $d(A) = \{x \in A | dx = x\}$ is closed under the pseudo-MV operations of $\mathbf{A}$, i.e., $\langle d(A); \oplus, ^-, ^\sim, 1\rangle$ is a pseudo-MV subalgebra of $\mathbf{A}$.*

Proof Let $x, y \in d(A)$. Then $d(x \oplus y) = (x \oplus y) \wedge G(x \oplus y) \wedge H(x \oplus y)$. Since $dx = x$ and $dy = y$, it turns out that $x \leq Gx$, $x \leq Hx$ and $y \leq Gy$, $y \leq Hy$. By Definition 2 (2), we have $x \oplus y \leq Gx \oplus Gy \leq G(x \oplus y)$. Similarly, $x \oplus y \leq H(x \oplus y)$, so $d(x \oplus y) = x \oplus y$ which implies that $d(A)$ is closed under operation $\oplus$. According to Lemma 1, $d(A)$ is closed under unary operations $^-$ and $^\sim$. Moreover, since $d(0) = 0$ and $d(1) = 1$, we have $0, 1 \in d(A)$. Hence $d(A)$ is a pseudo-MV subalgebra of $\mathbf{A}$. □

Definition 3 Let $(\mathbf{A}_1, G_1, H_1)$ and $(\mathbf{A}_2, G_2, H_2)$ be tense pseudo-MV algebras. A function $f : (\mathbf{A}_1, G_1, H_1) \to (\mathbf{A}_2, G_2, H_2)$ is called a *homomorphism*, if $f : \mathbf{A}_1 \to \mathbf{A}_2$ is a homomorphism of pseudo-MV algebras and $f(G_1 x) = G_2(f(x))$ and $f(H_1 x) = H_2(f(x))$ for all $x \in A$.

Remark 2 Let $f : (\mathbf{A}_1, G_1, H_1) \to (\mathbf{A}_2, G_2, H_2)$ be a homomorphism of tense pseudo-MV algebras. If $x \in d(A_1)$, then $f(x) \in d(A_2)$. Hence we can define a function $f^d = f|_{d(A_1)} : d(A_1) \to d(A_2)$, it follows that f^d is a homomorphism of pseudo-MV algebras. In fact, the assignment $A \mapsto d(A)$ and $f \mapsto f^d$ define a covariant functor $(.)^d$ from the category of tense pseudo-MV algebras to the category of pseudo-MV algebras

In the following, we characterize homomorphisms from an arbitrary tense pseudo-MV algebra to the tense pseudo-MV algebra in Example 3.

Lemma 2 *Let (A, G, H) be a tense pseudo-MV algebra, X a non-empty set, $\mathbf{L}$ a complete pseudo-MV chain and $f : \mathbf{A} \to \mathbf{L}^X$ a homomorphism of pseudo-MV algebras. For $x, y \in X$, let*
$$\alpha_{xy} = \bigwedge_{a \in A}(f(a)(x) \to^R f(Pa)(y));\ \beta_{xy} = \bigwedge_{b \in A}(f(b)(y) \to^L f(Fb)(x));$$
$$\gamma_{xy} = \bigwedge_{c \in A}(f(Gc)(y) \to^R f(c)(x));\ \delta_{xy} = \bigwedge_{d \in A}(f(Hd)(x) \to^L f(d)(y)).$$
Then $\alpha_{xy} = \beta_{xy} = \gamma_{xy} = \delta_{xy}$.

Proof We have

$$
\begin{aligned}
\beta_{xy} &= \bigwedge_{b\in A}(f(b)(y) \to^L f(Fb)(x)) = \bigwedge_{c\in A}(f(c^\sim)(y) \to^L f(Fc^\sim)(x)) \\
&= \bigwedge_{c\in A}((f(Fc^\sim)(x))^- \to^R (f(c^\sim)(y))^-) \\
&= \bigwedge_{c\in A}(f((Fc^\sim)^-)(x) \to^R f((c^\sim)^-)(y)) \\
&= \bigwedge_{c\in A}(f(Gc)(x) \to^R f(c)(y)) = \gamma_{xy},
\end{aligned}
$$

and

$$
\begin{aligned}
\alpha_{xy} &= \bigwedge_{a\in A}(f(a)(x) \to^R f(Pa)(y)) = \bigwedge_{d\in A}(f(d^-)(x) \to^R f(Pd^-)(y)) \\
&= \bigwedge_{d\in A}((f(Pd^-)(y))^\sim \to^L (f(d^-)(x))^\sim) \\
&= \bigwedge_{d\in A}(f((Pd^-)^\sim)(y) \to^L f((d^-)^\sim)(x)) \\
&= \bigwedge_{d\in A}(f(Hd)(y) \to^L f(d)(x)) = \delta_{xy}.
\end{aligned}
$$

For any $c \in A$, since $PGc^- \leq c^-$ by Proposition 4(9), it follows that

$$
\begin{aligned}
\alpha_{xy} &\leq f(Gc^-)(x) \to^R f(PGc^-)(y) \\
&\leq f(Gc^-)(x) \to^R f(c^-)(y) \\
&= (f(c^-)(y))^\sim \to^L (f(Gc^-))^\sim \\
&= f(c)(y) \to^L f(Fc)(x),
\end{aligned}
$$

therefore, $\alpha_{xy} \leq \wedge_{c\in A}(f(c)(y) \to^L f(Fc)(x)) = \beta_{xy}$.

On the other hand, for $b \in A$, since $FHb^\sim \leq b^\sim$ by Proposition 4(9) again, we have

$$
\begin{aligned}
\beta_{xy} &\leq f(Hb^\sim)(y) \to^L f(FHb^\sim)(y) \\
&\leq f(Hb^\sim)(y) \to^L f(b^\sim)(x) \\
&= (f(b^\sim)(x))^- \to^R (f(Hb^\sim)(y))^- \\
&= f(b)(x) \to^R f(Pb)(y),
\end{aligned}
$$

therefore, $\beta_{xy} \leq \wedge_{b\in A} f(b)(x) \to^R f(Pb)(y) = \alpha_{xy}$, it turns out that $\alpha_{xy} = \beta_{xy}$. Hence $\alpha_{xy} = \beta_{xy} = \gamma_{xy} = \delta_{xy}$. □

Given a homomorphism of pseudo-MV algebras $f : \mathbf{A} \to \mathbf{L}^X$, we define the following binary relation R_f on X: $R_f = \{(x, y) \in X^2 | \alpha_{xy} = 1\}$.

Proposition 7 *Let* $\mathbf{A}$ *and* $\mathbf{L}^X$ *be pseudo-MV algebras and* $f : \mathbf{A} \to \mathbf{L}^X$ *a homomorphism. Then the following assertions are equivalent:*
(1) $f : (\mathbf{A}, G, H) \to (\mathbf{L}^X, G^*, H^*)$ *is a homomorphism of tense pseudo-MV algebras;*
(2) For any $a \in A$ *and* $x \in X$*, the inequalities* $\bigwedge\{f(a)(y) | xR_f y\} \leq f(Ga)(x)$ *and* $\bigwedge\{f(a)(y) | yR_f x\} \leq f(Ha)(x)$ *hold;*
(3) For any $a \in A$ *and* $x \in X$*, the inequalities* $f(Fa)(x) \leq \bigvee\{f(a)(y) | xR_f y\}$ *and* $f(Pa)(x) \leq \bigvee\{f(a)(y) | yR_f x\}$ *hold.*

Proof The proof is similar to the case of tense MV-algebras. See [8]. □

Let $\mathbf{A}$ be a pseudo-MV algebra and $\mathfrak{F}$ a nonempty subset of A. $\mathfrak{F}$ is called a *filter* of $\mathbf{A}$ if the following conditions are satisfied: (F1) $1 \in \mathfrak{F}$; (F2) if $x, y \in \mathfrak{F}$, then $x \odot y \in \mathfrak{F}$ and $y \odot x \in \mathfrak{F}$; (F3) if $x \in \mathfrak{F}$ and $y \in A$ with $x \leq y$, then $y \in \mathfrak{F}$. Moreover, a filter $\mathfrak{F}$ with $x \odot \mathfrak{F} = \mathfrak{F} \odot x$ for $x \in A$ is called a *normal filter*. Any normal filter $\mathfrak{F}$ of $\mathbf{A}$ determines a congruence $\theta_{\mathfrak{F}}$ on $\mathbf{A}$ defined by $\langle x, y\rangle \in \theta_{\mathfrak{F}}$ iff $x \to^R y \in \mathfrak{F}$ and $y \to^R x \in \mathfrak{F}$ iff $x \to^L y \in \mathfrak{F}$ and $y \to^L x \in \mathfrak{F}$. Conversely, a congruence θ on $\mathbf{A}$ determines a normal filter $\mathfrak{F}_\theta$ of $\mathbf{A}$ defined by $\mathfrak{F}_\theta = \{x \in A | \langle x, 1\rangle \in \theta\}$. It is easy to show that the correspondence between the set of all normal filters and the set of all congruences is bijective.

Let $(\mathbf{A}, G, H)$ be a tense pseudo-MV algebra and $\mathfrak{F}$ a filter of $\mathbf{A}$. Then $\mathfrak{F}$ is called a *tense filter* of $(\mathbf{A}, G, H)$, if $\mathfrak{F}$ is closed under operators G and H. In other words, if $\mathfrak{F}$ is a tense filter of $\mathbf{A}$, then for all $x \in \mathfrak{F}$, we have $Gx \in \mathfrak{F}$ and $Hx \in \mathfrak{F}$. If a tense filter $\mathfrak{F}$ is normal, then $\mathfrak{F}$ is called a *normal tense filter*. On the other hand, let θ be a congruence on $\mathbf{A}$. If $\langle x, y\rangle \in \theta$ implies $\langle Gx, Gy\rangle \in \theta$ and $\langle Hx, Hy\rangle \in \theta$, then θ is called a *congruence* on $(\mathbf{A}, G, H)$.

Proposition 8 *Let* $(\mathbf{A}, G, H)$ *be a tense pseudo-MV algebra. Then there exists a bijective correspondence between the normal tense filters of* $(\mathbf{A}, G, H)$ *and the congruences on* $(\mathbf{A}, G, H)$.

Proof Straightforward. □

Definition 4 Let $(\mathbf{A}, G, H)$ be a tense pseudo-MV algebra. If the tense operators G and H satisfy $G^2x = Gx$, $H^2x = Hx$ and $GHx = HGx = x$ for all $x \in A$, we call $(\mathbf{A}, G, H)$ a *strong tense pseudo-MV algebra.*

Proposition 9 *Let* $(\mathbf{A}, G, H)$ *be a strong tense pseudo-MV algebra. The tense filter* $[a)$ *of* $(\mathbf{A}, G, H)$ *generated by* $\{a\}$ *has the following form* $[a) = \{x \in A | x \geq \mu_1^{n_1} \odot \mu_2^{n_2} \odot \cdots \odot \mu_t^{n_t}$*, where* $\mu_i \in \{a, Ga, Ha\}, n_i \in \mathbb{N}, i = 1, 2, \ldots, t\}$.

Proof Obviously, $1 \in [a)$. Let $x, y \in [a)$. Then $x \geq \mu_1^{n_1} \odot \mu_2^{n_2} \odot \cdots \odot \mu_t^{n_t}$ where $\mu_i \in \{a, Ga, Ha\}, n_i \in \mathbb{N}, i = 1, 2, \ldots, t$ and $y \geq \nu_1^{m_1} \odot \nu_2^{m_2} \odot \cdots \odot \nu_s^{m_s}$ where $\nu_j \in \{a, Ga, Ha\}, m_j \in \mathbb{N}, j = 1, 2, \ldots, s$. Then $x \odot y \geq \mu_1^{n_1} \odot \mu_2^{n_2} \odot \cdots \odot \mu_t^{n_t}$

$\odot \nu_1^{m_1} \odot \nu_2^{m_2} \odot \cdots \odot \nu_s^{m_s}$ where $\mu_i, \nu_j \in \{a, Ga, Ha\}$ and $n_i, m_j \in \mathbb{N}$, $i = 1, 2, \ldots, t$, $j = 1, 2, \ldots, s$. Hence $x \odot y \in [a)$. Similarly, $y \odot x \in [a)$. Suppose that $x \in [a)$ and $y \in A$ with $x \leq y$. Then $y \in [a)$. For all $x \in [a)$. Then $x \geq \mu_1^{n_1} \odot \mu_2^{n_2} \odot \cdots \odot \mu_t^{n_t}$ where $\mu_i \in \{a, Ga, Ha\}, n_i \in \mathbb{N}, i = 1, 2, \ldots, t$, it turns out that $Gx \geq G(\mu_1^{n_1}) \odot G(\mu_2^{n_2}) \odot \cdots \odot G(\mu_t^{n_t}) \geq (G\mu_1)^{n_1} \odot (G\mu_2)^{n_2} \odot \cdots \odot (G\mu_t)^{n_t}$ where $G\mu_i \in \{a, Ga\} \subseteq \{a, Ga, Ha\}$, thus $Gx \in [a)$. Similarly, $Hx \in [a)$. Hence $[a)$ is a tense filter of $(\mathbf{A}, G, H)$. □

Corollary 1 *Let $(\mathbf{A}, G, H)$ be a strong tense MV-algebra. Then the tense filter $[a)$ of $(\mathbf{A}, G, H)$ generated by $\{a\}$ has the following form: $[a) = \{x \in A | x \geq a^{n_1} \odot (Ga)^{n_2} \odot (Ha)^{n_3}, n_i \in \mathbb{N}, i = 1, 2, 3\}$. Moreover, $[a)$ is normal.*

Proposition 10 *Let $(\mathbf{A}, G, H)$ be a strong tense MV-algebra. Then the following assertions are equivalent:*
(1) $(\mathbf{A}, G, H)$ is a simple strong tense MV-algebra;
(2) For each $a \in A \backslash \{1\}$, there exist $n_1, n_2, n_3 \in \mathbb{N}$ such that $a^{n_1} \odot (Ga)^{n_2} \odot (Ha)^{n_3} = 1$.

Proposition 11 *Let $(\mathbf{A}, G, H)$ be a strong tense MV-algebra. Then the following assertions are equivalent:*
(1) $(\mathbf{A}, G, H)$ is a subdirectly irreducible strong tense MV-algebra;
(2) There exists $b \in A \backslash \{1\}$ such that for any $a \in A \backslash \{1\}$ there exist $n_1, n_2, n_3 \in \mathbb{N}$ such that $b \leq a^{n_1} \odot (Ga)^{n_2} \odot (Ha)^{n_3}$.

We denote $B(A) = \{x \in A | x \oplus x = x\}$.

Corollary 2 *Let $(\mathbf{A}, G, H)$ be a strong tense MV-algebra and $a \in B(A)$. Then the tense filter $[a)$ of $(\mathbf{A}, G, H)$ generated by $\{a\}$ is: $[a) = \{x \in A | x \geq a \odot Ga \odot Ha\}$. Moreover, $[a)$ is normal.*

4 Conclusions

In this paper, we investigate tense operators on pseudo-MV algebras which can be regraded as a non-commutative generalization of tense MV-algebras. In view of importance of filters in studying algebras of logics, we discuss properties of tense filters in tense pseudo-MV algebras. In our future work, we will consider characterization of special tense filters and focus on tense operators on other non-commutative logical algebras.

Acknowledgments I would like to show my sincere thanks to the referees for their valuable comments. This project is supported by the National Natural Science Foundation of China (Grant No. 11501245).

References

1. Botur, M., Paseka, J.: On tense MV-algebras. Fuzzy Sets Syst. **259**, 111–125 (2015)
2. Botur, M., Chajda, I., Halas, R., Kolarik, M.: Tense operators on basic algebras. Int. J. Theor. Phys. **50**, 3737–3749 (2011)
3. Burges, J.P.: Basic tense logic. In: Gabbay, D.M., Gunther, F. (eds.) Handbook of Philosophical Logic, vol. II, pp. 89–133. Springer, Netherlands (1984)
4. Chajda, I.: Algebraic axiomatization of tense intuitionistic logic. Cent. Eur. J. Math. **9**, 1185–1191 (2011)
5. Chajda, I., Kolarik, M.: Dynamic effect algebras. Math. Slovaca. **62**, 379–388 (2012)
6. Chajda, I., Paseka, J.: Dynamic effect algebras and their representation. Soft. Comput. **16**, 1733–1741 (2012)
7. Chajda, I., Paseka, J.: Tense operators in fuzzy logic. Fuzzy Sets Syst. **276**, 100–113 (2015)
8. Doacpmescu, D., Georgescu, G.: Tense operators on MV-algebras and Łukasiewicz-Moisil algebras. Fundam. Inform. **81**, 379–408 (2007)
9. Figallo, A.V., Pelaitay, G.: Tense operators on SH_n-algebras. Pioneer J. Algebra, Number Theory Appl. **1**, 33–41 (2011)
10. Figallo, A.V., Gallardo, G., Pelaitay, G.: Tense operators on m-symmetric algebras. Int. Math. Forum. **41**, 2007–2014 (2011)
11. Georgescu, G., Iorgulescu, A.: Pseudo MV algebras: a noncommutative extension of MV algebras. In: Proceedings of the Fourth International Symposium on Economic Informatics, pp. 961–968. Bucharest (1999)
12. Georgescu, G., Iorgulescu, A.: Pseudo MV algebras. Mult Valued Log **6**, 95–135 (2001)
13. Kowalski, T.: Varieties of tense algebras. Rep. Math. Logic **30**, 1–43 (1996)
14. Lemmon, E.J.: Algebraic semantics for modal logic. J. Symb. Logic **31**, 191–218 (1996)
15. Leustean, I.: Non-commutative Łukasiewicz propositional logic. Arch. Math. Logic **45**, 191–213 (2006)
16. Paseka, J.: Operators on MV-algebras and their representations. Fuzzy Sets Syst. **232**, 62–73 (2013)
17. Rachunek, J.: A non-commutative generalization of MV-algebras. Czechoslvak Math. J. **52**, 255–273 (2002)

(f, g)-Derivations on R_0-Algebras

Hua-Rong Zhang

Abstract In this paper, the notions of (f, g)-derivations and isotone (f, g)-derivations on R_0-algebras are introduced and related properties are discussed. Some characterization theorems of these derivations are derived. We give some equivalent characterizations for isotone (f, g)-derivations. Finally we investigate the properties of (f, g)-derivations on linearly-ordered R_0-algebras.

Keywords (f, g)-derivation · R_0-algebra · Boolean element · Isotone (f, g)-derivation

1 Introduction

In [25], Wang proposed the notion of R_0-algebras for the purpose of providing an algebraic structure for fuzzy logic system $\mathfrak{L}^*$. Recently, algebraic theory of R_0-algebras has been intensively explored [8, 11, 17, 21, 27]. The notion of derivation, introduced from analysis, is helpful to study the structure and property of algebraic systems. In recent years, some authors [1, 3, 4, 15, 16, 22–24] have investigated derivations in rings, near rings, prime rings and semiprime rings. After these studies, several authors [10, 24, 26] discussed the derivations of lattices. Subsequently, the research about generalized derivations on rings [9, 12] and lattices showed up. In [6], Ceven and Ozturk studied f-derivations of lattices. Ceven [5] also discussed symmetric bi-derivations of lattices. In [20], Ozbal and Firat introduced the notion of symmetric f-bi-derivations of a lattice. Generalized (f, g)-derivations on lattices were investigated by Mustafa and Sahin [18]. The derivations and f-derivations of BCI-algebras were defined and studied in [13, 14, 19]. In [2], Alshehri applied the notion of derivations to MV-algebras [7] and discussed some properties.

H.-R. Zhang (✉)
College of Sciences, China Jiliang University, Hangzhou 310018, China
e-mail: hrzhang2008@cjlu.edu.cn

T.-H. Fan et al. (eds.), *Quantitative Logic and Soft Computing 2016*,
Advances in Intelligent Systems and Computing 510,
DOI 10.1007/978-3-319-46206-6_35

Now, in this paper, we study the (f, g)-derivations on R_0-algebras. Since derivations and f-derivation of R_0-algebras are special (f, g)-derivations, we can know the corresponding properties of derivations and f-derivations from (f, g)-derivations.

The paper is organized as follows. In Sect. 2, basic definitions and results are given. In Sect. 3, we introduce the notion of (f, g)-derivations on R_0-algebras and study their properties. In Sect. 4, we study the properties of isotone (f, g)-derivations of R_0-algebras and give many equivalent characterizations for isotone (f, g)-derivations. In Sect. 5, we particularly discuss the (f, g)-derivations of linearly-ordered R_0-algebras.

2 Preliminaries

Definition 1 ([25]) Let L be a bounded distributive lattice with order-reversing involution $'$ and a binary operation $\to$, $(L, ', \vee, \to)$ is called an R_0-algebra if it satisfies the following axioms:
(1) $x \to y = y' \to x'$;
(2) $1 \to x = x$;
(3) $(y \to z) \wedge ((x \to y) \to (x \to z)) = y \to z$;
(4) $x \to (y \to z) = y \to (x \to z)$;
(5) $x \to (y \vee z) = (x \to y) \vee (x \to z)$;
(6) $(x \to y) \vee ((x \to y) \to (x' \vee y)) = 1$.

Now, if we define $x \leq y$ if and only if $x \vee y = y$ for all $x, y \in L$, then according to [25], $\leq$ is an order relation over L. If the order $\leq$ defined on L is total, then we say that L is a linearly ordered R_0-algebra.

Let L be an R_0-algebra, for $x, y \in L$, define $x \odot y = (x \to y')'$, $x \oplus y = x' \to y$. It is easily proved that $\odot$ and $\oplus$ are commutative, associative and $x \oplus y = (x' \odot y')'$. In [21], Pei proved that R_0-algebra is a particular MTL algebra and its t-norm $\odot$ is a nilpotent minimum t-norm.

In what follows, L will denote an R_0-algebra, unless otherwise specified.

Definition 2 ([11]) Let $a \in L$, a is called a Boolean element of L if $a \vee a' = 1$.

In this paper, we use B(L) to represent all Boolean elements of L. It is trivial that if $a \in B(L)$, then $a' \in B(L)$ and $0, 1 \in B(L)$.

Lemma 1 ([26]) *An R_0-algebra L has the following properties, for all $x, y, z \in L$:*
(1) $x' = x \to 0$, $x'' = x$;
(2) $x \odot y \leq x \wedge y$;
(3) $x \odot x' = 0$;
(4) $x \odot (y \vee z) = (x \odot y) \vee (x \odot z)$;
(5) $x \leq y$ if and only if $y' \leq x'$;
(6) If $x \leq y$, then $x \odot z \leq y \odot z$;
(7) If $x \odot y = 1$, then $x = y = 1$;
(8) $x \odot y \leq z$ if and only if $x \leq y \to z$;
(9) $a \vee (x \odot y) = (a \vee x) \odot (a \vee y)$, for $a \in B(L)$.

Lemma 2 ([10]) *Let L be an R_0-algebra. Then the following conditions are equivalent:*
(1) $a \in B(L)$;
(2) $a \vee a' = 1$;
(3) $a \wedge a' = 0$;
(4) $a \odot a = a$;
(5) $a \odot x = a \wedge x$ *for all* $x \in L$.

About R_0-algebras, and t-norms, there are lots of results. But there are few results about the operation $\oplus$. For this, we give the following results:

Lemma 3 *Let* $x, y, z \in L$. *Then the following hold:*
(1) $0 \oplus 0 = 0$;
(2) $x \oplus x' = 1$;
(3) $a \oplus a = a$ *if and only if* $a \in B(L)$;
(4) $a \oplus x = a \vee x$ *for* $a \in B(L)$;
(5) *If* $x \leq y$, *then* $x \oplus z \leq y \oplus z$;
(6) $a \wedge (x \oplus y) = (a \wedge x) \oplus (a \wedge y)$ *for* $a \in B(L)$;
(7) $x \vee y \leq x \oplus y$;
(8) *If* $x \oplus y = 0$, *then* $x = y = 0$.

Proof (1) $0 \oplus 0 = (0' \to 0) = 1 \to 0 = 0$.

(2) $x \oplus x' = (x' \to x') = 1$.

(3) If $a \in$ B(L), then $a \oplus a = (a' \odot a')' = a'' = a$. Conversely, $a \odot a = (a' \oplus a')' = a'' = a$. This shows that $a \in$ B(L).

(4) Assume that $a \in$ B(L), then $a \oplus x = (a' \odot x')' = (a' \wedge x')' = a \vee x$.

(5) Suppose $x \leq y$, then $y' \leq x'$. By Lemma 1(6), $y' \odot z' \leq x' \odot z'$. Thus $(x' \odot z')' \leq (y' \odot z')'$. That is, $x \oplus z \leq y \oplus z$.

(6) By Lemma 1(1) and Lemma 1(9), $a \wedge (x \oplus y) = a'' \wedge (x' \odot y')' = (a' \vee (x' \odot y'))' = ((a' \vee x') \odot (a' \vee y'))' = ((a \wedge x)' \odot (a \wedge y)')' = (a \wedge x) \oplus (a \wedge y)$.

(7) By Lemma 1(2), $x' \odot y' \leq x', y'$. Thus $x, y \leq (x' \odot y')'$. Hence $x \vee y \leq x \oplus y$.

(8) By (7), $x \vee y \leq x \odot y = 0$. Thus $x = y = 0$. □

Definition 3 ([25]) Let L and J be two R_0-algebras. The function $f : L \longrightarrow J$ is called a homomorphism if it satisfies the following conditions, for all $x, y \in$ L:
(1) $f(0_L) = 0_J$, $f(1_L) = 1_J$;
(2) $f(x') = (f(x))'$;
(3) $f(x \vee_L y) = f(x) \vee_J f(y)$; $f(x \to_L y) = f(x) \to_J f(y)$.

A homomorphism f: L $\longrightarrow$ L is called an endomorphism. A homomorphism f is called an isomorphism if it is bijective (see [25]).

Remark 1 [25] If f is a homomorphism, then $f(x \wedge_L y) = f(x) \wedge_J f(y)$, $f(x \odot_L y) = f(x) \odot_J f(y)$, $f(x \oplus_L y) = f(x) \oplus_J f(y)$.

Definition 4 ([8]) An ideal I of L is a subset of L satisfying the following conditions:
(1) $0 \in$ I;
(2) $\forall x, y \in$ I imply $x \oplus y \in$ I;
(3) $x \in$ I and $y \leq x$ imply $y \in$ I.

Lemma 4 *Let I be an ideal of L and f: L $\longrightarrow$ L an isomorphism. Then $f(I)$ is an ideal.*

Proof It is obvious. □

Lemma 5 *The following conditions are equivalent, for all $x, y \in L$*
(1) $x \leq y$;
(2) $x \odot y\prime = 0$;
(3) $x\prime \oplus y = 1$.

Proof (1)$\Longrightarrow$(2) Suppose $x \leq y$, then $y\prime \leq x\prime$. Thus $x \odot y\prime \leq x \odot x\prime = 0$. Hence $x \odot y\prime = 0$.

(2)$\Longrightarrow$(1) Given that $x \odot y\prime = 0$, then $x \odot y\prime \leq 0$. Hence $x \leq y\prime \rightarrow 0 = y\prime\prime = y$.

(2)$\Longleftrightarrow$(3) $x \odot y\prime = 0 \Longleftrightarrow (x \odot y\prime)\prime = 1 \Longleftrightarrow (x\prime\prime \odot y\prime)\prime{=}1 \Longleftrightarrow x\prime \oplus y = 1$. □

3 (f, g)-Derivations of R_0-Algebras

Definition 5 Let L be an R_0-algebras and d: L $\longrightarrow$ L a function. We call d a derivation on L, if $d(x \odot y) = (d(x) \odot y) \oplus (x \odot d(y))$ for all $x, y \in$ L.

Definition 6 Let L be an R_0-algebras and f: L $\longrightarrow$ L be a homomorphism. A function d : L $\longrightarrow$ L is called a f-derivation on L, if $d(x \odot y) = (d(x) \odot f(y)) \oplus (f(x) \odot d(y))$.

Definition 7 Let L be an R_0-algebras and f, g: L $\longrightarrow$ L homomorphisms. A function d : L $\longrightarrow$ L is called an (f, g)-derivation on L, if $d(x \odot y) = (d(x) \odot f(y)) \oplus (g(x) \odot d(y))$.

We often abbreviate $d(x), f(x), g(x)$ as dx, fx, gx.

Remark 2 In Definition 7, when $f = g$, d is an f-derivation; when $f = g = i$, i is the identity map, d is a derivation.

Example 1 Define d: $L \longrightarrow L$ by $dx = 0$, for all $x \in L$. For any homomorphism $f, g : L \longrightarrow L$, d is an (f, g)-derivation.

Example 2 Let $L = \{0, 1/3, 2/3, 1\}$ with Cayley tables as follows:

x	x'
0	1
1/3	2/3
2/3	1/3
1	0

$\rightarrow$	0	1/3	2/3	1
0	1	1	1	1
1/3	2/3	1	1	1
2/3	1/3	1/3	1	1
1	0	1/3	2/3	1

It can be easily checked that L is an R_0-algebra. Define $f, g : L \longrightarrow L$, by

$$f(x) = \begin{cases} 0, 0, 1/3; \\ 1, 2/3, 1. \end{cases} \text{ and } g = i, i \text{ is the identity map.}$$

Then f and g are homomorphisms. We define $d_1(0) = d_1(1) = d_1(1/3) = 0$, $d_1(2/3) = 1/3$, then d_1 is an (f, g)-derivation. But $d_2(0) = d_2(1) = 0, d_2(1/3) = d_2(2/3) = 1/3$ is not an (f, g)-derivation.

Lemma 6 *Let L be an R_0 algebra and d an (f, g)-derivation. Then $d1 \in B(L)$.*

Proof Assume that f, g are two homomorphisms, then $f1 = g1 = 1$. Thus $d1 = d(1 \odot 1) = (d1 \odot f1) \oplus (g1 \odot d1) = (d1 \odot 1) \oplus (1 \odot d1) = d1 \oplus d1$.
By Lemma 3(3), we know that $d1 \in B(L)$. □

Theorem 1 *Let d be an (f, g)-derivation on L. Then, for all $x \in L$, the following hold:*
(1) $d0 = 0$;
(2) $dx \odot f(x') = fx \odot d(x') = dx \odot g(x') = gx \odot d(x') = 0$;
(3) $dx \leq fx, gx$;
(4) $dx = (d1 \odot fx) \oplus dx = dx \oplus (gx \odot d1)$.

Proof (1) Suppose that f, g are two homomorphisms, then $f0 = g0 = 0$. Thus $d0 = d(0 \odot 0) = (d0 \odot f0) \oplus (g0 \odot d0) = (d0 \odot 0) \oplus (0 \odot d0) = 0 \oplus 0 = 0$.

(2) $0 = d0 = d(x \odot x') = (dx \odot f(x')) \oplus (gx \odot d(x'))$ and $0 = d0 = d(x' \odot x) = (d(x') \odot fx) \oplus (g(x') \odot dx)$. By Lemma 3(8), we have $dx \odot f(x') = gx \odot d(x') = fx \odot d(x') = dx \odot g(x') = 0$.

(3) By (2) and Remark 1, $0 = dx \odot f(x') = dx \odot (fx)'$. According to Lemma 5, we get $dx \leq fx$. Similarly $dx \leq gx$.

(4) $dx = d(1 \odot x) = (d1 \odot fx) \oplus (g1 \odot dx) = (d1 \odot fx) \oplus dx$. Similarly, we have $dx = dx \oplus (gx \odot d1)$. □

Proposition 1 *Let d be an (f, g)-derivation of L and $x, y \in L$. If $x \leq y$, then the following hold:*
(1) $d(x \odot y\prime) = dx \odot dy\prime = 0$. In particular, $dx \odot dx\prime = 0$.
(2) $dx \leq fy, gy$ and $d(y\prime) \leq (fx)\prime, (gx)\prime$.

Proof (1) Suppose that $x \leq y$, then $x \odot y\prime = 0$. Thus $d(x \odot y\prime) = 0$. On the other hand, since f is a homomorphism, $x \leq y$ implies that $fx \leq fy$. Hence we have $dx \leq fx \leq fy$. Then $dx \odot d(y\prime) \leq fy \odot d(y\prime) \leq fy \odot f(y\prime) = fy \odot (fy)\prime = 0$. Therefore, $dx \odot d(y\prime) = 0$. When $x = y$, we have $dx \odot d(x\prime) = 0$.

(2) By (1), we have $0 = d(x \odot y\prime) = (dx \odot f(y\prime)) \oplus (gx \odot d(y\prime))$. According to Lemma 3(8), we have $dx \odot f(y\prime) = gx \odot d(y\prime) = 0$. Then, by Lemma 5, $dx \leq fy, d(y\prime) \leq (gx)\prime$. On the other hand, $0 = d(y\prime \odot x) = (d(y\prime) \odot fx) \oplus (g(y\prime) \odot dx)$. Thus, we have $d(y\prime) \odot fx = g(y\prime) \odot dx = 0$. Therefore, by Lemma 5, we have $dx \leq gy, d(y\prime) \leq (fx)\prime$. □

Proposition 2 *Let d be an (f, g)-derivation of L. Then the following identities hold:*
(1) If $fx \leq d1$ and $gx \leq d1$, then $dx = fx \vee gx$;
(2) If $fx \geq d1$ and $gx \geq d1$, then $dx \geq d1$.

Proof (1) If $fx \leq d1$ and $gx \leq d1$. By Lemma 6 and Theorem 1(4), $dx = dx \oplus (fx \odot d1) = dx \oplus (fx \wedge d1) = dx \oplus fx \geq fx$. Similarly, we also have $dx \geq gx$. Thus $dx \geq fx \vee gx$. We know $dx \leq fx \vee gx$ from Theorem 1(3). This means $dx = fx \vee gx$.

(2) If $fx \geq d1$ and $gx \geq d1$, then $dx = (d1 \odot fx) \oplus dx = (d1 \wedge fx) \oplus dx = d1 \oplus dx \geq d1$. □

Proposition 3 *Let d be a (f, g)-derivation on L. Then $d(x\prime) = (dx)\prime$ if and only if either $dx = fx$ or $dx = gx$.*

Proof Suppose that $dx = fx$. Since f is a homomorphism, we get that, for all $x \in L$, $f(x\prime) = (fx)\prime$. Thus $d(x\prime) = (dx)\prime$. Conversely, Suppose that $d(x\prime) = (dx)\prime$. We have $0 = fx \odot d(x\prime) = fx \odot (dx)\prime$. Thus, by Lemma 5, $fx \leq dx$. And $dx \leq fx$. Hence $dx = fx$. Similarly, we can get that if $d(x\prime) = (dx)\prime$, then $dx = gx$. □

Theorem 2 *Let d be an (f, g)-derivation on L such that f, g are isomorphisms and I an ideal of L. Then $d(I) \subseteq f(I) \cap g(I)$.*

Proof If $y \in d$(I), then there exists $x \in$ I such that $y = dx$. By Theorem 1(3), we have $y = dx \leq fx \in f$(I) and $y = dx \leq gx \in g$(I). Since I is an ideal, by Lemma 4, we know that f(I) and g(I) are ideals. Thus $y \in f$(I) $\cap$ g(I). This means, d(I) $\subseteq f$(I) $\cap$ g(I). □

4 Isotone (f, g)-Derivations of R_0-Algebras

Definition 8 Let d be an (f, g)-derivation of L. If $x \leq y$ implies $dx \leq dy$ for all $x, y \in$ L, then d is called an isotone (f, g)-derivation.

Definition 9 Let d be an (f, g)-derivation of L. If $d(x \oplus y) = dx \oplus dy$ for all $x, y \in$ L, then d is called an additive (f, g)-derivation.

Example 3 Let L = $\{0, a, b, 1\}$. $0 \leq a \leq 1, 0 \leq b \leq 1$, a and b are not comparable, with the following Cayley tables:

x	x'
0	1
a	b
b	a
1	0

$\rightarrow$	0	a	b	1
0	1	1	1	1
a	b	1	b	1
b	a	a	1	1
1	0	a	b	1

It can be easily checked that L is an R_0-algebras. Define maps $f, g : L \longrightarrow L$ by

$$fx = \begin{cases} 0, \ x = 0; \\ b, \ x = a; \\ a, \ x = b; \\ 1, \ x = 1. \end{cases} \qquad gx = \begin{cases} 0, \ x = 0; \\ 0, \ x = a; \\ 1, \ x = b; \\ 1, \ x = 1. \end{cases}$$

Then f and g are homomorphisms. Now, we define $d : L \longrightarrow L$ by $d1 = db = a, d0 = da = 0$, then d is an isotone (f, g)-derivation.

Theorem 3 *Let d be an (f, g)-derivation on L. Then the following conditions are equivalent:*
(1) d is isotone;
(2) $dx \leq d1$;
(3) $dx = d1 \odot fx = d1 \odot gx$.

Proof (1)$\Longrightarrow$(2) Trivial.

(2)$\Longrightarrow$(3) Since $dx \leq d1$ and $dx \leq fx$, we have $dx \leq d1 \wedge fx = d1 \odot fx$. On the other hand, by Theorem 1(4), $dx = (d1 \odot fx) \oplus dx \geq d1 \odot fx$. Thus $dx = d1 \odot fx$. Similarly, $dx = d1 \odot gx$.

(3)$\Longrightarrow$(1) Suppose that $x \leq y$. Since f is a homomorphism, we have $fx \leq fy$. Thus $dx = d1 \odot fx \leq d1 \odot fy = dy$. When $dx = d1 \odot gx$, similarly we have $dx \leq dy$. □

Proposition 4 *Let d be an isotone (f, g)-derivation of L. Then $d(L) \subseteq B(L)$.*

Proof Since d is an isotone (f, g)-derivation, we get that, for $x \in L$, $dx = dx \oplus (d1 \odot fx) = dx \oplus dx$. This means that $d(L) \subseteq B(L)$. □

Proposition 5 *Let d be an (f, g)-derivation of L which is not a Boolean algebra. Then the identity map i is not an (f, g)-derivation.*

Proof Suppose i is an (f, g)-derivation. It is trivial that i is isotone. By Proposition 4, for any $x \in L$, we have $i(x) = x \in B(L)$. It is a contradiction. Thus i is not an (f, g)-derivation. □

Theorem 4 *Let d be an (f, g)-derivation on L. Then the following are equivalent:*
(1) d is isotone;
(2) $d(x \odot y) = dx \odot dy$;
(3) $d(x \oplus y) = dx \oplus dy$.

Proof (1)$\Longrightarrow$(2)

Suppose that d is isotone, then, by Theorem 3(3), $d(x \odot y) = d1 \odot f(x \odot y) = d1 \odot d1 \odot fx \odot fy = (d1 \odot fx) \odot (d1 \odot fy) = dx \odot dy$.

(2)$\Longrightarrow$(1) Since $dx = d(x \odot 1) = dx \odot d1 \leq d1$. By Theorem 3(2), we have that d is isotone.

(1)$\Longrightarrow$(3) $d(x \oplus y) = d1 \odot f(x \oplus y) = d1 \wedge (fx \oplus fy) = (d1 \wedge fx) \oplus (d1 \wedge fy) = (d1 \odot fx) \oplus (d1 \odot fy) = dx \oplus dy$.

(3)$\Longrightarrow$(1) Since $d1 = d(x \oplus 1) = dx \oplus d1 \geq dx$, by Theorem 3(2), we know that d is isotone. □

Corollary 1 *d is an isotone (f, g)-derivation of L if and only if d is an additive (f, g)-derivation.*

Proof Obvious. □

Theorem 5 *Let d be an (f, g)-derivation of L. Then the following are equivalent:*
(1) d is isotone;
(2) $d(x \odot y) = dx \odot fy = gx \odot dy$.

Proof (1)$\Longrightarrow$(2) Since $d(x \odot y) = (dx \odot fy) \oplus (gx \odot dy) \geq dx \odot fy$ and $d(x \odot y) = dx \odot dy \leq dx \odot fy$. We have $d(x \odot y) = dx \odot fy$. Similarly, $d(x \odot y) = gx \odot dy$ holds.

(2)$\Longrightarrow$(1) Assume $d(x \odot y) = dx \odot fy$, then $dx = d(1 \odot x) = d1 \odot fx$. By Theorem 3(3), we have that d is isotone. □

Theorem 6 *Let d be an (f, g)-derivation on L. Then the following are equivalent:*
(1) d is isotone;
(2) $d(x \wedge y) = dx \wedge dy$;
(3) $d(x \vee y) = dx \vee dy$.

Proof (1)$\Longrightarrow$(2) By Theorem 3(3) and Lemma 6, $d(x \wedge y) = d1 \odot f(x \wedge y) = d1 \wedge (fx \wedge fy) = (d1 \wedge fx) \wedge (d1 \wedge fy) = (d1 \odot fx) \wedge (d1 \odot fy) = dx \wedge dy$.

(2)$\Longrightarrow$(1) Suppose $dx = d(x \wedge 1) = dx \wedge d1$. Then $dx \leq d1$. By Theorem 3(2), we know that d is isotone.

(1)$\Longrightarrow$(3) $d(x \vee y) = d1 \odot f(x \vee y) = d1 \wedge (fx \vee fy) = (d1 \wedge fx) \vee (d1 \wedge fy) = (d1 \odot fx) \vee (d1 \odot fy) = dx \vee dy$.

(3)$\Longrightarrow$(1) It is similar to the proof of (2)$\Longrightarrow$(1). □

Theorem 7 *Let d be an isotone (f, g)-derivation of L. Then the following hold:*
(1) $d(x \wedge y) = d(x \odot y) = dx \wedge fy = gx \wedge dy$;
(2) $d(x \vee y) = d(x \oplus y)$.

Proof (1) Assume that d is isotone, then by Theorems 4 and 6, $d(x \wedge y) = dx \wedge dy = dx \odot dy = d(x \odot y) = dx \odot fy = dx \wedge fy = gx \odot dy = gx \wedge dy$.
(2) $d(x \vee y) = dx \vee dy = dx \oplus dy = d(x \oplus y)$. □

Proposition 6 *Let d be an isotone (f, g)-derivation of L. Then, if $fx \geq d1$ and $gx \geq d1$, then $dx = d1$.*

Proof By Proposition 2, if $fx \geq d1$ and $gx \geq d1$, then $dx \geq d1$. And d is isotone, thus $dx \leq d1$. Hence $dx = d1$. □

Let L be an R_0-algebras and d: $L \longrightarrow L$ a function. Define $\mathrm{Fix}_d(L) = \{x \in L\colon dx = x\}$.

Proposition 7 *Let d be an isotone (f, g)-derivation on L. Then the following hold:*
(1) $0 \in Fix_d(L)$;
(2) *If* $x, y \in Fix_d(L)$, $x \oplus y \in Fix_d(L)$.

Proof (1) It is trivial that $0 \in \mathrm{Fix}_d(L)$.
(2) Suppose $x, y \in \mathrm{Fix}_d(L)$. Then $d(x \oplus y) = dx \oplus dy = x \oplus y$. This shows that $x \oplus y \in \mathrm{Fix}_d(L)$. □

Theorem 8 *Let d be an (f, g)-derivation on L. If $d(x\prime) = dx$, for all $x \in L$, then the following hold:*
(1) $d1 = 0$;
(2) $dx \odot dx = 0$;
(3) *If d is isotone, then $d = 0$.*

Proof (1) By Lemma 1(1), we have $d1 = d(0\prime) = d0 = 0$;
(2) By Proposition 1(1), $0 = dx \odot d(x\prime) = dx \odot dx$;
(3) Suppose d is isotone, then, for $x \in$ L, $dx \leq d1 = 0$. Thus $d = 0$. □

5 (f, g)-Derivations of Linearly-Ordered R_0-Algebras

Lemma 7 ([11]) *If L is linearly ordered, then B(L) = {0, 1}.*

Proposition 8 *Let d be an (f, g)-derivation of a linearly ordered R_0-algebras L. Then $d1 = 0$ or $d1 = 1$.*

Proof By Proposition 4, $d1 \in \{0, 1\}$. Thus $d1 = 0$ or $d1 = 1$. □

Theorem 9 *Let d be an isotone (f, g)-derivation of a linearly ordered R_0-algebras L. Then, if $d1 = 0$, then $d = 0$.*

Proof If $d1 = 0$, then $dx \leq d1 = 0$, and so $d = 0$. □

Theorem 10 *Let d be an isotone (f, g)-derivation of a linearly ordered R_0-algebras L. Then $d^{-1}(0) = \{x \in L | d(x) = 0\}$ is an ideal of L.*

Proof Since $d0 = 0$, we have that $0 \in d^{-1}(0)$. Suppose that $x \in d^{-1}(0)$ and $y \leq x$. Then $dx = 0$. Since d is isotone, we have $dy \leq dx = 0$ which implies that $dy = 0$, Therefore $y \in d^{-1}(0)$. Now, Let $x, y \in d^{-1}(0)$. Then, $d(x \oplus y) = dx \oplus dy = 0 \oplus 0 = 0$. Thus $x \oplus y \in d^{-1}(0)$. This means $d^{-1}(0)$ is an ideal of L. □

Acknowledgments This work is supported by the National Natural Science Foundation of China (No. 11371130, 61273018).

References

1. Albas, E.: On ideals and orthogonal generalized derivations of semiprime ring. Math. J. Okayama Univ. **49**, 53–58 (2007)
2. Alshehri, N.O.: Derivations of MV-algebras. Int. J. Math. Sci. (2011)
3. Bell, H.E., Mason, G.: On derivations in near-rings and rings. Math. J. Okayama Univ. **34**, 135–144 (1994)
4. Bell, H.E., Kappe, L.C.: Rings in which derivations satisfy certain algebraic conditions. Acta Mathematica Hungarica **53**, 339–346 (1989)
5. Ceven, Y.: Symmetric *bi*-derivations of lattices. Quaest. Math. **32**, 241–245 (2009)
6. Ceven, Y., Ozturk, M.A.: On f-derivations of lattices. B. Korean Math. Soc. **45**, 701–707 (2008)
7. Chang, C.C.: Algebraic analysis of many-valud logic. Trans. Am. Math. Soc. **88**, 467–490 (1988)
8. Cheng, G.S.: The filters and ideals in R_0-algebras. Fuzzy Syst. Math. **15**, 58–61 (1988) (in Chinese)
9. Du, Y., Wang, Y.: A result on generalized derivations in prime rings. Hacet. J. Math. Stat. **42**, 81–85 (2013)
10. Ferrari, L.: On derivations of lattices. Pure Math. Appl. **12**, 365–382 (2001)
11. Han, C.: R_0-algebras and theory of similarity measures between vague sets. Shaanxi Normal University Dissertation (2006) (in Chinese)
12. Hvala, B.: Generalized derivations in rings*. Commun. Algebra. **26**, 1147–1166 (1998)
13. Javed, Y.B., Aslam, M.: A note on f-derivations of BCI-algebras. Commun. Korean Math. Soc. **24**, 321–331 (2009)
14. Jun, Y.B., Xin, X.L.: On derivations of BCI-algebras. Inf. Sci. **159**, 167–176 (2004)
15. Jung, Y.S., Chang, I.S.: On approximately higher ring derivations. J. Math. Anal. Appl. **343**, 636–643 (2008)
16. Kamal, A.A., Al-Shaalan, K.H.: Existence of derivations on near-rings. Math. Slovaca. **63**, 431–448 (2013)
17. Liu, L.Z., Li, K.T.: Fuzzy implicative and Boolean filters of R_0-algebras. Inf. Sci. **171**, 61–71 (2005)
18. Mustafa, A., Ceran, S.: Generalized (f, g)-derivations of lattices. Math. Sci. Appl. E-notes **1**, 56–62 (2013)
19. Nisar, F.: On f-derivations of BCI-algebras. J. Prime Res. Math. **5**, 176–191 (2009)

20. Ozbal, S.A., Firat, A.: Symmetric f-bi derivations of lattices. ARS. Combinatoria **97**, 471–477 (2010)
21. Pei, D.W.: On equivalent forms of fuzzy logic systems NM and IMTL. Fuzzy Set Syst. **138**, 187–195 (2003)
22. Posner, E.: Derivations in prime rings. P. Am. Math. Soc. **8**, 1093–1100 (1957)
23. Rehman, N.U.: Jordan triple $(\alpha, \beta)^*$-derivations on semiprime rings with involution. Hacet. J. Math. Stat. **42**, 641–651 (2013)
24. Szász, G.: Derivations of lattices. Acta Scientiarum Mathematicarum **37**, 149–154 (1975)
25. Wang, G.J.: Non-Classical Mathematical Logic and Approximate Reasoning. Science Press, Beijing (2000)
26. Xin, X.L., Li, T.Y., Lu, J.H.: On derivations of lattices. Inf. Sci. **178**, 307–316 (2008)
27. Zhou, X.N., Li, Q.G.: Boolean products of R_0-algebras. Math. Logic Quart. **56**, 289–298 (2010)

Parametrization Filters and Their Properties in Residuated Lattices

Lian-Zhen Liu and Xiang-Yang Zhang

Abstract Filters play a key role in studying algebraic structures of logics. Recently, various special filters in residuated lattices have been introduced. Hence it is very important to develop a general definition for special filters. In the paper, we introduce the notion of parametrization filters and study some of their properties. The relationship of Extension property (Triple of equivalent characteristics, and Quotient characteristics) between the set of parametrization filters and the set of $(\alpha, \beta]$-fuzzy parametrization filters is investigated.

Keywords Parametrization filter · $(\alpha, \beta]$-fuzzy parametrization filter · Extension property · Triple of equivalent characteristics · Quotient characteristics

1 Introduction

Filters play an important role in studying algebraic structures of logics. From logical point of view, filters correspond to sets of provable formulae. Based on this reason, various types of filters have been proposed and some results of them have been obtained.

The "Extension property", "Triple of equivalent characteristics" and "Quotient characteristics" were often studied for each special filter and its fuzzification. In order to illuminate the triviality of these properties, and to provide a tool for reviewers dealing with papers about new types of filters, Víta [10, 11] introduced t-filters and fuzzy t-filters in bounded commutative integral residuated lattices. He studied the "Extension property", "Triple of equivalent characteristics" and "Quotient characteristics" of t-filters and fuzzy t-filters, respectively, and demonstrated that the

L.-Z. Liu (✉) · X.-Y. Zhang
College of Science, Jiangnan University, Wuxi 214122, China
e-mail: lian712000@126.com

X.-Y. Zhang
e-mail: zhangxy@jiangnan.edu.cn

T.-H. Fan et al. (eds.), *Quantitative Logic and Soft Computing 2016*,
Advances in Intelligent Systems and Computing 510,
DOI 10.1007/978-3-319-46206-6_36

existing results in many papers can be seen as straightforward consequences of his theory. Then he left the following open problem:

Problem. Whether this theory holds for a special type of filter that is defined by a quasi-equation and it is not definable as a t-filter (all of special types of filters that the author has seen could be defined as t-filters for suitable t).

As Víta said, t-filters cover many special types of filters, such as Boolean filters, implicative filters, positive implicative filters, etc. However, as we can see, prime filters–the most important filters in logical algebras, are not included in t-filters. On the other hand, in [10, 11], the author investigated the "Extension property", "Triple of equivalent characteristics" and "Quotient characteristics" of t-filters and fuzzy t-filters, but he did not consider the intrinsical connection of the above three properties between a special filter and its fuzzification.

The motivations of this paper are to develop a general definition for special filters, and to study the intrinsical connection between special filters and their fuzzification.

As side effects, the results of this paper point out that for a special filter or its fuzzification, it is enough to study either one of them, the other is a straightforward consequence of the former. Also, these results provide a tool for reviewers dealing with papers about new types of filters or fuzzy filters.

2 Preliminaries

We recollect some definitions and results which will be used in the following and we shall not cite them every time they are used.

Definition 1 ([1, 12]) An integral residuated lattice (IRL for short) is a structure $L = (L, \vee, \wedge, \odot, \rightarrow, \rightsquigarrow, 1)$ of type (2,2,2,2,2,0) satisfying the following axioms:
(C1) $(L, \vee, \wedge)$ is a lattice;
(C2) $(L, \odot, 1)$ is a monoid, i.e. $\odot$ is associative and $x \odot 1 = 1 \odot x = x$;
(C3) $x \odot y \leq z$ if and only if $x \leq y \rightarrow z$ if and only if $y \leq x \rightsquigarrow z$ for all $x, y, z \in L$;
(C4) $x \leq 1$ for all $x \in L$.

Let L be an IRL. L is called commutative, if $x \odot y = y \odot x$ for any $x, y \in L$. L is called bounded if L has a bottom element 0.

Let L be a bounded IRL. For any $x \in L$, we define $\neg x = x \rightarrow 0$, $\sim x = x \rightsquigarrow 0$. If $\neg \sim x = \sim \neg x$, then L is call good. If $\neg \sim x = \sim \neg x = x$, then L is called involutive.

Definition 2 ([12]) Let L be an IRL, $\emptyset \neq F \subseteq L$. F is called a filter in L if the following conditions are fulfilled:
(F1) If $x, y \in F$, then $x \odot y \in F$;
(F2) If $x \leq y$ and $x \in F$, then $y \in F$.

The set of filters in L is denoted by $F(L)$

Let F be a filter, from Definition 2, it follows that F is closed under operations $\wedge, \vee, \rightarrow, \rightsquigarrow$. That is, if $x, y \in F$, then $x \vee y, x \wedge y, x \rightarrow y, x \rightsquigarrow y \in F$.

In the sequel, we will use $\bar{x}$ as an abbreviation of a finite sequence $x_1, x_2, \ldots$.

Let L be an IRL, $\alpha, \beta \in [0, 1]$, $\alpha < \beta$. A fuzzy subset of L is a mapping $\mu : L \to [0, 1]$. For $t \in [0, 1]$, the set $\mu^t = \{x \in L : \mu(x) \geq t\}$ is called level set of μ.

Let μ be a fuzzy subset of L. If $\mu(x) \leq \alpha$ for all $x \in L$, then $\mu^t = \emptyset$ for all $t > \alpha$. If $\mu(x) \geq \beta$ for all $x \in L$, then $\mu^t = L$ for all $t \leq \beta$. In what follows, we assume that μ satisfies $\mu(a) > \alpha$, $\mu(b) < \beta$ for some $a, b \in L$.

Definition 3 ([5–7]) Let μ be a fuzzy subset in L. μ is called an $(\alpha, \beta]$-fuzzy filter if for any $t \in (\alpha, \beta]$, every non-empty level set μ^t is a filter in L.

Remark 1 $(\alpha, \beta]$-fuzzy filters generalize many types of fuzzy filters in IRL. For example, fuzzy filters [3, 4, 8, 10] can be seen as (0,1]-fuzzy filters; $(\in, \in \vee q)$-fuzzy filters [6, 15] can be seen as (0, 0.5]-fuzzy filters; $(\overline{\in}, \overline{\in \vee q})$-fuzzy filters [6] can be seen as (0.5, 1]-fuzzy filters.

Corollary 1 *If μ is an $(\alpha, \beta]$-fuzzy filter, then $\mu(1) > \alpha$.*

Corollary 2 *If μ is an $(\alpha, \beta]$-fuzzy filter, then $\mu^{\mu(1)\wedge\beta}$ is a filter.*

Theorem 1 ([5–7]) *Let μ be a fuzzy subset of L. Then the following are equivalent:*
(1) *μ is an $(\alpha, \beta]$-fuzzy filter;*
(2) *μ satisfies the following:*
(FF1) *$\forall x, y \in L$, $\mu(x \odot y) \vee \alpha \geq \mu(x) \wedge \mu(y) \wedge \beta$;*
(FF2) *$\forall x, y \in L$, if $x \leq y$, then $\mu(y) \vee \alpha \geq \mu(x) \wedge \beta$.*

3 Parametrization Filters in Residuated Lattices

In [10], Víta showed that all of special types of filters that he has seen could be defined as t-filters for suitable t. However, as we can see, prime filters which are special type of filters are not included in t-filters. Hence there raises a natural question: how to define a general notion for types of filters which will cover many known special filters. In the following, we introduce parametrization filters with the intent on developing a unified definition for types of filters.

Definition 4 Let F be a filter in residuated lattice L, $s_{ij}(\bar{x})(1 \leq i \leq n, 1 \leq j \leq m)$, $t_{ij}(\bar{x})(1 \leq i \leq k, 1 \leq j \leq h)$ be terms on L. F is called a parametrization filter with parameter $(s_{11}(\bar{x}), \ldots, s_{nm}(\bar{x}); t_{11}(\bar{x}), \ldots, t_{kh}(\bar{x}))$, if it satisfies property P, where property P is:
(P1) If $s_{11}(\bar{x}) \in F, \ldots, s_{1m_1}(\bar{x}) \in F$, then $t_{11}(\bar{x}) \in F$ or $t_{12}(\bar{x}) \in F$ or $\cdots$ or $t_{1k_1}(\bar{x}) \in F$;
(P2) If $s_{21}(\bar{x}) \in F, \ldots, s_{2m_2}(\bar{x}) \in F$, then $t_{21}(\bar{x}) \in F$ or $t_{22}(\bar{x}) \in F$ or $\cdots$ or $t_{2k_2}(\bar{x}) \in F$;

$\vdots$

(Pn) If $s_{n1}(\bar{x}) \in F, \ldots, s_{nm_n}(\bar{x}) \in F$, then $t_{n1}(\bar{x}) \in F$ or $t_{n2}(\bar{x}) \in F$ or $\cdots$ or $t_{nk_n}(\bar{x}) \in F$.

The set of parametrization filters with parameter $(s_{11}(\bar{x}), \ldots, s_{nm}(\bar{x}); t_{11}(\bar{x}), \ldots, t_{kh}(\bar{x}))$ is denoted by $(PF(L), (s_{11}(\bar{x}), \ldots, s_{nm}(\bar{x}); t_{11}(\bar{x}), \ldots, t_{kh}(\bar{x})))$ ($PF(L)$ for short).

Remark 2 The notion of P-filters introduced in [2] is a special case of parametrization filters.

Example 1 (1) Let F be a filter in L, $t(\bar{x})$ a term, $s(\bar{x}) = \bar{x} \to \bar{x}$. If F satisfies the condition:

$$\text{If } s(\bar{x}) \in F, \text{ then } t(\bar{x}) \in F.$$

By Definition 4, F is a parametrization filter with parameter $(s(\bar{x}); t(\bar{x}))$. On the other hand, from [10, 11], F is a t-filter. This shows that t-filters are parametrization filters with parameter $(s(\bar{x}); t(\bar{x}))$.

(2) Let F be a filter in L such that $F \neq L$, and let $s_{11}(\bar{x}) = x \vee y, t_{11}(\bar{x}) = x, t_{12}(\bar{x}) = y$. If F satisfies the following:

$$\text{If } s_{11}(\bar{x}) \in F, \text{ then } t_{11}(\bar{x}) \in F \text{ or } t_{12}(\bar{x}) \in F.$$

By Definition 4, F is a parametrization filter with parameter $(s_{11}(\bar{x}); t_{11}(\bar{x}), t_{12}(\bar{x}))$. On the other hand, from [1, 12], F is a prime filter. This shows that prime filters are parametrization filters with parameter $(s_{11}(\bar{x}); t_{11}(\bar{x}), t_{12}(\bar{x}))$.

(3) Let L be a bounded commutative integral residuated lattice, F a filter in L. Let $s_{11}(\bar{x}) = \neg\neg x, t_{11}(\bar{x}) = x$. If F satisfies the following:

$$\text{If } s_{11}(\bar{x}) \in F, \text{ then } t_{11}(\bar{x}) \in F.$$

Then F is a parametrization filter with parameter $(s_{11}(\bar{x}); t_{11}(\bar{x}))$. From [13], F is an EIMTL filter. This shows that EIMTL filters are parametrization filters with parameter $(s_{11}(\bar{x}); t_{11}(\bar{x}))$.

(4) Let F be a filter in L and let $s_{11}(\bar{x}) = x \to y, t_{11}(\bar{x}) = x \rightsquigarrow y$. If F satisfies the following condition:

(i) if $s_{11}(\bar{x}) \in F$, then $t_{11}(\bar{x}) \in F$.

(ii) if $t_{11}(\bar{x}) \in F$, then $s_{11}(\bar{x}) \in F$.

Then from Definition 4, F is a parametrization filter with parameter $(s_{11}(\bar{x}), t_{11}(\bar{x}); t_{11}(\bar{x}), s_{11}(\bar{x}))$. By [1, 12], F is a normal filter in L. This shows that normal filters are parametrization filters with parameter $(s_{11}(\bar{x}), t_{11}(\bar{x}); t_{11}(\bar{x}), s_{11}(\bar{x}))$.

Transform Rule: For property P, we construct P^* as follows:

(1) If $s_{11}(\bar{x}) \geq 1, \ldots, s_{1m_1}(\bar{x}) \geq 1$, then $t_{11}(\bar{x}) \geq 1$ or $t_{12}(\bar{x}) \geq 1$ or $\cdots$ or $t_{1k_1}(\bar{x}) \geq 1$.

(2) If $s_{21}(\bar{x}) \geq 1, \ldots, s_{2m_2}(\bar{x}) \geq 1$, then $t_{21}(\bar{x}) \geq 1$ or $t_{22}(\bar{x}) \geq 1$ or $\cdots$ or $t_{2k_2}(\bar{x}) \geq 1$.

$$\vdots$$

(n) If $s_{n1}(\bar{x}) \geq 1, \ldots, s_{nm_n}(\bar{x}) \geq 1$, then $t_{n1}(\bar{x}) \geq 1$ or $t_{n2}(\bar{x}) \geq 1$ or $\cdots$ or $t_{nk_n}(\bar{x}) \geq 1$.

Given a class **B** of residuated lattices, let $\mathbf{B}(\mathbf{P}^*)$ denote its subclass given by P^*.

Example 2 (1) For t-filters, the corresponding P^* is: if $\bar{x} \to \bar{x} \geq 1$, then $t(\bar{x}) \geq 1$ for any $\bar{x} \in L$.

(2) For prime filters, the corresponding P^* is: if $x \vee y \geq 1$, then $x \geq 1$ or $y \geq 1$.

(3) For EIMTL-filters, the corresponding P^* is: if $\neg\neg x \geq 1$, then $x \geq 1$.

Definition 5 ([9, 10]) Let L be an IRL, $X \subseteq F(L)$. We say that

- X satisfies the intersection property, if X is closed under intersections.
- X satisfies the Extension property, if for any $X_1, X_2 \in F(L)$, $X_1 \in X$ and $X_1 \subseteq X_2$ imply $X_2 \in X$.
- X satisfies Triple of equivalent characteristics, if the following are equivalent:
  (TEC1) $F \in X$ for all $F \in F(L)$;
  (TEC2) $\{1\} \in X$;
  (TEC3) $L \in \mathcal{C}$, where $\mathcal{C}$ is a class of residuated lattices.

Definition 6 ([10]) Let L be an IRL. We say X satisfies Quotient characteristics, if for any normal filter F, $F \in X$ if and only if $L/F \in \mathcal{C}$.

Lemma 1 *Let L be an IRL, $X \subseteq F(L)$. If X satisfies Extension property, then X satisfies Triple of equivalent characteristics.*

Proof It is obvious. □

Lemma 2 *Let $F_i (i \in I) \in (PF(L), (s_{11}(\bar{x}), \ldots, s_{nm}(\bar{x}); t_{11}(\bar{x}), \ldots, t_{kh}(\bar{x})))$. If $t_{j1}(\bar{x}) = t_{j2}(\bar{x}) = \cdots = t_{jk_j}(\bar{x})$ $(1 \leq j \leq n)$, then $\cap_{i \in I} F_i \in (PF(L), (s_{11}(\bar{x}), \ldots, s_{nm}(\bar{x}); t_{11}(\bar{x}), \ldots, t_{kh}(\bar{x})))$.*

Proof It can be derived from Definition 4. □

Corollary 3 *Let $X \subseteq (PF(L), (s_{11}(\bar{x}), \ldots, s_{nm}(\bar{x}); t_{11}(\bar{x}), \ldots, t_{kh}(\bar{x})))$. If $t_{j1}(\bar{x}) = t_{j2}(\bar{x}) = \cdots = t_{jk_j}(\bar{x})$ $(1 \leq j \leq n)$, then X satisfies intersection property.*

Corollary 4 *Let L be an IRL. The following hold:*
(1) t-filters satisfy intersection property;
(2) EIMTL-filters satisfy intersection property;
(3) Normal filters satisfy intersection property.

In general, if there exist $t_{ij}(\bar{x}) \neq t_{ij'}(\bar{x})$ for some $i(1 \leq i \leq n), j, j'(1 \leq j, j' \leq k_i)$, then the result of Lemma 2 may not be true. Indeed, let L be the residuated lattice defined in Example 3. Routine calculation shows that $F = \{1, b, c, d\}$, $G = \{1, a, c, d\}$ are prime filters, but $F \cap G = \{1, c, d\}$ is not a prime filter. This shows that prime filters do not satisfy intersection property.

Lemma 3 *Let L be a bounded IRL, F be a filter in L. If $m(x_1, x_2, \ldots, x_n)$ be a 0-free term on L and $x_1, x_2, \ldots, x_n \in F$, then $m(x_1, x_2, \ldots, x_n) \in F$.*

Proof For $x, y \in L$, if $x, y \in F$, then $x \vee y, x \wedge y, x \rightarrow y, x \rightsquigarrow y \in F$ as F is a filter. Since $m(x_1, x_2, \ldots, x_n)$ is a 0-free term, and $x_1, x_2, \ldots, x_n \in F$, we have $m(x_1, x_2, \ldots, x_n) \in F$. □

Lemma 4 *Let* $G \in F(L), F \subseteq G, F \in (PF(L), (s_{11}(\bar{x}), \ldots, s_{nm}(\bar{x}); t_{11}(\bar{x}), \ldots, t_{kh}(\bar{x})))$. *If there exist* 0*-free terms* $m_i(\bar{x})(1 \leq i \leq n)$ *such that* $m_i(s_{i1}, s_{i2}, \ldots, s_{in_i}) \rightarrow t_{ij}(\bar{x}) \in F(1 \leq i \leq n, 1 \leq j \leq k_i)$, *then* $G \in (PF(L), (s_{11}(\bar{x}), \ldots, s_{nm}(\bar{x}); t_{11}(\bar{x}), \ldots, t_{kh}(\bar{x})))$.

Proof For each $i(1 \leq i \leq n)$, if $s_{i1}(\bar{x}), s_{i2}(\bar{x}), \ldots, s_{in_i}(\bar{x}) \in G$, then $m_i(s_{i1}(\bar{x}), s_{i2}(\bar{x}), \ldots, s_{in_i}(\bar{x})) \in G$. From $m_i(s_{i1}, s_{i2}, \ldots, s_{in_i}) \rightarrow t_{ij}(\bar{x}) \in F \subseteq G$, we have $t_{ij}(\bar{x}) \in G$. By Definition 4, $G \in (PF(L), (s_{11}(\bar{x}), \ldots, s_{nm}(\bar{x}); t_{11}(\bar{x}), \ldots, t_{kh}(\bar{x})))$. □

The following examples show that the Extension property and Triple of equivalent characteristics of some filters do not hold.

Example 3 Let $L = \{0, a, b, c, d, 1\}$ be a lattice whose Hasse diagram is below.

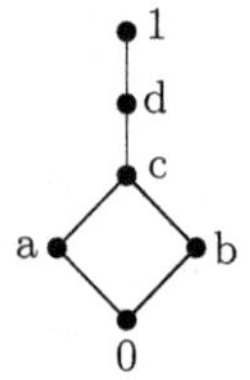

Define $\odot$ and $\rightarrow$ on L as follows:

$\odot$	0	a	b	c	d	1
0	0	0	0	0	0	0
a	0	a	0	a	a	a
b	0	0	b	b	b	b
c	0	a	b	c	c	c
d	0	a	b	c	c	d
1	0	a	b	c	d	1

$\rightarrow = \rightsquigarrow$	0	a	b	c	d	1
0	1	1	1	1	1	1
a	b	1	b	1	1	1
b	a	a	1	1	1	1
c	0	a	b	1	1	1
d	0	a	b	c	1	1
1	0	a	b	c	d	1

Routine calculation shows that $(L, \wedge, \vee, \odot, \rightarrow, \rightsquigarrow, 1)$ is an integral residuated lattice. Obviously, $\{1\}, \{1, c, d\}, \{1, b, c, d\}, \{1, a, c, d\}$ are filters in L. It is easily checked that the filter $\{1\}$ is a prime filter. But $\{1, c, d\}$ is not a prime filter. This shows that Extension property and Triple of equivalent characteristics of prime filters do not hold.

Example 4 Let $L = \{0, a, b, c, 1\}$ be a chain with Cayley tables as follows:

$\odot$	0	a	b	c	1
0	0	0	0	0	0
a	0	0	0	0	a
b	0	0	b	b	b
c	0	0	b	c	c
1	0	a	b	c	1

$\rightarrow = \rightsquigarrow$	0	a	b	c	1
0	1	1	1	1	1
a	c	1	1	1	1
b	a	a	1	1	1
c	a	a	b	1	1
1	0	a	b	c	1

Define $\wedge$ and $\vee$ operations on L as min and max, respectively. Routine calculation shows that $(L, \wedge, \vee, \odot \rightarrow, \rightsquigarrow, 1)$ is a residuated lattice. Obviously, $\{1\}$, $\{1, c\}$, $\{1, c, b\}$ are filters in L. It is easily checked that $\{1\}$ is an EIMTL-filter, $\{1, c\}$ is not an EIMTL-filter because $\neg\neg b = c \in \{1, c\}$, but $b \notin \{1, c\}$. This shows that Extension property and Triple of equivalent characteristics of EIMTL filters do not hold.

Lemma 5 *$PF(L)$ satisfies Quotient characteristics.*

4 $(\alpha, \beta]$-Fuzzy Parametrization Filters in Residuated Lattices

Inspired by [15], in this section, we introduce the notion of $(\alpha, \beta]$-fuzzy parametrization filters and investigate some of their properties.

Definition 7 Let μ be an $(\alpha, \beta]$-fuzzy filter. μ is called an $(\alpha, \beta]$-fuzzy parametrization filter with parameter $(s_{11}(\bar{x}), \ldots, s_{nm}(\bar{x}); t_{11}(\bar{x}), \ldots, t_{kh}(\bar{x}))$ ($(\alpha, \beta]$-fuzzy parametrization filter for short), if for each $t \in (\alpha, \beta]$, every non-empty level set μ^t is a parametrization filter with parameter $(s_{11}(\bar{x}), \ldots, s_{nm}(\bar{x}); t_{11}(\bar{x}), \ldots, t_{kh}(\bar{x}))$ in L.

Let $FPF(L)$ be the set of all $(\alpha, \beta]$-fuzzy parametrization filters with parameter $(s_{11}(\bar{x}), \ldots, s_{nm}(\bar{x}); t_{11}(\bar{x}), \ldots, t_{kh}(\bar{x}))$ in L.

From Definition 7, the following is clearly obtained.

Corollary 5 *Let F be a filter in L. Then F is a parametrization filter with parameter $(s_{11}(\bar{x}), \ldots, s_{nm}(\bar{x}); t_{11}(\bar{x}), \ldots, t_{kh}(\bar{x}))$ if and only if the characteristics function χ_F is an $(\alpha, \beta]$-fuzzy parametrization filter.*

Theorem 2 *Let μ be a fuzzy subset in L. Then the following are equivalent:*
(1) μ is an $(\alpha, \beta]$-fuzzy parametrization filter;
(2) μ is an $(\alpha, \beta]$-fuzzy filter and satisfies the following conditions:
(FP1) $\mu(t_{11}(\bar{x})) \vee \cdots \vee \mu(t_{1k_1}(\bar{x})) \vee \alpha \geq \mu(s_{11}(\bar{x})) \wedge \cdots \wedge \mu(s_{1m_1}(\bar{x})) \wedge \beta$ *for any* $\bar{x} \in L$;
(FP2) $\mu(t_{21}(\bar{x})) \vee \cdots \vee \mu(t_{2k_2}(\bar{x})) \vee \alpha \geq \mu(s_{21}(\bar{x})) \wedge \cdots \wedge \mu(s_{2m_2}(\bar{x})) \wedge \beta$ *for any* $\bar{x} \in L$;

$$\vdots$$

(FPn) $\mu(t_{n1}(\bar{x})) \vee \cdots \vee \mu(t_{nk_n}(\bar{x})) \vee \alpha \geq \mu(s_{n1}(\bar{x})) \wedge \cdots \wedge \mu(s_{nm_n}(\bar{x})) \wedge \beta$ *for any* $\bar{x} \in L$.

Proof (1) $\Rightarrow$ (2) Suppose μ is an $(\alpha, \beta]$-fuzzy parametrization filter, then μ is an $(\alpha, \beta]$-fuzzy filter by Definitions 3 and 7. For any $\bar{x} \in L$, let $t = \mu(s_{11}(\bar{x})) \wedge \cdots \wedge \mu(s_{1m_1}(\bar{x})) \wedge \beta$, we have $s_{11}(\bar{x}) \in \mu^t, \ldots, s_{1m_1}(\bar{x}) \in \mu^t$. If $t \leq \alpha$, then $\mu(t_{11}(\bar{x})) \vee \cdots \vee \mu(t_{1k_1}(\bar{x})) \vee \alpha \geq \alpha \geq t$. If $\alpha < t$, then from (1) and Definition 7, we get that

μ^t is a parametrization filter, thus $t_{11}(\bar{x}) \in \mu^t$, or $\cdots$, or $t_{1k_1}(\bar{x}) \in \mu^t$. This means that $\mu(t_{11}(\bar{x})) \vee \cdots \vee \mu(t_{1k_1}(\bar{x})) \vee \alpha \geq t$. Hence (FP1) holds. Similarly, we can prove (FP2)-(FPn)hold. This proves that (2) holds.

(2) $\Rightarrow$ (1) For any $t \in (\alpha, \beta]$, suppose $\mu^t \neq \emptyset$, then μ^t is a filter by (2) and Definition 7. If $s_{11}(\bar{x}) \in \mu^t, \ldots, s_{1m_1}(\bar{x}) \in \mu^t$, then $\mu(s_{11}(\bar{x})) \wedge \cdots \wedge \mu(s_{1m_1}(\bar{x})) \wedge \beta \geq t$. From (FP1), it follows that $\mu(t_{11}(\bar{x})) \vee \cdots \vee \mu(t_{1k_1}(\bar{x})) \vee \alpha \geq t$, and so $t_{11}(\bar{x}) \in \mu^t$ or $\cdots$ or $t_{1k_1}(\bar{x}) \in \mu^t$. This means that (P1) holds. Analogously, we can prove (P2)-(Pn) hold. By Definition 4, μ^t is a parametrization filter. Hence μ is an $(\alpha, \beta]$-fuzzy parametrization filter by Definition 7. □

Corollary 6 *Let μ be an $(\alpha, \beta]$-fuzzy filter. Then the following are equivalent:*
(1) μ is an $(\alpha, \beta]$-fuzzy normal filter;
(2) $\mu(x \to y) \vee \alpha \geq \mu(x \rightsquigarrow y) \wedge \beta$ and $\mu(x \rightsquigarrow y) \vee \alpha \geq \mu(x \to y) \wedge \beta$ for any $x, y \in L$.

Corollary 7 *If μ is an $(\alpha, \beta]$-fuzzy normal filter, then $(\mu(x \rightsquigarrow y) \vee \alpha) \wedge \beta = (\mu(x \to y) \vee \alpha) \wedge \beta$ for any $x, y \in L$.*

Next, we mainly discuss the relationship of Extension property (Triple of equivalent characteristics, Quotient characteristics) between parametrization filters and $(\alpha, \beta]$-fuzzy parametrization filters.

From [7, 9], the following theorem is obvious.

Theorem 3 *Let L be an IRL. Then the following are equivalent:*
(1) $PF(L)$ satisfies the intersection property;
(2) $FPF(L)$ satisfies the intersection property, that is, if $\mu_i \in FPF(L)(i \in \Gamma)$, then $\bigwedge_{i \in \Gamma} \in FPF(L)$.

Theorem 4 *Let L be an IRL. Then the following are equivalent:*
(1) $PF(L)$ satisfies Extension property; (2) $FPF(L)$ satisfies Extension property, i.e., if $\mu \in FPF(L)$, ν is an $(\alpha, \beta]$-fuzzy filter and $\mu \leq \nu$, $\mu(1) \vee \alpha = \nu(1) \vee \alpha$, then $\nu \in FPF(L)$.

Proof (1) $\Rightarrow$ (2) From Definition 7, it will suffice to prove that for any $t \in (\alpha, \beta]$, if $\nu^t \neq \emptyset$, then $\nu^t \in PF(L)$. It is clear that $1 \in \nu^t$. From $\mu(1) \vee \alpha = \nu(1) \vee \alpha$ and $\mu \leq \nu$, we have $\mu^t \subseteq \nu^t$ and $1 \in \mu^t$. Thus $\mu^t \in PF(L)$ as $\mu \in FPF(L)$. By (1), $\nu^t \in PF(L)$.

(2) $\Rightarrow$ (1) Let $F \subseteq G$ be two filters. Then χ_F and χ_G are $(\alpha, \beta]$-fuzzy filters and $\chi_F \leq \chi_G$, $\chi_F(1) \vee \alpha = \chi_G(1) \vee \alpha$. If $F \in PF(L)$, then Corollary 5 shows that $\chi_F \in FPF(L)$. By (2), $\chi_G \in FPF(L)$, so $G \in PF(L)$ by Corollary 5. From Definition 5, we get that $PF(L)$ satisfies Extension property. □

We call $FPF(L)$ satisfying Triple of equivalent characteristics, if the following are equivalent:
(FTEC1) $\mu \in FPF(L)$ for every $(\alpha, \beta]$-fuzzy filter μ in L;
(FTEC2) $\chi_{\{1\}} \in FPF(L)$;
(TEC3) $L \in \mathbf{B}(\mathbf{P}^*)$.

Theorem 5 *Let L be an IRL. Then the following are equivalent:*
(1) $PF(L)$ satisfies Triple of equivalent characteristics;
(2) $FPF(L)$ satisfies Triple of equivalent characteristics.

Proof (1) $\Rightarrow$ (2) We will prove (FTEC1) $\Rightarrow$ (FTEC2)$\Rightarrow$ (TEC3)$\Rightarrow$(FTEC1).

(FTEC1) $\Rightarrow$ (FTEC2) It is obvious.

(FTEC2)$\Rightarrow$ (TEC3) Suppose $\chi_{\{1\}} \in FPF(L)$. Then $\{1\} \in PF(L)$ by Definition 7. By (1), we get $L \in \mathbf{B}(\mathbf{P}^*)$, i.e., (TEC3) holds.

(TEC3)$\Rightarrow$(FTEC1) Let μ be an $(\alpha, \beta]$-fuzzy filter. By Definition 7, for $\alpha < t \leq \beta$, if $\mu^t \neq \emptyset$, then μ^t is a filter. From (TEC3), it follows that $\{1\} \in PF(L)$. This together with (1) leads to $\mu^t \in PF(L)$, and so $\mu \in FPF(L)$ by Definition 7.

(2) $\Rightarrow$ (1) It is clear that (TEC1) $\Rightarrow$ (TEC2)$\Rightarrow$ (TEC3) by (2). Now we prove (TEC3) $\Rightarrow$ (TEC1). Let F be a filter, then χ_F is an $(\alpha, \beta]$-fuzzy filter by Definition 3. From (TEC1) and (2), we get that $\chi_F \in FPF(L)$, and so $F \in PF(L)$ by Definition 7. This proves that (TEC1) holds. □

In view of Theorems 4, 5 and Lemma 1, the following is obvious.

Corollary 8 *If $FPF(L)$ satisfies Extension property, then it satisfies Triple of equivalent characteristics.*

Let μ be an $(\alpha, \beta]$-fuzzy filter. For any $x \in L$, we define

$\mu_x : L \to [0, 1]$ $\mu_x(y) = ((\mu(x \to y) \wedge \mu(y \to x)) \vee \alpha) \wedge \beta$.

$\mu_{[x]} : L \to [0, 1]$ $\mu_{[x]}(y) = ((\mu(x \rightsquigarrow y) \wedge \mu(y \rightsquigarrow x)) \vee \alpha) \wedge \beta$.

Lemma 6 *Let μ be an $(\alpha, \beta]$-fuzzy filter. Then for any $x, y \in L$, the following hold:*
(1) $\mu_x = \mu_y$ if and only if $(\mu(x \to y) \vee \alpha) \wedge \beta = (\mu(y \to x) \vee \alpha) \wedge \beta = \mu(1) \wedge \beta$;
(2) $\mu_{[x]} = \mu_{[y]}$ if and only if $(\mu(x \rightsquigarrow y) \vee \alpha) \wedge \beta = (\mu(y \rightsquigarrow x) \vee \alpha) \wedge \beta = \mu(1) \wedge \beta$.

Proof (1) Suppose $\mu_x = \mu_y$. Then $\mu_x(y) = \mu_y(y)$, i.e., $((\mu(x \to y) \wedge \mu(y \to x)) \vee \alpha) \wedge \beta = (\mu(1) \vee \alpha) \wedge \beta = \mu(1) \wedge \beta$. Thus $(\mu(x \to y) \vee \alpha) \wedge \beta \geq \mu(1) \wedge \beta$, $(\mu(y \to x) \vee \alpha) \wedge \beta \geq \mu(1) \wedge \beta$. From Theorem 1 and Corollary 1, we know that $\mu(1) = \mu(1) \vee \alpha \geq \mu(x \to y) \wedge \beta$, $\mu(1) = \mu(1) \vee \alpha \geq \mu(y \to x) \wedge \beta$, so $\mu(1) \wedge \beta = (\mu(1) \vee \alpha) \wedge \beta \geq (\mu(x \to y) \vee \alpha) \wedge \beta$, $\mu(1) \wedge \beta = (\mu(1) \vee \alpha) \wedge \beta \geq (\mu(y \to x) \vee \alpha) \wedge \beta$. Hence $(\mu(x \to y) \vee \alpha) \wedge \beta = (\mu(y \to x) \vee \alpha) \wedge \beta = \mu(1) \wedge \beta$.

Conversely, if $(\mu(x \to y) \vee \alpha) \wedge \beta = (\mu(y \to x) \vee \alpha) \wedge \beta = \mu(1) \wedge \beta$, then from $\mu(x \to z) \vee \alpha \geq \mu(x \to y) \wedge \mu(y \to z) \wedge \beta$, we have $(\mu(x \to z) \vee \alpha) \wedge \beta \geq (\mu(y \to z) \vee \alpha) \wedge \beta$. Similarly, we can prove $(\mu(y \to z) \vee \alpha) \wedge \beta \geq (\mu(x \to z) \vee \alpha) \wedge \beta$. Hence $(\mu(y \to z) \vee \alpha) \wedge \beta = (\mu(x \to z) \vee \alpha) \wedge \beta$. Analogously, $(\mu(z \to x) \vee \alpha) \wedge \beta = (\mu(z \to y) \vee \alpha) \wedge \beta$. Therefore $\mu_x(z) = ((\mu(x \to z) \wedge \mu(z \to x)) \vee \alpha) \wedge \beta = (\mu(y \to z) \wedge \mu(z \to y)) \vee \alpha) \wedge \beta = \mu_y(z)$.

(2) can be proved similarly. □

Corollary 9 *Let μ be an $(\alpha, \beta]$-fuzzy filter. Then for any $x, y \in L$, the following hold:*
(1) $\mu_x = \mu_y$ if and only if $x \to y \in \mu^{\mu(1) \wedge \beta}$, $y \to x \in \mu^{\mu(1) \wedge \beta}$;
(2) $\mu_{[x]} = \mu_{[y]}$ if and only if $x \rightsquigarrow y \in \mu^{\mu(1) \wedge \beta}$, $y \rightsquigarrow x \in \mu^{\mu(1) \wedge \beta}$.

Proof (1) $\mu_x = \mu_y$ if and only if $(\mu(x \to y) \vee \alpha) \wedge \beta = (\mu(y \to x) \vee \alpha) \wedge \beta = \mu(1) \wedge \beta$ if and only if $\mu(x \to y) \vee \alpha \geq \mu(1) \wedge \beta$, $\mu(y \to x) \vee \alpha \geq \mu(1) \wedge \beta$ if and only if $\mu(x \to y) \geq \mu(1) \wedge \beta$, $\mu(y \to x) \geq \mu(1) \wedge \beta$ if and only if $x \to y \in \mu^{\mu(1)\wedge\beta}$, $y \to x \in \mu^{\mu(1)\wedge\beta}$.

(2) is proved similarly. □

Lemma 7 *If μ is an $(\alpha, \beta]$-fuzzy normal filter, then $\mu_x = \mu_{[x]}$ for all $x \in L$.*

Proof By definitions of μ_x and $\mu_{[x]}$, we have $\mu_x(y) = ((\mu(x \to y) \vee \alpha) \wedge \beta) \wedge ((\mu(y \to x) \vee \alpha) \wedge \beta$, $\mu_{[x]}(y) = ((\mu(x \rightsquigarrow y) \vee \alpha) \wedge \beta) \wedge ((\mu(y \rightsquigarrow x) \vee \alpha) \wedge \beta$. If μ is an $(\alpha, \beta]$-fuzzy normal filter, then from Corollary 7, we know that $(\mu(x \rightsquigarrow y) \vee \alpha) \wedge \beta = (\mu(x \to y) \vee \alpha) \wedge \beta$, $(\mu(y \rightsquigarrow x) \vee \alpha) \wedge \beta = (\mu(y \to x) \vee \alpha) \wedge \beta$. Hence $\mu_x = \mu_{[x]}$. □

Corollary 10 *Let μ be an $(\alpha, \beta]$-fuzzy normal filter. Then for any $x, y \in L$, $\mu_x = \mu_y$ if and only if $x \equiv_{\mu^{\mu(1)\wedge\beta}} y$.*

Let μ be an $(\alpha, \beta]$-fuzzy normal filter. Let $L/\mu = \{\mu_x : x \in L\}$. Define
$\mu_x \vee_\mu \mu_y = \mu_{x\vee y}$, $\mu_x \wedge_\mu \mu_y = \mu_{x\wedge y}$, $\mu_x \odot_\mu \mu_y = \mu_{x\odot y}$,
$\mu_x \to_\mu \mu_y = \mu_{x\to y}$, $\mu_x \rightsquigarrow_\mu \mu_y = \mu_{x\rightsquigarrow y}$

Lemma 8 *If μ is an $(\alpha, \beta]$-fuzzy normal filter, then $L/\mu = (L/\mu, \wedge_\mu, \vee_\mu, \odot_\mu, \to_\mu, \rightsquigarrow_\mu, \mu_1)$ is an integral residuated lattice, and is isomorphic to $L/\mu^{\mu(1)\wedge\beta}$.*

Proof The proof is similar to that of [4]. □

We say that $FPF(L)$ satisfies Quotient characteristics, if for any $(\alpha, \beta]$-fuzzy normal filter μ, $\mu \in FPF(L)$ if and only if $L/\mu \in \mathbf{B}(\mathbf{P}^*)$.

Lemma 9 *Let μ be an $(\alpha, \beta]$-fuzzy normal filter. If $\mu \in FPF(L)$, then $L/\mu \in \mathbf{B}(\mathbf{P}^*)$.*

Proof Suppose μ is an $(\alpha, \beta]$-fuzzy normal filter. Then $\mu^{\mu(1)\wedge\beta}$ is a normal filter in L. If $\mu \in FPF(L)$, then $\mu^{\mu(1)\wedge\beta} \in PF(L)$ by Definition 7. Since $L/\mu^{\mu(1)\wedge\beta} \in \mathbf{B}(\mathbf{P}^*)$, we have $L/\mu \in \mathbf{B}(\mathbf{P}^*)$ follows from Lemma 8. □

The following example shows that the converse of Lemma 9 may not be true.

Example 5 Let L be the integral residuated lattice defined in Example 3. Define the fuzzy subset μ as $\mu(1) = 0.9$, $\mu(d) = 0.85$, $\mu(c) = 0.85$, $\mu(b) = 0.3$, $\mu(a) = 0.2$, $\mu(0) = 0.2$. Routine calculation shows that μ is a $(0.2, 0.92]$-fuzzy normal filter, hence by Lemma 8, $(L/\mu, \vee_\mu, \wedge_\mu, \odot_\mu, \to_\mu, \rightsquigarrow_\mu, \mu_1)$ is an integral residuated lattice. For any $x, y \in L$, if $\mu_x \vee_\mu \mu_y = \mu_1$, then $\mu_x = \mu_1$ or $\mu_y = 1$. This shows that L/μ satisfies condition P^*: $x \vee y = 1$ implies $x = 1$ or $y = 1$. But μ is not a $(0.2, 0.92]$-fuzzy prime filter because $\mu(a) \vee \mu(b) \vee 0.2 \ngeq \mu(a \vee b) \wedge 0.92$.

Lemma 10 *Let μ be an $(\alpha, \beta]$-fuzzy filter, $\alpha < t \leq \beta$. If $\mu^t \neq \emptyset$, then $\mu(1) \wedge \beta \geq t$ and $\mu^{\mu(1)\wedge\beta} \subseteq \mu^t$.*

Proof If $\mu^t \neq \emptyset$, then $1 \in \mu^t$, i.e., $\mu(1) \geq t$. Hence $\mu(1) \wedge \beta \geq t$ and $\mu^{\mu(1)\wedge\beta} \subseteq \mu^t$. □

Lemma 11 *Let L be an IRL. If $PF(L)$ satisfies Extension property, then $(\alpha, \beta]$-fuzzy normal filter $\mu \in FPF(L)$ if and only if $L/\mu \in B(P^*)$.*

Proof Suppose $L/\mu \in B(P^*)$. Then from Lemma 8, we know that $L/\mu^{\mu(1)\wedge\beta} \in B(P^*)$, so $\mu^{\mu(1)\wedge\beta} \in PF(L)$. Let $\alpha < t \leq \beta$, if $\mu^t \neq \emptyset$, then $\mu^{\mu(1)\wedge\beta} \subseteq \mu^t$. Since $PF(L)$ satisfies Extension property, we get that $\mu^t \in PF(L)$. Hence $\mu \in FPF(L)$. The necessity is followed by Lemma 9. □

The following theorem is a consequence of the preceding results.

Theorem 6 *If $PF(L)$ satisfies Extension property, then the following are equivalent:*
(1) $PF(L)$ satisfies Quotient characteristics;
(2) $FPF(L)$ satisfies Quotient characteristics.

Acknowledgments This work is supported by NSFC (Nos. 60875084, 61273017), by the Fundamental Research Funds for the Central Universities (JUSRP21118, JUSRP211A24), by Jiangsu Overseas Research & Training Program for University Prominent Young & Middle-aged Teachers and Presidents and the Project-sponsored by SRF for ROCS, SEM.

References

1. Blount, K., Tsinakis, C.: The structure of residuated lattices. Int. J. Algebra Comput. **13**, 437–461 (2003)
2. Liu, L.Z.: Generalized intuitionistic fuzzy filters on residuated lattices. J. Intell. Fuzzy Syst. **28**, 1545–1552 (2015)
3. Liu, L.Z., Li, K.T.: Fuzzy filters of BL-algebras. Inform. Sci. **173**, 141–154 (2005)
4. Liu, L.Z., Li, K.T.: Fuzzy Boolean and positive implicative filters of BL-algebras. Fuzzy Sets Syst. **152**, 333–348 (2005)
5. Liu, L.Z., Zhang, X.Y.: Applications of transfer principle to fuzzy theory with thresholds (α, β). In: Proceedings of International Conference on Quantitative Logic and Quantification of Software. pp. 125–129 (2009)
6. Ma, X.L., Zhan, J.M., Dudek, W.A.: Some kinds of $(\bar{\in}, \bar{\in} \vee \bar{q})$-fuzzy filters of BL-algebras. Comput. Math. Appl. **58**, 248–256 (2009)
7. Ma, Z.M.: Lattices of (generalized) fuzzy filters in residuated lattices. J. Intell. Fuzzy Syst. **27**, 2281–2287 (2014)
8. Rachůnek, J., Šalounova, D.: Fuzzy filters and fuzzy prime filters of bounded *RL*-monoids and pseudo BL-algebras. Inform. Sci. **178**, 3474–3481 (2008)
9. Swamy, U.M., Viswanadha Raju, D.: Algebraic fuzzy systems. Fuzzy Sets Syst. **41**, 187–194 (1991)
10. Víta, M.: Fuzzy t-filters and their properties. Fuzzy Sets Syst. **247**, 127–134 (2014)
11. Víta, M.: Why are papers about filters on residuated structures (usually) trivial? Inform. Sci. **276**, 387–391 (2014)
12. van Alten, C.J.: Representable Biresiduated Lattices. J. Algebra. **247**, 672–691 (2002)

13. Xiao, L., Liu, L.Z.: A note on some filters in residuated lattices. J. Intell. Fuzzy Syst. **30**, 493–500 (2016)
14. Yuan, X.H., Zhang, C., Ren, Y.H.: Generalized fuzzy groups and many-valued implications. Fuzzy Sets Syst. **138**, 205–211 (2003)
15. Zhang, H.R., Li, Q.G.: $(\in, \in \vee q)$-fuzzy t-filters on residuated lattices. J. Intell. Fuzzy Syst. **29**, 1521–1526 (2015)

Weak Pseudo-Quasi-Wajsberg Algebras

Wen-Jun Liu and Wen-Juan Chen

Abstract In this paper we introduce a generalization of pseudo-quasi-Wajsberg algebras, called weak pseudo-quasi-Wajsberg algebras (weak PQW-algebras, for short). And then some properties of weak PQW-algebras are investigated. Finally, we define weak pseudo-quasi-MV algebras and a related categorical equivalence is established.

Keywords Wajsberg algebras · Quasi-Wajsberg algebras · Weak pseudo-quasi-Wajsberg algebras · Weak pseudo-quasi-MV algebras

1 Introduction

It is known that quasi-MV algebra arising from the quantum computational logic is an algebraic model for describing the set of all density operators of the Hilbert space $\mathbb{C}^2$, endowed with a suitable stock of quantum logical gates. In the past decade, many properties and their associated logics of quasi-MV algebras have been investigated [1, 8–11]. Recently, pseudo-quasi-MV algebras (PQMV-algebras, for short) as a non-commutative generalization of quasi-MV algebras were introduced in [4, 5]. Meanwhile, the variety of PQMV-algebras as a subvariety of quasi-pseudo-MV algebras (QPMV-algebras, for short) introduced in [2, 3] plays an important role in studying QPMV-algebras.

In [5], the authors showed that PQMV-algebras are term equivalent to pseudo-quasi-Wajsberg algebras which are non-commutative generalization of quasi-Wajsberg algebras. In 2010, in order to study the logical aspects of quasi-MV algebras, Bou et. al. introduced quasi-Wajsberg algebras in [1]. Thus pseudo-quasi-

W.-J. Liu · W.-J. Chen (✉)
School of Mathematical Sciences, University of Jinan, No. 336, West Road of Nan Xinzhuang, Jinan 250022, Shandong, China
e-mail: wjchenmath@gmail.com

W.-J. Liu
e-mail: liuwenjun@mail.ujn.edu.cn

T.-H. Fan et al. (eds.), *Quantitative Logic and Soft Computing 2016*, Advances in Intelligent Systems and Computing 510,
DOI 10.1007/978-3-319-46206-6_37

Wajsberg algebras are related to the non-commutative quantum computational logic. In this paper we want to continue and generalize this study. We present and discuss structures which are weaker than those of pseudo-quasi-Wajsberg algebras.

The paper is organized as follows. In Sect. 2, we recall some definitions and results which is used in thesequel. In Sect. 3, we introduce weak pseudo-quasi-Wajsberg algebras (weak PQW-algebras, for short) and investigate properties of weak PQW-algebras. In Sect. 4, we define weak pseudo-quasi-MV algebras and establish a related categorical equivalence.

2 Preliminary

In this section, we recall some definitions and results which will be used in what follows.

Wajsberg algebras were defined and investigated by Font et al. in [7]. Among other results, the categorical equivalence between Wajsberg algebras and MV algebras was established.

Definition 1 ([7]) An algebra $A = \langle A; \to, ', 1\rangle$ of type $\langle 2, 1, 0\rangle$ is called a Wajsberg algebra, if it satisfies the following axioms for all $x, y, z \in A$:
(W1) $1 \to x = x$;
(W2) $(x \to y) \to ((y \to z) \to (x \to z)) = 1$;
(W3) $(x \to y) \to y = (y \to x) \to x$;
(W4) $(x' \to y') \to (y \to x) = 1$.

Quasi-Wajsberg algebras were introduced as a generalization of Wajsberg algebras. In fact, it is a term equivalent version of quasi-MV algebras.

Definition 2 ([1]) An algebra $A = \langle A; \to, ', 1\rangle$ of type $\langle 2, 1, 0\rangle$ is called a quasi-Wajsberg algebra, if it satisfies the following axioms for all $x, y, z \in A$:
(QW1) $1 \to (x \to y) = x \to y$;
(QW2) $(x \to y) \to ((y \to z) \to (x \to z)) = 1$;
(QW3) $(x \to y) \to y = (y \to x) \to x$;
(QW4) $(x' \to y') \to (y \to x) = 1$;
(QW5) $x'' = x$;
(QW6) $1 \to (1 \to x)' = (1 \to x)'$.

According to the definition, any Wajsberg algebra is a quasi-Wajsberg algebra. Conversely, if a quasi-Wajsberg algebra satisfies the condition $1 \to x = x$, then it is a Wajsberg algebra.

In [5], Liu and Chen introduced pseudo-quasi-MV algebras as non-commutative generalization of quasi-MV algebras. At the same time, they defined and studied pseudo-quasi-Wajsberg algebras which were categorically equivalent to pseudo-quasi-MV algebras.

Definition 3 ([5]) An algebra $A = \langle A; \to, \rightsquigarrow, ', 1\rangle$ of type $\langle 2, 2, 1, 0\rangle$ is called a pseudo-quasi-Wajsberg algebra, if it satisfies the following axioms for all $x, y, z \in A$:
(PQW1) $1 \to x = 1 \rightsquigarrow x$;
(PQW2) $1 \to (x \to y) = x \to y$, $1 \rightsquigarrow (x \rightsquigarrow y) = x \rightsquigarrow y$;
(PQW3) $(x \rightsquigarrow y) \to y = (y \rightsquigarrow x) \to x = (y \to x) \rightsquigarrow x = (x \to y) \rightsquigarrow y$;
(PQW4) $(x \to y) \to ((y \to z) \rightsquigarrow (x \to z)) = 1$,
$(x \rightsquigarrow y) \rightsquigarrow ((y \rightsquigarrow z) \to (x \rightsquigarrow z)) = 1$;
(PQW5) $(x' \rightsquigarrow y') \to (y \to x) = 1$, $(x' \to y') \to (y \rightsquigarrow x) = 1$;
(PQW6) $x \to y' = y \rightsquigarrow x'$;
(PQW7) $1 \to (1 \to x)' = (1 \to x)'$, $1 \rightsquigarrow (1 \rightsquigarrow x)' = (1 \rightsquigarrow x)'$;
(PQW8) $x'' = x$.

Following from the definition, quasi-Wajsberg algebras are pseudo-quasi-Wajsberg algebras. On the other hand, a pseudo-quasi-Wajsberg algebra in which the two implications coincide is a quasi-Wajsberg algebra.

Definition 4 ([5]) An algebra $A = \langle A; \oplus, ', 0\rangle$ of type $\langle 2, 1, 0\rangle$ is called a pseudo-quasi-MV algebra, if it satisfies the following axioms for all $x, y, z \in A$:
(PQMV1) $(x \oplus y) \oplus z = x \oplus (y \oplus z)$;
(PQMV2) $x \oplus y \oplus 0 = x \oplus y$;
(PQMV3) $x \oplus 0 = 0 \oplus x$;
(PQMV4) $x \oplus 0' = 0' = 0' \oplus x$;
(PQMV5) $(x \oplus 0)' = x' \oplus 0$;
(PQMV6) $y \oplus (x' \oplus y)' = (y \oplus x')' \oplus y = x \oplus (y' \oplus x)' = (x \oplus y')' \oplus x$;
(PQMV7) $x'' = x$.

Theorem 1 ([5]) *Let $A = \langle A; \to, \rightsquigarrow, ', 1\rangle$ be a pseudo-quasi-Wajsberg algebra. Define a binary operation $\oplus$ and a constant 0 by $x \oplus y = x' \to y = y' \rightsquigarrow x$ and $0 = 1'$. Then $f(A) = \langle A; \oplus, ', 0\rangle$ is a pseudo-quasi-MV algebra.*

Theorem 2 ([5]) *Let $A = \langle A; \oplus, ', 0\rangle$ be a pseudo-quasi-MV algebra. Define the binary operations by $x \to y = x' \oplus y$, $x \rightsquigarrow y = y \oplus x'$ and a constant 1 by $1 = 0'$. Then $g(A) = \langle A; \to, \rightsquigarrow, ', 1\rangle$ is a pseudo-quasi-Wajsberg algebra.*

3 Weak Pseudo-quasi-Wajsberg Algebras

In this section, we introduce the concepts of weak pseudo-quasi-Wajsberg algebras and study the related properties between them.

Definition 5 An algebra $A = \langle A; \to, \rightsquigarrow, ', 1\rangle$ of type $\langle 2, 2, 1, 0\rangle$ is called a weak pseudo-quasi-Wajsberg algebra (weak PQW-algebra, for short), if it satisfies the following axioms for all $x, y, z \in A$:
(WW1) $1 \to x = 1 \rightsquigarrow x$;
(WW2) $1 \to (x \to y) = x \to y$, $1 \rightsquigarrow (x \rightsquigarrow y) = x \rightsquigarrow y$;

(WW3) $(x \to y) \rightsquigarrow y = (y \to x) \rightsquigarrow x$,
$(x \rightsquigarrow y) \to y = (y \rightsquigarrow x) \to x$;
(WW4) $(x \to y) \to ((y \to z) \rightsquigarrow (x \to z)) = 1$,
$(x \rightsquigarrow y) \rightsquigarrow ((y \rightsquigarrow z) \to (x \rightsquigarrow z) = 1$;
(WW5) $(x' \rightsquigarrow y') \to (y \to x) = 1$, $(x' \to y') \to (y \rightsquigarrow x) = 1$;
(WW6) $x \to y' = y \rightsquigarrow x'$;
(WW7) $1 \to (1 \to x)' = (1 \to x)'$;
(WW8) $x'' = x$;
(WW9) $x \to (y \rightsquigarrow z) = y \rightsquigarrow (x \to z)$.

Obviously, pseudo-quasi-Wajsberg algebra is a weak PQW-algebra. Conversely, a weak PQW algebra which satisfies the axiom $(x \rightsquigarrow y) \to y = (y \rightsquigarrow x) \to x = (y \to x) \rightsquigarrow x = (x \to y) \rightsquigarrow y$ is a pseudo-quasi-Wajsberg algebra.

Now we list some properties of weak PQW-algebras.

Proposition 1 *In a weak PQW algebra $A = \langle A; \to, \rightsquigarrow, ', 1\rangle$, the following equalities and implications hold for all $x, y \in A$:*
(P1) *If* $x \to y = 1$ *and* $y \to x = 1$, *then* $1 \to x = 1 \to y$,
If $x \rightsquigarrow y = 1$ *and* $y \rightsquigarrow x = 1$, *then* $1 \rightsquigarrow x = 1 \rightsquigarrow y$;
(P2) *If* $x \to y = 1$ *and* $y \to z = 1$, *then* $x \to z = 1$,
If $x \rightsquigarrow y = 1$ *and* $y \rightsquigarrow z = 1$, *then* $x \rightsquigarrow z = 1$;
(P3) $1 \to 1 = 1$, $1 \rightsquigarrow 1 = 1$;
(P4) $(1 \to x) \to (1 \to x) = 1$, $(1 \to x) \rightsquigarrow (1 \to x) = 1$;
(P5) $(x \to y) \to (x \to y) = 1$, $(x \to y) \rightsquigarrow (x \to y) = 1$,
$(x \rightsquigarrow y) \rightsquigarrow (x \rightsquigarrow y) = 1$, $(x \rightsquigarrow y) \to (x \rightsquigarrow y) = 1$;
(P6) $(1 \to x) \to x = 1$, $(1 \rightsquigarrow x) \rightsquigarrow x = 1$;
(P7) $(1 \to x) \to 1 = 1$, $(1 \rightsquigarrow x) \rightsquigarrow 1 = 1$;
(P8) $(x \to y) \rightsquigarrow 1 = 1$, $(x \rightsquigarrow y) \to 1 = 1$,
$(x \to y) \to 1 = 1$, $(x \rightsquigarrow y) \rightsquigarrow 1 = 1$;
(P9) $((x \rightsquigarrow y) \to y) \rightsquigarrow x = (x \to (y \rightsquigarrow x)) \rightsquigarrow (y \rightsquigarrow x)$,
$((x \to y) \rightsquigarrow y) \to x = (x \rightsquigarrow (y \to x)) \to (y \to x)$;
(P10) $(x \to (1 \to y)) \to (1 \rightsquigarrow (x \to y)) = 1$,
$(x \rightsquigarrow (1 \rightsquigarrow y)) \rightsquigarrow (1 \to (x \rightsquigarrow y)) = 1$;
(P11) $(x \rightsquigarrow y)' = (x \rightsquigarrow y) \to 1'$, $(x \to y)' = (x \to y) \rightsquigarrow 1'$;
(P12) $((x \to y)' \rightsquigarrow 1') \to (x \to y) = 1$, $((x \to y)' \to 1') \rightsquigarrow (x \to y) = 1$,
$((x \rightsquigarrow y)' \to 1') \rightsquigarrow (x \rightsquigarrow y) = 1$, $((x \rightsquigarrow y)' \rightsquigarrow 1') \to (x \rightsquigarrow y) = 1$;
(P13) $((x \to y) \to (1' \rightsquigarrow (x \to y))) \rightsquigarrow (1' \rightsquigarrow (x \to y)) = 1$,
$((x \rightsquigarrow y) \rightsquigarrow (1' \to (x \rightsquigarrow y))) \to (1' \to (x \rightsquigarrow y)) = 1$;
(P14) $1' \rightsquigarrow (x \to y) = 1$, $1' \to (x \rightsquigarrow y) = 1$;
(P15) $x \to 1 = 1$, $x \rightsquigarrow 1 = 1$;
(P16) $x \rightsquigarrow (1 \to x) = 1$, $x \to (1 \rightsquigarrow x) = 1$;
(P17) $(1 \to x) \rightsquigarrow y = x \rightsquigarrow y$, $(1 \rightsquigarrow x) \to y = x \to y$;
(P18) $x \rightsquigarrow (1 \to y) = x \rightsquigarrow y$, $x \to (1 \rightsquigarrow y) = x \to y$;
(P19) $(1 \to x) \rightsquigarrow (1 \to y) = x \rightsquigarrow y$, $(1 \rightsquigarrow x) \to (1 \rightsquigarrow y) = x \to y$;
(P20) $1 \to 1' = 1'$, $1 \rightsquigarrow 1' = 1'$.

Proof (P1) If $x \to y = 1$ and $y \to x = 1$, we have $1 \to y = 1 \rightsquigarrow y = (x \to y) \rightsquigarrow y = (y \to x) \rightsquigarrow x = 1 \rightsquigarrow x = 1 \to x$ by (WW3) and (WW1). If $x \rightsquigarrow y = 1$ and $y \rightsquigarrow x = 1$, we have $1 \rightsquigarrow y = 1 \to y = (x \rightsquigarrow y) \to y = (y \rightsquigarrow x) \to x = 1 \to x = 1 \rightsquigarrow x$ by (WW3) and (WW1).

(P2) If $x \to y = 1$ and $y \to z = 1$, we have $1 = (x \to y) \to ((y \to z) \rightsquigarrow (x \to z)) = 1 \to (1 \rightsquigarrow (x \to z)) = 1 \rightsquigarrow (x \to z) = x \to z$ by (WW4) and (WW2). If $x \rightsquigarrow y = 1$ and $y \rightsquigarrow z = 1$, we have $1 = (x \rightsquigarrow y) \rightsquigarrow ((y \rightsquigarrow z) \to (x \rightsquigarrow z)) = 1 \rightsquigarrow (1 \to (x \rightsquigarrow z)) = 1 \to (x \rightsquigarrow z) = x \rightsquigarrow z$ by (WW4) and (WW2).

(P3) Applying (WW4) twice and (WW2), we have $1 \to 1 = 1 \to ((x \to y) \to ((y \to z) \rightsquigarrow (x \to z))) = (x \to y) \to ((y \to z) \rightsquigarrow (x \to z)) = 1$. Moreover, $1 \rightsquigarrow 1 = 1 \to 1 = 1$.

(P4) We have $(1 \to x) \to (1 \to x) = (1 \rightsquigarrow x) \to (1 \rightsquigarrow x) = 1 \to ((1 \rightsquigarrow x) \to (1 \rightsquigarrow x)) = (1 \rightsquigarrow 1) \rightsquigarrow ((1 \rightsquigarrow x) \to (1 \rightsquigarrow x)) = 1$ by (WW1), (WW2), (P3) and (WW4). The second equation can be proved similarly.

(P5) We have $(x \to y) \to (x \to y) = (1 \to (x \to y)) \to (1 \to (x \to y)) = 1$ by (WW2) and (P4). The rest can be proved similarly.

(P6) We have $(1 \to x) \to x = (1 \rightsquigarrow x) \to x = (x \rightsquigarrow 1) \to 1 = (x \rightsquigarrow 1) \to ((x \rightsquigarrow 1) \rightsquigarrow (x \rightsquigarrow 1)) = (x \rightsquigarrow 1) \to ((1 \to (x \rightsquigarrow 1)) \rightsquigarrow (x \rightsquigarrow 1)) = (x \rightsquigarrow 1) \to (((x \rightsquigarrow 1) \to 1) \rightsquigarrow 1) = (1 \to (x \rightsquigarrow 1)) \to (((x \rightsquigarrow 1) \to 1) \rightsquigarrow (1 \to 1)) = 1$ by (WW1), (WW3), (P5), (WW2), (WW3), (WW2), (P3) and (WW4). The second equation can be proved similarly.

(P7) We have $(1 \to x) \to 1 = (1 \to x) \to ((1 \to x) \rightsquigarrow (1 \to x)) = (1 \to x) \to ((1 \to (1 \to x)) \rightsquigarrow (1 \to x)) = (1 \to x) \to (((1 \to x) \to 1) \rightsquigarrow 1) = (1 \to (1 \to x)) \to (((1 \to x) \to 1) \rightsquigarrow (1 \to 1)) = 1$ by (P5), (WW2), (WW3), (P3) and (WW4). The second equation can be proved similarly.

(P8) We have $(x \to y) \rightsquigarrow 1 = (1 \rightsquigarrow (x \to y)) \rightsquigarrow 1 = 1$ by (WW1), (WW2) and (P7). The rest can be proved similarly.

(P9) We have $((x \rightsquigarrow y) \to y) \rightsquigarrow x = ((y \rightsquigarrow x) \to x) \rightsquigarrow x = (x \to (y \rightsquigarrow x)) \rightsquigarrow (y \rightsquigarrow x)$ by (WW3). The second equation can be proved similarly.

(P10) We have $(x \to (1 \to y)) \to (1 \rightsquigarrow (x \to y)) = (x \to (1 \to y)) \to (((1 \to y) \to y) \rightsquigarrow (x \to y)) = 1$ by (P6) and (WW4). The second equation can be proved similarly.

(P11) We have $(x \rightsquigarrow y) \to 1' = 1 \rightsquigarrow (x \rightsquigarrow y)' = 1 \rightsquigarrow (1 \rightsquigarrow (x \rightsquigarrow y))' = 1 \to (1 \rightsquigarrow (x \rightsquigarrow y))' = (1 \rightsquigarrow (x \rightsquigarrow y))' = (x \rightsquigarrow y)'$ by (WW6), (WW2), (WW1), (WW7) and (WW2). The second equation can be proved similarly.

(P12) We have $((x \to y)' \rightsquigarrow 1') \to (x \to y) = (1 \to (x \to y)) \to (x \to y) = 1$ by (WW6) and (P6). The rest can be proved similarly.

(P13) We have $((x \to y) \to (1' \rightsquigarrow (x \to y))) \rightsquigarrow (1' \rightsquigarrow (x \to y)) = (((x \to y) \rightsquigarrow 1') \to 1') \rightsquigarrow (x \to y) = ((x \to y)' \to 1') \rightsquigarrow (x \to y) = 1$ by (P7), (P11) and (P12). The second equation can be proved similarly.

(P14) We have $1' \rightsquigarrow (x \to y) = 1 \rightsquigarrow (1' \rightsquigarrow (x \to y)) = ((x \to y) \to 1) \rightsquigarrow (1' \rightsquigarrow (x \to y)) = ((x \to y) \to ((1 \to (x \to y)') \to 1)) \rightsquigarrow (1' \rightsquigarrow (x \to y)) = ((x \to y) \to (((x \to y) \rightsquigarrow 1') \to 1)) \rightsquigarrow (1' \rightsquigarrow (x \to y)) = ((x \to y) \to ((x \to y)' \to 1)) \rightsquigarrow (1' \rightsquigarrow (x \to y)) = ((x \to y) \to (1' \rightsquigarrow (x \to y))) \rightsquigarrow (1' \rightsquigarrow (x \to$

$y)) = 1$ by (WW2), (P8), (P7), (WW6), (P11) and (P13). It is similar to the second equation.

(P15) We have $x \to 1 = 1' \rightsquigarrow x' = 1 \rightsquigarrow (1' \rightsquigarrow x') = (1' \rightsquigarrow (1 \to x')) \rightsquigarrow (1' \rightsquigarrow x') = (1' \rightsquigarrow (1 \rightsquigarrow x')) \rightsquigarrow (1 \to (1' \rightsquigarrow x')) = 1$ by (WW6), (P14), (WW2), (WW1), (WW6) and (P10). The second equation can be proved similarly.

(P16) We have $x \rightsquigarrow (1 \to x) = x \rightsquigarrow ((x' \to 1) \to x) = x \rightsquigarrow ((1' \rightsquigarrow x) \to x) = x \rightsquigarrow ((x \rightsquigarrow 1') \to 1') = x \rightsquigarrow (1 \rightsquigarrow (x \rightsquigarrow 1')') = x \rightsquigarrow (1 \to (x \rightsquigarrow 1')') = x \rightsquigarrow (1 \to (1 \to x')') = x \rightsquigarrow (1 \to x')' = (1 \to x') \to x' = 1$ by (P15), (WW6), (WW1), (WW6), (WW7), (WW6) and (P6). The second equation can be proved similarly.

(P17) On the one hand, we have $1 = (x \rightsquigarrow (1 \to x)) \rightsquigarrow (((1 \to x) \rightsquigarrow y) \to (x \rightsquigarrow y)) = 1 \to (((1 \to x) \rightsquigarrow y) \to (x \rightsquigarrow y)) = ((1 \to x) \rightsquigarrow y) \to (x \rightsquigarrow y)$ by (WW4), (P16), (WW1) and (WW2). On the other hand, we have $1 = ((1 \rightsquigarrow x) \rightsquigarrow x) \rightsquigarrow ((x \rightsquigarrow y) \to ((1 \to x) \rightsquigarrow y)) = 1 \to ((x \rightsquigarrow y) \to ((1 \to x) \rightsquigarrow y)) = (x \rightsquigarrow y) \to ((1 \to x) \rightsquigarrow y)$ by (WW4), (WW1), (P6) and (WW2). Then, by (WW3) and (WW2), we obtain $(1 \to x) \rightsquigarrow y = x \rightsquigarrow y$. The second equation can be proved similarly.

(P18) The proof is similar to (P17).

(P19) We have $(1 \to x) \rightsquigarrow (1 \to y) = x \rightsquigarrow (1 \to y) = x \rightsquigarrow y$ by (P17) and (P18). The second equation can be proved similarly.

(P20) We have $1 \rightsquigarrow 1' = 1 \to (1 \to 1)' = (1 \to 1)' = 1'$ by (WW1), (P3) and (WW7). The second equation can be proved similarly. □

In any weak PQW-algebra, we can define two relations $x \leq_1 y \Leftrightarrow x \to y = 1$ and $x \leq_2 y \Leftrightarrow x \rightsquigarrow y = 1$. Let us mention that in a pseudo-quasi-Wajsberg algebra, it is easy to see that $x \to y = 1$ if and only if $x \rightsquigarrow y = 1$. Thus the relations $\leq_1$ and $\leq_2$ coincide in a pseudo-quasi-Wajsberg algebra.

Proposition 2 *Let A be a weak PQW-algebra. Then the relation* $\leq_1$ *with the lowest element* 0 *and the greatest element* 1 *has the following properties for any* $x, y \in A$*:*
(1) $x \leq_1 x$*;*
(2) *if* $x \leq_1 y$ *and* $y \leq_1 x$*, then* $1 \to x = 1 \to y$*;*
(3) *if* $x \leq_1 y$ *and* $y \leq_1 z$*, then* $x \leq_1 z$*.*

Proof (1) Since $x \to x = (1 \rightsquigarrow x) \to (1 \rightsquigarrow x) = 1$ by (P19), (WW1) and (P4), we have $x \leq_1 x$.

(2) If $x \leq_1 y$ and $y \leq_1 x$, then $x \to y = 1$ and $y \to x = 1$, it follows that $1 \to x = 1 \to y$ by (P1).

(3) If $x \leq_1 y$ and $y \leq_1 z$, then $x \to y = 1$ and $y \to z = 1$, it follows that $x \to z = 1$ by (P2), so $x \leq_1 z$. Moreover, since $x \to 1 = 1$ by (P15), we have $x \leq_1 1$. Also $0 \leq_1 1 \to x$ by (P14) and $1 \to x \leq_1 x$ by (P6), we have $0 \leq_1 x$ by (2). □

Similarly, we have

Proposition 3 *Let A be a weak PQW-algebra. Then the relation* $\leq_2$ *with the lowest element* 0 *and the greatest element* 1 *has the following properties for all* $x, y \in A$*:*

(1) $x \leq_2 x$;
(2) *if* $x \leq_2 y$ *and* $y \leq_2 x$, *then* $1 \rightsquigarrow x = 1 \rightsquigarrow y$;
(3) *if* $x \leq_2 y$ *and* $y \leq_2 z$, *then* $x \leq_1 z$.

Remark 1 Let A be a weak PQW-algebra. Let $R(A) = \{x \in A | 1 \to x = x\}$. Then the relations $\leq_1$ and $\leq_2$ restricted to $R(A)$ are partial orderings.

The following proposition is easy to prove. The proof is omitted.

Proposition 4 *In a weak PQW-algebra A, the following are true for all* $x, y, z \in A$:
(1) $x \leq_1 1 \rightsquigarrow x$ *and* $x \leq_2 1 \to x$;
(2) $1 \rightsquigarrow x \leq_1 x$ *and* $1 \to x \leq_2$;
(3) $x \to y \leq_1 (y \to z) \rightsquigarrow (x \to z)$, $x \rightsquigarrow y \leq_2 (y \rightsquigarrow z) \to (x \rightsquigarrow z)$;
(4) $x \rightsquigarrow y \leq_1 (z \rightsquigarrow x) \to (z \rightsquigarrow y)$, $x \to y \leq_2 (z \to x) \rightsquigarrow (z \to y)$;
(5) $x \leq_1 (y \rightsquigarrow x)$, $x \leq_2 (y \to x)$;
(6) $x \leq_1 y \rightsquigarrow z \Leftrightarrow y \leq_2 x \to z$, $x \leq_2 y \to z \Leftrightarrow y \leq_1 x \rightsquigarrow z$;
(7) $x \leq_1 y \Rightarrow y \to z \leq_2 x \to z$, $x \leq_2 y \Rightarrow y \rightsquigarrow z \leq_1 x \rightsquigarrow z$;
(8) $x \leq_1 y \Rightarrow z \to x \leq_1 z \to y$, $x \leq_2 y \Rightarrow z \rightsquigarrow x \leq_2 z \rightsquigarrow y$;
(9) $x \leq_1 y \Leftrightarrow y' \leq_2 x'$, $x \leq_2 y \Leftrightarrow y' \leq_1 x'$.

Now, we define two binary operations $x \vee_1 y = (x \to y) \rightsquigarrow y = (y \to x) \rightsquigarrow x$ and $x \vee_2 y = (x \rightsquigarrow y) \to y = (y \rightsquigarrow x) \to x$.

Proposition 5 *Let A be a weak PQW-algebra. Then the following are true for all* $x, y \in A$:
(1) $x \vee_1 y$ *is a supremum for x and y with respect to* $\leq_1$;
(2) $x \vee_2 y$ *is a supremum for x and y with respect to* $\leq_2$.

Proof (1) By definition of $\vee_1$, (WW3) and (WW9), we have $x \to (x \vee_1 y) = x \to ((x \to y) \rightsquigarrow y) = x \to ((y \to x) \rightsquigarrow x) = 1$, thus $x \leq_1 x \vee_1 y$. Similarly, we have $y \leq_1 x \vee_1 y$. Let $z \in A$ such that $x \leq_1 z$ and $y \leq_1 z$. Then $x \to z = 1$ and $y \to z = 1$. By (WW4), (WW1) and (WW2), we have $1 = (y \to z) \to ((z \to x) \rightsquigarrow (y \to x)) = 1 \to ((z \to x) \rightsquigarrow (y \to x)) = (z \to x) \rightsquigarrow (y \to x)$. Then we obtain $1 = ((z \to x) \rightsquigarrow (y \to x)) \rightsquigarrow (((y \to x) \rightsquigarrow x) \to ((z \to x) \rightsquigarrow x)) = ((y \to x) \rightsquigarrow x) \to ((z \to x) \rightsquigarrow x) = (x \vee_1 y) \to (x \vee_1 z) = (x \vee_1 y) \to (1 \to z)$. Thus we have $(x \vee_1 y) \leq_1 (1 \to z)$.

(2) The proof is similar to (1). □

Dually, we can define the binary operations $x \wedge_1 y = (x' \vee_2 y')'$ and $x \wedge_2 y = (x' \vee_1 y')'$.

Proposition 6 *Let A be a weak PQW-algebra. Then the following are true for all* $x, y \in A$:
(1) $x \wedge_1 y$ *is an infimum for x and y with respect to* $\leq_1$;
(2) $x \wedge_2 y$ *is an infimum for x and y with respect to* $\leq_2$.

Proposition 7 *Let A be a weak PQW-algebra. Then*
(1) $x \vee_1 y = y \vee_1 x$ *and* $x \wedge_1 y = y \wedge_1 x$;
(2) $(x \vee_1 y) \vee_1 z = x \vee_1 (y \vee_1 z)$ *and* $(x \wedge_1 y) \wedge_1 z = x \wedge_1 (y \wedge_1 z)$;
(3) $x \vee_1 x = 1 \to x$ *and* $x \wedge_1 x = 1 \to x$;
(4) $(x \vee_1 y) \wedge_1 x = 1 \to x$ *and* $(x \wedge_1 y) \vee_1 x = 1 \to x$.

Proof We only check the case of $\vee_1$. The case of $\wedge_1$ can be proved dually.

(1) Directly from the definition of $\vee_1$.

(2) Since $y \leq_1 y \vee_1 z$, we have $x \vee_1 y \leq_1 x \vee_1 (y \vee_1 z)$ and $z \leq_1 y \vee_1 z \leq_1 x \vee_1 (y \vee_1 z)$, it turns out that $(x \vee_1 y) \vee_1 z \leq_1 x \vee_1 (y \vee_1 z)$. Similarly, $x \vee_1 (y \vee_1 z) \leq_1 (x \vee_1 y) \vee_1 z$. Thus $(x \vee_1 y) \vee_1 z = x \vee_1 (y \vee_1 z)$.

(3) We have $x \vee_1 x = (x \rightsquigarrow x) \to x = ((1 \to x) \rightsquigarrow (1 \to x)) \to x = 1 \to x$ by (P19) and (P4).

(4) Since $1 \to x \leq_1 x \vee_1 y$ and $1 \to x \leq_1 x$ by Proposition 4 and (WW1), we have $1 \to x \leq_1 (x \vee_1 y) \wedge_1 x$. On the other hand, $(x \vee_1 y) \wedge_1 x \leq_1 x \leq_1 1 \to x$. Thus $(x \vee_1 y) \wedge_1 x = 1 \to x$. □

Similarly, we have

Proposition 8 *Let A be a weak PQW-algebra. Then*
(1) $x \vee_2 y = y \vee_2 x$ *and* $x \wedge_2 y = y \wedge_2 x$;
(2) $(x \vee_2 y) \vee_2 z = x \vee_2 (y \vee_2 z)$ *and* $(x \wedge_2 y) \wedge_2 z = x \wedge_2 (y \wedge_2 z)$;
(3) $x \vee_2 x = 1 \rightsquigarrow x$ *and* $x \wedge_2 x = 1 \rightsquigarrow x$;
(4) $(x \vee_2 y) \wedge_2 x = 1 \rightsquigarrow x$ *and* $(x \wedge_2 y) \vee_1 x = 1 \rightsquigarrow x$.

Proposition 9 *In a weak PQW algebra* $A = \langle A; \to, \rightsquigarrow, 0, 1\rangle$, *the following are true for all* $x, y, z \in A$:
(1) $(x \vee_1 y) \to z = (x \to z) \wedge_2 (y \to z)$, $(x \vee_2 y) \rightsquigarrow z = (x \rightsquigarrow z) \wedge_1 (y \rightsquigarrow z)$;
(2) $z \to (x \wedge_1 y) = (z \to x) \wedge_1 (z \to y)$, $z \rightsquigarrow (x \wedge_2 y) = (z \rightsquigarrow x) \wedge_2 (z \rightsquigarrow y)$;
(3) $(x \vee_1 y) \to y = x \to y$, $(x \vee_2 y) \rightsquigarrow y = x \rightsquigarrow y$;
(4) $x \to (x \wedge_1 y) = x \to y$, $x \rightsquigarrow (x \wedge_2 y) = x \rightsquigarrow y$.

Proof The proof is similar to [6]. □

4 Weak Pseudo-Quasi-MV Algebras and a Categorical Equivalence

In this section we introduce the concept of weak pseudo-quasi-MV algebras and prove that they are term equivalent to weak PQW-algebras.

Definition 6 An algerbra $A = \langle A; \oplus, ', 0\rangle$ of type $\langle 2, 1, 0\rangle$ is called a weak pseudo-quasi-MV algebra, if it satisfies the following axioms for all $x, y, z \in A$:
(WPQ1) $(x \oplus y) \oplus z = x \oplus (y \oplus z)$;
(WPQ2) $x \oplus y \oplus 0 = x \oplus y$;

(WPQ3) $x \oplus 0 = 0 \oplus x$;
(WPQ4) $x \oplus 0' = 0' = 0' \oplus x$;
(WPQ5) $(x \oplus 0)' = x' \oplus 0$;
(WPQ6) $y \oplus (x' \oplus y)' = x \oplus (y' \oplus x)'$, $(y \oplus x')' \oplus y = (x \oplus y')' \oplus x$;
(WPQ7) $x'' = x$.

Remark 2 In any weak pseudo-quasi-MV algebra, we denote $1 = 0'$ and have $x' \oplus x = x \oplus x' = 1$. Indeed, $x' \oplus x = 0 \oplus x' \oplus x = (0 \oplus x') \oplus x = (x \oplus 0)' \oplus x = (x \oplus 1')' \oplus x = (1 \oplus x')' \oplus 1 = 1' \oplus 1 = 0 \oplus 1 = 1$.

On any weak pseudo-quasi-MV algebra, we define:
(1) $x \vee_1 y = y \oplus (x' \oplus y)'$, $x \vee_2 y = (y \oplus x')' \oplus y$;
(2) $x \wedge_1 y = (x' \vee_2 y')'$, $x \wedge_2 y = (x' \vee_1 y')'$;
(3) $x \leq_1 y = x \vee_1 y = y \oplus 0$, $x \leq_2 y = x \vee_2 y = y \oplus 0$.

Proposition 10 *Let A be a weak pseudo-quasi-MV algebra. For $x, y \in A$, the following conditions are equivalent:*
(1) $x \leq_1 y$,
(2) $x' \oplus y = 1$;

Proof (1) ⇒ (2) Since $x \leq_1 y$, we have $x \vee_1 y = y \oplus 0$. Thus $x' \oplus y = (x' \oplus y) \oplus 0 = x' \oplus (y \oplus 0) = x' \oplus (x \vee_1 y) = x' \oplus (y \oplus (x' \oplus y)') = (x' \oplus y) \oplus (x' \oplus y)' = 1$.

(2) ⇒ (1) Since $x \vee_1 y = y \oplus (x' \oplus y)' = y \oplus 1' = y \oplus 0$, we have$x \leq_1 y$. □

Similarly, we have

Proposition 11 *Let A be a weak pseudo-quasi-MV algebra. For $x, y \in A$, the following conditions are equivalent:*
(1) $x \leq_2 y$;
(2) $y \oplus x' = 1$.

Remark 3 Based on Propositions 10 and 11, it is easy to see that $0 \leq_1 x$ and $x \leq_1 1$ for each $x \in A$. Similarly, we have $0 \leq_2 x$ and $x \leq_2 1$.

Theorem 3 *Let $A = \langle A; \oplus, ', 0\rangle$ be a weak pseudo-quasi-MV algebra. Define the binary operations $x \to y = x' \oplus y$, $x \rightsquigarrow y = y \oplus x'$ and a constant $1 = 0'$. Then $f(A) = \langle A; \to, \rightsquigarrow, ', 1\rangle$ is a weak PQW-algebra.*

Proof We check the conditions in Definition 5 consecutively.

(WW1) $1 \to x = 1' \oplus x = 0 \oplus x = x \oplus 0 = x \oplus 1' = 1 \rightsquigarrow x$.

(WW2) $1 \to (x \to y) = 1' \oplus (x' \oplus y) = 0 \oplus (x' \oplus y) = x' \oplus y = x \to y$. Similarly, we have $1 \rightsquigarrow (x \rightsquigarrow y) = x \rightsquigarrow y$.

(WW3) Since $(x \rightsquigarrow y) \to y = (y \oplus x')' \oplus y$ and $(y \rightsquigarrow x) \to x = (x \oplus y')' \oplus x$, we have $(x \rightsquigarrow y) \to y = (y \rightsquigarrow x) \to x$ by (WPQ6). Similarly, we can prove $(x \to y) \rightsquigarrow y = (y \to x) \rightsquigarrow x$.

(WW4) Obviously, $(x \to y) \to ((y \to z) \rightsquigarrow (x \to z)) \leq 1$. Since $(x \to y) \to ((y \to z) \rightsquigarrow (x \to z)) = (x' \oplus y) \to ((y' \oplus z) \rightsquigarrow (x' \oplus z)) = (x' \oplus y) \to ((x' \oplus z) \oplus (y' \oplus z)') = (x' \oplus y)' \oplus ((x' \oplus z) \oplus (y' \oplus z)') = ((x' \oplus y)' \oplus x') \oplus (z \oplus (y' \oplus z)') = (x' \vee_2 y') \oplus (z \vee_1 y) \geq y' \oplus y = 1$, we have $(x \to y) \to ((y \to z) \rightsquigarrow (x \to z)) = 1$. Similarly, we have $(x \rightsquigarrow y) \rightsquigarrow ((y \rightsquigarrow z) \to (x \rightsquigarrow z)) = 1$.

(WW5) We have $(x' \rightsquigarrow y') \to (y \to x) = (y' \oplus x) \to (y' \oplus x) = (y' \oplus x)' \oplus (y' \oplus x) = 1$. Similarly, we have $(x' \to y') \to (y \rightsquigarrow x) = 1$.

(WW6) We have $x \to y' = x' \oplus y' = y \rightsquigarrow x'$.

(WW7) We have $1 \to (1 \to x)' = 1' \oplus (1 \to x)' = 0 \oplus (0 \oplus x)' = 0 \oplus x' = (0 \oplus x)' = (1 \to x)'$ by (WPQ5). Similarly, we have $1 \rightsquigarrow (1 \rightsquigarrow x)' = (1 \rightsquigarrow x)'$.

(WW8) From(WPQ7).

(WW9) We have $x \to (y \rightsquigarrow z) = x' \oplus (y \rightsquigarrow z) = x' \oplus (z \oplus y') = (x' \oplus z) \oplus y' = (x \to z) \oplus y' = y \rightsquigarrow (x \to z)$. □

Conversely, we have

Theorem 4 *Let* $A = \langle A; \to, \rightsquigarrow, ', 1 \rangle$ *be a weak PQW-algebra. Define a binary operation* $\oplus$ *and a constant 0 by* $x \oplus y = x' \to y = y' \rightsquigarrow x$ *and* $0 = 1'$. *Then* $g(A) = \langle A; \oplus, ', 0 \rangle$ *is a weak pseudo-quasi-MV algebra.*

Proof We check the conditions in Definition 6 consecutively.

(WPQ1) Since $(x \oplus y) \oplus z = (x' \to y) \oplus z = z' \rightsquigarrow (x' \to y)$ and $x \oplus (y \oplus z) = x \oplus (y' \to z) = x' \to (y' \to z) = x' \to (z' \rightsquigarrow y)$, we have $(x \oplus y) \oplus z = x \oplus (y \oplus z)$ by (WW9).

(WPQ2) Since $x \oplus y \oplus 0 = (x' \to y) \oplus 0 = (x' \to y) \oplus 1' = 1 \rightsquigarrow (x' \to y) = 1 \to (x' \to y) = x' \to y$ by (WW1) and (WW2), we have $x \oplus y \oplus 0 = x \oplus y$.

(WPQ3) We have $x \oplus 0 = 0' \rightsquigarrow x = 1 \rightsquigarrow x = 1 \to x = 0' \to x = 0 \oplus x$.

(WPQ4) We have $0' \oplus x = x' \rightsquigarrow 0' = x' \rightsquigarrow 1 = 1 = 0'$ by (L13). Similarly, $x \oplus 0' = 0'$.

(WPQ5) Since $(x \oplus 0)' = (0' \rightsquigarrow x)' = (1 \rightsquigarrow x)' = (1 \rightsquigarrow x) \to 0 = x \to 0 = 1 \rightsquigarrow x'$ by (WW6), (L9), (L15) and (WW6), and $x' \oplus 0 = 0' \rightsquigarrow x' = 1 \rightsquigarrow x'$, we have $(x \oplus 0)' = x' \oplus 0$.

(WPQ6) We have $y \oplus (x' \oplus y)' = y \oplus (x \to y)' = (x \to y) \rightsquigarrow y = (y \to x) \rightsquigarrow x = (y' \oplus x) \rightsquigarrow x = x \oplus (y' \oplus x)'$ by (WW3). Similarly, $(y \oplus x')' \oplus y = (x \rightsquigarrow y)' \oplus y = (x \rightsquigarrow y) \to y = (y \rightsquigarrow x) \to x = (x \oplus y') \to x = (x \oplus y')' \oplus x$.

(WPQ7) From (WW8). □

Based on Theorems 3 and 4, we can easily check that

Theorem 5 *The mappings f and g defined in Theorems 3 and 4 are mutually inverse correspondences.*

Acknowledgments We would like to show our sincere thanks to the referees for their valuable comments. This project is supported by the National Natural Science Foundation of China (Grant No. 11501245).

References

1. Bou, F., Giuntini, R., Ledda, A., Paoli, F.: The logic of quasi-MV algebras. J. Log. Comput. **20**, 619–643 (2010)
2. Chen, W.J., Dudek, W.A.: Quantum computational algebra with a non-commutative generalization. Math. Slovaca. **66**(1), 19–34 (2016)
3. Chen, W.J., Dudek, W.A.: The representation of square root quasi-pseudo-MV algebras. Soft Comput. **19**(2), 269–282 (2015)
4. Chen, W.J., Bavvaz, D.: Some classes of quasi-pseudo-MV algebras. Log. J. IGPL. (2016). doi:10.1093/jigpal/jzw034
5. Liu, J.M., Chen, W.J.: A non-communicative generalization of quasi-MV algebras. In: IEEE International Conference on Fuzzy Systems, in press (2016)
6. Ceterchi, R.: Weak pseudo-Wajsberg and weak pseudo-MV algebras. Soft Comput. **5**, 334–346 (2001)
7. Font, J.M., Rodriguez, A.J., Torrens, A.: Wajsberg Algebras. Stochastica. VII **I**(1), 5–31 (1984)
8. Jipsen, P., Ledda, A., Paoli, F.: On some properties of quasi-MV algebras and $\sqrt{'}$quasi-MV algebras Part IV. Rep. Math. Log. **48**, 3–36 (2013)
9. Kowalski, T., Paoli, F.: On some properties of quasi-MV algebras and $\sqrt{'}$quasi-MV algebras Part III. Rep. Math. Log. **45**, 161–199 (2010)
10. Giuntini, R., Konig, M., Ledda, A., Paoli, F.: MV algebras and quantum computation. Stud. Log. **82**, 245–270 (2006)
11. Freytes, H., Giuntini, R., Ledda, A., Paoli, F.: On some properties of quasi-MV algebras and $\sqrt{'}$quasi-MV algebras Part I. Rep. Math. Log. **44**, 31–63 (2009)

Characterizations of a Class of Commutative Algebras of Logic Systems

Xue-Min Ling and Luo-Shan Xu

Abstract This paper gives new properties of WBR_0-algebras and commutative WBR_0-algebras. A comprehensive characterization for commutative WBR_0-algebras by various other logic algebras is obtained. Three characterizations with simpler types and fewer axioms of commutative WBR_0-algebras are also given.

Keywords WBR_0-algebra · CFI-algebra · MV-algebra · NLI-algebra · Residuated lattice · Lattice implication algebra

1 Introduction

The study of non-classical logics includes syntax, semantics and algebras of logic calculus of logical systems. In the study of logic systems, one found that most logic systems are related to some logic algebras. For example, the Łukasiewicz continuous-valued logic system is matched with MV-algebras, the intuitionistic propositional logic system is matched with Heyting algebras. The formal logic system $\mathcal{L}^*$ posed by Wang is matched with R_0-algebras [8]. So, researches on various logic algebras are important in the study of non-classical logics. Afterwards, a new algebraic structure of WBR_0-algebras [10] was posed according to characterizations of BR_0-algebras. Significant results [1–3, 9] were obtained.

The purpose of this paper is to study commutative WBR_0-algebras. Many properties of (commutative) WBR_0-algebras are obtained. It is proved that a commutative WBR_0-algebra is actually equivalent to a (2, 0)-type algebra expressed by an "$\rightarrow$" operator. Several characterizations with simpler types and fewer axioms of commutative WBR_0-algebras are given.

X.-M. Ling
Department of Common Courses, Anhui Xinhua University, Hefei 230031, China
e-mail: kikifariy@126.com

L.-S. Xu (✉)
Department of Mathematics, Yangzhou University, Yangzhou 225002, China
e-mail: luoshanxu@hotmail.com

T.-H. Fan et al. (eds.), *Quantitative Logic and Soft Computing 2016*,
Advances in Intelligent Systems and Computing 510,
DOI 10.1007/978-3-319-46206-6_38

2 Preliminaries

In the study of various kinds of logic systems, one put forward many logic algebras. On the basis of these logic algebras, a number of related stronger or weaker algebras are proposed. Basic concepts and related results of WBR$_0$-algebras, FI-algebras and NLI-algebras are given below. For other concepts and results not clearly specified please refer to [2, 3, 6, 7, 9, 14].

Definition 1 ([10]) A (2, 2, 0, 0)-type algebra $(M, \oplus, \rightarrow, 0, 1)$ is called a WBR$_0$-algebra, if $\forall a,\ b,\ c \in M$, the following statements hold:
(WB1) $a \oplus 0 = a$;
(WB2) $a \oplus b = b \oplus a$;
(WB3) $a \rightarrow b = (b \rightarrow 0) \rightarrow (a \rightarrow 0)$;
(WB4) $a \rightarrow (b \rightarrow c) = b \rightarrow (a \rightarrow c)$;
(WB5) $(b \rightarrow c) \rightarrow ((a \rightarrow b) \rightarrow (a \rightarrow c)) = 1$;
(WB6) $a \rightarrow (a \oplus b) = 1$;
(WB7) $(a \oplus b) \rightarrow c = (((a \rightarrow c) \rightarrow 0) \oplus ((b \rightarrow c) \rightarrow 0)) \rightarrow 0$;
(WB8) $1 \rightarrow a = a$;
(WB9) If $a \rightarrow b = b \rightarrow a = 1$, then $a = b$.

Definition 2 ([7]) A (2, 0)-type algebra $(M, \rightarrow, 0)$ is called an FI-algebra if $\forall a,\ b,\ c \in M$ and $1 = 0 \rightarrow 0$, the following statements hold:
(I1) $a \rightarrow (b \rightarrow c) = b \rightarrow (a \rightarrow c)$;
(I2) $(a \rightarrow b) \rightarrow ((b \rightarrow c) \rightarrow (a \rightarrow c)) = 1$;
(I3) $a \rightarrow a = 1$;
(I4) If $a \rightarrow b = b \rightarrow a = 1$, then $a = b$;
(I5) $0 \rightarrow a = 1$.
If M also satisfies the following condition:

$$(Com) : (a \rightarrow b) \rightarrow b = (b \rightarrow a) \rightarrow a,\ \ \forall a, b \in M,$$

then M is said to be a commutative FI-algebra, briefly, CFI-algebra.

Remark 1 (1) (WB4) in Definition 1 is not independent by [9]. (2) By [1, 10], a WBR$_0$-algebra is an FI-algebra and a BR$_0$-algebra is a WBR$_0$-algebra.

Lemma 1 ([2, 9]) *In a WBR$_0$-algebra,* $\forall a,\ b,\ c \in M$*:*
(1) $a \rightarrow a = 1$, *specially* $0 \rightarrow 0 = 1$*;*
(2) *If* $a \rightarrow c = 1,\ b \rightarrow c = 1$*, then* $(a \oplus b) \rightarrow c = 1$*;*
(3) $(a \rightarrow 0) \rightarrow 0 = a$*;*
(4) $0 \rightarrow a = 1$.

Definition 3 ([2]) A WBR$_0$-algebra M is called a commutative WBR$_0$-algebra, briefly, CWBR$_0$-algebra, if M satisfies the following condition:

$$\text{(Com)}: (a \rightarrow b) \rightarrow b = (b \rightarrow a) \rightarrow a,\ \ \forall a, b \in M.$$

Definition 4 ([6]) A (2, 0)-type algebra $(L, \to, 0)$ is called an NFI-algebra, if $\forall a,\ b,\ c \in L$, one has:
(NLI-1) $0 \to a = 1$, where $1 = 0 \to 0$;
(NLI-2) $(a \to 0) \to 0 = a$;
(NLI-3) $a \to b = (b \to 0) \to (a \to 0)$;
(NLI-4) $(a \to b) \to (a \to c) = (b \to a) \to (b \to c)$.

3 Further Properties of (Commutative) WBR$_0$-Algebras

New properties not expressed or not explicitly expressed in the literature of WBR$_0$-algebras and CWBR$_0$-algebras are given in this section.

Proposition 1 *If M is a WBR$_0$-algebra, then $\forall a,\ b\ c,\ d \in M$, one has,*
(W1) $a \to 1 = 1$;
(W2) *If* $1 \to a = 1$, *then* $a = 1$;
(W3) $a \to (b \to a) = 1$;
(W4) *If* $a \to 0 = 0$, *then* $a = 1$;
(W5) *If* $a \to b = 1$, $b \to c = 1$, *then* $a \to c = 1$;
(W6) *If* $a \to b = 1$, $c \to d = 1$, *then* $(a \oplus c) \to (b \oplus d) = 1$.

Proof (W1) By Definition 1 (WB4) and Lemma 1 (1, 4), we have

$$a \to 1 = a \to (0 \to a) = 0 \to (a \to a) = 0 \to 1 = 1.$$

(W2) It follows from (W1) and Definition 1 (WB9).
(W3) By Definition 1 (WB4), Lemma 1 (1) and (W1), we know

$$a \to (b \to a) = b \to (a \to a) = b \to 1 = 1.$$

(W4) By Lemma 1 (1, 3), $a = (a \to 0) \to 0 = 0 \to 0 = 1$.
(W5) By Definition 1 (WB5) and (WB8), we have

$$1 = (a \to b) \to ((b \to c) \to (a \to c)) = 1 \to (1 \to (a \to c)) = a \to c.$$

(W6) By Definition 1 (WB6), we have $b \to (b \oplus d) = 1$. By (W5), we have $a \to (b \oplus d) = 1$. Similarly, $c \to (b \oplus d) = 1$. By Lemma 1 (2), we have $(a \oplus c) \to (b \oplus d) = 1$. □

Proposition 2 *If M is a CWBR$_0$-algebra, then $\forall a,\ b \in M$,*
(CBR1) $((a \to b) \to b) \to b = a \to b$;
(CBR2) $a \oplus b = (a \to b) \to b$.

Proof (CBR1) On one hand, by Definition 1 (WB4), Lemma 1 (1), we have

$$(a \to b) \to (((a \to b) \to b) \to b)$$
$$= ((a \to b) \to b) \to ((a \to b) \to b) = 1.$$

On the other hand, we have

$$(((a \to b) \to b) \to b) \to (a \to b)$$
$$= (((b \to a) \to a) \to b) \to (a \to b) \quad \text{(by Definition 3 (Com))}$$
$$= 1 \to ((((b \to a) \to a) \to b) \to (a \to b)) \quad \text{(by (WB8))}$$
$$= (a \to ((b \to a) \to a)) \to ((((b \to a) \to a) \to b) \to (a \to b)) \quad \text{(by (W3))}$$
$$= 1. \quad \text{(by (WB4), (WB8))}$$

(CBR2) On one hand,we have

$$a \to ((a \to b) \to b) = (a \to b) \to (a \to b) = 1, \quad \text{(by (WB4), Lemma 1 (1))}$$
$$b \to ((a \to b) \to b) = (a \to b) \to (b \to b) = 1. \quad \text{(by (WB4), Lemma 1 (1), (W1))}$$

By Lemma 1 (2), we have $(a \oplus b) \to ((a \to b) \to b) = 1$. Further, we have

$$((a \to b) \to b) \to (a \oplus b) = ((a \to b) \to b) \to (1 \to (a \oplus b)) \quad \text{(by (WB8))}$$
$$= ((a \to b) \to b) \to ((b \to (a \oplus b)) \to (a \oplus b)) \quad \text{(by (WB6))}$$
$$= ((a \to b) \to b) \to (((a \oplus b) \to b) \to b) \quad \text{(by Definition 3 (Com))}$$
$$= ((a \oplus b) \to b) \to (((a \to b) \to b) \to b) \quad \text{(by (WB4))}$$
$$= ((a \oplus b) \to b) \to (a \to b) \quad \text{(by (CBR1))}$$
$$= (a \to (a \oplus b)) \to (((a \oplus b) \to b) \to (a \to b)) \quad \text{(by (WB6), (WB8))}$$
$$= 1. \quad \text{(by (WB4), (WB5))}$$

So, by Definition 1 (WB9), $a \oplus b = (a \to b) \to b$. □

Proposition 3 *Let M be a WBR$_0$-algebra. Then M is commutative iff*

$$\forall a, b \in L, \qquad a \oplus b = (a \to b) \to b.$$

Proof Sufficiency: Suppose that M is a WBR$_0$-algebra and satisfies condition $a \oplus b = (a \to b) \to b$. By Definition 1 (WB2), we have $a \oplus b = b \oplus a$. So, (Com) in Definition 3 holds and M is a CWBR$_0$-algebra.

Necessity: It follows from (CBR2) that $a \oplus b = (a \to b) \to b$. □

Remark 2 By Propositions 2 and 3, in CWBR$_0$-algebras, "$\oplus$" can be expressed by "$\to$". So a CWBR$_0$-algebra is actually equivalent to a (2, 0)-type algebra which is expressed by an "$\to$" operator.

Proposition 4 *If M is a CWBR$_0$-algebra, then* $\forall a, b, c \in M$
(CBR3) $(a \to c) \to (b \to c) = b \to (a \oplus c)$;
(CBR4) $(a \to b) \to (a \to c) = (b \to a) \to (b \to c)$;

(CBR5) *If* $a \to b = 1$, *then* $(b \to c) \to (a \to c) = (c \to a) \to (c \to b) = 1$;
(CBR6) $a \oplus (a \oplus b) = a \oplus b$;
(CBR7) $((a \oplus b) \to a) \oplus ((a \oplus b) \to b) = 1$;
(CBR8) *If* $a \to b = 1$, *then* $a \to (b \oplus c) = 1$;
(CBR9) $(a \to b) \oplus (b \to a) = 1$;
(CBR10) $(a \oplus b) \to a = b \to a$, $(a \oplus b) \to b = a \to b$.

Proof (CBR3) By Definition 1 (WB4) and Proposition 2 (CBR2), one has

$$\begin{aligned} &(a \to c) \to (b \to c) = b \to ((a \to c) \to c) && \text{(by (WB4))} \\ &= b \to (a \oplus c). && \text{(by (CBR2))} \end{aligned}$$

(CBR4) We have that

$$\begin{aligned} &(a \to b) \to (a \to c) \\ &= ((b \to 0) \to (a \to 0)) \to ((c \to 0) \to (a \to 0)) && \text{(by (WB3))} \\ &= (c \to 0) \to ((b \to 0) \oplus (a \to 0)) && \text{(by (CBR3))} \\ &= (c \to 0) \to ((a \to 0) \oplus (b \to 0)) && \text{(by (WB2))} \\ &= (c \to 0) \to (((a \to 0) \to (b \to 0)) \to (b \to 0)) && \text{(by (CBR2))} \\ &= ((a \to 0) \to (b \to 0)) \to ((c \to 0) \to (b \to 0)) && \text{(by (WB4))} \\ &= (b \to a) \to (b \to c). && \text{(by (WB3))} \end{aligned}$$

(CBR5) Set $a \to b = 1$. Then we have

$$\begin{aligned} &(b \to c) \to (a \to c) = a \to ((b \to c) \to c) && \text{(by (WB4))} \\ &= a \to ((c \to b) \to b) && \text{(by Definition 3 (Com))} \\ &= (c \to b) \to (a \to b) && \text{(by (WB4))} \\ &= (c \to b) \to 1 = 1. && \text{(by the Assumption, (W1))} \end{aligned}$$

$$\begin{aligned} &(a \to b) \to ((c \to a) \to (c \to b)) \\ &= 1 \to ((c \to a) \to (c \to b) = 1. && \text{(by the Assumption, (WB5))} \end{aligned}$$

So by Proposition 1 (W2), we have $(c \to a) \to (c \to b) = 1$.

(CBR6) On one hand, by Definition 1 (WB6), we have $(a \oplus b) \to (a \oplus (a \oplus b)) = 1$. On the other hand, we have

$$\begin{aligned} &(a \oplus (a \oplus b)) \to (a \oplus b) \\ &= ((a \to (a \oplus b)) \to (a \oplus b)) \to (a \oplus b) && \text{(by (CBR2))} \\ &= (1 \to (a \oplus b)) \to (a \oplus b) = 1. && \text{(by (WB6), (WB8), Lemma 1 (1))} \end{aligned}$$

By Definition 1 (WB9), we see $a \oplus (a \oplus b) = a \oplus b$.

(CBR7) We have

$$
\begin{aligned}
&((a \oplus b) \to a) \oplus ((a \oplus b) \to b) \\
&= (((a \oplus b) \to a) \to ((a \oplus b) \to b)) \to ((a \oplus b) \to b) && \text{(by (CBR2))} \\
&= ((b \to 0) \to ((a \to 0) \oplus ((a \oplus b) \to 0))) \to ((a \oplus b) \to b) && \text{(by (WB3))} \\
&= ((((a \to 0) \oplus ((a \oplus b) \to 0)) \to 0) \to b) \to ((a \oplus b) \to b) && \text{(by (WB3), (CBR3))} \\
&= ((a \oplus (a \oplus b)) \to b) \to ((a \oplus b) \to b) && \text{(by Lemma 1 (3), (WB7))} \\
&= ((a \oplus b) \to b) \to ((a \oplus b) \to b) = 1. && \text{(by Lemma 1 (1), (CBR6))}
\end{aligned}
$$

(CBR8) Set $a \to b = 1$. Then we have

$$
\begin{aligned}
&a \to (b \oplus c) = a \to ((b \to c) \to c)) && \text{(by (CBR2))} \\
&= a \to (((c \to b) \to b)) && \text{(by Definition 3 (Com))} \\
&= (c \to b) \to (a \to b) && \text{(by (WB4))} \\
&= (c \to b) \to 1 = 1. && \text{(by the Assumption, (W1))}
\end{aligned}
$$

(CBR9) By Definition 1 (WB6), we have $a \to (a \oplus b) = 1$ and $b \to (a \oplus b) = 1$. By (CBR5), we have $((a \oplus b) \to b) \to (a \to b) = 1$ and $((a \oplus b) \to a) \to (b \to a) = 1$. By Proposition 1 (W6) and (CBR7), we have

$$
\begin{aligned}
1 &= (((a \oplus b) \to b) \oplus ((a \oplus b) \to a) \to ((a \to b) \oplus (b \to a)) && \text{(by (W6))} \\
&= 1 \to ((a \to b) \oplus (b \to a)). && \text{(by (CBR7))}
\end{aligned}
$$

By Proposition 1 (W2), we see that $(a \to b) \oplus (b \to a) = 1$ holds.

(CBR10) On one hand, we have

$$
\begin{aligned}
&((a \oplus b) \to a) \to (b \to a) \\
&= (((a \to b) \to b) \to a) \to (b \to a) && \text{(by (CBR2))} \\
&= (b \to ((a \to b) \to b)) \to ((((a \to b) \to b) \to a) \to (b \to a)) && \text{(by (W3), (WB8))} \\
&= 1. && \text{(by (WB4), (WB5))}
\end{aligned}
$$

On the other hand, we have

$$
\begin{aligned}
&(b \to a) \to ((a \oplus b) \to a) \\
&= (b \to a) \to (((a \to b) \to b) \to a) && \text{(by (CBR2))} \\
&= ((a \to b) \to b) \to ((b \to a) \to a) = 1. && \text{(by (WB4), Lemma 1 (1), Definition 3 (Com))}
\end{aligned}
$$

By (WB9), we see $(a \oplus b) \to a = b \to a$. Similarly, $(a \oplus b) \to b = a \to b$. □

4 Characterization of Commutative WBR$_0$-Algebras

Firstly, a comprehensive characterization for CWBR$_0$-algebras by various logic algebras are given. For some involved algebras please refer to [11] for commutative BR$_0$-algebras, [8] for MV-algebras and residuated lattices, [13] for lattice implication algebras, [14] for regular BL-algebras and [5] for bounded commutative BCK-algebras. A commutative residuated lattice is a residuated lattice with (Com).

Theorem 1 *The following algebraic structures are equivalent to each other:*
(1) *CWBR$_0$-algebras;*
(2) *CFI-algebras;*
(3) *Commutative BR$_0$-algebras;*
(4) *MV-algebras;*
(5) *lattice implication algebras;*
(6) *NLI-algebras;*
(7) *regular BL-algebras;*
(8) *bounded commutative BCK-algebras;*
(9) *commutative residual lattices.*

Proof It follows from Theorem 3.4.2 in [2] that (1), (2), (3) and (5) are equivalent. By Theorem 3.7 in [7], (2), (3), (4) and (5) are equivalent. By Theorem 2.1 in [6], (5) and (6) are equivalent. By Theorem 3.2 in [14], (5) and (7) are equivalent. By Theorem 3.3 in [5], (3) and (8) are equivalent. The equivalence of (1) and (9) appeared as Theorem 1 in [3]. So, all the nine algebraic systems are equivalent to each other. □

The following lemma is useful in characterizing CWBR$_0$-algebras.

Lemma 2 *In a (2, 0)-type algebra* $(M, \to, 0)$, *let* $1 = 0 \to 0$. *Then*
(1) *The conditions* (WB4) *and* (WB5) *imply the following condition:*

$$(\mathrm{WB5}^*)(a \to b) \to ((b \to c) \to (a \to c)) = 1.$$

(2) *Conditions* (WB8) *and* (Com) *imply* (I4) *in* Definition 2.

Proof (1) By Definition 1 (WB4, WB5), we have

$$(a \to b) \to ((b \to c) \to (a \to c)) = (b \to c) \to ((a \to b) \to (a \to c)) = 1.$$

So, (WB5*) holds.

(2) If $a \to b = b \to a = 1$, then by (WB8) and (Com), we have

$$a = 1 \to a = (b \to a) \to a = (a \to b) \to b = 1 \to b = b.$$

So, (I4) in Definition 2 holds. □

The following theorem gives a simpler form of CWBR$_0$-algebras.

Theorem 2 *A (2, 0)-type algebra* $(M, \to, 0)$ *is equivalent to a CWBR*$_0$*-algebra iff* $\forall a, b, c \in M$ *and* $1 = 0 \to 0$*, the following holds:*
(WB3) $a \to b = (b \to 0) \to (a \to 0)$;
(WB4) $a \to (b \to c) = b \to (a \to c)$;
(WB5*) $(a \to b) \to ((b \to c) \to (a \to c)) = 1$;
(WB8) $1 \to a = a$;
(Com) $(a \to b) \to b = (b \to a) \to a$.

Proof Necessity can be obtained by Definition 3, Proposition 3 and Lemma 2. To show the sufficiency, note that (WB4) = (I1) and (WB5*) = (I2). So, by Theorem 1, it suffices to verify (I3), (I4) and (I5) in Definition 2.

Verification of (I3): By (WB5*) and (WB8), we have

$$a \to a = 1 \to (a \to a) = (1 \to 1) \to ((1 \to a) \to (1 \to a)) = 1. \tag{4-1}$$

Verification of (I4): It follows from Lemma 2 (2).

Verification of (I5): By condition (Com), (WB8) and (4-1), we have

$$(a \to 1) \to 1 = (1 \to a) \to a = a \to a = 1. \tag{4-2}$$

By (WB8), (4-2), (4-1) and (WB5*), we have

$$\begin{aligned} a \to 1 &= (1 \to a) \to ((a \to 1) \to 1) \\ &= (1 \to a) \to ((a \to 1) \to (1 \to 1)) = 1. \end{aligned} \tag{4-3}$$

By (WB3), (4-1) and (4-3), we have

$$0 \to a = (a \to 0) \to (0 \to 0) = (a \to 0) \to 1 = 1.$$

Summing up the above, we see that sufficiency holds. □

Theorem 3 *A (2, 0)-type algebra* $(M, \to, 0)$ *equivalent to a CWBR*$_0$*-algebra iff* $\forall a, b, c \in M$ *and* $1 = 0 \to 0$*, the following holds:*
(WB3) $a \to b = (b \to 0) \to (a \to 0)$;
(WB5) $(b \to c) \to ((a \to b) \to (a \to c)) = 1$;
(WB8) $1 \to a = a$;
(Com) $(a \to b) \to b = (b \to a) \to a$.

Proof Necessity can be obtained by Definitions 1 and 3.
To show sufficiency, it suffices to verify (WB4) and (WB5*) in Theorem 2.

Verification of (WB5*): By (WB3) and (WB5), we have

$$\begin{aligned} &(a \to b) \to ((b \to c) \to (a \to c)) \\ &= ((b \to 0) \to (a \to 0)) \to (((c \to 0) \to (b \to 0)) \to ((c \to 0) \to (a \to 0))) \\ &= 1. \end{aligned} \tag{4-4}$$

Then (WB5*) holds.

Verification of (WB4): By (WB8) and (4-4), we have

$$a \to ((a \to c) \to c) = (1 \to a) \to ((a \to c) \to (1 \to c)) = 1. \tag{4-5}$$

By (4-4), we have

$$(b \to (a \to c)) \to (((a \to c) \to c) \to (b \to c)) = 1. \tag{4-6}$$

If $a \to b = 1$, then by (WB8) and (4-4), we have

$$\begin{aligned} &(b \to c) \to (a \to c) \\ &= 1 \to ((b \to c) \to (a \to c)) \\ &= (a \to b) \to ((b \to c) \to (a \to c)) = 1. \end{aligned} \tag{4-7}$$

By (4-5), we can use $(a \to c) \to c$ to replace b which in the premise condition in (4-7) that $a \to b = 1$, use $b \to c$ to replace c which in the $(b \to c) \to (a \to c) = 1$ in (4-7), we have

$$(((a \to c) \to c) \to (b \to c)) \to (a \to (b \to c)) = 1. \tag{4-8}$$

If $a \to b = 1$ and $b \to c = 1$, by (WB8) and (4-4), we have

$$\begin{aligned} &a \to c = 1 \to (a \to c) \\ &= (b \to c) \to (a \to c) \\ &= (a \to b) \to ((b \to c) \to (a \to c)) = 1. \end{aligned} \tag{4-9}$$

By (4-6), (4-8), (4-9), we have $(b \to (a \to c)) \to (a \to (b \to c)) = 1$. Similarly, $(a \to (b \to c)) \to (b \to (a \to c)) = 1$. By Lemma 2 (2), we have $a \to (b \to c) = b \to (a \to c)$.

Summing up the above, we see that sufficiency holds. □

Since by Theorem 1 that CWBR_0-algebras are equivalent to NLI-algebras, we give a characterization of CWBR_0-algebra related to Definition 4.

Theorem 4 *A (2, 0)-type algebra* $(M, \to, 0)$ *is equivalent to a* $CWBR_0$*-algebra iff* $\forall a, b, c \in M$, *let* $1 = 0 \to 0$, *the following conditions are satisfied:*

(CWB1) $(a \to b) \to (a \to c) = (b \to a) \to (b \to c)$;
(CWB2) $a \to (b \to 0) = b \to (a \to 0)$;
(CWB3) $(a \to 0) \to 0 = a$;
(CWB4) $a \to 1 = 1$.

Proof Necessity: Note that (CWB1) = (NLI-4), (CWB3) = (NLI-2), (CWB4) = (W1) and (CWB2) is a special case of (WB4). So, the necessity holds by Theorem 1.

To show sufficiency, by Theorem 1, it suffices to verify conditions (NLI-1) and (NLI-3) in the definition of NLI-algebras.

Verification of (NLI-3): By (CWB2) and (CWB3), we have

$$(b \to 0) \to (a \to 0) = a \to ((b \to 0) \to 0) = a \to b. \tag{4-10}$$

Verification of (NLI-1): By (CWB3), (4-10) and (CWB4), we have

$$\begin{aligned} 0 \to a &= ((0 \to 0) \to 0) \to ((a \to 0) \to 0) \\ &= (a \to 0) \to (0 \to 0) = (a \to 0) \to 1 = 1. \end{aligned}$$

Summing up the above, we see that sufficiency holds. □

Theorems 3 and 4 are characterizations of $CWBR_0$-algebras with simpler types and fewer axioms. By Theorem 1, they can also be viewed as characterizations of MV-algebras, lattice implication algebras and CFI-algebras.

Acknowledgments Supported by the NSF of China (61472343, 61300153) and University Science Research Projects (KJ2016A310, 15KJD110006) of Anhui and Jiangsu Provinces.

References

1. Chen, D.Q., Wu, H.B.: Relations of regular FI-algebras and WBR_0-algebras. J. Shaanxi Norm. Univ. (Nat. Sci.) **39**(5), 7–10 (2011)
2. Chen, D.Q.: The researches about BR_0-algebras and WBR_0-algebras. Dissertation of Shanxi Normal University (2011)
3. Fan, X., Wang, G.J.: Equivalent characterizations of WBR_0-algebras. J. Xi'an Inst. Lit. Sci. **13**(3), 1–3 (2010)
4. Ling, X.M., Xu, L.S.: Conditions for BR_0-algebras to be Boolean algebras. Int. J. Contemp. Math. Sci. **5**(54), 2663–2671 (2010)
5. Ling, X.M., Xu, L.S.: Characterizations and properties of commutative BR_0-algebras. J. Yangzhou Univ. (Nat. Sci.) **15**(1), 1–4 (2012)
6. Liu, C.H., Xu, L.S.: The representative theorems of lattice implication algebras by implication operators. Fuzzy Syst. Math. **24**(4), 27–32 (2014)
7. Liu, C.H., Wu, H.X., Xu, L.S.: On CFI-algebras. J. Yangzhou Univ. Lit. Sci. **10**(4), 1–4 (2007)
8. Wang, G.J.: Non-classical Mathematical Logic and Approximate Reasoning, 2nd edn. Science Press, Beijing (2008)
9. Wang, Z.M., Wu, H.B.: An construction and properties of WBR_0-algebras. Fuzzy Syst. Math. **25**(4), 5–61 (2011)
10. Wu, H.B., Wang, Z.H.: The non-ordered form of BR_0-algebras and properties of WBR_0-algebras. J. Eng. Math. **26**(3), 456–460 (2009)
11. Wu, S.P., Wang, G.J.: Commutative weak R_0-algebras. J. Yunnan Norm. Univ. **27**(1), 1–4 (2007)
12. Wu, W.M.: Fuzzy implication algebras. Fuzzy Syst. Math. **4**(1), 56–64 (1990)
13. Xu, Y.: Lattice implication algebras. J. Southwest Jiao Tong Univ. **28**(1), 20–27 (1993)
14. Zhu, X., Zhan, X.Q., Hu, M.L.: Study of two groups of addtional conditions of BL-algebras. J. Jilin Inst. Chem. Tech. **30**(7), 78–81 (2013)

Ideals in Residuated Lattices

Qing-Jun Luo

Abstract In this paper, we first introduce the notion of ideals in residuated lattices as a natural generalization of the concept of ideals in BL-algebras. Then we give several equivalent characterizations of ideals in residuated lattices. Finally, the congruence induced by ideals in residuated lattices is also obtained.

Keywords Residuated lattice · Ideal · Congruence

1 Introduction

Ideal theory is a very effective tool for studying various algebraic structures. As in rings, the notion of ideals is at the center in the theory of MV-algebras, while in BL-algebras, MTL-algebras, resiudated lattices and the corresponding non-commutative version, the focus is turned to filter theory [1, 3, 6, 10]. In meantime, some authors introduced the notion of ideals in BL-algebras as a natural generalization of that of ideals in MV-algebras (see, e.g., [4, 5, 9]). The ideal in residuated lattices has not been studied. Thus, the purpose of this paper is to introduce the notion of ideals in residuated lattices as a framework studying ideal theory in logical algebras. Then we give several equivalent characterizations of ideals in residuated lattices and obtain congruences via these ideals.

2 Ideals and Congruences in Residuated Lattices

In this section, we first recall some basic notions and results relative to residuated lattices. Then, we introduce the concept of ideals in residuated lattices and investigate some of their properties. Based on ideals in a residuated lattice, equivalence relations are obtained. Let us start with the following definition.

Q.-J. Luo (✉)
Institute of Applied Mathematics, Xi'an University of Finance and Economics, Xi'an 710100, China
e-mail: qingjunlou@163.com

T.-H. Fan et al. (eds.), *Quantitative Logic and Soft Computing 2016*, Advances in Intelligent Systems and Computing 510,
DOI 10.1007/978-3-319-46206-6_39

Definition 1 ([2, 4]) A residuated lattice is an algebra $\mathscr{A} = (A, \wedge, \vee, \otimes, \rightarrow, 0, 1)$ of type (2, 2, 2, 2, 0, 0) satisfying the following axioms:
(RL1) $(A, \wedge, \vee, 0, 1)$ is a bounded lattice;
(RL2) $(A, \otimes, 1)$ is a commutative monoid;
(RL3) $\otimes$ and $\rightarrow$ form an adjoint pair, that is, $a \otimes b \leq c$ if and only if $a \leq b \rightarrow c$.

In what follows, by A we denote the universe of a residuated lattice. For $a \in A$, we denote $a' = a \rightarrow 0$, $a'' = (a \rightarrow 0) \rightarrow 0$. A residuated lattice A is called a regular residuated lattice [10] if for all $a \in A, a'' = a$. Next, we recall some important classes of residuated lattices.

• A residuated lattice A is called an MTL-algebra if it satisfies the following equation for all $a, b \in A$:

$$\text{(MTL)}\ (a \rightarrow b) \vee (b \rightarrow a) = 1\ \text{(prelinearity)}.$$

• An MTL-algebra A is called a BL-algebra if it satisfies the following equation for all $a, b \in A$:

$$\text{(BL)}\ a \wedge b = a \otimes (a \rightarrow b)\ \text{(divisibility)}.$$

• A BL-algebra A is call an MV-algebra if it satisfies the following condition for all $a, b \in A$:

$$\text{(MV)}\ (a \rightarrow b) \rightarrow b = (b \rightarrow a) \rightarrow a,$$

or equivalently, $a'' = a$ for all $a \in A$.

• A BL-algebra A is call a Gödel-algebra if $a \otimes a = a$ for all $a \in A$.

Some basic properties of residuated lattices and regular residuated lattices that also will be used later, are listed in the following proposition.

Proposition 1 ([2, 10]) *Let A be a residuated lattice. For all $a, b, c \in A$, we have:*

(i) $1 \rightarrow a = a$;
(ii) $a \leq b$ *if and only if* $a \rightarrow b = 1$;
(iii) *if* $a \leq b$, *then* $c \rightarrow a \leq c \rightarrow b$ *and* $b \rightarrow c \leq a \rightarrow c$;
(iv) *if* $a \leq c$ *and* $b \leq d$, *then* $a \otimes b \leq c \otimes d$;
(v) $a \rightarrow (b \rightarrow c) = a \otimes b \rightarrow c = b \rightarrow (a \rightarrow c)$;
(vi) $a \leq a''$ *and* $a''' = a'$;
(vii) $a \rightarrow b \leq b' \rightarrow a'$;
(viii) $(a \rightarrow b'')'' = a \rightarrow b''$;
(ix) $(a \vee b)' = a' \wedge b'$;
(x) $a \otimes (b \vee c) = (a \otimes b) \vee (a \otimes c)$;
(xi) $a \otimes b = 0$ *if and only if* $a \leq b'$ *if and only if* $b \leq a'$.

Furthermore, if A is regular, then we also have the following:

(xii) $a' \rightarrow b' = b \rightarrow a$, $a' \rightarrow b = b' \rightarrow a$;
(xiii) $a \otimes b = (a \rightarrow b')'$;
(xiv) $a \rightarrow b = (a \otimes b')'$.

Proposition 2 *Let A be a residuated lattice. Define a binary operation* $\oplus$ *on A as follows:*

$$a \oplus b = a' \to b, \quad a, b \in A.$$

Then the following properties hold for all $a, b, c \in A$*:*

(i) $a \vee b \leq a \oplus b$;
(ii) *if* $a \leq b$, *then* $a \oplus c \leq b \oplus c$ *and* $c \oplus a \leq c \oplus b$;
(iii) $0 \oplus a = a$, $a \oplus 0 = a''$;
(iv) $a \oplus 1 = 1 \oplus a = 1$;
(v) $a' \oplus a'' = a'' \oplus a' = 1$;
(vi) $a \oplus b = 1$ *if and only if* $a' \leq b$;
(vii) $a \oplus b'' = a'' \oplus b'' = b \oplus a''$;
(viii) $a' \oplus b' = b' \oplus a'$;
(ix) $a' \oplus 0 = 0 \oplus a' = a'$.

Proof The proposition follows immediately from Proposition 1. □

It should be noted that the operation $\oplus$ is not commutative in general, since the equation $a' \to b = b' \to a$ is not necessarily true in an arbitrary residuated lattice.

According to Propositions 1 and 2, we give a list of equivalent conditions for $\oplus$ to be commutative below.

Proposition 3 *Let A be a residuated lattice. Then the following conditions are equivalent to each other:*

(i) *A is a regular residuated lattice;*
(ii) *for all* $a, b \in A$, $a \oplus b = b \oplus a$;
(iii) *for all* $a \in A$, $a \oplus 0 = a$;
(iv) *for all* $a \in A$, $a' \oplus a = 1$.

Lele and Nganou proved that the operation $\oplus$ defined in Proposition 2, is associative in BL-algebras [5]. However, in a general residuated lattice, $\oplus$ is not necessarily associative, as shown in the following example.

Example 1 Let $A = \{0, a, b, c, d, 1\}$ with $0 < a < c < d < 1, 0 < b < c < d < 1$, where a and b are incomparable. The operations $\otimes$ and $\to$ are defined respectively in the following tables.

$\otimes$	0	a	b	c	d	1
0	0	0	0	0	0	0
a	0	0	0	0	a	a
b	0	0	0	0	b	b
c	0	0	0	0	c	c
d	0	a	b	c	d	d
1	0	a	b	c	d	1

$\to$	0	a	b	c	d	1
0	1	1	1	1	1	1
a	c	1	c	1	1	1
b	c	c	1	1	1	1
c	c	c	c	1	1	1
d	0	a	b	c	1	1
1	0	a	b	c	d	1

Then it is easily verified that $(A, \wedge, \vee, \otimes, \rightarrow, 0, 1)$ is a residuated lattice. Since $(a' \oplus a) \oplus 0 = (a'' \rightarrow a)' \rightarrow 0 = (c \rightarrow a)' \rightarrow 0 = c' \rightarrow 0 = c \rightarrow 0 = c$, $a' \oplus (a \oplus 0) = a' \oplus a'' = a'' \rightarrow a'' = 1 \neq c$, we have that the operation $\oplus$ is not associative in A.

We can show that in any regular residuated lattice A, the operation $\oplus$ is associative, since $(a \oplus b) \oplus c = (a' \rightarrow b)' \rightarrow c = c' \rightarrow (a' \rightarrow b) = a' \rightarrow (c' \rightarrow b) = a' \rightarrow (b' \rightarrow c) = a \oplus (b \oplus c)$ for all $a, b, c \in A$. The converse, however, is not true in general, as is shown in the following example.

Example 2 Let $A = [0, 1]$. Define two binary operations $\otimes$ and $\rightarrow$ on A as follows:

$$x \rightarrow y = \begin{cases} 1, & x \leqslant y, \\ y, & x > y, \end{cases} \quad x \otimes y = \min\{x, y\}.$$

Then it is easy to verify that $(A, \max, \min, \otimes, \rightarrow, 0, 1)$ is a Gödel-algebra, and the equality $(a \oplus b) \oplus c = a \oplus (b \oplus c)$ holds for all $a, b, c \in A$. However, $(\frac{1}{2})'' = (\frac{1}{2} \rightarrow 0) \rightarrow 0 = 0 \rightarrow 0 = 1 \neq \frac{1}{2}$. This shows that A is not a regular residuated lattice.

In the sequel, by using the operation $\oplus$ defined in Proposition 2, we introduce the concept of ideals in residuated lattices.

Definition 2 Let A be a residuated lattice and I a non-empty subset of A. Then I is called an ideal of A if for all $a, b \in A$:
(I1) $a \leq b$ and $b \in I$ imply $a \in I$;
(I2) $a, b \in I$ implies $a \oplus b \in I$.

It is clear that $\{0\}$ and A are ideals of A, and the least element 0 belongs to every ideal of A. Moreover, the set of all ideals of A is closed under arbitrary intersections.

Let I be an ideal of A. By Definition 2, it is easy to prove that for all $a \in A$, $a \in I$ if and only if $a'' \in I$.

For any $B \subseteq A$, the smallest ideal containing B is called the ideal generated by B, and is denoted by $\langle B \rangle$. It is evident that $\langle \emptyset \rangle = \{0\}$. If $B \neq \emptyset$, then one can easily check that $\langle B \rangle = \{a \in A \mid a \leq b_1 \oplus b_2 \oplus \cdots \oplus b_n$ for some $n \in N$ and $b_1, b_2, \ldots, b_n \in B\}$.

Remark 1 Given a residuated lattice A. If A is a BL-algebra, the concept of ideals introduced in Definition 2 coincides with that of ideals given by Lele and Nganou in [5]. In particular, if A is an MV-algebra, then the notion of ideals here is identical to the well-known concept of ideal MV-algebra (see, e.g., [8]).

Next, we present equivalent characterizations of ideals in residuated lattices.

Proposition 4 *Let I be a subset containing 0 of a residuated lattice A. Then I is an ideal of A if and only if for all $a, b \in A$, $a \in I$ and $a' \otimes b \in I$ imply $b \in I$.*

Proof Assume that I is an ideal of A, $a \in I$ and $a' \otimes b \in I$. By Definition 2, we have that $a \oplus (a' \otimes b) \in I$, i.e., $a' \to a' \otimes b \in I$. Since $b \leq a' \to a' \otimes b$, we obtain $b \in I$.

Conversely, suppose that $a \leq b$ and $b \in I$. Then $b' \leq a'$, and thus, $b' \otimes a \leq a' \otimes a = 0 \in I$, which yields $b' \otimes a \in I$. By the hypothesis, we get $a \in I$. Let $a, b \in I$. Since $a' \otimes (a \oplus b) = a' \otimes (a' \to b) \leq b \in I$, we have $a' \otimes (a \oplus b) \in I$, it then follows that $a \oplus b \in I$. Therefore, I is an ideal of A. □

Proposition 5 *A subset I containing 0 of a residuated lattice A is an ideal if and only if for all $a, b \in A$, $a \in I$ and $(a' \to b')' \in I$ imply $b \in I$.*

Proof Suppose that I is an ideal of A, $a \in I$ and $(a' \to b')' \in I$. Since $a' \otimes b \leq (a' \otimes b)'' = (a' \to b')'$, by Definition 2, we have $a' \otimes b \in I$. Then it follows from Proposition 4 that $b \in I$.

Conversely, assume that $a \leq b$ and $b \in I$. Then $b' \leq a'$, and thus, $(b' \to a')' = 1' = 0 \in I$. By the hypothesis, we obtain $a \in I$. Note that, $a'' \in I$ whenever $a \in I$. Let us now suppose that $a, b \in I$. Since $a' \otimes (a \oplus b) = a' \otimes (a' \to b) \leq b$ and $b \in I$, we have $a' \otimes (a \oplus b) \in I$. Thus, $(a' \to (a \oplus b)')' = (a' \otimes (a \oplus b))'' \in I$, which yields $a \oplus b \in I$. This shows that I is an ideal of A. □

In what follows, we will see that every ideal of a residuated lattice can induce an equivalence relation. Furthermore, this equivalence relation is a congruence if the residuated lattice is a regular MTL-algebra (i.e., an MTL-algebra A which satisfies the identity $(a \to 0) \to 0 = a$ for all $a \in A$).

Proposition 6 *Let I be an ideal of a residuated lattice A. Define a binary operation θ_I on A as follows:*

$$a\theta_I b \text{ if and only if } a' \otimes b,\ b' \otimes a \in I.$$

Then θ_I is an equivalence relation (θ_I is also called the equivalence relation induced by I).

Proof For any $a \in A$, $a' \otimes a = 0 \in I$, we have $a\theta_I a$, i.e., θ_I is reflexive. By the definition of θ_I, it is clear that θ_I is symmetric. Suppose now that $a\theta_I b$ and $b\theta_I c$. Then $a' \otimes b, b' \otimes c \in I$. Since $(a' \otimes b)' \otimes (a' \otimes c) = (a' \to b') \otimes (a' \otimes c) \leq b' \otimes c \in I$, by Definition 2 and Proposition 4, we have $a' \otimes c \in I$. A similar argument gives $c' \otimes a \in I$, thus, $a\theta_I c$. This shows that θ_I is transitive. As a consequence, θ_I is an equivalence relation. □

In the rest of this paper, under the equivalence relation θ_I, by $[a]_{\theta_I}$ we mean the equivalent class of a for all $a \in A$.

Proposition 7 *Let A be a residuated lattice and I an ideal of A. Then the following properties hold for all $a \in A$:*

(i) $[a]_{\theta_I} = I$ *if and only if* $a \in I$*;*
(ii) $[a]_{\theta_I} = [a'']_{\theta_I}$*, i.e.,* $a\theta_I a''$*;*
(iii) $[a]_{\theta_I} = [1]_{\theta_I}$ *if and only if* $a' \in I$.

In order to prove Theorem 1 below, we need the following lemma.

Lemma 1 *Let* I *be an ideal of a residuated lattice* A *and* $a, b \in A$*. If* $a\theta_I b$*, then for all* $x \in A$*,* $(a \vee x)\theta_I(b \vee x)$ *and* $(a \otimes x)\theta_I(b \otimes x)$*.*

Proof Assume $a\theta_I b$. For any $x \in A$, since

$$\begin{aligned}(a \vee x)' \otimes (b \vee x) &= (a' \wedge x') \otimes (b \vee x)\\ &= ((a' \wedge x') \otimes b) \vee ((a' \wedge x') \otimes x)\\ &= ((a' \wedge x') \otimes b) \vee 0\\ &\leq a' \otimes b,\end{aligned}$$

and $a' \otimes b \in I$, we obtain that $(a \vee x)' \otimes (b \vee x) \in I$. Similarly, we can prove that $(b \vee x)' \otimes (a \vee x) \in I$. Therefore, $(a \vee x)\theta_I(b \vee x)$ for all $x \in A$.

Suppose $a\theta_I b$. For any $x \in A$, since

$$\begin{aligned}(a \otimes x)' \otimes (b \otimes x) &= (x \to a') \otimes (b \otimes x) \leq a' \otimes b \in I,\\ (b \otimes x)' \otimes (a \otimes x) &= (x \to b') \otimes (a \otimes x) \leq b' \otimes a \in I,\end{aligned}$$

we have that $(a \otimes x)' \otimes (b \otimes x) \in I$, $(b \otimes x)' \otimes (a \otimes x) \in I$. Thus, $(a \otimes x)\theta_I (b \otimes x)$ for all $x \in A$. □

Theorem 1 *Let* I *be an ideal of a residuated lattice* A*. Then the following properties hold for all* $a, b, c, d \in A$*:*

(i) *if* $a\theta_I b$ *and* $c\theta_I d$*, then* $(a \vee c)\theta_I(b \vee d)$*,* $(a \otimes c)\theta_I(b \otimes d)$*;*
(ii) *if* $a\theta_I b$*, then* $a'\theta_I b'$*;*
(iii) *if* $a\theta_I b$*, then* $(a \to c')\theta_I(b \to c')$*;*
(iv) *if* A *is regular,* $a\theta_I b$*,* $c\theta_I d$*, then* $(a \to c)\theta_I(b \to d)$*;*
(v) *if* A *is an MTL-algebra,* $a\theta_I b$*,* $c\theta_I d$*, then* $(a \wedge c)\theta_I(b \wedge d)$*.*

Proof (i) Let $a\theta_I b$, $c\theta_I d$. It then follows from Lemma 1 that $(a \vee c)\theta_I(b \vee c)$ and $(b \vee c)\theta_I(b \vee d)$. By the transitivity of θ_I, we obtain $(a \vee c)\theta_I(b \vee d)$. It can be proved that $(a \otimes c)\theta_I(b \otimes d)$ in a similar manner.

(ii) Assume $a\theta_I b$. Then $b' \otimes a \in I$. Since $(b' \otimes a)' \otimes (a'' \otimes b') = (b' \to a') \otimes (a'' \otimes b') \leq a' \otimes a'' = 0 \in I$, by Proposition 6, we get $a'' \otimes b' \in I$. Similarly, we can prove that $b'' \otimes a' \in I$, and thus, $a'\theta_I b'$.

(iii) Suppose $a\theta_I b$. Since $a \to c' = (a \otimes c)'$, $b \to c' = (b \otimes c)'$. Then by (i) and (ii), we obtain that $(a \to c')\theta_I(b \to c')$ for all $c \in A$.

(iv) Let $a\theta_I b$ and $c\theta_I d$. Then $c'\theta_I d'$, and by (iii), $(c' \to a')\theta_I(d' \to a')$. It follows from Proposition 1 that $(a \to c)\theta_I(a \to d)$. Furthermore, $d = d''$ gives $(a \to d)\theta_I(b \to d)$. According to the transitivity of θ_I, we conclude that $(a \to c)\theta_I(b \to d)$.

(v) Suppose A is an MTL-algebra. Then it follows from Proposition 1 of that [1] that $(a \wedge b)' = a' \vee b'$ for all $a, b \in A$. Thus, the proof of (v) is similar to that of Lemma 1. □

By Theorem 1, we have the following result.

Corollary 1 *Let A be a regular MTL-algebra and I an ideal of A. Then θ_I is a congruence on A, i.e., for all $a, b, c, d \in A$, if $a\theta_I b, c\theta_I d$, then $(a \wedge c)\theta_I(b \wedge d)$, $(a \vee c)\theta_I(b \vee d)$, $(a \otimes c)\theta_I(b \otimes d)$ and $(a \to c)\theta_I(b \to d)$.*

Note that in Corollary 1, the condition that the residuated lattice is a regular MTL-algebra is indispensable.

In Example 1, it is easy to check that $b\theta_I b$, $a\theta_I b$ and $a\theta_I c$ with respect to the ideal $I = \{0\}$. However, $(b \to a)\theta_I(b \to b)$ does not hold, since $b \to a = c, b \to b = 1$, but c and 1 are not equivalent with respect to θ_I. Also, $(a \wedge b)\theta_I(c \wedge b)$ does not hold, since $a \wedge b = 0, c \wedge b = b, 0' \otimes b = b \notin I$.

3 Conclusions

In the present paper, we introduced the notion of ideals in residuated lattices, and gave several equivalent characterizations of these ideals. As applications, the congruence induced by ideals in residuated lattices was also obtained. All results obtained in this paper also hold in special subclasses of residuated lattices such as MTL-algebras, BL-algebras and MV-algebras. For future work, on one hand, we could study other types of ideals in residuated lattices such as prime ideals, maximal ideals and primary ideals, and examine the relationships between these ideals and the corresponding filters in residuated lattices. On the other hand, we could define rough ideals in residuated lattices by using rough sets of Pawlak [7].

Acknowledgments This work described here is partially supported by the grant from the National Science Foundation of China (No. 61473336), Science Research Program Funded by Shaanxi Provincial Education Department (No. 16JK1302), and Foundation of Xi'an University of Finance and Economics (No. 15JD01).

References

1. Borzooei, R.A., Shoar, S.K., Ameri, R.: Some Types of Filters in MTL-algebras. Fuzzy Sets Syst. **187**, 92–102 (2012)
2. Dilworth, R.P., Ward, M.: Residuated lattices. Trans. Am. Math. Soc. **45**, 335–354 (1939)
3. Gasse, B.V., Deschrijver, G., Cornelis, C., Kerre, E.E.: Filters of residuated lattices and triangle algebras. Inform. Sci. **180**, 3006–3020 (2010)
4. Hájek, P.: Metamathematics of Fuzzy Logic. Kluwer Academic Publishers, Dordrecht (1998)
5. Lele, C., Nganou, J.B.: MV-algebras derived from ideals in BL-algebras. Fuzzy Sets Syst. **218**, 103–113 (2013)

6. Liu, L.Z., Li, K.T.: Fuzzy Boolean and positive implicative filters of BL-algebra. Fuzzy Sets Syst. **152**, 333–348 (2005)
7. Pawlak, Z.: Rough Sets. Int. J. Comput. Inform. Sci. **11**, 341–356 (1982)
8. Rasouli, S., Davvaz, B.: Roughness in MV-algebras. Inform. Sci. **180**, 737–747 (2010)
9. Turunen, E., Sessa, S., Architettura, F.: Local BL-algebras. Multiple-Valued Log. **6**, 229–249 (2001)
10. Zhu, Y.Q., Xu, Y.: Filter theory of residuated lattices. Inform. Sci. **180**, 3614–3632 (2010)

Hyper Equality Algebras

Xiao-Yun Cheng, Xiao-Long Xin and Young-Bae Jun

Abstract In this paper, we introduce a new structure, called hyper equality algebras which are a generalization of equality algebras, and investigate some related properties. Then we define (weak, strong) hyper filters and (weak, strong) hyper deductive systems, and give relations between them. Moreover we discuss relations between hyper equality algebras and other hyper structures, such as hyper EQ-algebras, hyper BCK-algebras and weak hyper residuated lattices. Finally, we also obtain quotient hyper equality algebras via regular hyper congruence relations.

Keywords Hyper equality algebra · Hyper filter · Hyper deductive system · Hyper congruence relation · Quotient hyper equality algebra

1 Introduction

EQ-algebras were proposed by Nov*á*k in [1] which generalizes commutative residuated lattices. One of the motivations was to introduce a special algebra as the correspondence of truth values for high-order fuzzy type theory(FTT). Another motivation is from the equational style of proof in logic. An EQ-algebra has three connectives: meet $\wedge$, product $\otimes$ and fuzzy equality $\sim$. The product in an EQ-algebra is quite

X.-Y. Cheng · X.-L. Xin (✉)
School of Mathematics, Northwest University, Xi'an 710127,
People's Republic of China
e-mail: xlxin@nwu.edu.cn

X.-Y. Cheng
e-mail: chengxiaoyun2004@163.com

X.-Y. Cheng
General Education Center, Xi'an Peihua University, Xi'an 710065, China

Y.-B. Jun
Department of Mathematics Education, Gyeongsang National University,
Chinju 660-701, Korea
e-mail: skywine@gmail.com

T.-H. Fan et al. (eds.), *Quantitative Logic and Soft Computing 2016*,
Advances in Intelligent Systems and Computing 510,
DOI 10.1007/978-3-319-46206-6_40

loose which can be replaced by any other smaller binary operation, but still obtains an EQ-algebra. Based on the above reasons, Jenei [2] introduced equality algebras, as a candidate for a possible algebraic semantics of fuzzy type theory similar to EQ-algebras but without a product. An equality algebra in [2] is an algebra $(E; \sim, \wedge, 1)$ of type $(2, 2, 0)$ such that for all $x, y, z \in E$:

(E1) $(E; \wedge, 1)$ is a meet-semilattice with top element 1;
(E2) $x \sim y = y \sim x$;
(E3) $x \sim x = 1$;
(E4) $x \sim 1 = x$;
(E5) $x \leq y \leq z$ implies $x \sim z \leq y \sim z$ and $x \sim z \leq x \sim y$;
(E6) $x \sim y \leq (x \wedge z) \sim (y \wedge z)$;
(E7) $x \sim y \leq (x \sim z) \sim (y \sim z)$.

And the author proved the term equivalence of equivalential equality algebras to BCK-meet-semilattices.

Hyper structure theory was introduced by Marty [3], at the 8th Congress of Scandinavian Mathematicians. In an algebraic hyper structure, the composition of two elements is not an element but a set. Since then hyper structure theory has been intensively researched in [4–9]. Recently, Borzooei has applied hyper theory to EQ-algebras to introduce hyper EQ-algebras [10] which are a generalization of EQ-algebras. Now hyper structure theory has been applied to many disciplines such as geometry, graph theory, automata, cryptography, artificial intelligence and probability theory, dismutation reactions, inheritance, etc. Hyper filters (or hyper deductive systems) are important tools in studying hyper structures [4, 11, 12]. The above are the motivation of introducing and studying hyper equality algebras.

This paper is organized as follows: in Sect. 2, we introduce the concept of hyper equality algebras, give some examples and investigate related properties. In Sect. 3, we introduce the notion of (weak, strong) hyper filters and (weak, strong) hyper deductive systems in hyper EQ-algebras and focus on discussing relations between them. In Sect. 4, we discuss relations between hyper equality algebras and other hyper structures. Moreover we construct quotient hyper equality algebras via regular hyper congruence relations.

2 Hyper Equality Algebras

Let H be a nonempty set and $\circ$ be a function from $H \times H$ to the nonempty power set of H, $P(H)$. Then $\circ$ is said to be a hyper operation on H.

Definition 1 A hyper equality algebra $(H; \sim, \wedge, 1)$ is a nonempty set H endowed with a binary operation $\wedge$, a binary hyper operation $\sim$ and a top element 1 satisfying the following axioms, for all $x, y, z \in H$:

(HE1) $(H; \wedge, 1)$ is a meet-semilattice with top element 1;
(HE2) $x \sim y \ll y \sim x$;

(HE3) $1 \in x \sim x$;
(HE4) $x \in 1 \sim x$;
(HE5) $x \le y \le z$ implies $x \sim z \ll y \sim z$ and $x \sim z \ll x \sim y$;
(HE6) $x \sim y \ll (x \wedge z) \sim (y \wedge z)$;
(HE7) $x \sim y \ll (x \sim z) \sim (y \sim z)$.
where $x \le y$ iff $x \wedge y = x$; $A \ll B$ is defined by for all $x \in A$, there exists $y \in B$ such that $x \le y$.

In any hyper equality algebra $(H; \sim, \wedge, 1)$, define operations $\rightarrow, \leftrightarrow$ by $x \rightarrow y := x \sim (x \wedge y)$, $x \leftrightarrow y := (x \rightarrow y) \wedge (y \rightarrow x)$ for all $x, y \in H$, respectively. Moreover, for any nonempty subsets $A, B \subseteq H$, we write $A \wedge B = \{a \wedge b : a \in A, b \in B\}$ and $A \circ B = \bigcup_{a \in A, b \in B} a \circ b$, where $\circ \in \{\sim, \rightarrow, \leftrightarrow\}$. Here after we denote $x \circ y$ instead of $x \circ \{y\}$, $\{x\} \circ y$ or $\{x\} \circ \{y\}$.

Example 1 Let $(H; \sim, \wedge, 1)$ be an equality algebra and define $\circ$ on H by $x \circ y = \{x \sim y\}$ for all $x, y \in H$. Then $(H; \circ, \wedge, 1)$ is a hyper equality algebra.

Example 2 Let $H = [0, 1]$. Define $\wedge$ and $\sim$ on H as follows: for all $x, y \in H$, $x \wedge y = min\{x, y\}$ and $x \sim y = \{1- \mid x - y \mid, 1\}$. Then $(H; \sim, \wedge, 1)$ is a hyper equality algebra where

$$x \rightarrow y = \begin{cases} \{1 - x + y, 1\}, & y < x. \\ \{1\}, & x \le y. \end{cases}$$

Example 3 Let $H = (0, 1)$. Define $\wedge$ and $\sim$ on H as follows: for all $x, y \in H$, $x \wedge y = min\{x, y\}$ and

$$x \sim y = \begin{cases} [y, 1], & x = 1. \\ H, & otherwise. \end{cases}$$

Then $(H; \sim, \wedge, 1)$ is a hyper equality algebra.

Example 4 Let $H = \{0, a, 1\}$ with $0 < a < 1$. Define operations $\wedge$, $\sim_1$ and $\sim_2$ on H as follows: $x \wedge y = min\{x, y\}$ and

$\sim_1$	0	a	1
0	$\{1\}$	$\{0\}$	$\{0\}$
a	$\{0\}$	$\{1\}$	$\{a, 1\}$
1	$\{0\}$	$\{a, 1\}$	$\{1\}$

$\sim_2$	0	a	1
0	$\{1\}$	$\{a, 1\}$	$\{a, 1\}$
a	$\{a, 1\}$	$\{0, a, 1\}$	$\{a, 1\}$
1	$\{0, 1\}$	$\{a, 1\}$	$\{1\}$

Then one can check that $(H; \sim_1, \wedge, 1)$ and $(H; \sim_2, \wedge, 1)$ are hyper equality algebras.

Example 5 Let $H = \{0, a, 1\}$ with $0 < a < 1$. Define operations $\wedge$, $\sim_1$ and $\sim_2$ on H as follows: $x \wedge y = min\{x, y\}$ and

$\sim_1$	0	a	1
0	$\{1\}$	$\{0, a\}$	$\{0, a\}$
a	$\{0, a\}$	$\{1\}$	$\{a\}$
1	$\{0, a\}$	$\{0, a\}$	$\{1\}$

$\sim_2$	0	a	1
0	$\{0, 1\}$	$\{0, a\}$	$\{0\}$
a	$\{0, a\}$	$\{1\}$	$\{a\}$
1	$\{0\}$	$\{a\}$	$\{1\}$

Then it is easily calculated that $(H; \sim_1, \wedge, 1)$ and $(H; \sim_2, \wedge, 1)$ are hyper equality algebras.

Example 6 Let $H = \{0, a, b, 1\}$ in which the Hasse diagram and operations $\wedge, \sim$ on H are as follows:

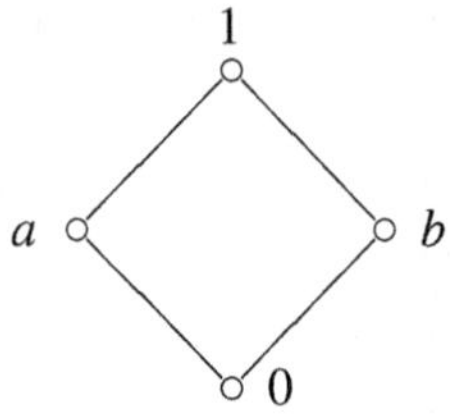

$\wedge$	0	a	b	1
0	0	0	0	0
a	0	a	0	a
b	0	0	b	b
1	0	a	b	1

$\sim$	0	a	b	1
0	$\{1\}$	$\{1\}$	$\{b, 1\}$	$\{0, a\}$
a	$\{1\}$	$\{1\}$	$\{a, 1\}$	$\{a\}$
b	$\{b, 1\}$	$\{a, 1\}$	$\{1\}$	$\{b, 1\}$
1	$\{0, a\}$	$\{a\}$	$\{b, 1\}$	$\{1\}$

Then $(H; \wedge, \sim, 1)$ is a hyper equality algebra.

Proposition 1 *Let $(H; \sim, \wedge, 1)$ be a hyper equality algebra. Then for all $x, y, z \in H$ the following are equivalent:*
(HE5) $x \leq y \leq z$ *implies* $x \sim z \ll y \sim z$ *and* $x \sim z \ll x \sim y$;
(HE5a) $x \sim (x \wedge y \wedge z) \ll x \sim (x \wedge y)$;
(HE5b) $x \to (y \wedge z) \ll x \to y$.

Proof By (HE2) and the proof of Proposition 1 in [2]. □

Proposition 2 *Let $(H; \sim, \wedge, 1)$ be a hyper equality algebra. Then for all $x, y, z \in H, A, B, C \subseteq H$:*
(P1) $A \sim B \ll B \sim A$;
(P2) $A \ll B$ *and* $B \ll C$ *imply* $A \ll C$;
(P3) $1 \ll A$ *iff* $1 \in A$;
(P4) $A \subseteq B$ *implies* $A \ll B$;
(P5) $x \leq y$ *implies* $y \to x = y \sim x$;
(P6) $x \leq y$ *and* $y \leq x$ *imply* $x = y$;
(P7) $1 \in x \to x,\ 1 \in x \to 1,\ x \ll x \sim 1,\ x \in 1 \to x,\ 1 \in x \leftrightarrow x$;
(P8) $x \sim y \ll x \to y,\ x \sim y \ll y \to x$;
(P9) $x \sim y \ll x \leftrightarrow y,\ x \leftrightarrow y \ll x \to y$ *and* $x \leftrightarrow y \ll y \to x$;
(P10) $x \leq y$ *implies* $1 \in x \to y$;
(P11) $x \leq y$ *implies* $x \sim 1 \ll y \sim 1,\ x \sim 1 \ll x \sim y$;
(P12) $x \to y \ll (x \wedge z) \to y$;
(P13) $x \leq y \leq z$ *implies* $z \sim x \ll z \sim y$ *and* $z \sim x \ll y \sim x$;
(P14) $x \ll y \to x,\ A \ll B \to A$;
(P15) $x \leq y$ *implies* $z \to x \ll z \to y$ *and* $y \to z \ll x \to z$;

(P16) $A \ll B$ *implies* $C \rightarrow A \ll C \rightarrow B$ *and* $B \rightarrow C \ll A \rightarrow C$;
(P17) $x \rightarrow (y \wedge z) \ll (x \wedge z) \rightarrow y$;
(P18) $x \rightarrow y = x \rightarrow x \wedge y$;
(P19) $x \leq y$ *implies* $x \ll y \sim x$;
(P20) $y \ll (x \rightarrow y) \rightarrow y$;
(P21) $x \rightarrow y \ll (y \rightarrow z) \rightarrow (x \rightarrow z)$.

Proof (P1)–(P6) are straightforward.

(P7) This is easy to verify by (HE2), (HE3) and (HE4).

(P8) $x \sim y \ll x \rightarrow y$ holds by replacing z by x in (HE6). Replacing z by y in (HE6) and using (HE2), we get that $x \sim y \ll (x \wedge y) \sim y \ll y \sim (x \wedge y) = y \rightarrow x$. Again considering (P2), we show $x \sim y \ll y \rightarrow x$.

(P9) By (P8), $x \sim y \ll x \rightarrow y$, $x \sim y \ll y \rightarrow x$. That is, for any $a \in x \sim y$, there exist $b \in x \rightarrow y$ and $c \in y \rightarrow x$ such that $a \leq b$ and $a \leq c$. Hence $a \leq b \wedge c$. This shows that $x \sim y \ll (x \rightarrow y) \wedge (y \rightarrow x) = x \leftrightarrow y$. The rest are obvious.

(P10) Let $x \leq y$. Then $1 \in x \sim x = x \sim (x \wedge y) = x \rightarrow y$.

(P11) By (HE5).

(P12) By (HE6), we have $x \sim (x \wedge y) \ll (x \wedge z) \sim (x \wedge y \wedge z)$. This implies that $x \rightarrow y \ll (x \wedge z) \rightarrow y$.

(P13) By (HE2), (HE5) and (P2).

(P14) It follows from (P7) and (P12) that $x \ll 1 \rightarrow x \ll (1 \wedge y) \rightarrow x = y \rightarrow x$. The other part is obvious.

(P15) Let $x \leq y$. Then by (HE5a) we have $z \rightarrow x = z \sim (z \wedge x) = z \sim (z \wedge x \wedge y) \ll z \sim (z \wedge y)$. This shows that $z \rightarrow x \ll z \rightarrow y$. The other part can be verified by (P12).

(P16) By (P15).

(P17) By (HE5b), (P12) and (P3).

(P18) From (P5) $x \rightarrow y = x \sim x \wedge y = x \rightarrow x \wedge y$.

(P19) Let $x \leq y$. Then by (P14) $x \ll y \rightarrow x = y \sim x$.

(P20) Since $y \ll x \rightarrow y$ and (P19) we have $y \ll (x \rightarrow y) \sim y = (x \rightarrow y) \rightarrow y$.

(P21) First, by (P12) and (HE5b) we can get that $y \rightarrow z \ll (y \wedge x) \rightarrow z$ and $x \rightarrow (z \wedge y) \ll x \rightarrow z$ respectively. Hence by (P15) $((y \wedge x) \rightarrow z) \rightarrow (x \rightarrow (z \wedge y)) \ll (y \rightarrow z) \rightarrow (x \rightarrow (z \wedge y)) \ll (y \rightarrow z) \rightarrow (x \rightarrow z)$. This implies that $((y \wedge x) \rightarrow z) \rightarrow (x \rightarrow (z \wedge y)) \ll (y \rightarrow z) \rightarrow (x \rightarrow z)$.

On the other hand, it follows from (P1) and (HE7) that $x \rightarrow y = x \sim (x \wedge y) \ll (x \wedge y) \sim x \ll ((y \wedge x) \sim (y \wedge z \wedge x)) \sim (x \sim (x \wedge z \wedge y))$. That is, $x \rightarrow y \ll ((y \wedge x) \rightarrow z) \rightarrow (x \rightarrow (z \wedge y))$.

Combing the above two parts, we obtain $x \rightarrow y \ll (y \rightarrow z) \rightarrow (x \rightarrow z)$. □

Proposition 3 *Let* $(H; \sim, \wedge, 1)$ *be a hyper equality algebra and let* $y \in H$. *If* $1 \sim y = y$ *or* $x \leq y$ *for all* $x, z \in H$, *then*
(1) $x \ll (x \sim y) \sim y$;
(2) $x \ll (x \rightarrow y) \rightarrow y$;
(3) $x \ll y \rightarrow z$ *iff* $y \ll x \rightarrow z$;
(4) $x \rightarrow (y \rightarrow z) \ll y \rightarrow (x \rightarrow z)$.

Proof Let $x, y, z \in H$

(1) If $1 \sim y = y$, then by (P7) and (HE7), we have $x \ll x \sim 1 \ll (x \sim y) \sim (1 \sim y) = (x \sim y) \sim y$. Hence from (P2) $x \ll (x \sim y) \sim y$.

If $x \leq y$, $x \ll (x \sim y) \sim y$ holds by (P20).

(2) It follows from (1) that $x \ll (x \sim (x \wedge y)) \sim (x \wedge y) \ll (x \to y) \sim (x \wedge y)$. Since $x \wedge y \leq y \ll x \to y$, then by (P8) and (P15) $(x \to y) \sim (x \wedge y) \ll (x \to y) \to (x \wedge y) \ll (x \to y) \to y$. Hence $x \leq (x \to y) \to y$.

(3) Let $x \ll y \to z$. Then by (2) and (P15) we get $y \ll (y \to z) \to z \ll x \to z$. Thus $y \ll x \to z$. By the symmetry of x and y, the converse is true.

(4) Since from (2) $y \ll (y \to z) \to z$, then by (P21) and (P15) we can get that $x \to (y \to z) \ll ((y \to z) \to z) \to (x \to z) \ll y \to (x \to z)$. Therefore $x \to (y \to z) \ll y \to (x \to z)$. □

Definition 2 A hyper equality algebra $(H; \sim, \wedge, 1)$ is called good if $x = 1 \sim x$ for all $x \in H$.

Example 7 Let $H = \{0, a, 1\}$. Define operations $\wedge$ and $\sim$ on H as follows: $x \wedge y = min\{x, y\}$ and

$\sim$	0	a	1
0	$\{1\}$	$\{0, a\}$	$\{0\}$
a	$\{0, a\}$	$\{1\}$	$\{a\}$
1	$\{0\}$	$\{a\}$	$\{1\}$

Then $(H; \wedge, \sim, 1)$ is a good hyper equality algebra.

Using Proposition 3 we can get the following corollary immediately.

Corollary 1 *Let $(H; \sim, \wedge, 1)$ be a good hyper equality algebra. Then*
(1) $x \ll (x \sim y) \sim y$;
(2) $x \ll (x \to y) \to y$;
(3) $x \ll y \to z$ *iff* $y \ll x \to z$;
(4) $x \to (y \to z) \ll y \to (x \to z)$.

A hyper equality algebra H is called bounded, if there is bottom element 0 in H. In this case, denote $x^* = x \sim 0$. Then $x^* = x \to 0$.

Proposition 4 *Let $(H; \sim, \wedge, 1)$ be a bounded hyper equality algebra. Then for all $x, y, z \in H$,*
(1) $0 \in 1^*$ *and* $1 \in 1^{**}$;
(2) $1 \in 0^*$ *and* $0 \in 0^{**}$;
(3) $x \sim y \ll x^* \sim y^*$;
(4) $x \to y \ll y^* \to x^*$;
(5) $x \leq y$ *implies* $y^* \ll x^*$.

Proof (1) Since $0 \in 1 \to 0$, we have $0 \in 1^*$ and $1 \in 0 \to 0 \in (1 \to 0) \to 0$. Hence $1 \in 1^{**}$.

(2) From $1 \in 0 \to 0$, $1 \in 0^*$ follows. Also $0 \in 1 \to 0 \in (0 \to 0) \to 0$. That is $0 \in 0^{**}$.

(3) Taking $z = 0$ in (HE7).

(4) Taking $z = 0$ in (P21).

(5) By (P15) $x \leq y$ implies $y \to 0 \ll x \to 0$. □

3 Hyper Filters and Hyper Deductive Systems

From now on, unless otherwise stated we assume that H is a hyper equality algebra.

Definition 3 A nonempty subset S containing 1 of H is said to be a hyper subalgebra, if S is a hyper equality algebra with respect to the hyper operation $\sim$ and the binary operation $\wedge$ on H.

Theorem 1 *A nonempty subset S of H is a hyper subalgebra if and only if $x \sim y \subseteq S$ and $x \wedge y \in S$ for all $x, y \in S$.*

Proof Necessity is obvious. Now assume for any $x, y \in S, x \sim y \in S$ and $x \wedge y \in S$. By (HE3), we have $1 \in x \sim x \subseteq S$ and so $1 \in S$. The rest proof are easy. □

Example 8 (1) In Example 3, $S = \{0.5, 1\}$ is not a hyper subalgebra, since $1 \sim 0.5 = [0.5, 1] \nsubseteq S$.
(2) In Example 2, $S = [0.5, 1]$ and $\{1\}$ are hyper subalgebras of H.

Definition 4 Let F be a nonempty subset of H satisfying
(F) $x \in F$ and $x \leq y$ imply $y \in F$ for all $x, y \in H$.
F is called a
• weak hyper filter, if
(WHF) $x \in F$ and $x \sim y \subseteq F$ imply $y \in F$ for all $x, y \in H$.
• hyper filter, if
(HF) $x \in F$ and $F \ll x \sim y$ imply $y \in F$ for all $x, y \in H$.
• strong hyper filter, if
(SHF) $x \in F$ and $x \sim y \cap F \neq \emptyset$ imply $y \in F$ for all $x, y \in H$.

Remark 1 (1) If F is a (weak, strong) hyper filter of H, then $1 \in F$.
(2) Every strong hyper filter of H is a (weak) hyper filter.

Example 9 (1) Let $(H; \sim, \wedge, 1)$ be the hyper equality algebra defined in Example 6. Then one can see that $F = \{b, 1\}$ is not a hyper filter of H, since $b \in F$ and $F \ll b \sim 0$, but $0 \notin F$. Furthermore, we can check that $F = \{b, 1\}$ is not a weak hyper filter of H, since $b \in F$ and $b \sim 0 \subseteq F$, but $0 \notin F$. So we know that $F = \{b, 1\}$ is not a strong hyper filter of H.
(2) Let $(H; \sim_1, \wedge, 1)$ be the hyper equality algebra defined in Example 4. Then one can see that $F = \{a, 1\}$ is a (weak, strong) hyper filter of H.

The following example indicates that a (weak) hyper filter of H may not be a strong hyper filter of H.

Example 10 Let $(H; \sim_1, \wedge, 1)$ be the hyper equality algebra defined in Example 5. Then one can see that $F = \{a, 1\}$ is a (weak) hyper filter of H. But $F = \{a, 1\}$ is not a strong hyper filter of H, since $a \in F$ and $a \sim_1 0 \cap F \neq \emptyset$, but $0 \notin F$.

The following example shows that a weak hyper filter of H may not be a hyper filter of H.

Example 11 Let $(H; \sim_1, \wedge, 1)$ be the hyper equality algebra defined in Example 6. Then one can see that $F = \{1\}$ is a weak hyper filter of H. But $F = \{1\}$ is not a hyper filter of H, since $1 \in F$ and $F \ll 1 \sim b$, but $b \notin F$.

Definition 5 Let D be a nonempty subset of H. D is called a
- weak hyper deductive system, if D satisfies (F) and
(WHD) $x \in D$ and $x \to y \subseteq D$ imply $y \in D$ for all $x, y \in H$.
- hyper deductive system, if D satisfies (F) and
(HD) $x \in D$ and $D \ll x \to y$ imply $y \in D$ for all $x, y \in H$.
- strong hyper deductive system, if $1 \in D$ and D satisfies
(SHD) $x \in D$ and $x \to y \cap D \neq \emptyset$ imply $y \in D$ for all $x, y \in H$.

Remark 2 (1) If D is a hyper deductive system of H, then $1 \in D$.
(2) If D is a strong hyper deductive system of H, then D satisfies (F). In fact, let $x \in D$ and $x \leq y$. Then $1 \in x \to y$ by (P10). Hence $x \to y \cap D \neq \emptyset$. This implies $y \in D$.
(3) Every strong hyper deductive system of H is a (weak) hyper deductive system of H.

Example 12 (1) Let $(H; \sim, \wedge, 1)$ be the hyper equality algebra defined in Example 6. Then one can see that $D = \{a, b, 1\}$ is not a hyper deductive system of H, since $a \in D$ and $D \ll a \to 0$, but $0 \notin D$. Furthermore, we can see that $D = \{a, b, 1\}$ is not a strong hyper deductive system of H.
(2) Let $(H; \sim_1, \wedge, 1)$ be the hyper equality algebra defined in Example 4. Then one can see that $D = \{a, 1\}$ is a (strong) hyper deductive system of H.

The following example indicate that a hyper deductive system of H may not be a strong hyper deductive system of H.

Example 13 Let $(H; \sim_2, \wedge, 1)$ be the hyper equality algebra defined in Example 5. Then one can see that $D = \{a, 1\}$ is a hyper deductive system of H. But $D = \{a, 1\}$ is not a strong hyper deductive system of H, since $a \in D$ and $a \to_2 0 \cap D \neq \emptyset$, but $0 \notin D$.

Lemma 1 *Let D be a nonempty subset satisfying (F) of H. Then for any nonempty subset A, B of H, $A \cap D \neq \emptyset$ and $A \ll B$ imply $B \cap D \neq \emptyset$.*

Proof Since $A \cap D \neq \emptyset$, then there is $a \in A$ such that $a \in D$. For the above $a \in A$, it follows from $A \ll B$ that there exists $b \in B$ such that $a \leq b$. Again since $a \in D$ and D satisfies (F), we get $b \in D$. This shows that $B \cap D \neq \emptyset$. □

Proposition 5 (1) *Every hyper deductive system of H is a hyper filter of H.*
(2) *Every strong hyper deductive system of H is a strong hyper filter of H.*

Proof (1) Assume that D is a hyper deductive system of H and $x \in D$, $D \ll x \sim y$. Since $x \sim y \ll x \rightarrow y$ by (P8), using (P2) we have $D \ll x \rightarrow y$. Hence $y \in D$, which implies that D is a hyper filter of H.

(2) Assume that D is a strong hyper deductive system of H and $x \in D$, $x \sim y \cap D \neq \emptyset$. Since $x \sim y \ll x \rightarrow y$ by (P8), then by Lemma 1 we obtain $x \rightarrow y \cap D \neq \emptyset$. Therefore $y \in D$, which shows that D is a strong hyper filter of H. □

Let S be a nonempty subset of H. Denote by $[S)$ the least strong hyper deductive system of H containing S, called the strong hyper deductive system generated by S. In particular, if $S = \{a\}$, we write $[\{a\}) = [a)$, called the principal strong hyper deductive system generated by element a in H. In addition, we use $[D \cup \{x\})$ to denote the strong hyper deductive system generated by D and x, where $x \in H - D$. The following are some results about the generated strong hyper deductive system.

Theorem 2 *Let S be a nonempty subset of H. Then $[S) \supseteq \{x \in H : 1 \in a_n \rightarrow (\cdots (a_2 \rightarrow (a_1 \rightarrow x)) \cdots)$ for some $a_1, a_2, \cdots, a_n \in S\}$.*

Proof Similar to the proof of Theorem 3.9 in [11]. □

Corollary 2 *Let $a \in H$. Then $[a) \supseteq \{x \in H : 1 \in a \circ x\}$.*

Theorem 3 *Let D be a strong hyper deductive system of H and $a \in H - D$. Then $[D \cup \{a\}) \supseteq \{x \in H : a \rightarrow x \cap D \neq \emptyset\}$.*

Proof Similar to the proof of Theorem 3.9 in [11]. □

4 Hyper Equality Algebras and Other Hyper Structures

The following are the relations between hyper equality algebras and hyper EQ-algebras, hyper BCK-algebras and weak hyper residuated lattices.

Definition 6 ([10]) A hyper EQ-algebra $(H; \wedge, \otimes, \sim, 1)$ is a nonempty set H endowed with a binary operation $\wedge$, two binary hyper operations $\otimes, \sim$ and a top element 1 satisfying the following axioms, for all $x, y, z, t \in H$:
(HEQ1) $(H; \wedge, 1)$ is a commutative idempotent monoid with a top element 1;
(HEQ2) $(H; \otimes, 1)$ is a commutative semihypergroup with 1 as an identity and $\otimes$ is isotone w.r.t. $\leq$, i.e., if $x \leq y$, then $x \otimes z \ll y \otimes z$(where $x \leq y$ if and only if $x \wedge y = x$);

(HEQ3) $((x \wedge y) \sim z) \otimes (t \sim x) \ll z \sim (t \wedge y)$;
(HEQ4) $(x \sim y) \otimes (z \sim t) \ll (x \sim z) \sim (y \sim t)$;
(HEQ5) $(x \wedge y \wedge z) \sim x \ll (x \wedge y) \sim x$;
(HEQ6) $(x \wedge y) \sim x \ll (x \wedge y \wedge z) \sim (x \wedge z)$;
(HEQ7) $x \otimes y \ll x \sim y$;
where $A \ll B$ means that for all $x \in A$, there exists $y \in B$ such that $x \leq y$.

Proposition 6 ([10]) *Let $(H; \wedge, \otimes, \sim, 1)$ be a hyper EQ-algebra. Then the following properties hold for all $x, y, z, t \in H$:*
(QP1) $1 \in x \sim x$;
(QP2) $A \ll B$ *and* $B \ll C$ *imply* $A \ll C$;
(QP3) $x \sim y \ll y \sim x$;
(QP4) $x \ll x \sim 1$;
(QP5) $x \leq y \leq z$ *implies* $z \sim x \ll z \sim y$ *and* $x \sim z \ll x \sim y$;
(QP6) $x \sim y \ll (x \wedge z) \sim (y \wedge z)$;
(QP7) $x \sim y \ll (x \sim z) \sim (y \sim z)$.

Theorem 4 *Let $(H; \wedge, \otimes, \sim, 1)$ be a hyper EQ-algebra. If $x \in 1 \sim x$ for all $x \in H$, then $(H; \sim, \wedge, 1)$ is a hyper equality algebra.*

Proof By hypothesis (HE4) is true. (HE1), (HE2), (HE3), (HE6) and (HE7) follow from (HEQ1), (QP3), (QP1), (QP6) and (QP7) respectively. (HE5) can be obtained by (QP2), (QP3) and (QP5). □

Definition 7 ([6]) A weak hyper residuated lattice $(H; \vee, \wedge, \otimes, \rightarrow, 0, 1)$ is a non-empty set H endowed with two binary operations $\vee, \wedge$, two binary hyper operations $\otimes, \rightarrow$ and two constant element 0, 1 satisfying the following axioms, for all $x, y, z \in H$:
(WHR1) $(H; \leq, \wedge, \vee, 0, 1)$ is a bounded lattice;
(WHR2) $(H, \otimes, 1)$ is a commutative semihypergroup with 1 as the identity;
(WHR3) $x \otimes y \leq z$ if and only if $y \leq x \rightarrow z$.
where $A \leq B$ means that for there exist $x \in A$, $y \in B$ such that $x \leq y$ and $A \ll B$ means that for all $x \in A$ there exist $y \in B$ such that $x \leq y$.

Definition 8 A hyper equality algebra $(H; \sim, \wedge, 1)$ is called
(1) separated if $1 \in x \sim y$ implies $x = y$;
(2) symmetric if $x \sim y = y \sim x$.
for all $x, y \in H$.

Example 14 In Example 7, one can see that $(H; \wedge, \sim, 1)$ is a separated and symmetric hyper equality algebra.

Theorem 5 *Let $(H; \vee, \wedge, \otimes, \rightarrow, 0, 1)$ be a weak hyper residuated lattice, where 1 is a scalar element of H (i.e. $1 \otimes x$ has only one element). Then $(H; \sim, \wedge, 1)$ is a separated and symmetric hyper equality algebra, where $x \sim y = x \leftrightarrow y = (x \rightarrow y) \wedge (y \rightarrow x)$.*

Proof By [10] Theorem 5.11 and $x \sim y = y \sim x$, we know that $(H; \wedge, \otimes, \sim, 1)$ is a separated hyper EQ-algebra. Using [6] Proposition 2.5 (v) and Theorem 3.13, this proof is completed. □

Definition 9 ([8]) By a hyper BCK-algebra we mean a nonempty set H endowed with a binary hyper operation $\rightarrow$ and a constant 1 satisfying the following axioms, for all $x, y, z \in H$:
(HK1) $x \rightarrow y \ll (z \rightarrow x) \rightarrow (z \rightarrow y)$;
(HK2) $x \rightarrow (y \rightarrow z) = y \rightarrow (x \rightarrow z)$;
(HK3) $x \ll y \rightarrow x$;
(HK4) $x \leq y$ and $y \leq x$ imply $x = y$.
where $x \leq y$ is defined by $1 \in x \rightarrow y$ and for every $A, B \subseteq H$, $A \ll B$ is defined as for all $x \in A$, there exists $y \in B$ such that $x \leq y$.

Definition 10 A hyper BCK-algebra $(H; \rightarrow, 1)$ is called a hyper BCK-meet-semilattice if $(H; \leq)$ is a meet $(\wedge)$-semilattice.

Theorem 6 *Let $(H; \sim, \wedge, 1)$ be a hyper equality algebra such that $x \rightarrow (y \rightarrow z) = y \rightarrow (x \rightarrow z)$ and $x \rightarrow y \ll (z \rightarrow x) \rightarrow (z \rightarrow y)$ for all $x, y, z \in H$. Then $(H; \rightarrow, \wedge, 1)$ is a hyper BCK-meet-semilattice, where $x \rightarrow y = x \sim x \wedge y$.*

Proof From (HE1) we know that $(H, \leq)$ is a meet-semilattice. (HK1) and (HK2) are true by hypothesis. (HK3) and (HK4) follow from (P14) and (P6) respectively. □

Theorem 7 *Let $(H; \rightarrow, \wedge, 1)$ be a linearly ordered hyper BCK-meet-semilattice. If $x \rightarrow y \ll (y \rightarrow z) \rightarrow (x \rightarrow z)$, and $x \leq y$ implies that $z \rightarrow x \ll z \rightarrow y$, for all $x, y, z \in H$, then $(H; \sim, \wedge, 1)$ is a symmetric hyper equality algebra, where $x \sim y = (x \rightarrow y) \wedge (y \rightarrow x)$.*

Proof Let $(H; \rightarrow, \wedge, 1)$ be a hyper BCK-meet-semilattice and let $x, y, z \in H$. Then it is obvious that (HE1) and (HE2) hold. (HE3) follows from $1 = 1 \wedge 1 \in (x \rightarrow x) \wedge (x \rightarrow x) = x \sim x$ and (HE4) follows from $x = x \wedge 1 \in (1 \rightarrow x) \wedge (x \rightarrow 1) = 1 \sim x$. Let $x \leq y \leq z$ and now we will prove (HE5). Since $1 \in x \rightarrow z, y \rightarrow z$ and $x \leq y$ implies that $z \rightarrow x \ll z \rightarrow y$, we have $(x \rightarrow z) \wedge (z \rightarrow x) \ll z \rightarrow x \ll z \rightarrow y \ll (z \rightarrow y) \wedge (y \rightarrow z)$. That is $x \sim z \ll y \sim z$. The other part is similar. Since H is linearly ordered, then one can easily check that (HE6) holds. In the following we show (HE7) and we let $x \leq y \leq z$ without loss of generality. Since $1 \in x \rightarrow z, 1 \in y \rightarrow z$ and thus $x \rightarrow y \ll (z \rightarrow x) \rightarrow (z \rightarrow y) \ll ((x \rightarrow z) \wedge (z \rightarrow x)) \rightarrow ((y \rightarrow z) \wedge (z \rightarrow y)) = (x \sim z) \rightarrow (y \sim z)$ and $x \rightarrow y \ll (y \rightarrow z) \rightarrow (x \rightarrow z) \ll (((y \rightarrow z) \wedge (z \rightarrow y)) \rightarrow ((x \rightarrow z) \rightarrow (z \rightarrow x)) = (y \sim z) \rightarrow (x \sim z)$, we can obtain that $x \rightarrow y \ll (x \sim z) \rightarrow (y \sim z)$ and $y \rightarrow x \ll (y \sim z) \rightarrow (x \sim z)$. Hence $x \sim y = (x \rightarrow y) \wedge (y \rightarrow x) \ll (((x \sim z) \rightarrow (y \sim z)) \wedge ((y \sim z) \rightarrow (x \sim z)) = (x \sim z) \sim (y \sim z)$. Therefore $(H; \sim, \wedge, 1)$ is a symmetric hyper equality algebra. □

In the following, we construct quotient hyper equality algebras via regular hyper congruence relations. To do this we give some related concepts and results.

Definition 11 Let θ be an equivalence relation on H.
(1) For any $A, B \subseteq H$, $A\overline{\theta}B$ means for any $a \in A$ there exists $b \in B$ such that $a\theta b$ and for any $b \in B$ there exists $a \in A$ such that $a\theta b$;
(2) For any $A, B \subseteq H$, $A\overline{\overline{\theta}}B$ means for any $a \in A$ and any $b \in B$ such that $a\theta b$;
(3) θ is called a hyper congruence relation if for all $x, y, u, v \in H$, $x\theta y$ and $u\theta v$ imply $x \sim u\overline{\theta}y \sim v$ and $x \wedge u\theta y \wedge v$;
(4) θ is called a strong hyper congruence relation if for all $x, y, u, v \in H$, $x\theta y$ and $u\theta v$ imply $x \sim u\overline{\overline{\theta}}y \sim v$ and $x \wedge u\theta y \wedge v$.

Let θ be a hyper congruence relation on H. Denote $H/\theta = \{[x]_\theta : x \in H\}$, where $[x]_\theta = \{y \in H : y\theta x\}$ and denote $[A]_\theta = \{[a]_\theta : a \in A\}$. $\overline{\sim}$, $\overline{\rightarrow}$ and $\overline{\wedge}$ on H/θ are defined by $[x]_\theta\overline{\sim}[y]_\theta = [x \sim y] = \{[a]_\theta : a \in x \sim y\}$, $[x]_\theta\overline{\rightarrow}[y]_\theta = [x \rightarrow y]_\theta = \{[a]_\theta : a \in x \rightarrow y\}$ and $[x]_\theta\overline{\wedge}[y]_\theta = [x \wedge y]_\theta$, respectively. $[x]_\theta \leq_\theta [y]_\theta$ iff $[x]_\theta = [x]_\theta\overline{\wedge}[y]_\theta$ iff $x \wedge y\theta x$ for any $[x]_\theta, [y]_\theta \in H/\theta$. For all $A, B \subseteq H/\theta, A \ll_\theta B$ means that for any $[a]_\theta \in A$, there exists $[b]_\theta \in B$ such that $[a]_\theta \leq_\theta [b]_\theta$. Clearly, $x \leq y$ implies that $[x]_\theta \leq_\theta [y]_\theta$.

Definition 12 A hyper congruence relation θ on H is called regular if for all $x, y \in H$, $[x]_\theta \leq_\theta [y]_\theta$ implies $[x]_\theta \ll [y]_\theta$.

Proposition 7 *Let θ be a regular hyper congruence relation on H. Then for any $x, y \in H$, $[x]_\theta \leq_\theta [y]_\theta$ implies that $[1]_\theta \in [x]_\theta\overline{\rightarrow}[y]_\theta$.*

Proof Let $[x]_\theta \leq_\theta [y]_\theta$ for any $x, y \in H$. Then for any $t \in [x]_\theta$ there exists $s \in [y]_\theta$ such that $t \leq s$. Hence $1 \in t \rightarrow s$ and so $[1]_\theta \in [t \rightarrow s]_\theta = [t]_\theta\overline{\rightarrow}[s]_\theta = [x]_\theta\overline{\rightarrow}[y]_\theta\theta$. □

Lemma 2 *Let θ be a regular hyper congruence relation on H. Then for any $A, B \subseteq H$, $A \ll B$ implies $[A]_\theta \ll_\theta [B]_\theta$.*

Proof Let $A \ll B$ for any $A, B \subseteq H$. Then for any $a \in A$ there exists $b \in B$ such that $a \leq b$. Hence $[a]_0 \leq_0 [b]_0$ and therefore $[A] \ll_0 [B]$. □

Theorem 8 *Let θ be a regular hyper congruence relation on $(H; \sim, \wedge, 1)$. Then $(H/\theta; \overline{\sim}, \overline{\wedge}, [1])$ is a hyper equality algebra, which is called a quotient hyper equality algebra with respect to θ.*

Proof Clearly, $\overline{\wedge}$ is well defined. Assume that $[x_1]_\theta = [x_2]_\theta$ and $[y_1]_\theta = [y_2]_\theta$, $x_1, x_2, y_1, y_2 \in H$. Then $x_1\theta x_2$ and $y_1\theta y_2$. Thus $x_1 \sim y_1\theta x_2 \sim y_2$. Let $[a]_\theta \in [x_1]_\theta\overline{\sim}[y_1]_\theta$ where $a \in x_1 \sim y_1$. Since $x_1 \sim y_1\theta x_2 \sim y_2$, then there exists $c \in x_2 \sim y_2$ such that $a\theta c$. Hence $[a]_\theta = [c]_\theta \in [x_2]_\theta\overline{\sim}[y_2]_\theta$, which shows that $[x_1]_\theta\overline{\sim}[y_1]_\theta \subseteq [x_2]_\theta\overline{\sim}[y_2]_\theta$. Similarly, we have $[x_2]_\theta\overline{\sim}[y_2]_\theta \subseteq [x_1]_\theta\overline{\sim}[y_1]_\theta$. It follows that $[x_1]_\theta\overline{\sim}[y_1]_\theta = [x_2]_\theta\overline{\sim}[y_2]_\theta$. Therefore $\overline{\sim}$ is well defined. Similarly, $\overline{\otimes}$ is well defined.
(HE1) $(H/\theta; \overline{\wedge}, [1]_\theta)$ is a meet-semilattice with top element $[1]_\theta$. Since for any $x \in H, x \leq 1$ implies that $[x]_\theta \leq_\theta [1]_\theta$.
(HE2) $[x]_\theta\overline{\sim}[y]_\theta \ll_\theta [y]_\theta\overline{\sim}[x]_\theta$. By Lemma 2 and $x \sim y \ll y \sim x$.

(HE3) $[1]_\theta \in [x]_\theta \widetilde{\sim} [x]_\theta$. Since $1 \in x \sim x$, we have $[1]_\theta \in [x \sim x]_\theta = [x]_\theta \widetilde{\sim} [x]_\theta \theta$.
(HE4) $[x]_\theta \in [0]_\theta \widetilde{\sim} [x]_\theta$. Since $x \in 1 \sim x$, then $[x]_\theta \in [1 \sim x]_\theta = [1]_\theta \widetilde{\sim} [x]_\theta$.
(HE5) $[x]_\theta \leq_\theta [y]_\theta \leq_\theta [z]_\theta$ implies $[x]_\theta \widetilde{\sim} [z]_\theta \ll_\theta [y]_\theta \widetilde{\sim} [z]_\theta$ and $[x]_\theta \widetilde{\sim} [z]_\theta \ll_\theta [x]_\theta \widetilde{\sim} [y]_\theta$. Indeed, if $[x]_\theta \leq_\theta [y]_\theta \leq_\theta [z]_\theta$ for any $x, y, z \in H$, then by Proposition 7 for any $t \in [x]_\theta$, there exists $s \in [y]_\theta$ and $u \in [z]_\theta$ such that $t \leq s \leq u$. Hence $t \sim u \ll s \sim u$ and $t \sim u \ll t \sim s$. According to Lemma 2 $[t]_\theta \widetilde{\sim} [u]_\theta \ll_\theta [s]_\theta \widetilde{\sim} [u]_\theta$ and $[t]_\theta \widetilde{\sim} [u]_\theta \ll_\theta [t]_\theta \widetilde{\sim} [s]_\theta$. This proves that $[x]_\theta \widetilde{\sim} [z]_\theta \ll_\theta [y]_\theta \widetilde{\sim} [z]_\theta$ and $[x]_\theta \widetilde{\sim} [z]_\theta \ll_\theta [x]_\theta \widetilde{\sim} [y]_\theta$.
(HE6) $[x]_\theta \widetilde{\sim} [y]_\theta \ll_\theta ([x]_\theta \widetilde{\wedge} [z]_\theta) \widetilde{\sim} ([y]_\theta \widetilde{\wedge} [z]_\theta)$. By Lemma 2 and $x \sim y \ll (x \wedge z) \sim (y \wedge z)$.
(HE7) $[x]_\theta \widetilde{\sim} [y]_\theta \ll_\theta ([x]_\theta \widetilde{\sim} [z]_\theta)[x] \widetilde{\sim} ([y]_\theta \widetilde{\sim} [z]_\theta)$. By Lemma 2 and $x \sim y \ll (x \sim z) \sim (y \sim z)$. □

5 Conclusions

In this paper, we introduce hyper equality algebras which is a generalization of equality algebras. We investigate some types of hyper filters and hyper deductive systems. Also we give relations between hyper equality algebras and other hyper structures, and moreover construct quotient hyper equality algebras via regular hyper congruence relations. Next we will further study quotient hyper equality algebras and construct quotient hyper equality algebras via hyper filters or hyper deductive systems.

Acknowledgments This research is supported by a grant of National Natural Science Foundation of China (11571281).

References

1. Novák, V., Baets, B.D.: EQ-algebras. Fuzzy Sets Syst. **160**, 2956–2978 (2009)
2. Jenei, S.: Equality algebras. Stud. Log. **100**, 1201–1209 (2012)
3. Marty, F.: Surune generalization de la notion de group. The 8th Congress Math. Scandinaves, Stockholm, pp. 45–49 (1934)
4. Borzooei, R.A., Bakhshi, M., Zahiri, O.: Filter theory on hyper residuated lattice. Quasigroups Relat. Syst. **22**, 33–50 (2014)
5. Borzooei, R.A., Hasankhani, A., Zahedi, M.M., JUN, Y.B.: On hyper K-algebras. Math. Japonicae **52**, 113–121 (2000)
6. Borzooei, R.A., Niazian, S.: Weak hyper residuated lattics. Quasigroups Relat. Syst. **21**, 29–42 (2013)
7. Ghorbani, S., Hasankhani, A., Eslami, E.: Hyper MV-algebras. Set-Valued Math. Appl. **1**, 205–222 (2008)
8. Jun, Y.B., Zahedi, M.M., Xin, X.L., Borzooei, R.A.: On hyper BCK-algebras. Italian J. Pure Appl. Mathematics-N **8**, 127–136 (2000)
9. Xin, X.L.: Hyper BCI-algebras. Discuss. Math. General Algebra Appl. **26**, 5–19 (2006)

10. Borzooei, R.A., Saffar, B.G., Ameri, R.: On hyper EQ-algebras. Italian J. Pure Appl. Mathematics-N. **31**, 77–96 (2013)
11. Cheng, X.Y., Xin, X.L.: Filter theory On hyper BE-algebras. Italian J. Pure Appl. Mathematics-N. **35**, 509–526 (2015)
12. Cheng, X.Y., Xin, X.L.: Deductive systems in hyper EQ-algebras. J. Math. Res. Appl. (submitted)

A Special Sub-algebras of $N(2, 2, 0)$-Algebras

Fang-An Deng, Lu Chen, Shou-Heng Tuo and Sheng-Zhang Ren

Abstract In this paper, we introduce a subalgebra of $N(2, 2, 0)$-algebras, investigate the relations between $N(2, 2, 0)$-algebras and other algebras, such as G-algebra, B-algebra, Q-algebra and CI-algebra. In particular, we find out a class of subalgebras of $N(2, 2, 0)$-algebras, show some properties of those subalgebras and prove that the subalgebras is G-algebra, B-algebra, Q-algebra and CI-algebra. Finally, we give an important result on $N(2, 2, 0)$-algebras.

Keywords $N(2, 2, 0)$-algebra · G-algebra · CI-algebra · Q-algebra

1 Introduction

Some algebras with one nulary operations were introduced to set up an algebraic counterpart of implication reduct of classical or non-classical propositional logics. In 1966 [5], Imai and Iseki introduced two classes of abstract algebras: BCK-algebras and BCI-algebras.

In 1996 [2, 3], we introduced $N(2, 2, 0)$-algebra, showed some basic properties $(S, *, \triangle, 0)$ of $N(2, 2, 0)$-algebra and proved:

1. If the operations $*$ is idempotent, then $(S, *, \triangle, 0)$ is a rewriting systems and
2. If the operation $\triangle$ is nilpotent, then $(S, \triangle, 0)$ is a associated BCI-algebra.

Recently, H.S. Kim and Y.H. Kim defined a BE-algebra in [7], Meng defined the notion of CI-algebra as a generalization of the BE-algebra in [8] and Bandru and Rafi introduced a new notion, called G-algebras in [1]. In 2001 [9], Neggers introduced the notion of Q-algebras which is a generalization of BCI/BCK-algebras. Negger and Kim introduced an important class of logical algebras, called B-algebras in 2002 [10].

F.-A. Deng (✉) · L. Chen · S.-H. Tuo · S.-Z. Ren
School of Mathematics and Computer Science, Shaanxi SCI-TECH University, Hanzhong 723000, Shaanxi, People's Republic of China
e-mail: dengfangans@126.com

T.-H. Fan et al. (eds.), *Quantitative Logic and Soft Computing 2016*,
Advances in Intelligent Systems and Computing 510,
DOI 10.1007/978-3-319-46206-6_41

In this paper, we recall the basic definitions and some elementary aspects which are necessary for the sequel in Sect. 2 and investigate relations among the semigroup of $N(2,2,0)$-algebras, G-algebras, B-algebras, Q-algebras and CI-algebras in Sect. 3.

2 Paper Preparation

Let $N(2,2,0)$-algebra be an algebra of type $(2,2,0)$. The notion was formulated firstly by Deng [3] in 1996, and some properties were obtained in [5], which was inspired by the fuzzy implication algebra introduced by Wu in [14]. Wu proved that in a fuzzy implication algebra $(X,\to,0)$, the order relation $\leq$ satisfying $x \leq y$ iff $x \to y = 1$ is a partial order. Deng in [3] introduced a binary operation $*$ which was defined on fuzzy implication algebra $(X,\to,0)$ such that for all $a, b, u \in X$,

$$u \leq a \to b \Leftrightarrow a * u \leq b.$$

where $(*,\to)$ is an adjoint pair on X.

In the corresponding fuzzy logic, the operation $*$ is recognized as logic connective "conjunction" and $\to$ is considered as "implication". If the above expression holds for a product $*$, then $\to$ is the residunm of $*$. For a product $*$, the corresponding residunm $\to$ is uniquely defined by a $\to b = \vee\{x|a * x \leq b\}$.

Let us note that $a \to b$ is the greatest element of the set $\{u|a * u \leq b\}$. We proved that if for all $a, b, u \in X$, the following formulas hold:

$$u \to (a * b) = b \to (u \to a) \tag{1}$$

$$(a * u) \to b = u \to (a \to b), \tag{2}$$

then $(X,*)$ is a semigroup.

In fact, the multiplication defined as above is associative.

Let $(X,\to,0)$ be a fuzzy implication algebra. It was shown in [14] that for every $a, b, c \in X$, there are $1 \to a = a$,

$$a * (b * c) = 1 \to (a * (b * c)) = (b * c) \to (1 \to a) = (b * c) \to a = c \to (b \to a) \text{ and}$$

$$(a * b) * c = 1 \to ((a * b) * c) = c \to (1 \to (a * b)) = c \to (a * b) = b \to (c \to a).$$

Note $b \to (c \to a) = c \to (b \to a)$. Then we have $(a * b) * c = a * (b * c)$. Then $(X,*)$ is a semigroup.

By generalizing the expressions (1) and (2), we obtain the basic equations of $N(2,2,0)$-algebra. Recall the following definition in Deng [3].

Definition 1 An $N(2,2,0)$-algebra is a non-empty set S with a constant 0 and two binary operations $*$, Δ such that for all $a, b, c \in S$:

(N_1) $a * (b \bigtriangleup c) = c * (a * b)$;
(N_2) $(a \bigtriangleup b) * c = b * (a * c)$ and
(N_3) $0 * a = a$.

By substituting $*$ and $\bigtriangleup$ in expressions (N_1) and (N_2) with $\rightarrow$ and $*$, respectively, we arrive at the expressions (2) and (3). Recall the following theorem and corollary in Dong [3].

Theorem 1 *Let $(S, *, \bigtriangleup, 0)$ be a $N(2, 2, 0)$ algebra. Then, for all $a, b, c \in S$:*
(1) $a * b = b \bigtriangleup a$;
(2) $(a * b) * c = a * (b * c)$, $(a \bigtriangleup b) \bigtriangleup c = a \bigtriangleup (b \bigtriangleup c)$;
(3) $a * (b * c) = b * (a * c)$, $(a \bigtriangleup b) \bigtriangleup c = (a \bigtriangleup c) \bigtriangleup b$.

Corollary 1 *If $(S, *, \bigtriangleup, 0)$ is a $N(2, 2, 0)$-algebra, then both $(S, *, 0)$ and $(S, \bigtriangleup, 0)$ are semigroups.*

So $N(2, 2, 0)$-algebra is an algebra system with a pair of dual semigroups. Some interesting properties of $N(2, 2, 0)$-algebra have been discussed earlier in Deng [3]. Recall the following definition in Deng [2].

Definition 2 A residuated poset is a structure $(A; \leq, \rightarrow, ., 0, 1)$ such that
(R_1) $(A; \leq, 0, 1)$ is a bounded poset;
(R_2) $(A; ., 1)$ is a commutative monoid;
(R_3) It satisfies the adjoint property, i.e., $x \cdot y \leq z \Longleftrightarrow x \leq y \rightarrow z$.

Then, by the definitions and Theorem 2 for $N(2, 2, 0)$-algebra, we have the following results:

Remark 1 Let $(S, *, \bigtriangleup, 0)$ be a $N(2, 2, 0)$-algebra. In every semigroup $(S, *, 0)$ of $(S, *, \bigtriangleup, 0)$, one can define a binary relation $\leq$. Then $x \leq y \Leftrightarrow x \rightarrow y = 1$ for all $x, y \in S$, where $1 = 0 \rightarrow 0$ and $x * 0 = x$ for all $x \in S$. It is easy to check that semigroup $(S, *, 0)$ is a residuated poset.

Remark 2 Let fuzzy implication algebra $(X, \rightarrow, 0)$ with a partial order "$\leq$" satisfy $a \leq b \Leftrightarrow a \rightarrow b = 1$ and $u \leq a \rightarrow b \Leftrightarrow a * u \leq b$ for all $a, b, u \in X$. If $u \rightarrow (a * b) = b \rightarrow (u \rightarrow a)$ and $(a * u) \rightarrow b = u \rightarrow (a \rightarrow b)$ for all $a, b, u \in X$, then $(X, \rightarrow, *, 0)$ is a $N(2, 2, 0)$-algebra.

Remark 3 Let $(S, *, \bigtriangleup, 0)$ be a $N(2, 2, 0)$-algebra. Then semigroups $(S, *, 0)$ and $(S, \bigtriangleup, 0)$ are a pair of dual semigroup. A pair of dual operations $(*, \bigtriangleup)$ forms an adjoint pair $(\rightarrow, *)$. Then $u \leq a \rightarrow b \Leftrightarrow a * u \leq b$ for every $a, b, u \in S$.

Recall the following theorem in Deng [3].

Theorem 2 *Let $(S, *, \bigtriangleup, 0)$ be a $N(2, 2, 0)$ algebra. Then $x \bigtriangleup (y * z) = y * (x \bigtriangleup z)$ and $x * (y \bigtriangleup z) = y \bigtriangleup (x * z)$ for every x, y, z in S.*

3 Main Results

Theorem 3 *Let* $(S, *, \triangle, 0)$ *be an* $N(2, 2, 0)$*-algebra. If* $x * x = 0$ *for every* $x \in S$*, then, for all* $a, x, y \in S$*,*
(1) $x * 0 = 0 * x = x$;
(2) $x * y = y * x$;
(3) $x * y = 0$ *implies* $x = y$;
(4) $a * x = a * y$ *implies* $x = y$.

Proof (1) Let $(S, *, \triangle, 0)$ be a $N(2, 2, 0)$ algebra. For every $x \in S$, if $x * x = 0$, then $x * 0 = x * (x * x) = (x * x) * x = 0 * x = x$ by (N_3).

(2) $x * y = x * (y * 0) = y * (x * 0) = y * x$ by (1).

(3) Let $x * y = 0$. Then $y = 0 * y = (x * y) * y = x * 0 = x$ by (N_3) and (1).

(4) Let $a, x, y \in S$, then $a * x = a * y \Rightarrow a * (a * x) = a * (a * y) \Rightarrow x = 0 * x = (a * a) * x = (a * a) * y = 0 * y = y \Rightarrow x = y$. □

Definition 3 Let $(S, *, \triangle, 0)$ and $(\overline{S}, \overline{*}, \overline{\triangle}, \overline{0})$ be $N(2, 2, 0)$-algebras. A mapping $f : S \to \overline{S}$ is called a homomorphism if $f(x * y) = f(x)\overline{*}f(y), f(x \triangle y) = f(x)\overline{\triangle} f(y)$ for all $x, y \in S$.

A homomorphism f is called a monomorphism (resp., epimorphism) if it is injective (resp., surjective). A bijective homomorphism is called an isomorphism. Two $N(2, 2, 0)$-algebras $(S, *, \triangle, 0)$ and $(\overline{S}, \overline{*}, \overline{\triangle}, \overline{0})$ are said to be isomorphic, denoted by $S \cong \overline{S}$, if there exists an isomorphism $f : S \to \overline{S}$. For every homomorphism $f : S \to \overline{S}$, the set $\{x \in S | f(x) = 0\}$ is called kernel of f, denoted by $Ker(f)$ and the set $\{f(x) | x \in S\}$ is called image of f, denoted by $Im(f)$. We denote by $Hom(S, \overline{S})$ the set of all homomorphisms of $N(2, 2, 0)$-algebras from S to $\overline{S}$.

Suppose $f : S \to \overline{S}$ is a homomorphism of $N(2, 2, 0)$-algebras. Then:

(1) $f(0) = \overline{0}$;
(2) f is an isomorphism, i.e., if $x * y = 0, x, y \in S$, then $f(x) * f(y) = \overline{0}$.

In $N(2, 2, 0)$-algebra $(S, *, \triangle, 0)$, a endomorphism on semigroup $(S, *, 0)$ is a anti-homomorphism f from semigroup $(S, *, 0)$ to dual semigroup $(S, \triangle, 0)$.

Note that, in $N(2, 2, 0)$-algebra $(S, *, \triangle, 0)$, if $x * x = 0$ for every $x \in S$, then $(S, *, 0)$ is isomorphic to $(S, \triangle, 0)$. Then we define $x \sim y$ if and only if $x * y = 0$ for every $x, y \in S$. Now we prove that $\sim$ is an equivalence relation on S. Note $x * x = 0$ and $x \sim x$. Then $\sim$ is reflexive. By Theorem 3, if $x * y = 0$, then $y * x = 0$. Then $x \sim y \Rightarrow y \sim x$. Then $\sim$ is symmetric. If $x \sim y$ and $y \sim z$, then $x * z = x * (0 * z) = x * ((x * y) * z) = ((x * x) * y) * z = (0 * y) * z = y * z = 0$. Then $x \sim z$. Then $\sim$ is transitive. Then $\sim$ is an equivalence relation on S. Furthermore we have the following theorem.

Theorem 4 *Let* $(S, *, \triangle, 0)$ *be a* $N(2, 2, 0)$ *algebra. If* $x * x = 0$, $x \sim y$ *and* $u \sim v$ *for every* $x, y, u, v \in S$*, then* $x * u \sim y * v$*. Then* $\sim$ *is a congruence relation on semigroup* $(S, *, 0)$.

Proof Note $x \sim y$ and $u \sim v$. Then $x * y = 0$, $u * v = 0$. So $0 = 0 * 0 = (x * y) * (u * v) = u * ((x * y) * v) = u * (x * (y * v)) = x * (u * (y * v)) = (x * u) * (y * v)$. Then $x * u \sim y * v$. This completes the proof. □

Definition 4 A non-empty set X with a constant 0 and a binary operation $*$ is said to be G-algebra if it satisfies the following axioms: $(G_1)\, x * x = 0$; $(G_2)\, x * (x * y) = y$, for all $x, y \in X$. A G-algebra is denoted by $(X, *, 0)$.

Theorem 5 *Let $(S, *, \triangle, 0)$ be a $N(2, 2, 0)$-algebra. If $x * x = 0$ for every $x \in S$, then $(S, *, 0)$ is a G-algebra.*

Proof Let $(S, *, \triangle, 0)$ be a $N(2, 2, 0)$-algebra. If $x * x = 0$ for every $x \in S$, then $x * (x * y) = (x * x) * y = 0 * y = y$ for all $x, y \in S$. Hence $(S, *, 0)$ is a G-algebra. □

Recall the following example in Dong [1].

Example 1 Let $X = \{0, 1, 2, 3, 4, 5, 6, 7\}$ be a set with the following table:

$*$	0	1	2	3	4	5	6	7
0	0	2	1	3	4	5	6	7
1	1	0	3	2	5	4	7	6
2	2	3	0	1	6	7	4	5
3	3	2	1	0	7	6	5	4
4	4	5	6	7	0	2	1	3
5	5	4	7	6	1	0	3	2
6	6	7	4	5	2	3	0	1
7	7	6	5	4	3	2	1	0

Then $(X, *, 0)$ is a G algebra but is not a semigroup $(S, *, 0)$ of $N(2, 2, 0)$-algebra $(S, *, \triangle, 0)$ since $0 * x \neq x$.

Definition 5 A non-empty subset S of a G-algebra X is called a G-subalgebra of X if $x * y \in S$ for all $x, y \in S$.

In the following, we suppose that S is an $N(2, 2, 0)$-algebra which satisfies: $x \triangle x = 0$ for every $x \in S$. Denote $[x] = \{y \in S | x \sim y\} = \{y \in S | x * y = 0\}$ by the equivalence class of x. Note $0 * x = x$. Then $[0] = \{0\}$.

Denote $S/\sim = \{[x] | x \in S\}$ and define that $[x]♣[y] = [x * y]$.

Note $\sim$ is a congruence relation on S. Then the operation ♣ is well-defined. In the following, we will prove that $([x]; ♣, [0])$ is a G-algebra.

Let $[x], [y], [z]$ and $[0]$ be in $S/\sim$. Then we have the following properties:

(G_1). $[x]♣[x] = [0]$,

(G_2). $[x]♣([x]♣[y]) = [y]$ for all $x, y \in S$.

Then $S/\sim$ is a G-algebra by the above facts. Then $S/\sim$ is a quotient G-algebra too.

B-algebras is an important class of logical algebras which is introduced by Neggers and Kim in [10] and is extensively investigated by some researchers. Recall the definition of B-algebras in [10].

Definition 6 A non-empty set X with a constant 0 and a binary operation $\triangle$ is called a B-algebra if it satisfies the following axioms: for all $x, y, z \in X$,
(B_1) $x \triangle x = 0$;
(B_2) $x \triangle 0 = x$;
(B_3) $(x \triangle y) \triangle z = x \triangle (z \triangle (0 \triangle y))$.

Theorem 6 *Let $(S, *, \triangle, 0)$ be an $N(2, 2, 0)$-algebra. If $x \triangle x = 0$ for every $x \in S$, then $(S, \triangle, 0)$ is a B-algebra but is not conversely.*

Proof Let $(S, *, \triangle, 0)$ be an $N(2, 2, 0)$-algebra. If $x \triangle x = 0$ for every $x \in S$, then

$$(x \triangle y) \triangle z = (x \triangle z) \triangle y = x \triangle (z \triangle y) = x \triangle ((z \triangle 0) \triangle y) = x \triangle (z \triangle (0 \triangle y))$$

for all $x, y, z \in S$. This implies B_3. Then $(S, \triangle, 0)$ is a B-algebra. □

Recall the following example in [10]
Let $X = \{0, 1, 2, 3\}$ be a set with the following table:

$\triangle$	0	1	2	3
0	0	3	2	1
1	1	0	3	2
2	2	1	0	3
3	3	2	1	0

Then $(X, \triangle, 0)$ is a B-algebra, but is not a semigroup $(S, \triangle, 0)$ of $N(2, 2, 0)$-algebra $(S, *, \triangle, 0)$ since $x \triangle y \neq y \triangle x$.
Recall the following theorem in [13].

Theorem 7 *$(X, *, 0)$ is a B-algebra if and only if:*
(B_4) $x * x = 0$;
(B_5) $0 * (0 * x) = x$;
(B_6) $(x * z) * (y * z) = x * y$ for any $x, y, z \in X$.

Then we have the following theorem.

Theorem 8 *Let $(S, *, 0)$ be a semigroup of $N(2, 2, 0)$-algebra $(S, *, \triangle, 0)$. Then $(S, *, 0)$ is a B-algebra if and only if $(S, *, 0)$ satisfies $x * x = 0$ for every $x \in S$.*

Proof Suppose that $(S, *, 0)$ is a B-algebra. By (N_3), for each $x \in S$, we have $0 * (0 * x) = 0 * x = x$. Consequently, (B_5) is valid in $(S, *, 0)$. Then, by Theorem 3, we have $x * 0 = x$. Then we obtain

$$(x * y) * (y * z) = y * ((x * z) * z) = y * (x * (z * z)) = x * (y * 0) = x * y.$$

This implies (B_6). Then we complete the proof by Theorem 10. □

Recall the following definition in [6].

Definition 7 A BCH-algebra is an algebra $(X, *, 0)$ of type (2, 0) satisfying the following axioms:
(BCH_1) $x * x = 0$;
(BCH_2) $x * y = y * x = 0 \Rightarrow x = y$;
(BCH_3) $(x * y) * z = (x * z) * y$.

We have the following theorem.

Theorem 9 *Let $(S, *, \Delta, 0)$ be a $N(2, 2, 0)$-algebra. If $x \Delta x = 0$ for every $x \in S$, then $(S, \Delta, 0)$ is a BCH-algebra but not conversely.*

Proof Note Theorems 1 and 3 Then $(S, \Delta, 0)$ is a BCH-algebra. □

Recall the following definition in [6].

Definition 8 A BCI-algebra is an algebra $(X, *, 0)$ of type $(2, 0)$ satisfying the following axioms:
(BCI_1) $((x * y) * (x * z)) * (z * y) = 0$;
(BCI_2) $(x * (x * y)) * y = 0$;
(BCI_3) $x * x = 0$;
(BCI_4) $x * y = y * x = 0 \Rightarrow x = y$ for all $x, y, z \in X$.

Hu and Li shew that the BCI-algebras is a proper subclass of the BCH-algebras in [4]. Then, by Theorem 9 and Definition 8, we have:

Theorem 10 *Let $(S, *, \Delta, 0)$ be an $N(2, 2, 0)$-algebra. If $x \Delta x = 0$ for every $x \in S$, then $(S, *, 0)$ is a associative BCI-algebra but not conversely.*

Recall the following definitions and example in [9].

Definition 9 A Q-algebra is a non-empty set X with a constant 0 and a binary operation Δ satisfying the following axioms:
(Q_1) $x \Delta x = 0$;
(Q_2) $x \Delta 0 = x$;
(Q_3) $(x \Delta y) \Delta z = (x \Delta z) \Delta y$ for all $x, y, z \in X$.

Example 2 Let $X = \{0, 1, 2\}$ be a set with the following table:

Δ	0	1	2
0	0	2	1
1	1	0	2
2	2	1	0

Then $(X, \Delta, 0)$ is a Q-algebra.

Definition 10 A non-empty subset S of a Q-algebra X is called a Q-subalgebra of X if $x \Delta y \in S$, for all $x, y \in S$.

Recall the following definition and example in [8].

Definition 11 A CI-algebra is an algebra $(X, *, 0)$ of type $(2, 0)$ satisfying the following axioms for all $x, y, z \in X$,
(CI_1) $x * x = 0$;
(CI_2) $0 * x = x$;
(CI_3) $x * (y * z) = y * (x * z)$.

Example 3 Let $X = \{0, a, b, c\}$ be a set with the following the table:

$*$	0	a	b	c
0	0	a	b	c
a	0	0	b	b
b	0	a	0	a
c	0	0	0	0

Then $(X, *, 0)$ is a CI-algebra.

Then, by Theorem 1, we have

Theorem 11 *Let* $(S, *, \Delta, 0)$ *be an* $N(2, 2, 0)$*-algebra. If* $x * x = 0$ *every* $x \in S$, *then:*
(1) $(S, *, 0)$ *is a* Q*-algebra;*
(2) $(S, \Delta, 0)$ *is a* CI*-algebra.*

Theorem 12 *Let* $(S, *, 0)$ *be a semigroup of* $N(2, 2, 0)$*-algebra* $(S, *, \Delta, 0)$ *and* $x * x = 0$ *for every* $x \in S$. *Then:*
(1) *The order of semigroup* $(S, *, 0)$ *is* 2, *Denoted by* $\| S \| = 2$.
(2) $(S, *, 0)$ *is* G*-algebras,* B*-algebras,* Q*-algebras and* CI*-algebras.*

Proof Suppose $S = \{0, a, b\}, a \neq 0$ and $b \neq 0$. Then, by Theorem 3, we have $a * 0 = a$ for every $a \in S$ if $a * a = 0$. Consider the following two cases:

Case 1. If $a * b = a$, then $a * (a * b) = a * a$. Then $(a * a) * b = 0 * b = b = 0$. This contradicts to $b \neq 0$.

Case 2. As similar as the above Case 1, let $a * b = b$. Then $a * (b * b) = b * b$. Then $a * (b * b) = a * 0 = a = 0$. This contradicts to $a \neq 0$.

Then $a * b = b * a = 0$. Note Theorem 3 If $a * b = b * a = 0$, Then $a = b$. Then S contains only two elements.

Let $S = \{0, 1\}$. Define a binary operation "$*$" on S by the following table:

$*$	0	1
0	0	1
1	1	0

If $\| S \| > 3$, then the proof is similar to the case of $\| S \| = 3$. We obtain that there is not $N(2, 2, 0)$ algebra $(S, *, \Delta, 0)$ of order $\| S \| \geq 3$ if $x * x = 0$ for every $x \in S$. It is easy to demonstrate that $(S, *, 0)$ is G-algebra, B-albebra, Q-algebra and CI-algebra. □

Acknowledgments The work Partially supported by Qinba Mountains of Bio-Resource Collaborative Innovation Center of Southern Shaanxi province of China (QBXT-Z(Y)-15-4).

References

1. Bandru, R.K., Rafi, N.: On G-algebras. Sci. Manga. **8**(3), 1–7 (2012)
2. Deng, F.A., Chen, L., Tuo, S.H., Ren, S.Z.: Characterizations of N(2; 2; 0) algebras. Algebra, **2016**, Article ID 2752681, 1–7 (2016). http://dx.doi.org/10.1155/2016/2752681
3. Deng, F.A., Xu, Y.: On $N(2, 2, 0)$-algebra. J. Southwest Jiaotong Univ. **131**, 457–463 (1996)
4. Hu, Q.P., Li, X.: On BCH-algebras. Math. Semin. Notes **11**, 313–320 (1983)
5. Imai, Y., Iseki, K.: On axiom systems of propositional calculi. XIV Proc. Jpn. Acad. **42**, 19–22 (1966)
6. Iseki, I.: On BCI-algebras. Math. Semin. Notes **8**, 125 130 (1980)
7. Kim, H.S., Kim, Y.H.: On BE-algebras. Scientiae Mathematicae Japonica Online. pp. 1299–1302(2006)
8. Meng, B.L.: CI-algebra. Scientiae Mathematicae Japonica Online. pp. 695–701 (2009)
9. Neggers, J., Ahn, S.S.: On Q-algebras. Int. J. Math. Math. Sci. **27**, 749–757 (2007)
10. Neggers, J., Kim, H.S.: On B-algebras. Mathematichki Vesnik **54**, 21–29 (2002)
11. Saeid, A.B.: CI-algebra is equivalent to dual Q-algebra. J. Egypt. Math. Soc. **21**, 1–2 (2013)
12. Senapati, T.: Translations of Intuitionistic Fuzzy B-algebras. Fuzzy Inf. Eng. **7**, 389–404 (2015)
13. Walendziak, A.: Some axiomtizations of B-algebras. Math. Aslovaca. **56**(3), 301–306 (2006)
14. Wu, W.M.: Fuzzy implication algebra. Fuzzy Syst. Math. **1**, 56–64 (1990)

Generalized Fuzzy Filters of R_0-Algebras

Long-Chun Wang, Xiang-Nan Zhou and Hua-Rong Zhang

Abstract In R_0-algebras, the notions of (α, β)-fuzzy (implicative, positive implicative, fantastic) filters where (α, β) are any two of $\{\in, q, \in \vee q, \in \wedge q\}$ with $\alpha \neq \in \wedge q$ are introduced and related properties are discussed. Some characterization theorems of these generalized fuzzy filters are derived. In particular, we prove that a fuzzy set is an $(\in, \in)$-fuzzy (implicative, positive implicative, fantastic) filter if and only if it is a fuzzy (implicative, positive implicative, fantastic) filter. Moreover, we also give the conditions for an $(\in, \in \vee q)$-fuzzy (implicative, positive implicative, fantastic) filter to be an $(\in, \in)$-fuzzy (implicative, positive implicative, fantastic) filter, and the conditions for a fuzzy set to be a $(q, \in \vee q)$-fuzzy (implicative, positive implicative, fantastic) filter.

Keywords R_0-algebra · Belong to · Quasi-coincident with · (α, β)-fuzzy (implicative, positive implicative, fantastic) filter

1 Introduction

Non-classical logic has become a considerable formal tool for computer science and artificial intelligence to deal with fuzzy information and uncertain information. Many-valued logic, a great extension and development of classical logic [1] has always been a crucial direction in non-classical logic. In order to research the many-valued logical system whose propositional value is given in a lattice, Xu [2] proposed the concept of lattice implication algebras and discussed some of their properties. Later on, Xu and Qin [3, 4] proposed the concept of implicative filters in lattice implication algebras and discussed some of their properties. On the other hand, the notion of Boolean deductive system, or equivalently, the Boolean filter in BL-algebras [5] was introduced by Turunen [6, 7]. The concept of R_0-algebras was first introduced

L.-C. Wang · X.-N. Zhou (✉) · H.-R. Zhang
College of Mathematics and Econometrics, Hunan University,
Changsha 410082, China
e-mail: xnzhou81026@163.com

T.-H. Fan et al. (eds.), *Quantitative Logic and Soft Computing 2016*,
Advances in Intelligent Systems and Computing 510,
DOI 10.1007/978-3-319-46206-6_42

by Wang [8] by providing an algebraic proof of the completeness theorem of a formal deductive system (see Wang [9]). R_0-algebras are essentially different from BL algebras and lattice implication algebras. They all have the implication operator $\rightarrow$. Therefore, it is meaningful to generalize the lattice implication algebras and BL algebras to R_0-algebras. The theory of fuzzy sets was first introduced by Zadeh [10] and has been applied to many branches. Zadeh [11] introduced the concept of interval valued fuzzy subset. The interval valued fuzzy subgroups were first defined and studied by Biswas [12] which are the subgroups of the same nature of the fuzzy subgroups defined by Rosenfeld (see Biswas [12]). The $(\in, \in \vee q)$-fuzzy subgroups was introduced in an earlier paper of Bhakat and Das [13] by using the combined notions of "belongingness" and "quasi-coincidence" of fuzzy points and fuzzy sets, which was introduced by Pu and Liu [14]. The $(\in, \in \vee q)$-fuzzy subgroup is an important generalization of Rosenfeld's fuzzy subgroup. Recently, Liu and Li [2] discussed the fuzzy implicative and Boolean filters of R_0-algebras, Ma and Zhan et al. [15] studies the $(\in, \in \vee q)$-fuzzy filters of R_0-algebras. As a generalization of those papers, we introduce the concepts of (α, β)-fuzzy filters in R_0-algebras and investigate related properties. In Sect. 3, we discuss the properties of (α, β)-fuzzy (implicative, positive implicative, fantastic) filters and describe the relationships among these fuzzy filters. In Sect. 4, we study the properties of $(\in, \in)$-fuzzy (implicative, positive implicative, fantastic) filters and discuss the relationships between $(\in, \in)$-fuzzy filters and fuzzy filters. In Sect. 5, we discuss the properties of $(\in, \in \vee q)$-fuzzy (implicative, positive implicative, fantastic) filter. In Sect. 6, we give the conditions for a fuzzy set to be a $(q, \in \vee q)$-fuzzy filter. In Sect. 7, we discuss the relations among $(\alpha, \in \wedge q)$-fuzzy filters where $\alpha \in \{\in, q, \in \vee q\}$.

2 Preliminaries

Definition 1 Let L be a bounded distributive lattice with order-reversing involution $\prime$ and a binary operation $\rightarrow$, $(L, \prime, \vee, \rightarrow)$ is called an R_0 -algebra if it satisfies the following axioms (see [8]):
(1) $x \rightarrow y = y\prime \rightarrow x\prime$;
(2) $1 \rightarrow x = x$;
(3) $(y \rightarrow z) \wedge ((x \rightarrow y) \rightarrow (x \rightarrow z)) = y \rightarrow z$;
(4) $x \rightarrow (y \rightarrow z) = y \rightarrow (x \rightarrow z)$;
(5) $x \rightarrow (y \vee z) = (x \rightarrow y) \vee (x \rightarrow z)$;
(6) $(x \rightarrow y) \wedge ((x \rightarrow y) \rightarrow (x\prime \vee y)) = 1$.

Let L be an R_0-algebra, for any $x, y \in L$, define $x \odot y = (x \rightarrow y\prime)\prime$. It is proved that $\odot$ is commutative, associative and $(L, \prime, \vee, \odot, \rightarrow, 0, 1)$ is a residuated lattice. In the following, let x^n denote $\underbrace{x \odot x \cdots \odot x}$ for $n \geq 1$.

In the follows, L will denote an R_0-algebra, unless otherwise specified.

Lemma 1 *For all $x, y \in L$, the following properties hold (see [16, 17]):*
(1) $x \leq y$ *if and only if* $x \to y = 1$;
(2) $x \leq y \to x$;
(3) $x\prime = x \to 0$;
(4) $x \vee y = ((x \to y) \to y) \wedge ((y \to x) \to x)$;
(5) *if* $x \leq y$, *then* $x \to z \geq y \to z$;
(6) *if* $x \leq y$,*then* $z \to x \leq z \to y$;
(7) $x \odot y \leq x \wedge y, x \odot (x \to y) \leq x \wedge y$;
(8) $x \odot y \leq z$ *if and only if* $x \leq y \to z$.

A non-empty subset F of L is called a filter of L if it satisfies the following conditions (see [16]):
(i) $1 \in F$;
(ii) $\forall x \in F, y \in L, x \to y \in F \Rightarrow y \in F$.
It is easy to check that a non-empty subset F of L is a filter if and only if it satisfies (see [16]):
(i) $\forall x, y \in F, x \odot y \in F$;
(ii) $\forall x \in F, x \leq y \Rightarrow y \in F$.
A subset F of L is called an implicative filter of L if it satisfies the following conditions (see [17]):
(i) $1 \in F$;
(ii) for all $x, y, z \in L, x \to (y \to z) \in F, x \to y \in F \Rightarrow x \to z \in F$.
A subset F of L is called a positive implicative filter of L if it satisfies the following conditions:
(i) $1 \in F$;
(ii) for all $x, y, z \in L, x \in F, x \to ((y \to z) \to y) \in \mathrm{F} \Rightarrow y \in F$.
A subset F of L is called a fantastic filter of L if it satisfies the following conditions:
(i) $1 \in F$;
(ii) for all $x, y, z \in L, z \in F, z \to (y \to x) \in F \Rightarrow ((x \to y) \to y) \to x \in F$.
A fuzzy set of L is a function $f\colon L \to [0, 1]$. For a fuzzy set f of L and $t \in [0, 1]$, the crisp set $f_t = \{x \in L \mid f(x) \geq t\}$ is called a level subset of f (see [10]).
A fuzzy set of f of L is called a fuzzy filter of L if it satisfies (see [17]):
$(F1) f(1) \geq f(x)$ for all $x \in L$.
$(F2) f(y) \geq f(x \to y) \wedge f(x)$ for all $x, y \in L$.

Lemma 2 *Let f be a fuzzy filter of L, for any $x, y \in L$, the following hold (see [17]):*
(1) *If* $x \leq y$, *then* $f(x) \leq f(y)$;
(2) *If* $f(x \to y) = f(1)$, *then* $f(x) \leq f(y)$.

A fuzzy set f of L having the form

$$f(y) = \begin{cases} t(\neq 0), & \text{if } y = x, \\ 0, & \text{if } y \neq x. \end{cases} \tag{1}$$

is said to be fuzzy point with support x and value t and is denoted by $U(x; t)$. A fuzzy point $U(x; t)$ is said to belong to (resp. be quasi-coincident with) a fuzzy set f, written as $U(x;t) \in f$ (resp. $U(x;t)qf$) if $f(x) \geq t$ (resp. $f(x)+t > 1$). If $U(x;t) \in f$ or (resp. and) $U(x;t)qf$, then we write $U(x;t) \in \vee q$ (resp. $\in \wedge q$). The symbol $\overline{\in \vee q}$ means that $\in \vee q$ does not hold. Using the notion of "membership ($\in$)" and "quasi-coincidence (q)" of fuzzy points with fuzzy subsets, we obtain the concepts of (α, β)-fuzzy subsemigroup, where α and β are any two of $(\in, q, \in \vee q, \in \wedge q)$ with $\alpha \neq \in \wedge q$. As a generalization of (α, β)-fuzzy subsemigroup, we introduce the concepts positive implicative filter of (α, β)-fuzzy filters of L.

3 (α, β)-Fuzzy Filter

Definition 2 A fuzzy subset f of L is called an (α, β)-fuzzy filters of L if for all $r, t \in (0, 1]$ and $x, y, z \in L$,
(F3) $U(x;r)\alpha f \Rightarrow U(1;r)\beta f$;
(F4) $U(x;r)\alpha f$, $U(x \rightarrow y)\alpha f \Rightarrow U(y; \min\{r,t\})\beta f$. Where $\alpha, \beta \in \{\in, q, \in \vee q, \in \wedge q\}$ with $\alpha \neq \in \wedge q$.

Note that if f is a fuzzy set in L defined by $f(x) \leq 0.5$ for all $x \in L$, the set $\{U(x;t) \mid U(x;t) \in \wedge qf\}$ is empty. (i) (α, β)-fuzzy implicative filter of L if it satisfies $(F3)$ and $(F5) U(x \rightarrow (y \rightarrow z); r)\alpha f$ and $U(x \rightarrow y; t)\alpha f \Rightarrow U(x \rightarrow z; \min\{r,t\})\beta f$.
(ii) (α, β)-fuzzy positive implicative filter of L if it satisfies $(F3)$ and $(F6) U(x;r)\alpha f$ and $U(x \rightarrow ((y \rightarrow z) \rightarrow y); t)\alpha f \Rightarrow U(y; \min\{r,t\})\beta f$.
(iii) (α, β)-fuzzy fantastic filter of L if it satisfies $(F3)$ and $(F7) U(z \rightarrow (y \rightarrow x); r)\alpha f$ and $U(z;t)\alpha f \Rightarrow U(((x \rightarrow y) \rightarrow y) \rightarrow x; \min\{r,t\})\beta f$

Theorem 1 *Any (α, β)-fuzzy implicative filter of L is an (α, β)-fuzzy filter.*

Proof In $(F5)$, Let $x = 1, y = x, z = y$. By $U(1 \rightarrow (x \rightarrow y); r)\alpha f$ and $U(1 \rightarrow x; t)\alpha f$, we have $U(1 \rightarrow y; \min\{r,t\})\beta f$. That is, $U(x \rightarrow y; r)\alpha f$ and $U(x;t)\alpha f$, we have $U(y; \min\{r,t\})\beta f$. Hence f is an (α, β)-fuzzy filter. □

Similarly, we have the following result.

Theorem 2 *Every (α, β)-fuzzy positive implicative (fantastic) filter of L is an (α, β)-fuzzy filter.*

Theorem 3 *If f is an (α, β)-fuzzy filter of L, then f is an $(\alpha, \in \vee q)$-fuzzy filter where $\alpha, \beta \in \{\in, q, \in \vee q, \in \wedge q\}$ and $\alpha \neq \in \wedge q$.*

Proof Assume that f is an (α, β)-fuzzy filter of L. Let $x, y \in L$ and $r, t \in (0, 1]$ satisfy $U(x;r)\alpha f$ and $U(x \rightarrow y; t)\alpha f$. Then we have the following cases:

Case 1: $\beta = \in$. Then $U(1;r) \in f$ and $U(y; \min\{r,t\}) \in f$. Thus $U(1;r) \in \vee qf$ and $U(y; \min\{r,t\}) \in \vee qf$.

Case 2: $\beta = q$. Then $U(1;r)qf$ and $U(y; \min\{r,t\})qf$. Thus $U(1;r) \in \vee qf$ and $U(y; \min\{r,t\}) \in \vee qf$.

Case 3: $\beta = \in \vee q$. This completes the proof.

Case 4: $\beta = \in \wedge q$. Then $U(1; r) \in \wedge qf$ and $U(y; \min\{r, t\}) \in \wedge qf$. Thus $U(1; r) \in \vee qf$ and $U(y; \min\{r, t\}) \in \vee qf$. The proof is complete. □

Similarly, we have:

Theorem 4 *Every* (α, β)*-fuzzy (implicative, positive implicative, fantastic) filter of* L *is an* $(\alpha, \in \vee q)$*- fuzzy (implicative, positive implicative, fantastic) filter.*

Theorem 5 *Every* $(\in \vee q, \beta)$*-fuzzy filter is an* (α, β)*-fuzzy filter.*

Proof Suppose that f is an $(\in \vee q, \beta)$-fuzzy filter and $U(x; r)\alpha f, U(x \to y; t)\alpha f$. Then $U(x; r) \in \vee qf, U(x \to y; t) \in \vee qf$. Thus $U(1; r)\beta f, U(y; \min\{r, t\})\beta f$. This shows that f is an (α, β)-fuzzy filter. □

Theorem 6 *Every* $(\in \vee q, \beta)$*-fuzzy (implicative, positive implicative, fantastic) filter is an* (α, β)*-fuzzy (implicative, positive implicative, fantastic) filter.*

Proof The proof is similar to that of Theorem 5. □

Theorem 7 *Every* $(\alpha, \in \wedge q)$*-fuzzy filter of* L *is an* (α, β)*- fuzzy filter.*

Proof Suppose that f is an $(\alpha, in \wedge q)$-fuzzy filter and $U(x; r)\alpha f, U(x \to y; t)\alpha f$. Then $U(x; r) \in \wedge qf, U(y; \min\{r, t\}) \in \wedge qf$. Thus $U(1; r)\beta f, U(y; \min\{r, t\})\beta f$. This shows f is an (α, β)-fuzzy filter. □

Similarly, we have the following result.

Theorem 8 *Every* $(\alpha, \in \wedge q)$*-fuzzy (implicative, positive implicative, fantastic) filter is an* (α, β)*-fuzzy (implicative, positive implicative, fantastic) filter.*

4 $(\in, \in)$-Fuzzy Filters

In this section, let $\alpha = \in, \beta = \in$ in Definition 2. Firstly, we introduce some concepts of fuzzy filters.

Definition 3 A fuzzy set f of L is called a fuzzy implicative filter of L if it satisfies $(F1)$ and (see [17])
(F8) $f(x \to z) \geq f(x \to (y \to z)) \wedge f(x \to y)$ for all $x, y, z \in L$.

Definition 4 A fuzzy set f of L is called a fuzzy positive implicative filter of L if it satisfies $(F1)$ and (F9) $f(y) \geq f(x) \wedge f(x \to ((y \to z) \to y))$ for all $x, y, z \in L$.

Definition 5 A fuzzy set f of L is called a fuzzy fantastic filter of L if it satisfies $(F1)$ and (F10) $f(((x \to y) \to y) \to x) \geq f(z \to (y \to x)) \wedge f(z)$ for all $x, y, z \in L$.

Example 1 Let $L = \{0, a, b, c, 1\}$ be a chain with the following Cayley table:

x	x'	$\to$	0	a	b	c	1
0	1	0	1	1	1	1	1
a	c	a	c	1	1	1	1
b	b	b	b	b	1	1	1
c	a	c	a	a	b	1	1
1	0	1	0	a	b	c	1

Define the $\vee$ and $\wedge$ operations on L as min and max respectively. Then $(L, \vee, \wedge, \odot, \to)$ is an R_0-algebra. Now, define a fuzzy set f in L by $f(1) = f(c) = 0.4$, $f(0) = f(a) = f(b) = 0.2$. It is easy to check that f is a fuzzy fantastic filter.

Example 2 Suppose L$= \{0, a, b, c, d, 1\}$ is a chain with the following Cayley table:

x	x'	$\to$	0	a	b	c	d	1
0	1	0	1	1	1	1	1	1
a	d	a	d	1	1	1	1	1
b	c	b	c	c	1	1	1	1
c	b	c	b	b	b	1	1	1
d	a	d	a	a	b	c	1	1
1	0	1	0	a	b	c	d	1

Define the $\vee$ and $\wedge$ operations on L as min and max respectively. Then L is an R_0-algebra. Define a fuzzy subset f on L by $f(1) = f(d) = f(c) = 0.8$, $f(0) = f(a) = f(b) = 0.3$. Then f is not only a fuzzy implicative but also a fuzzy positive implicative filter.

Theorem 9 *Every $(\in, \in)$-fuzzy filter is a fuzzy filter and vice versa.*

Proof Let f be a fuzzy filter of L and $U(x; t) \in f$. Then $f(x) \geq r$. By $(F1)$, we have $f(1) \geq r$, thus $U(1; r) \in f$. Suppose $U(x; r) \in f$ and $U(x \to y; t) \in f$. Then $f(x) \geq r$ and $f(x \to y) \geq t$. It follows from $(F2)$ that: $f(y) \geq f(x) \wedge f(x \to y) \geq \min\{r, t\}$. Thus $U(y; \min\{r, t\}) \in f$.

Conversely, let f be an $(\in, \in)$-fuzzy filter of L and $f(1) < f(x)$. Then $\exists s \in (0, 1]$ such that $f(1) < s < f(x)$. This shows that $f(x) \geq s$ but $f(1) < s$. This contradicts with $(F3)$. So we have $f(1) \geq f(x)$. Note that $U(x; r) \in f$ and $U(x \to y; t) \in f$ for all $x, y \in L$, where $r = f(x), t = f(x \to y)$. It follows from $(F4)$ that $U(y; \min\{r, t\}) \in f)$. That is, $f(y) \geq \min\{r, t\} = f(x) \wedge f(x \to y)$. □

Theorem 10 *Every $(\in, \in)$-fuzzy implicative filter of L is a fuzzy implicative filter and vice versa.*

Proof $(F1) \Leftrightarrow (F3)$ See Theorem 9.

$(F2) \Leftrightarrow (F4)$ Assume that f satisfies $(F2)$. Let $x, y, z \in L$ and $r, t \in (0, 1]$ satisfy $U(x \to (y \to z); r) \in f$ and $U(x \to y; t) \in f$. Then $f(x \to (y \to z)) \geq r$

and $f(x \to y) \geq t$. It follows from $(F2)$ that $f(x \to z) \geq f(x \to (y \to z)) \wedge f(x \to y) \geq \min\{r, t\}$. This shows that $U(x \to z; \min\{r, t\}) \in f$. Now suppose that $(F4)$ is valid. Note that $U(x \to (y \to z); r) \in f$ and $U(x \to y; t) \in f$ for all $x, y, z \in L$ where $r = f(x \to (y \to z)), t = f(x \to y)$. It follows from $(F4)$ that $U(x \to z; \min\{r, t\}) \in f$. This shows that $f(x \to z) \geq \min\{r, t\} = f(x \to (y \to z)) \wedge f(x \to y)$. This completes the proof. □

Similarly, we have,

Theorem 11 *Every $(\in, \in)$-fuzzy positive implicative (fantastic) filter is a fuzzy positive implicative (fantastic) filter and vice versa.*

By Theorems 1, 10 and 11, we have the following result.

Theorem 12 *Every fuzzy (implicative, positive implicative, fantastic) filter of L is a fuzzy filter.*

Theorem 13 *f is a fuzzy implicative filter of L if and only if f is a fuzzy positive implicative filter (see [17]).*

Corollary 1 *f is an $(\in, \in)$-fuzzy implicative filter if and only if f is an $(\in, \in)$-fuzzy positive implicative filter.*

Theorem 14 *Let f be a fuzzy filter of L, f is a fuzzy fantastic filter if and only if it satisfies: $(F11) f(((x \to y) \to y) \to x) \geq f(y \to x)$ for all $x, y \in L$.*

Proof Let f be a fuzzy filter. Taking $z = 1$ in $(F10)$, we have $f(((x \to y) \to y) \to x) \geq f(1 \to (y \to x)) \wedge f(1) = f(y \to x) \wedge f(1) = f(y \to x)$.

Conversely, since f is a fuzzy filter, then $f(z \to (y \to x)) \wedge f(z) \leq f(y \to x)$ for all $x, y, z \in L$. By $(F11)$, we have $f(z \to (y \to x)) \wedge f(z) \leq f(((x \to y) \to y) \to x)$. The proof is completed. □

Theorem 15 *Let f be a fuzzy filter of L, f is a fuzzy positive implicative filter if and only if it satisfies:*
(F12) $f(y) = f((y \to z) \to y)$ *for all* $y, z \in L$.

Proof Assume that f is a fuzzy positive implicative filter, we have $f(y) \geq f(1) \wedge f(1 \to ((y \to z) \to y)) = f((y \to z) \to y)$. On the other hand, $y \leq (y \to z) \to y$, therefore $(F12)$ holds.

Conversely, suppose that f is a fuzzy filter. We have $f(x) \wedge f(x \to ((y \to z) \to y)) \leq f((y \to z) \to y) \leq f(y)$. This shows that f is a fuzzy positive implicative filter. □

Theorem 16 *Every fuzzy positive implicative filter of L is a fuzzy fantastic filter of L.*

Proof Suppose that f is a fuzzy positive implicative filter of L. Since $x \odot ((x \to y) \to y) \leq x$, then $x \leq (((x \to y) \to y) \to x)$. By Lemma 1, we have $(((x \to y) \to y) \to x) \to y \leq x \to y$, thus $((((x \to y) \to y) \to x) \to y) \to (((xy) \to y) \to x) \geq (x \to y) \to (((x \to y) \to y) \to x) = ((x \to y) \to y) \to ((x \to y) \to x) \geq y \to x$. Thus $f((((x \to y) \to y) \to x) \to y) \to (((x \to y) \to y) \to x) \geq f(y \to x)$. By hypothesis and Theorem 15, we have $f(((x \to y) \to y) \to x) \geq f(y \to x)$. By Theorem 14, f is a fuzzy fantastic filter of L. □

The converse of Theorem 16 may not be true. In Example 1, $f((b \to 0) \to b) = f(b \to b) = f(1) \neq f(b)$, By Theorem 15, we have that f is not a fuzzy positive implicative filter.

Corollary 2 *Every* $(\in, \in)$*-fuzzy positive implicative filter of L is an* $(\in, \in)$*-fuzzy fantastic filter.*

Lemma 3 *f is a fuzzy filter of L if and only if for each $t \in (0, 1]$, f_t is either empty or a filter of L (see [17]).*

Theorem 17 *Let f be a fuzzy filter in L, f is a fuzzy positive implicative filter if and only if for each $t \in (0, 1]$, f_t is either empty or a positive implicative filter of L.*

Proof Suppose that f is a fuzzy positive implicative filter and for each $t \in (0, 1]$, $f_t \neq \emptyset$. Then f_t is a filter by Lemma 3. Thus $1 \in f_t$. Suppose $x \in f_t, x \to ((y \to z) \to y) \in f_t$. That is $f(x) \geq t, f(x \to ((y \to z) \to y)) \geq t$. Hence $f(y) \geq f(x) \wedge f(x \to ((y \to z) \to y)) \geq t$, That is, $y \in f_t$. This shows that f_t is a positive implicative filter. Conversely, suppose that f is a fuzzy filter and for each $t \in (0, 1]$, f_t is either empty or a positive implicative filter of L. Let $t = f(x) \wedge f(x \to ((y \to z) \to y))$, then $x \in f_t, x \to ((y \to z) \to y) \in f_t$. Since f_t is a positive implicative filter, we have $y \in f_t$, and so $f(y) \geq t = f(x) \wedge f(x \to ((y \to z) \to y))$. The proof is complete. □

Theorem 18 *Let f be a fuzzy filter in L, f is a fuzzy fantastic filter if and only if for each $t \in (0, 1]$, f_t is either empty or a fantastic filter of L.*

Proof The proof is similar to the proof of Theorem 17. □

Theorem 19 *Let f be a fuzzy filter of L. f is an $(\in, \in)$-fuzzy (positive implicative, fantastic) filter if and only if for each $t \in (0, 1]$, f_t is either empty or a (positive implicative, fantastic) filter of L.*

5 $(\in, \in \vee Q)$-Fuzzy Filters

In this section, let $\alpha = \in$, $\beta = \in \vee q$ in Definition 2.

Example 3 In Example 1, define a fuzzy set f by $f(1) = 0.6$, $f(c) = 0.7$, $f(0) = f(a) = f(b) = 0.3$. It is routine to verify that F is an $(\in, \in \vee q)$-fuzzy fantastic filter.

Example 4 In Example 2, define a fuzzy set f by $f(1) = 0.6$, $f(d) = f(c) = 0.8$, $f(0) = f(a) = f(b) = 0.4$. It is easy to check that F is an $(\in, \in \vee q)$-fuzzy positive implicative filter.

By Theorems 3 and 9, we have

Theorem 20 *Every fuzzy filter is an* $(\in, \in \vee q)$*-fuzzy filter.*

Theorem 21 f *is an* $(\in, \in \vee q)$*-fuzzy filter of L if and only if:*
(F13) $f(1) \geq \min\{f(x), 0.5\}$;
(F14) $f(y) \geq \min\{f(x), f(x \to y), 0.5\}$.

Proof $(F3) \Rightarrow (F13)$

Suppose $f(1) < \min\{f(x), 0.5\}$. There have two cases:

Case 1 $f(x) < 0.5$, then $\min\{f(x), 0.5\} = f(x)$, thus $f(1) < f(x)$ and $f(1) + f(x) < 1$. By $f(x) \geq f(x)$ and $(F3)$. We should have $U(1; f(x)) \in \vee qf$. i.e., $f(1) \geq f(x)$ and $f(1) + f(x) > 1$. A contradiction.

Case 2 $f(x) \geq 0.5$. Then $\min\{f(x), 0.5\} = 0.5$. Thus $f(1) < 0.5$ and $f(1) + 0.5 < 1$. By $(F3)$, we should have $U(1; 0.5) \in \vee qf$. i.e., $f(1) \geq 0.5$ and $f(1) + 0.5 > 1$. A contradiction.

$(F13) \Rightarrow (F3)$

Assume $U(x; r) \in f$. Then $f(x) \geq r$. By $(F13)$, we have $f(1) \geq \min\{f(x), 0.5\} \geq \min\{r, 0.5\}$. If $r > 0.5$, then $f(1) + r > 1$; If $r \leq 0.5$, then $f(1) \geq r$. This shows $U(1; r) \in \vee qf$.

$(F4) \Rightarrow (F14)$

Let $t_0 = \min\{f(x), f(x \to y), 0.5\}$. We have $f(x) \geq t_0$, $f(x \to y) \geq t_0$. That is, $U(x; t_0) \in f$ and $U(x \to y; t_0) \in f$. By $(F4)$, we have $U(y; \min\{t_0, t_0\}) \in \vee qf$. Then $f(y) \geq t_0$ or $f(y) + t_0 > 1$. If $f(y) + t_0 > 1$, note that $t_0 \leq 0.5$. Then $f(y) > 1 - t_0 \geq t_0$. Thus $f(y) \geq t_0$. This shows $f(y) \geq t_0 = \min\{f(x), f(x \to y), 0.5\}$.

$(F14) \Rightarrow (F4)$

Suppose $f(x) \geq r$, $f(x \to y) \geq t$. By $(F14)$, $f(y) \geq \min\{f(x), f(x \to y), 0.5\} \geq \min\{r, t, 0.5\}$. If $\min\{r, t\} > 0.5$, then $f(y) \geq 0.5$. Thus $f(y) + \min\{r, t\} > 1$; If $\min\{r, t\} \leq 0.5$, then $f(y) \geq \min\{r, t\}$. This shows $U(y; \min\{r, t\}) \in \vee qf$. □

Similarly, we have,

Theorem 22 *Let* f *be an* $(\in, \in \vee q)$*-fuzzy (implicative, positive implicative, fantastic) filter of L if and only if it satisfies* $(F13)$ *and* (F15) $f(x \to z) \geq \min\{f(x \to (y \to z)), f(x \to y), 0.5\}$; (F16) $f(y) \geq \min\{f(x), f(x \to ((y \to z) \to y)), k\ 0.5\}$; (F17) $f(((x \to y) \to y) \to x) \geq \min\{f(z \to (y \to x)), f(z), 0.5\}$.

Theorem 23 f *is an* $(\in, \in \vee q)$*-fuzzy implicative filter of L if and only if* f *is an* $(\in, \in \vee q)$*-fuzzy positive implicative filter (see* [15]*).*

Theorem 24 *Let* f *be an* $(\in, \in \vee q)$*-fuzzy implicative filter of L, then* f *is an* $(\in, \in \vee q)$*-fuzzy fantastic filter.*

Proof Suppose that f is an $(\in, \in \vee q)$-fuzzy implicative filter. Then f is an $(\in, \in \vee q)$-fuzzy filter. By $(F14)$, we have $\min\{f(z), f(z \to (y \to x)), 0.5\} = \min\{f(z), f(z \to (y \to x)), 0.5, 0.5\} \le \min\{f(y \to x), 0.5\}$. By the proof of Theorem 16, we have $y \to x \le ((x \to y) \to y) \to x$, Then $\min\{f(y \to x), 0.5\} \le f(((x \to y) \to y) \to x)$ (see Theorems 3.3 and 3.6 in [15]). Thus $\min\{f(z), f(z \to (y \to x)), 0.5\} \le f(((x \to y) \to y) \to x)$. □

Theorem 25 *A fuzzy set of L is an $(\in, \in \vee q)$-fuzzy (positive implicative, fantastic) filter of L if and only if f_t is either empty or a (positive implicative, fantastic) filter for all $t \in (0, 0.5]$.*

Proof The proof is similar to that of Theorem 3.10 in [15]. □

Theorem 26 *If f is an $(\in, \in \vee q)$-fuzzy positive implicative filter of L with $f(1) < 0.5$, then f is a fuzzy positive implicative filter.*

Proof Suppose that f is an $(\in, \in \vee q)$-fuzzy positive implicative filter with $f(1) < 0.5$. Then we have $f(1) \ge f(x)$. If $f(1) < f(x)$, then $f(1) < \min\{f(x), 0.5\}$. This contradicts with $(F13)$. On the other hand, $f(y) \ge \min\{f(x), f(x \to ((y \to z) \to y)), 0.5\} \ge \min\{f(x), f(x \to ((y \to z) \to y))\}$ by $f(x) \le f(1) < 0.5$. This shows that f is a fuzzy positive implicative filter. □

Theorem 27 *If f is an $(\in, \in \vee q)$-fuzzy fantastic filter of L with $f(1) < 0.5$, then f is a fuzzy fantastic filter.*

6 $(q, \in \vee Q)$-Fuzzy Filters

In this section, let $\alpha = q, \beta = \in \vee q$ in Definition 2.

Theorem 28 *Let F be a filter of L and f a fuzzy set in L, such that*
(i) $\forall x \in L \setminus F,\ f(x) = 0$;
(ii) $\forall x \in F,\ f(x) \ge 0.5$.
Then f is a $(q, \in \vee q)$-fuzzy filter.

Proof Let $x \in L$ and $r \in (0, 1]$ satisfy $U(x; r)qf$. Then $f(x) + r > 1$ and so $x \in F$. Since $1 \in F$, Thus $f(1) \ge 0.5$. There have two cases. If $r > 0.5$, then $f(1) + r > 0.5 + 0.5 = 1$; If $r \le 0.5$, then $f(1) \ge 0.5 \ge r$. So that $U(1; r) \in \vee qf$. Let $x, y \in L$ and $r, t \in (0, 1]$ satisfy $U(x; r)qf$ and $U(x \to y; t)qf$. i.e., $f(x) + r > 1$ and $f(x \to y) + t > 1$. Then $x \in F$ and $x \to y \in F$. Since F is a filter, it follows that $y \in F$. So that $f(y) \ge 0.5$. If $r \le 0.5$ or $t \le 0.5$, then $f(y) \ge 0.5 \ge \min\{r, t\}$. Hence $U(y; \min\{r, t\}) \in f$. If $r > 0.5$ and $t > 0.5$, then $f(y) + \min\{r, t\} > 0.5 + 0.5 = 1$ and so $U(y; \min\{r, t\})qf$. Consequently $U(y; \min\{r, t\}) \in \vee qf$. Therefore f is a $(q, \in \vee q)$-fuzzy filter of L. □

Theorem 29 *In Theorem 28, if F is an implicative (positive implicative, fantastic) filter of L, then f is a $(q, \in \vee q)$-fuzzy (implicative, positive implicative, fantastic) filter of L.*

7 $(\alpha, \in \wedge Q)$-Fuzzy Filters

In this section, let $\beta = \in \wedge q$ in Definition 2.

Theorem 30 *f is an $(\in, \in \wedge q)$-fuzzy filter if and only if f is an $(\in, \in)$-fuzzy filter and $(\in, q)$-fuzzy filter.*

Proof Assume that f is an $(\in, \in \wedge q)$-fuzzy filter. By Theorem 7, we have that f is an $(\in, \in)$-fuzzy filter and $(\in, q)$-fuzzy filter. On the other hand, suppose that f is not only an $(\in, \in)$-fuzzy filter but also an $(\in, q)$-fuzzy filter and $U(x; r) \in f$, $U(x \rightarrow y; t) \in f$. Then $U(1; r) \in f$ and $U(1; r)qf$. Thus $U(1; r) \in \wedge q$, $U(y; \min\{r, t\}) \in f$ and $U(y; \min\{r, t\})qf$. Hence $U(y; \min\{r, t\}) \in \wedge q$. This shows that f is an $(\in, \in \wedge q)$-fuzzy filter. □

Similarly, we have,

Theorem 31 *f is an $(\in, \in \wedge q)$-fuzzy (implicative, positive implicative, fantastic) filter if and only if f is an $(\in, \in)$-fuzzy (implicative, positive implicative, fantastic) filter and $(\in, q)$-fuzzy (implicative, positive implicative, fantastic) filter.*

Theorem 32 *f is an $(q, \in \wedge q)$-fuzzy (implicative, positive implicative, fantastic) filter if and only if f is an $(q, \in)$-fuzzy (implicative, positive implicative, fantastic) filter and (q, q)-fuzzy (implicative, positive implicative, fantastic) filter.*

Theorem 33 *f is an $(\in \vee q, \in \wedge q)$-fuzzy (implicative, positive implicative, fantastic) filter if and only if f is an $(\in \vee q, \in)$-fuzzy (implicative, positive implicative, fantastic) filter and $(\in \vee q, q)$-fuzzy (implicative, positive implicative, fantastic) filter.*

8 Conclusions

We introduced the concepts of (α, β)-fuzzy (implicative, positive implicative, fantastic) filter in R_0- algebras, where α, β are any two of $\{\in, q, \in \vee q, \in \wedge q\}$ with $\alpha \neq \in \wedge q$. We investigated relations between $(\in, \in)$-fuzzy (implicative, positive implicative, fantastic) filter and fuzzy (implicative, positive implicative, fantastic) filter. We established characterizations of $(\in, \in \vee q)$-fuzzy (implicative, positive implicative, fantastic) filter and provide conditions for an $(\in, \in \vee q)$-fuzzy (implicative, positive implicative, fantastic) filter to be an $(\in, \in)$-fuzzy (implicative, positive implicative, fantastic) filter. We give the conditions for a fuzzy set to be a $(q, \in \vee q)$-fuzzy (implicative, positive implicative, fantastic) filter. And we discuss the relations of $(\alpha, \in \wedge q)$-fuzzy filters where $\alpha \in \{\in, q, \in \vee q\}$.

Acknowledgments This work is supported by the National Natural Science Foundation of China, Grant No. 11101135, 11371130.

References

1. Borns, D.W., Mack, J.M.: An Algebraic Introduction to Mathematical Logic. Springer, Berlin (1975)
2. Xu, Y.: Lattice implication algebras. J. Southwest Jiaotong Univ. **1**, 20–27 (1993)
3. Xu, Y., Qin, K.Y.: On filters of lattice implication algebras. J. Fuzzy Math. **2**, 251–260 (1993)
4. Xu, Y., Qin, K.Y.: Fuzzy lattice implication algebras. J. Southwest Jiaotong Univ. **2**, 121–127 (1995)
5. ájek, P.H.: Metamathematics of Fuzzy Logic. Kluwer Academic Publishers, Dordrecht (1998)
6. Turunen, E.: BL-algebras of basic fuzzy logic. Math. Soft. Comput. **6**, 49–61 (1999)
7. Turunen, E.: Boolean deductive systems of BL algebras. Arch. Math. Logic **40**, 467–473 (2001)
8. Wang, G.J.: Non-classical Mathematical Logic and Approximate Reasoning. Science Press, Beijing (2000)
9. Wang, G.J.: On the logic foundation of fuzzy reasoning. Inf. Sci. **117**, 47–88 (1999)
10. Zageh, L.A.: Fuzzy Sets Inf. Control **8**, 338–353 (1965)
11. Zadeh, L.A.: The concept of a lingistic variable and its application to approximate reason. Inf. Control **18**, 199–249 (1975)
12. Biswas, R.: Rosenfeld's fuzzy subgroups with interval valued membership functions. Fuzzy Sets Syst. **63**, 87–90 (1994)
13. Biswas, R., Das, P.: $(\in, \in \vee q)$-fuzzy subgroups. Fuzzy Sets Syst. **80**, 359–368 (1996)
14. Pu, P.M., Liu, Y.M.: Fuzzy topology I: neighourhood structure of a fuzzy point and Moore–Smith convergence. J. Math. Anal. Appl. **76**, 571–599 (1980)
15. Ma, X.L., Zhan, J.M., Jun, Y.B.: On $(\in, \in \vee q)$-fuzzy filters of R0-algebras. Math. Log. Quart. **55**, 493–508 (2009)
16. Pei, D.W., Wang, G.J.: The completeness and application of the formal system L^*. Sci. China, Series F **45**, 40–50 (2002)
17. Liu, L.Z., Li, K.T.: Fuzzy implicative and Boolean filters of R0-algebras. Inf. Sci. **171**, 61–71 (2005)

On Generalized Annihilators in BL-Algebras

Yu-Xi Zou, Xiao-Long Xin and Young-Bae Jun

Abstract The theory of generalized annihilators on BL-algebras are developed in this paper. Firstly, some properties of generalized annihilators on BL-algebras are supplemented. Secondly, we introduce the notion of involutory ideals relative to an ideal I and denote the set of all of them by $S_I(L)$. Then $S_I(L)$ can be made into a complete Boolean lattice and a BL-algebra with respect to the suit operations, respectively. Finally, the prime ideals can be characterized by the generalized annihilators, and the generalized annihilators of the quotient algebra induced by an ideal I in a BL-algebra L are studied.

Keywords BL-algebra · Generalized annihilator · Involutory ideal · Prime ideal

1 Introduction

In order to study the basic logic framework of fuzzy set system, based on continuous triangle norm and under the theoretical framework of residuated lattices theory, Hǻjek [1] proposed a new fuzzy logic system−BL-system and the corresponding logic algebraic system−BL-algebra. And MV-algebras were introduced by Chang [2] to give an algebraic proof of the completeness theorem of Lukasiewice system of many valued logic.

Y.-X. Zou · X.-L. Xin (✉) · Y.-B. Jun
School of Mathematics, Northwest University, Xi'an 710127,
People's Republic of China
e-mail: xlxin@nwu.edu.cn

Y.-X. Zou
e-mail: 616298751@qq.com

Y.-B. Jun
Department of Mathematics Education, Gyeongsang National University,
Chinju 660-701, Korea
e-mail: skywine@gmail.com

T.-H. Fan et al. (eds.), *Quantitative Logic and Soft Computing 2016*,
Advances in Intelligent Systems and Computing 510,
DOI 10.1007/978-3-319-46206-6_43

Ideals are a very effective tool for studying algebraic and logical systems. In the theory of MV-algebras the notion of ideals is central one and deductive systems and ideals are dual notions, while in BL-algebra, due to lack of suitable algebraic addition, the focus is shifted to deductive systems also called filters [3–9]. So the notion of ideals is missing in BL-algebras. To fill this gap [10] introduced the notion of ideals in BL-algebras, which generalized in a natural sense the existing notion in MV-algebras and subsequently all results about ideals in MV-algebras. The paper also constructed some examples to show that ideals and filters behave quite differently in BL-algebra. So the notion of ideal from a view of purely algebraic point has a proper meaning in BL-algebras.

Much work has been done with respect to annihilators and co-annihilators. For example, in [11], Davery studied the relationship between minimal prime ideals conditions and annihilators conditions on distributive lattices. Turunen [12] defined co-annihilator of a non-empty set X of L and proved some of its properties on BL-algebras, they got that $A^{\perp}$ is a prime filter if and only if A is linear and $A \neq \{1\}$. Also, in [13] B.A. Laurentiu Leustean introduced the notion of co-annihilator of A relative to F on pseudo-BL-algebras, which is a generalization of co-annihilator, and they also extended some results obtained in [11]. Moreover, in [14], Zou Y.X. et al. introduced the notion of annihilator and generalized annihilator on BL-algebras. Now, we further study generalized annihilator of BL-algebras based on [14].

2 Preliminaries

Definition 1 ([1]) An algebra structure $(L, \wedge, \vee, \odot, \rightarrow, 0, 1)$ of type $(2, 2, 2, 2, 0, 0)$ is called a BL-algebra, if it satisfies the following conditions: for all $x, y, z \in L$

$(BL1)$ $(L, \wedge, \vee, 0, 1)$ is a bounded lattice relative to the order $\leq$;
$(BL2)$ $(L, \odot, 1)$ is a commutative monoid;
$(BL3)$ $x \odot y \leq z$ if and only if $x \leq y \rightarrow z$;
$(BL4)$ $x \wedge y = x \odot (x \rightarrow y)$;
$(BL5)$ $(x \rightarrow y) \vee (y \rightarrow x) = 1$.

For each $x \in L$ and a natural number n, we define $\overline{x} = x \rightarrow 0$, $\overline{\overline{x}} = \overline{(\overline{x})}$, $x^0 = 1$ and $x^n = x^{n-1} \odot x$ for $n \geq 1$. For every $x, y \in L$, we adopt the notation: $x \oslash y = \overline{x} \rightarrow y$.

Proposition 1 ([10]) *In every BL-algebra L, the following hold:*
(1) *the operation $\oslash$ is associative, that is, for every $x, y, z \in L$, $(x \oslash y) \oslash z = x \oslash (y \oslash z)$;*
(2) *the operation $\oslash$ is compatible with the order, that is, for every $x, y, z, t \in L$, such that $x \leq y$ and $z \leq t$, then $x \oslash z \leq y \oslash t$.*

Definition 2 ([10]) Let L be a BL-algebra and I be a nonempty subset of L. We say that I is an ideal of L if it satisfies

(I1) for all $x, y \in L$, if $x \leq y$ and $y \in I$, then $x \in I$;
(I2) for all $x, y \in I$, $x \oslash y \in I$.

Proposition 2 ([10]) *An ideal P of a BL-algebra L is prime if and only if for all $x, y \in L$, $x \wedge y \in P$ implies that $x \in P$ or $y \in P$.*

Let L be a BL-algebra, $\emptyset \neq A \subseteq L$ and I be an ideal of L. Denote $A^{\perp} = \{x \in L \mid a \wedge x = 0, \text{for all } a \in A\}$, which is called an annihilator of A and $A_I^{\perp} = \{x \in L \mid a \wedge x \in I, \text{for all } a \in A\}$ in [14]. We call $A_I^{\perp}$ a generalized annihilator of A relative to I in this paper. Moreover, we denote $\{a\}_I^{\perp}$ by $a_I^{\perp}$.

Proposition 3 *Let L be a BL-algebra, $I, J \in I(L)$ and $\emptyset \neq X, X' \subseteq L$. Then we have*
(1) $I \subseteq J$ *implies* $X_I^{\perp} \subseteq X_J^{\perp}$;
(2) $X \subseteq X'$ *implies* $(X')_I^{\perp} \subseteq X_I^{\perp}$;
(3) $(\cup_{\lambda \in \Lambda} X_\lambda)_I^{\perp} = \cap_{\lambda \in \Lambda} (X_\lambda)_I^{\perp}$;
(4) $X_I^{\perp} = \cap_{x \in X} x_I^{\perp}$;
(5) $X_{\cap_{\lambda \in \Lambda} I_\lambda}^{\perp} = \cap_{\lambda \in \Lambda} X_{I_\lambda}^{\perp}$;
(6) $(X]_I^{\perp} = X_I^{\perp}$.

Proof Similar to Proposition 3.24 in [14] □

Proposition 4 ([14]) *Let L be a BL-algebra, $\emptyset \neq X \subseteq L$ and I be an ideal of L. Then*
(1) $I \subseteq X_I^{\perp}$;
(2) $X_I^{\perp} = L$ *if and only if* $X \subseteq I$.

Proposition 5 ([14]) *Let L be a BL-algebra, I, J and $H \in I(L)$. Then we have*
(1) $J_I^{\perp} \cap J \subseteq I$;
(2) $J \cap H \subseteq I$ *if and only if* $H \subseteq J_I^{\perp}$.

3 Generalized Annihilators

In this section, we focus on generalized annihilators of BL-algebras. Using generalized annihilators we define involutory ideals relative to an ideal I and research the structure of the set of all involutory ideals relative to an ideal I.

Proposition 6 *Let L be a BL-algebra, $\emptyset \neq A \subseteq L$, I be an ideal of L. Then the following hold:*
(1) $A \subseteq (A_I^{\perp})_I^{\perp}$;
(2) $A_I^{\perp} = ((A_I^{\perp})_I^{\perp})_I^{\perp}$;
(3) $(A_I^{\perp}) \cap A \subseteq I$;
(4) *if A is also an ideal of L with $I \subseteq A$, then* $(A_I^{\perp}) \cap A = I$;
(5) $(A_I^{\perp})_I^{\perp} \cap A_I^{\perp} = I$;

(6) $L_I^{\perp} = I$;
(7) $I_I^{\perp} = L$;
(8) $(I_I^{\perp})_I^{\perp} = I$;
(9) $(L_I^{\perp})_I^{\perp} = L$;
(10) $A_I^{\perp} = \{x \in L \mid (x] \cap (A] \subseteq I\}$.
(11) $(a \wedge b)_I^{\perp} = a_{b_I^{\perp}}^{\perp} = b_{a_I^{\perp}}^{\perp}$, *for all* $a, b \in L$. *In particular,* $a_I^{\perp} = a_{a_I^{\perp}}^{\perp}$, *for all* $a \in L$;
(12) $a_I^{\perp} \cap b_I^{\perp} = (a \vee b)_I^{\perp} = (a \oslash b)_I^{\perp}$, *for all* $a, b \in L$.

Proof (1) For any $x \in A$, by the definition of $A_I^{\perp}$, we have $x \wedge y \in I$ for all $y \in A_I^{\perp}$, which implies that $x \in (A_I^{\perp})_I^{\perp}$. Therefore, $A \subseteq (A_I^{\perp})_I^{\perp}$.

(2) Since $A \subseteq (A_I^{\perp})_I^{\perp}$, by Proposition 3(2), we have $((A_I^{\perp})_I^{\perp})_I^{\perp} \subseteq A_I^{\perp}$. Conversely, taking $A_I^{\perp}$ as A, using (1), we have $A_I^{\perp} \subseteq ((A_I^{\perp})_I^{\perp})_I^{\perp}$. Therefore, $A_I^{\perp} = ((A_I^{\perp})_I^{\perp})_I^{\perp}$.

(3) For any $x \in (A_I^{\perp}) \cap A$, then $x \in A_I^{\perp}$ and $x \in A$, so we have that $x = x \wedge x \in I$. That is, $(A_I^{\perp}) \cap A \subseteq I$.

(4) By (3), we have $(A_I^{\perp}) \cap A \subseteq I$. Conversely, if $I \subseteq A$, then $I = I \cap A \subseteq A_I^{\perp} \cap A$, by Proposition 4(1). Therefore, $(A_I^{\perp}) \cap A = I$.

(5) Since $A_I^{\perp}$ is an ideal and $I \subseteq A_I^{\perp}$. By (4), we have $(A_I^{\perp})_I^{\perp} \cap A_I^{\perp} = I$.

(6) Taking L as A in (4), we obtain $L_I^{\perp} = L_I^{\perp} \cap L = I$, which implies that $L_I^{\perp} = I$.

(7) By Proposition 4(2), we have $I_I^{\perp} = L$.

(8) By (6) and (7), we have $(I_I^{\perp})_I^{\perp} = I$.

(9) By (6) and (7), we have $(L_I^{\perp})_I^{\perp} = L$.

(10) For any $x \in L$, satisfied that $(x] \cap (A] \subseteq I$. Then $x \in (x] \subseteq (A]_I^{\perp} = A_I^{\perp}$, that is, $\{x \in L \mid (x] \cap (A] \subseteq I\} \subseteq A_I^{\perp}$. Conversely, for any $x \in A_I^{\perp}$, we have $x \in A_I^{\perp} = (A]_I^{\perp}$, which implies that $(x] \subseteq (A]_I^{\perp}$, that is, $(x] \cap (A] \subseteq I$. Therefore, $A_I^{\perp} = \{x \in L \mid (x] \cap (A] \subseteq I\}$.

(11) $x \in (a \wedge b)_I^{\perp} \Leftrightarrow x \wedge a \wedge b \in I \Leftrightarrow x \wedge a \in b_I^{\perp} \Leftrightarrow x \in a_{b_I^{\perp}}^{\perp}$, so we have $(a \wedge b)_I^{\perp} = a_{b_I^{\perp}}^{\perp}$. Similarly, we get that $(a \wedge b)_I^{\perp} = b_{a_I^{\perp}}^{\perp}$. If $a = b$, then $a_I^{\perp} = a_{a_I^{\perp}}^{\perp}$.

(12) Since $a, b \le a \vee b \le a \oslash b$, then we have $(a \oslash b)_I^{\perp} \subseteq (a \vee b)_I^{\perp} \subseteq a_I^{\perp}, b_I^{\perp}$, it follows that $(a \oslash b)_I^{\perp} \subseteq (a \vee b)_I^{\perp} \subseteq a_I^{\perp} \cap b_I^{\perp}$. Conversely, if $x \in a_I^{\perp} \cap b_I^{\perp}$, then $x \wedge a \in I$ and $x \wedge b \in I$, it follows that $(x \wedge a) \oslash (x \wedge b) \in I$ as I is an ideal. Since $x \wedge (a \oslash b) \le (x \wedge a) \oslash (x \wedge b)$, it follows that $x \in (a \oslash b)_I^{\perp}$. This shows that $a_I^{\perp} \cap b_I^{\perp} \subseteq (a \vee b)_I^{\perp} \subseteq (a \oslash b)_I^{\perp}$. Therefore, $a_I^{\perp} \cap b_I^{\perp} = (a \vee b)_I^{\perp} = (a \oslash b)_I^{\perp}$. □

Proposition 7 *Let* A, B, *and* I *be ideals of a BL-algebra* L. *Then* $((A \cap B)_I^{\perp})_I^{\perp} = (A_I^{\perp})_I^{\perp} \cap (B_I^{\perp})_I^{\perp}$.

Proof Since $A \cap B \subseteq A, B$, we have $A_I^{\perp}, B_I^{\perp} \subseteq (A \cap B)_I^{\perp}$, and then $((A \cap B)_I^{\perp})_I^{\perp} \subseteq (A_I^{\perp})_I^{\perp}, (B_I^{\perp})_I^{\perp}$, which implies that $((A \cap B)_I^{\perp})_I^{\perp} \subseteq (A_I^{\perp})_I^{\perp} \cap (B_I^{\perp})_I^{\perp}$. Conversely, let $z \in (A_I^{\perp})_I^{\perp} \cap (B_I^{\perp})_I^{\perp}$. For any $x \in A$, $y \in B$, we have $x \wedge y \in A \cap B$. For all $u \in (A \cap B)_I^{\perp}$, we have $u \wedge x \wedge y \in I$, then $z \wedge u \wedge x \wedge y \in I$, so we get $z \wedge u \wedge x \in B_I^{\perp}$. Moreover, since $z \wedge u \wedge x \le z$ and $z \in (B_I^{\perp})_I^{\perp}$, we have $z \wedge u \wedge x \in (B_I^{\perp})_I^{\perp}$, it follows that $z \wedge u \wedge x \in B_I^{\perp} \cap (B_I^{\perp})_I^{\perp} = I$, which implies that $z \wedge u \in A_I^{\perp}$. Since $z \in (A_I^{\perp})_I^{\perp}$, then we have $z \wedge u \in (A_I^{\perp})_I^{\perp}$, it follows that $z \wedge u \in A_I^{\perp} \cap (A_I^{\perp})_I^{\perp} = I$, which implies that $z \in ((A \cap B)_I^{\perp})_I^{\perp}$. Therefore, $((A \cap B)_I^{\perp})_I^{\perp} = (A_I^{\perp})_I^{\perp} \cap (B_I^{\perp})_I^{\perp}$. □

Proposition 8 *Let A, B be ideals of a BL-algebra L with $A \subseteq B$ and $B = B^{\perp\perp}$. Then $B = (B_A^{\perp})_A^{\perp}$.*

Proof By Proposition 6(1), we have that $B \subseteq (B_A^{\perp})_A^{\perp}$. To prove the converse it suffices to show that if $x \notin B$, then we have $x \notin (B_A^{\perp})_A^{\perp}$. Now let $x \notin B = B^{\perp\perp}$, then $x \wedge y \neq 0$ for some $y \in B^{\perp}$. But $y \in B^{\perp}$ implies $x \wedge y \in B^{\perp}$, since $B^{\perp} \subseteq B_A^{\perp}$, then we have $x \wedge y \in B_A^{\perp}$. On the other hand, $x \wedge y \neq 0$, $x \wedge y \in B^{\perp}$ and $B \cap B^{\perp} = \{0\}$ implies $x \wedge y \notin B$. By $A \subseteq B$, we have $x \wedge y \notin A$. Since $B_A^{\perp} \cap (B_A^{\perp})_A^{\perp} = A$, then we have $x \wedge y \notin (B_A^{\perp})_A^{\perp}$, which means that $x \notin (B_A^{\perp})_A^{\perp}$. Therefore, $B = (B_A^{\perp})_A^{\perp}$. □

Definition 3 Let L be a BL-algebra, A and I ideals of L, I is said to be involutory relative to A if $I = (I_A^{\perp})_A^{\perp}$. If every ideal of L is involutory relative to A, then L is called an involutory BL-algebra relative to A. If $A = \{0\}$, we will simply call I to be an involutory ideal, i.e., I is an involutory ideal if $I = I^{\perp\perp}$. If every ideal of L is involutory, then L is called an involutory BL-algebra. The set of all involutory ideals relative to A of L is denoted by $S_A(L)$. The set of all involutory ideals of L is denoted by $S(L)$.

Example 1 Let $L = \{0, a, b, 1\}$ be a set, where $0 \leq a, b \leq 1$. The Cayley tables are as follows.

$\odot$	0	a	b	1
0	0	0	0	0
a	0	a	0	a
b	0	0	b	b
1	0	a	b	1

$\to$	0	a	b	1
0	1	1	1	1
a	b	1	b	1
b	a	a	1	1
1	0	a	b	1

Then $(L, \wedge, \vee, \odot, \to, 0, 1)$ is a BL-algebra. It is easy to check that $I_0 = \{0\}$, $I_a = \{0, a\}$, $I_b = \{0, b\}$, L are all ideals of L.

(i) $I_0^{\perp} = L$, $(I_0)_{I_a}^{\perp} = L$, $(I_0)_{I_b}^{\perp} = L$, $(I_0)_L^{\perp} = L$.
$I_a^{\perp} = I_b$, $(I_a)_{I_a}^{\perp} = L$, $(I_a)_{I_b}^{\perp} = I_b$, $(I_a)_L^{\perp} = L$.
$I_b^{\perp} = I_a$, $(I_b)_{I_a}^{\perp} = I_a$, $(I_b)_{I_b}^{\perp} = L$, $(I_b)_L^{\perp} = L$.
$L^{\perp} = I_0$, $L_{I_a}^{\perp} = I_a$, $L_{I_b}^{\perp} = I_b$, $L_L^{\perp} = L$.

(ii) $I_0^{\perp\perp} = \{0\} = I_0$, so I_0 is an involutory ideal.
$I_a^{\perp\perp} = I_b^{\perp} = I_a$, so I_a is an involutory ideal.
$I_b^{\perp\perp} = I_a^{\perp} = I_b$, so I_b is an involutory ideal.
$L^{\perp\perp} = I_0^{\perp} = L$, so L is an involutory ideal.
Therefore, L is an involutory BL-algebra.

(iii) $((I_0)_{I_a}^{\perp})_{I_a}^{\perp} = L_{I_a}^{\perp} = I_a$, so I_0 is not involutory relative to I_a.
$((I_a)_{I_a}^{\perp})_{I_a}^{\perp} = L_{I_a}^{\perp} = I_a$, so I_a is involutory relative to I_a.
$((I_b)_{I_a}^{\perp})_{I_a}^{\perp} = (I_a)_{I_a}^{\perp} = L$, so I_a is not involutory relative to I_a.
$(L_{I_a}^{\perp})_{I_a}^{\perp} = (I_a)_{I_a}^{\perp} = L$, so L is involutory relative to I_a.
Similarly, we can check that I_b and L are involutory relative to I_b, I_0 and I_a are not involutory relative to I_a.

$((I_0)_L^{\perp})_L^{\perp} = ((I_a)_L^{\perp})_L^{\perp} = ((I_b)_L^{\perp})_L^{\perp} = (L_L^{\perp})_L^{\perp} = L$, so L is involutory relative to L. I_0, I_a, I_b are not involutory relative to L.

Proposition 9 *Let L be a BL-algebra and A be an ideal of L. Then $S_A(L) = \{B_A^{\perp} \mid B \in I(L), A \subseteq B\}$.*

Proof Let $I \in S_A(L)$, then we have $I = (I_A^{\perp})_A^{\perp}$. Considering $B = I_A^{\perp}$, then $B \in I(L)$ and $I = B_A^{\perp}$. Therefore, $S_A(L) \subseteq \{B_A^{\perp} \mid B \in I(L), A \subseteq B\}$. Conversely, since $B_A^{\perp} = ((B_A^{\perp})_A^{\perp})_A^{\perp}$, then we have $B_A^{\perp} \in S_A(L)$. Therefore, $S_A(L) = \{B_A^{\perp} \mid B \in I(L), A \subseteq B\}$. □

Proposition 10 *Let I be an ideal of a BL-algebra L, $\emptyset \neq A \subseteq L$ such that $(A] \in S_I(L)$. Then $(A] = (A_I^{\perp})_I^{\perp}$.*

Proof Since $(A]_I^{\perp} = A_I^{\perp}$ and $((A]_I^{\perp})_I^{\perp} = (A]$, it follows that $(A] = ((A]_I^{\perp})_I^{\perp} = (A_I^{\perp})_I^{\perp}$. □

Proposition 11 *Let I be an ideal of a BL-algebra L. If $A, B \in S_I(L)$, then the following hold:*
(1) $I \subseteq A$;
(2) $A_I^{\perp} \cap A = I$;
(3) $A \cap B \subseteq I$ implies $B \subseteq A_I^{\perp}$;
(4) $I, L \in S_I(L)$.

Proof (1) By Proposition 4(1), we have $I \subseteq A_I^{\perp}$, so $I \subseteq (A_I^{\perp})_I^{\perp} = A$.
(2) By Proposition 6(4), it is clear.
(3) By Proposition 5(2), it is clear.
(4) By Proposition 6(8) and (9), it is clear. □

The above Proposition shows that I and L are the least and the largest elements in the $S_I(L)$ with respect to the set-theoretic inclusion, respectively.

Proposition 12 *Let I be an ideal of a BL-algebra L, and $\{A_\lambda \mid \lambda \in \Lambda\} \subseteq S_I(L)$; where $\Lambda \neq \emptyset$. Then $\cap_{\lambda\in\Lambda} A_\lambda \in S_I(L)$. Hence $\cap_{\lambda\in\Lambda} A_\lambda$ is the infimum of the set $\{A_\lambda \mid \lambda \in \Lambda\}$ in $S_I(L)$ with respect to the set-theoretic inclusion.*

Proof By Proposition 3(3), we have $\cap_{\lambda\in\Lambda} A_\lambda = \cap_{\lambda\in\Lambda}((A_\lambda)_I^{\perp})_I^{\perp} = (\cup_{\lambda\in\Lambda}(A_\lambda)_I^{\perp})_I^{\perp}$, then it follows that $((\cap_{\lambda\in\Lambda} A_\lambda)_I^{\perp})_I^{\perp} = ((((\cup_{\lambda\in\Lambda}(A_\lambda)_I^{\perp})_I^{\perp})_I^{\perp})_I^{\perp} = (\cup_{\lambda\in\Lambda}(A_\lambda)_I^{\perp})_I^{\perp} = \cap_{\lambda\in\Lambda} A_\lambda$. Therefore, $\cap_{\lambda\in\Lambda} A_\lambda \in S_I(L)$. □

Let I be an ideal of a BL-algebra L, for any nonempty subsets $A_\lambda(\lambda \in \Lambda \neq \emptyset)$ of L, define $\sqcup_{\lambda\in\Lambda} A_\lambda := ((\cup_{\lambda\in\Lambda} A_\lambda)_I^{\perp})_I^{\perp}$. Then by Proposition 3(3), we have $\sqcup_{\lambda\in\Lambda} A_\lambda = (\cap_{\lambda\in\Lambda}(A_\lambda)_I^{\perp})_I^{\perp}$.

Proposition 13 *Let I be an ideal of a BL-algebra L, and $\{A_\lambda \mid \lambda \in \Lambda\} \subseteq S_I(L)$; where $\Lambda \neq \emptyset$. Then $((\cup_{\lambda\in\Lambda} A_\lambda)_I^{\perp})_I^{\perp} \in S_I(L)$ and $\sqcup_{\lambda\in\Lambda} A_\lambda := ((\cup_{\lambda\in\Lambda} A_\lambda)_I^{\perp})_I^{\perp}$ is the supremum of the set $\{A_\lambda \mid \lambda \in \Lambda\}$ in $S_I(L)$ with respect to the set-theoretic inclusion.*

Proof Clearly, $((\cup_{\lambda\in\Lambda} A_\lambda)_I^{\perp})_I^{\perp} \in S_I(L)$. Since $A_\lambda \subseteq \cup_{\lambda\in\Lambda} A_\lambda \subseteq ((\cup_{\lambda\in\Lambda} A_\lambda)_I^{\perp})_I^{\perp}$ for all $\lambda \in \Lambda$, that is, $((\cup_{\lambda\in\Lambda} A_\lambda)_I^{\perp})_I^{\perp}$ is an upper bound of the set $\{A_\lambda \mid \lambda \in \Lambda\}$. Now we take any $A \in S_I(L)$ with $A_\lambda \subseteq A$ for all $\lambda \in \Lambda$. Then $A_I^{\perp} \subseteq (A_\lambda)_I^{\perp}$, and so $A_I^{\perp} \subseteq \cap_{\lambda\in\Lambda}(A_\lambda)_I^{\perp}$. So it follows that $((\cup_{\lambda\in\Lambda} A_\lambda)_I^{\perp})_I^{\perp} = (\cap_{\lambda\in\Lambda}(A_\lambda)_I^{\perp})_I^{\perp} \subseteq (A_I^{\perp})_I^{\perp} = A$. Therefore, $\sqcup_{\lambda\in\Lambda} A_\lambda := ((\cup_{\lambda\in\Lambda} A_\lambda)_I^{\perp})_I^{\perp}$ is the supremum of the set $\{A_\lambda \mid \lambda \in \Lambda\}$ in $S_I(L)$ with respect to the set-theoretic inclusion. □

Theorem 1 *Let I be an ideal of a BL-algebra L, then $(S_I(L), \cap, \sqcup, I, L)$ is a complete Boolean lattice, in which I is the least element and L is the largest element, respectively.*

Proof By Propositions 12 and 13, we have that $(S_I(L), \cap, \sqcup, I, L)$ is a complete lattice. And by Proposition 11, we know that I and L are the least and the largest elements in the $S_I(L)$, respectively. For any $A \in S_I(L)$, by Proposition 11(2), we have $A_I^{\perp} \cap A = I$, $A \sqcup A_I^{\perp} = ((A \cup A_I^{\perp})_I^{\perp})_I^{\perp} = (A_I^{\perp} \cap A)_I^{\perp} = I_I^{\perp} = L$, which implies that the lattice $(S_I(L), \cap, \sqcup, I, L)$ is a complemented lattice.

In what follows, we prove the distributive law. For any $A, B, C \in S_I(L)$, denote $D = (A \cap B) \sqcup (A \cap C)$, then $A \cap B \subseteq D$ and $A \cap C \subseteq D$, it follows that $B \cap (A \cap D_I^{\perp}) = I$ and $C \cap (A \cap D_I^{\perp}) = I$, which implies that $B \subseteq (A \cap D_I^{\perp})_I^{\perp}$ and $C \subseteq (A \cap D_I^{\perp})_I^{\perp}$. Hence $B \sqcup C \subseteq (A \cap D_I^{\perp})_I^{\perp}$, so $(B \sqcup C) \cap A \cap D_I^{\perp} = A$ and $X \cap (B \sqcup C) \subseteq (D_I^{\perp})_I^{\perp} = D = (A \cap B) \sqcup (A \cap C)$. The converse inclusion is clear. So the lattice $(S_I(L), \cap, \sqcup, I, L)$ is distributive, moreover, $(S_I(L), \cap, \sqcup, I, L)$ is a Boolean lattice. □

Corollary 1 *Let L be a BL-algebra. Then $(S(L), \cap, \sqcup, \{1\}, L)$ is a complete Boolean lattice, in which $\{0\}$ is the least element and L is the largest element, respectively.*

Theorem 2 *Let I be an ideal of a BL-algebra L, for any $A, B \in S_I(L)$, define*

$$A \leq B \; if \; and \; only \; if \; A \subseteq B;$$
$$A \to B = B \sqcup A_I^{\perp} \in S_I(L);$$
$$A \odot B = A \cap B \in S_I(L).$$

Then $(S_I(L), \cap, \sqcup, \odot, \to, I, L)$ is a BL-algebra; where I is the least element and L is the largest element, respectively.

Proof (BL1): By Theorem 1, we have known that $(S_I(L), \cap, \sqcup, I, L)$ is a lattice with the least element I and the largest element L, where the order relation is the set-theoretic inclusion.

(BL2): Clearly.

(BL3): Let $A, B, C \in S_I(L)$. If $A \leq B \to C$, then $A \subseteq C \sqcup B_I^{\perp}$, and $A \odot B = A \cap B \subseteq C \sqcup B_I^{\perp} \cap B = (C \cap B) \sqcup (B_I^{\perp} \cap B) = (C \cap B) \sqcup I = C \cap B \subseteq C$. Therefore, $A \leq B \to C$ implies $A \odot B \leq C$. Conversely, if $A \odot B \leq C$, then $A = A \cap (B \sqcup B_I^{\perp}) = (A \cap B) \sqcup (A \cap B_I^{\perp}) \subseteq C \sqcup B_I^{\perp} = B \to C$. Therefore, $A \odot B \leq C$ implies $A \leq B \to C$.

(BL4): Let $A, B \in S_I(L)$. $A \odot (A \to C) = A \cap (B \sqcup A_I^{\perp}) = (A \cap B) \sqcup (A \cap A_I^{\perp}) = (A \cap B) \sqcup I = A \cap B$.

(BL5): Let $A, B \in S_I(L)$. Then $(A \to B) \sqcup (B \to A) = (B \sqcup A_I^{\perp}) \sqcup (A \sqcup B_I^{\perp}) = (A \sqcup A_I^{\perp}) \sqcup (B \sqcup B_I^{\perp}) = L \sqcup L = L$.

Therefore, $(S_I(L), \cap, \sqcup, \odot, \to, I, L)$ is a BL-algebra. □

Corollary 2 *Let L be a BL-algebra. For any $A, B \in S(L)$, define*

$$A \leq B \; if \; and \; only \; if \; A \subseteq B;$$
$$A \rightarrow B = B \sqcup A^{\perp} \in S(L);$$
$$A \odot B = A \cap B \in S(L).$$

Then $(S(L), \cap, \sqcup, \odot, \rightarrow, I, L)$ is a BL-algebra where $\{0\}$ is the least element and L is the largest element, respectively.

Proposition 14 *Let L be a BL-algebra, I an ideal and P be a prime ideal of L with $I \subseteq P$. Then the following hold:*
(1) for any nonempty subset A of L with $A \nsubseteq P$, we have $A_I^{\perp} \subseteq P$;
(2) $x_I^{\perp} \subseteq P$, for all $x \notin P$.

Proof (1) Since $A \nsubseteq P$, there exists $a \in A$ but $a \notin P$. For any $x \in A_I^{\perp}$, then we have $x \wedge a \in I \subseteq P$. Since P is prime, it follows that $x \in P$. Therefore, $A_I^{\perp} \subseteq P$.
(2) This is a special case of (1). □

Corollary 3 *Let P be a prime ideal of a BL-algebra L. Then the following hold:*
(1) for any nonempty subset A of L with $A \nsubseteq P$, we have $A_P^{\perp} = P$;
(2) $x_P^{\perp} = P$, for all $x \notin P$.

Proof Since $P \subseteq A_P^{\perp}$, it is clear by taking P as I in Proposition 14. □

Proposition 15 *Let P be an ideal of a BL-algebra L. Then P is prime if and only if for any nonempty subset A of L with $A_P^{\perp} \neq L$, $A_P^{\perp} = P$.*

Proof Let P be a prime ideal of L, A a nonempty subset of L such that $A_P^{\perp} \neq L$. Then by Proposition 4(2), we have $A \nsubseteq P$, and by Corollary 3, we have $A_P^{\perp} = P$. Conversely, let A be a nonempty subset of L with $A_P^{\perp} \neq L$, then $A_P^{\perp} = P$. Suppose that $a \wedge b \in P$ and $b \notin P$, since $b \wedge b = b \notin P$, we have $b \notin b_P^{\perp}$, it follows that $b_P^{\perp} \neq L$, which implies that $b_P^{\perp} = P$. Moreover, since $a \wedge b \in P$, we have $a \in b_P^{\perp} = P$. Therefore, P is a prime ideal. □

Proposition 16 *Let P be an ideal of a BL-algebra L. Then P is prime if and only if $x_P^{\perp} \subseteq P$, for all $x \notin P$.*

Proof Let P is prime, then by Corollary 3, we have $x_P^{\perp} \subseteq P$, for all $x \notin P$. Conversely, suppose that $a \wedge b \in P$ and $b \notin P$, since $b \wedge b = b \notin P$, we have $b \notin b_P^{\perp}$, it follows that $b_P^{\perp} \neq L$, which implies that $b_P^{\perp} = P$. Since $a \wedge b \in P$, we have $a \in b_P^{\perp} = P$. Therefore, P is a prime ideal. □

For an ideal I and a prime ideal P of L, denote $I_P := \{x \in L \mid x_I^{\perp} \nsubseteq P\}$. If $I = \{0\}$, then $\{0\}_P = \{x \in L \mid x^{\perp} \nsubseteq P\}$, we simply write 0_P instead of $\{0\}_P$.

Proposition 17 *Let I be an ideal of a BL-algebra L, and P, Q be two prime ideals with $I \subseteq P, Q$. Then the following hold:*
(1) if $P \subseteq Q$, then $I_Q \subseteq I_P$;
(2) $I \subseteq I_P$;
(3) $x \in I_P$ if and only if there exists $a \notin P$ such that $x \in a_I^{\perp}$;
(4) $I_P \subseteq P$;
(5) I_P is an ideal of L.

Proof (1) Let $x \in I_Q$, we have $x_I^{\perp} \nsubseteq Q$, since $P \subseteq Q$, it follows that $x_I^{\perp} \nsubseteq P$, i.e., $x \in I_P$. Therefore, $I_Q \subseteq I_P$.

(2) For any $x \in I$, we have $x_I^{\perp} = L \nsubseteq P$, so $x \in I_P$, i.e., $I \subseteq I_P$.

(3) From $x \in I_P$, we have $x_I^{\perp} \nsubseteq P$, i.e., there exist $a \in x_I^{\perp}$ but $a \notin P$, so $a \wedge x \in I$, that is $x \in a_I^{\perp}$. Conversely, if there is $a \notin P$ such that $x \in a_I^{\perp}$, then $a \wedge x \in I$, it follows that $a \in x_I^{\perp}$ and $a \notin P$, so $x_I^{\perp} \nsubseteq P$, which means that $x \in I_P$.

(4) Let $x \in I_P$, by (3), there is $a \notin P$ such that $x \wedge a \in I$. Since P is prime, we have $x \in P$.

(5) Since $0_I^{\perp} = L \nsubseteq P$, we have $0 \in I_P$. If $a \in I_P$ and $b \leq a$, then $a_I^{\perp} \nsubseteq P$ and $a_I^{\perp} \subseteq b_I^{\perp}$, which implies that $b_I^{\perp} \nsubseteq P$, i.e., $b \in I_P$. If $a \in I_P$ and $b \in I_P$, then $a_I^{\perp} \nsubseteq P$ and $b_I^{\perp} \nsubseteq P$, so there exist $x, y \in L$ such that $x \in a_I^{\perp}$, but $x \notin P$, and $y \in b_I^{\perp}$, but $y \notin P$. Since P is prime, we have $x \wedge y \notin P$. $(x \wedge y) \wedge (a \oslash b) \leq (x \wedge y \wedge a) \oslash (x \wedge y \wedge b) \in I$, which implies that $x \wedge y \in (a \oslash b)_I^{\perp}$, but we have already known that $x \wedge y \notin P$, so $(a \oslash b)_I^{\perp} \nsubseteq P$, i.e., $a \oslash b \in I_P$. Therefore, I_P is an ideal of L. □

Proposition 18 *Let L be a BL-algebra, I an ideal of L and $A \neq \emptyset \subseteq L$. Then $A_I^{\perp} = \cap\{P$ is a prime ideal of $L \mid F \subseteq P$ and $A \nsubseteq P\}$.*

Proof Denote $T := \cap\{P$ is a prime ideal a of $L \mid F \subseteq P$ and $A \nsubseteq P\}$. Let $x \in A_I^{\perp}$, for any $P \in T$, we have $x \wedge A \subseteq I \subseteq P$, it follows that $x \in A_P^{\perp} = P$ by Corollary 3, i.e., $x \in P$, which implies that $A_I^{\perp} \subseteq T$. On the other hand, let $x \in T$, then for any prime ideal P with $I \subseteq P$ and $A \nsubseteq P\}$, we have $x \in P = A_P^{\perp}$. And since $I \subseteq P$, then $P = A_P^{\perp} \subseteq A_I^{\perp}$. Thus $x \in A_I^{\perp}$, which implies that $T \subseteq A_I^{\perp}$. Therefore, $A_I^{\perp} = T$. □

Let I be an ideal of a BL-algebra L. Define $x \sim_I y$ if and only if $\bar{x} \odot y \in I$ and $\bar{y} \odot x \in I$, for any $x, y \in L$. Then we have that $\sim$ is a congruence on L. The set of all congruence classes is defined by L/I, i.e., $L/I = \{[x] \mid x \in L\}$, where $[x] = \{y \in L \mid x \sim_I y\}$. Define $[x] \odot [y] = [x \odot y]$; $[x] \rightarrow [y] = [x \rightarrow y]$; $[x] \wedge [y] = [x \wedge y]$; $[x] \vee [y] = [x \vee y]$. Then in [10] it has been proved that $(L/I, \wedge, \vee, \odot, \rightarrow, [0], [1])$ is an MV-algebra and $[0] = I$.

Proposition 19 *Let P be an ideal of a BL-algebra L. Then P is prime if and only if $[x] \wedge [y] = [0]$ implies $[x] = [0]$ or $[y] = [0]$ in L/P.*

Proof Let P be a prime ideal and $[x] \wedge [y] = [0]$, then we have $[x \wedge y] = [0] = P$, it follows that $x \wedge y \in P$. Since P is prime, then $x \in P$ or $y \in P$, which implies that $[x] = [0]$ or $[y] = [0]$. Conversely, suppose that $x \wedge y \in P$, then we have $[x \wedge y] =$

$P = [0]$, it follows that either $[x] = [0]$ or $[y] = [0]$. Hence $x \in P$ or $y \in P$, which means that P is prime. □

Proposition 20 *Let I be an ideal of a BL-algebra L and A be a nonempty subset of L. Then the following hold:*
(1) $(A/I)^{\perp} = A_I^{\perp}/I$;
(2) $[x]^{\perp} = x_I^{\perp}/I$;
(3) *If I is a prime ideal and* $[x] \neq I$, *then* $[x]^{\perp} = I$;
(4) $(A/I)^{\perp\perp} = (A_I^{\perp})_I^{\perp}/I$.

Proof (1) $(A/I)^{\perp} = \{[y] \in L/I \mid [x] \wedge [y] = [0],\ \text{for all } x \in A\} = \{[y] \in L/I \mid [x \wedge y] = [0], \text{for all } x \in A\} = \{[y] \in L/I \mid x \wedge y \in I, \text{for all } x \in A\} = \{[y] \in L/I \mid y \in A_I^{\perp}\} = A_I^{\perp}/I$.

(2) This is a special case of (1).

(3) If I is a prime ideal of L and $[x] \neq I$, then $x \notin I$, by Corollary 3, we have $x_I^{\perp} = I$. Therefore, by (2) we have $[x]^{\perp} = I$.

(4) $(A/I)^{\perp\perp} = (A_I^{\perp}/I)_I^{\perp}/I = (A_I^{\perp})_I^{\perp}/I$. □

Corollary 4 *Let I, J be two ideals of a BL-algebra L. If J is an involutory ideal relative to I, then J/I is an involutory ideal of L/I.*

Proof It is clear by Proposition 20(4). □

Acknowledgments This research is supported by a grant of National Natural Science Foundation of China (11571281).

References

1. Hájek, P.: Metamathematics of Fuzzy Logic. Kluwer Academic Publishers, Dordrecht (1998)
2. Chang, C.C.: Algebraic analysis of many valued logics. Trans Am Math Soc **88**, 467–490 (1958)
3. Blyth, T.S.: Lattices and Ordered Algebraic Structures. Springer, London (2005)
4. Busneag, D., Piciu, D.: On the lattice of deductive systems of a BL-algebras. Central Eur J Math **1**, 221–238 (2003)
5. Saeid, A.B., Motamed, S.: Some results in BL-algebras. Math Logic Q **55**, 649–658 (2009)
6. Liu, L.Z., Li, K.T.: Fuzzy Boolean and positive implicative filters of BL-algebras. Fuzzy Sets Syst **152**, 333–348 (2005)
7. Turunen, E.: Boolean deductive systems of BL-algebras. Archive Math Logic **40**, 467–473 (2000)
8. Zhang, X.H., Jun, Y.B., Im, D.M.: On fuzzy filters and fuzzy ideals of BL-algebras. Fuzzy Syst Math **20**, 8–12 (2006)
9. Zhan, J.M., Jun, Y.B., Kim, H.S.: Some types of falling fuzzy filters of BL-algebras and its application. J Intell Fuzzy Syst **26**, 1675–1685 (2014)
10. Lele, C., Nganou, J.B.: MV-algebras derived from ideals in BL-algebras. Fuzzy Sets Syst **218**, 103–113 (2013)
11. Davey, B.A.: Some annihilator conditions on distributive lattices. Algebra Univ **4**, 316–322 (1974)
12. Turunen, E.: BL-algebras of basic fuzzy logic. Mathware Soft Comput **6**, 49–61 (1999)

13. Leustean, L.: Some algebraic properties of non-commutative fuzzy stuctures. Stud Inf Control **9**, 365–370 (2000)
14. Zou, Y.X., Xin, X.L., He, P.F.: On annihilators in BL-algebras. Open Math **14**, 324–337 (2016)

Part VI
Fuzzy Set Theory and Applications

On the Equivalence of Convergence of Fuzzy Number Series with Respect to Different Metrics

Hong-Mei Wang and Tai-He Fan

Abstract In this paper, we discuss the equivalence of convergence of fuzzy number series under different metrics. It is proved that the convergence of a series of fuzzy numbers with respect to most of the metrics can be converted into the convergence of the corresponding remainder sequence, in the case of convergence the limit of the latter must be 0. Also, the levelwise convergence of a series of fuzzy numbers is equivalent to the convergence of its remainder. It is proved that the convergence of a series of fuzzy numbers with respect to the sendograph metric D, the supremum metric d_∞ and the convergence of series of support sets with Hausdorff metric are equivalent to each other.

Keywords Series of fuzzy numbers · Supremum metric · Sendograph metric · Endograph metric · Convergence

1 Introduction

Fuzzy number is the basic concept of fuzzy mathematics. One of the main aspects of fuzzy number theory is the study of metrics for fuzzy numbers. The most commonly used metrics for fuzzy numbers are the supremum metric d_∞, the sendograph metric, the endograph metric, and the d_p metric and so on.

Convergence of series is an important topic in classical real analysis. It is an effective way to study known functions and plays an important role in approximate calculation.

Since fuzzy number series is a generalization of the ordinary series, so the convergence of series of fuzzy numbers is a basic problem in fuzzy set theory. In recent years, much study has been made on metrics on fuzzy numbers. In this paper, we

H.-M. Wang · T.-H. Fan (✉)
Department of Mathematics Science, Zhejiang Sci-Tech University, Hangzhou 310027, China
e-mail: Taihefan@163.com

H.-M. Wang
e-mail: 876858031@qq.com

T.-H. Fan et al. (eds.), *Quantitative Logic and Soft Computing 2016*,
Advances in Intelligent Systems and Computing 510,
DOI 10.1007/978-3-319-46206-6_44

study the convergence of fuzzy number series and the relations of convergence of fuzzy number series with respect to different metrics.

2 Preliminaries

Definition 1 ([1]) Let X be a set, a mapping $u : X \to [0, 1]$ is called a fuzzy set on X, $u_\alpha = \{x : u(x) \geq \alpha\}$ is called the $\alpha-$cut of u, where $\alpha \in (0, 1]$.

Definition 2 A fuzzy set u on the real line R is called a fuzzy number if the following conditions are satisfied:
(1) u is normal, i.e., $u(x_0) = 1$ for some $x_0 \in R$;
(2) u is fuzzy convex, i.e., $u(rx + (1 - r)y) \geq \min\{u(x), u(y)\}$ for all $x, y \in R$ and $r \in I$;
(3) u is upper semicontinuous, i.e., u_α is a closed set, $u_\alpha = \{x : u(x) \geq \alpha\}$, $\alpha \in (0, 1]$;
(4) The topological support of u is compact: $u_0 = cl\{x|x \in R, u(x) > 0\}$ is compact.

Let E^1 denote the set of all fuzzy numbers. For $u \in E^1$, each $\alpha \in I$, let $u_\alpha = [u_L(\alpha), u_R(\alpha)]$.

Definition 3 Let $u \in E^1$, the family of closed intervals $\{\{|\lambda| : \lambda \in u_\alpha\} : \alpha \in [0, 1]\}$ determine a unique fuzzy number, denoted by $|u|$, which is called the absolute value of u.

Definition 4 The sum and difference of two fuzzy sets u, v on R are defined by the Zadeh's extension principle as follows:

$$(u + v)(x) = \sup\{\min\{u(a), v(b)\} : x = a + b\}$$

$$(u - v)(x) = \sup\{\min\{u(a), v(b)\} : x = a - b\}$$

It should be noted that the addition and subtraction of fuzzy numbers are not inverse to each other, this is inconvenient in practice. Therefore, in this paper, we use the Hukuhara difference (H difference for short) as the subtraction operation for fuzzy numbers.

Definition 5 ([9]) Let $u, v \in E^1$, if there exists a fuzzy number w, such that $w + v = u$, we say that the H difference of u, v exists and w is called the H difference of u, v, simply denoted by $u -_h v$.

Lemma 1 ([9]) *If $u, v \in E^1$, $\lambda \in R$, then*
(1) $u -_h v$ exists if and only if $u_L(\alpha) - v_L(\alpha) \leq u_R(\alpha) - v_R(\alpha)$ and $[u_L(\alpha) - v_L(\alpha), u_R(\alpha) - v_R(\alpha)] \subseteq [u_L(\beta) - v_L(\beta), u_R(\beta) - v_R(\beta)]$, $0 \leq \beta \leq \alpha \leq 1$;
(2) If $u -_h v$ exists, then $(u -_h v)(\alpha) = u(\alpha) - v(\alpha)$, $\alpha \in [0, 1]$.

Definition 6 ([1]) Let (X, d) be a metric space, A, B are two nonempty compact sets in X, we call $H(A, B) = \max\{d_H^*(A, B), d_H^*(B, A)\}$ the Hausdorff distance between A and B, where $d_H^*(A, B) = \sup_{a\in A} \inf_{b\in B} d(a, b)$.

In the following we give definition and some basic properties of the supremum metric for fuzzy numbers:

Definition 7 ([2]) For $u, v \in E^1$, the supremum metric of u and v is defined as follows:

$$d_\infty(u, v) = \sup_{\alpha\in(0,1]} H(u_\alpha, v_\alpha)$$

It is well known that (E^1, d_∞) is a complete but not separable metric space.

Definition 8 ([8]) For $u \in E^1$, Let

$$send(u) = \{(x, y)|x \in u_0, 0 \le y \le u(x)\}$$

$$end(u) = \{(x, y)|x \in R, 0 \le y \le u(x)\}$$

$send(u)$ and $end(u)$ are called the sendograph and the endograph of u respectively. For $u, v \in E^1$

$$D(u, v) = H(send(u), send(v))$$

$$D'(u, v) = H(end(u), end(v))$$

$$d_p = (\int_0^1 H(u_\alpha, v_\alpha)^p d\alpha)^{1/p}, (1 \le p < +\infty)$$

are three types of metrics on E^1, called the sendograph metric, the endograph metric and the d_p metrics respectively.

Lemma 2 ([2]) *From the properties of Hausdorff metric, we have*

$$D(u + w, v + w') \le D(u, v) + D(w, w'), u, v, w, w' \in E^1,$$

$$D(u + w, v + w) \le D(u, v), u, v, w \in E^1.$$

(E^1, D) is not a complete metric space.

Lemma 3 ([8]) *For $u, v, w \in E^1$, if $\rho = D(u, v) > 0$ and $H(w_0, w_1) \in A_m$, then*

$$D(u + w, v + w) \le D(u, v) \le mD(u + w, v + w)$$

$$A_m = \begin{cases} \{0\}, & m=1 \\ (0, \frac{1}{2}]\rho, & m=2 \\ (\frac{1}{2}, \frac{2+\sqrt{3}}{3}]\rho, & m=3 \\ (\sum_{i=0}^{m-2} \frac{\sqrt{(m-i-2)(m+i-2)}}{m-1}, \sum_{i=0}^{m-1} \frac{\sqrt{(m-i-1)(m+i-1)}}{m}]\rho, & m>3 \end{cases}$$

(As $A_i \cap A_j = \emptyset, \cup_{m=1}^{\infty} = [0, \infty)$, when $i \neq j$, then for each w, there exists a unique m such that $H(w_0, w_1) \in A_m$.)

Lemma 4 ([2]) *For $c \in R \backslash \{0\}, u, v, w, w' \in E^1, 1 \le p \le \infty$*
(1) $d_p(cu, cv) = |c|\, d_p(u, v)$;
(2) $d_p(u + w, v + w) = d_p(u, v)$;
(3) $d_p(u + w, v + w') \le d_p(u, v) + d_p(w, w')$.

It's easy to show that the supremum metric d_∞ also has the above three properties.

In [2] completeness and relations between convergences with respect to different metrics on E^1 are studied; In [3, 6] detailed characterizations on relations and convergences between different metrics are listed; In [2] compactness of fuzzy number space with respect to sendograph metric and supremum metric are given.

Remark 1 ([4]) It's obvious from the definition that a fuzzy number u is equi-right-continuous at $\alpha = 0$. u is equi-right-continuous at $\alpha \in (0, 1]$ if and only if the mapping $\alpha \to u_\alpha$ is right-continuous at α. u is equi-left-continuous at $\alpha \in (0, 1]$.

Next, we give definition on the convergence of fuzzy number sequence and related properties:

Definition 9 ([10]) Let $u_n (n = 1, 2, \ldots), u \in E^1$, d a metric on E^1,
(1) We say that a fuzzy number sequence $\{u_n\}$ converges to u with respect to d if $d(u_n, u) \to 0 (n \to \infty)$, denoted by $u_n \xrightarrow{d} u$;
(2) We say that a fuzzy number sequence $\{u_n\}$ converges levelwise to u if $H((u_n)_\alpha, u_\alpha) \to 0 (n \to \infty)$, for all $\alpha \in (0, 1]$, denoted by $u_n \xrightarrow{H} u$.

From the definition, we have

Proposition 1 (1) *If $u_n \xrightarrow{d_\infty} u$, then $u_n \xrightarrow{H} u$;*
(2) $u_n \xrightarrow{d_\infty} u \Leftrightarrow \{(u_n)_L(\alpha)\}$ converge uniformly to $u_L(\alpha)$ and $\{(u_n)_R(\alpha)\}$ converge uniformly to $u_R(\alpha)$, for all $\alpha \in I$.

Proposition 2 *For $a_n, a \in R, u_n, v_n, u, v \in E^1$*
(1) If $u_n \xrightarrow{d_\infty} u, v_n \xrightarrow{d_\infty} v$, then $u_n + v_n \xrightarrow{d_\infty} u + v$;
(2) If $a_n \to a, u_n \xrightarrow{d_\infty} u$, then $a_n u_n \xrightarrow{d_\infty} au$. where $u_L(\alpha), u_R(\alpha)$ are the left and right endpoints of u_α respectively, i.e., $u_\alpha = [u_L(\alpha), u_R(\alpha)]$.

Proof (1) and (2) can be easily obtained by the definition of the supremum metric d_∞ and the Hausdorff metric H. □

Lemma 5 ([7]) *Suppose that* $u_n(n = 1, 2, 3, \ldots), u \in E^1$. *Then* $u_n \xrightarrow{D} u$ *if and only if*
(1) $H((u_n)_0, u_0) \to 0 (n \to \infty)$;
(2) *For all* $\epsilon > 0$, *there exists a natural number* $N(\epsilon)$ *such that for all* $n > N(\epsilon)$:
(i) *For all* $x \in R$, *there exists* $\{x_n\} \in R$ *such that* $|x_n - x| < \epsilon, u_n(x) < u(x_n) + \epsilon$;
(ii) *For all* $x \in u_0$, *there exists* $\{x_n\} \in (u_n)_0$ *such that* $|x_n - x| < \epsilon, u(x) < u_n(x_n) + \epsilon$.

Definition 10 ([5]) Suppose that $u_n \in E^1$, $(n = 1, 2, 3, \ldots)$. The expression

$$\sum_{n=1}^{\infty} u_n = u_1 + u_2 + \cdots + u_n + \cdots \tag{1}$$

is called a fuzzy number series, and u_n the general term of the series.

According to the Zadeh's extension principle, the membership function of the fuzzy number series (1) is as follows:

$$(\sum_{i=1}^{\infty} u_i)(x) = \sup\{\inf\{u_i(x_i)\}_{i \in N}, x \in \sum_{i=1}^{\infty} x_i\}.$$

It should be noted that the membership function of a fuzzy number series may not be the membership function of any fuzzy number.

Let

$$S_1 = u_1, S_2 = u_1 + u_2, S_3 = u_1 + u_2 + u_3, \ldots$$

$$S_n = u_1 + u_2 + \cdots + u_n = \sum_{k=1}^{n} u_k, \ldots$$

Then we get a fuzzy number sequence $S_n = \sum_{k=1}^{n} u_k (n = 1, 2, 3, \ldots)$ for a fuzzy number series $\sum_{n=1}^{\infty} u_n$, S_n is called the partial sum of $\sum_{n=1}^{\infty} u_n$, and $\{S_n\}$ is the partial sum sequence of the fuzzy number series.

On the contrary, for a fuzzy number sequence $\{S_n\}$, there may not exist fuzzy number series $\sum_{n=1}^{\infty} u_n$ such that $\{S_n\}$ is the partial sum sequence of $\sum_{n=1}^{\infty} u_n$. The reason is that the H difference between two fuzzy numbers may not exist [1] (See Example 1 below).

By the definition of convergence of sequence in metric spaces, we give the concept of convergence of fuzzy number series with respect to metrics as follows:

Definition 11 Let d be a metric on E^1, a fuzzy number series $\sum_{n=1}^{\infty} u_n$ is said to be convergent with respect to metric d if the partial sum sequence $\{S_n\}$ of the fuzzy number series converges to some $S \in E^1$ with respect to metric d, i.e.

$$S_n \xrightarrow{d} S$$

denoted by

$$\sum_{n=1}^{\infty} u_n = S(d)$$

S is also called the sum of the fuzzy number series. Fuzzy numbers series $\sum_{n=1}^{\infty} u_n$ is said to be divergent if the partial sum sequence $\{S_n\}$ of fuzzy numbers series is divergent, i.e., if $\{S_n\}$ does not converge in E^1.

The sequence

$$r_n = S -_h S_n = \sum_{k=n+1}^{\infty} u_k = u_{n+1} + u_{n+2} + u_{n+3} + \cdots$$

is called the remainder of fuzzy number series.

Thus, the study on the convergence of fuzzy number series is transformed into the study of convergence of the partial sum sequence, which enables us to apply the knowledge about fuzzy number sequence to the study of fuzzy number series.

3 Main Results

In this section, we first study the convergence of fuzzy number series with respect to different metrics. Then we study the fuzzy number series by using the remainder of the series.

Proposition 3 *Let* $\{u_n\}, \{v_n\} \subset E^1, \lambda \in R^+$*:*
(1) *If the fuzzy number series* $\sum_{n=1}^{\infty} u_n$ *is convergent with respect to the supremm metric* d_∞*, then* $u_n \xrightarrow{d_\infty} 0$*;*
(2) *If* $\sum_{n=1}^{\infty} u_n$ *and* $\sum_{n=1}^{\infty} v_n$ *are convergent with respect to the supreme metric* d_∞*, then* $\sum_{n=1}^{\infty}(u_n + v_n) = \sum_{n=1}^{\infty} u_n + \sum_{n=1}^{\infty} v_n$*;*
(3) *If* $\sum_{n=1}^{\infty} u_n$ *is convergent with respect to the supremm metric* d_∞*, then* $\sum_{n=1}^{\infty} \lambda u_n = \lambda \sum_{n=1}^{\infty} u_n$*.*

Proof (1) If $\sum_{n=1}^{\infty} u_n$ is convergent with respect to the supremm metric d_∞, i.e., for each $\epsilon > 0$, there exists N, such that $m > n > N$, $d_\infty(S_m, S_n) < \epsilon$, let $m = n + 1$, then we have $d_\infty(u_{n+1}, 0) < \epsilon$. Thus $u_n \xrightarrow{d_\infty} 0$.

(2) and (3) are obvious. □

Example 1 Let

$$S(x) = \begin{cases} 1, \ x = 0 \\ 0, \ x \neq 0 \end{cases}$$

$$S_n(x) = \begin{cases} (1-x)^n, & 0 \le x \le 1 \\ 0, & x \ne 0 \end{cases}$$

then

$$S_\alpha = \{0\}, (S_n)_\alpha = [0, 1 - \sqrt[n]{\alpha}], \text{ for all } \alpha \in (0, 1]$$

and

$$(S_n)_L(\alpha) - (S_{n-1})_L(\alpha) = 0$$

$$(S_n)_R(\alpha) \quad (S_{n-1})_R(\alpha) = 1 - \sqrt[n]{\alpha} - 1 + \sqrt[n-1]{\alpha} = \sqrt[n]{\alpha} - \sqrt[n-1]{\alpha} \le 0$$

From Lemma 1(1), we know that $S_n -_h S_{n-1}$ does not exist. Thus the fuzzy number sequence $\{S_n\}$ is not the partial sum sequence of any fuzzy number series. Which shows that fuzzy number sequence and fuzzy number series can not be converted to each other.

Now, we discuss the problem whether the convergence of a fuzzy number series can be transformed into the convergence of its remainder, which enables us apply the knowledge about the limit of fuzzy number sequence to fuzzy number series.

Proposition 4 (1) *The convergence of a fuzzy number series* $\sum_{n=1}^{\infty} u_n$ *is equivalent to the fact that its remainder converges to 0 with respect to the supremum metric. i.e.,* $r_n \xrightarrow{d_\infty} 0 (n \to \infty)$;
(2) *The convergence of a fuzzy number series* $\sum_{n=1}^{\infty} u_n$ *is equivalent to the fact that its remainder converges to 0 with respect to* d_p *metrics. i.e.,* $r_n \xrightarrow{d_p} 0 (n \to \infty)$.

Proof (1) Obviously, $d_\infty(u + w, v + w) = d_\infty(u, v)$. Since

$$S = S_n + r_n, S_n = S_n + 0,$$

we have

$$d_\infty(S, S_n) = d_\infty(r_n, 0).$$

We conclude that the partial sum sequence of fuzzy number series is convergent with respect to supremum metric d_∞ if and only if its remainder converges to 0.

The proof of (2) is similiar to (1). □

Proposition 5 *The convergence of a fuzzy number series* $\sum_{n=1}^{\infty} u_n$ *is equivalent to the fact that its remainder converges to 0 with respect to the sendograph metric, i.e.,* $r_n \xrightarrow{D} 0 (n \to \infty)$.

Proof From Lemma 3, we have

$$D(S, S_n) \to 0 (n \to \infty) \Leftrightarrow D(r_n, 0) \to 0 (n \to \infty)$$

Thus the partial sum sequence of the fuzzy number series converges with respect to the sendograph metric D if and only if its remainder converges to 0. □

Proposition 6 *A fuzzy number series* $\sum_{n=1}^{\infty} u_n$ *converges levelwise if and only if* $(r_n)_\alpha \xrightarrow{H} 0(n \to \infty), \alpha \in (0, 1]$.

Proof Since

$$H((r_n)_\alpha, 0) = H((S -_h S_n)_\alpha, 0) = H(S_\alpha - (S_n)_\alpha, 0)$$

thus

$$(r_n)_\alpha \xrightarrow{H} 0 \Leftrightarrow (S_n)_\alpha \xrightarrow{H} S_\alpha$$

□

Example 2 That the partial sum sequence of a fuzzy number series $\sum_{n=1}^{\infty} u_n$ converges with respect to the endograph metric D' is not equal to the fact that its remainder converges to 0. The counterexample is as follows:

Let $\{a_n\}$ be a monotone decreasing real number sequence which converges to 0. The general term of the fuzzy number series is

$$u_n(x) = \begin{cases} 1, & x = 0 \\ a_n, & x \in (0, n] \\ 0, & otherwise \end{cases}$$

It follows that the remainder $r_n = \sum_{k=n+1}^{\infty} u_k \xrightarrow{D'} 0(n \to \infty)$. Obviously, $end(S_n) = end(\sum_{k=1}^{n} u_k)$ is not convergent with respect to the Hausdorff metric, thus $\sum_{n=1}^{\infty} u_n$ is not convergent with respect to the endograph metric D'.

Proposition 7 *If the partial sum sequence of a fuzzy number series* $\sum_{n=1}^{\infty} u_n$ *converges levelwise to a fuzzy number u, then* $\max\limits_{a \in (r_n)_\alpha} |a| \to 0(n \to \infty), \alpha \in (0, 1]$.

Proof Obvious. □

Next, based on the existing results, we discuss the relations among different kind of convergence of fuzzy number series with respect to different metrics.

Theorem 1 *Let* $S, S_n, u_n (n = 1, 2, 3, \ldots) \in E^1$, *then the following conditions are equivalent:*

(1) $\sum_{n=1}^{\infty} u_n$ *converges to S with respect to the supremum metric* d_∞, *i.e.,* $S_n \xrightarrow{d_\infty} S(n \to \infty)$;

(2) *The 0 cut set sequence* $(S_n)_0$ *of the partial sum* S_n *of the fuzzy number series* $\sum_{n=1}^{\infty} u_n$ *converges to* S_0 *with respect to the Hausdorff metric, i.e.,* $(S_n)_0 \xrightarrow{H} S_0$ $(n \to \infty)$;

(3) $\sum_{n=1}^{\infty} u_n$ *converges to S with respect to the sendograph metric D, i.e.,* $S_n \xrightarrow{D} S(n \to \infty)$.

Proof By virtue of Propositions 4, 5 and 6, the convergence of a fuzzy number series can be transformed into the convergence of its remainder to 0. Let r_n be the remainder, we need to show the following relations:

$$r_n \xrightarrow{d_\infty} 0(n \to \infty) \Leftrightarrow (r_n)_0 \xrightarrow{H} 0(n \to \infty) \Leftrightarrow r_n \xrightarrow{D} 0(n \to \infty)$$

(1) $\Rightarrow$ (2): Since $\sup_{\alpha>0} H((r_n)_\alpha, 0) \to 0(n \to \infty)$, thus $H((r_n)_\alpha, 0) \to 0(n \to \infty)$, $\forall \alpha \in (0, 1]$, i.e., for each $\epsilon > 0$, there exists $N > 0$, such that whenever $n > N$, then $H((r_n)_\alpha, 0) < \epsilon/2$, for all $\alpha \in (0, 1]$.

As H is right continuous at $\alpha = 0$, pick a sequence $\alpha_k \to 0$, $\alpha_k \in (0, 1]$, $k = 1, 2, 3, \ldots$, thus $H((r_n)_{\alpha_k}, (r_n)_0) < \epsilon/2$, for $n > N$. $H((r_n)_0, 0) \le H((r_n)_{\alpha_k}, (r_n)_0) + H((r_n)_{\alpha_k}, 0) < \epsilon$, for $n > N$.

We conclude that the 0-cut sequence $(S_n)_0$ of the partial sum S_n of $\sum_{n=1}^{\infty} u_n$ converges to S_0 with respect to the Hausdorff metric.

(2) $\Rightarrow$ (1): Since $H((r_n)_0, 0) \to 0(n \to \infty)$, i.e., $\max_{a\in(r_n)_0} |a| \to 0(n \to \infty)$.

As $(r_n)_\alpha \subset (r_n)_0$, hence, $\max_{a\in(r_n)_\alpha} |a| \to 0(n \to \infty)$, $\alpha \in (0, 1]$. Thus $\sup_{\alpha>0} H((r_n)_\alpha, 0) \to 0(n \to \infty)$. This shows that (2) implies (1).

(2) $\Rightarrow$ (3): Since $H((r_n)_0, 0) \to 0(n \to \infty)$, $\max_{a\in(r_n)_0} |a| \to 0(n \to \infty)$, $send(r_n) = \{(x, \alpha)|(x, \alpha) \in (r_n)_0 \times I, 0 \le \alpha \le u(x)\}$. Thus $send(r_n) \xrightarrow{H} send(0)$ $(n \to \infty)$.

(3) $\Rightarrow$ (2), Since $send(r_n) \xrightarrow{H} send(0)(n \to \infty)$, from Lemma 5, we have $(r_n)_0 \xrightarrow{H} 0(n \to \infty)$. This completes the proof of the theorem. □

Proposition 8 *If the partial sum sequence of a fuzzy number series $\sum_{n=1}^{\infty} u_n$ is a Cauchy sequence with respect to the supremum metric d_∞, then its sum S is a fuzzy number.*

Proof This is simply because that (E^1, d_∞) is a complete metric space, thus S is a fuzzy number by uniqueness of the limit. □

Proposition 9 *If $u_n + v_n = w_n$, $u_n, v_n, w_n (n = 1, 2, 3, \ldots) \in E^1$ and $\sum_{n=1}^{\infty} u_n$ is convergent with respect to the supremum metric, then $\sum_{n=1}^{\infty} w_n$ is convergent if and only if $\sum_{n=1}^{\infty} v_n$ is convergent with respect to the supremum metric.*

Proof Sufficiency: Since $v_n = w_n -_h u_n$, i.e., v_n is the H difference between u_n and w_n. Suppose that the fuzzy number series $\sum_{n=1}^{\infty} u_n$ converges with respect to supremum metric d_∞, denote the partial sum sequence of $\sum_{n=1}^{\infty} u_n, \sum_{n=1}^{\infty} v_n, \sum_{n=1}^{\infty} w_n$ by S_n^u, S_n^v and S_n^w respectively, then $S_n^v = S_n^w -_h S_n^u$.

Since $d_\infty(S_n^w, S^w) \to 0(n \to \infty)$, $d_\infty(S_n^u, S^u) \to 0(n \to \infty)$, where S^u and S^w are the sum of fuzzy numbers series $\sum_{n=1}^{\infty} u_n$ and $\sum_{n=1}^{\infty} w_n$ respectively. Since $S_n^v = S_n^w -_h S_n^u$, $S^w -_h S^u$ exists and $S^w -_h S^u \in E^1$.

Clearly

$$(S^w -_h S^u)_\alpha = (S^w)_\alpha -_h (S^u)_\alpha, (S_n^w -_h S_n^u)_\alpha = (S_n^w)_\alpha - (S_n^u)_\alpha$$

for all $\alpha \in (0, 1]$. From the definition of Hausdorff distance, we have,

$$H((S_n^w)_\alpha, (S^w)_\alpha) = H((S_n^u)_\alpha, (S^u)_\alpha) + H((S_n^w -_h S^u)_\alpha, (S^w -_h S^u)_\alpha)$$

for all $\alpha \in (0, 1]$.

Since $d_\infty(S_n^w, S^w) \to 0 (n \to \infty), d_\infty(S_n^u, S^u) \to 0 (n \to \infty)$, we have $H((S_n^w -_h S^u)_\alpha, (S^w -_h S^u)_\alpha) \to 0$ holds uniformly for all $\alpha \in (0, 1]$. Thus $d_\infty(S_n^v, S^w -_h S^u) \to 0 (n \to \infty)$. Let $S^v = S^w -_h S^u$, then $\sum_{n=1}^\infty v_n = S^v$.

Necessity: This is a part of Proposition 2. □

Lemma 6 *If $\sum_{n=1}^\infty a_n$ is an interval numbers series, then $\sum_{n=1}^\infty a_n$ is convergent with respect to the Hausdorff metric if and only if its left and right endpoints series $\sum_{n=1}^\infty (a_n)_L$ and $\sum_{n=1}^\infty (a_n)_R$ are all convergent as ordinary series.*

Proof This is a direct corollary of Proposition 1. □

Proposition 10 *Suppose that $u_n = a_n + v_n, a_n = \chi_{(u_n)_1}$, then v_n is a unimodal fuzzy number whose normal point is 0. Moreover, the convergence of $\sum_{n=1}^\infty u_n$ is equivalent to the convergence of both $\sum_{n=1}^\infty a_n$ and $\sum_{n=1}^\infty v_n$.*

Proof Sufficiency: Suppose that $\sum_{n=1}^\infty u_n$ converges with respect to the supremum metric d_∞, then $\sum_{n=1}^\infty (u_n)_1$ is also convergent. i.e., $\sum_{n=1}^\infty a_n$ is convergent. By Proposition 9, we have that $\sum_{n=1}^\infty v_n$ is convergent with respect to the supremum metric d_∞.

Necessity: By Proposition 9. □

Proposition 11 *If the fuzzy number series $\sum_{n=1}^\infty |u_n|$ is convergent, then $\sum_{n=1}^\infty u_n$ is convergent. The opposite, however, is not necessarily true.*

Proof Suppose that the support series of $\sum_{n=1}^\infty |u_n|$ is convergent, we have, for all $\epsilon > 0$, there exists N, whenever $m > n > N$, we have $|u_{n+1}|_0 + |u_{n+2}|_0 + \cdots + |u_m|_0 < \epsilon$. By definition of the absolute value of fuzzy number, we have $|(u_{n+1})_0 + (u_{n+2})_0 + \cdots + (u_m)_0| < \epsilon$, thus the support series of $\sum_{n=1}^\infty u_n$ is convergent. This shows that $\sum_{n=1}^\infty u_n$ is convergent.

As an ordinary convergent series may not be absolutely convergent, and fuzzy number series is a generalization of the ordinary series, thus the inverse proposition is not necessarily true. □

Acknowledgments Project Supported by the National Natural Science Foundation of China (61170110, 11171308, 61379018).

References

1. Chen, M.H.: New fuzzy analysis theory. Sciences Press, Beijing (2009). (in Chinese)
2. Diamond, P., Kloeden, P.: Metric Spaces of Fuzzy Sets-theory and Applications. World Scientific, Singapore (1994)
3. Fan, T.H.: Equivalence of weak converence and endograph metric converence for fuzzy number spaces. In: Proceedings of IFSA 2005, Beijing, Part I: Mathematical Foundations of Fuzzy Set Theory, pp. 41–43 (2005)
4. Fan, T.H.: On the compactness of fuzzy number spaces with Sendograph metric. Fuzzy Sets Syst **143**(3), 471–477 (2004)
5. Fan, T.H., Wang, G.J.: Endographic approach on supremum and infimum of fuzzy numbers. Inf Sci **159**(3–4), 221–231 (2004)
6. Huang, H., Wu, C.X.: Characterizations of Γ convergence on fuzzy number space. In: Proceedings of IFSA 2005, Beijing, Part I: Mathematical Foundations of Fuzzy Sets Theory, pp. 66–70 (2005)
7. Kloeden, P.E.: Compact supported sendgraphs and fuzzy sets. Fuzzy Sets Syst **4**(2), 193–201 (1980)
8. Li, Y.: Properties of the sendograph metric of fuzzy sets. Master's Dissertation, Zhejiang: Zhejiang Sci-Tech University **06**, A002–79 (2012). (in Chinese)
9. Li, H.L., Pei, H.L., He, Q., Xi, Z.M.: The existence of H difference of fuzzy numbers. Fuzzy Syst Math **29**(3), 119–122 (2015). (in Chinese)
10. Zhang, R.Y., Li, L.T.: The convergence of fuzzy series. J Shanxi Univ **2**(23), 221–224 (1999)

Multicriteria Decision Making Based on Interval-Valued Intuitionistic Fuzzy Sets with a New Kind of Accuracy Function

Bei Liu and Min-Xia Luo

Abstract In this paper, we propose a new accuracy function based on interval-valued intuitionistic fuzzy sets, then use this new accuracy function to multicriteria decision making method. By comparing the new accuracy function with other accuracy function, some examples are given. While aggregating fuzzy information, we use the interval-valued intuitionistic fuzzy weighted aggregation operators, and rank the fuzzy information by the proposed accuracy function, which overcomes some difficulties arising in some existing accuracy functions for determining rank of interval-valued intuitionistic fuzzy information. Finally, the effectiveness and practicability of the proposed method are illustrated by examples.

Keywords Interval-valued intuitionistic fuzzy sets · Multicriteria decision making · Ranking of interval-valued intuitionistic fuzzy numbers · Accuracy function

1 Introduction

Since Zadeh introduced fuzzy sets in [15], many new approaches and theories treating vagueness and uncertainty have been proposed. As a generalization of the ordinary fuzzy set, interval-valued fuzzy set was first introduced by Zadeh [16–18]. Intuitionistic fuzzy set was introduced by Atanassov in [1], which is another extension of the classical fuzzy set. Atanassov and Gargov proposed the concept of interval-valued intuitionistic fuzzy set in [2], which is a further generalization of intuitionistic fuzzy set. When the membership function and non-membership function of an

B. Liu
Department of Mathematics, China Jiliang University, Hangzhou 310018, People's Republic of China

M.-X. Luo (✉)
School of Sciences, China Jiliang University, Hangzhou 310018, People's Republic of China
e-mail: minxialuo@163.com

T.-H. Fan et al. (eds.), *Quantitative Logic and Soft Computing 2016*,
Advances in Intelligent Systems and Computing 510,
DOI 10.1007/978-3-319-46206-6_45

interval-valued intuitionistic fuzzy set are exact numbers rather than intervals, the interval-valued intuitionistic fuzzy set reduces to an intuitionistic fuzzy set.

Interval-valued intuitionistic fuzzy set has received more and more attention since its appearance [8, 9]. It is a very useful tool to deal with uncertainty and become a popular topic in multicriteria decision making. Some researchers have established some aggregation operators based on interval-valued intuitionistic fuzzy set. To aggregate interval-valued intuitionistic fuzzy information, Xu and Chen [12] extended arithmetic aggregation operators (see [11]). Then Xu [13] developed interval-valued intuitionistic fuzzy weighted averaging operators and interval-valued intuitionistic fuzzy weighted geometric aggregation operators. Aggregated interval-valued intuitionistic fuzzy information can be ranked by accuracy functions. Hong and Choi [4] indicated that score function cannot discriminate some alternatives although they are apparently different, then they proposed accuracy functions and they are extended to interval-valued fuzzy set in [10, 11, 13]. Some other accuracy functions based on interval-valued intuitionistic fuzzy sets are studied in [5–7, 14]. However, in some cases, those existing accuracy functions do not rank correctly even in some comparable interval-valued intuitionistic fuzzy numbers, which may be troublesome for decision maker to make choices. To solve cases like this, we propose a new accuracy functions.

The rest of this paper is organized as follows. In Sect. 2 we review some basic definitions of interval-valued intuitionistic fuzzy sets. In Sect. 3 we introduce a new kind of accuracy functions and give some examples. In Sect. 4, two illustrative examples are given to demonstrate the validity of the proposed accuracy function. And finally, conclusions are stated in Sect. 5.

2 Preliminaries

Throughout this paper, let X be the universe of discourse, and $D[0, 1]$ the set of all closed subinterval of the unit interval $[0, 1]$.

Definition 1 ([2]) An interval-valued intuitionistic fuzzy set on X can be expressed as $A = \{(x, [\mu_A^-(x), \mu_A^+(x)], [\nu_A^-(x), \nu_A^+(x)])|x \in X\}$, where $[\mu_A^-(x), \mu_A^+(x)] \in D[0, 1]$, $[\nu_A^-(x), \nu_A^+(x)] \in D[0, 1]$ with the condition $\mu_A^+(x) + \nu_A^+(x) \leq 1$ for all $x \in X$.

For an interval-valued intuitionistic fuzzy set A based on X, the pair $([\mu_A^-(x), \mu_A^+(x)], [\nu_A^-(x), \nu_A^+(x)])$ is called an interval-valued intuitionistic fuzzy number [13] and is denoted by $\tilde{\alpha} = ([a, b], [c, d])$ for convenience.

In the following, we give some operations and relations on interval-valued intuitionistic fuzzy sets.

Definition 2 ([2]) Let X be a universe of discourse. For two interval-valued intuitionistic fuzzy sets $A = \{(x, [\mu_A^-(x), \mu_A^+(x)], [\nu_A^-(x), \nu_A^+(x)])|x \in X\}$, $B = \{(x, [\mu_B^-(x), \mu_B^+(x)], [\nu_B^-(x), \nu_B^+(x)])|x \in X\}$, the following relations and operations can be defined:

(1) $A \subseteq B$ iff $\mu_A^-(x) \leq \mu_B^-(x), \mu_A^+(x) \leq \mu_B^+(x), \nu_A^-(x) \geq \nu_B^-(x)$, and $\nu_A^+(x) \geq \nu_B^+(x)$ for each $x \in X$,
(2) $A = B$ iff $A \subseteq B$ and $B \subseteq A$,
(3) $A^c = \{[\nu_A^-(x), \nu_A^+(x)], [\mu_A^-(x), \mu_A^+(x)] | x \in X\}$.

Now, we introduce two weighted aggregation operators related to interval-valued intuitionistic fuzzy sets. We denoted by IVIFS(X) the set of all interval-valued intuitionistic fuzzy sets in X.

Definition 3 ([13]) Let $A_j \in$ IVIFS(X)$(j = 1, 2, \ldots, n)$. Weighted arithmetic average operators are defined as follows:

$$\begin{aligned} &F_\omega(A_1, A_2, \ldots, A_n) \\ &= \sum_{j=1}^n \omega_j A_j \\ &= \left(\left[1 - \prod_{j=1}^n (1 - u_{A_j}^-(x))^{\omega_j}, 1 - \prod_{j=1}^n (1 - u_{A_j}^+(x))^{\omega_j} \right], \right. \\ &\left. \left[\prod_{j=1}^n (v_{A_j}^-(x))^{\omega_j}, \prod_{j=1}^n (v_{A_j}^+(x))^{\omega_j}) \right] \right). \end{aligned} \tag{1}$$

where ω_j $(j = 1, 2, \ldots, n)$ are such that $\omega_j \in [0, 1]$ and $\sum_{j=1}^n \omega_j = 1$. They are called the weight of F_ω. Especially, assume $\omega_j = \frac{1}{n}$ $(j = 1, 2, \ldots, n)$, then F_ω is called the arithmetic average operator for interval-valued intuitionistic fuzzy sets.

Definition 4 ([13]) Let $A_j \in$ IVIFS(X)$(j = 1, 2, \ldots, n)$. The weighted geometric average operators is defined by

$$\begin{aligned} &G_\omega(A_1, A_2, \ldots, A_n) \\ &= \sum_{j=1}^n A_j^{\omega_j} \\ &= \left(\left[\prod_{j=1}^n (u_{A_j}^-(x))^{\omega_j}, \prod_{j=1}^n (u_{A_j}^+(x))^{\omega_j}) \right], \right. \\ &\left. \left[1 - \prod_{j=1}^n (1 - v_{A_j}^-(x))^{\omega_j}, 1 - \prod_{j=1}^n (1 - v_{A_j}^+(x))^{\omega_j}) \right] \right). \end{aligned} \tag{2}$$

where ω_j is the weight of A_j $(j = 1, 2, \ldots, n)$, $\omega_j \in [0, 1]$ and $\sum_{j=1}^{n} \omega_j = 1$. Especially, assume $\omega_j = \frac{1}{n}$ $(j = 1, 2, \ldots, n)$, then G_ω is called the geometric average operator for interval-valued intuitionistic fuzzy sets.

Next, we review the concepts of score functions and accuracy functions proposed for ranking interval-valued intuitionistic fuzzy numbers (see [5–7, 13, 14]).

Definition 5 ([13]) Let $\tilde{\alpha} = ([a, b], [c, d])$ be an interval-valued intuitionistic fuzzy number, then the score the value of it is defined as:

$$S(\tilde{\alpha}) = \frac{a + b - c - d}{2} \tag{3}$$

while the accuracy value is defined as:

$$H(\tilde{\alpha}) = \frac{a + b + c + d}{2} \tag{4}$$

Definition 6 ([14]) Let $\tilde{\alpha} = ([a, b], [c, d])$ be an interval-valued intuitionistic fuzzy number. Novel accuracy value $M(\tilde{\alpha})$ of the interval-valued intuitionistic fuzzy number is defined as follow:

$$M(\tilde{\alpha}) = \frac{a - (1 - a - c) + b - (1 - b - d)}{2} = a + b - 1 + \frac{c + d}{2} \tag{5}$$

where $M(\tilde{\alpha}) \in [-1, 1]$.

Definition 7 ([7]) Let $\tilde{\alpha} = ([a, b], [c, d])$ be an interval-valued intuitionistic fuzzy number. An improved accuracy value $K(\tilde{\alpha})$ of the interval-valued intuitionistic fuzzy number, including hesitancy degree of IVIFSs, is defined by:

$$K(\tilde{\alpha}) = \frac{a + b(1 - a - c) + b + a(1 - b - d)}{2} \tag{6}$$

where $K(\tilde{\alpha}) \in [0, 1]$.

Definition 8 ([5]) Let $\tilde{\alpha} = ([a, b], [c, d])$ be an interval-valued intuitionistic fuzzy number. A new novel accuracy value L of the interval-valued intuitionistic fuzzy number is defined as follows:

$$L(\tilde{\alpha}) = \frac{a - d(1 - b) + b - c(1 - a)}{2} \tag{7}$$

where $L(\tilde{\alpha}) \in [-1, 1]$.

Definition 9 ([6]) Let $\tilde{\alpha} = ([a, b], [c, d])$ be an interval-valued intuitionistic fuzzy number, then the general accuracy value of $\tilde{\alpha}$ is defined as

$$LG(\tilde{\alpha}) = \frac{a + b + \delta(2 - a - b - c - d)}{2} \tag{8}$$

where $\delta \in [0, 1]$ is a parameter depending on the individual's intention.

3 Multicriteria Decision Making Method Based on a New Accuracy Function

3.1 Ranking by a New Accuracy Function

Definition 10 Let $\tilde{\alpha} = ([a, b], [c, d])$ be an interval-valued intuitionistic fuzzy number. The new accuracy value of the interval-valued intuitionistic fuzzy number is defined by:

$$A(\tilde{\alpha}) = \frac{a + \delta_1(1 - a - c) + b + \delta_2(1 - b - d)}{2}, \tag{9}$$

where $\delta_1, \delta_2 \in [-1, 1]$ is a parameter depending on the individual's intention.

Remark 1 Let $\delta_1 = \delta_2 = -1$, then we have $A(\tilde{\alpha}) = \frac{a-(1-a-c)+b-(1-b-d)}{2} = M(\tilde{\alpha})$.
Let $\delta_1 = b, \delta_2 = a$, then we have $A(\tilde{\alpha}) = \frac{a+b(1-a-c)+b+a(1-b-d)}{2} = K(\tilde{\alpha})$.
Let $\delta_1 = \delta_2 = \delta \in [0, 1]$, then we have $A(\tilde{\alpha}) = \frac{a+b+\delta(2-a-b-c-d)}{2} = LG(\tilde{\alpha})$
Accuracy function M was proposed in [14], K was provided in [7] and LG was presented in [6], accuracy functions M, K and LG are special kinds of Definition 10.

Example 1 ([7]) Let $\tilde{A_1} = ([0.2, 0.4], [0.3, 0.4])$, $\tilde{A_2} = ([0.1, 0.5], [0.2, 0.5])$ be interval-valued intuitionistic fuzzy numbers for two alternatives. If we want to determine the desirable alternative according to the accuracy function K, we have $M(\tilde{A_1}) = M(\tilde{A_2}) = -0.05$. In this case, we do not know which alternative is better.

Since $A(\tilde{A_1}) = 0.3 + 0.25\delta_1 + 0.1\delta_2$, $A(\tilde{A_2}) = 0.3 + 0.35\delta_1$. When $\delta_1 > \delta_2$, $A(\tilde{A_1}) < A(\tilde{A_2})$, and when $\delta_1 < \delta_2$, $A(\tilde{A_1}) > A(\tilde{A_2})$. So according to the values of δ_1 and δ_2, we can know which one is better.

Example 2 Let $\tilde{\alpha_1} = ([0.2, 0.6], [0.2, 0.3])$, $\tilde{\alpha_2} = ([0.3, 0.5], [0, 0.4])$ be interval-valued intuitionistic fuzzy numbers for two alternatives. If we want to determine the desirable alternative according to the accuracy function K, we have $K(\tilde{\alpha_1}) = K(\tilde{\alpha_2}) = 0.59$. In this case, we do not know which alternative is better.

Since $A(\tilde{\alpha_1}) = 0.4 + 0.3\delta_1 + 0.05\delta_2$, $A(\tilde{\alpha_2}) = 0.15 + 0.35\delta_1 + 0.05\delta_2$, it's obvious that $A(\tilde{\alpha_1}) > A(\tilde{\alpha_2})$. Then $\tilde{\alpha_1}$ is better than $\tilde{\alpha_2}$.

Example 3 Let $\tilde{\alpha_1} = ([0.2, 0.2], [0.2, 0.7])$, $\tilde{\alpha_2} = ([0.1, 0.3], [0.4, 0.5])$ be interval-valued intuitionistic fuzzy numbers for two alternatives. If we want to determine the desirable alternative according to the accuracy function LG, we have $LG(\tilde{\alpha_1}) = LG(\tilde{\alpha_2}) = 0.2 + 0.35\delta$. In this case, we do not know which alternative is better.

But if $\delta_1 \neq \delta_2$, we have $A(\tilde{\alpha_1}) = 0.2 + 0.3\delta_1 + 0.05\delta_2$, $A(\tilde{\alpha_2}) = 0.2 + 0.25\delta_1 + 0.1\delta_2$. If $\delta_1 > \delta_2$, $A(\tilde{\alpha_1}) > A(\tilde{\alpha_2})$, and when $\delta_1 < \delta_2$, $A(\tilde{\alpha_1}) < A(\tilde{\alpha_2})$. So according to the values of δ_1 and δ_2, we can know which one is better.

The following example can illustrate the effectiveness of the proposed new accuracy function.

Example 4 Let $\tilde{\alpha_1} = ([\frac{1}{2}, \frac{1}{2}], [\frac{1}{16}, \frac{1}{4}])$, $\tilde{\alpha_2} = ([\frac{1}{2}, \frac{1}{2}], [\frac{1}{32}, \frac{9}{32}])$ be interval-valued intuitionistic fuzzy numbers for two alternatives. We have $H(\tilde{\alpha_1}) = H(\tilde{\alpha_2}) = \frac{21}{32}$, $M(\tilde{\alpha_1}) = M(\tilde{\alpha_2}) = \frac{5}{32}$, $K(\tilde{\alpha_1}) = K(\tilde{\alpha_2}) = \frac{43}{64}$, $L(\tilde{\alpha_1}) = L(\tilde{\alpha_2}) = \frac{27}{64}$, $LG(\tilde{\alpha_1}) = LG(\tilde{\alpha_2}) = \frac{1}{2} + \frac{11}{32}\delta$. If we want to determine the desirable alternative according to the accuracy function A, we have $A(\tilde{\alpha_1}) = \frac{1}{2} + \frac{7}{32}\delta_1 + \frac{1}{8}\delta_2$, $A(\tilde{\alpha_2}) = \frac{1}{2} + \frac{15}{64}\delta_1 + \frac{7}{64}\delta_2$. If we take $\delta_1 \neq \delta_2$, we have $A(\tilde{\alpha_1}) = 0.2 + 0.3\delta_1 + 0.05\delta_2$, $A(\tilde{\alpha_2}) = 0.2 + 0.25\delta_1 + 0.1\delta_2$. If we take $\delta_1 > \delta_2$, $A(\tilde{\alpha_1}) < A(\tilde{\alpha_2})$, and when $\delta_1 < \delta_2$, $A(\tilde{\alpha_1}) > A(\tilde{\alpha_2})$. So according to the values of δ_1 and δ_2, we can know which one is better.

Theorem 1 *Let* $\tilde{\alpha_1} = ([a_1, b_1], [c_1, d_1])$ *and* $\tilde{\alpha_2} = ([a_2, b_2], [c_2, d_2])$ *be two interval-valued intuitionistic fuzzy numbers for two alternatives. If* $\tilde{\alpha_1} \subseteq \tilde{\alpha_2}$, *then* $A(\tilde{\alpha_1}) \leq A(\tilde{\alpha_2})$.

Proof As we know, $\tilde{\alpha_1} \subseteq \tilde{\alpha_2}$ if and only if $a_1 \leq a_2$, $b_1 \leq b_2$, $c_1 \geq c_2$ and $d_1 \geq d_2$, then

$$
\begin{aligned}
&2A(\tilde{\alpha_1}) - 2A(\tilde{\alpha_2}) \\
&= [a_1 + \delta_1(1 - a_1 - c_1) + b_1 + \delta_2(1 - b_1 - d_1)] - \\
&\quad [a_2 + \delta_1(1 - a_2 - c_2) + b_2 + \delta_2(1 - b_2 - d_2)] \\
&= (1 - \delta_1)(a_1 - a_2) + (1 - \delta_2)(b_1 - b_2) + \delta_1(c_2 - c_1) + \\
&\quad \delta_2(d_2 - d_1) \\
&\leq 0.
\end{aligned}
$$

□

3.2 Multicriteria Decision Process

Mathematically speaking, the multicriteria decision making problem about m alternatives with n criteria can be expressed as

$$
\begin{array}{c} \\ u_1 \\ u_2 \\ \vdots \\ u_m \end{array}
\begin{array}{c} \begin{array}{cccc} c_1 & c_2 & \cdots & c_n \end{array} \\
\begin{pmatrix} r_{11} & r_{12} & \cdots & r_{1n} \\ r_{21} & r_{22} & \cdots & r_{2n} \\ \vdots & \vdots & \ddots & \vdots \\ r_{m1} & r_{m2} & \cdots & r_{mn} \end{pmatrix} \end{array}
\tag{10}
$$

where $U = \{u_1, u_2, \ldots, u_m\}$ is the set of alternatives; $C = \{c_1, c_2, \ldots, c_n\}$ is the set of criteria, and the weight of the criterion $c_j (j = 1, 2, \ldots, n)$ is ω_j, $\omega_j \in [0, 1]$

and $\sum_{j=1}^{n} \omega_j = 1$; r_{ij} $(i = 1, 2, \ldots, m,\ j = 1, 2, \ldots, n)$ is the evaluation information of alternatives u_i under c_j provided by experts. Let $\beta_i = ([a_i, b_i], [c_i, d_i])$ be the aggregating interval-valued intuitionistic fuzzy number for u_i $(i = 1, 2, \ldots, m)$, then $\beta_i = ([a_i, b_i], [c_i, d_i]) = F_\omega(r_{i1}, r_{i2}, \ldots, r_{in})$ or $\beta_i = ([a_i, b_i], [c_i, d_i]) = G_\omega(r_{i1}, r_{i2}, \ldots, r_{in})$.

In summary, the decision procedure for the proposed method can be summarized as follows:

(i) Obtain the weighted arithmetic average values using Eq. (1) or the weighted geometric average values using Eq. (2) for u_i $(i = 1, 2, \ldots, m)$.

(ii) Compute the score value of r_{ij} by Eq. (3) for $i = 1, 2, \ldots, m,\ j = 1, 2, \ldots, n$. Let a_i be the number of r_{ij} whose order are not changed by their score values, and we take $\delta_1 = \max_{1 \le i \le m} \{\frac{a_i}{n}\}$ and $\delta_2 = \min_{1 \le i \le m} \{\frac{a_i}{n}\}$. Then compute the accuracy values $A(\beta_i)$ $(i = 1, 2, \ldots, m)$ using Eq. (9).

(iii) Ranking the alternatives u_i $(i = 1, 2, \ldots, m)$ and choose the best one according to $A(\beta_i)$ $(i = 1, 2, \ldots, m)$.

4 Multicriteria Decision Making Method Based on the New Accuracy Function

The following are two examples of multicriteria decision making problems.

Example 5 ([3]) Assume a invest panel has four possible alternatives to invest money, they are a car company u_1, a food company u_2, a computer company u_3 and an arms company u_4. The investment company wants to decide a decision according to three criteria given by the risk analysis c_1, the growth analysis c_2 and the environmental impact analysis c_3. Using the interval-valued intuitionistic fuzzy information provided by the decision maker under the above three criteria, the four possible alternatives are to be evaluated and listed in Table 1:

Suppose that the weights of c_1, c_2 and c_3 are $\omega_1 = 0.35$, $\omega_2 = 0.25$ and $\omega_3 = 0.4$, respectively.

(i) We can compute the weighted arithmetic average value β_i for u_i $(i = 1, 2, 3, 4)$ as follows:

Table 1 Information for example 5

	c_1	c_2	c_3
u_1	([0.4, 0.5], [0.3, 0.4])	([0.4, 0.6], [0.2, 0.4])	([0.1, 0.3], [0.5, 0.6])
u_2	([0.6, 0.7], [0.2, 0.3])	([0.6, 0.7], [0.2, 0.3])	([0.4, 0.6], [0.1, 0.2])
u_3	([0.3, 0.6], [0.3, 0.4])	([0.5, 0.6], [0.3, 0.4])	([0.3, 0.6], [0.1, 0.3])
u_4	([0.7, 0.8], [0.1, 0.2])	([0.6, 0.7], [0.1, 0.3])	([0.3, 0.4], [0.1, 0.2])

$\beta_1 = ([0.2943, 0.4996], [0.3225, 0.5030]), \beta_2 = ([0.5025, 0.7449], [0.1515, 0.2550]),$

$\beta_3 = ([0.3949, 0.5626], [0.1677, 0.3565]), \beta_4 = ([0.5227, 0.6565], [0.1189, 0.2213]).$

(ii) We obtain $A(\beta_i)$ $(i = 1, 2, 3, 4)$ as follows:

$$A(\beta_1) = 0.5836, A(\beta_2) = 0.7967, A(\beta_3) = 0.6975, A(\beta_4) = 0.7688.$$

(iii) Ranking all alternatives according to the accuracy values of $A(\beta_i)$ $(i = 1, 2, 3, 4)$: $A_2 > A_4 > A_3 > A_1$.

Thus the alternative u_2 is the most desirable alternative according to weighted arithmetic average operator.

Next we use the weighted geometric average operator. By calculating, we can obtain that $A_2 > A_4 > A_3 > A_1$.

Thus the alternative u_2 is the most desirable alternative according to weighted geometric average operator. In this example, our approach produces the same ranking as reference ([7]).

Example 6 Given that there are four possible alternatives for an enterprise to select: a car company (u_1), a clothes company (u_2), a software company (u_3) and a domestic company (u_4). The factors that must be considered before gathering evaluation information conclude risk (c_1), benefit (c_2), social and political response (c_3). Once the possible alternatives and corresponding factors are determined, the concrete evaluation information can be estimated by experts who are commissioned by the investor. Finally, the assessment report about this problem is listed in Table 2.

Suppose that the weights of c_1, c_2 and c_3 are equal, i.e. $\omega_1 = \omega_2 = \omega_3 = \frac{1}{3}$.

(i) We can compute the weighted arithmetic average value β_i for u_i $(i = 1, 2, 3, 4)$ as follows:

$\beta_1 = ([0.2440, 0.6417], [0.0000, 0.2282]), \beta_2 = ([0.3803, 0.5632], [0.0000, 0.3497]),$

$\beta_3 = ([0.2000, 0.6000], [0.2000, 0.3000]), \beta_4 = ([0.3000, 0.5000], [0.0000, 0.4000]).$

(ii) Obtain $K(\beta_i)$ $(i = 1, 2, 3, 4)$ as follows:

$$K(\beta_1) = 0.7012, K(\beta_2) = 0.6628, K(\beta_3) = 0.5900, K(\beta_4) = 0.5900.$$

Table 2 Information for example 6

	c_1	c_2	c_3
u_1	([0.20, 0.5000], [0.36, 0.5000])	([0.10, 0.8000], [0.00, 0.1000])	([0.40, 0.5400], [0.20, 0.2378])
u_2	([0.30, 0.7500], [0.00, 0.2400])	([0.15, 0.5000], [0.00, 0.3000])	([0.60, 0.3333], [0.00, 0.5940])
u_3	([0.00, 0.2000], [0.40, 0.7500])	([0.20, 0.8000], [0.10, 0.1800])	([0.36, 0.6000], [0.20, 0.2000])
u_4	([0.30, 0.4792], [0.00, 0.4000])	([0.51, 0.7000], [0.00, 0.2000])	([0.00, 0.2000], [0.00, 0.8000])

From values of $K(\beta_i)$ $(i = 1, 2, 3, 4)$, we can not sure weather u_3 or u_4 is better.

Next, we use the weighted geometric average operator, and also take $\omega_1 = \omega_2 = \omega_3 = \frac{1}{3}$. By calculating, we can obtain that $K(\beta_1) = 0.5900$, $K(\beta_2) = 0.5900$, $K(\beta_3) = 0.4020$, $K(\beta_4) = 0.4063$.

We could not well either u_1 or u_2 is better. And the result of two aggregation operators are different, so use the accuracy function K could not make decision. Next, we use the accuracy function A instead of K.

The weighted arithmetic average value β_i for u_i $(i = 1, 2, 3, 4)$ are as follows:

$$\beta_1 = ([0.2440, 0.6417], [0.0000, 0.2282]),\ \beta_2 = ([0.3803, 0.5632], [0.0000, 0.3497]),$$

$$\beta_3 = ([0.2000, 0.6000], [0.2000, 0.3000]),\ \beta_4 = ([0.3000, 0.5000], [0.0000, 0.4000]).$$

Then, we obtain $A(\beta_i)$ $(i = 1, 2, 3, 4)$ as follows:

$$A(\beta_1) = 0.5689,\ A(\beta_2) = 0.5750,\ A(\beta_3) = 0.5000,\ A(\beta_4) = 0.5167.$$

Ranking all alternatives according to the accuracy values of $A(\beta_i)$ $(i = 1, 2, 3, 4)$: $A_2 > A_1 > A_4 > A_3$.

Next, the weighted geometric average value β_i for u_i $(i = 1, 2, 3, 4)$ are as follows:

$$\beta_1 = ([0.2000, 0.6000], [0.2000, 0.3000]),\ \beta_2 = ([0.3000, 0.5000], [0.0000, 0.4000]),$$

$$\beta_3 = ([0.0000, 0.4579], [0.2440, 0.4526]),\ \beta_4 = ([0.0000, 0.4063], [0.0000, 0.5421]).$$

We obtain $A(\beta_i)$ $(i = 1, 2, 3, 4)$ as follows:

$$A(\beta_1) = 0.5000,\ A(\beta_2) = 0.5167,\ A(\beta_3) = 0.3550,\ A(\beta_4) = 0.3698.$$

Ranking all alternatives according to the accuracy values of $A(\beta_i)$ $(i = 1, 2, 3, 4)$: $A_2 > A_1 > A_4 > A_3$.

Thus the alternative u_2 is the most desirable alternative according to weighted arithmetic and the weighted geometric average operator.

5 Conclusion

In this paper, we propose a new accuracy function for interval-valued intuitionistic fuzzy numbers, which can be used to rank interval-valued intuitionistic fuzzy numbers more correctly than the existing accuracy functions. In addition, we utilize the interval-valued intuitionistic fuzzy weighted aggregation operators to aggregate the interval-valued intuitionistic fuzzy information when dealing with multicriteria decision making problems, and rank all aggregated interval-valued intuitionistic fuzzy information according to the accuracy values. Finally, we give two examples, the first is used to illustrate the validity of the proposed new accuracy function, and the second a decision making problem, which could not be solved if not use the accuracy function K.

Acknowledgments This work is supported by the National Natural Science Foundation of China (Nos. 61273018 and 61302190).

References

1. Atanassov, K.: Intuitionistic fuzzy sets. Fuzzy Sets Syst. **20**, 87–96 (1986)
2. Atanassov, K., Gargov, G.: Interval-valued intuitionistic fuzzy sets. Fuzzy Sets Syst. **31**, 343–349 (1989)
3. Herrera, F., Herrera-Viedma, E.: Linguistic decision analysis: Steps for solving decision problems under linguistic information. Fuzzy Sets Syst. **115**, 67–82 (2000)
4. Hong, D.H., Choi, C.H.: Multi criteria fuzzy decision-making problems based on vague set theory. Fuzzy Sets Syst. **114**, 103–113 (2000)
5. Nayagam, V., Muralikrish, S., Sivaraman, G.: Multicriteria decision-making method based on interval-valued intuitionistic fuzzy set. Expert Syst. Appl. **38**(3), 1464–1467 (2011)
6. Nayagam, V.G., Sivaraman, G.: Ranking of interval-valued intuitionistic fuzzy sets. Appl. Soft Comput. **11**, 3368–3372 (2011)
7. Sahin, R.: Fuzzy multicriteria decision making method based on the improved accuracy function for interval-valued intuitionistic fuzzy sets. Soft Comput. 1–7 (2015)
8. Xu, Z.S.: A method based on diatance measure for interval-valued intuitionistic fuzzy group decision making. Inf. Sci. **180**(1), 181–190 (2010)
9. Xu, Z.S., Cai, X.Q.: Incomplete interval-valued intuitionistic fuzzy preference relations. Int. J. Gen. Syst. **38**(8), 871–886 (2009)
10. Xu, Z.S., Chen, J.: On geometric aggregation over interval valued intuitionistic fuzzy information. In: FSKD, Fourth International Conference on Fuzzy Systems and Knowledge Discovery, vol. 2, pp. 466–471 (2007)
11. Xu, Z.S.: Intuitionistic fuzzy aggregation operators. IEEE Trans. Fuzzy Syst. **15**(6), 1179–1187 (2007)
12. Xu, Z.S.: Intuitionistic preference relations and their application in group decision making. Inf. Sci. **177**(11), 2363–2379 (2007)
13. Xu, Z.S.: Methods for aggregating interval-valued intuitionistic fuzzy information and their application to decision making. Control Decis. **22**, 215–219 (2007)
14. Ye, J.: Multicriteria fuzzy decision-making methods based on a novel accuracy function under interval-valued intuitionistic fuzzy environment. Expert Syst. Appl. **36**, 6899–6902 (2009)
15. Zadeh, L.A.: Fuzzy sets. Inf. Control **8**, 338–356 (1965)
16. Zadeh, L.A.: The concept of a linguistic variable and its application to approximate reasoning (I). Inf. Sci. **8**, 199–249 (1975)
17. Zadeh, L.A.: The concept of a linguistic variable and its application to approximate reasoning (II). Inf. Sci. **8**, 301–357 (1975)
18. Zadeh, L.A.: The concept of a linguistic variable and its application to approximate reasoning (III). Inf. Sci. **9**, 43–80 (1975)

Regression Analysis Model Based on Normal Fuzzy Numbers

Cui-Ling Gu, Wei Wang and Han-Yu Wei

Abstract Fuzzy regression analysis plays an important role in analyzing the correlation between the dependent and explanatory variables in the fuzzy system. This paper put forward the FLS (Fuzzy Least Squares) method for parameter estimating of the fuzzy linear regression model with input, output variables and regression coefficients that are normal fuzzy numbers. Our improved method proves the statistical properties, i.e., linearity and unbiasedness of the fuzzy least square estimators. Residuals, residual sum of squares and coefficient of determination are given to illustrate the fitting degree of the regression model. Finally, the method is validated in both rationality and validity by solving a practical parameter estimation problem.

Keywords Normal fuzzy numbers · Fuzzy regression analysis · Fuzzy least squares · Coefficient of determination

1 Introduction

The term regression was introduced by Francis Galton. Now, regression analysis is a fundamental analytic tool in many research fields. The method gives a crisp relationship between the dependent and explanatory variables with an estimated variance of measurement errors. Fuzzy regression [1] techniques provide a useful means to model the functional relationships between the dependent variable and independent variables in a fuzzy environment. After the introduction of fuzzy linear

C.-L. Gu · H.-Y. Wei
Mathematics and Statistical Institute, Zhoukou Normal University, Zhoukou 466001, Henan, China
e-mail: 20121024@zknu.edu.cn

H.-Y. Wei
e-mail: weihanyu8207@163.com

W. Wang (✉)
School of Network Engineering, Zhoukou Normal University, Zhoukou 466001, Henan, China
e-mail: wangwei@zknu.edu.cn

T.-H. Fan et al. (eds.), *Quantitative Logic and Soft Computing 2016*, Advances in Intelligent Systems and Computing 510, DOI 10.1007/978-3-319-46206-6_46

regression by Tanaka et al. [2], there has been a great deal of literatures on this topic [3–13]. Diamond [3] defined the distance between two fuzzy numbers and the estimated fuzzy regression parameters by minimizing the sum of the squares of the deviation. Chang [4] summarized three kinds of fuzzy regression methods from existing regression models: minimum fuzzy rule, the rule of least squares fitting and interval regression analysis method. For the purpose of integration of fuzziness and randomness, mixed regression model is put forward in [5]. Chang proposed the triangular fuzzy regression parameters least squares estimation by using the weighted fuzzy arithmetic and least-square fitting criterion. Sakawa and Yano [6] studied the fuzzy linear regression relation between the dependent variable and the fuzzy explanatory variable based on three given linear programming methods. In order to estimate the parameters of fuzzy linear regression model with input, output variables and regression coefficients are LR typed fuzzy numbers, Zhang [7] first represented the observed fuzzy data by using intervals, and then used the left, right point and the midpoint data sets of intervals to derive the corresponding regression coefficients of conventional linear regression models. Zhang [8] discussed the least squares estimation and the error estimate of the fuzzy regression analysis when the coefficient is described by trapezoidal fuzzy numbers depicting the fuzzy concept by using the gaussian membership function corresponding to human mind. To our knowledge, few researches are conducted on fuzzy regression analysis based on normal fuzzy numbers. Therefore, in this paper, we first calculate the least squares estimator of the fuzzy linear regression model, and then discuss statistical properties of the fuzzy least squares (FLS) estimator. Then, we give residuals, residual sum of squares and coefficient of determination and illustrate the fitting degree of the regression model. Last, we also verify the rationality and validity of the parameter estimation method by a numerical example (Fig. 1).

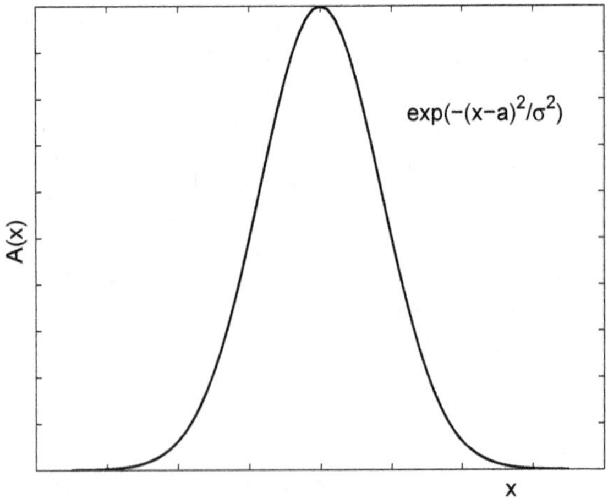

Fig. 1 The schematic diagram of fuzzy normal numbers

2 Preliminaries

Definition 1 ([14]) If fuzzy number $\tilde{A}$ has the following membership function

$$\tilde{A}(x) = exp\left\{-\frac{(x-a)^2}{\sigma^2}\right\}, x, a \in R, \sigma > 0$$

where R is a set of real numbers, then $\tilde{A}$ is called a normal fuzzy number determined by a and σ^2, and thus denoted by $\tilde{A} = (a, \sigma^2)$.

Let $\tilde{A} = (a, \sigma_a^2)$ and $\tilde{B} = (b, \sigma_b^2)$, then three operations of the normal fuzzy numbers are defined as follows: (1) $\tilde{A} + \tilde{B} = (a + b, \sigma_a^2 + \sigma_b^2)$; (2) $\lambda\tilde{A} = (\lambda a, \lambda\sigma_a^2)$; (3)$\frac{1}{\tilde{A}} = (\frac{1}{a}, \frac{1}{\sigma_a^2})$, where $a \neq 0$.

Definition 2 ([15]) The expectation of fuzzy number $\tilde{A}$ is

$$E(\tilde{A}) \triangleq \frac{\int_{-\infty}^{+\infty} x\tilde{A}(x)dx}{\int_{-\infty}^{+\infty} \tilde{A}(x)dx} \tag{1}$$

where $\int_{-\infty}^{+\infty} \tilde{A}(x)dx > 0$. The average of $\tilde{A}$ is denoted by the expectation $E(\tilde{A})$ of fuzzy number $\tilde{A}$. In particular, when $\tilde{A} = (a, \sigma_a^2)$, $E(\tilde{A}) = a$.

Definition 3 ([15]) The variance of fuzzy number $\tilde{A}$ is

$$D(\tilde{A}) \triangleq \frac{\int_{-\infty}^{+\infty} \tilde{A}(x)(x - E(\tilde{A}))^2 dx}{\int_{-\infty}^{+\infty} \tilde{A}(x)dx} \tag{2}$$

where $\int_{-\infty}^{+\infty} \tilde{A}(x)dx > 0$. The spread of $\tilde{A}$ is denoted by the variance $D(\tilde{A})$ of fuzzy number $\tilde{A}$. In particular, when $\tilde{A} = (a, \sigma_a^2)$, $D(\tilde{A}) = \frac{\sigma^2}{2}$.

Definition 4 ([15]) Multiplication between fuzzy numbers $\tilde{A}$ and $\tilde{B}$ is defined as:

$$\tilde{A} \otimes \tilde{B} \triangleq \int_{-\infty}^{+\infty} \tilde{A}(x)dx \int_{-\infty}^{+\infty} \tilde{B}(y)dy \tag{3}$$

when $\tilde{A} = \tilde{B}$, and $\tilde{A} \otimes \tilde{B} = \tilde{A} \otimes \tilde{A} = [\int_{\infty}^{\infty} \tilde{A}(x)dx]^2$, $\tilde{A} \otimes \tilde{A} = ||\tilde{A}||^2$ is called the module of $\tilde{A}$.

Let $\tilde{A}$ and $\tilde{B}$ denote the fuzzy numbers $\tilde{A} = (a, \sigma_a^2)$, and $\tilde{B} = (b, \sigma_b^2)$ respectively, then

$$\tilde{A} \otimes \tilde{B} \triangleq \int_{\infty}^{\infty} \tilde{A}(x)dx \int_{\infty}^{\infty} \tilde{B}(y)dy = \int_{\infty}^{\infty} e^{-\frac{(x-a)^2}{\sigma_a^2}} dx \int_{\infty}^{\infty} e^{-\frac{(y-b)^2}{\sigma_b^2}} dy = \pi\sigma_a\sigma_b,$$

Specifically, when $\tilde{A} = \tilde{B}, \tilde{A} \otimes \tilde{A} = ||\tilde{A}||^2$.

Definition 5 ([16]) Let $\tilde{A} = (a, \sigma_a^2), \tilde{B} = (b, \sigma_b^2)$, then the distance between $\tilde{A}$ and $\tilde{B}$ is defined as:

$$d^2(\tilde{A}, \tilde{B}) = (a-b)^2 + \frac{1}{2}(\sigma_a^2 - \sigma_b^2)^2 \tag{4}$$

3 The Least Squares Estimator of Fuzzy Linear Regression Model

The classical linear regression model is as follows:

$$Y = \beta_0 + \beta_1 X_1 + \beta_2 X_2 + \cdots + \beta_k X_k \tag{5}$$

where Y is explained as variable and $X_1, X_2, \ldots, X_k$ are explanatory variables, β_0, $\beta_1, \ldots, \beta_k$ are regression coefficients. Let$\{(X_i, Y_i) : i = 1, 2, \ldots, n\}$ be a set of sample observations, ordinary least squares estimation is frequently based on the fact that the overall error between the estimated $\hat{Y}_i$ and the observations Y_i should be as small as possible. That is, the corresponding Q residual between the estimated $\hat{Y}_i$ and the observations Y_i should be as small as possible. Symbolically,

$$Q = \sum_{i=1}^{n}(Y_i - \hat{Y}_i)^2 = \sum_{i=1}^{n}(Y_i - (\hat{\beta}_0 + \hat{\beta}_1 X_{1i} + \cdots + \hat{\beta}_k X_{ki}))^2 \tag{6}$$

According to the principle of differential and integral calculus, Q will be the minimum value when the first order partial derivative of Q about $\beta_0, \beta_1, \ldots, \beta_k$ is equal to zero.

However, in many cases, the fuzzy relations in formula (5) must be considered. In general, there are the following three conditions [9]:

(a) $\tilde{Y}_i = \beta_0 + \beta_1 \tilde{X}_{1i} + \beta_2 \tilde{X}_{2i} + \cdots + \beta_k \tilde{X}_{ki}, \beta_0, \beta_1, \ldots, \beta_k \in R, \tilde{X}_1, \ldots, \tilde{X}_k, \tilde{Y}_i \in \widetilde{F}(R), i = 1, 2, \ldots, n$;

(b) $\tilde{Y}_i = \tilde{\beta}_0 + \tilde{\beta}_1 X_{1i} + \tilde{\beta}_2 X_{2i} + \cdots + \tilde{\beta}_k X_{ki}, \tilde{\beta}_0, \tilde{\beta}_1, \ldots, \tilde{\beta}_k, \tilde{Y}_i \in \widetilde{F}(R), X_1, \ldots, X_k \in R, i = 1, 2, \ldots, n$;

(c) $\tilde{Y}_i = \tilde{\beta}_0 + \tilde{\beta}_1 \tilde{X}_{1i} + \tilde{\beta}_2 \tilde{X}_{2i} + \cdots \tilde{\beta}_k \tilde{X}_{ki}, \tilde{\beta}_0, \tilde{\beta}_1, \ldots, \tilde{\beta}_k, \tilde{X}_1, \ldots, \tilde{X}_k, \tilde{Y}_i \in \widetilde{F}(R), i = 1, 2, \ldots, n$.

In fact, (b) is the most common conditions. For (b), we focus on the fuzzy linear regression model in which dependent variables are the form of real numbers and explanatory variables and the regression coefficients are the form of normal fuzzy numbers.

Theorem 1 *Assume the fuzzy multiple linear regression model is as follows:*

$$\tilde{Y}_i = \tilde{\beta}_0 + \tilde{\beta}_1 X_{1i} + \tilde{\beta}_2 X_{2i} + \cdots + \tilde{\beta}_k X_{ki}$$

then

$$\begin{aligned}\tilde{Y}_i = (a_i, \sigma_i^2) &= (a_{\tilde{\beta}_0}, \sigma^2_{\tilde{\beta}_0}) + (a_{\tilde{\beta}_1}, \sigma^2_{\tilde{\beta}_1})X_{1i} + (a_{\tilde{\beta}_2}, \sigma^2_{\tilde{\beta}_2})X_{2i} + \cdots + (a_{\tilde{\beta}_k}, \sigma^2_{\tilde{\beta}_k})X_{ki} \\ &= (a_{\tilde{\beta}_0} + a_{\tilde{\beta}_1}X_{1i} + \cdots + a_{\tilde{\beta}_k}X_{ki}) + (\sigma^2_{\tilde{\beta}_0} + \sigma^2_{\tilde{\beta}_1}X_{1i}^2 + \cdots + \sigma^2_{\tilde{\beta}_k}X_{ki}^2)\end{aligned}$$

Let $\boldsymbol{a} = \boldsymbol{X}\boldsymbol{\psi}$, $\boldsymbol{b} = \boldsymbol{X_1}\boldsymbol{\zeta}$, *where*

$$\boldsymbol{\psi} = [a_{\tilde{\beta}_0}, a_{\tilde{\beta}_1}, \ldots, a_{\tilde{\beta}_k}]^T, \boldsymbol{\zeta} = [\sigma^2_{\tilde{\beta}_0}, \sigma^2_{\tilde{\beta}_1}, \ldots, \sigma^2_{\tilde{\beta}_k}]^T, \boldsymbol{b} = [\sigma_1^2, \sigma_2^2, \ldots, \sigma_n^2]^T$$

$$\boldsymbol{a} = [a_1, a_2, \ldots, a_n]^T, \boldsymbol{A} = [\hat{a}_{\tilde{\beta}_0}, \hat{a}_{\tilde{\beta}_1}, \ldots, \hat{a}_{\tilde{\beta}_k}]^T, \boldsymbol{\sigma} = [\hat{\sigma}^2_{\tilde{\beta}_0}, \hat{\sigma}^2_{\tilde{\beta}_1}, \ldots, \hat{\sigma}^2_{\tilde{\beta}_k}]^T$$

$$\boldsymbol{X} = \begin{pmatrix} 1 & X_{11} & X_{21} & \cdots & X_{k1} \\ 1 & X_{12} & X_{22} & \cdots & X_{k2} \\ \vdots & \vdots & \vdots & \ddots & \vdots \\ 1 & X_{1n} & X_{2n} & \cdots & X_{kn} \end{pmatrix}$$

$$\boldsymbol{X_1} = \begin{pmatrix} 1 & X_{11}^2 & X_{21}^2 & \cdots & X_{k1}^2 \\ 1 & X_{12}^2 & X_{22}^2 & \cdots & X_{k2}^2 \\ \vdots & \vdots & \vdots & \ddots & \vdots \\ 1 & X_{1n}^2 & X_{2n}^2 & \cdots & X_{kn}^2 \end{pmatrix}$$

where $i = 1, 2, \ldots, n$; $\tilde{\beta}_0, \tilde{\beta}_1, \ldots, \tilde{\beta}_k, \tilde{Y} \in \tilde{F}(R)$; $X_1, X_2, \ldots, X_k \in R$. *Then, the FLS of* $\tilde{\beta}_0, \tilde{\beta}_1, \ldots, \tilde{\beta}_k$ *are defined as:*

$$\begin{cases} \boldsymbol{A} = (\boldsymbol{X}'\boldsymbol{X})^{-1}\boldsymbol{X}'\boldsymbol{a} \\ \boldsymbol{\sigma} = (\boldsymbol{X_1}'\boldsymbol{X_1})^{-1}\boldsymbol{X_1}'\boldsymbol{b} \end{cases}$$

Proof Assuming that $\{(X_i, \tilde{Y}_i), i = 1, 2, \ldots, n\}$ are the set of known samples, and $\tilde{Y}_i = (a_i, \sigma_i^2)$, the sum Q of the squares of the dispersion between the estimated $\hat{\tilde{Y}}_i$ and the observations $\tilde{Y}_i$ should be minimized. That is,

$$\begin{aligned}
\tilde{Q} &= \sum_{i=1}^{n}(\tilde{Y}_i - \hat{\tilde{Y}}_i)^2 = \sum_{i=1}^{n}\{(a_i, \sigma_i^2) - [(\hat{a}_{\tilde{\beta}_0}, \hat{\sigma}^2_{\tilde{\beta}_0}) + (\hat{a}_{\tilde{\beta}_1}, \hat{\sigma}^2_{\tilde{\beta}_1})X_{1i} + \cdots \\
&+ (\hat{a}_{\tilde{\beta}_k}, \hat{\sigma}^2_{\tilde{\beta}_k})X_{ki}]\} \\
&= \sum_{i=1}^{n}[(a_i, \sigma_i^2) - (\hat{a}_{\tilde{\beta}_0} + \hat{a}_{\tilde{\beta}_1}X_{1i} + \cdots + \hat{a}_{\tilde{\beta}_k}X_{ki}, \\
&\quad \hat{\sigma}^2_{\tilde{\beta}_0} + \hat{\sigma}^2_{\tilde{\beta}_1}X_{1i}^2 + \cdots + \hat{\sigma}^2_{\tilde{\beta}_1}X_{ki}^2)] \\
&= \sum_{i=1}^{n}[(a_i - \hat{a}_{\tilde{\beta}_0} - \hat{a}_{\tilde{\beta}_1}X_{1i} - \cdots - \hat{a}_{\tilde{\beta}_k}X_{ki})^2 \\
&+ \frac{1}{2}(\sigma_i^2 - \hat{\sigma}^2_{\tilde{\beta}_0} - \hat{\sigma}^2_{\tilde{\beta}_1}X_{1i}^2 - \cdots - \hat{\sigma}^2_{\tilde{\beta}_k}X_{ki}^2)^2]
\end{aligned}$$

should be minimized. $\tilde{Q}$ will be the minimum value when the first order partial derivatives of Q about $\hat{\tilde{\beta}}_0, \hat{\tilde{\beta}}_1, \ldots, \hat{\tilde{\beta}}_k$ are equal to zero. In this case, fuzzy ordinary least squares estimator can be calculated.

$$\begin{cases}
\dfrac{\partial \tilde{Q}}{\partial \hat{a}_{\tilde{\beta}_0}} = -2\sum_{i=1}^{n}(a_i - \hat{a}_{\tilde{\beta}_0} - \hat{a}_{\tilde{\beta}_1}X_{1i} - \cdots - \hat{a}_{\tilde{\beta}_k}X_{ki}) = 0 \\
\dfrac{\partial \tilde{Q}}{\partial \hat{a}_{\tilde{\beta}_1}} = -2\sum_{i=1}^{n}(a_i - \hat{a}_{\tilde{\beta}_0} - \hat{a}_{\tilde{\beta}_1}X_{1i} - \cdots - \hat{a}_{\tilde{\beta}_k}X_{ki})X_{1i} = 0 \\
\cdots\cdots \\
\dfrac{\partial \tilde{Q}}{\partial \hat{a}_{\tilde{\beta}_k}} = -2\sum_{i=1}^{n}(a_i - \hat{a}_{\tilde{\beta}_0} - \hat{a}_{\tilde{\beta}_1}X_{1i} - \cdots - \hat{a}_{\tilde{\beta}_k}X_{ki})X_{ki} = 0 \\
\dfrac{\partial \tilde{Q}}{\partial \hat{\sigma}^2_{\tilde{\beta}_0}} = -\sum_{i=1}^{n}(\sigma_i^2 - \hat{\sigma}^2_{\tilde{\beta}_0} - \hat{\sigma}^2_{\tilde{\beta}_1}X_{1i}^2 - \cdots - \hat{\sigma}^2_{\tilde{\beta}_k}X_{ki}^2) = 0 \\
\dfrac{\partial \tilde{Q}}{\partial \hat{\sigma}^2_{\tilde{\beta}_1}} = -\sum_{i=1}^{n}(\sigma_i^2 - \hat{\sigma}^2_{\tilde{\beta}_0} - \hat{\sigma}^2_{\tilde{\beta}_1}X_{1i}^2 - \cdots - \hat{\sigma}^2_{\tilde{\beta}_k}X_{ki}^2)X_{1i}^2 = 0 \\
\cdots\cdots \\
\dfrac{\partial \tilde{Q}}{\partial \hat{\sigma}^2_{\tilde{\beta}_k}} = -\sum_{i=1}^{n}(\sigma_i^2 - \hat{\sigma}^2_{\tilde{\beta}_0} - \hat{\sigma}^2_{\tilde{\beta}_1}X_{1i}^2 - \cdots - \hat{\sigma}^2_{\tilde{\beta}_k}X_{ki}^2)X_{ki}^2 = 0
\end{cases}$$

Then, the above equations can be simplified as

$$\begin{cases} \sum_{i=1}^{n} a_i &= n\hat{a}_{\tilde{\beta}_0} + \hat{a}_{\tilde{\beta}_1} \sum_{i=1}^{n} X_{1i} + \cdots + \hat{a}_{\tilde{\beta}_k} \sum_{i=1}^{n} X_{ki} \\ \sum_{i=1}^{n} a_i X_{1i} &= \hat{a}_{\tilde{\beta}_0} \sum_{i=1}^{n} X_{1i} + \hat{a}_{\tilde{\beta}_1} \sum_{i=1}^{n} X_{1i}^2 + \cdots + \hat{a}_{\tilde{\beta}_k} \sum_{i=1}^{n} X_{ki} X_{1i} \\ \cdots\cdots & \\ \sum_{i=1}^{n} a_i X_{ki} &= \hat{a}_{\tilde{\beta}_0} \sum_{i=1}^{n} X_{ki} + \hat{a}_{\tilde{\beta}_1} \sum_{i=1}^{n} X_{ki} X_{ki} + \cdots + \hat{a}_{\tilde{\beta}_k} \sum_{i=1}^{n} X_{ki}^2 \\ -\sum_{i=1}^{n} \sigma_i^2 &= n\hat{\sigma}_{\tilde{\beta}_0}^2 + \hat{\sigma}_{\tilde{\beta}_1}^2 \sum_{i=1}^{n} X_{1i}^2 + \cdots + \hat{\sigma}_{\tilde{\beta}_k}^2 \sum_{i=1}^{n} X_{ki}^2 \\ -\sum_{i=1}^{n} \sigma_i^2 X_{1i}^2 &= \hat{\sigma}_{\tilde{\beta}_0}^2 \sum_{i=1}^{n} X_{1i}^2 + \hat{\sigma}_{\tilde{\beta}_1}^2 \sum_{i=1}^{n} X_{1i}^4 + \cdots + \hat{\sigma}_{\tilde{\beta}_k}^2 \sum_{i=1}^{n} X_{ki}^2 X_{1i}^2 \\ \cdots\cdots & \\ -\sum_{i=1}^{n} \sigma_i^2 X_{ki}^2 &= \hat{\sigma}_{\tilde{\beta}_0}^2 \sum_{i=1}^{n} X_{ki}^2 + \hat{\sigma}_{\tilde{\beta}_1}^2 \sum_{i=1}^{n} X_{ki}^2 X_{1i}^2 + \cdots + \hat{\sigma}_{\tilde{\beta}_k}^2 \sum_{i=1}^{n} X_{ki}^4 \end{cases}$$

The matrix expression of the normal equations is as follows

$$\begin{cases} (\boldsymbol{X}'\boldsymbol{X})\boldsymbol{A} = \boldsymbol{X}'\boldsymbol{a} \\ (\boldsymbol{X_1}'\boldsymbol{X_1})\boldsymbol{\sigma} = \boldsymbol{X_1}'\boldsymbol{b} \end{cases}$$

And least squares estimator of parameters are as follows

$$\begin{cases} \boldsymbol{A} = (\boldsymbol{X}'\boldsymbol{X})^{-1}\boldsymbol{X}'\boldsymbol{a} \\ \boldsymbol{\sigma} = (\boldsymbol{X_1}'\boldsymbol{X_1})^{-1}\boldsymbol{X_1}'\boldsymbol{b} \end{cases}$$

□

Corollary 1 *Assume that the fuzzy simple linear regression model is as follows*

$$\tilde{Y}_i = \tilde{\beta}_0 + \tilde{\beta}_1 X_i, i = 1, 2, \ldots, n, \tilde{\beta}_0, \tilde{\beta}_1, \tilde{Y}_i \in \tilde{F}(R), X_i \in R, \tilde{Y}_i = (a_i, \sigma_i^2),$$

that is

$$\tilde{Y}_i = (a_i, \sigma_i^2) = (a_{\tilde{\beta}_0}, \sigma_{\tilde{\beta}_0}^2) + (a_{\tilde{\beta}_1}, \sigma_{\tilde{\beta}_1}^2)X_i, i = 1, 2, \ldots, n, \tilde{\beta}_0, \tilde{\beta}_1 \tilde{Y}_i \in \tilde{F}(R), X_i \in R$$

where $\hat{\tilde{\beta}}_0 = (\hat{a}_{\tilde{\beta}_0}, \hat{\sigma}_{\tilde{\beta}_0}^2)$ *and* $\hat{\tilde{\beta}}_1 = (\hat{a}_{\tilde{\beta}_1}, \hat{\sigma}_{\tilde{\beta}_1}^2)$ *are respectively the FLS of* $\tilde{\beta}_0$ *and* $\tilde{\beta}_1$. *then*

$$\begin{cases} \hat{a}_{\tilde{\beta}_0} = \dfrac{\sum_{i=1}^{n} a_i \sum_{i=1}^{n} X_i^2 - \sum_{i=1}^{n} X_i \sum_{i=1}^{n} a_i X_i}{n\sum_{i=1}^{n} X_i^2 - (\sum_{i=1}^{n} X_i)^2}, & \hat{\sigma}_{\tilde{\beta}_0}^2 = \dfrac{\sum_{i=1}^{n} \sigma_i^2 \sum_{i=1}^{n} X_i^4 - \sum_{i=1}^{n} X_i^2 \sum_{i=1}^{n} X_i^2 \sigma_i^2}{-n\sum_{i=1}^{n} X_i^4 + (\sum_{i=1}^{n} X_i^2)^2} \\ \hat{a}_{\tilde{\beta}_1} = \dfrac{n\sum_{i=1}^{n} a_i X_i - \sum_{i=1}^{n} X_i \sum_{i=1}^{n} a_i}{n\sum_{i=1}^{n} X_i^2 - (\sum_{i=1}^{n} X_i)^2}, & \hat{\sigma}_{\tilde{\beta}_1}^2 = \dfrac{-n\sum_{i=1}^{n} X_i^2 \sigma_i^2 - \sum_{i=1}^{n} X_i^2 \sum_{i=1}^{n} \sigma_i^2}{-n\sum_{i=1}^{n} X_i^4 + (\sum_{i=1}^{n} X_i^2)^2} \end{cases}$$

Proof Let $X_i, \tilde{Y}_i, i = 1, 2, \ldots, n$ be a set of sample observations and $\tilde{Y}_i = (a_i, \sigma_i^2)$, according to Theorem 1, is revised as follows:

$$\tilde{Q} = \sum_{i=1}^{n} (\tilde{Y}_i - \hat{\tilde{Y}}_i)^2$$

$$= \sum_{i=1}^{n} \{(a_i, \sigma_i^2) - [(\hat{a}_{\tilde{\beta}_0}, \hat{\sigma}^2_{\tilde{\beta}_0}) + (\hat{a}_{\tilde{\beta}_1}, \hat{\sigma}^2_{\tilde{\beta}_1})X_i]\}^2$$

$$= \sum_{i=1}^{n} [(a_i, \sigma_i^2) - (\hat{a}_{\tilde{\beta}_0} + \hat{a}_{\tilde{\beta}_1} X_i, \hat{\sigma}^2_{\tilde{\beta}_0} + \hat{\sigma}^2_{\tilde{\beta}_1} X_i^2)]^2$$

$$= \sum_{i=1}^{n} \left[(a - \hat{a}_{\tilde{\beta}_0} - \hat{a}_{\tilde{\beta}_1} X_i)^2 + \frac{1}{2} (\sigma_i^2 - \hat{\sigma}^2_{\tilde{\beta}_0} + \hat{\sigma}^2_{\tilde{\beta}_1} X_i^2)^2 \right]$$

Obviously, $\tilde{Q}$ will be minimized when the first order partial derivatives of $\tilde{Q}$ about $\hat{\tilde{\beta}}_0$, $\hat{\tilde{\beta}}_1$ and are equal to zero. That is, we can solve the question by making the first order partial derivatives of $\tilde{Q}$ about $\hat{a}_{\tilde{\beta}_0}, \hat{a}_{\tilde{\beta}_1}, \hat{\sigma}^2_{\tilde{\beta}_0}, \hat{\sigma}^2_{\tilde{\beta}_1}$ respectively equal to zero.

$$\begin{cases} \dfrac{\partial \tilde{Q}}{\partial \hat{a}_{\tilde{\beta}_0}} & = -2 \sum\limits_{i=1}^{n} (a_i - \hat{a}_{\tilde{\beta}_0} - \hat{a}_{\tilde{\beta}_1} X_i) = 0 \\ \dfrac{\partial \tilde{Q}}{\partial \hat{a}_{\tilde{\beta}_1}} & = -2 \sum\limits_{i=1}^{n} (a_i - \hat{a}_{\tilde{\beta}_0} - \hat{a}_{\tilde{\beta}_1} X_i) X_i = 0 \\ \dfrac{\partial \tilde{Q}}{\partial \hat{\sigma}^2_{\tilde{\beta}_0}} & = - \sum\limits_{i=1}^{n} (\sigma_i^2 - \hat{\sigma}^2_{\tilde{\beta}_0} + \hat{\sigma}^2_{\tilde{\beta}_1} X_i^2) = 0 \\ \dfrac{\partial \tilde{Q}}{\partial \hat{\sigma}^2_{\tilde{\beta}_1}} & = - \sum\limits_{i=1}^{n} (\sigma_i^2 - \hat{\sigma}^2_{\tilde{\beta}_0} + \hat{\sigma}^2_{\tilde{\beta}_1} X_i^2) X_i^2 = 0 \end{cases}$$

The above equations may be written as

$$\begin{cases} n\hat{a}_{\tilde{\beta}_0} + \hat{a}_{\tilde{\beta}1} \sum\limits_{i=1}^{n} X_i = \sum\limits_{i=1}^{n} a_i \\ \hat{a}_{\tilde{\beta}_0} \sum\limits_{i=1}^{n} X_i + \hat{a}_{\tilde{\beta}_1} \sum\limits_{i=1}^{n} X_i^2 = \sum\limits_{i=1}^{n} a_i X_i \\ n\hat{\sigma}^2_{\tilde{\beta}_0} - \hat{\sigma}^2_{\tilde{\beta}_1} \sum\limits_{i=1}^{n} X_i^2 = - \sum\limits_{i=1}^{n} \sigma_i^2 \\ \hat{\sigma}^2_{\tilde{\beta}_0} \sum\limits_{i=1}^{n} X_i^2 - \hat{\sigma}^2_{\tilde{\beta}_1} \sum\limits_{i=1}^{n} X_i^4 = - \sum\limits_{i=1}^{n} \sigma_i^2 X_i^2 \end{cases}$$

Then, in terms of Cramer's rule, we can obtain the linear fuzzy least squares estimator of the simple linear regression model by solving the above equations. □

4 The Statistical Properties of Fuzzy Least Squares Estimator

Theorem 2 *Fuzzy least squares estimator*

$$\begin{cases} \boldsymbol{A} = (\boldsymbol{X}'\boldsymbol{X})^{-1}\boldsymbol{X}'\boldsymbol{a} \\ \boldsymbol{\sigma} = (\boldsymbol{X_1}'\boldsymbol{X_1})^{-1}\boldsymbol{X_1}'\boldsymbol{b} \end{cases}$$

is a linear estimator.

Proof Since

$$\begin{cases} \boldsymbol{A} = (\boldsymbol{X}'\boldsymbol{X})^{-1}\boldsymbol{X}'\boldsymbol{a} = \boldsymbol{Ca} \\ \boldsymbol{\sigma} = (\boldsymbol{X_1}'\boldsymbol{X_1})^{-1}\boldsymbol{X_1}'\boldsymbol{b} = \boldsymbol{Db} \end{cases}$$

where $\boldsymbol{C} = (\boldsymbol{X}'\boldsymbol{X})^{-1}\boldsymbol{X}', \boldsymbol{D} = (\boldsymbol{X}'\boldsymbol{X})^{-1}\boldsymbol{X}'$, the parameter estimator is a linear combination of explanatory variables. □

In order to know statistic properties of the parameter estimator in simple fuzzy regression mode1, let $x_i = X_i - \bar{X}$, where $\bar{X} = \frac{1}{n}\sum_{i=1}^{n} X_i$. When $\bar{\tilde{Y}} = (\bar{a}, \bar{\sigma}^2,) y_i = (\check{a}, \check{\sigma}^2) = \tilde{Y}_i - \bar{\tilde{Y}} = (a_i, \sigma_i^2) - (\bar{a}, \bar{\sigma}^2) = (a_i - \bar{a}, \sigma_i^2 - \bar{\sigma}^2)$, where $\bar{a} = E(\frac{1}{n}\sum_{i=1}^{n}\tilde{Y}_i) = \frac{1}{n}\sum_{i=1}^{n} E(\tilde{Y}_i) = \frac{1}{n}\sum_{i=1}^{n} a_i, \bar{\sigma^2} = Var(\frac{1}{n}\sum_{i=1}^{n}\tilde{Y}_i) = \frac{1}{n^2}\sum_{i=1}^{n} Var(\tilde{Y}_i) = \frac{1}{n^2}\sum_{i=1}^{n}\sigma_i^2$, then

$$\sum_{i=1}^{n} x_i^2 = \sum_{i=1}^{n} (X_i - \bar{X})^2 = \sum_{i=1}^{n} X_i^2 - \frac{1}{n}\left(\sum_{i=1}^{n} X_i\right)^2$$

that is

$$n\sum_{i=1}^{n} x_i^2 = n\sum_{i=1}^{n} (X_i - \bar{X})^2 = \sum_{i=1}^{n} X_i^2 - \left(\sum_{i=1}^{n} X_i\right)^2$$

so

$$\begin{cases} \hat{a}_{\tilde{\beta}_0} = \bar{a} - \hat{a}_{\tilde{\beta}_1}\bar{X} \\ \hat{a}_{\tilde{\beta}_1} = \dfrac{\sum_{i=1}^{n} \check{a}_1 x_i}{\sum_{i=1}^{n} x_i^2} \end{cases}$$

Corollary 2 *Expectations* $\hat{a}_{\tilde{\beta}_0}$ *and* $\hat{a}_{\tilde{\beta}_1}$ *of fuzzy least squares estimator* $\hat{\tilde{\beta}}_0 = (\hat{a}_{\tilde{\beta}_0}, \hat{\sigma}^2_{\tilde{\beta}_0})$ *and* $\hat{\tilde{\beta}}_1 = (\hat{a}_{\tilde{\beta}_1}, \hat{\sigma}^2_{\tilde{\beta}_1})$ *are linear estimators.*

Proof

$$\hat{a}_{\tilde{\beta}_1} = \frac{\sum_{i=1}^{n} \breve{a}_i x_i}{\sum_{i=1}^{n} x_i^2} = \frac{\sum_{i=1}^{n} (a_i - \bar{a}) x_i}{\sum_{i=1}^{n} x_i^2} = \frac{\sum_{i=1}^{n} a_i x_i}{\sum_{i=1}^{n} x_i^2} - \frac{\sum_{i=1}^{n} \bar{a} x_i}{\sum_{i=1}^{n} x_i^2}$$

$$= \frac{\sum_{i=1}^{n} a_i x_i}{\sum_{i=1}^{n} x_i^2} - \frac{\bar{a} \sum_{i=1}^{n} x_i}{\sum_{i=1}^{n} x_i^2} = \sum_{i=1}^{n} k_i a_i$$

where $k_i = \frac{x_i}{\sum_{i=1}^{n} x_i^2}$, $\sum_{i=1}^{n} x_i = 0$;

$$\hat{a}_{\tilde{\beta}_0} = \bar{a} - \hat{a}_{\tilde{\beta}_1} \bar{X} = \frac{1}{n} \sum_{i=1}^{n} a_i - \sum_{i=1}^{n} k_i a_i \bar{X} = \sum_{i=1}^{n} \left(\frac{1}{n} - \bar{X} k_i \right) a_i = \sum_{i=1}^{n} w_i a_i$$

where $w_i = \frac{1}{n} - \bar{X} k_i$. □

Theorem 3 *Fuzzy least squares estimator*

$$\begin{cases} \boldsymbol{A} = (\boldsymbol{X}'\boldsymbol{X})^{-1} \boldsymbol{X}' \boldsymbol{a} \\ \boldsymbol{\sigma} = (\boldsymbol{X_1}'\boldsymbol{X_1})^{-1} \boldsymbol{X_1}' \boldsymbol{b} \end{cases}$$

are unbiased estimators.

Proof

$$E(\boldsymbol{A}) = E[(\boldsymbol{X}'\boldsymbol{X})^{-1} \boldsymbol{X}' \boldsymbol{a})] = E[(\boldsymbol{X}'\boldsymbol{X})^{-1} \boldsymbol{X}' \boldsymbol{X} \boldsymbol{\psi}] = E(\boldsymbol{\psi}) = \boldsymbol{\psi}$$
$$E(\boldsymbol{\sigma}) = E[(\boldsymbol{X_1}'\boldsymbol{X_1})^{-1} \boldsymbol{X_1}' \boldsymbol{b}] = E[(\boldsymbol{X_1}'\boldsymbol{X_1})^{-1} \boldsymbol{X_1}' (\boldsymbol{X_1} \boldsymbol{\zeta})] = \boldsymbol{\zeta}$$

So fuzzy least squares estimators are unbiased. □

Corollary 3 *Expectations* $a_{\tilde{\beta}_0}$ *and* $a_{\tilde{\beta}_1}$ *of fuzzy least squares estimator* $\hat{\tilde{\beta}}_0 = (\hat{a}_{\tilde{\beta}_0}, \hat{\sigma}^2_{\tilde{\beta}_0})$ *and* $\hat{\tilde{\beta}}_1 = (\hat{a}_{\tilde{\beta}_1}, \hat{\sigma}^2_{\tilde{\beta}_1})$ *are unbiased estimators of the parameters* $\tilde{\beta}_0$, $\tilde{\beta}_1$.

Proof

$$\hat{a}_{\tilde{\beta}_1} = \sum_{i=1}^{n} k_i a_i = \sum_{i=1}^{n} k_i (a_{\tilde{\beta}_0} + a_{\tilde{\beta}_1} X_i) = a_{\tilde{\beta}_0} \sum_{i=1}^{n} k_i + a_{\tilde{\beta}_1} \sum_{i=1}^{n} k_i X_i$$

where $k_i = \frac{x_i}{\sum\limits_{i=1}^{n} x_i^2}$, $\sum\limits_{i=1}^{n} k_i = \frac{\sum\limits_{i=1}^{n} x_i}{\sum\limits_{i=1}^{n} x_i^2} = \frac{\sum\limits_{i=1}^{n} x_i}{\sum\limits_{i=1}^{n} x_i^2} = 0$

$$\sum_{i=1}^{n} k_i X_i = \frac{\sum\limits_{i=1}^{n} x_i X_i}{\sum\limits_{i=1}^{n} x_i^2} = \frac{\sum\limits_{i=1}^{n} x_i(x_i + \bar{X})}{\sum\limits_{i=1}^{n} x_i^2} = \frac{\bar{X}\sum\limits_{i=1}^{n} x_i}{\sum\limits_{i=1}^{n} x_i^2} + \frac{\sum\limits_{i=1}^{n} x_i^2}{\sum\limits_{i=1}^{n} x_i^2} = 1$$

so $E(\hat{a}_{\tilde{\beta}_1}) = a_{\tilde{\beta}_1}$;

$$\hat{a}_{\tilde{\beta}_0} = \sum_{i=1}^{n} w_i a_i = \sum_{i=1}^{n} [w_i(a_{\tilde{\beta}_0} + a_{\tilde{\beta}_1} X_i)] = a_{\tilde{\beta}_0} \sum_{i=1}^{n} w_i + a_{\tilde{\beta}_1} \sum_{i=1}^{n} w_i X_i$$

where, $w_i = \frac{1}{n} - \bar{X}k_i$,$E\left(\sum\limits_{i=1}^{n} w_i\right) = E\left(\sum\limits_{i=1}^{n} \left(\frac{1}{n} - \bar{X}k_i\right)\right) = 1$,

$$\sum_{i=1}^{n} w_i X_i = \sum_{i=1}^{n} \left(\frac{1}{n} - \bar{X}k_i\right) X_i = \frac{1}{n}\sum_{i=1}^{n} X_i - \bar{X}\sum_{i=1}^{n} k_i X_i = \bar{X} - \bar{X} = 0$$

so $E(\hat{a}_{\tilde{\beta}_0}) = a_{\tilde{\beta}_0}$. □

5 Assessment on Fuzzy Multiple Linear Regression Model

Regression analysis is a useful statistical method for analyzing quantitative relationships between two or more variables. It is important for the regression analysis to assess the performance of fitting regression model. That is to say, after estimating parameter of fuzzy liner regression model, how far is it from the parameter estimation to the true value? In fuzzy regression analysis, the simplest method evaluating the fuzzy regression model is to take the residual and the Coefficient of Determination as metrics. According to Classical Regression Mode [17], we can calculate the residual and the Coefficient of Determination about Fuzzy Regression Model by using fuzzy calculation rule which listed previously.

Theorem 4 give the module formula of residual $|\check{e}_i|$ and require that it is as small as possible. The fuzzy total sum of squares(FTSS) and the fuzzy explained sum of squares(FESS) are given in Theorem 5, and we obtain fuzzy coefficient of determination $\tilde{R}^2$ in Theorem 6, $\tilde{R}$ is bigger, and more better.

Theorem 4 *The residual produced by the fuzzy multiple linear regression model based on normal fuzzy numbers is defined as*

$$|\check{e}_i| = \sqrt{\pi}\sqrt{\hat{\sigma}^2_{\tilde{\beta}_0} + \sigma_i^2} + \sqrt{\pi}\sigma_{\tilde{\beta}_1}|X_{1i}| + \sqrt{\pi}\sigma_{\tilde{\beta}_2}|X_{2i}| + \cdots + \sqrt{\pi}\sigma_{\tilde{\beta}_k}|X_{ki}|$$

Proof

$$\begin{aligned}
|\check{e}_i| &= |\hat{\tilde{Y}}_i - \tilde{Y}_i| \\
&= |(\hat{a}_{\tilde{\beta}_0}, \hat{\sigma}^2_{\tilde{\beta}_0}) + (\hat{a}_{\tilde{\beta}_1}, \hat{\sigma}^2_{\tilde{\beta}_1})X_{1i} + (\hat{a}_{\tilde{\beta}_1}, \hat{\sigma}^2_{\tilde{\beta}_2})X_{2i} + \cdots + (\hat{a}_{\tilde{\beta}_k}, \hat{\sigma}^2_{\tilde{\beta}_k})X_{ki} - (a_i, \sigma_i^2)| \\
&= |(\hat{a}_{\tilde{\beta}_0} - a_i, \hat{\sigma}^2_{\tilde{\beta}_0} + \sigma_i^2) + (\hat{a}_{\tilde{\beta}_1}, \hat{\sigma}^2_{\tilde{\beta}_1})X_{1i} + (\hat{a}_{\tilde{\beta}_1}, \hat{\sigma}^2_{\tilde{\beta}_2})X_{2i} + \cdots + (\hat{a}_{\tilde{\beta}_k}, \hat{\sigma}^2_{\tilde{\beta}_k})X_{ki}| \\
&= |(\hat{a}_{\tilde{\beta}_0} - a_i, \hat{\sigma}^2_{\tilde{\beta}_0} + \sigma_i^2| + |(\hat{a}_{\tilde{\beta}_1}, \hat{\sigma}^2_{\tilde{\beta}_1})||X_{1i}| + \cdots + |(\hat{a}_{\tilde{\beta}_k}, \hat{\sigma}^2_{\tilde{\beta}_k})||X_{ki}| \\
&= \sqrt{\pi}\sqrt{\hat{\sigma}^2_{\tilde{\beta}_0} + \sigma_i^2} + \sqrt{\pi}\sigma_{\tilde{\beta}_1}|X_{1i}| + \sqrt{\pi}\sigma_{\tilde{\beta}_2}|X_{2i}| + \cdots + \sqrt{\pi}\sigma_{\tilde{\beta}_k}|X_{ki}| \qquad \square
\end{aligned}$$

Corollary 4 *The residual produced by the fuzzy simple linear regression model based on normal fuzzy numbers is expressed as*

$$\check{e}_i = \sqrt{\pi}\sqrt{\hat{\sigma}^2_{\tilde{\beta}_0} + \sigma_i^2} + \sqrt{\pi}\hat{\sigma}_{\tilde{\beta}_1}|X_i|$$

Proof

$$\begin{aligned}
\check{e}_i &= |\hat{\tilde{Y}}_i - \tilde{Y}_i| = |(\hat{a}_{\tilde{\beta}_0}, \hat{\sigma}^2_{\tilde{\beta}_0}) + (\hat{a}_{\tilde{\beta}_1}, \hat{\sigma}^2_{\tilde{\beta}_1})X_i - (a_i, \sigma_i^2)| \\
&= |(\hat{a}_{\tilde{\beta}_0} - a_i, \hat{\sigma}^2_{\tilde{\beta}_0} + \sigma_i^2) + (\hat{a}_{\tilde{\beta}_1}, \hat{\sigma}^2_{\tilde{\beta}_1})X_i| \\
&= |(\hat{a}_{\tilde{\beta}_0} - a_i, \hat{\sigma}^2_{\tilde{\beta}_0} + \sigma_i^2)| + |(\hat{a}_{\tilde{\beta}_1}, \hat{\sigma}^2_{\tilde{\beta}_1})||X_i| \\
&= \sqrt{\pi}\sqrt{\hat{\sigma}^2_{\tilde{\beta}_0} + \sigma_i^2} + \sqrt{\pi}\hat{\sigma}_{\tilde{\beta}_1}|X_i| \qquad \square
\end{aligned}$$

Theorem 5 *The residual sum of squares produced by the fuzzy multiple linear regression model based on normal fuzzy numbers is defined as*

$$\begin{aligned}
FTSS &= \pi\sum_{i=1}^{n}(\hat{\sigma}^2_{\tilde{\beta}_0} + \sigma_i^2) + \pi\sum_{i=1}^{n}\sum_{j=1}^{k}\hat{\sigma}^2_{\tilde{\beta}_j}X_{ji}^2 \\
&+ 2\pi\sum_{i=1}^{n}\sum_{j=1}^{k}\hat{\sigma}_{\tilde{\beta}_j}X_{ji}\sqrt{(\hat{\sigma}^2_{\tilde{\beta}_0} + \sigma_i^2)} + \pi\sum_{i=1}^{n}\sum_{j\neq r}^{k}\hat{\sigma}_{\tilde{\beta}_j}\hat{\sigma}_{\tilde{\beta}_r}X_{ri}X_{ji}
\end{aligned}$$

The explained sum of squares produced by the fuzzy multiple linear regression model based on normal fuzzy numbers is defined as

$$FESS = n\pi(\hat{\sigma}^2_{\tilde{\beta}_0} + \bar{\sigma}^2) + \pi \sum_{i=1}^{n}\sum_{j=1}^{k} \hat{\sigma}^2_{\tilde{\beta}_j} X^2_{ji}$$

$$+ 2\pi\sqrt{\hat{\sigma}^2_{\tilde{\beta}_0} + \bar{\sigma}^2} \sum_{i=1}^{n}\sum_{j-1}^{k} \hat{\sigma}_{\tilde{\beta}_j} X_{ji} + \pi \sum_{i=1}^{n}\sum_{j\neq r}^{k} \hat{\sigma}_{\tilde{\beta}_j}\hat{\sigma}_{\tilde{\beta}_r} X_{ri}X_{ji}$$

Proof

$$\begin{aligned}
FTSS &= \textstyle\sum_{i=1}^{n} (\hat{\tilde{Y}}_i - \tilde{Y}_i)^2 \\
&= \textstyle\sum_{i=1}^{n} [(\hat{a}_{\tilde{\beta}_0}, \hat{\sigma}^2_{\tilde{\beta}_0}) + (\hat{a}_{\tilde{\beta}_1}, \hat{\sigma}^2_{\tilde{\beta}_1})X_{1i} + \cdots + (\hat{a}_{\tilde{\beta}_k}, \hat{\sigma}^2_{\tilde{\beta}_k})X_{ki} - (a_i, \sigma_i^2)]^2 \\
&= \textstyle\sum_{i=1}^{n} [(\hat{a}_{\tilde{\beta}_0} - a_i, \hat{\sigma}^2_{\tilde{\beta}_0} + \sigma_i^2) + (\hat{a}_{\tilde{\beta}_1}, \hat{\sigma}^2_{\tilde{\beta}_1})X_{1i} + \cdots + (\hat{a}_{\tilde{\beta}_k}, \hat{\sigma}^2_{\tilde{\beta}_k})X_{ki}]^2 \\
&= \textstyle\sum_{i=1}^{n} \Big[(\hat{a}_{\tilde{\beta}_0} - a_i, \hat{\sigma}^2_{\tilde{\beta}_0} + \sigma_i^2)^2 + \sum_{j=1}^{k} (\hat{a}_{\tilde{\beta}_j}, \hat{\sigma}^2_{\tilde{\beta}_1})^2 X^2_{ji} \\
&\quad + 2\textstyle\sum_{j=1}^{k} (\hat{a}_{\tilde{\beta}_0} - a_i, \hat{\sigma}^2_{\tilde{\beta}_0} + \sigma_i^2)(\hat{a}_{\tilde{\beta}_j}, \hat{\sigma}^2_{\tilde{\beta}_j})X_{ji} \\
&\quad + \textstyle\sum_{r\neq j}^{k} (\hat{a}_{\tilde{\beta}_j}, \hat{\sigma}^2_{\tilde{\beta}_j})(\hat{a}_{\tilde{\beta}_r}, \hat{\sigma}^2_{\tilde{\beta}_r})X_{ji}X_{ri}\Big] \\
&= \textstyle\sum_{i=1}^{n} \Big[(\hat{a}_{\tilde{\beta}_0} - a_i, \hat{\sigma}^2_{\tilde{\beta}_0} + \sigma_i^2)^2 + \sum_{i=1}^{n}\sum_{j=1}^{k} (\hat{a}_{\tilde{\beta}_j}, \hat{\sigma}^2_{\tilde{\beta}_1})^2 X^2_{ji} \\
&\quad + 2\textstyle\sum_{i=1}^{n}\sum_{j=1}^{k} (\hat{a}_{\tilde{\beta}_j}, \hat{\sigma}^2_{\tilde{\beta}_j})(\hat{a}_{\tilde{\beta}_0} - a_i, \hat{\sigma}^2_{\tilde{\beta}_0} + \sigma_i^2)X_{ji} \\
&\quad + \textstyle\sum_{i=1}^{n}\sum_{r\neq j}^{k} (\hat{a}_{\tilde{\beta}_j}, \hat{\sigma}^2_{\tilde{\beta}_j})(\hat{a}_{\tilde{\beta}_r}, \hat{\sigma}^2_{\tilde{\beta}_r})X_{ji}X_{ri}\Big] \\
&= \pi\textstyle\sum_{i=1}^{n} (\hat{\sigma}^2_{\tilde{\beta}_0} + \sigma_i^2) + \pi\sum_{i=1}^{n}\sum_{j=1}^{k} \hat{\sigma}^2_{\tilde{\beta}_j} X^2_{ji} \\
&\quad + 2\textstyle\sum_{i=1}^{n}\sum_{j=1}^{k} [(\hat{a}_{\tilde{\beta}_0} X_{ji}, \hat{\sigma}^2_{\tilde{\beta}_0} X^2_{ji})(\hat{a}_{\tilde{\beta}_0} - a_i, \hat{\sigma}^2_{\tilde{\beta}_0} + \sigma_i^2)] \\
&\quad + \textstyle\sum_{i=1}^{n}\sum_{r\neq j}^{k} (\hat{a}_{\tilde{\beta}_j} X_{ji}, \hat{\sigma}^2_{\tilde{\beta}_j} X^2_{ji})(\hat{a}_{\tilde{\beta}_r} X_{ri}, \hat{\sigma}^2_{\tilde{\beta}_r} X^2_{ri}) \\
&= \pi\textstyle\sum_{i=1}^{n} (\hat{\sigma}^2_{\tilde{\beta}_0} + \sigma_i^2) + \pi\sum_{i=1}^{n}\sum_{j=1}^{k} \hat{\sigma}^2_{\tilde{\beta}_j} X^2_{ji} \\
&\quad + 2\pi\textstyle\sum_{i=1}^{n}\sum_{j=1}^{k} \hat{\sigma}_{\tilde{\beta}_j} X_{ji}\sqrt{(\hat{\sigma}^2_{\tilde{\beta}_0} + \sigma_i^2)} + \pi\sum_{i=1}^{n}\sum_{j\neq r}^{k} \hat{\sigma}_{\tilde{\beta}_j}\hat{\sigma}_{\tilde{\beta}_r} X_{ri}X_{ji}
\end{aligned}$$

$$
\begin{aligned}
FESS &= \textstyle\sum_{i=1}^{n}(\hat{\tilde{Y}}_i - \bar{\tilde{Y}})^2 \\
&= \textstyle\sum_{i=1}^{n}[(\hat{a}_{\tilde{\beta}_0}, \hat{\sigma}^2_{\tilde{\beta}_0}) + (\hat{a}_{\tilde{\beta}_1}, \hat{\sigma}^2_{\tilde{\beta}_1})X_{1i} + \cdots + (\hat{a}_{\tilde{\beta}_k}, \hat{\sigma}^2_{\tilde{\beta}_k})X_{ki} - (\bar{a}, \bar{\sigma}^2)]^2 \\
&= \textstyle\sum_{i=1}^{n}[(\hat{a}_{\tilde{\beta}_0} - \bar{a}, \hat{\sigma}^2_{\tilde{\beta}_0} + \bar{\sigma}^2) + (\hat{a}_{\tilde{\beta}_1}, \hat{\sigma}^2_{\tilde{\beta}_1})X_{1i} + \cdots + (\hat{a}_{\tilde{\beta}_k}, \hat{\sigma}^2_{\tilde{\beta}_k})X_{ki}]^2 \\
&= \textstyle\sum_{i=1}^{n}\Big[(\hat{a}_{\tilde{\beta}_0} - \bar{a}, \hat{\sigma}^2_{\tilde{\beta}_0} + \bar{\sigma}^2)^2 + \sum_{j=1}^{k}(\hat{a}_{\tilde{\beta}_j}, \hat{\sigma}^2_{\tilde{\beta}_j})^2 X^2_{ji} \\
&\quad +2(\hat{a}_{\tilde{\beta}_0} - \bar{a}, \hat{\sigma}^2_{\tilde{\beta}_0} + \bar{\sigma}^2)\textstyle\sum_{j=1}^{k}(\hat{a}_{\tilde{\beta}_j}, \hat{\sigma}^2_{\tilde{\beta}_j})X_{ji} \\
&\quad + \textstyle\sum_{r\neq j}^{k}(\hat{a}_{\tilde{\beta}_j}, \hat{\sigma}^2_{\tilde{\beta}_j})(\hat{a}_{\tilde{\beta}_r}, \hat{\sigma}^2_{\tilde{\beta}_r})X_{ji}X_{ri}\Big] \\
&= n(\hat{a}_{\tilde{\beta}_0} - \bar{a}, \hat{\sigma}^2_{\tilde{\beta}_0} + \bar{\sigma}^2)^2 + \textstyle\sum_{i=1}^{n}\sum_{j=1}^{k}(\hat{a}_{\tilde{\beta}_1}, \hat{\sigma}^2_{\tilde{\beta}_1})^2 X^2_{ji} \\
&\quad +2(\hat{a}_{\tilde{\beta}_0} - \bar{a}, \hat{\sigma}^2_{\tilde{\beta}_0} + \bar{\sigma}^2)\textstyle\sum_{i=1}^{n}\sum_{j=1}^{k}(\hat{a}_{\tilde{\beta}_j}, \hat{\sigma}^2_{\tilde{\beta}_j})X_{ji} \\
&\quad + \textstyle\sum_{i=1}^{n}\sum_{r\neq j}^{k}(\hat{a}_{\tilde{\beta}_j}, \hat{\sigma}^2_{\tilde{\beta}_j})(\hat{a}_{\tilde{\beta}_r}, \hat{\sigma}^2_{\tilde{\beta}_r})X_{ji}X_{ri} \\
&= n\pi(\hat{\sigma}^2_{\tilde{\beta}_0} + \bar{\sigma}^2) + \pi\textstyle\sum_{j=1}^{k}\sum_{i=1}^{n}\hat{\sigma}^2_{\tilde{\beta}_j}X^2_{ji} \\
&\quad +2\sqrt{\pi}\sqrt{\hat{\sigma}_{\tilde{\beta}_0} + \bar{\sigma}^2}\textstyle\sum_{i=1}^{n}\sum_{j=1}^{k}\sqrt{\pi}\hat{\sigma}_{\tilde{\beta}_j}X_{ji} \\
&\quad + \textstyle\sum_{i=1}^{n}\sum_{r\neq j}^{k}(\hat{a}_{\tilde{\beta}_j}X_{ji}, \hat{\sigma}^2_{\tilde{\beta}_j}X^2_{ji})(\hat{a}_{\tilde{\beta}_r}X_{ri}, \hat{\sigma}^2_{\tilde{\beta}_r}X^2_{ri}) \\
&= n\pi(\hat{\sigma}^2_{\tilde{\beta}_0} + \bar{\sigma}^2) + \pi\textstyle\sum_{i=1}^{n}\sum_{j=1}^{k}\hat{\sigma}^2_{\tilde{\beta}_j}X^2_{ji} \\
&\quad +2\pi\sqrt{\hat{\sigma}^2_{\tilde{\beta}_0} + \bar{\sigma}^2}\textstyle\sum_{i=1}^{n}\sum_{j=1}^{k}\hat{\sigma}_{\tilde{\beta}_j}X_{ji} + \pi\sum_{i=1}^{n}\sum_{j\neq r}^{k}\hat{\sigma}_{\tilde{\beta}_j}\hat{\sigma}_{\tilde{\beta}_r}X_{ri}X_{ji}
\end{aligned}
$$

□

Corollary 5 *The residual produced by the fuzzy simple linear regression model based on normal fuzzy numbers is expressed as*

$$FTSS = \pi\sum_{i=1}^{n}(\hat{\sigma}^2_{\tilde{\beta}_0} + \sigma_i^2) + 2\pi\hat{\sigma}^2_{\tilde{\beta}_1}\sum_{i=1}^{n}\sqrt{(\hat{\sigma}^2_{\tilde{\beta}_0} + \sigma_i^2)X_i^2} + \pi\hat{\sigma}^2_{\tilde{\beta}_1}\sum_{i=1}^{n}X_i^2$$

The explained sum of squares produced by fuzzy simple linear regression model based on normal fuzzy numbers is defined as

$$FESS = n\pi(\hat{\sigma}^2_{\tilde{\beta}_0} + \bar{\sigma}^2) + 2\sqrt{\pi}\sqrt{\hat{\sigma}^2_{\tilde{\beta}_0} + \bar{\sigma}^2}\sum_{i=1}^{n}X_i + \pi\hat{\sigma}^2_{\tilde{\beta}_1}\sum_{i=1}^{n}X_i^2$$

Proof

$$
\begin{aligned}
FTSS &= \sum_{i=1}^{n}(\hat{\tilde{Y}}_i - \tilde{Y}_i)^2 \\
&= \sum_{i=1}^{n}[(\hat{a}_{\tilde{\beta}_0}, \hat{\sigma}^2_{\tilde{\beta}_0}) + (\hat{a}_{\tilde{\beta}_1}, \hat{\sigma}^2_{\tilde{\beta}_1})X_i - (a_i, \sigma_i^2)]^2 \\
&= \sum_{i=1}^{n}[(\hat{a}_{\tilde{\beta}_0} - a_i, \hat{\sigma}^2_{\tilde{\beta}_0} + \sigma_i^2) + (\hat{a}_{\tilde{\beta}_1}, \hat{\sigma}^2_{\tilde{\beta}_1})X_i]^2 \\
&= \sum_{i=1}^{n}[(\hat{a}_{\tilde{\beta}_0} - a_i, \hat{\sigma}^2_{\tilde{\beta}_0} + \sigma_i^2)^2 + 2(\hat{a}_{\tilde{\beta}_0} - a_i, \hat{\sigma}^2_{\tilde{\beta}_0} + \sigma_i^2)(\hat{a}_{\tilde{\beta}_1}, \hat{\sigma}^2_{\tilde{\beta}_1})X_i \\
&\quad + (\hat{a}_{\tilde{\beta}_1}, \hat{\sigma}^2_{\tilde{\beta}_1})^2 X_i^2] \\
&= \sum_{i=1}^{n}(\hat{a}_{\tilde{\beta}_0} - a_i, \hat{\sigma}^2_{\tilde{\beta}_0} + \sigma_i^2) + 2(\hat{a}_{\tilde{\beta}_1}, \hat{\sigma}^2_{\tilde{\beta}_1})\sum_{i=1}^{n}(\hat{a}_{\tilde{\beta}_0} - a_i, \hat{\sigma}^2_{\tilde{\beta}_0} + \sigma_i^2)X_i \\
&\quad + (\hat{a}_{\tilde{\beta}_1}, \hat{\sigma}^2_{\tilde{\beta}_1})^2 \sum_{i=1}^{n} X_i^2 \\
&= \pi\sum_{i=1}^{n}(\hat{\sigma}^2_{\tilde{\beta}_0} + \sigma_i^2) + 2\sqrt{\pi}\hat{\sigma}_{\tilde{\beta}_1}\sum_{i=1}^{n}[(\hat{a}_{\tilde{\beta}_0} - a_i)X_i, (\hat{\sigma}^2_{\tilde{\beta}_0} + \sigma_i^2)X_i^2] \\
&\quad + \pi\hat{\sigma}^2_{\tilde{\beta}_1}\sum_{i=1}^{n} X_i^2 \\
&= \pi\sum_{i=1}^{n}(\hat{\sigma}^2_{\tilde{\beta}_0} + \sigma_i^2) + 2\pi\hat{\sigma}^2_{\tilde{\beta}_1}\sum_{i=1}^{n}\sqrt{(\hat{\sigma}^2_{\tilde{\beta}_0} + \sigma_i^2)X_i^2} + \pi\hat{\sigma}^2_{\tilde{\beta}_1}\sum_{i=1}^{n} X_i^2
\end{aligned}
$$

$$
\begin{aligned}
FESS &= \sum_{i=1}^{n}(\hat{\tilde{Y}}_i - \bar{\tilde{Y}}_i)^2 \\
&= \sum_{i=1}^{n}[(\hat{a}_{\tilde{\beta}_0}, \hat{\sigma}^2_{\tilde{\beta}_0}) + (\hat{a}_{\tilde{\beta}_1}, \hat{\sigma}^2_{\tilde{\beta}_1})X_i - (\bar{a}, \bar{\sigma}^2)]^2 \\
&= \sum_{i=1}^{n}[(\hat{a}_{\tilde{\beta}_0} - \bar{a}, \hat{\sigma}^2_{\tilde{\beta}_0} + \bar{\sigma}^2) + (\hat{a}_{\tilde{\beta}_1}, \hat{\sigma}^2_{\tilde{\beta}_1})X_i]^2 \\
&= \sum_{i=1}^{n}[(\hat{a}_{\tilde{\beta}_0} - \bar{a}, \hat{\sigma}^2_{\tilde{\beta}_0} + \bar{\sigma}^2)^2 + 2(\hat{a}_{\tilde{\beta}_0} - \bar{a}, \hat{\sigma}^2_{\tilde{\beta}_0} \\
&\quad + \bar{\sigma}^2)(\hat{a}_{\tilde{\beta}_1}, \hat{\sigma}^2_{\tilde{\beta}_1})X_i + (\hat{a}_{\tilde{\beta}_1}, \hat{\sigma}^2_{\tilde{\beta}_1})^2 X_i^2] \\
&= n(\hat{a}_{\tilde{\beta}_0} - \bar{a}, \hat{\sigma}^2_{\tilde{\beta}_0} + \bar{\sigma}^2)^2 \\
&\quad + 2(\hat{a}_{\tilde{\beta}_0} - \bar{a}, \hat{\sigma}^2_{\tilde{\beta}_0} + \bar{\sigma}^2)(\hat{a}_{\tilde{\beta}_1}, \hat{\sigma}^2_{\tilde{\beta}_1})\sum_{i=1}^{n} X_i + (\hat{a}_{\tilde{\beta}_1}, \hat{\sigma}^2_{\tilde{\beta}_1})^2 \sum_{i=1}^{n} X_i^2 \\
&= n\pi(\hat{\sigma}^2_{\tilde{\beta}_0} + \bar{\sigma}^2) + 2\sqrt{\pi}\sqrt{\hat{\sigma}^2_{\tilde{\beta}_0} + \bar{\sigma}^2}\sum_{i=1}^{n} X_i + \pi\hat{\sigma}^2_{\tilde{\beta}_1}\sum_{i=1}^{n} X_i^2
\end{aligned}
$$

□

The greater the regression sum of squares, the smaller the sum of squared residuals, and the better the fitting between regression line and the sample points.

Theorem 6 *The coefficient of determination of the fuzzy multiple linear regression model based on normal fuzzy numbers is defined as*

$$\tilde{R}^2 = \frac{FESS}{FTSS}$$
$$= \frac{n\pi(\hat{\sigma}^2_{\tilde{\beta}_0} + \bar{\sigma}^2) + \pi\sum_{i=1}^{n}\sum_{j=1}^{k}\hat{\sigma}^2_{\tilde{\beta}_j}X_{ji}^2 + 2\pi\sqrt{\hat{\sigma}^2_{\tilde{\beta}_0} + \bar{\sigma}^2}\sum_{i=1}^{n}\sum_{j=1}^{k}\hat{\sigma}_{\tilde{\beta}_j}X_{ji} + \pi\sum_{i=1}^{n}\sum_{j\neq r}^{k}\hat{\sigma}_{\tilde{\beta}_j}\hat{\sigma}_{\tilde{\beta}_r}X_{ri}X_{ji}}{\pi\sum_{i=1}^{n}(\hat{\sigma}^2_{\tilde{\beta}_0} + \sigma_i^2) + \pi\sum_{i=1}^{n}\sum_{j=1}^{k}\hat{\sigma}^2_{\tilde{\beta}_j}X_{ji}^2 + 2\pi\sum_{i=1}^{n}\sum_{j=1}^{k}\hat{\sigma}_{\tilde{\beta}_j}X_{ji}\sqrt{(\hat{\sigma}^2_{\tilde{\beta}_0} + \sigma_i^2)} + \pi\sum_{i=1}^{n}\sum_{j\neq r}^{k}\hat{\sigma}_{\tilde{\beta}_j}\hat{\sigma}_{\tilde{\beta}_r}X_{ri}X_{ji}}$$

Proof It is easy to prove Theorem 6 by using Theorem 5. □

Corollary 6 *The coefficient of the determination of fuzzy simple linear regression model based on normal fuzzy numbers is expressed as*

$$\tilde{R}^2 = \frac{FESS}{FTSS}$$
$$= \frac{n\pi(\hat{\sigma}^2_{\tilde{\beta}_0} + \bar{\sigma}^2) + 2\sqrt{\pi}\sqrt{\hat{\sigma}^2_{\tilde{\beta}_0} + \bar{\sigma}^2}\sum_{i=1}^{n}X_i + \pi\hat{\sigma}^2_{\tilde{\beta}_1}\sum_{i=1}^{n}X_i^2}{\pi\sum_{i=1}^{n}(\hat{\sigma}^2_{\tilde{\beta}_0} + \sigma_i^2) + 2\pi\hat{\sigma}^2_{\tilde{\beta}_1}\sum_{i=1}^{n}\sqrt{(\hat{\sigma}^2_{\tilde{\beta}_0} + \sigma_i^2)X_i^2} + \pi\hat{\sigma}^2_{\tilde{\beta}_1}\sum_{i=1}^{n}X_i^2}$$

6 Numerical Example

Assume that the fuzzy linear regression model is as follows:

$$\tilde{Y}_i = \tilde{\beta}_0 + \tilde{\beta}_1 X_{1i} + \tilde{\beta}_2 X_{2i}$$

where, $\tilde{Y}$ is the dependent variable, X_1 and X_2 the explanatory variables, and $(X_{1i}, X_{2i}, \tilde{Y}_i), i = 1, 2, \ldots, n, X_1, X_2 \in R, \tilde{Y} \in \tilde{F}(R)$. Now, our goal is to solve the fuzzy regression and evaluate the model with the observed data shown in Table 1.

Then, the fuzzy regression mode can be obtained by our proposed method.

$$\tilde{Y}_i = (20.5371, 0.0542^2) + (41.5827, 0.0087^2)X_{1i} + (14.5884, 0.0035^2)X_{2i}$$

Residual series of the regression model are shown in Table 2. According to the formulas in Theorem 5, we can calculate the evaluation indexes of the fuzzy model i.e., $FTSS = 7.2531$, $FESS = 6.9836$, and $\tilde{R}^2 = 0.9628$. Clearly, the uncertainty of the practical problem is better considered by the fuzzy linear regression analysis. Using fuzzy numbers to represent the observation data makes it more effective to

Table 1 The observed data

Order	X_1	X_2	$\tilde{Y}$	Order	X_1	X_2	$\tilde{Y}$
1	0.16	0.86	$(40.0, 0.31^2)$	7	0.28	1.15	$(48.6, 0.18^2)$
2	0.18	0.89	$(41.0, 0.22^2)$	8	0.29	1.18	$(49.4, 0.19^2)$
3	0.23	0.94	$(42.0, 0.25^2)$	9	0.32	1.25	$(50.8, 0.23^2)$
4	0.24	0.96	$(43.0, 0.16^2)$	10	0.35	1.29	$(54.3, 0.24^2)$
5	0.22	0.98	$(46.5, 0.17^2)$	11	0.39	1.33	$(57.0, 0.25^2)$
6	0.26	0.99	$(47.2, 0.20^2)$	12	0.45	1.37	$(59.2, 0.21^2)$

Table 2 Residual series of the regression model

Order	Fuzzy residual	Order	Fuzzy residual	Order	Fuzzy residual
1	0.1833	5	0.0659	9	0.1160
2	0.0993	6	0.0862	10	0.1207
3	0.1253	7	0.0741	11	0.1302
4	0.0602	8	0.0810	12	0.0988

resolve the problem. In this example, the residual sequence of the regression model and the coefficient of determination help to understand how well the regression model can fit the sample points. The coefficient of determination 96.28 % implies that the change 96.28 % of the explained variable can be explained by the change of explanatory variables.

7 Conclusions

The paper proposes an improved FLS method for parameter estimating of the fuzzy linear regression model when the explanatory variables are precise and the explained variables and regression parameters are normal fuzzy numbers. Specifically, the paper figures out the fuzzy least squares estimation of multivariate linear regression analysis and gets some statistical properties, i.e., linearity and unbiasedness, of the fuzzy least square estimators. Finally, it illustrates the feasibility and effectiveness of the proposed method by the numerical example.

Acknowledgments This work is supported in part by the Science and Technology Department of Henan Province (Grant No. 152300410230) and the Key Scientific Research Projects of Henan Province (Grant No. 17A110040).

References

1. Zhang, A.W.: Statistical analysis of fuzzy linear regression model based on centroid method. J. Fuzzy Syst. Math. **5**, 172–177 (2012)
2. Tanaka, H., Uejima, S., Asai, K.: Linear regression analysis with fuzzy model. J. Syst. Man Cyber. **12**, 903–907 (1982)
3. Diamond, P.: Fuzzy least squares. J. Inf. Sci. **46**, 141–157 (1988)
4. Chang, Y.H., Ayyub, B.M.: Fuzzy regression methods–a comparative assessment. J. Fuzzy Sets Syst. **119**, 187–203 (2011)
5. Chang, Y.H.: Hybrid fuzzy least-squares regression analysis and its reliability measures. J. Fuzzy Sets Syst. **119**, 225–246 (2011)
6. Sakawa, M., Yano, H.: Multiobjective fuzzy liear regression analysis for fuzzy input-out data. J. Fuzzy Set Syst. **157**, 137–181 (1992)
7. Zhang, A.W.: Parameter estimation of the fuzzy regressive model with lr typed fuzzy coefficients. J. Fuzzy Syst. Math. **27**, 140–147 (2013)
8. Zhang, A.W.: A least-squares approach to fuzzy regression analysis with trapezoidal fuzzy number. J. Math. Pract. Theory **42**, 235–244 (2012)
9. Parvathi, R., Malathi, C., Akram, M.: Intuitionistic fuzzy linear regression analysis. J. Fuzzy Optim. Decis. Mak. **2**, 215–229 (2013)
10. Xu, R.N., Li, C.L.: Multidimensional least squares fitting with a fuzzy model. J. Fuzzy Sets Syst. **119**, 215–223 (2001)
11. Cope, R., D'Urso, P., Giordani, P., et al.: Least squares estimation of a linear regression model with LR fuzzy response. J. Comput. Stat. Data Anal. **51**, 267–286 (2006)
12. Bisserier, A., Boukezzoula, B., Galichet, S.: Linear Fuzzy Regression Using Trapezoidal Fuzzy Intervals. I PMU, Malaga (2008)
13. Liang, Y., Wei, L.L.: Fuzzy linear least-squares regression with LR-type fuzzy coefficients. J. Fuzzy Syst. Math. **3**, 112–117 (2007)
14. Peng, Z.Y., Sun, Y.Y.: The Fuzzy Mathematics and Its Application. Wuhan University Press, Wuhan (2007)
15. Li, A.G., Zhang, Z.H., Meng, Y., et al.: Fuzzy Mathematics and Its Application. Metallurgical Industry Press, Beijing (2005)
16. Xu, R.N.: Research on time series prediction problem with the normal fuzzy number. J. J. Guangzhou Univ. **12**, 82–89 (1988)
17. Li, Z.N.: Econmetrics. Higher Education Press, Beijing (2012)

Weighted L_p Metric on Fuzzy Numbers

Hao-Yue Liu and Tai-He Fan

Abstract In order to reflect the difference of membership degree on metric of fuzzy numbers, we introduce a weighted metric L_ω for fuzzy numbers in this paper. First, the condition on weighted function under which the expression L_ω is a metric is given. Then the topological properties such as completeness and separability of the weighted fuzzy number metric spaces (E^1, L_ω) are discussed. Finally, relations between the weighted metric and the corresponding unweighted one is discussed briefly.

Keywords Fuzzy number · Weighted function · Weighted metric · Completeness · Separability

1 Introduction

In practical issues, metrics of fuzzy numbers reflects the depth of relation between fuzzy numbers. Metrics of fuzzy numbers have been applied in data analysis in various areas such as fuzzy decision, fuzzy structural analysis, fuzzy clustering and so on. The most often used metrics are the Hausdorff metric, supremum metric, L_p metrics, sendograph metric, endograph metric and so on.

In [1] Diamond and Kloeden introduced L_p metrics for fuzzy numbers and discussed completeness and separability of fuzzy number set with respect to L_p metrics, where the difference between membership function values were not considered, this may not be quite rational since at least syntactically higher membership values might be more important than the lower values, the metrics might better reflect such phenomena. Therefore, in [2] the author considered the influence of membership functions on the distance between fuzzy numbers and gave the definition of δ_{p-q} metric. In [3] by using weighted function, considering the influence of cut set on the

H.-Y. Liu · T.-H. Fan (✉)
Department of Mathematical Science, Zhejiang Sci-Tech University, Hangzhou 310018, China
e-mail: Taihefan@163.com

H.-Y. Liu
e-mail: Lhyty7@163.com

T.-H. Fan et al. (eds.), *Quantitative Logic and Soft Computing 2016*,
Advances in Intelligent Systems and Computing 510,
DOI 10.1007/978-3-319-46206-6_47

distance between fuzzy numbers in terms of the new metric of mid and spread. In [4] essentially a weighted average of distances which is the convex combinations of the infimum and supremum of fuzzy numbers was proposed. Such idea were also reflected by using concept like reducing function by Voxman [5], Tan [6], R.N. Xu [7] and so on. In this paper, we study the weighted L_p metric on fuzzy number space. We give condition on the weighted function $\omega(\alpha)$ under which the expression L_ω is a metric on E^1. Then we take one-dimensional space as an example to discuss completeness and separability of the weighted metric space (E^1, L_ω). Finally, relations between the weighted metric and the original metric are discussed briefly.

2 Basic Notation and Preliminaries

Throughout the whole paper R denotes the set of all real numbers and I denotes the closed unit interval [0, 1], $K_c(R)$ denotes the family of non-empty compact convex sets of R.

First, we recall the basics of fuzzy numbers.

Definition 1 Let $E^n = \{u \mid u : R^n \to [0,1],\ u\, has\, the\, following\, properties\, (i){-}(iv)\}$

(i) u is normal i.e. $u(x_0) = 1$ for some $x_0 \in R^n$,
(ii) u is quasiconvex (fuzzy convex),
(iii) u is upper semicontinuous,
(iv) $[u]^0 = \{x \in R^n \mid u(x) > 0\}$ is compact.

Elements in E^n are called n-dimension fuzzy numbers.

Definition 2 For $u \in E^n$ and $\alpha \in I$, let $[u(x)]^\alpha = \{x \in R | u(x) \geq \alpha\}$, $[u]^\alpha$ is called the $\alpha - cut$ of u. Then all cut sets of u are nonempty closed intervals.

Definition 3 The Hausdorff metric d_H on $K_c(R^n)$ is defined as follows:

$$d_H(A, B) = \max\{\rho(A, B), \rho(B, A)\}$$

For $A, B \in K_c(R^n)$, where $\rho(A, B) = \max\limits_{a \in A} dist(a, B)$, $dist(x, A) = \min\limits_{a \in A} \|x - a\| (x \in R^n)$.

Definition 4 For each $u, v \in E^n$ and $\alpha \in I$, $1 \leq p \leq \infty$, the $L_p - metric$ between u, v is defined as follows:

$$L_p(u, v) = \left(\int_0^1 [d_H([u]^\alpha, [v]^\alpha)]^p d\alpha \right)^{1/p}.$$

3 Main Result

In this section we introduce weighted function $\omega(\alpha)$ and discuss the condition that weighted function should satisfy. Then give the definition of weighted metric $L_\omega(u, v)$. Moreover, completeness and separability of the weighted metric (E^1, L_ω) are proved, and briefly explain the influence of membership degree on the distance between two fuzzy numbers.

First, weighted function $\omega(\alpha) : I \to R$ and weighted metric $L_p\omega$ are characterized.

Definition 5 A function $\omega : I \to R$ is a weighted function if ω is a nonnegative function.

In order to define weighted metric, we need to restrict the weight function. For example, to define weighted L_p metric, we must require that the corresponding integral exist at least.

Definition 6 If the weight function $\omega(\alpha)$ is integrable on I, then for $u, v \in E^n$ and $0 \leq p < \infty$, we define the weighted L_p metric between u, v as follows:

$$L_{p\omega} = \left(\int_0^1 \omega(\alpha)[d_H([u]^\alpha), [v]^\alpha]^p d\alpha \right)^{1/p} \tag{1}$$

Taking one-dimensional setting as an example we discuss the restriction on the weight function $\omega(\alpha)$ such that $L_\omega = L_{1\omega}$ is indeed a metric. The discussion of n-dimensional space and other weighted metric is similar.

We need to state that the above weighted expression may not satisfy the condition of metric of fuzzy numbers. We study the conditions such that (1) is indeed a metric of fuzzy numbers. We describe as follows:

Theorem 1 *$L_\omega = \int_0^1 \omega(\alpha)[d_H([u]^\alpha, [v]^\alpha)]d\alpha$ is a metric on E^1 if and only if for each subinterval I_1 of I, $\int_{I_1} \omega(\alpha)d\alpha > 0$.*

Proof If L_ω is a metric on E^1, then for all $u, v, u_1 \in E^1$, the following condition (i)–(iii) on a metric are satisfied:

(i) Positivity, i.e., $\int_0^1 \omega(\alpha)d_H([u]^\alpha, [v]^\alpha)d\alpha > 0$;
(ii) Symmetry, i.e., $\int_0^1 \omega(\alpha)d_H([u]^\alpha, [v]^\alpha)d\alpha = \int_0^1 \omega(\alpha)d_H([v]^\alpha, [u]^\alpha)d\alpha$;
(iii) Triangle inequality, i.e., $\int_0^1 \omega(\alpha)d_H([u]^\alpha, [v]^\alpha)d\alpha \leq \int_0^1 \omega(\alpha)d_H([u]^\alpha, [u_1]^\alpha)d\alpha + \int_0^1 \omega(\alpha)d_H([u_1]^\alpha, [v]^\alpha)d\alpha$

By the basic properties of the integral, $L_\omega(u, v)$ always satisfies symmetry and triangle inequality. Hence, we need to prove that (i) is equivalent to the condition in the theorem.

Sufficiency. For $u, v \in E^1$ such that $u \neq v$, by the left-continuity of the cut set function, there exists $\alpha_0 \in (0, 1)$ and $\varepsilon_0 > 0$ such that $d_H([u]^\alpha, [v]^\alpha) \geq \varepsilon_0$ for $\alpha \in [\alpha_0 - \varepsilon, \alpha_0]$, since $\int_{\alpha_0-\varepsilon}^{\alpha_0} \omega(\alpha)d\alpha > 0$ and

$$\begin{aligned} L_\omega(u, v) &= \int_0^1 \omega(\alpha)d_H([u]^\alpha, [v]^\alpha)d\alpha \\ &\geq \int_{\alpha_0-\varepsilon}^{\alpha_0} \omega(\alpha) \cdot \varepsilon_0 d\alpha = \varepsilon_0 \int_{\alpha_0-\varepsilon}^{\alpha_0} \omega(\alpha)d\alpha > 0 \end{aligned}$$

Necessity. If $\omega(\alpha)$ does not satisfy the condition, there are $\alpha_1, \alpha_2 \in I$ and $\alpha_1 < \alpha_2$ such that $\int_{\alpha_1}^{\alpha_2} \omega(\alpha)d\alpha = 0$. Define fuzzy numbers u, v as follows:

$$u(x) = \begin{cases} \alpha_1, & 1 \leq x < 4, \\ 1, & x = 4, \\ 0, & else. \end{cases}$$

and

$$v(x) = \begin{cases} \alpha_2, & 1 \leq x < 4, \\ 1, & x = 4, \\ 0, & else. \end{cases}$$

Obviously,

$$d_H([u]^\alpha, [v]^\alpha) = \begin{cases} 3, & \alpha \in (\alpha_1, \alpha_2], \\ 0, & otherwise. \end{cases}$$

Then $\int_0^1 \omega(\alpha)d_H([u]^\alpha, [v]^\alpha)d\alpha = \int_{\alpha_1}^{\alpha_2} \omega(\alpha)d_H([u]^\alpha, [v]^\alpha)d\alpha = \int_{\alpha_1}^{\alpha_2} \omega(\alpha) \cdot 3d\alpha = 0$. Thus the positivity does not hold. The proof is thus completed. □

The following example shows the condition that $\omega(\alpha) \neq 0$ a.e. on I is not equivalent to the condition in Theorem 1.

Example 1 Fix $\alpha \in (0, 1)$. First divided I into three sub-intervals, and we remove the open interval of length $\frac{1}{3} \cdot \alpha$ right on the middle. Then divided the remaining two intervals similarly, and remove the open intervals right on the middle of length $\frac{1}{3^2} \cdot \alpha$. The remaining four sub-intervals are dealt with similarly. Generally, in the nth step, we remove the middle 2^{n-1} open sub-intervals from the remaining 2^{n-1} sub-intervals right in the middle, whose length are all $\frac{1}{3^n} \cdot \alpha$. This procedure is similar to the construction of the Cantor set, the only difference is that the length of the removed open sub-intervals are smaller. The union of the removed open sub-intervals is denoted by G_0. Let G_1 be the open sub-interval first removed, its length is $\frac{1}{3} \cdot \alpha$. Let G_2 be the union of the two sub-intervals removed the second time, their length are all $\frac{1}{3^2} \cdot \alpha$. Similarly, the union of the open sub-intervals removed in the nth step is denoted by G_n, they are all of length $\frac{1}{3^n} \cdot \alpha$, and the total number of sub-intervals is 2^{n-1}. The remaining elements of I constitute a Cantor positive measure set denoted by P_0, i.e., $G_0 = \bigcup_{n=1}^{\infty} G_n$, $P_0 = I \backslash G_0$.

Obviously, $mG_0 = \alpha$ and $mP_0 = 1 - \alpha > 0$. Define $\omega(\alpha)$ as follows:

$$\omega(\alpha) = \begin{cases} 1 & \alpha \in G_1, \\ 1/2 & \alpha \in G_2, \\ 1/2^2 & \alpha \in G_3, \\ \dots & \dots \\ 1/2^{n-1} & \alpha \in G_n, \\ 0 & otherwise. \end{cases}$$

Then we have
(1) $\omega(\alpha) \equiv 0$ on P_0, but $\omega(\alpha) \neq 0$ a.e. on I.
(2) If I_1 is an arbitrary sub-interval on I, then $\int_{I_1} \omega(\alpha)d\alpha > 0$.

Proof (1) Obvious.

(2) Clearly, $\omega(\alpha)$ is integrable on I. From the structure of G_0, G_0 is an open set. Let $I_1 \subset I$ be an open interval (if I_1 is a closed interval, we just remove the endpoints of the interval), then from the construction G_0 we have $I_1 \cap G_0 \neq \varnothing$, then there exists $x \in I_1 \cap G_0$, since $I_1 \cap G_0$ is open, there exists a neighborhood $\delta(x, \varepsilon) \subset I_1 \cap G_0$, thus $\int_{I_1} \omega(\alpha)d\alpha > \int_{\delta(x,\varepsilon)} \omega(\alpha)d\alpha > 0$.

This completes the proof. □

In practice, when we studied distance of fuzzy numbers, the differences on the larger cut set are more important than that on the lower cut set. Therefore, it should be a natural condition that the weighted function is monotone. Now we take one-dimensional Euclidean space as an example to discuss the completeness and separability of the weighted metric fuzzy number space. Based on application consideration we require that the weighted function $\omega(\alpha)$ be monotonically increasing. The case of n-dimensional space can be discussed similarly.

Definition 7 ([7]) For $u \in E^n$, the support function s of u is define as follows:

$$s(\alpha, x) = \sup_{p \in [u]^\alpha} < p, x >, (\alpha, x) \in I \times S^{n-1}$$

where S^{n-1} is the unit sphere on R^n, $<, >$ is the inner product on R^n.

For all $u, v \in E^n$, we have the following relation:

$$d_H(u]^\alpha, [v]^\alpha) = \sup_{p \in S^{n-1}} |s([u]^\alpha, p) - s([v]^\alpha, p)|. \quad (2)$$

Theorem 2 *Let $\omega(\alpha)$ be a positive monotone weighted function on I such that $\int_I \omega(\alpha)d\alpha > 0$. Then $(E^1(K), L_\omega)$ is a complete metric space for each nonempty compact subset K of R. Where $E^1(K) = \{u \in E^1 : [u]^0 \subseteq K\}$.*

Proof For $u \in E^1(K)$, let s_u be the support function of u. Let $\{u_n | n = 1, 2, \dots\}$ be a Cauchy sequence in $(E^1(K), L_\omega)$. From [1] it follows that the support function $\omega(\alpha)s_{u_n}$ is a Cauchy sequence in the Banach space of all L_1 integral function $L_1(I, C(S^{n-1}))$ form I to the set of all continuous function defined on S^{n-1}.

Hence there exists $\bar{s} \in L_1(I, C(S^{n-1}))$ such that $\omega(\alpha)s_{u_n} \to \bar{s}(n \to \infty)$ with respect to the L_1 metric and hence there exists a subsequence $\omega(\alpha)s_{u_{n(l)}}$ of $\omega(\alpha)s_{u_n}$ such that $\omega(\alpha)s_{u_{n(l)}}(\alpha, \cdot) \to \bar{s}(\alpha, \cdot)$ a.e. on I. Thus there exists $A \subset I$ with Lebesgue measure 0 and $I \setminus A$ is dense in I such that for each $\alpha \in I \setminus A$, $\{\omega(\alpha) \cdot s_{u_{n(i)}(\alpha,\cdot)}\}$ is Cauchy sequence in $L_1(I, C(S^{n-1}))$. Since

$$\omega(\alpha)d_H([u_{n(j)}]^\alpha, [u_{n(i)}]^\alpha) = \omega(\alpha) \cdot \| s_{u_{n(j)}(\alpha,\cdot) - s_{u_{n(i)}}(\alpha,\cdot)} \|,$$

then, $\| s_{u_{n(j)}} - s_{u_{n(i)}}(\alpha, \cdot) \| \to 0$, by formula (2). Hence for each $\alpha \in I \setminus A$,

$$d_H([u_{n(j)}]^\alpha, [u_{n(i)}]^\alpha) \to 0(i, j \to 0).$$

Thus, $\{[u_{n(j)}]^\alpha\}$ is a Cauchy sequence in (K_C, d_H).

For $\alpha \in I \setminus A$. This shows that there exists $A_\alpha \in K_C$ such that

$$d_H([u_{n(j)}]^\alpha, A_\alpha) \to 0(j \to \infty).$$

Let $d_H^*(A, B) = \inf\{\varepsilon | N(A, \varepsilon) \supset B\}$. For $\alpha_1, \alpha_2 \in I \setminus A$ with $\alpha_2 \le \alpha_1$, since $[u_{n(j)}]^{\alpha_1} \subseteq [u_{n(j)}]^{\alpha_2}$, we have

$$\begin{aligned} d_H^*(A_{\alpha_1}, A_{\alpha_2}) &\le d_H^*(A_{\alpha_1}, [u_{n(j)}]^{\alpha_1}) + d_H^*([u_{n(j)}]^{\alpha_1}, [u_{n(j)}]^{\alpha_2}) \\ &\quad + d_H^*([u_{n(j)}]^{\alpha_2}, A_{\alpha_2}) \to \infty(j \to \infty). \end{aligned}$$

Hence $d_H^*([u_n]^{\alpha_1}, [u_n]^{\alpha_2}) = 0$ and $A_{\alpha_1} \subseteq A_{\alpha_2}$.

For $\alpha \in A \setminus \{0\}$, define $A_\alpha = \cap\{A_\beta : \beta \in (0, \alpha) \setminus A\}$, so $A_\alpha \subseteq A_\beta \subseteq A_{\alpha'}$, for all $0 < \alpha' \le \beta \le \alpha < 1$, where $\alpha, \alpha' \in A \setminus \{0\}$, $\beta \in I \setminus A$. In order to complete the definition of the family $\{A_\alpha\}$, for $\alpha = 0$ we define

$$A_0 = \overline{\cup\{u_\alpha : \alpha \in (0, 1]\}}.$$

So $A_0 \subseteq K$. For all $\alpha \in (0, 1]$, clearly, $A_\alpha \in K_C$. It remains to show that if $\{\alpha_n\}$ is a nondecreasing sequence in I and α_n converges to $\alpha \in I$, then $A_\alpha = \cap_{n \ge 0} A_{\alpha_n}$.

Let $\varepsilon > 0$. For $l > 0$, in (K_C, d_H), since $[u_{n(j)}]^{\alpha_n} \to A_{\alpha_n}$, choose a sequence of cut sets $\{[u_{n(jl)}]^{\alpha_n}\} \subset \{[u_{n(j)}]^{\alpha_n}\}$ such that $d_H([u_{n(jl)}]^{\alpha_n}, A_{\alpha_n}) < \frac{\varepsilon}{2^n}$. Since all these cut sets are nonempty, convex and compact, there is a convergent subsequence $\{[u_{n(jl_p)}]^{\alpha_{l_p}}\}$ of $\{[u_{n(jl)}]^{\alpha_p}\}$, let $\lambda_{l(p)} = A_{k(jl_p)}$, let $d_H([\lambda_{l_p}]^{l_p}, T_\alpha) \to 0$ for some $T_\alpha \in K_C$, as $l_p \to \infty$.

Then for $l_p \ge N(\varepsilon, \alpha)$,

$$\begin{aligned} d_H\left(\bigcap_{n=1}^{l_p} A_{\alpha_n}, T_\alpha\right) &= d_H(A_{\alpha_{l_p}}, A_\alpha) \\ &\le d_H(A_{\alpha_{l_p}}, [\lambda_{l_p}]^{\alpha_{nl_p}}) + d_H([\lambda_{l_p}]^{\alpha_{l_p}}, T_\alpha) \\ &\le \tfrac{\varepsilon}{2^{l_p}} + \tfrac{\varepsilon}{2} \end{aligned}$$

thus $d_H(\bigcap_{n=1}^{\infty} A_{\alpha_n}, T_\alpha) \leq \frac{\varepsilon}{2}$. Since ε is arbitrary, $T_\alpha = \bigcap_{n\geq 0} A_{\alpha_n}$. Now we show that $T_\alpha = A_\alpha$. First, we have $A_\alpha \subseteq T_\alpha$. Assuming $T_\alpha \neq A_\alpha$, choose $x \in T_\alpha \setminus A_\alpha$. Since A_α is compact, there exists a neighborhood V of x and a neighborhood W of A_α such that $V \bigcap W = \varnothing$. By the convergence of $[u_j]^\alpha$, $[u_j]^\alpha \subset W$ for all sufficiently large j, which means that for all $y \in V$, $u_j(y) < \alpha$. Then there exists $j \geq j_0(\varepsilon, \alpha)$, $y \in V$ such that

$$u_j(y) < \alpha - \frac{\varepsilon}{2}. \tag{3}$$

On the other hand, x is a limit point of $x_{l_p} \in [u_{l_p}]^{\alpha_{l_p}}$ and $\lambda_{l_p}(x_{l_p}) \geq \alpha_{l_p}$, when l_p sufficiently large, $j_{l_p} \geq j_1$, $x_{l_p} \in V$, $X \notin W$, $j_{l_p} \geq \max\{j_0, j_1\}$, thus

$$\lambda_{l_p}(x_{l_p}) < \alpha - \frac{\varepsilon}{2}. \tag{4}$$

However, since α_n is monotonic increasing and converges to α, we have $\alpha_{l_p} > \alpha - \frac{\varepsilon}{2}$ hence (3) and (4) are contradictory. Hence $T_\alpha = A_\alpha$. So there is a fuzzy set $u \in E^1$ such that $[u]^\alpha = A_\alpha$.

Next we show that $L_\omega(u_n, u) \to 0$ by showing that $\omega(\alpha) d_H([u_{n(j)}]^\alpha, [u]^\alpha) \to 0$ a.e.

Define $\varphi_n(\alpha) = \omega(\alpha) d_H([u_n]^\alpha, [u]^\alpha)$, the

$$| \varphi_n(\alpha) | \leq 2 \parallel K \parallel \cdot \omega(1) < \infty$$

and

$$| \varphi_k(\alpha) - \varphi_l(\alpha) | \leq \omega(\alpha) d_H([u_k]^\alpha, [u_l]^\alpha).$$

Hence, $\{\varphi_n\}$ is a Cauchy sequence in $L_1(I)$, thus there exists $\overline{\varphi} \in L_1(I)$ such that $\varphi_n \to \overline{\varphi}$, and subsequence $\{\varphi_{n(j)}\}$ converges to $\overline{\varphi}$ almost everywhere. Clearly, $\overline{\varphi} \equiv 0$ a.e. on I, so $\{\varphi_n\}$ converges to 0 function in $L_1(I)$, i.e., $L_\omega(u_n, u) \to 0$, thus $(E^1(K), L_\omega)$ is complete metric space. And the proof is completed. □

Theorem 3 *Let $\omega(\alpha)$ be a positive monotone weight function on I such that $\int_I \omega(\alpha) d\alpha > 0$. Then (E^1, L_ω) is separable.*

Proof Let $u \in E^1$ and $\varepsilon > 0$ be arbitrary. Since $[u]^0$ is compact, take a closed interval $[a_0, b_0]$ such that $[u]^0 \subset [a_0, b_0]$ and a_0, b_0 are rational number, take a division $a_0 = x_0 < x_1 < x_1 < \ldots < x_n = b_0$ of $[a_0, b_0]$ such that $x_{i+1} - x_i = \frac{b_0 - a_0}{n} < \frac{1}{\int_0^1 \omega(\alpha) d\alpha} \cdot \frac{\varepsilon}{2}$. For each $\alpha \in I$, let $V_\alpha = \{\bigcup [x_i - 1, x_i] | [u]^\alpha \bigcap [x_i - 1, x_i] \neq \varnothing\}$. Since $[u]^\alpha$ is an interval, there exists unique $i_1(\alpha)$ and $i_n(\alpha)$ such that

$$[x_{i_1(\alpha)-1}, x_{i_1(\alpha)}] \cap [u]^\alpha \neq \varnothing \neq [x_{i_n(\alpha)-1}, x_{i_n(\alpha)}] \cap [u]^\alpha,$$

and

$$V_\alpha = [x_{i_1(\alpha)-1}, x_{i_1(\alpha)}] \cup [x_{i_1(\alpha)}, x_{i_1(\alpha)+1}] \cup \cdots [x_{i_n(\alpha)-1}, x_{i_n(\alpha)}] \supset u_\alpha.$$

Obviously, $i_1(\alpha)$ is increasing about α and $i_n(\alpha)$ is decreasing about α. Since there are only finitely many number of $i_1(\alpha)$ and $i_n(\alpha)$, and $[u]^\alpha$ is left continuous about α, there are finitely many α_i, $i = 1, 2, \ldots, m$ such that $\alpha_1 = 0, \alpha_m = 1$ and for $\alpha \in (\alpha_{i-1}, \alpha_i]$ $(i = 1, 2, \ldots, m)$, $i_1(\alpha) = i_n(\alpha)$. Clearly, $\nu = \bigcup_{i=1}^{m} \alpha_i V_{\alpha_i}$ determines a fuzzy number $\nu \in E^1$, and

$$d_H([u]^\alpha, [\nu]^\alpha)d\alpha < \frac{1}{\int_0^1 \omega(\alpha)d\alpha} \cdot \frac{\varepsilon}{2},$$

hence

$$\int_0^1 \omega(\alpha)d_H([u]^\alpha, [\nu]^\alpha)d\alpha < \frac{\varepsilon}{2}.$$

Choose rational $\beta_i \in I$ and $\beta_m = 1$ such that $0 < \beta_i - \alpha_i < \frac{1}{m} \cdot \frac{1}{b_0 - a_0} \cdot \frac{1}{\int_0^1 \omega(\alpha)d\alpha} \cdot \frac{\varepsilon}{2}$, $\alpha_i < \beta_i < \alpha_{i+1}$ $(i = 1, 2, \ldots, m-1)$, define fuzzy number ν_1 as follows:

$$\nu_1 = \bigcup_{i=1}^{m} \beta_i V_{\alpha_i}.$$

Clearly, $\nu_1 \in E^1$. Only on $[\alpha_i, \beta_i]$ $(i = 1, 2, \ldots, m-1)$ the cut sets of ν and ν_1 are difference, and the union of these interval are denoted by Ω, $\Omega = \bigcup_{i=1}^{m} [\alpha_i, \beta_i]$. Since $\int_0^1 \omega(\alpha)d\alpha > 0$, and

$$\begin{aligned}\int_0^1 \omega(\alpha)d_H([\nu]^\alpha, [\nu_1]^\alpha)d\alpha &= \int_\Omega \omega(\alpha)d_H([\nu]^\alpha, [\nu_1]^\alpha)d\alpha \\ &< \int_\Omega \omega(\alpha) \cdot m \cdot (b_0 - a_0) \cdot \frac{1}{m} \cdot \frac{1}{b_0 - a_0} \cdot \frac{1}{\int_0^1 \omega(\alpha)d\alpha} \cdot \frac{\varepsilon}{2} d\alpha \\ &< \frac{\varepsilon}{2}.\end{aligned}$$

Hence,

$$\begin{aligned}\int_0^1 \omega(\alpha)d_H([u]^\alpha, [\nu_1]^\alpha)d\alpha \leq &\int_0^1 \omega(\alpha)d_H([u]^\alpha, [\nu]^\alpha)d\alpha \\ &+ \int_0^1 \omega(\alpha)d_H([\nu]^\alpha, [\nu_1]^\alpha)d\alpha < \varepsilon.\end{aligned}$$

Clearly, the set of all fuzzy numbers of form ν_1 is countable. Thus (E^1, L_ω) is separable. □

Remark 1 Form the definition of L_ω metric, when the weight function is a monotonic increasing function, the influence of the higher cut sets on the distance between two fuzzy numbers u, ν is greater than the lower cut sets on $L_\omega(u, \nu)$. This is quite different from the classical L_1 metric ($\omega(\alpha) \equiv 1$) which treats the distance of all cut sets as the same. Thus the L_ω metric can reflect the difference between the fuzzy numbers much better.

Example 2 Let $\omega(\alpha) = \alpha$, $\alpha \in I$, and $0 < \delta < 1$, define a family of fuzzy numbers $\{u_t\}_{t\in[0,1-\delta]}$ by the cut set function as follows:

$$[u_t]^\alpha = \begin{cases} [0, 1], & \alpha \in [0, t], \\ 1, & otherwise. \end{cases}$$

Clearly, $L_1(u_t, u_{t+\delta}) = \delta$ for each t. But

$$L_\omega(u_t, u_{t+\delta}) = \int_t^{t+\delta} \alpha \cdot 1 = \frac{1}{2}[(t+\delta)^2 - t^2] = t\delta + \frac{\delta^2}{2}$$

Obviously, the smaller t is, the smaller the distance between u_t and $u_{t+\delta}$.

Remark 2 In this paper, the discuss is about the weighted metric spaces (E^1, L_ω). It follows that all conclusions can be easily generalized to the higher dimensional cases and we just give the most basic properties of the metric. Further research on the topological properties and the study of other forms of weighted metric spaces on the fuzzy number space will be the subject of our future research.

References

1. Diamond, P., Kloeden, P.: Metric spaces of fuzzy sets: Theory and Applications. World Scientific Publishing, Singapore (1994)
2. Grzegorzewski, P.: Metric and orders in space of fuzzy numbers. Fuzzy Sets Syst. **97**(1), 83–94 (1998)
3. Trutschnig, W., Gonzalez-Rodrigue, G., Colubi, A.: A new family of metric for compact, convex fuzzy sets based on a generalized concept of mid and spread. Inf. Sci. **179**, 3964–3972 (2009)
4. Bertoluzza, C.N., Corrdal, N., Salas, A.: On a new class of distances between fuzzy numbers. Mathw. Soft Comput. **2**, 71–84 (1995)
5. Voxman, W.: Some remarks on distances between fuzzy numbers. Fuzzy Sets Syst. **100**(1–3), 353–365 (1998)
6. Tran, L., Duckstein, L.: Comparison of fuzzy numbers using a fuzzy distance measure. Fuzzy Sets Syst. **130**, 331–341 (2002)
7. Chen, M.L.: A New Fuzzy Analysis Theory. Scientific Publishing, New York (2009)
8. Rao, M.M.: Measure Theory and Integration. Wiley, Hoboken New Jersy (1987)
9. Chen, S.L., Li, X.G., Wang, X.G.: Fuzzy Set and Application. Scientific Publishing, New York (2013)

Multi-variable-term Latticized Linear Programming with Addition-Min Fuzzy Relation Inequalities Constraint

Hai-Tao Lin and Xiao-Peng Yang

Abstract P2P network can be reduced into a system of fuzzy relation inequalities with addition-min composition. In this paper we introduce multi-variable-term latticized linear programming subject to this system. Firstly, we introduce some properties on the minimal solution of the system. Next we define the minimal intervals of the system. Meanwhile, We prove that the optimal solution of the programming is the minimal solution of the system. Finally, we get algorithm for the programming by translating it into some linear programming problems with minimal intervals constraint. An example is given to show the efficiency and feasibility of the algorithm.

Keywords Fuzzy relation inequality · Latticized linear programming · Addition-min composition · Minimal interval

1 Introduction

Fuzzy relation equation was first proposed by E. Sanchez [1, 2] in 1976 and has played an important role in fuzzy logic, fuzzy implication, engineering management, image processing and other application fields.

Optimization problem with fuzzy relation equation or inequality constraint is an research topic. P.-Z. Wang [3] studied the resolution of max-min fuzzy relation inequalities and introduced the corresponding latticized linear programming problem. After then, optimal solution(s) was/were selected from the minimal solution set of the constraint by comparing their objective function values. S.-C. Fang [4] was the first researcher who considered the linear programming problem subject to fuzzy relation equations. Due to the special structure of max-min fuzzy relation equation, the proposed problem was equivalently converted into a 0–1 integer programming

H.-T. Lin · X.-P. Yang (✉)
School of Mathematics and Statistics, Hanshan Normal University, Guangdong 521041, Chaozhou, China
e-mail: happyyangxp@163.com

T.-H. Fan et al. (eds.), *Quantitative Logic and Soft Computing 2016*,
Advances in Intelligent Systems and Computing 510,
DOI 10.1007/978-3-319-46206-6_48

and then solved by the branch-and-bound method. For more works on minimizing a linear function under the fuzzy relation equation constraint, the readers may refer to [5–12].

Recently, S.-J. Yang et al. [13, 14] introduced the concept of fuzzy relation inequality with addition-min composition with application in bittorrent-like peer-to-peer (P2P) file sharing system and investigated the corresponding fuzzy relation linear programming problem. To present some further results on the P2P file sharing system, X.-P. Yang et al. proposed the fuzzy relation multi-level linear programming and min-max programming problems with effective solution algorithms. As pointed out in [14], the file sharing system, under bittorrent-like peer-to-peer transmission mechanism, could be reduced into a system of addition-min fuzzy relation inequalities. The optimization model established and studied in [13] as follows:

$$\begin{aligned} &\min\ z(x) = c_1x_1 + c_2x_2 + \cdots + c_nx_n \\ &\text{s.t. } A \circ x^T \geq b^T. \end{aligned} \tag{1}$$

where $\circ$ is the composition operator of addition-min. The author aimed at minimizing the linear sum of the quality levels, i.e. $x_1, x_2, \ldots, x_n$. The coefficient c_j the weighted factor x_j, $j = 1, 2, \ldots, n$, and the objective function $z(x)$ reflected the network congestion in the P2P file sharing system. However, the coefficients $c_1, c_2, \ldots, c_n$ are usually objective and are given by some specific experts. Thus, it is necessary to consider more than one group of values of the coefficients. Suppose that there are p experts providing p groups of values of the coefficients, i.e.

$$\{C_{tj} > 0 | t = 1, 2, \ldots, p, j = 1, 2, \ldots, n\},$$

and the tth function $z_j(x) = \sum_{j=1}^{n} C_{tj}x_j$ still reflects the network congestion, $t = 1, 2, \ldots, p$. In this paper, we aim at minimizing the maximum network congestions described by p experts. We established the following optimization model:

$$\begin{aligned} &\min\ z(x) = \sum_{j=1}^{n} C_{1j}x_j \vee \sum_{j=1}^{n} C_{2j}x_j \cdots \vee \sum_{j=1}^{n} C_{pj}x_j \\ &\text{s.t. } A \circ x^T \geq b^T, \end{aligned} \tag{2}$$

where "$\vee$" is the maximum value and $A \circ x^T \geq b^T$ is a system of addition-min fuzzy relation inequalities.

In this paper, we aim at obtainning an algorithm for model (2). The remaining content is organized as follows. In Sect. 2 we introduce some concepts and results of the system of addition-min fuzzy relation inequalities. In Sect. 3, we discuss the optimal solution of problem (2) by converting it into other models. We get an algorithm for our problem in Sect. 4. An example is cited to explain the notations and the algorithm for problem (2) in Sect. 5 and the conclusion is in Sect. 6.

2 Preliminaries

Fuzzy relation inequalities with addition-min composition can be written as:

$$\begin{cases} a_{11} \wedge x_1 + a_{12} \wedge x_2 + \cdots a_{1n} \wedge x_n \geq b_1 \\ a_{21} \wedge x_1 + a_{22} \wedge x_2 + \cdots + a_{2n} \wedge x_n \geq b_2 \\ \cdots \\ a_{m1} \wedge x_1 + a_{m2} \wedge x_2 + \cdots + a_{mn} \wedge x_n \geq b_m. \end{cases} \tag{3}$$

where "$\wedge$" is the minimum value.

Let $A = (a_{ij})_{m \times n} \in [0, 1]^{m \times n}$, $x = (x_1, x_2, \ldots, x_n) \in [0, 1]^n$, $b = (b_1, b_2, \ldots, b_m) \in [0, 1]^m$, and $\circ$ addition-min composition. $I = \{1, 2, \ldots, m\}$ and $J = \{1, 2, \ldots, n\}$ two index sets, then the system (3) can be reduced to

$$A \circ x^T \geq b^T, \tag{3'}$$

Denote $X = [0, 1]^n$ and $X(A, b) = \{x \in X | A \circ x^T \geq b^T\}$, where $X(A, b)$ is the solution set of system (3).

Definition 1 For $x = (x_1, x_2, \ldots, x_n)$, $y = (y_1, y_2, \ldots, y_n) \in X$, $x \geq y$ $(x \leq y)$ if $x_j \geq y_j$ $(x_j \leq y_j)$ for all $j \in J$.

Definition 2 System (3) is said to be consistent if $X(A, b) \neq \emptyset$. Otherwise, it is said to be inconsistent.

Definition 3 A solution $\hat{x} \in X(A, b)$ is said to be the maximum solution of system (3) when $x \leq \hat{x}$ for all $x \in X(A, b)$. A solution $\check{x} \in X(A, b)$ is said to be a minimal solution of system (3) when $x \leq \check{x}$ implies $x = \check{x}$ for any $x \in X(A, b)$.

Lemma 1 (i) *Let $x' \in X(A, b)$, $x \in X$. For any $x \geq x'$, then $x \in X(A, b)$;*
(ii) *Let $x', x \in X$ and $x' \notin X(A, b)$. For any $x \leq x'$, then $x \notin X(A, b)$.*

Proof (i) Let $x' = (x'_1, x'_2, \ldots, x'_n)$ and $x = (x_1, x_2, \ldots, x_n)$. Since $x \geq x'$ and $x' \in X(A, b)$, then for any $i \in I$,

$$a_{i1} \wedge x_1 + a_{i2} \wedge x_2 + \cdots + a_{in} \wedge x_n \geq a_{i1} \wedge x'_1 + a_{i2} \wedge x'_2 + \cdots + a_{in} \wedge x'_n \geq b_i,$$

which implies that x is a feasible solution of system (3).

The proof of (ii) is similar to (i). □

Theorem 1 *For system* (3), *we have equivalent conditions:*
(i) *system* (3) *is consistent;*
(ii) $\sum_{j=1}^{n} a_{ij} \geq b_i$ *for all* $i \in I$;
(iii) $\hat{x} = (1, 1, \ldots, 1)$ *is the maximum solution of* (3).

Proof (i)⇒(ii) If system (3) is consistent, then there exist $x = (x_1, x_2, \ldots, x_n)$, which satisfies $a_{i1} \wedge x_1 + a_{i2} \wedge x_2 + \cdots + a_{in} \wedge x_n \geq b_i$ for any $i \in I$. Thus $\sum_{j=1}^{n} a_{ij} \geq b_i \geq a_{i1} \wedge x_1 + a_{i2} \wedge x_2 + \cdots + a_{in} \wedge x_n \geq b_i$ for any $i \in I$.

(ii)⇒(iii) Since $\sum_{j=1}^{n} a_{ij} = a_{i1} \wedge 1 + a_{i2} \wedge 1 + \cdots + a_{in} \wedge 1 \geq b_i$ for any $i \in I$, then $\hat{x} = (1, 1, \ldots, 1)$ is the maximum solution of (3).

(iii)⇒(i) Obviously, $\hat{x} = (1, 1, \ldots, 1) \in X(A, b)$ implies that system (3) is consistent. □

Theorem 2 *If system* (3) *is consistent, then the feasible solutions set is*

$$X(A, b) = \bigcup_{\check{x} \in \check{X}(A,b)} \{x \in X \mid \check{x} \leq x \leq \hat{x}\},$$

where $\hat{x} = (1, 1, \ldots, 1)$ *and* $\check{X}(A, b)$ *is the set of all the minimal solutions of* (3).

Proof The proof is trivial by Lemma 1 and Theorem 1. □

Theorem 3 *Suppose* $x = (x_1, x_2, \ldots, x_n) \in X(A, b)$. *Then for any* $i \in I$, $j \in J$,

$$x_j \geq b_i - \sum_{k \in J - \{j\}} a_{ik}.$$

Proof Since $x = (x_1, x_2, \ldots, x_n) \in X(A, b)$, then

$$\sum_{k \in J} a_{ik} \wedge x_k = a_{ij} \wedge x_j + \sum_{k \in J - \{j\}} a_{ik} \wedge x_k \geq b_i,$$

or

$$x_j \geq a_{ij} \wedge x_j = b_i - \sum_{k \in J - \{j\}} a_{ik} \wedge x_k \geq b_i - \sum_{k \in J - \{j\}} a_{ik},$$

and this completes the proof. □

Denote $\check{a}_{ij} = max\{0, b_i - \sum_{k \in J - \{j\}} a_{ik}\}$, $\check{a}_j = max\{\check{a}_{ij} \mid i \in I\}$.

Theorem 4 *If* $x \in X(A, b)$ *of system* (3), *then* $x \geq \check{a}$, *where* $\check{a} = (\check{a}_1, \check{a}_2, \ldots, \check{a}_n)$.

Proof For $x = (x_1, x_2, \ldots, x_n) \in X(A, b)$. $x_j \geq b_i - \sum_{k \in J - \{j\}} a_{ik}$ for any $i \in I$, $j \in J$. Since $x_j \geq 0$, then $x_j \geq max\{0, b_i - \sum_{k \in J - \{j\}} a_{ik}\} = \check{a}_{ij}$ for all $i \in I$, $j \in J$. Hence $x_j \geq max\{\check{a}_{ij} \mid i \in I\} = \check{a}_j$ for all $j \in J$. Thus $x \geq \check{a}$. □

Denote $\hat{a}_j = max\{a_{ij} \mid i \in I\}$, and $\hat{a} = (\hat{a}_1, \hat{a}_2, \ldots, \hat{a}_n)$.

Theorem 5 *If* $\check{x}$ *is a minimal solution of system* (3), *then* $\check{x} \leq \hat{a}$.

Proof For all $i \in I$, $a_{i1} \wedge \hat{a}_1 + a_{i2} \wedge \hat{a}_2 + \cdots + a_{in} \wedge \hat{a}_n \geq a_{i1} \wedge a_{i1} + a_{i2} \wedge a_{i2} + \cdots + a_{in} \wedge a_{in} = \sum_{j=1}^{n} a_{ij} \geq b_i$, then $\hat{a} \in X(A, b)$ if system (3) is consistent. Thus for all $x > \hat{a}$, x is a solution of (3) by (i) of Lemma 1, which means each $x > \hat{a}$ is not a minimal solution of (3). Hence, $\check{x} \leq \hat{a}$. □

Corollary 1 *Let $\check{x}$ be any minimal solution of system* (3), *i.e., $\check{x} \in \check{X}(A, b)$, then $\check{a} \leq \check{x} \leq \hat{a}$.*

Proof The proof is trivial by Theorems 4 and 5. □

System (3) always has minimal solution if it is consistent. And all the minimal solutions are between $\check{a}$ and $\hat{a}$. Next, we want to find all the minimal solutions of system (3) and introduce some definitions.

Definition 4 For $j \in J$, denote $D_j = \{d_{0j}, d_{1j}, \ldots, d_{s_j j}\}$ and $D = D_1 \times D_2 \cdots \times D_n$, where D_j satisfy:
(i) $\check{a}_j = d_{0j} < d_{1j} <, \cdots, < d_{s_j j} = \hat{a}_j$, where $d_{kj} \in \{a_{ij} | i \in I\}$, $k = 1, 2, \ldots, s_j - 1$;
(ii) For any $j \in J$, if $a_{ij} \geq \check{a}_j$, there exists unique $i' \in \{0, 1, 2, \ldots, s_j\}$, such that $a_{ij} = d_{i'j}$.

Definition 5 A interval $[\bar{x}^-, \bar{x}]$ is a minimal interval if it satisfies:
(i) $\bar{x} = (d_{k_1 1}, d_{k_2 2}, \ldots, d_{k_n n}) \in D$ and $\bar{x}^- = (d_{(k_1-1)1}, d_{(k_2-1)2}, \ldots, d_{(k_n-1)n}) \in D$;
(ii) $\bar{x}$ is a solution of system (3);
(iii) $\bar{x}^-$ is not a solution of system (3).

Example 1 Consider the following systems:

$$\begin{cases} 0.6 \wedge x_1 + 0.45 \wedge x_2 + 0.55 \wedge x_3 \geq 1.55 \\ 0.4 \wedge x_1 + 0.9 \wedge x_2 + 0.55 \wedge x_3 \geq 1.65 \\ 0.75 \wedge x_1 + 0.5 \wedge x_2 + 0.85 \wedge x_3 \geq 1.6 \end{cases} \tag{4}$$

Solution:

By definition, it is easy to get the results as following:

$$(\check{a}_{ij}) = \begin{pmatrix} 0.55 & 0.4 & 0.5 \\ 0.2 & 0.7 & 0.35 \\ 0.25 & 0 & 0.35 \end{pmatrix}, \check{a} = (0.55, 0.7, 0.5), \hat{a} = (0.75, 0.9, 0.85),$$

$$D_1 = \{0.55, 0.6, 0.75\}, D_2 = \{0.7, 0.9\}, D_3 = \{0.5, 0.55, 0.85\}.$$

Then, all the intervals which satisfied (i) of Definition 5 are:
$d_1 = [(0.55, 0.7, 0.5)^T, (0.6, 0.9, 0.55)^T]$, $d_2 = [(0.55, 0.7, 0.55)^T, (0.6, 0.9, 0.85)^T]$, $d_3 = [(0.6, 0.7, 0.5)^T, (0.75, 0.9, 0.55)^T]$, $d_4 = [(0.6, 0.7, 0.55)^T, (0.75, 0.9, 0.85)^T]$.

Now check (ii) and (iii) of Definition 5 to get the minimal intervals.

Since $(0.6, 0.9, 0.55)$ is a solution of (4) and $(0.55, 0.7, 0.5)$ is not a solution of (4), then d_1 is a minimal interval.

Since $(0.6, 0.9, 0.85)$ is a solution of (4) and $(0.55, 0.7, 0.55)$ is a solution of (4), then d_2 is not a minimal interval.

Since $(0.75, 0.9, 0.55)$ is a solution of (4) and $(0.6, 0.7, 0.5)$ is not a solution of (4), then d_3 is a minimal interval.

Since $(0.75, 0.9, 0.85)$ is a solution of (4) and $(0.6, 0.7, 0.55)$ is a solution of (4), then d_1 is not a minimal interval.

We get all the minimal intervals of system (4) are d_1 and d_3.

The following theorem will show that all the minimal solutions of system (3) can be found in minimal intervals.

Theorem 6 *If $\check{x}$ is a minimal solution of system* (3), *then there exist some minimal interval* $[\bar{x}^-, \bar{x}]$ *of system* (3), *s.t.* $\check{x} \in [\bar{x}^-, \bar{x}]$.

Proof If $\check{x} = (\check{x}_1, \check{x}_2, \ldots, \check{x}_n)$ is a minimal solution of system (3), according to Corollary 1, then $\check{a} \le \check{x} \le \hat{a}$.

For any $j \in J, \check{a}_j = d_{0j} < d_{1j} <, \cdots, < d_{s_j j} = \hat{a}_j$, there exist some $k_j \in \{1, 2, \ldots, s_j\}$, s.t. $d_{(k_j-1)j} < \check{x}_j \le d_{(k_j)j}$. Let $\bar{x}^- = (d_{(k_1-1)1}, d_{(k_2-1)2}, \ldots, d_{(k_n-1)n})$ and $\bar{x} = (d_{k_1 1}, d_{k_2 2}, \ldots, d_{k_n n})$, then $\bar{x}^- < \check{x} \le \bar{x}$.

Since $\bar{x} \ge \check{x}$, then $\bar{x}$ is a solution of system (3). Since $\bar{x}^- < \check{x}$ and $\check{x}$ is a minimal solution of system (3), then $\bar{x}^-$ is not a solution of system (3). □

3 The Optimal Solution of (2)

In this subsection, we aim to solve the optimal solution of problem (2). First, we prove that the optimal solution of (2) is a minimal solution of (3) by Theorem 7. Then, to seek the optimal solution of (2), we set up an optimization problem (5), which will prove to be equivalent to problem (2) by Theorem 8. Furthermore, we construct a range of optimization problem (P_t) to deal with problem (5) by Theorem 9. Finally, we solve problem (2) by Theorem 10.

Then optimal problem (2) can be rewritten as:

$$\min \quad z(x) = \bigvee_{i=1}^{p} \sum_{j=1}^{n} C_{ij} x_j \tag{2'}$$

$$s.t. \quad A \circ x^T \ge b^T$$

Theorem 7 *If x^* is the optimal solution of problem* (2), *then x^* is a minimal solution of* (3).

Proof Let $x^* = (x_1^*, x_2^*, \ldots, x_n^*)$ be an optimal solution of system (2), $x = (x_1, x_2, \ldots, x_1)$ be any feasible solution of (3) and $x \le x^*$. We will prove $x = x^*$ which means that x^* is a minimal solution of system (3).

Since x^* is an optimal solution, then $z(x^*) \leq z(x)$, i.e.,

$$\sum_{j=1}^{n} C_{1j}x_j^* \vee \sum_{j=1}^{n} C_{2j}x_j^* \cdots \vee \sum_{j=1}^{n} C_{pj}x_j^* \leq \sum_{j=1}^{n} C_{1j}x_j \vee \sum_{j=1}^{n} C_{2j}x_j \cdots \vee \sum_{j=1}^{n} C_{pj}x_j \ .$$

There exists $i_0 \in \{1, 2, \ldots, p\}$, such that

$$\sum_{j=1}^{n} C_{i_0 j}x_j = \sum_{j=1}^{n} C_{1j}x_j \vee \sum_{j=1}^{n} C_{2j}x_j \cdots \vee \sum_{j=1}^{n} C_{pj}x_j \ .$$

Then

$$\sum_{j=1}^{n} C_{i_0 j}x_j \geq \sum_{j=1}^{n} C_{1j}x_j^* \vee \sum_{j=1}^{n} C_{2j}x_j^* \cdots \vee \sum_{j=1}^{n} C_{pj}x_j^* \ ,$$

which implies that

$$\sum_{j=1}^{n} C_{i_0 j}x_j \geq \sum_{j=1}^{n} C_{i_0 j}x_j^* \ ,$$

or

$$\sum_{j=1}^{n} C_{i_0 j}(x_j - x_j^*) \geq 0.$$

Since for $j \in J$, $C_{i_0 j} \geq 0$ and $x_j \leq x_j^*$, then $x_j = x_j^*$. Thus $x = x^*$. □

For seeking the optimal solution of (2), we construct a optimization model as follows:

$$\min y$$

$$\text{s.t.} \begin{cases} y \geq \sum_{j=1}^{n} C_{ij}x_j, i = 1, 2, \ldots, p. \\ A \circ x^T \geq b^T \end{cases} \quad (5)$$

Theorem 8 (i) *The optimal value (objective function) of problem* (2) *is equal to the optimal value (objective function) of problem* (5).
(ii) *The complete optimal solution set of* (2) *is exactly the complete optimal solution of* (5)*;*

Proof Let $x^{1*} = (x_1^{1*}, x_2^{1*}, \ldots, x_n^{1*})$ be any optimal solution of (2) and z^* is the optimal value (objective function) of (2). Let $x^{2*} = (x_1^{2*}, x_2^{2*}, \ldots, x_n^{2*})$ be any optimal solution of (5) and y^* is the optimal value of (5). We will prove that:

(i) $y^* = z^*$, which implies that the optimal value (objective function) of problem (2) is equal to that of (5);

(ii) x^{1*} is an optimal solution of (5) and x^{2*} is an optimal solution of (2), which implies that the complete optimal optimal solution set of problem (2) is exactly the same as that of (5).

On one hand, since x^{1*} is an optimal solution of (2), then x^{1*} satisfies $A \circ (x^{1*})^T \geq b^T$ in problem (2). Furthermore, $z^* = z(x^{1*}) = \sum_{j=1}^{n} C_{1j}x_j^{1*} \vee \sum_{j=1}^{n} C_{2j}x_j^{1*} \cdots \vee \sum_{j=1}^{n} C_{pj}x_j^{1*} \geq \sum_{j=1}^{n} C_{ij}x_j^{1*}$ for any $i = 1, 2, \ldots, p$, which indicates that x^{1*} is a feasible solution of problem (5). Hence $y^* \leq z(x^{1*}) = z^*$.

On the other hand, since x^{2*} is an optimal solution of (5) and y^* is the optimal value of (5), then x^{2*} satisfies the constraints of problem (5), i.e., $y^* \geq \sum_{j=1}^{n} C_{ij}x_j^{2*}$ for $i = 1, 2, \ldots, p$, which means $y^* \geq \sum_{j=1}^{n} C_{1j}x_j^{2*} \vee \sum_{j=1}^{n} C_{2j}x_j^{2*} \cdots \vee \sum_{j=1}^{n} C_{pj}x_j^{2*} = z(x^{2*})$. Meanwhile $A \circ (x^{2*})^T \geq b^T$. Then, x^{2*} is a feasible solution of (2), which means $z(x^{2*}) \geq z(x^{1*}) = z^*$. Hence $y^* \geq z(x^{2*}) \geq z(x^{1*}) = z^*$.

Therefore, $y^* = z^*$.

From the proof above, x^{2*} is an feasible solution of (5) and $z(x^{2*}) = z^*$. Then x^{2*} is an optimal solution of (5). Since x^{1*} is an feasible solution of (2) and its subject value is y^*, then x^{1*} is an optimal solution of (2). □

Suppose that the minimal intervals of system (3) are $d_1, d_2, \ldots, d_s$. We call the following models as problem (P_t), where $t = 1, 2, \ldots, s$.

$$(P_t) \min y$$

$$\text{s.t.} \begin{cases} y \geq \sum_{j=1}^{n} C_{ij}x_j, i = 1, 2, \ldots, p \\ A \circ x^T \geq b^T \\ x \in d_t. \end{cases} \tag{6}$$

Theorem 9 *Suppose that the optimal value (objective function) of problem (P_t) is y^{t*} for $t = 1, 2, \ldots, s$. Then the optimal value (objective function) of problem (5) is $y^* = \bigwedge_{t=1}^{s} y^{t*}$.*

Proof Suppose that $x^* = (x_1^*, x_2^*, \ldots, x_n^*)$ is an optimal solution of problem (5) and the optimal value (objective function) is y^*. Then x^* is an optimal solution of problem (2) by Theorem 8. Hence, x^* is a minimal solution of (3) by Theorem 7. Then by Theorem 6, there exist some minimal interval d_{t_0}, s.t. $x^* \in d_{t_0}$, $t_0 \in \{1, 2, \ldots, s\}$. Therefore, x^* is a feasible solution of problem (P_{t_0}), which implies $y^* \geq y^{t_0 *}$.

For $t \in \{1, 2, \ldots, s\}$, $y^* \leq y^{t*}$ since problem (P_t) is problem (5) limited to the minimal interval d_t. Then $y^* \leq \bigwedge_{t=1}^{s} y^{t*} \leq y^{t_0 *}$.

Therefore, $y^* = \bigwedge_{t=1}^{s} y^{t*}$. □

Theorem 10 *Suppose that the optimal value (objective function) of problem* (P_t) *is* y^{t*} *for* $t = 1, 2, \ldots, t$. *Then the optimal value (objective function) of problem* (2) *is* $y^* = \bigwedge_{t=1}^{s} y^{t*}$.

Proof The proof is trivial by Theorems 8 and 9. □

4 Algorithm for Solving Problem (2)

Based on the concepts and results above, we obtain an algorithm for the optimal solution of (2):

Step 1. Check the consistency of system (3) by Theorem 1. If it is consistent, go to Step 2.

Step 2. Compute $\check{a}$, $\hat{a}$ of problem (2) and $D_1, D_2, \ldots, D_n$ by Definition 4.

Step 3. Find the minimal intervals d_t of (2) by Definition 5, where $t = 1, 2, \ldots, s$.

Step 4. Solve the linear programming problem (P_t), $t = 1, 2, \ldots, s$.

Step 5. Obtain the optimal value of problem (2) by $y^* = \bigwedge_{t=1}^{s} y^{t*}$, where y^{t*} is the optimal value of (P_t).

Step 6. Obtain the optimal solutions of problem (2) by finding $t_0 \in \{1, 2, \ldots, s\}$, s.t. $y^{t_0*} = y^*$.

5 Numerical Example

Example 2 Solve the fuzzy optimal problem:

$$\begin{aligned} \min g(x) &= (0.7x_1 + 0.25x_2 + 0.9x_3) \vee (0.8x_1 + 0.2x_2 + 0.65x_3) \\ &\vee (0.45x_1 + 0.8x_2 + 0.35x_3) \\ \text{s.t.} &\begin{cases} 0.6 \wedge x_1 + 0.45 \wedge x_2 + 0.55 \wedge x_3 \geq 1.55, \\ 0.4 \wedge x_1 + 0.9 \wedge x_2 + 0.55 \wedge x_3 \geq 1.65, \\ 0.75 \wedge x_1 + 0.5 \wedge x_2 + 0.85 \wedge x_3 \geq 1.6. \end{cases} \end{aligned} \tag{7}$$

Solution:

Step 1–3. we have checked and computed in Example 1.

Step 4. There are only two minimal intervals, we will compute them one by one. Choose $[(0.55, 0.7, 0.5)^T, (0.6, 0.9, 0.55)^T]$ as an minimal interval and solve (P_1):

$$\min y$$

$$\text{s.t.} \begin{cases} y \geq 0.7x_1 + 0.25x_2 + 0.9x_3, \\ y \geq 0.8x_1 + 0.2x_2 + 0.65x_3, \\ y \geq 0.45x_1 + 0.8x_2 + 0.35x_3, \\ x_1 + 0.45 + x_3 \geq 1.55, \\ 0.4 + x_2 + x_3 \geq 1.65, \\ x_1 + 0.5 + x_3 \geq 1.6, \\ 0.55 \leq x_1 \leq 0.6, 0.7 \leq x_2 \leq 0.9, 0.5 \leq x_3 \leq 0.55. \end{cases} \tag{8}$$

We get the optimal solution is y^{1*}=1.055 and the optimal solution is (0.55, 0.7, 0.55).

Choose $[(0.6, 0.7, 0.5)^T, (0.75, 0.9, 0.55)^T]$ as an minimal interval and solve (P_2):

$$\min y$$

$$\text{s.t.} \begin{cases} y \geq 0.7x_1 + 0.25x_2 + 0.9x_3, \\ y \geq 0.8x_1 + 0.2x_2 + 0.65x_3, \\ y \geq 0.45x_1 + 0.8x_2 + 0.35x_3, \\ 0.6 + 0.45 + 0.55 \geq 1.55, \\ 0.4 + x_2 + 0.55 \geq 1.65, \\ x_1 + 0.5 + x_3 \geq 1.6, \\ 0.6 \leq x_1 \leq 0.75, 0.7 \leq x_2 \leq 0.9, 0.5 \leq x_3 \leq 0.55. \end{cases} \tag{9}$$

We get the optimal solution is $y^{2*} = 1.045$ and the optimal solution is (0.6, 0.7, 0.5).

Step 5. The optimal value of problem (7) is $y^* = 1.055 \wedge 1.045 = 1.045$.

Step 6. The optimal solutions of problem (7) is (0.6, 0.7, 0.5) since $y^{2*} = y^*$=1.045.

6 Conclusion

In this paper, we provide an algorithm for multi-variable-term latticized linear programming with addition-min fuzzy relation inequalities constraint. First, the optimal solution of problem (2) is a minimal solution of system (3) (Theorem 7) and, a minimal solution of system (3) is in some minimal interval (Theorem 6). Second, problem (2) is equivalent to (5), which can be decomposed into some linear subproblems (P_t). Relation among problem (2), problem (5) and subproblems (P_t) is discussed in Theorem 8–10. The key technology of the algorithm is that we decompose problem (2) into several subproblems (P_t) by defining the minimal intervals.

According to the analysis and examples, the problem proposed can be convert into several linear programming problems. The computational complexity of the problem depends on the minimal intervals, which reduced much work. As we have shown in

Example 2, it just needs to solve 2 linear programming problems and check which one is better(smaller).

Acknowledgments This work is supported by the Natural Science Foundation of Guangdong Province (2016A030307037; 2016A030313552) and the Innovation and Building Strong School Project of Colleges of Guangdong Province (2015KQNCX094).

References

1. Sanchez, E.: Resolution of composite fuzzy relation equations. In. Control **30**, 38–48 (1976)
2. Sanchez, E.: Solutions in composite fuzzy relation equations: Application to medical diagnosis in Brouwerian logic, In: M. M. Gupta, G. N. Saridis, & B. R. Gaines (Ed.). Fuzzy automata and decision processes. pp. 221-234. Amsterdam: North-Holland(1977)
3. Wang, P.Z., Zhang, D.Z., Sanchez, E., Lee, E.S.: Latticized linear programming and fuzzy relation inequalities. Jo. Math. Anal. Appl. **159**(1), 72–87 (1991)
4. Fang, S.C., Li, G.: Solving fuzzy relation equations with a linear objective function. Fuzzy Sets and Syst. **103**, 107–113 (1999)
5. Loetamonphong, J., Fang, S.-C.: Optimization of fuzzy relation equations with max-product composition. Fuzzy Sets Syst. **118**, 509–517 (2001)
6. Wu, Y.K., Guu, S.M., Liu, J.Y.C.: An accelerated approach for solving fuzzy relation equations with a linear objective function. IEEE Trans. Fuzzy Syst. **10**(4), 552–558 (2002)
7. Ghodousian, A., Khorram, E.: Solving a linear programming problem with the convex combination of the max-min and the max-average fuzzy relation equations. Applied. Math. Comput. **180**, 411–418 (2006)
8. Wu, Y.K., Guu, S.M.: Minimizing a linear function under a fuzzy max-min relational equation constraint. Fuzzy Sets Syst. **150**, 147–162 (2005)
9. Guo, F.F., Pang, L.P., Meng, D., Xia, Z.Q.: An algorithm for solving optimization problems with fuzzy relational inequality constraints. Inf. Sci. **252**, 20–31 (2013)
10. Chang, C.W., Shieh, B.S.: Linear optimization problem constrained by fuzzy max-min relation equations. Inf. Sci. **234**, 71–79 (2013)
11. Li, P., Liu, Y.: Linear optimization with bipolar fuzzy relational equation constraints using the Łukasiewicz triangular norm. Soft Comput. **18**(7), 1399–1404 (2014)
12. Qu, X.B., Wang, X.P.: Minimization of linear objective functions under the constraints expressed by a system of fuzzy relation equations. Inf. Sci. **178**, 3482–3490 (2008)
13. Yang, S.J.: An algorithm for minimizing a linear objective function subject to the fuzzy relationin inequalities with addition-min composition. Fuzzy Sets Syst. **255**, 41–51 (2014)
14. Li, J.X., Yang, S.J.: Fuzzy relation equalities about the data transmission mechanism in bittorrent-like peer-to-peer file sharing systems, In: Proceedings of the 2012 9th International Conference on Fuzzy Systems and Knowledge Discovery, FSKD, pp. 452-456 (2012)

Possibility Interval-Valued Multi-fuzzy Soft Sets and Theirs Applications

Dong-Xue Li, Jian-Hua Jin and Wei Ran

Abstract The notions of interval-valued multi-fuzzy soft set and possibility interval-valued multi-fuzzy soft set are proposed in this paper. Several interesting algebraic properties of them are then investigated. In particular, both interval-valued multi-fuzzy soft set and possibility interval-valued multi-fuzzy soft set with union and intersection operators turn out to be distributive lattices. Finally, possibility interval-valued multi-fuzzy soft sets are applied to decision making and an illustrated example is given.

Keywords Multi-fuzzy set · Interval-valued fuzzy set · Interval-valued multi-fuzzy soft set · Possibility interval-valued multi-fuzzy soft set · Lattice

1 Introduction

Since Zadel [12] introduced fuzzy set in 1965, extensions on fuzzy set have been widely discussed. Researchers expand fuzzy sets from different viewpoints like interval-valued fuzzy set [1], intuitionistic fuzzy set [2] and multi-fuzzy set [8] and so on. Motodtsov proposed soft set theory [6], which is a useful tool to solve uncertainty. Combining soft set theory with fuzzy set theory, many scholars investigate generalized soft sets such as interval-valued fuzzy soft set [9], intuitionistic fuzzy soft set [5], multi-fuzzy soft set [10], interval-valued intuitionistic fuzzy soft set [3], bipolar multi-fuzzy soft set theory [11] and possibility multi-fuzzy soft set [13].

D.-X. Li · J.-H. Jin (✉) · W. Ran
School of Sciences, Southwest Petroleum University, Chengdu 610500, China
e-mail: jinjianhua@swpu.edu.cn
URL: http://www.swpu.edu.cn

D.-X. Li
e-mail: lidongxueever@163.com

W. Ran
e-mail: jjh2006ok@aliyun.com

T.-H. Fan et al. (eds.), *Quantitative Logic and Soft Computing 2016*,
Advances in Intelligent Systems and Computing 510,
DOI 10.1007/978-3-319-46206-6_49

In our real world, much information involving different parameters is uncertain and vague. And it is also difficult for people to characterize it by a precise number. To more accurately express the information, we propose interval-valued multi-fuzzy soft set, combined interval-valued fuzzy set and multi-fuzzy soft set. Considering each parameters' possibility degree, we propose possibility interval-valued multi-fuzzy soft set. Some important algebraic properties on these system are investigated.

The rest of the paper is arranged as follows. Section 2 briefly reviews some backgrounds on soft sets, interval-valued fuzzy sets, intuitionistic fuzzy sets and multi-fuzzy sets. Section 3 gives the notion of interval-valued multi-fuzzy soft sets based on interval-valued fuzzy sets and multi-fuzzy soft sets, some operators on them are then investigated in detail. Section 4 gives the notion of possibility interval-valued multi-fuzzy soft sets and investigated some operations. An example is given to illustrate application of possibility interval-valued multi-fuzzy soft sets to decision making in Sect. 5. Section 6 summarizes the conclusion.

2 Preliminaries

In the section we present some definitions and preliminaries in brief which are required in the sequel of our work.

Definition 1 ([6]) Let U be an initial universe set and E a set of parameters. $A \subseteq E$. A pair (F, A) is called a soft set over U if F: $A \rightarrow P(U)$ is a mapping, where $P(U)$ is the set of all subsets of U.

Definition 2 ([4]) Let U be an initial universe set and E be a set of parameters. A pair (F, A) is called a fuzzy soft set over (U, E) if $A \subseteq E$ and F is a mapping given by $F : A \rightarrow f(U)$, where $f(U)$ is the set of all fuzzy sets over U.

Definition 3 ([1]) Let U be an initial universe set. A set X is called an interval-valued fuzzy set over U if $X : U \rightarrow Int([0, 1])$ is a mapping, where $Int([0, 1])$ stands for the set of all closed subintervals of [0,1].

Let $IVF(U)$ denote the set of all interval-valued fuzzy sets on U and E a set of parameters. A pair (F, A) is called an interval-valued fuzzy soft set [9] over (U, E) if $A \subseteq E$ and $F : A \rightarrow IVF(U)$ is a mapping.

Definition 4 ([2]) Let X be an initial universe set. An intuitionistic fuzzy set A on X is defined as an object of the following form: $A = \{< x, \mu_A(x), \nu_A(x) >: x \in X\}$. The function $\mu_A : X \rightarrow [0, 1]$ and $\nu_A : X \rightarrow [0, 1]$ define the degree of membership and the degree of nonmembership of element $x \in X$ to set A. For all $x \in X$, $0 \leq \mu_A(x) + \nu_A(x) \leq 1$.

Let $IF(U)$ denote the set of all intuitionistic fuzzy sets over U and E a set of parameters. A pair (F, A) is called an intuitionistic fuzzy soft set [5] over (U, E) if $A \subseteq E$ and F is a mapping given by $F : A \rightarrow IF(U)$.

Definition 5 ([7]) Let X be an initial universe, an interval-valued intuitionistic fuzzy set on X is an object of the form $A = \{< x, \mu_A(x), \nu_A(x) >: x \in X\}$, where $\mu_A(x) = [\mu_A^-(x), \mu_A^+(x)]$, $\nu_A(x) = [\nu_A^-(x), \nu_A^+(x)], \mu_A(x) : X \to Int[0, 1], \nu_A(x) : X \to Int[0, 1]$, and satisfied $\mu_A^+(x) + \nu_A^+(x) \leq 1$.

Let U be an initial universe set and E be set of parameters. *IVIFS* denotes the set of all interval-valued intuitionistic fuzzy sets of U. A pair (F, A) is called an interval-valued intuitionistic fuzzy soft set [3] over (U, E) if $A \subseteq E$ and F is a mapping given by $F : A \to IVIFS$.

Definition 6 ([8]) Let U be an initial universe set and k be positive integer. A multi-fuzzy set A over U is a set of ordered sequences denoted by: $A = \{u/(\mu_1(u), \mu_2(u), \ldots, \mu_k(u)) : u \in U\}$, where $\mu_i \in F(U), i = 1, 2, \ldots, k$.

Function $\mu_A = (\mu_1, \mu_1, \ldots, \mu_k)$ is called the multi-membership function of multi-fuzzy set A, and k is called the dimension of A. The set of all multi-fuzzy sets of dimension k in U is denoted by $M^k FS$.

If $\sum_{i=1}^{k} \mu_i(u) \leq 1$, $\forall u \in U$, then the multi-fuzzy set of dimension k is called a normalized multi-fuzzy set. If $\sum_{i=1}^{k} \mu_i(u) = l > 1$, for some $u \in U$, we redefine the multi-member degree $\{u/(\mu_1(u), \mu_2(u), \ldots, \mu_k(u)) : u \in U\}$ as $\{\frac{1}{l}(u/(\mu_1(u), \mu_2(u), \ldots, \mu_k(u))) : u \in U\}$, then a non-normalized multi-fuzzy set can be changed into a normalized multi-fuzzy set.

Let U be an initial universe set and E be set of parameters. A pair (F, A) is called a multi-fuzzy soft set [10] of dimension k over U, where $F : A \to M^k FS$ is a mapping.

3 Interval-Valued Multi-fuzzy Soft Set

Combining interval-valued fuzzy set with multi-fuzzy soft set, we define an interval-valued multi-fuzzy soft set as an extension of multi-fuzzy soft set.

Definition 7 Let U be an initial universe and E a set of parameters. A pair $(\widetilde{F}, A)$ is called an interval-valued multi-fuzzy soft set (for short *IVM*^{k}*FSS*) over (U, E) if $A \subseteq E$ and $\widetilde{F}$ is a mapping given by $\widetilde{F} : A \to IVM^k FS$, where *IVM*$^k FS$ is a special multi-fuzzy set denoted by *IVM*$^k FS = \{u/(\mu_1(u), \mu_2(u), \ldots, \mu_k(u)) : u \in U\}$, $\mu_i(u) = [\mu_i^-(u), \mu_i^+(u)]$, $0 \leq \mu_i^-(u) \leq \mu_i^+(u) \leq 1, i = 1, 2, \ldots, k$. The form of $\widetilde{F}$ is $\widetilde{F}(e) = \{\widetilde{F}(e_i), i = 1, 2, 3\}$.

If $\sum_{i=1}^{k} \mu_i^+(u) \leq 1, \forall u \in U$, then the interval-valued multi-fuzzy soft set of dimension k is called a normalized interval-valued multi-fuzzy soft set. If $\sum_{i=1}^{k} \mu_i^+(u) = l > 1$ for some $u \in U$, we redefine the interval-valued multi-member degree $\{u/([\mu_1^-(u), \mu_1^+(u)], [\mu_2^-(u), \mu_2^+(u)], \ldots, [\mu_k^-(u), \mu_k^+(u)])\}$ as $\{\frac{1}{l}(u/([\mu_1^-(u), \mu_1^+(u)], [\mu_2^-(u),$

$\mu_2^+(u)], \ldots, [\mu_k^-(u), \mu_k^+(u)]))\}$, for $\forall u \in U, \forall e \in E$. Then a non-normalized interval-valued multi-fuzzy soft set can be changed into a normalized interval-valued multi-fuzzy soft set.

If $\mu_i^-(u) = \mu_i^+(u)$, $\forall e \in E, u \in U, i = 1, 2, \ldots, k$, *IVM*^{k}*FSS* degenerates to multi-fuzzy soft set. If $k = 2$, *IVM*^{k}*FSS* degenerates to interval-valued intuitionistic fuzzy soft set. If $\mu_i^-(u) = \mu_i^+(u)$ and $k = 2$, *IVM*^{k}*FSS* degenerates to intuitionistic fuzzy soft set.

Definition 8 Let U be a finite initial universe and E a finite set of parameters. F and G are two *IVM*^{k}*FSS* over U. $F = \{u/(\mu_1(u), \mu_2(u), \ldots, \mu_k(u)) : u \in U\}$, $G = \{u/(\nu_1(u), \nu_2(u), \ldots, \nu_k(u)) : u \in U\}$, where $\mu_i(u) = [\mu_i^-(u), \mu_i^+(u)]$, $\nu_i(u) = [\nu_i^-(u), \nu_i^+(u)], i = 1, 2, \ldots k$. Then the following relations and operations on them are defined:

(1) $F \subseteq G$ if $\mu_i(u) \le \nu_i(u)$, i.e., $\mu_i^-(u) \le \nu_i^-(u)$ and $\mu_i^+(u) \le \nu_i^+(u), \forall u \in U,\ i = 1, 2, \ldots k$.

(2) $F = G$ if $\mu_i(u) = \nu_i(u)$, i.e., $\mu_i^-(u) = \nu_i^-(u)$ and $\mu_i^+(u) = \nu_i^+(u), \forall u \in U,\ i = 1, 2, \ldots k$.

(3) $F \cup G = \{u/(\mu_1(u) \vee \nu_1(u), \mu_2(u) \vee \nu_2(u), \ldots, \mu_k(u) \vee \nu_k(u)) : u \in U\}$, where $\mu_i(u) \vee \nu_i(u) = [\mu_i^-(u) \vee \nu_i^-(u), \mu_i^+(u) \vee \nu_i^+(u)], i = 1, 2, \ldots k$.

(4) $F \cap G = \{u/(\mu_1(u) \wedge \nu_1(u), \mu_2(u) \wedge \nu_2(u), \ldots, \mu_k(u) \wedge \nu_k(u)) : u \in U\}$, where $\mu_i(u) \wedge \nu_i(u) = [\mu_i^-(u) \wedge \nu_i^-(u), \mu_i^+(u) \wedge \nu_i^+(u)], i = 1, 2, \ldots k$.

(5)$F^c = \{u/(1 - \mu_1(u), 1 - \mu_2(u), \ldots, 1 - \mu_k(u)) : u \in U\}$, where $1 - \mu_i(u) = [1 - \mu_i^+(u), 1 - \mu_i^-(u)], i = 1, 2, \ldots k$.

Definition 9 Let $U = \{u_1, u_2, \ldots, u_n\}$ be a finite initial universe and E a finite set of parameters, $E = \{e_1, e_2, \ldots, e_m\}$. An *IVM*^{k}*FSS* $(\widetilde{F}, A) = \widetilde{F}(e) = \{\widetilde{F}(e_i) | i = 1, 2, \ldots, m\}$, $\widetilde{F}(e_i) = \{u/(\mu_1(u), \mu_2(u), \ldots, \mu_k(u))\}$ has a complement set given by $(\widetilde{F}^c, A)$, denoted by $(\widetilde{F}^c, A) = (\widetilde{G}, A)$, where $\widetilde{G}$ is a mapping denoted by $\widetilde{G} : A \to IVM^kFS$ and $\widetilde{G}(e) = 1 - \widetilde{F}(e)$, where $\widetilde{G}(e) = \{\widetilde{G}(e_i) | i = 1, 2, \ldots, m\}$, $\widetilde{G}(e_i) = \{u/(1 - \mu_1(u), 1 - \mu_2(u), \ldots, 1 - \mu_k(u)) : u \in U\}$, $1 - \mu_j(u) = [1 - \mu_j^+(u), 1 - \mu_j^-(u)]$, $j = 1, 2, \ldots, k$.

Definition 10 Let $U = \{u_1, u_2, \ldots, u_n\}$ an initial universe and E be a set of parameters, $E = \{e_1, e_2, \ldots, e_m\}$. $(\widetilde{F}, A)$ and $(\widetilde{G}, A)$ are two *IVM*^{k}*FSSs* over (U, E). $(\widetilde{F}, A)$ is said to be an interval-valued multi-fuzzy soft subset of $(\widetilde{G}, A)$ denoted by $\widetilde{F} \subseteq \widetilde{G}$ if $\widetilde{F}(e) \subseteq \widetilde{G}(e)$. i.e., $\widetilde{F}(e_i) \subseteq \widetilde{G}(e_i), i = 1, 2, \ldots, m$.

Definition 11 Let $U = \{u_1, u_2, \ldots, u_n\}$ be an initial universe and E a set of parameters, $E = \{e_1, e_2, \ldots, e_m\}$. $(\widetilde{F}, A)$ and $(\widetilde{G}, A)$ two *IVM*^{k}*FSSs* over (U, E). $(\widetilde{F}, A) = \widetilde{F}(e) = \{\widetilde{F}(e_i)\}$, $\widetilde{F}(e_i) = \{u/(\mu_1(u), \mu_2(u), \ldots, \mu_k(u))\}$. $(\widetilde{G}, A) = \widetilde{G}(e) = \{\widetilde{G}(e_i\}$, $\widetilde{G}(e_i) = \{u/(\nu_1(u), \nu_2(u), \ldots, \nu_k(u))\}$, $i = 1, 2, \ldots, m$, $u \in U$. An union operation on $\widetilde{F}$ and $\widetilde{G}$ denoted by $\widetilde{F} \cup \widetilde{G}$ is defined by a mapping given by $\widetilde{H} : A \to IVM^kFS$, $(\widetilde{H}, A) = \widetilde{H}(e) = \{\widetilde{H}(e_i),\ i = 1, 2, \ldots, m\}$, $\widetilde{H}(e_i) = \{u/(\xi_1(u), \xi_2(u), \ldots, \xi_k(u)) : u \in U\}$, where $\widetilde{H}(e) = \widetilde{F}(e) \cup \widetilde{G}(e)$.

Definition 12 Let $U = \{u_1, u_2, \ldots, u_n\}$ be an initial universe and E be a set of parameters, $E = \{e_1, e_2, \ldots, e_m\}$. $(\widetilde{F}, A)$ and $(\widetilde{G}, A)$ be two $IVM^k FSSs$ over (U, E). $(\widetilde{F}, A) = \widetilde{F}(e) = \{\widetilde{F}(e_i)\}$, $\widetilde{F}(e_i) = \{u/(\mu_1(u), \mu_2(u), \ldots, \mu_k(u))\}$. $(\widetilde{G}, A) = \widetilde{G}(e) = \{\widetilde{G}(e_i)\}$, $\widetilde{G}(e_i) = \{u/(\nu_1(u), \nu_2(u), \ldots, \nu_k(u))\}$. $i = 1, 2, \ldots, m$, $u \in U$. An intersection operation on $\widetilde{F}$ and $\widetilde{G}$ denoted by $\widetilde{F} \cap \widetilde{G}$ is defined by a mapping $\widetilde{H} : A \to IVM^k FS$, $(\widetilde{H}, A) = \widetilde{H}(e) = \{\widetilde{H}(e_i),\ i = 1, 2, \ldots, m\}$, $\widetilde{H}(e_i) = \{u/(\xi_1(u), \xi_2(u), \ldots, \xi_k(u)) : u \in U\}$, where $\widetilde{H}(e) = \widetilde{F}(e) \cap \widetilde{G}(e)$.

Proposition 1 *Let $\widetilde{F}, \widetilde{G}$ and $\widetilde{H}$ be three $IVM^k FSSs$ over (U, E). Then*
(1) $\widetilde{F} \cup \widetilde{F} = \widetilde{F}$; (2) $\widetilde{F} \cap \widetilde{F} = \widetilde{F}$;
(3) $\widetilde{F} \cup \widetilde{G} = \widetilde{G} \cup \widetilde{F}$, (4) $\widetilde{F} \cap \widetilde{G} = \widetilde{G} \cap \widetilde{F}$;
(5) $(\widetilde{F} \cup \widetilde{G}) \cup \widetilde{H} = \widetilde{F} \cup (\widetilde{G} \cup \widetilde{H})$, (6) $(\widetilde{F} \cap \widetilde{G}) \cap \widetilde{H} = \widetilde{F} \cap (\widetilde{G} \cap \widetilde{H})$.

Proof It can be easily proved by Definitions 11 and 12. □

Proposition 2 *Let $\widetilde{F}$ and $\widetilde{G}$ be two $IVM^k FSSs$ over (U, E). Then*
(1) $(\widetilde{F} \cup \widetilde{G})^c = \widetilde{F}^c \cap \widetilde{G}^c$; (2) $(\widetilde{F} \cap \widetilde{G})^c = \widetilde{F}^c \cup \widetilde{G}^c$.

Proof We just prove the first one, the second can be proved similarly. Let $\widetilde{F} = \widetilde{F}(e) = \{\widetilde{F}(e_i) | i = 1, 2, \ldots, m\}$, $\widetilde{F}(e_i) = \{u/(\mu_1(u), \mu_2(u), \ldots, \mu_k(u)) : u \in U\}$, $\mu_j(u) = [\mu_j^-(u), \mu_j^+(u)]$, $j = 1, 2, \ldots, k$.

$\widetilde{G} = \widetilde{G}(e) = \{\widetilde{G}(e_i) | i = 1, 2, \ldots, m\}$, $\widetilde{G}(e_i) = \{u/(\nu_1(u), \nu_2(u), \ldots, \nu_k(u)) : u \in U\}$, $\nu_j(u) = [\nu_j^-(u), \nu_j^+(u)]$, $j = 1, 2, \ldots, k$.

$\widetilde{F}(e_i) \cup \widetilde{G}(e_i) = \{u/(\mu_1(u) \vee \nu_1(u), \mu_2(u) \vee \nu_2(u), \ldots, \mu_k(u) \vee \nu_k(u)) : u \in U\}$, $(\widetilde{F}(e_i) \cup \widetilde{G}(e_i))^c = \{u/(1 - (\mu_1(u) \vee \nu_1(u)), 1 - (\mu_2(u) \vee \nu_2(u)), \ldots, 1 - (\mu_k(u) \vee \nu_k(u))) : u \in U\} = \{u/((1 - \mu_1(u)) \wedge (1 - \nu_1(u)), (1 - \mu_2(u)) \wedge (1 - \nu_2(u)), \ldots, (1 - \mu_k(u)) \wedge (1 - \nu_k(u))) : u \in U\} = \widetilde{F}^c(e_i) \cap \widetilde{G}^c(e_i)$. So $(\widetilde{F} \cup \widetilde{G})^c = \widetilde{F}^c \cap \widetilde{G}^c$. □

Proposition 3 *Let $\widetilde{F}$ and $\widetilde{G}$ be two $IVM^k FSSs$ over (U, E). Then*
(1) $(\widetilde{F} \cup \widetilde{G}) \cap \widetilde{F} = \widetilde{F}$; (2) $(\widetilde{F} \cap \widetilde{G}) \cup \widetilde{F} = \widetilde{F}$.

Proof We just prove the first one, the second can be proved similarly. Let $\widetilde{F} = \{\widetilde{F}(e_i)\}$, $\widetilde{F}(e_i) = \{u/(\mu_1(u), \mu_2(u), \ldots, \mu_k(u))\}$, $\mu_j(u) = [\mu_j^-(u), \mu_j^+(u)]$. $\widetilde{G} = \{\widetilde{G}(e_i)\}$, $\widetilde{G}(e_i) = \{u/(\nu_1(u), \nu_2(u), \ldots, \nu_k(u))\}$, $\nu_j(u) = [\nu_j^-(u), \nu_j^+(u)]$, $u \in U$, $i = 1, 2, \ldots, m$, $j = 1, 2, \ldots, k$.

$(\widetilde{F}(e_i) \cup \widetilde{G}(e_i)) \cap \widetilde{F}(e_i) = \{u/((\mu_1(u) \vee \nu_1(u)) \wedge \mu_1(u), (\mu_2(u) \vee \nu_2(u)) \wedge \mu_2(u), \ldots, (\mu_k(u) \vee \nu_k(u)) \wedge \mu_k(u)) : u \in U\}$. $(\mu_j(u) \vee \nu_j(u)) \wedge \mu_j(u) = \mu_j(u)$, $j = 1, 2, \ldots, k$, i.e., $(\widetilde{F}(e_i) \cup \widetilde{G}(e_i)) \cap \widetilde{F}(e_i) = \widetilde{F}(e_i)$. So $(\widetilde{F} \cup \widetilde{G}) \cap \widetilde{F} = \widetilde{F}$. □

Proposition 4 *Let $\widetilde{F}, \widetilde{G}$ and $\widetilde{H}$ be three $IVM^k FSSs$ over (U, E). Then*
(1) $\widetilde{F} \cup (\widetilde{G} \cap \widetilde{H}) = (\widetilde{F} \cup \widetilde{G}) \cap (\widetilde{F} \cup \widetilde{H})$, (2) $\widetilde{F} \cap (\widetilde{G} \cup \widetilde{H}) = (\widetilde{F} \cap \widetilde{G}) \cup (\widetilde{F} \cap \widetilde{H})$.

Proof We just prove the first one, the second can be proved similarly. Let $\widetilde{F} = \{\widetilde{F}(e_i)\}$, $\widetilde{F}(e_i) = \{u/(\mu_1(u), \mu_2(u), \ldots, \mu_k(u))\}$, $\mu_j(u) = [\mu_j^-(u), \mu_j^+(u)]$. $\widetilde{G} = \{\widetilde{G}(e_i)\}$, $\widetilde{G}(e_i) = \{u/(\nu_1(u), \nu_2(u), \ldots, \nu_k(u))\}$, $\nu_j(u) = [\nu_j^-(u), \nu_j^+(u)]$.

$\widetilde{H} = \{\widetilde{H}(e_i)\}$, $\widetilde{H}(e_i) = \{u/(\xi_1(u), \xi_2(u), \ldots, \xi_k(u))\}$, $\xi_j(u) = [\xi_j^-(u), \xi_j^+(u)]$, $u \in U, i = 1, 2, \ldots, m, j = 1, 2, \ldots, k$. $\widetilde{F}(e_i) \cup (\widetilde{G}(e_i) \cap \widetilde{H}(e_i)) = \{u/(\mu_1(u) \vee (\nu_1(u) \wedge \xi_1(u)), \mu_2(u) \vee (\nu_2(u) \wedge \xi_2(u)), \ldots, \mu_k(u) \vee (\nu_k(u) \wedge \xi_k(u))) : u \in U\}$, $(\widetilde{F}(e_i) \cup \widetilde{G}(e_i)) \cap (\widetilde{F}(e)_i \cup \widetilde{H}(e_i)) = \{u/((\mu_1(u) \vee \nu_1(u)) \wedge (\mu_1(u) \vee \xi_1(u)), (\mu_2(u) \vee \nu_2(u)) \wedge (\mu_2(u) \vee \xi_2(u)), \ldots, (\mu_k(u) \vee \nu_k(u)) \wedge (\mu_k(u) \vee \xi_k(u)))\}$. $\mu_j(u) \vee (\nu_j(u) \wedge \xi_j(u)) = (\mu_j(u) \vee \nu_j(u)) \wedge (\mu_j(u) \vee \xi_j(u))$, $j = 1, 2, \ldots, k$. $\widetilde{F}(e_i) \cup (\widetilde{G}(e_i) \cap \widetilde{H}(e_i)) = (\widetilde{F}(e_i) \cup \widetilde{G}(e_i)) \cap (\widetilde{F}(e_i) \cup \widetilde{H}(e_i))$. So $\widetilde{F} \cup (\widetilde{G} \cap \widetilde{H}) = (\widetilde{F} \cup \widetilde{G}) \cap (\widetilde{F} \cup \widetilde{H})$. □

Theorem 1 *Let $\Gamma(U, E)$ be the set of all interval-valued multi-fuzzy soft sets over (U, E), i.e., $\Gamma(U, E) = \{\widetilde{F} | \widetilde{F}$ is an IVM^{k}FSS over $(U, E)\}$. Then the algebraic system $Q = (\Gamma(U, E), \cup, \cap)$ is a distributive lattice.*

Proof It can be easily proved by Propositions 1, 3 and 4. □

4 Possibility Interval-Valued Multi-fuzzy Soft Sets

In this section, we generalized the concept of possibility fuzzy soft sets to possibility interval-valued multi-fuzzy soft set.

Definition 13 ([13]) Let U be an initial universe and E a set of parameters. A pair (F_α, E) is called a possibility multi-fuzzy soft set of dimension k if F_α is a mapping given by $F_\alpha : E \to M^kFS \times M^kFS$, where $F_\alpha(e) = (F(e)(u), \alpha(e)(u))$, $\forall u \in U,\ e \in E$.

Definition 14 Let U be a finite initial universe and E a finite set of parameters. Suppose that $\widetilde{F} : E \to IVM^kFS$, and f is an interval-valued multi-fuzzy subset of E, i.e., $f : E \to IVM^kFS$. $\widetilde{F}_f$ is called a possibility interval-valued multi-fuzzy soft set of dimension k (for short $PIVM^k$FSS) over (U, E) if $\widetilde{F}_f$ is a mapping given by

$$\widetilde{F}_f : E \to IVM^kFS \times IVM^kFS,$$

where $\widetilde{F}_f(e) = (\widetilde{F}(e)(u), f(e)(u))$, $\forall u \in U,\ e \in E$.

For each parameter e_i, $\widetilde{F}_f(e_i) = (\widetilde{F}(e_i)(u), f(e_i)(u))$ indicates not only the multi-membership degree of elements in U belonging to $\widetilde{F}(e_i)$ but also include multi-membership degree of possibility of elements in U belonging to $\widetilde{F}(e_i)$, which is represented by $f(e_i)$. So we can write $\widetilde{F}_f$ as follows.

$$\widetilde{F}_f(e_i) = \{(u/\mu_{\widetilde{F}(e_i)}(u), \mu_{f(e_i)}(u)) : u \in U\},$$

where $\mu_{\widetilde{F}(e_i)}(u) = (\mu^1_{\widetilde{F}(e_i)}(u), \mu^2_{\widetilde{F}(e_i)}(u), \ldots, \mu^k_{\widetilde{F}(e_i)}(u))$, and
$\mu_{f(e_i)}(u) = (\mu^1_{f(e_i)}(u), \mu^2_{f(e_i)}(u), \ldots, \mu^k_{f(e_i)}(u))$,
$\mu^k_{\widetilde{F}(e_i)}(u) = [\mu^{k-}_{\widetilde{F}(e_i)}(u), \mu^{k+}_{\widetilde{F}(e_i)}(u)]$,
$\mu^k_{f(e_i)}(u) = [\mu^{k-}_{f(e_i)}(u), \mu^{k+}_{f(e_i)}(u)]$.

Sometimes we write $\widetilde{F}_f$ as $(\widetilde{F}_f, E)$. If $A \subseteq E$, we have $PIVM^kFSS\ (\widetilde{F}_f, A)$.

A possibility interval-valued multi-fuzzy soft set of dimension k is also a special case of a soft set. If $\mu^-(u) = \mu^+(u)$, i.e., $\forall e \in E,\ \forall u \in U, \mu^-_{\widetilde{F}(e_i)}(u) = \mu^+_{\widetilde{F}(e_i)}(u)$ and $\mu^-_{f(e_i)}(u) = \mu^+_{f(e_i)}(u)$, a $PIVM^kFSS$ will be degenerates to a possibility multi-fuzzy soft set. If $k = 1$ and $\mu^-(u) = \mu^+(u)$, a $PIVM^kFSS$ will degenerates to a possibility fuzzy soft set.

Definition 15 Let U be a finite initial universe and E a finite set of parameters. $\widetilde{F}_f$ is a $PIVM^kFSS$ over (U, E). The complement set of $\widetilde{F}_f$ is also a $PIVM^kFSS$ denoted by $\widetilde{F}^c_f$, where $\widetilde{F}^c_f = \widetilde{G}_g$ is a mapping given by $\widetilde{G}_g : E \to IVM^kFS \times IVM^kFS$. $\widetilde{G}_g(e) = (\widetilde{G}(e), g(e)), \widetilde{G}(e) = \widetilde{F}^c(e),\ g(e) = f^c(e)$.

From the above definition, we have $(\widetilde{F}^c_f)^c = \widetilde{F}_f$.

Definition 16 Let $\widetilde{F}_f$ and $\widetilde{G}_g$ be two $PIVM^kFSSs$ over (U, E). $\widetilde{F}_f$ is said to be a possibility interval-valued multi-fuzzy soft subset of $\widetilde{G}_g$ denoted by $\widetilde{F}_f \subseteq \widetilde{G}_g$ if and only if for any $e \in E$, and $u \in U$, $\widetilde{F}(e)$ is a interval-valued multi-fuzzy subset of $\widetilde{G}(e)$ and $f(e)$ is a interval-valued multi-fuzzy subset of $g(e)$.

Definition 17 Let $\widetilde{F}_f$ and $\widetilde{G}_g$ be two $PIVM^kFSSs$ over (U, E). $\widetilde{F}_f$ is said to be a possibility interval-valued multi-fuzzy soft equal of $\widetilde{G}_g$ denoted by $\widetilde{F}_f = \widetilde{G}_g$ if and only if $\widetilde{F}_f$ is a possibility interval-valued multi-fuzzy soft subset of $\widetilde{G}_g$ and $\widetilde{G}_g$ is a possibility interval-valued multi-fuzzy soft subset of $\widetilde{F}_f$.

Definition 18 Let $\widetilde{F}_f$ and $\widetilde{G}_g$ be two $PIVM^kFSSs$ over (U, E). The union and intersection operations on $\widetilde{F}_f$ and $\widetilde{G}_g$, denoted by $\widetilde{F}_f \cup \widetilde{G}_g$ and $\widetilde{F}_f \cap \widetilde{G}_g$ respectively, are defined by mappings $\widetilde{H}_h, \widetilde{K}_k : E \to IVM^kFS \times IVM^kFS$, where $\widetilde{H}_h(e) = (\widetilde{H}(e), h(e)),\ \ \widetilde{K}_k(e) = (\widetilde{K}(e), k(e)),\ \ \widetilde{H}(e) = \widetilde{F}(e) \cup \widetilde{G}(e),\ \ h(e) = f(e) \cup g(e)$, $\widetilde{K}(e) = \widetilde{F}(e) \cap \widetilde{G}(e)$ and $k(e) = f(e) \cap g(e)$. $\forall u \in U,\ e \in E$.

Proposition 5 *Let $\widetilde{F}_f$, $\widetilde{G}_g$ and $\widetilde{H}_h$ be three $PIVM^kFSSs$ over (U, E). Then we have*

(1) $\widetilde{F}_f \cup \widetilde{F}_f = \widetilde{F}_f$; (2) $\widetilde{F}_f \cap \widetilde{F}_f = \widetilde{F}_f$;

(3) $\widetilde{F}_f \cup \widetilde{G}_g = \widetilde{G}_g \cup \widetilde{F}_f$; (4) $\widetilde{F}_f \cap \widetilde{G}_g = \widetilde{G}_g \cap \widetilde{F}_f$;

(5) $(\widetilde{F}_f \cup \widetilde{G}_g) \cup \widetilde{H}_h = \widetilde{F}_f \cup (\widetilde{G}_g \cup \widetilde{H}_h)$; (6) $(\widetilde{F}_f \cap \widetilde{G}_g) \cap \widetilde{H}_h = \widetilde{F}_f \cap (\widetilde{G}_g \cap \widetilde{H}_h)$.

Proposition 6 *Let $\widetilde{F}_f$ and $\widetilde{G}_g$ be two $PIVM^kFSSs$ over (U, E). Then*

(1) $(\widetilde{F}_f \cup \widetilde{G}_g)^c = \widetilde{F}^c_f \cap \widetilde{G}^c_g$; (2) $(\widetilde{F}_f \cap \widetilde{G}_g)^c = \widetilde{F}^c_f \cup \widetilde{G}^c_g$;

(3) $(\widetilde{F}_f \cup \widetilde{G}_g) \cap \widetilde{F}_f = \widetilde{F}_f$; (4) $(\widetilde{F}_f \cap \widetilde{G}_g) \cup \widetilde{F}_f = \widetilde{F}_f$;

(5) $\widetilde{F}_f \cup (\widetilde{G}_g \cap \widetilde{H}_h) = (\widetilde{F}_f \cup \widetilde{G}_g) \cap (\widetilde{F}_f \cup \widetilde{H}_h)$;

(6) $\widetilde{F}_f \cap (\widetilde{G}_g \cup \widetilde{H}_h) = (\widetilde{F}_f \cap \widetilde{G}_g) \cup (\widetilde{F}_f \cap \widetilde{H}_h)$.

Theorem 2 *Let $\Theta(U, E)$ be the set of all possibility interval-valued multi-fuzzy soft sets over (U, E), i.e., $\Theta(U, E) = \{\widetilde{F}_f | \widetilde{F}_f$ is a $PIVM^kFSS$ over $(U, E)\}$. Then the algebraic system $Q = (\Theta(U, E), \cup, \cap)$ is a distributive lattice.*

Proof It can be easily proved by Propositions 5 and 6. □

Definition 19 Let $(\widetilde{F}_f, A)$ and $(\widetilde{G}_g, B)$ be two $PIVM^kFSSs$ over (U, E). The operation "$(\widetilde{F}_f, A)\ AND\ (\widetilde{G}_g, B)$" denoted by $\widetilde{F}_f \wedge \widetilde{G}_g$ is defined as $(\widetilde{F}_f, A) \wedge (\widetilde{G}_g, B) = (\widetilde{H}_h, A \times B)$, where $\widetilde{H}_h(\alpha, \beta) = (\widetilde{H}(\alpha, \beta)(u), h(\alpha, \beta)(u))$. For all $(\alpha, \beta) \in A \times B$, $\widetilde{H}(\alpha, \beta) = \widetilde{F}(\alpha) \cap \widetilde{G}(\beta)$ and $h(\alpha, \beta) = f(\alpha) \cap g(\beta)$.

Definition 20 Let $(\widetilde{F}_f, A)$ and $(\widetilde{G}_g, B)$ be two $PIVM^kFSSs$ over (U, E). The operation "$(\widetilde{F}_f, A)\ OR\ (\widetilde{G}_g, B)$" denoted by $\widetilde{F}_f \vee \widetilde{G}_g$ is defined as $(\widetilde{F}_f, A) \vee (\widetilde{G}_g, B) = (\widetilde{H}_h, A \times B)$, where $\widetilde{H}_h(\alpha, \beta) = (\widetilde{H}(\alpha, \beta)(u), h(\alpha, \beta)(u))$. For all $(\alpha, \beta) \in A \times B$, $\widetilde{H}(\alpha, \beta) = \widetilde{F}(\alpha) \cup \widetilde{G}(\beta)$ and $h(\alpha, \beta) = f(\alpha) \cup g(\beta)$.

Remark 1 Let $(\widetilde{F}_f, A)$ and $(\widetilde{G}_g, B)$ be two $PIVM^kFSSs$ over (U, E). For all $(\alpha, \beta) \in (A \times B)$, if $\alpha \neq \beta$, then $(\widetilde{F}_f, A) \wedge (\widetilde{G}_g, B) \neq (\widetilde{G}_g, B) \wedge (\widetilde{F}_f, A)$ and $(\widetilde{F}_f, A) \vee (\widetilde{G}_g, B) \neq (\widetilde{G}_g, B) \vee (\widetilde{F}_f, A)$.

Theorem 3 *Let $(\widetilde{F}_f, A)$ and $(\widetilde{G}_g, B)$ be two $PIVM^kFSSs$ over (U, E). Then*
(1) $((\widetilde{F}_f, A) \vee (\widetilde{G}_g, B))^c = (\widetilde{F}_f, A)^c \wedge (\widetilde{G}_g, B)^c$;
(2) $((\widetilde{F}_f, A) \wedge (\widetilde{G}_g, B))^c = (\widetilde{F}_f, A)^c \vee (\widetilde{G}_g, B)^c$.

Proof We just prove the first one, the second can be proved similarly. Suppose $(\widetilde{F}_f, A) \vee (\widetilde{G}_g, B) = (\widetilde{H}_h, A \times B)$. Where $\widetilde{H}_h^c(\alpha, \beta) = (\widetilde{H}^c(\alpha, \beta), h^c(\alpha, \beta))$. From Definition 18 and Proposition 6, for all $u \in U$, $(\alpha, \beta) \in (A \times B)$, we have $\widetilde{H}^c(\alpha, \beta) = (\widetilde{F}(\alpha) \cup \widetilde{G}(\beta))^c = \widetilde{F}^c(\alpha) \cap \widetilde{G}^c(\beta)$ and $h^c(\alpha, \beta) = (f(\alpha) \cup g(\beta))^c = f^c(\alpha) \cap g^c(\beta)$. Also suppose that $(\widetilde{F}_f, A)^c \wedge (\widetilde{G}_g, B)^c = (\widetilde{L}_l, A \times B)$, where $\widetilde{L}_l(\alpha, \beta) = (\widetilde{L}(\alpha, \beta), l(\alpha, \beta))$. For all $u \in U$, $(\alpha, \beta) \in (A \times B)$, $\widetilde{L}(\alpha, \beta) = \widetilde{F}^c(\alpha) \cap \widetilde{G}^c(\beta)$, and $l(\alpha, \beta) = f^c(\alpha) \cap g^c(\beta)$. $\widetilde{H}_h^c = \widetilde{L}_l$. So $((\widetilde{F}_f, A) \vee (\widetilde{G}_g, B))^c = (\widetilde{F}_f, A)^c \wedge (\widetilde{G}_g, B)^c$. □

Proposition 7 *Let $(\widetilde{F}_f, A)$, $(\widetilde{G}_g, B)$ and $(\widetilde{H}_h, C)$ be three $PIVM^kFSSs$ over (U, E). Then we have*
(1) $((\widetilde{F}_f, A) \vee (\widetilde{G}_g, B)) \vee (\widetilde{H}_h, C) = (\widetilde{F}_f, A) \vee ((\widetilde{G}_g, B) \vee (\widetilde{H}_h, C))$;
(2) $((\widetilde{F}_f, A) \wedge (\widetilde{G}_g, B)) \wedge (\widetilde{H}_h, C) = (\widetilde{F}_f, A) \wedge ((\widetilde{G}_g, B) \wedge (\widetilde{H}_h, C))$;
(3) $(\widetilde{F}_f, A) \vee ((\widetilde{G}_g, B) \wedge (\widetilde{H}_h, C)) = ((\widetilde{F}_f, A) \vee (\widetilde{G}_g, B)) \wedge ((\widetilde{F}_f, A) \vee (\widetilde{H}_h, C))$;
(4) $(\widetilde{F}_f, A) \wedge ((\widetilde{G}_g, B) \vee (\widetilde{H}_h, C)) = ((\widetilde{F}_f, A) \wedge (\widetilde{G}_g, B)) \vee ((\widetilde{F}_f, A) \wedge (\widetilde{H}_h, C))$.

5 Decision-Making

In this section, we give an example to illustrated application of possibility interval-valued multi-fuzzy soft sets in decision-making.

Assume that a company want to select a manager from three candidates $U = \{u_1, u_2, u_3\}$. The set of parameters $E = \{e_1, e_2, e_3\}$, where e_1 represent "experience" which includes three levels: rich, average, poor, e_2 stands for "computer skills" which contains three levels: skilled, average, poor, e_3 express "young age" involving three levels: old, medium, young. Two experts gives two $PIVM^3FSSs$ $\widetilde{F}_f$ and $\widetilde{G}_g$ over (U, E). $\widetilde{F}_f$ and $\widetilde{G}_g$ are given as follows.

$$\begin{aligned}
\widetilde{F}_f(e_1) = \{&(u_1/([0.05, 0.1], [0.4, 0.45], [0.35, 0.45]), ([0.3, 0.4], [0.4, 0.5], [0.05, 0.1])),\\
&(u_2/([0.05, 0.1], [0.2, 0.3], [0.5, 0.6]), ([0.1, 0.2], [0.2, 0.3], [0.3, 0.4])),\\
&(u_3/([0.35, 0.4], [0.4, 0.5], [0.05, 0.1]), ([0.2, 0.3], [0.3, 0.4], [0.25, 0.3]))\};\\
\widetilde{F}_f(e_2) = \{&(u_1/([0.05, 0.1], [0.3, 0.4], [0.4, 0.5]), ([0.2, 0.3], [0.35, 0.45], [0.2, 0.25])),\\
&(u_2/([0.15, 0.2], [0.3, 0.4], [0.3, 0.4]), ([0, 0.05], [0.05, 0.2], [0.6, 0.7])),\\
&(u_3/([0.2, 0.3], [0.3, 0.4], [0.1, 0.2]), ([0.2, 0.3], [0.3, 0.4], [0.25, 0.3]))\};\\
\widetilde{F}_f(e_3) = \{&(u_1/([0.25, 0.3], [0.4, 0.5], [0.1, 0.15]), ([0.2, 0.3], [0.3, 0.4], [0.15, 0.2])),\\
&(u_2/([0.25, 0.3], [0.4, 0.5], [0.15, 0.2]), ([0.1, 0.2], [0.4, 0.45], [0.3, 0.35])),\\
&(u_3/([0.1, 0.2], [0.25, 0.3], [0.3, 0.4]), ([0.1, 0.2], [0.3, 0.4], [0.3, 0.4]))\};\\
\widetilde{G}_g(e_1) = \{&(u_1/([0.1, 0.15], [0.2, 0.3], [0.4, 0.5]), ([0.25, 0.3], [0.3, 0.4], [0.15, 0.2])),\\
&(u_2/([0.1, 0.2], [0.3, 0.4], [0.3, 0.4]), ([0, 0.05], [0.1, 0.2], [0.5, 0.6])),\\
&(u_3/([0.15, 0.2], [0.3, 0.4], [0.05, 0.1]), ([0.25, 0.3], [0.25, 0.3], [0.3, 0.35]))\};\\
\widetilde{G}_g(e_2) = \{&(u_1/([0.2, 0.3], [0.4, 0.5], [0.05, 0.1]), ([0.1, 0.2], [0.2, 0.3], [0.4, 0.5])),\\
&(u_2/([0.25, 0.3], [0.3, 0.4], [0.25, 0.3]), ([0.2, 0.3], [0.3, 0.35], [0.25, 0.35])),\\
&(u_3/([0.2, 0.25], [0.3, 0.35], [0.3, 0.4]), ([0.2, 0.3], [0.35, 0.4], [0.25, 0.3]))\};\\
\widetilde{G}_g(e_3) = \{&(u_1/([0.3, 0.4], [0.4, 0.5], [0.05, 0.1]), ([0.2, 0.3], [0.45, 0.5], [0.1, 0.2])),\\
&(u_2/([0.2, 0.3], [0.35, 0.4], [0.2, 0.25]), ([0.25, 0.3], [0.3, 0.35], [0.2, 0.3])),\\
&(u_3/([0.15, 0.2], [0.2, 0.3], [0.2, 0.25]), ([0.3, 0.35], [0.4, 0.45], [0.1, 0.2]))\}.
\end{aligned}$$

Here, we use AND operation to consider the two experts' opinions. We have $(\widetilde{F}_f, A)$ AND $(\widetilde{G}_g, B) = (\widetilde{H}_h, A \times B)$, where, for example, $\widetilde{H}_h(e_1, e_1) = \{(u_1/$ $([0.05, 0.1], [0.2, 0.3], [0.35, 0.45]), ([0.25, 0.3], [0.3, 0.4], [0.05, 0.1])),$ $(u_2/([0.05, 0.1], [0.2, 0.3], [0.3, 0.4]), ([0, 0.05], [0.1, 0.2], [0.3, 0.4])),$ $(u_3/([0.15, 0.2], [0.3, 0.4], [0.05, 0.1]), ([0.2, 0.3], [0.25, 0.3], [0.25, 0.3]))\}$.

To choose the best candidate, we have to compute the numerical grade $t_{ij}(u_k)$ and the corresponding grade $s_{ij}(u_k)$ for each (e_i, e_j).

Table 1 The numerical grades and corresponding possibility grade

	(e_1, e_1)	(e_1, e_2)	(e_1, e_3)	(e_2, e_1)	(e_2, e_2)	(e_2, e_3)	(e_3, e_1)	(e_3, e_2)	(e_3, e_3)
$t(u_1)$	$[-0.8, 0.65]$	$[-0.4, 0.25]$	$[-0.25, 0.45]$	$[-0.4, 0.6]$	$[-0.9, -0.1]$	$[-0.75, 0.1]$	$[-0.6, 0.2]$	$[-0.5, 0.45]$	$[-0.25, 0.6]$
$s(u_1)$	$[-0.35, 0.5]$	$[-1.15, -0.1]$	$[-0.4, 0.7]$	$[0.45, 0.55]$	$[-0.6, 0.45]$	$[-0.15, 1.05]$	$[-0.15, 0.75]$	$[-0.9, 0.1]$	$[-0.45, 0.7]$
$t(u_2)$	$[-0.4, 0.5]$	$[-0.25, 0.25]$	$[-0.35, 0.4]$	$[-0.2, 0.75]$	$[-0.1, 0.8]$	$[0, 0.85]$	$[-0.1, 0.8]$	$[-0.4, 0.4]$	$[-0.35, 0.6]$
$s(u_2)$	$[-0.9, 0]$	$[-0.5, 0.6]$	$[-0.8, 0.35]$	$[-0.7, 0.35]$	$[-1.15, -0.05]$	$[-1.35, -0.15]$	$[-0.95, -0.1]$	$[-0.3, 0.7]$	$[-0.5, 0.6]$
$t(u_3)$	$[-0.65, 0.25]$	$[-0.25, 0.4]$	$[-0.6, 0.25]$	$[-0.9, 0.05]$	$[-0.3, 0.5]$	$[-0.55, 0.35]$	$[-0.6, 0.25]$	$[-0.5, 0.45]$	$[-0.8, 0.1]$
$s(u_3)$	$[-0.05, 0.8]$	$[0.05, 1.1]$	$[-0.5, 0.65]$	$[-0.35, 0.6]$	$[0.15, 1.2]$	$[-0.3, 0.9]$	$[-0.2, 0.65]$	$[-0.3, 0.7]$	$[-0.75, 0.4]$

Table 2 Grade table

	(e_1, e_1)	(e_1, e_2)	(e_1, e_3)	(e_2, e_1)	(e_2, e_2)	(e_2, e_3)	(e_3, e_1)	(e_3, e_2)	(e_3, e_3)
u_i	u_1	u_3	u_1	u_2	u_2	u_2	u_2	u_2	u_1
Highest grade	×	0.075	0.1	0.275	×	0.425	0.35	0	×
possibility grade		0.575	0.15	−0.175		−0.75	−0.525	0.2	

$t_{ij}(u_k) = \sum\limits_{u\in U}((\mu^1_{\widetilde{H}(e_i,e_j)}(u_k) - \mu^1_{\widetilde{H}(e_i,e_j)}(u)) + (\mu^2_{\widetilde{H}(e_i,e_j)}u_k) - \mu^2_{\widetilde{H}(e_i,e_j)}(u)) +$
$(\mu^3_{\widetilde{H}(e_i,e_j)}(u_k) - \mu^3_{\widetilde{H}(e_i,e_j)}(u))),$

$s_{ij}(u_k) = \sum\limits_{u\in U}((\mu^1_{h(e_i,e_j)}(u_k) - \mu^1_{h(e_i,e_j)}(u)) + (\mu^2_{h(e_i,e_j)}(u_k) - \mu^2_{h(e_i,e_j)}(u)) +$
$(\mu^3_{h(e_i,e_j)}(u_k) - \mu^3_{h(e_i,e_j)}(u))).$

For example, $t_{12}(u_1) = [-0.4, 0.25]$, and $s_{12}(u_1) = [-1.15, -0.1]$.

All the $t_{ij}(u_k)$ and $s_{ij}(u_k)$ are shown in Table 1.

Next we transform interval-number $[a, b]$ to $\frac{a+b}{2}$ and mark the highest numerical grade in each column excluding the columns which are the possibility grade of such belongingness of a candidate against each pair of parameters (see Table 2). By taking the sum of these numerical grades with the corresponding possibility $s_{ij}(u_k)$, we calculate the score of each candidate. The candidate who has the highest score is the best one. This method is similar to Ref. [13].

$Score(u_1){=}0.1 \times 0.15{=}0.015$, $Score(u_2) = 0.275 \times (-0.175) + 0.425{\times}(-0.75) + 0.35 \times (-0.525) + 0 \times 0.2 = -0.550625$, $Score(u_3) = 0.075 \times 0.575 = 0.043125$. The company will select the candidate u_3 with the highest score.

6 Conclusions

Soft set theory is a hot topic and an effective mathematical tool to deal with uncertainty. The paper extends the soft notion and proposes the concepts of interval-valued multi-fuzzy soft sets and possibility interval-valued multi-fuzzy soft sets which can more precisely describe the uncertainty in real life. We investigate some based operations and relations on them and discuss the possibility interval-valued fuzzy soft sets with "AND" operator which is successfully applied to solve a decision-making problem in this paper. In the future work, the new decision-making methods about possibility interval-valued multi-fuzzy soft sets should be an important issue to be discussed.

Acknowledgments This work is supported by National Natural Science Foundation of China (Grant No. 11401495) and Students' Extracurricular Open Experiment Project of Southwest Petroleum University (Grant No. KSZ15149).

References

1. Gorzalzany, M.B.: A method of inference in approximate reasoning based on interval-valued fuzzy sets. Fuzzy Sets Syst. **21**, 1–17 (1987)
2. Hur, K., Jang, S.Y., Ahn, Y.S.: Intuitionistic fuzzy equivalence relations. Honam Math. J. **27**(2), 163–181 (2005)
3. Jiang, Y.C., Tang, Y., Chen, Q.M., Liu, H., Tang, T.C.: Interval-valued intuitionistic fuzzy soft sets and their properties. Comput. Math. Appl. **60**, 906–918 (2010)
4. Maji, P.K., Biswas, R., Roy, A.R.: Fuzzy Soft Set. J. Fuzzy Math. **9**(3), 589–602 (2001)
5. Maji, P.K., Biswas, R., Roy, A.R.: On intuitionistic fuzzy soft sets. J. Fuzzy Math. **12**, 669–683 (2004)
6. Molodtsov, D.: Soft set theory-First results. Comput. Math. Appl. **37**, 19–31 (1999)
7. Muharrem, D.: A new distance measure for interval valued intuitionistic fuzzy sets and its application to group decision making problems with incomplete weights information. Appl. Soft Comput. **41**, 120–134 (2016)
8. Sebastian, S., Ramakrishnan, T.V.: Multi-fuzzy sets: an extension of fuzzy sets. Fuzzy Inf. Eng. **1**, 35–43 (2011)
9. Xiao, Z., Chen, W.J., Li, L.L.: A method based on interval-valued fuzzy soft set for multi-attribute group decision-making problems under uncertain environment. Knowl. Inf. Syst. **34**, 653–669 (2013)
10. Yang, Y., Tan, X., Meng, C.C.: The multi-fuzzy soft set and its application in decision making. Appl. Math. Model. **37**, 4915–4923 (2013)
11. Yang, Y., Peng, X.D., Chen, H., Zeng, L.: A decision making approach baesd on bipolar multi-fuzzy soft set theory. J. Intell. Fuzzy Syst. **27**, 1861–1872 (2014)
12. Zadeh, L.A.: Fuzzy sets. Inf. Control. 338–353 (1965)
13. Zhang, H.D., Shu, L.: Possibility multi-fuzzy soft set and its application in decision making. J. Intell. Fuzzy Syst. **27**(4), 2115–2125 (2014)

Weighted Interval-Valued Belief Structures on Atanassov's Intuitionistic Fuzzy Sets

Xin-Hong Xu, De-Chao Li and Zhi-Song Liu

Abstract The Dempster–Shafer (D–S) theory of evidence provides a powerful tool for combination of uncertainty information, and has been extensively applied to deal with uncertainty and vagueness. This paper shows a new approach to combine weighted interval-valued belief structures based on Atanassov's intuitionistic fuzzy sets (A-IF) theory. Two numerical examples are provided to illustrate the rule of combination of weighted interval-valued belief structures. It is then found that this combination can lead to a good result in dealing with the corresponding belief structures in accordance with people's cognitive thinking.

Keywords Evidence · Intuitionistic fuzzy sets · Weighted interval-valued belief structures · Combination

1 Introduction

Since it was originally investigated by Dempster and Shafer [1, 2], D–S theory of belief functions have been a practically usable tool to model and manipulate uncertain, imprecise, incomplete and even vague information. The basic representational structure in D–S theory is a belief structure. The fundamental numeric measures derived from the belief structure are a dual pair of belief and plausibility functions.

X.-H. Xu · D.-C. Li (✉) · Z.-S. Liu
School of Mathematics, Physics and Information Science, Zhejiang Ocean University, Zhoushan 316022, Zhejiang, China
e-mail: dch1831@163.com

X.-H. Xu
e-mail: 1170825404@qq.com

Z.-S. Liu
e-mail: 455597137@qq.com

X.-H. Xu · D.-C. Li · Z.-S. Liu
Key Laboratory of Oceanographic Big Data Mining and Application of Zhejiang Province, Zhoushan 316022, Zhejiang, China

T.-H. Fan et al. (eds.), *Quantitative Logic and Soft Computing 2016*,
Advances in Intelligent Systems and Computing 510,
DOI 10.1007/978-3-319-46206-6_50

Since its inception, evidential reasoning has emerged as a powerful methodology for pattern recognition, image analysis, diagnosis, knowledge discovery, information fusion and decisionmaking [3–12].

In practice, there are many problems, like incompleteness, lack of information and linguistic ambiguity. This results in which probability masses may be uncertain or imprecise in belief structures. Therefore, interval-valued belief degree is more suitable than precise one. Interval-valued belief structure assigns belief degree to each individual hypothesis lies within a certain interval. Lee and Zhu handled interval-valued basic probability assignments [13]. Denoeux extended main concepts of D–S theory to the case where degrees of belief in various propositions are only known to lie within certain intervals [5]. Wang et. al. investigated combination and normalization of interval-valued belief structures within the framework of D-S theory of evidence [14]. Yager considered the situation in which our knowledge of weights associated with the focal elements is that they lie in some known intervals [15]. Su et al. discussed the existence of credible and maximal interval-valued beliefs assigned to focal elements [16].

In addition, different experts may give different degrees of belief in group decisionmaking. In order to prevent loss of some important information, an A-IF set is more objective than a fuzzy set to describe the vagueness of data or information. Therefore, it is arisen to consider the problem of extending D–S theory to interval-valued belief functions on intuitionistic fuzzy setting. Such extension allows us to use directly Dempster's rule of combination to aggregate local criteria presented by intuitionistic fuzzy values in decisionmaking problem. Dymova and Sevastjanov defined value of belief function of an A-IF set [17]. Feng et. al. studied probability problems of intuitionistic fuzzy sets and belief structures of general intuitionistic fuzzy information systems [18]. Song et al. developed a new approach for combining interval-valued belief structures based on intuitionistic fuzzy set [19].

It is well known that the kernel of D–S theory is Dempster's rule. However, as it stands Dempster's rule cannot be applied in this case where two pieces of evidence are in complete conflict. In order to overcome this drawback, many alternative evidence combination rules have been developed. Three typical strategies can be found in the literature: (1) allocating conflicting beliefs to the frame of discernment as global ignorance, (2) allocating conflicting beliefs to a subset of relevant focal propositions as local ignorance or redistributing it among focal propositions locally, and (3) modifying initial belief functions to better represent original information without modifying Dempster's rule [20].

In many practical applications, the decision makers or experts can only intuitively assign their assessments with their expertise. This results in the fact that interval-valued probability masses are usually irrational and invalid assignments. Therefore, it is crucial whether it is sufficient to identify rules to combine pieces of highly or completely conflicting evidence. In other words, how to combine pieces of evidence with various weights and reliabilities that have different meaning? The importance of a piece of evidence depends on the decision maker's judgment. This means that it is independent of the fact that who may use the evidence. This paper is aimed to combine multiple-pieces of independent evidence for intuitionistic fuzzy setting.

This research is also motivated to investigate the rationale and foundation of the evidence reason approach based on A-IF. Having this in mind, the rest of this paper is organized as follows. In Sect. 2, we give some definitions of basic notions and notations. In Sect. 3, we present a new approach to combine the weighted interval-valued belief structures based on synthesis of A-IF sets and D–S theory. Section 4 provides two examples to illustrate the performance of the new combination.

2 Preliminaries

First, we briefly summarize some basic concepts and results that are needed for further study.

Definition 1 ([19]) Let Θ be a frame of discernment. A basic probability assignment (bpa) is a function $m : 2^{\Theta} \rightarrow [0, 1]$, satisfying: $m(\emptyset) = 0$, $\sum_{A \subseteq \Theta} m(A) = 1$.

Associated with each bpa is a belief measure, denoted by $Bel(A)$, and a plausibility measure, denoted by $Pl(A)$, which are defined by the following equations [19]:

Definition 2 ([19]) Given a belief structure m on Θ, the belief function and plausibility function can be defined as: $Bel(A) = \sum_{B \subseteq A} m(B)$, $Pl(A) = \sum_{B \cap A \neq \emptyset} m(B) = 1 - \sum_{B \cap A = \emptyset} m(B)$, $\forall A \subseteq \Theta$.

Definition 3 ([19]) Let Bel_1 and Bel_2 be two belief functions of Θ, of which m_1 and m_2 are the corresponding basic probability assignments. The belief structure that results from application of Dempster's combination rule is given by

$$m_1 \oplus m_2(A) = \begin{cases} \dfrac{\sum\limits_{B \cap C = A} m_1(B) m_2(C)}{1 - \sum\limits_{B \cap C = \emptyset} m_1(B) m_2(C)} & \forall A \subseteq \Theta,\ A \neq \emptyset \\ 0 & A = \emptyset \end{cases}. \quad (1)$$

Definition 4 ([19]) Let $m(A_i) = [a_i, b_i](i = 1, \ldots, n)$ be the support interval of to A_i, where a_i and b_i can be respectively seen as the lower and upper bounds of the probability to which A_i is supposed, then $m(A_i) = [a_i, b_i]$ denotes the interval-valued belief structure.

Definition 5 ([19]) Let $\Theta = \{A_1, \ldots, A_n\}$ be a frame of discernment, $F_1, \ldots, F_N$ be N subsets of Θ, and $[a_i, b_i]$ be N intervals with $0 \leq a_i \leq b_i \leq 1 (i = 1, \ldots, N)$. Then an interval-valued belief structure is a belief structure on Θ meeting the following conditions:

(1) $a_i \leq m(F_i) \leq b_i, 0 \leq a_i \leq b_i \leq 1 (i = 1, \ldots, N)$;

(2) $\sum_{i=1}^{N} a_i \leq 1, \sum_{i=1}^{N} b_i \geq 1$;

(3) $m(H) = 0, \forall H \not\subseteq \{F_1, \ldots, F_N\}$.

Definition 6 ([14]) Let m be an interval-valued belief function with $a_i \leq m(F_i) \leq b_i (i = 1, \ldots, N)$. If a_i and b_i satisfy $\sum_{i=1}^{N} b_i - (b_k - a_k) \geq 1$, $\sum_{i=1}^{N} a_i + (b_k - a_k) \leq 1$, $\forall k \in \{1, \ldots, N\}$, then m is called a normalized interval-valued belief structure.

There are two kinds of non-normalized interval-valued belief functions [14]. The first contradicts the conditions that $\sum_{i=1}^{N} a_i \leq 1$ and $\sum_{i=1}^{N} b_i \geq 1$, which can be normalized by

$$\hat{a}_i = \frac{a_i}{a_i + \sum_{j=1, j \neq i}^{N} b_j}, \quad \hat{b}_i = \frac{b_i}{b_i + \sum_{j=1, j \neq i}^{N} a_j}, \quad i = 1, \ldots, N. \tag{2}$$

The second satisfies $\sum_{i=1}^{N} a_i \leq 1$ and $\sum_{i=1}^{N} b_i \geq 1$, but does not satisfy the conditions in Definition 5, i.e., $\sum_{i=1}^{N} b_i - (b_k - a_k) \geq 1$ and $\sum_{i=1}^{N} a_i + (b_k - a_k) \leq 1$, which can be normalized by

$$\hat{a}_i = \max\left\{a_i, 1 - \sum_{j=1, j \neq i}^{N} b_j\right\}, \quad \hat{b}_i = \max\left\{b_i, 1 - \sum_{j=1, j \neq i}^{N} a_j\right\}, \quad i = 1, \ldots, N. \tag{3}$$

Definition 7 ([19]) Let m be a normalized interval-valued belief structure on $\Theta = \{A_1, \ldots, A_n\}$ with interval-valued probability mass $a_i \leq m(F_i) \leq b_i$, $i = 1, \ldots, N$. m' denotes a Bayesian belief structure in correspondence with m. The basic probability assignment m' is given by

$$m'(A_j) = BetP(A_j) = [BetP^-(A_j), BetP^+(A_j)], \tag{4}$$

where $BetP^-(A_j) = \sum_{A_j \in F_i} \frac{a_i}{|F_i|}$, $BetP^+(A_j) = \min\left\{1, \sum_{A_j \in F_i} \frac{b_i}{|F_i|}\right\}$, $i = 1, \ldots, N, j = 1, \ldots, n$.

Definition 8 ([19]) Let X be a universe of discourse, then an intuitionistic fuzzy set (A-IF) A in X is defined as $A = \{(x, \mu_A(x), \gamma_A(x)) | x \in X\}$, in which $\mu_A(x) : X \to [0, 1]$ and $\gamma_A(x) : X \to [0, 1]$ are respectively called the membership degree and non-membership degree with the condition that $0 \leq \mu_A(x) + \gamma_A(x) \leq 1, \forall x \in X$.

Song et. al. discussed interval-valued belief structure based on A-IF set [19]. Suppose that $\mu(A_i)$, $\gamma(A_i)$ and $\pi(A_i)$ are membership degree function, non-membership degree function and intuition index. Let $M = \{\langle A_i, \mu(A_i), \gamma(A_i)\rangle | A_i \in \Theta, i = 1, \ldots, n\}$. The following relations can be effortlessly obtained:

$$\mu(A_i) = a_i, \tag{5}$$

$$\gamma(A_i) = 1 - b_i, \tag{6}$$

$$\pi(A_i) = b_i - a_i. \tag{7}$$

For two belief structures $m_1^{A_i}$ and $m_2^{A_i}$, we have

$$m_1^{A_i}(Yes) = \mu_1(A_i),\ m_1^{A_i}(No) = \gamma_1(A_i),\ m_1^{A_i}(Yes, No) = \pi_1(A_i), \tag{8}$$

$$m_2^{A_i}(Yes) = \mu_2(A_i),\ m_2^{A_i}(No) = \gamma_2(A_i),\ m_2^{A_i}(Yes, No) = \pi_2(A_i), \tag{9}$$

where $\{Yes\}$, $\{No\}$, $\{Yes, No\}$ are the answers of "Does the unknown object belong to A_i?".

Combining them by Dempster's rule of combination, we get

$$m_1^{A_i} \oplus m_2^{A_i}(H_1) = \frac{\sum\limits_{H_2 \cap H_3 = H_1} m_1^{A_i}(H_2) m_2^{A_i}(H_3)}{1 - \sum\limits_{H_2 \cap H_3 = \emptyset} m_1^{A_i}(H_2) m_2^{A_i}(H_3)}, \tag{10}$$

where $H_1, H_2, H_3 \in \{\{Yes\}, \{No\}, \{Yes, No\}\}$.

Definition 9 ([16]) The overall conflict of the two sources of evidence κ_{12} is defined as

$$\kappa_{12} = \sum_{\substack{A_1, A_2 \in 2^{\Theta} \\ A_1 \cap A_2 = \emptyset}} m_1(A_1) m_2(A_2). \tag{11}$$

Definition 10 ([16]) The similarity of s belief functions with regard to an non-empty subset A on Θ is given by

$$d_{12\ldots s}(A) = 1 - [\max\{m_i(A)\} - avg\{m_i(A)\}][\min\{m_i(A)\} - avg\{m_i(A)\}], \tag{12}$$

where $i = 1, \ldots, s$, max, min and avg represent the maximum, minimum and average value, respectively.

Definition 11 ([16]) The weight matrix W of s weight belief functions $s \times 2^{\Theta}$, which every line corresponds to a weight function and every column corresponds to a non-empty subset, is defined as

$$W_{i,j} = \begin{cases} \omega_i \ m_i(A) > 0 \\ 0 \ \ m_i(A) = 0 \end{cases}. \tag{13}$$

3 Weighted Interval-Valued Belief Structures on A-IF

In this section, we present a method to combine interval-valued belief structures with weight based on A-IF set. We firstly consider the combination of two belief structures. Let m_1^A and m_2^A be two belief structures, with weights of ω_1 and ω_2. According to Eq. (10), we have

$$\begin{cases} m_1^A \oplus^\omega m_2^A(Yes) = \frac{\mu_1(A)+\mu_2(A)(1-\mu_1(A))}{1+\mu_1(A)\gamma_2(A)+\mu_2(A)\gamma_1(A)} + \\ \kappa_{12} \frac{w_{12}(A)d_{12}^-(A)[m_{12}^-(A)+c_{12}^-(A)]}{\sum\limits_{\substack{A\in 2^\Theta-\{\emptyset\} \\ B\in 2^\Theta-\{\emptyset\} \\ A\cap B=\emptyset \\ m_{12}(A\cap B)>0}} w_{12}(A)d_{12}(A)[m_{12}(A)+c_{12}(A)]} \\ m_1^A \oplus^\omega m_2^A(No) = \frac{\gamma_1(A)+\gamma_2(A)(1-\gamma_1(A))}{1+\mu_1(A)\gamma_2(A)+\mu_2(A)\gamma_1(A)} - \\ \kappa_{12} \frac{w_{12}(A)d_{12}^-(A)[m_{12}^-(A)+c_{12}^-(A)]}{\sum\limits_{\substack{A\in 2^\Theta-\{\emptyset\} \\ B\in 2^\Theta-\{\emptyset\} \\ A\cap B=\emptyset \\ m_{12}(A\cap B)>0}} w_{12}(A)d_{12}(A)[m_{12}(A)+c_{12}(A)]} \\ m_1^A \oplus^\omega m_2^A(Yes, No) = \frac{\pi_1(A)\pi_2(A)}{1+\mu_1(A)\gamma_2(A)+\mu_2(A)\gamma_1(A)} \end{cases} \tag{14}$$

where $w_{12}(A)$ represents the sum of columns which correspond to the non-empty set A on weight matrix W and

$$m_{12}^-(A) = \frac{\mu_1(A) + \mu_2(A)(1 - \mu_1(A))}{1 + \mu_1(A)\gamma_2(A) + \mu_2(A)\gamma_1(A)}, \tag{15}$$

$$m_{12}^+(A) = 1 - \frac{\gamma_1(A) + \gamma_2(A)(1 - \gamma_1(A))}{1 + \mu_1(A)\gamma_2(A) + \mu_2(A)\gamma_1(A)}, \tag{16}$$

$$\kappa_{12} = 1 - \frac{1 + \mu_1(A)\gamma_2(A) + \mu_2(A)\gamma_1(A)}{2}, \tag{17}$$

$$d_{12}^-(A) = 1 - [max\{m_i^-(A)\} - avg\{m_i^-(A)\}][min\{m_i^-(A)\} - avg\{m_i^-(A)\}], \tag{18}$$

$$d_{12}^+(A) = 1 - [max\{m_i^+(A)\} - avg\{m_i^+(A)\}][min\{m_i^+(A)\} - avg\{m_i^+(A)\}], \tag{19}$$

$$c_{12}^-(A) = m_1^-(A) + m_2^-(A) \tag{20}$$

$$and \quad c_{12}^+(A) = m_1^+(A) + m_2^+(A). \tag{21}$$

Consequently, we come to define a new operation on A-IF set as follows.

Definition 12 Let $M_1 = \{\langle A_i, \mu_1(A_i), \gamma_1(A_i)\rangle | A_i \in \Theta\}$ and $M_2 = \{\langle A_i, \mu_2(A_i), \gamma_2(A_i)\rangle | A_i \in \Theta\}$ be two A-IF sets on $\Theta = \{A_1, \ldots, A_n\}$ with weights ω_1 and ω_2. The combination operation of M_1 and M_2, denoted by $M_1 \odot^\omega M_2$, is defined by

$$M_1 \odot^{\omega} M_2 = \left\{ \left\langle A_i, \frac{\mu_1(A_i) + \mu_2(A_i)(1 - \mu_1(A_i))}{1 + \mu_1(A_i)\gamma_2(A_i) + \mu_2(A_i)\gamma_1(A_i)} + \Psi_i, \right. \right.$$
$$\left. \left. \frac{\gamma_1(A_i) + \gamma_2(A_i)(1 - \gamma_1(A_i))}{1 + \mu_1(A_i)\gamma_2(A_i) + \mu_2(A_i)\gamma_1(A_i)} - \Psi_i \right\rangle | A_i \in \Theta, i = 1, \ldots, n \right\}, \tag{22}$$

where $\Psi_i = \kappa_{12} \frac{w_{12}(A_i) d_{12}^-(A_i)[m_{12}^-(A_i) + c_{12}^-(A_i)]}{\sum\limits_{\substack{A_i \in 2^{\Theta} - \{\emptyset\} \\ A_j \in 2^{\Theta} - \{\emptyset\} \\ A_i \cap A_j = \emptyset \\ m_{12}(A_i \cap A_j) > 0}} w_{12}(A_i) d_{12}(A_i)[m_{12}(A_i) + c_{12}(A_i)]}$.

Since the Dempster's rule can be applied to combine more than two belief structures, the combination $\odot^{\omega}$ can be then extended to N A-IF sets with their weights, which is denoted by $\odot^{\omega}_{j=1,\ldots,N} M_j$.

Theorem 1 *With the commutativity and associativity of Dempster's rule, it is easy to verify that the following properties hold:*

(1) $M_1 \odot^{\omega} M_2 = M_2 \odot^{\omega} M_1$;

(2) $M_1 \odot^{\omega} M_2 \odot^{\omega} M_3 = (M_1 \odot^{\omega} M_2) \odot^{\omega} M_3 = M_1 \odot^{\omega} (M_2 \odot^{\omega} M_3)$.

Proof Obviously.

Let m_1 and m_2 be two normalized interval-valued belief functions on the discernment frame $\Theta = \{A_1, A_2, \ldots, A_n\}$, their normalized interval-valued Bayesian probability are respectively denoted by $m_1(A_i) = [a_{1i}, b_{1i}]$ and $m_2(A_i) = [a_{2i}, b_{2i}]$, $i = 1, \ldots, n$. By Eqs. (5)–(7), the following two A-IF sets can then be introduced

$$M_1 = \{\langle A_i, a_{1i}, 1 - b_{1i}\rangle | A_i \in \Theta, i = 1, \ldots, n\}, \tag{23}$$

$$M_2 = \{\langle A_i, a_{2i}, 1 - b_{2i}\rangle | A_i \in \Theta, i = 1, \ldots, n\}. \tag{24}$$

According to Eq. (22), we can further define $M_1 \odot^{\omega} M_2$ as

$$M_1 \odot^{\omega} M_2 = \left\{ \left\langle A_i, \frac{a_{1i} + a_{2i}(1 - a_{1i})}{1 + a_{1i} + a_{2i} - a_{1i}b_{2i} - a_{2i}b_{1i}} + \Psi_i, \right. \right.$$
$$\left. \left. \frac{1 - b_{1i}b_{2i}}{1 + a_{1i} + a_{2i} - a_{1i}b_{2i} - a_{2i}b_{1i}} - \Psi_i \right\rangle | x \in E \right\}. \tag{25}$$

Considering the relationship between A-IF sets and belief structures by Eqs. (5)–(7), the interval-valued probability assignment of the combination of m_1 and m_2 with weights ω_1 and ω_2, denoted by $m_1 \oplus^{\omega} m_2$, is defined as follows:

$$m_1 \oplus^{\omega} m_2(A_i) = \left(\frac{a_{1i} + a_{2i}(1 - a_{1i})}{1 + a_{1i} + a_{2i} - a_{1i}b_{2i} - a_{2i}b_{1i}} + \Psi_i, \right.$$
$$\left. 1 - \frac{1 - b_{1i}b_{2i}}{1 + a_{1i} + a_{2i} - a_{1i}b_{2i} - a_{2i}b_{1i}} + \Psi_i \right), \tag{26}$$

where

$$\Psi_i = \kappa_{12i} \frac{w_{12}(A_i)d^-_{12}(A_i)[m^-_{12}(A_i) + c^-_{12}(A_i)]}{\sum\limits_{\substack{A_i \in 2^{\Theta} - \{\emptyset\} \\ A_j \in 2^{\Theta} - \{\emptyset\} \\ A_i \cap A_j = \emptyset \\ m_{12}(A_i \cap A_j) > 0}} w_{12}(A_i)d_{12}(A_i)[m_{12}(A_i) + c_{12}(A_i)]} \tag{27}$$

$$m^-_{12}(A_i) = \frac{a_{1i} + a_{2i}(1 - a_{1i})}{1 + a_{1i} + a_{2i} - a_{1i}b_{2i} - a_{2i}b_{1i}}, \tag{28}$$

$$m^+_{12}(A_i) = 1 - \frac{1 - b_{1i}b_{2i}}{1 + a_{1i} + a_{2i} - a_{1i}b_{2i} - a_{2i}b_{1i}}, \tag{29}$$

$$\kappa_{12i} = 1 - \frac{1 + a_{1i} + a_{2i} - a_{1i}b_{2i} - a_{2i}b_{1i}}{2}, \tag{30}$$

$$d^-_{12}(A_i) = 1 - [max\{m^-(A_i)\} - avg\{m^-(A_i)\}][min\{m^-(A_i)\} - avg\{m^-(A_i)\}], \tag{31}$$

$$d^+_{12}(A_i) = 1 - [max\{m^+(A_i)\} - avg\{m^+(A_i)\}][min\{m^+(A_i)\} - avg\{m^+(A_i)\}], \tag{32}$$

$$c^-_{12}(A_i) = m^-_1(A_i) + m^-_2(A_i) \tag{33}$$

$$and \quad c^+_{12}(A_i) = m^+_1(A_i) + m^+_2(A_i). \tag{34}$$

Finally, after $m_1 \oplus^{\omega} m_2$ is normalized according to Eq. (2) or (3), we obtain combination of weighted interval-valued belief functions.

Apparently, this combination operator $\oplus^{\omega}$ can also be extended to combine N interval-valued belief functions with their weights, denoted by $\oplus^{\omega}_{i=1,\ldots,N} m_i$. Similarly, the combination rule of weighted interval-valued belief functions is commutative and associative owning to the commutativity and associativity of combination operation on A-IF set. And then the following equations hold:

(1) $m_1 \oplus^{\omega} m_2 = m_2 \oplus^{\omega} m_1$;
(2) $m_1 \oplus^{\omega} m_2 \oplus^{\omega} m_3 = (m_1 \oplus^{\omega} m_2) \oplus^{\omega} m_3 = m_1 \oplus^{\omega} (m_2 \oplus^{\omega} m_3)$.

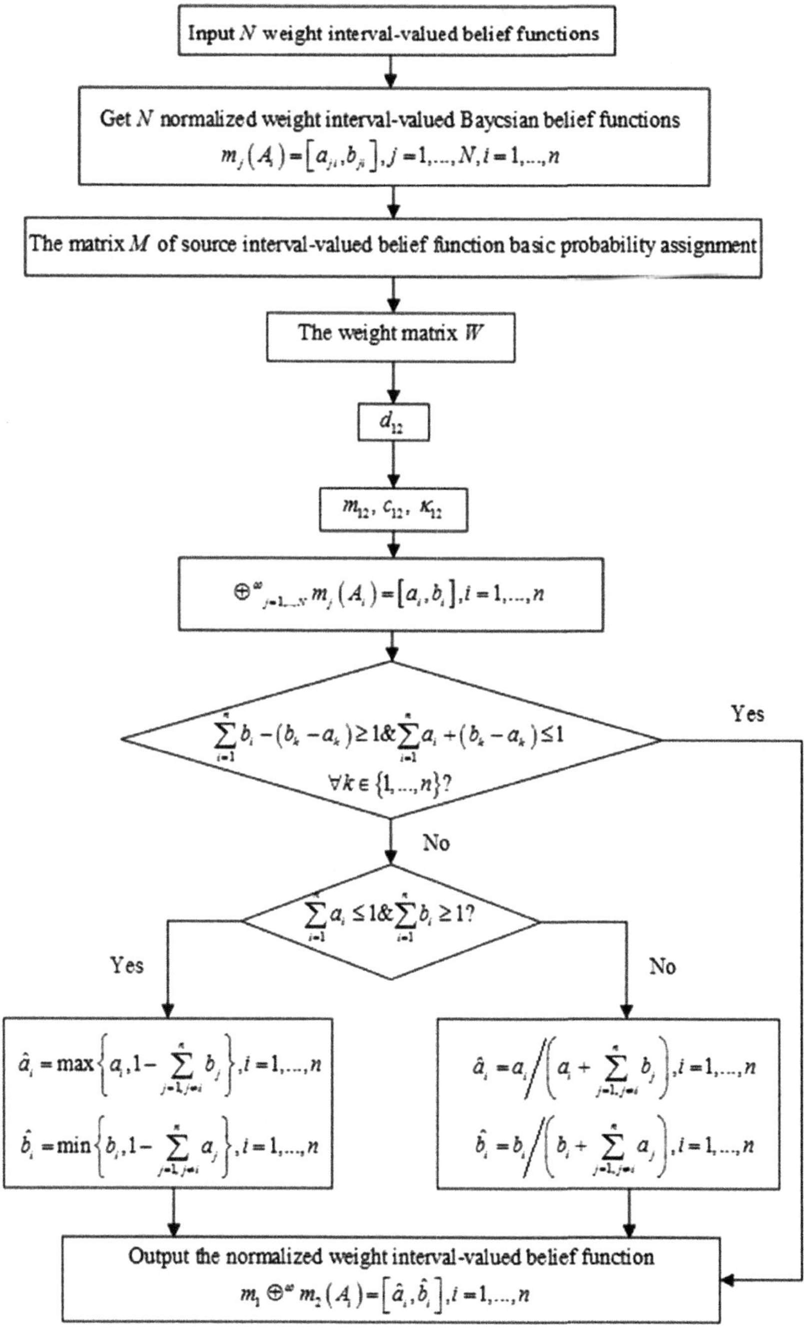

Fig. 1 The flow chart to combine weighted interval-valued belief structures

This implies that we can combine many weighted interval-valued belief functions one by one. The result does not depend on the order of combination. Of course, it does not change when normalization is postponed to a later point. In order to achieve associativity, the normalization process in combination should be postponed to the end of combination. For instance, if we need to combine N weighted interval-valued structures in the frame of discernment Θ, the final result is supposed got by firstly combining them and then normalizing the result. Eventually, Fig. 1 shows the flow of combining N weighted interval-valued belief structures on $\Theta = \{A_1, \ldots, A_n\}$. □

4 Numerical Examples

This subsection presents two illustrative examples of the proposed method.

Example 1 Two interval-valued belief structures on $\Theta = \{A_1, A_2, A_3\}$ are given as follows:

$m_1(A_1) = [0.2, 0.4]$, $m_1(A_2) = [0.3, 0.5]$, $m_1(A_3) = [0.1, 0.3]$;

$m_2(A_1) = [0.3, 0.4]$, $m_2(A_2) = [0.4, 0.5]$, $m_2(A_3) = [0.2, 0.3]$ with their weights $\omega_1 = 0.6$ and $\omega_2 = 0.4$.

It is no difficult to verify that these two weighted interval-valued belief structures are both normalized weighted interval-valued Bayesian belief structures. We can intuitively judge that the belief function is obtained after combining more belief to $\{A_2\}$, medium belief to $\{A_1\}$ and less to $\{A_3\}$. In order to get $m_1 \oplus^{\omega} m_2$, we therefore carry out the following procedure.

Step 1 Compute the matrix M of source interval-valued belief function basic probability assignment. M is given in Table 1.

Step 2 Compute the weight matrix W. W can be obtained by Definition 11 as shown in Table 2.

Step 3 Compute d_{12} by Eq. (2). The value of d_{12} is shown in Table 3.

Step 4 Compute m_{12}, c_{12}, κ_{12} by Eqs. (28)–(34), and we can get them as shown in Table 4.

Eventually, the result is obtained by Eqs. (26) and (27):

$m_1 \oplus^{\omega} m_2(A_1) = [0.388, 0.422]$, $m_1 \oplus^{\omega} m_2(A_2) = [0.491, 0.523]$, $m_1 \oplus^{\omega} m_2(A_3) = [0.267, 0.304]$.

It is easy to see that the result is normalized. Namely, it is a desired result.

Table 1 The matrix M of source interval-valued belief function basic probability assignment

	$\{A_1\}$		$\{A_2\}$		$\{A_3\}$		$\{A_1, A_2\}$		$\{A_1, A_3\}$		$\{A_2, A_3\}$		$\{A_1, A_2, A_3\}$	
	−	+	−	+	−	+	−	+	−	+	−	+	−	+
m_1	0.2	0.4	0.3	0.5	0.1	0.3	0	0	0	0	0	0	0	0
m_2	0.3	0.4	0.4	0.5	0.2	0.3	0	0	0	0	0	0	0	0

Table 2 The weight matrix W

	$\{A_1\}$		$\{A_2\}$		$\{A_3\}$		$\{A_1, A_2\}$		$\{A_1, A_3\}$		$\{A_2, A_3\}$		$\{A_1, A_2, A_3\}$	
	−	+	−	+	−	+	−	+	−	+	−	+	−	+
ω_1	0.6	0.6	0.6	0.6	0.6	0.6	0	0	0	0	0	0	0	0
ω_2	0.4	0.4	0.4	0.4	0.4	0.4	0	0	0	0	0	0	0	0
w_{12}	1	1	1	1	1	1	0	0	0	0	0	0	0	0

Table 3 The value of d_{12}

	$\{A_1\}$		$\{A_2\}$		$\{A_3\}$		$\{A_1, A_2\}$		$\{A_1, A_3\}$		$\{A_2, A_3\}$		$\{A_1, A_2, A_3\}$	
	−	+	−	+	−	+	−	+	−	+	−	+	−	+
max	0.2	0.4	0.3	0.5	0.1	0.3	0	0	0	0	0	0	0	0
min	0.3	0.4	0.4	0.5	0.2	0.3	0	0	0	0	0	0	0	0
avg	0.25	0.4	0.35	0.5	0.15	0.3	0	0	0	0	0	0	0	0
d_{12}	0.9975	1	0.9975	1	0.9975	1	1	1	1	1	1	1	1	1

Table 4 The results of m_{12}, c_{12} and κ_{12}

	$\{A_1\}$		$\{A_2\}$		$\{A_3\}$	
	−	+	−	+	−	+
m_{12}	0.3385	0.3538	0.4296	0.4444	0.2314	0.2479
c_{12}	0.5	0.8	0.7	1	0.3	0.6
κ_{12}	0.35		0.325		0.395	

As we can see, this result reveals that $\{A_2\}$ makes the greatest distribution to belief while $\{A_3\}$ makes the least distribution to belief. Obviously, the result is concordant with our previous intuitive feeling. Therefore, the combined result is satisfactory.

Example 2 Three interval-valued belief structures on $\Theta = \{A_1, A_2, A_3\}$ are given as follows:

$m_1(A_1) = [0.1, 0.4]$, $m_1(A_2) = [0.3, 0.6]$, $m_1(A_3) = [0, 0.3]$;
$m_2(A_1) = [0.2, 0.5]$, $m_2(A_2) = [0.4, 0.6]$, $m_2(A_3) = [0.1, 0.4]$;
$m_3(A_1) = [0.2, 0.4]$, $m_3(A_2) = [0.3, 0.7]$, $m_3(A_3) = [0.1, 0.3]$ with their weights $\omega_1 = 0.4$, $\omega_2 = 0.2$ and $\omega_3 = 0.4$.

We firstly combine m_1 and m_2. It is necessary to normalize the weights of m_1 and m_2. Let $\omega_1' = \frac{\omega_1}{\omega_1+\omega_2}$ and $\omega_2' = \frac{\omega_2}{\omega_1+\omega_2}$. The weights of m_1 and m_2 are respectively replaced by ω_1' and ω_2'. We then have the combination result of $m_1 \oplus^{\omega} m_2$ as

$m_1 \oplus^{\omega} m_2(A_1) = [0.279, 0.405]$, $m_1 \oplus^{\omega} m_2(A_2) = [0.526, 0.608]$, $m_1 \oplus^{\omega} m_2(A_3) = [0.109, 0.249]$.

Next, we combine $m_1 \oplus^{\omega} m_2$ and m_3 with the weights $\omega_{12} = \omega_1 + \omega_2$ and ω_3 similarly. Then, the final combination result of $(m_1 \oplus^{\omega} m_2) \oplus^{\omega} m_3$ is obtained:

$(m_1 \oplus^\omega m_2) \oplus^\omega m_3(A_1) = [0.375, 0.414]$,
$(m_1 \oplus^\omega m_2) \oplus^\omega m_3(A_2) = [0.601, 0.657]$,
$(m_1 \oplus^\omega m_2) \oplus^\omega m_3(A_3) = [0.198, 0.247]$.
Similarly, we can obtain the combination result of $m_1 \oplus^\omega (m_2 \oplus^\omega m_3)$:
$m_1 \oplus^\omega (m_2 \oplus^\omega m_3)(A_1) = [0.375, 0.414]$,
$m_1 \oplus^\omega (m_2 \oplus^\omega m_3)(A_2) = [0.601, 0.657]$,
$m_1 \oplus^\omega (m_2 \oplus^\omega m_3)(A_3) = [0.198, 0.247]$.
Obviously, the associativity $(m_1 \oplus^\omega m_2) \oplus^\omega m_3 = m_1 \oplus^\omega (m_2 \oplus^\omega m_3)$ holds.

5 Conclusions

In this paper, we have provided a novel combination of weighted interval-valued belief structures based on A-IF set. In our method, the evidence combination rule is associative, and then it is convenient to combine a lot of independent interval-valued belief structures. By two numerical examples, the accomplishment of our new method has been illustrated. Indeed, the combination of weighted interval-valued belief structures has a good result in the treatment of the corresponding belief structures.

Acknowledgments We thank all referees for their helpful comments. This work was supported by the Zhejiang Provincial Natural Science Foundation (Grant No: LY12A01009) and the Open Foundation from Marine Sciences in the Most Important Subject of Zhejiang (Grant No: 111040602136).

References

1. Dempster, A.P.: Upper and lower probabilities induced by a multivalued mapping. Ann. Math. Stat. **38**, 325–339 (1967)
2. Shafer, G.: A Mathematical Theory of Evidence. Princeton University Press, Princeton, NJ (1976)
3. Aggarwal, P., Bhatt, D., Devabhaktuni, V., Bhattacharya, P.: Dempster Shafer neural network algorithm for land vehicle navigation application. Inf. Sci. **253**, 26–33 (2013)
4. Benferhat, S., Saffiotti, A., Smets, P.: Belief functions and default reasoning. Artif. Intell. **122**, 1–69 (2000)
5. Denoeux, T.: Reasoning with imprecise belief structures. Int. J. Approx. Reason. **20**, 79–111 (1999)
6. Denoeux, T., Smets, P.: Classification using belief functions: relationship between case-based and model-based approaches. IEEE Trans. Syst. Man Cybern. -Part B: Cybern. **36**(6), 1395–1406 (2006)
7. Denoeux, T.: Maximum likelihood estimation from uncertain data in the belief function framework. IEEE Trans. Knowl. Data Eng. **25**(1), 119–130 (2013)
8. Jones, R.W., Lowe, A., Harrison, M.J.: A framework for intelligent medical diagnosis using the theory of evidence. Knowl. Based Syst. **15**, 77–84 (2002)
9. Laha, A., Pal, N.R., Das, J.: Land cover classification using fuzzy rules and aggregation of contextual information through evidence theory. Trans. Geosci. Remote Sensi. **44**(6), 1633–1641 (2006)
10. Liu, Z.G., Pan, Q., Dezert, J.: Evidential classifier for imprecise data based on belief functions. Knowl. -Based Syst. **52**, 246–257 (2013)

11. Sua, Z.G., Wang, Y.F., Wang, P.H.: Parametric regression analysis of imprecise and uncertain data in the fuzzy belief function framework. Int. J. Approx. Reason. **54**, 1217–1242 (2013)
12. Telmoudi, A., Chakhar, S.: Data fusion application from evidential databases as a support for decision making. Inform. Softw. Technol. **46**(8), 547–555 (2004)
13. Lee, E.S., Zhu, Q.: An interval Dempster-Shafer approach. Comput. Math. Appl. **24**(7), 89–95 (1992)
14. Wang, Y.M., Yang, J.B., Xu, D.L., Chin, K.S.: On the combination and normalization of interval-valued belief structures. Inf. Sci. **177**, 1230–1247 (2007)
15. Yager, R.R.: Dempster-Shafer belief structures with interval valued focal weights. Int. J. Intell. Syst. **16**(4), 497–512 (2001)
16. Su, Z.G., Wang, P.H., Yu, X.J., Lv, Z.Z.: Maximal confidence intervals of the interval-valued belief structure and applications. Inf. Sci. **181**, 1700–1721 (2001)
17. Dymova, L., Sevastjanov, P.: An interpretation of intuitionistic fuzzy sets in terms of evidence theory: decision making aspect. Knowl. -Based Syst. **23**(8), 772–782 (2010)
18. Feng, T., M, J.-S., Zhang, S.-P.: Belief functions on general intuitionistic fuzzy information systems. Inf. Sci. **271**, 143–158 (2014)
19. Song, Y.F., Wang, X.D., Lei, L., Xue, A.J.: Combination of interval-valued belief structures based on intuitionistic fuzzy set. Knowl. -Based Syst. **67**, 61–70 (2014)
20. Yang, J.B., Dong, L.X.: Evidential reasoning rule for evidence combination. Artif. Intell. **205**, 1–29 (2013)
21. Atanassov, K.T.: Intuitionistic fuzzy sets. Fuzzy Sets Syst. **20**, 87–96 (1986)
22. Atanassov, K.T.: Intuit. Fuzzy Sets. Springer Physica, Berlin (1999)

Determinations of Soft Topologies

Lu-Lu Liu, Sheng-Gang Li, Wen-Qing Fu, Sheng-Quan Ma and Xiao-Fei Yang

Abstract Cryptomorphic properties of soft topologies are studied in this paper. For an arbitrary set X, appropriate order relations $\leq$ on $\mathbb{SWCL}(X, I)$ (the set of all soft weak closure operators on X indexed by I), $\mathbb{SWIN}(X, I)$ (the set of all soft weak interior operators on X indexed by I), $\mathbb{SWOU}(X, I)$ (the set of all soft weak exterior operators on X indexed by I), and $\mathbb{SWB}(X, I)$ (the set of all soft weak boundary operators on X indexed by I) are defined respectively, which make $(\mathbb{SWCL}(X, I), \leq)$, $(\mathbb{SWIN}(X, I), \leq)$, $(\mathbb{SWOU}(X, I), \leq)$ and $(\mathbb{SWB}(X, I), \leq)$ complete lattices that are isomorphic to $(\mathbb{ST}(X, I), \subseteq)$, where $\mathbb{ST}(X, I)$ is the set of all soft topologies on X indexed by I.

Keywords Soft topology · Soft weak closure operator · Soft weak interior operator · Soft weak exterior operator · Soft weak boundary operator

L.-L. Liu · S.-G. Li
College of Mathematics and Information Science, Shaanxi Normal University, Xi'an 710062, People's Republic of China
e-mail: topologygroup1@163.com

S.-G. Li
e-mail: shengganglinew@126.com

W.-Q. Fu
School of Science, Xi'an Technological University, Xi'an 710032, People's Republic of China
e-mail: palace_2000@163.com

S.-Q. Ma (✉)
College of Information Science and Technology, Hainan Normal University, Haikou 571158, People's Republic of China
e-mail: mashengquan@163.com

X.-F. Yang
School of Science, Xi'an Polytechnic University, Xi'an 710048, People's Republic of China
e-mail: yangxiaofei2002@163.com

T.-H. Fan et al. (eds.), *Quantitative Logic and Soft Computing 2016*, Advances in Intelligent Systems and Computing 510,
DOI 10.1007/978-3-319-46206-6_51

1 Introduction

Soft set theory, which was firstly proposed by Molodtsov [1] in 1999, has received much attention (see [2–5] and reference in) for its application background (it provides a new mathematical tool for dealing with some uncertainties that traditional mathematical tools cannot handle effectively). Research work on soft set theory and its applications in various fields are progressing rapidly, some of which principally work out soft mathematical concepts and structures that are based on soft set-theoretic operations. For examples, Aktaş and Çağman [6] defined soft group, Jun [7] defined soft BCK/BCI-algebra, Feng et al. [8] defined soft semiring, Zhan and Jun [9] defined soft BL-algebra, Acar et. al. [10] defined soft ring, Atagün and Sezgin [11] defined soft subrings, soft ideals, soft subfields and soft submodules, Sezgin and Atagün [12] defined normalistic soft group and normalistic soft group homomorphism.

We notice that the notions above have something in common: A soft set (F, I) on a group (resp., on a ring, on a semiring, on a BCK-algebra, on a BCI-algebra, on a BL-algebra) X is said to be a soft group (resp., a soft ring, a soft semiring, a soft BCK-algebra, a soft BCI-algebra, a soft BL-algebra) if $F(i)$ is a subgroup (resp., a subring, a subsemiring, a subBCK-algebra, a subBCI-algebra, a subBL-algebra) of X $(\forall i \in I)$. This is partially true for the notion of soft topological space which is defined by Shabir and Naz [13] and Çağman et. al. [14] in a slightly different manner (the former is a special case of the latter). Thus it is necessary to consider the harmony between these two definitions of soft topologies and other soft mathematical concepts which have already been defined.

Many real-world problems can be tentatively described by a soft set (or a soft set with a structure), but effective descriptions to such problem (even efficient methods to solve such problem) at present almost surely rely on our understanding of the structure of these soft sets. This and the striking similarities and connections between soft mathematical concepts and structures and corresponding crisp mathematical concepts and structures urge us to continue to study soft topological spaces (precisely, cryptomorphic properties of soft topologies). We define appropriate order relations $\leq$ on $\mathbb{SWCL}(X, I)$ (the set of all soft weak closure operators on X indexed by I), $\mathbb{SWIN}(X, I)$ (the set of all soft weak interior operators on X indexed by I), $\mathbb{SWOU}(X, I)$ (the set of all soft weak exterior operators on X indexed by I), and $\mathbb{SWB}(X, I)$ (the set of all soft weak boundary operators on X indexed by I), and then we show that $(\mathbb{SWCL}(X, I), \leq)$, $(\mathbb{SWIN}(X, I), \leq)$, $(\mathbb{SWOU}(X, I), \leq)$ and $(\mathbb{SWB}(X, I), \leq)$ are all complete lattices that are isomorphic to $(\mathbb{ST}(X, I), \subseteq)$ (particularly, we give out these isomorphisms in detail). This implies that a soft topological structure on a given set can be displayed or shown in at least five equivalent forms (i.e. we can work out or compute the others from any of them). Thus one can choose the relatively convenient form or approach in related study and applications. We also give examples and algorithms to indicate possible applications of our results. Actually, the idea and technique of the present paper may also be used in simplification of soft mathematical concepts (even of soft set theory and rough set theory) if one aware the covered similarities and connections between soft mathematical concepts (resp., rough mathematical concepts) and some crisp mathematical concepts.

Now we give some definitions and preliminary results which will be used in this paper.

Definition 1 ([1]) A soft set on a set X is a pair (F, I), where I is a nonempty index set (called also a parameter set), and $F : I \longrightarrow 2^X$ (the set of all subset of X) is a mapping; The set of all soft sets on X indexed by I is denoted by $\mathfrak{S}(X, I)$.

Remark 1 We use $\langle F(1), F(2), \ldots, F(n)\rangle$ (or $F(1), F(2), \ldots, F(n)$) to replace (F, I) for convenience and intuitiveness when $I = \{1, 2, \ldots, n\}$ and n is a natural number. For each $A \in 2^X$, $\widetilde{A} \in \mathfrak{S}(X, I)$ is defined by $\widetilde{A}(i) = A$ $(\forall i \in I)$; we identify $\widetilde{\{x\}}$ with $\widetilde{x}$ $(\forall x \in X)$. For each $(F, I) \in \mathfrak{S}(X, I)$, $(F', I) \in \mathfrak{S}(X, I)$ is defined by $F'(i) = X - F(i)$ $(\forall i \in I)$; sometimes we use $(F, I)'$ to replace (F', I) for convenience.

Definition 2 ([15]) Let $\{(F, I)\}_{F \in \mathfrak{F}} \subseteq \mathfrak{S}(X, I)$.
(1) The member $(H, I) = \widetilde{\bigcup}_{F \in \mathfrak{F}}(F, I)$ (or written $(F, I)\widetilde{\cup}\,(H, I)\widetilde{\cup}\cdots\widetilde{\cup}(K, I)$ if $\mathfrak{F} = \{F, H, \ldots, K\}$ is a finite set) of $\mathfrak{S}(X, I)$ is called the union of the family $\{(F, I)\}_{F \in \mathfrak{F}}$, which is defined by $H(i) = \bigcup_{F \in \mathfrak{F}} F(i)$ $(\forall i \in I)$.
(2) The member $(G, I) = \widetilde{\bigcap}_{F \in \mathfrak{F}}(F, I)$ (or written $(F, I)\widetilde{\cap}\,(H, I)\widetilde{\cap}\cdots\widetilde{\cap}(K, I)$ if $\mathfrak{F} = \{F, H, \ldots, K\}$ is a finite set) of $\mathfrak{S}(X, I)$ is called the intersection of the family $\{(F, I)\}_{F \in \mathfrak{F}}$, which is defined by $G(i) = \bigcap_{F \in \mathfrak{F}} F(i)$ $(\forall i \in I)$.

Theorem 1 *For any members (F, I) and (H, I) in $\mathfrak{S}(X, I)$, define $(F, I) \le (H, I)$ if and only if $F(i) \subseteq H(i)$ $(\forall i \in I)$. Then $(\mathfrak{S}(X, I), \le)$ is a powerset lattice, and the least element and the greatest element of $(\mathfrak{S}(X, I), \le)$ are $(\widetilde{\emptyset}, I)$ and $(\widetilde{X}, I)$, respectively.*

Proof Actually, the mapping $\varphi : (\mathfrak{S}(X, I), \le) \longrightarrow (2^{X\times I}, \subseteq)$, defined by $\varphi((F, I)) = \bigcup_{i\in I}(F(i) \times \{i\})$ $(\forall (F, I) \in \mathfrak{S}(X, I))$, is an isomorphism, whose inverse mapping $\psi : (2^{X\times I}, \subseteq) \longrightarrow (\mathfrak{S}(X, I), \le)$ is given by $\psi(A) = (F_A, I)$ $(\forall A \in 2^{X\times I})$, where $F_A : I \longrightarrow 2^X$ $(\forall A \in 2^{X\times I})$ is defined by $F_A(i) = \{x \in X \mid (x, i) \in A\}$ $(\forall i \in I)$. □

Definition 3 Let $\mathfrak{T} \subseteq \mathfrak{S}(X, I)$. $(X, \mathfrak{T}, I)$ is called a
(1) Soft topological space indexed by I if $\mathfrak{T}$ is closed under the operations of arbitrary unions and nonempty finite intersections (it thus contains $(\widetilde{\emptyset}, I)$); $\mathfrak{T}$ is called a soft topology on X indexed by I, members of $\mathfrak{T}$ are called soft open sets, (F', I) is called a soft closed set for each $(F, I) \in \mathfrak{T}$. The set of all soft topologies on X indexed by I is denoted by $\mathbb{ST}(X, I)$.
(2) 2^X-topological space indexed by I if $\mathfrak{T}$ is a soft topology on X indexed by I which satisfies $(\widetilde{X}, I) \in \mathfrak{T}$ (in this case $\mathfrak{T}$ is called a 2^X-topology on X indexed by I). The set of all 2^X-topologies on X indexed by I is denoted by 2^X-$\mathbb{T}(X, I)$.
(3) Deranged soft topological space indexed by I (in this case $\mathfrak{T}$ is called a deranged soft topology on X indexed by I) if, for each $n \in I$, $(X_n, \mathfrak{T}(n))$ is a topological space, where $\mathfrak{T}(n) = \{F(n) \mid (F, I) \in \mathfrak{T}\}$ and $X_n = \bigcup \mathfrak{T}(n)$.

Remark 2 (1) It can be easily seen that $\mathfrak{T} \subseteq \mathfrak{S}(U, I)$ is a soft topology (resp., a 2^X-topology) on X indexed by I if and only if it is a soft topology in the sense of [14] (resp., [13]) on X. Thus every soft topology in the sense of [13] is a soft topology in the sense of [14], and every soft topology is a deranged soft topology (see also Remark 3 below).

(2) For a deranged soft topological space $(X, \mathfrak{T}, I)$ indexed by I, let $\mathcal{J} = \{U \in 2^X \mid U \cap X_n \in \mathfrak{T}(n)\ (\forall n \in I)\}$. Then $\mathcal{J}$ is the biggest topology (called ground topology of $\mathfrak{T}$) on X such that $((X_n, \mathfrak{T}(n)), i_{X_n})$ is a **Top**-subobject of $(X, \mathcal{J})$ (i.e. the inclusion mapping or embedding mapping $i_{X_n} : (X_n, \mathfrak{T}(n)) \longrightarrow (X, \mathcal{J})$ is continuous) for each $n \in I$, where **Top** is the category [16] of topological spaces and continuous mappings (compare also to the second paragraph of this section). This conclusion still holds if $(X, \mathfrak{T}, I)$ is a soft topological space (particularly, a 2^X-topological space) indexed by I. It is now plausible to call (contrast to definitions of soft group [6], soft ring [10], soft semiring [8], soft BCK/BCI-algebra [7], and soft BL-algebra [9] all "spaces" in Definition 3(3) (not just all "spaces" in Definition 3(2)) soft topological spaces. But we do not do this because deranged soft topological spaces (which are actually derangements of soft topological spaces) are not well-behaved (see Remark 3 below). Moreover we do not take (2) in Definition 3 to be the definition of soft topological space because the "spaces" in Definition 3(2) are too special (they are actually very special L-topological spaces [17] which are so special that fuzzy topologists look them to be ordinary topological spaces). However, research works towards applications of these two kind of "spaces" may be meaningful.

Remark 3 (1) $(\mathbb{ST}(X, I), \subseteq)$ and $(2^X\text{-}\mathbb{T}(X, I), \subseteq)$ are complete lattices, and $(\mathbb{DRST}(X, I), \subseteq)$ is a poset (but not necessarily a lattice[1]) with a greatest element. $\mathfrak{S}(X, I) = \{(F, I) \mid F \in (2^X)^I\}$ is the greatest element of $(\mathbb{ST}(X, I), \subseteq)$ (resp., $(2^X\text{-}\mathbb{T}(X, I), \subseteq)$, $(\mathbb{DRST}(X, I), \subseteq)$), $\{(F, I) \mid F \in (\{\emptyset\})^I\} = \{(\widetilde{\emptyset}, I)\}$ is the least element of $(\mathbb{ST}(X, I), \subseteq)$, and $\{(\widetilde{\emptyset}, I), (\widetilde{X}, I)\}$ is the least element of $(2^X\text{-}\mathbb{T}(X, I), \subseteq)$. For a subset $\{\mathfrak{T}_j\}_{j\in J}$ of $\mathbb{A}$ (where $J \neq \emptyset$ and $\mathbb{A} = (\mathbb{ST}(X, I), \subseteq)$ or $(2^X - \mathbb{T}(X, I), \subseteq)$), $\bigcap_{j\in J} \mathfrak{T}_j$ is the infimum of $\{\mathfrak{T}_j\}_{j\in J}$ in $\mathbb{A}$, and $\bigcap\{\mathfrak{T} \subseteq \mathfrak{S}(U, I) \mid \mathfrak{T}$ is a upper bound of $\{\mathfrak{T}_j\}_{j\in J}$ in $\mathbb{A}\}$ is the supremum of $\{\mathfrak{T}_j\}_{j\in J}$ in $\mathbb{A}$.

(2) If $X = \emptyset$, then $2^X\text{-}\mathbb{T}(X, I) = \mathbb{ST}(X, I) = \mathbb{DRST}(X, I) = \{(\widetilde{\emptyset}, I)\}$.

(3) If $X \neq \emptyset$, then $2^X\text{-}\mathbb{T}(X, I) \subset \mathbb{ST}(X, I)$. Obviously, $2^X\text{-}\mathbb{T}(X, I) \subseteq \mathbb{ST}(X, I)$. As $\{(\widetilde{\emptyset}, I)\} \in \mathbb{ST}(X, I)$ but $\{(\widetilde{\emptyset}, I)\} \notin 2^X\text{-}\mathbb{T}(X, I)$, $2^X\text{-}\mathbb{T}(X, I) \subset \mathbb{ST}(X, I)$.

(4) If $X \neq \emptyset$ and $|I| = 1$, then $\mathbb{ST}(X, I) = \mathbb{DRST}(X, I)$.

(5) If $X \neq \emptyset$ and $|I| > 1$, then $\mathbb{ST}(X, I) \subset \mathbb{DRST}(X, I)$.

For each $\mathfrak{T} \in \mathbb{ST}(X, I)$, one can easily verified that $(X_i, \mathfrak{T}(i))$ is a topological space $(\forall i \in I)$, therefore $\mathbb{ST}(X, I) \subseteq \mathbb{DRST}(X, I)$. In addition, $\mathcal{J} = \{U \subseteq X \mid f_i^{-1}(U) \in \mathfrak{T}_i\ (\forall i \in I)\}$ is a topology on X, and $f_i : (X_i, \mathfrak{T}(i)) \longrightarrow (X, \mathcal{J})$ is a continuous mapping $(\forall i \in I)$, where $f_i : X_i \longrightarrow X$ is the embedding mapping $(\forall i \in I)$. Without loss of generality we assume that $I = \{a, b\}$ $(a \neq b)$. Take

[1]Let $\mathfrak{T}_1 = \{(F_1, I), (H_1, I)\}$, $\mathfrak{T}_2 = \{(F_2, I), (H_2, I)\}$, where $I = \{a, b\}$ $(a \neq b)$, $F_1(a) = \emptyset$, $F_1(b) = X$, $H_1(a) = X$, $H_1(b) = \emptyset$, $F_2(a) = \emptyset$, $F_2(b) = \emptyset$, $H_2(a) = X$, $H_2(b) = X$. Then $\mathfrak{T}_1, \mathfrak{T}_2 \in \mathbb{DRST}(X, I)$, but $\mathfrak{T}_1$ and $\mathfrak{T}_2$ have no infimum in $(\mathbb{DRST}(X, I), \subseteq)$ since $\mathfrak{T}_1 \cap \mathfrak{T}_2 = \emptyset$.

$\mathfrak{T} = \{(F, I), (H, I)\}$, where $F(a) = \emptyset$, $F(b) = X$, $H(a) = X$, $H(b) = \emptyset$. Then $\mathfrak{T} \in \mathbb{DRST}(X, I) - \mathbb{ST}(X, I)$, which means $\mathbb{ST}(X, I) \subset \mathbb{DRST}(X, I)$.

For other undefined lattice-theoretical notions and symbols, please refer to [18].

2 Main Results

Definition 4 A mapping $c : \mathfrak{S}(X, I) \longrightarrow \mathfrak{S}(X, I)$ is called a soft weak closure operator on X indexed by I if it satisfies the following conditions:
(SWCL1) $(F, I) \leq c((F, I)) \quad (\forall (F, I) \in \mathfrak{S}(X, I))$.
(SWCL2) $c((F, I)\widetilde{\cup}(G, I)) = c((F, I))\widetilde{\cup}c((G, I)) \quad (\forall (F, I), (G, I) \in \mathfrak{S}(X, I))$.
(SWCL3) $c(c((F, I))) = c((F, I)) \quad (\forall (F, I) \in \mathfrak{S}(X, I))$.

The set of all soft weak closure operators on X indexed by I is denoted by $\mathbb{SWCL}(X, I)$.

Theorem 2 (1) *Define a relation* $\leq$ *on* $\mathbb{SWCL}(X, I)$ *by putting* $c_1 \leq c_2$ *iff* $c_1((F, I)) \geq c_2((F, I)) \quad (\forall (F, I) \in \mathfrak{S}(X, I))$, *then* $(\mathbb{SWCL}(X, I), \leq)$ *is a complete lattice.*
(2) *For each* $\mathfrak{T} \in \mathbb{ST}(X, I)$, *define a mapping* $\varphi_{1,2}(\mathfrak{T}) = c_{\mathfrak{T}} : \mathfrak{S}(X, I) \longrightarrow \mathfrak{S}(X, I)$ *by putting* $c_{\mathfrak{T}}((F, I)) = \bigcap\{(G, I) \in \mathfrak{S}(X, I) \mid (F, I) \leq (G, I), \quad (G, I)' \in \mathfrak{T}\}$ $(\forall (F, I) \in \mathfrak{S}(X, I))$. *Then* $c_{\mathfrak{T}}$ *is a soft weak closure operator on* X *indexed by* I, *called the soft weak closure operator on* X *indexed by* I *corresponding to* $\mathfrak{T}$. *Thus we obtain a mapping* $\varphi_{1,2} : (\mathbb{ST}(X, I), \subseteq) \longrightarrow (\mathbb{SWCL}(X, I), \leq)$, *which is an isomorphism between complete lattices, and whose inverse mapping* $\varphi_{2,1} : (\mathbb{SWCL}(X, I), \leq) \longrightarrow (\mathbb{ST}(X, I), \subseteq)$ *is given by* $\varphi_{2,1}(c) = \mathfrak{T}_c = \{(F, I) \in \mathfrak{S}(X, I) \mid c((F, I)') = (F, I)'\} \quad (\forall c \in \mathbb{SWCL}(X, I))$.
(3) *Given a* $c \in \mathbb{SWCL}(X, I)$, $\varphi_{2,1}(c) = \mathfrak{T}_c \in 2^X\text{-}\mathbb{T}(X, I)$ *if and only if* c *satisfies* $c((\widetilde{\emptyset}, I)) = (\widetilde{\emptyset}, I)$.

Definition 5 A mapping $i : \mathfrak{S}(X, I) \longrightarrow \mathfrak{S}(X, I)$ is called a soft weak interior operator on X indexed by I if it satisfies the following conditions:
(SWIN1) $(F, I) \geq i((F, I)) \quad (\forall (F, I) \in \mathfrak{S}(X, I))$.
(SWIN2) $i((F, I)\widetilde{\cap}(G, I)) = i((F, I))\widetilde{\cap}i((G, I)) \quad (\forall (F, I), (G, I) \in \mathfrak{S}(X, I))$.
(SWIN3) $i(i((F, I))) = i((F, I)) \quad (\forall (F, I) \in \mathfrak{S}(X, I))$.

The set of all soft interior operators on X indexed by I is denoted by $\mathbb{SWIN}(X, I)$.

Theorem 3 (1) *Define a relation* $\leq$ *on* $\mathbb{SWIN}(X, I)$ *by putting* $i_1 \leq i_2$ *iff* $i_1((F, I)) \leq i_2((F, I)) \quad (\forall (F, I) \in \mathfrak{S}(X, I))$, *then* $(\mathbb{SWIN}(X, I), \leq)$ *is a complete lattice.*
(2) *For each* $c \in \mathbb{SWCL}(X, I)$, *define a mapping* $\varphi_{2,3}(c) = i_c : \mathfrak{S}(X, I) \longrightarrow \mathfrak{S}(X, I)$ *by putting* $i_c((F, I)) = (c((F, I)'))' \quad (\forall (F, I) \in \mathfrak{S}(X, I))$. *Then* i_c *is a soft weak interior operator on* X *indexed by* I, *called the soft weak interior operator on* X *indexed by* I *corresponding to* c. *Thus we obtain a mapping* $\varphi_{2,3} : (\mathbb{SWCL}(X, I), \leq) \longrightarrow (\mathbb{SWIN}(X, I), \leq)$, *which is an isomorphism between complete lattices, and whose inverse mapping* $\varphi_{3,2} : (\mathbb{SWIN}(X, I), \leq) \longrightarrow (\mathbb{SWCL}(X, I), \leq)$ *is given by*

$\varphi_{3,2}(i)((F,I)) = c_i((F,I)) = (i((F,I)'))'$ $(\forall i \in \mathbb{SWCL}(X,I))$.
(3) *Given an* $i \in \mathbb{SWIN}(X,I)$, $\varphi_{3,2}(i) = c_i$ *satisfies* $c_i((\widetilde{\emptyset},I)) = (\widetilde{\emptyset},I)$ *if and only if* i *satisfies* $i((\widetilde{X},I)) = (\widetilde{X},I)$.

Remark 4 By Theorems 2 and 3, we obtain an isomorphism $\varphi_{1,3} = \varphi_{2,3} \circ \varphi_{1,2} : (\mathbb{ST}(X,I),\subseteq) \longrightarrow (\mathbb{SWIN}(X,I),\leq)$, which is defined by $\varphi_{1,3}(\mathfrak{T}) = i_{\mathfrak{T}} : \mathfrak{S}(X,I) \longrightarrow \mathfrak{S}(X,I)$ $(\forall \mathfrak{T} \in \mathbb{ST}(X,I))$ and $i_{\mathfrak{T}}((F,I)) = \widetilde{\bigcup}\{(G,I) \in \mathfrak{S}(X,I) \mid (F,I) \geq (G,I), (G,I) \in \mathfrak{T}\}$ $(\forall (F,I) \in \mathfrak{S}(X,I))$. The inverse mapping of $\varphi_{1,3}$ is $\varphi_{3,1} = \varphi_{2,1} \circ \varphi_{3,2}$, which is defined by $\varphi_{3,1}(i) = \mathfrak{T}_i = \{(F,I) \in \mathfrak{S}(X,I) \mid i((F,I)) = (F,I)\}$ $(\forall i \in \mathbb{SWIN}(X,I))$. Furthermore, given an $i \in \mathbb{SWIN}(X,I)$, $\varphi_{3,1}(i) = \mathfrak{T}_i \in 2^X\text{-}\mathbb{T}(X,I)$ if and only if i satisfies $i((\widetilde{X},I)) = (\widetilde{X},I)$.

Definition 6 A mapping $o : \mathfrak{S}(X,I) \longrightarrow \mathfrak{S}(X,I)$ is called a soft weak exterior operator on X indexed by I if it satisfies the following conditions:
(SWOU1) $o((F,I)) \leq (F,I)'$ $(\forall (F,I) \in \mathfrak{S}(X,I))$.
(SWOU2) $o((F,I)\widetilde{\cup}(G,I)) = o((F,I))\widetilde{\cap}o((G,I))$ $(\forall (F,I),(G,I) \in \mathfrak{S}(X,I))$.
(SWOU3) $o((o((F,I)))') = o((F,I))$ $(\forall (F,I) \in \mathfrak{S}(X,I))$.

The set of all soft weak exterior operators on X indexed by I is denoted by $\mathbb{SWOU}(X,I)$.

Theorem 4 (1) *Define a relation* $\leq$ *on* $\mathbb{SWOU}(X,I)$ *by putting* $o_1 \leq o_2$ *iff* $o_1((F,I)) \leq o_2((F,I))$ $(\forall (F,I) \in \mathfrak{S}(X,I))$, *then* $(\mathbb{SWOU}(X,I),\leq)$ *is a complete lattice.*
(2) *For each* $c \in \mathbb{SWCL}(X,I)$, *define a mapping* $\varphi_{2,4}(c) = o_c : \mathfrak{S}(X,I) \longrightarrow \mathfrak{S}(X,I)$ *by putting* $o_c((F,I)) = (c((F,I)))'$ $(\forall (F,I) \in \mathfrak{S}(X,I))$. *Then* o_c *is a soft weak exterior operator on X indexed by I, called the soft weak exterior operator on X indexed by I corresponding to c. Thus we obtain a mapping* $\varphi_{2,4} : (\mathbb{SWCL}(X,I),\leq) \longrightarrow (\mathbb{SWOU}(X,I),\leq)$, *which is an isomorphism between complete lattices, and whose inverse mapping* $\varphi_{4,2} : (\mathbb{SWOU}(X,I),\leq) \longrightarrow (\mathbb{SWCL}(X,I),\leq)$ *is given by* $\varphi_{4,2}(o)((F,I)) = c_o((F,I)) = (o((F,I)))'$ $(\forall o \in \mathbb{SWOU}(X,I))$.
(3) *Given an* $o \in \mathbb{SWOU}(X,I)$, $\varphi_{4,2}(o) = c_o$ *satisfies* $c_o((\widetilde{\emptyset},I)) = (\widetilde{\emptyset},I)$ *if and only if* o *satisfies* $o((\widetilde{\emptyset},I)) = (\widetilde{X},I)$.

Remark 5 (1) By Theorems 2 and 4, we obtain an isomorphism $\varphi_{1,4} = \varphi_{2,4} \circ \varphi_{1,2} : (\mathbb{ST}(X,I),\subseteq) \longrightarrow (\mathbb{SWOU}(X,I),\leq)$, which is defined by $\varphi_{1,4}(\mathfrak{T}) = o_{\mathfrak{T}} : \mathfrak{S}(X,I) \longrightarrow \mathfrak{S}(X,I)$ $(\forall \mathfrak{T} \in \mathbb{ST}(X,I))$ and $o_{\mathfrak{T}}((F,I)) = \widetilde{\bigcup}\{(G,I) \in \mathfrak{S}(X,I) \mid (F,I)' \geq (G,I), (G,I) \in \mathfrak{T}\}$ $(\forall (F,I) \in \mathfrak{S}(X,I))$. The inverse mapping of $\varphi_{1,4}$ is $\varphi_{4,1} = \varphi_{2,1} \circ \varphi_{4,2}$, which is defined by $\varphi_{4,1}(o) = \mathfrak{T}_o = \{(F,I) \in \mathfrak{S}(X,I) \mid o((F,I)') = (F,I)\}$ $(\forall o \in \mathbb{SWOU}(X,I))$. Furthermore, given an $o \in \mathbb{SWOU}(X,I)$, $\varphi_{4,1}(o) = \mathfrak{T}_o \in 2^X\text{-}\mathbb{T}(X,I)$ if and only if o satisfies $o((\widetilde{\emptyset},I)) = (\widetilde{X},I)$.

(2) By Theorems 3 and 4, we obtain an isomorphism $\varphi_{3,4} = \varphi_{2,4} \circ \varphi_{3,2} : (\mathbb{SWIN}(X,I),\leq) \longrightarrow (\mathbb{SWOU}(X,I),\leq)$, which is defined by $\varphi_{3,4}(i) = o_i : \mathfrak{S}(X,I) \longrightarrow \mathfrak{S}(X,I)$ $(\forall i \in \mathbb{SWIN}(X,I))$ and $o_i((F,I)) = i((F,I)')$ $(\forall (F,I) \in \mathfrak{S}(X,I))$. The inverse mapping of $\varphi_{3,4}$ is $\varphi_{4,3} = \varphi_{2,3} \circ \varphi_{4,2}$, which is defined by $\varphi_{4,3}(o) = i_o : \mathfrak{S}(X,I) \longrightarrow \mathfrak{S}(X,I)$ $(\forall i \in \mathbb{SWOU}(X,I))$ and $i_o((F,I)) = o((F,I)')$ $(\forall (F,I) \in \mathfrak{S}(X,I))$.

Definition 7 A mapping $b : \mathfrak{S}(X, I) \longrightarrow \mathfrak{S}(X, I)$ is called a soft weak boundary operator on X indexed by I if it satisfies the following conditions:
(SWB1) $b((F, I)) = b((F, I)')$ $(\forall (F, I) \in \mathfrak{S}(X, I))$.
(SWB2) $b((F, I)\widetilde{\cup}(G, I)) \leq b((F, I))\widetilde{\cup}b((G, I))$ $(\forall (F, I), (G, I) \in \mathfrak{S}(X, I))$.
(SWB3) If $(F, I), (G, I) \in \mathfrak{S}(X, I)$ and $(F, I) \leq (G, I)$, then $(F, I)\widetilde{\cup}b((F, I)) \leq (G, I)\widetilde{\cup}b((G, I))$.
(SWB4) $(F, I)\widetilde{\cup}b((F, I))\widetilde{\cup}b((F, I)\widetilde{\cup}b((F, I))) = (F, I)\widetilde{\cup}b((F, I))$ $(\forall (F, I) \in \mathfrak{S}(X, I))$.

The set of all soft weak boundary operators on X indexed by I is denoted by $\mathbb{SWB}(X, I)$.

Theorem 5 (1) *Define a relation $\leq$ on $\mathbb{SWB}(X, I)$ by putting $b_1 \leq b_2$ iff $b_1((F, I)) \geq b_2((F, I))$ $(\forall (F, I) \in \mathfrak{S}(X, I))$, then $(\mathbb{SWB}(X, I), \leq)$ is a complete lattice.*
(2) *For each $c \in \mathbb{SWCL}(X, I)$, define a mapping $\varphi_{2,5}(c) = b_c : \mathfrak{S}(X, I) \longrightarrow \mathfrak{S}(X, I)$ by putting $b_c((F, I)) = c((F, I))\widetilde{\cap}c((F, I)')$ $(\forall (F, I) \in \mathfrak{S}(X, I))$. Then b_c is a soft weak boundary operator on X indexed by I, called the soft weak boundary operator on X indexed by I corresponding to c. Thus we obtain a mapping $\varphi_{2,5} : (\mathbb{SWCL}(X, I), \leq) \longrightarrow (\mathbb{SWB}(X, I), \leq)$, which is an isomorphism between complete lattices, and whose inverse mapping $\varphi_{5,2} : (\mathbb{SWB}(X, I), \leq) \longrightarrow (\mathbb{SWCL}(X, I), \leq)$ is given by $\varphi_{5,2}(b)((F, I)) = c_b((F, I)) = (F, I)\widetilde{\cup}b((F, I))$ $(\forall b \in \mathbb{SWB}(X, I))$.*
(3) *Given a $b \in \mathbb{SWB}(X, I)$, $\varphi_{5,2}(b) = c_b$ satisfies $c_b((\widetilde{\emptyset}, I)) = (\widetilde{\emptyset}, I)$ if and only if b satisfies $b((\widetilde{\emptyset}, I)) = (\widetilde{\emptyset}, I)$.*

Remark 6 (1) By Theorems 2 and 5, we obtain an isomorphism $\varphi_{1,5} = \varphi_{2,5} \circ \varphi_{1,2} : (\mathbb{ST}(X, I), \subseteq) \longrightarrow (\mathbb{SWB}(X, I), \leq)$, which is defined by $\varphi_{1,5}(\mathfrak{T}) = b_{\mathfrak{T}} : \mathfrak{S}(X, I) \longrightarrow \mathfrak{S}(X, I)$ $(\forall \mathfrak{T} \in \mathbb{ST}(X, I))$ and

$$b_{\mathfrak{T}}((F, I)) = \left(\widetilde{\bigcap}\{(G, I) \in \mathfrak{S}(X, I) \mid (F, I) \leq (G, I), (G, I)' \in \mathfrak{T}\}\right)$$

$$\widetilde{\cap}\left(\widetilde{\bigcap}\{(G, I) \in \mathfrak{S}(X, I) \mid (F, I)' \leq (G, I), (G, I)' \in \mathfrak{T}\}\right)$$

$$(\forall (F, I) \in \mathfrak{S}(X, I)).$$

The inverse mapping of $\varphi_{1,5}$ is $\varphi_{5,1} = \varphi_{2,1} \circ \varphi_{5,2}$, which is defined by $\varphi_{5,1}(b) = \mathfrak{T}_b = \{(F, I) \in \mathfrak{S}(X, I) \mid (F, I)'\widetilde{\cup}b((F, I)') = (F, I)'\}$ $(\forall b \in \mathbb{SWB}(X, I))$. Furthermore, given a $b \in \mathbb{SWB}(X, I)$, $\varphi_{5,1}(b) = \mathfrak{T}_b \in 2^X\text{-}\mathbb{T}(X, I)$ if and only if b satisfies $b((\widetilde{\emptyset}, I)) = (\widetilde{\emptyset}, I)$.

(2) By Theorems 3 and 5, we obtain an isomorphism $\varphi_{3,5} = \varphi_{2,5} \circ \varphi_{3,2} : (\mathbb{SWIN}(X, I), \leq) \longrightarrow (\mathbb{SWB}(X, I), \leq)$, which is defined by $\varphi_{3,5}(i) = b_i : \mathfrak{S}(X, I) \longrightarrow \mathfrak{S}(X, I)$ $(\forall i \in \mathbb{SWIN}(X, I))$ and $b_i((F, I)) = (i((F, I))\widetilde{\cup}i((F, I)'))'$ $(\forall (F, I) \in \mathfrak{S}(X, I))$. The inverse mapping of $\varphi_{3,5}$ is $\varphi_{5,3} = \varphi_{2,3} \circ \varphi_{5,2}$, which is defined by $\varphi_{5,3}(b) = i_b : \mathfrak{S}(X, I) \longrightarrow \mathfrak{S}(X, I)$ $(\forall b \in \mathbb{SWB}(X, I))$ and $i_b((F, I)) = (F, I) -$

$b((F, I))$ $(\forall (F, I) \in \mathfrak{S}(X, I))$. Furthermore, given an $i \in \mathbb{SWIN}(X, I)$, $\varphi_{3,5}(i) = b_i$ satisfies $b_i((\widetilde{\emptyset}, I)) = (\widetilde{\emptyset}, I)$ if and only if i satisfies $i((\widetilde{X}, I)) = (\widetilde{X}, I)$.

(3) By Theorems 4 and 5, we obtain an isomorphism $\varphi_{4,5} = \varphi_{2,5} \circ \varphi_{4,2}$: $(\mathbb{SWOU}(X, I), \leq) \longrightarrow (\mathbb{SWB}(X, I), \leq)$, which is defined by $\varphi_{4,5}(o) = b_o : \mathfrak{S}(X, I) \longrightarrow \mathfrak{S}(X, I)$ $(\forall o \in \mathbb{SWOU}(X, I))$ and $b_o((F, I)) = (o((F, I))\widetilde{\cup} o((F, I)'))'$ $(\forall (F, I) \in \mathfrak{S}(X, I))$. The inverse mapping of $\varphi_{4,5}$ is $\varphi_{5,4} = \varphi_{2,4} \circ \varphi_{5,2}$, which is defined by $\varphi_{5,4}(b) = o_b : \mathfrak{S}(X, I) \longrightarrow \mathfrak{S}(X, I)$ $(\forall b \in \mathbb{SWB}(X, I))$ and $o_b((F, I)) = (F, I)' - b((F, I))$ $(\forall (F, I) \in \mathfrak{S}(X, I))$. Furthermore, given an $o \in \mathbb{SWOU}(X, I)$, $\varphi_{4,5}(o) = b_o$ satisfies $b_o((\widetilde{\emptyset}, I)) = (\widetilde{\emptyset}, I)$ if and only if o satisfies $o((\widetilde{\emptyset}, I)) = (\widetilde{X}, I)$.

From above we can see that: For a given set X, if we know one of a soft topology $\mathfrak{T}$ on X indexed by I, a soft weak closure operator on X indexed by I, a soft weak interior operator on X indexed by I, a soft weak exterior operator on X indexed by I, a soft weak boundary operator on X indexed by I, then we can work out or compute the others. This can be realized in computers when X and I are both finite sets. We will illustrate this by examples in the following (we may also give the corresponding algorithms).

Example 1 Let $X = \{x, y\}$ be a two-element set, and $I = \{1, 2\}$. Then $\mathfrak{T} = \{\langle\emptyset, \emptyset\rangle, \langle\emptyset, X\rangle, \langle X, \emptyset\rangle, \langle X, X\rangle\}$ is a soft topology on X indexed by I and $\mathfrak{S}(X, I)$ consists exactly of sixteen members: $\langle\emptyset, \emptyset\rangle$, $\langle\emptyset, \{x\}\rangle$, $\langle\emptyset, \{y\}\rangle$, $\langle\emptyset, X\rangle$, $\langle\{x\}, \emptyset\rangle$, $\langle\{x\}, \{x\}\rangle$, $\langle\{x\}, \{y\}\rangle$, $\langle\{x\}, X\rangle$, $\langle\{y\}, \emptyset\rangle$, $\langle\{y\}, \{x\}\rangle$, $\langle\{y\}, \{y\}\rangle$, $\langle\{y\}, X\rangle$, $\langle X, \emptyset\rangle$, $\langle X, \{x\}\rangle$, $\langle X, \{y\}\rangle$, $\langle X, X\rangle$. Next we compute the rest four: $c_{\mathfrak{T}}$, $i_{\mathfrak{T}}$, $o_{\mathfrak{T}}$, and $b_{\mathfrak{T}}$.

By Theorem 2 we know $c_{\mathfrak{T}} = \varphi_{1,2}(\mathfrak{T}) : \mathfrak{S}(X, I) \longrightarrow \mathfrak{S}(X, I)$ is defined by $c_{\mathfrak{T}}(\langle\emptyset, \emptyset\rangle) = \langle\emptyset, \emptyset\rangle$, $c_{\mathfrak{T}}(\langle\emptyset, \{x\}\rangle) = \langle\emptyset, X\rangle$, $c_{\mathfrak{T}}(\langle\emptyset, \{y\}\rangle) = \langle\emptyset, X\rangle$, $c_{\mathfrak{T}}(\langle\emptyset, X\rangle) = \langle\emptyset, X\rangle$, $c_{\mathfrak{T}}(\langle\{x\}, \emptyset\rangle) = \langle X, \emptyset\rangle$, $c_{\mathfrak{T}}(\langle\{x\}, \{x\}\rangle) = \langle X, X\rangle$, $c_{\mathfrak{T}}(\langle\{x\}, \{y\}\rangle) = \langle X, X\rangle$, $c_{\mathfrak{T}}(\langle\{x\}, X\rangle) = \langle X, X\rangle$, $c_{\mathfrak{T}}(\langle\{y\}, \emptyset\rangle) = \langle X, \emptyset\rangle$, $c_{\mathfrak{T}}(\langle\{y\}, \{x\}\rangle) = \langle X, X\rangle$, $c_{\mathfrak{T}}(\langle\{y\}, \{y\}\rangle) = \langle X, X\rangle$, $c_{\mathfrak{T}}(\langle\{y\}, X\rangle) = \langle X, X\rangle$, $c_{\mathfrak{T}}(\langle X, \emptyset\rangle) = \langle X, \emptyset\rangle$, $c_{\mathfrak{T}}(\langle X, \{x\}\rangle) = \langle X, X\rangle$, $c_{\mathfrak{T}}(\langle X, \{y\}\rangle) = \langle X, X\rangle$, $c_{\mathfrak{T}}(\langle X, X\rangle) = \langle X, X\rangle$.

By Remark 4 we know $i_{\mathfrak{T}} = \varphi_{1,3}(\mathfrak{T}) : \mathfrak{S}(X, I) \longrightarrow \mathfrak{S}(X, I)$ is defined by $i_{\mathfrak{T}}(\langle\emptyset, \emptyset\rangle) = \langle\emptyset, \emptyset\rangle$, $i_{\mathfrak{T}}(\langle\emptyset, \{x\}\rangle) = \langle\emptyset, \emptyset\rangle$, $i_{\mathfrak{T}}(\langle\emptyset, \{y\}\rangle) = \langle\emptyset, \emptyset\rangle$, $i_{\mathfrak{T}}(\langle\emptyset, X\rangle) = \langle\emptyset, X\rangle$, $i_{\mathfrak{T}}(\langle\{x\}, \emptyset\rangle) = \langle\emptyset, \emptyset\rangle$, $i_{\mathfrak{T}}(\langle\{x\}, \{x\}\rangle) = \langle\emptyset, \emptyset\rangle$, $i_{\mathfrak{T}}(\langle\{x\}, \{y\}\rangle) = \langle\emptyset, \emptyset\rangle$, $i_{\mathfrak{T}}(\langle\{x\}, X\rangle) = \langle\emptyset, X\rangle$, $i_{\mathfrak{T}}(\langle\{y\}, \emptyset\rangle) = \langle\emptyset, \emptyset\rangle$, $i_{\mathfrak{T}}(\langle\{y\}, \{x\}\rangle) = \langle\emptyset, \emptyset\rangle$, $i_{\mathfrak{T}}(\langle\{y\}, \{y\}\rangle) = \langle\emptyset, \emptyset\rangle$, $i_{\mathfrak{T}}(\langle\{y\}, X\rangle) = \langle\emptyset, X\rangle$, $i_{\mathfrak{T}}(\langle X, \emptyset\rangle) = \langle X, \emptyset\rangle$, $i_{\mathfrak{T}}(\langle X, \{x\}\rangle) = \langle X, \emptyset\rangle$, $i_{\mathfrak{T}}(\langle X, \{y\}\rangle) = \langle X, \emptyset\rangle$, $i_{\mathfrak{T}}(\langle X, X\rangle) = \langle X, X\rangle$.

By Remark 5 we know $o_{\mathfrak{T}} = \varphi_{1,4}(\mathfrak{T}) : \mathfrak{S}(X, I) \longrightarrow \mathfrak{S}(X, I)$ is defined by $o_{\mathfrak{T}}(\langle\emptyset, \emptyset\rangle) = \langle X, X\rangle$, $o_{\mathfrak{T}}(\langle\emptyset, \{x\}\rangle) = \langle X, \emptyset\rangle$, $o_{\mathfrak{T}}(\langle\emptyset, \{y\}\rangle) = \langle X, \emptyset\rangle$, $o_{\mathfrak{T}}(\langle\emptyset, X\rangle) = \langle X, \emptyset\rangle$, $o_{\mathfrak{T}}(\langle\{x\}, \emptyset\rangle) = \langle\emptyset, X\rangle$, $o_{\mathfrak{T}}(\langle\{x\}, \{x\}\rangle) = \langle\emptyset, \emptyset\rangle$, $o_{\mathfrak{T}}(\langle\{x\}, \{y\}\rangle) = \langle\emptyset, \emptyset\rangle$, $o_{\mathfrak{T}}(\langle\{x\}, X\rangle) = \langle\emptyset, \emptyset\rangle$, $o_{\mathfrak{T}}(\langle\{y\}, \emptyset\rangle) = \langle\emptyset, X\rangle$, $o_{\mathfrak{T}}(\langle\{y\}, \{x\}\rangle) = \langle\emptyset, \emptyset\rangle$, $o_{\mathfrak{T}}(\langle\{y\}, \{y\}\rangle) = \langle\emptyset, \emptyset\rangle$, $o_{\mathfrak{T}}(\langle\{y\}, X\rangle) = \langle\emptyset, \emptyset\rangle$, $o_{\mathfrak{T}}(\langle X, \emptyset\rangle) = \langle\emptyset, X\rangle$, $o_{\mathfrak{T}}(\langle X, \{x\}\rangle) = \langle\emptyset, \emptyset\rangle$, $o_{\mathfrak{T}}(\langle X, \{y\}\rangle) = \langle\emptyset, \emptyset\rangle$, $o_{\mathfrak{T}}(\langle X, X\rangle) = \langle\emptyset, \emptyset\rangle$.

By Remark 6 we know $b_{\mathfrak{T}} = \varphi_{1,5}(\mathfrak{T}) : \mathfrak{S}(X, I) \longrightarrow \mathfrak{S}(X, I)$ is defined by $b_{\mathfrak{T}}(\langle\emptyset, \emptyset\rangle) = \langle\emptyset, \emptyset\rangle$, $b_{\mathfrak{T}}(\langle\emptyset, \{x\}\rangle) = \langle\emptyset, X\rangle$, $b_{\mathfrak{T}}(\langle\emptyset, \{y\}\rangle) = \langle\emptyset, X\rangle$, $b_{\mathfrak{T}}(\langle\emptyset, X\rangle) = \langle\emptyset, \emptyset\rangle$, $b_{\mathfrak{T}}(\langle\{x\}, \emptyset\rangle) = \langle X, \emptyset\rangle$, $b_{\mathfrak{T}}(\langle\{x\}, \{x\}\rangle) = \langle X, X\rangle$, $b_{\mathfrak{T}}(\langle\{x\}, \{y\}\rangle) = \langle X, X\rangle$,

$b_{\mathfrak{T}}(\langle\{x\}, X\rangle) = \langle X, \emptyset\rangle$, $b_{\mathfrak{T}}(\langle\{y\}, \emptyset\rangle) = \langle X, \emptyset\rangle$, $b_{\mathfrak{T}}(\langle\{y\}, \{x\}\rangle) = \langle X, X\rangle$, $b_{\mathfrak{T}}(\langle\{y\}, \{y\}\rangle) = \langle X, X\rangle$, $b_{\mathfrak{T}}(\langle\{y\}, X\rangle) = \langle X, \emptyset\rangle$, $b_{\mathfrak{T}}(\langle X, \emptyset\rangle) = \langle\emptyset, \emptyset\rangle$, $b_{\mathfrak{T}}(\langle X, \{x\}\rangle) = \langle\emptyset, X\rangle$, $b_{\mathfrak{T}}(\langle X, \{y\}\rangle) = \langle\emptyset, X\rangle$, $b_{\mathfrak{T}}(\langle X, X\rangle) = \langle\emptyset, \emptyset\rangle$.

Example 2 Let $X = \{x, y\}$ be a two-element set, and $I = \{1, 2\}$. Then $\mathfrak{S}(X, I)$ has sixteen members (as in Example 1) and $c : \mathfrak{S}(X, I) \longrightarrow \mathfrak{S}(X, I)$, defined by $c(\langle\emptyset, \emptyset\rangle) = \langle\emptyset, \emptyset\rangle$ and $c(\langle F_1, F_2\rangle) = \langle X, X\rangle$ $(\forall\langle F_1, F_2\rangle \in (\mathfrak{S}(X, I) - \langle\emptyset, \emptyset\rangle))$, is a soft weak closure operator on X indexed by I. Next we compute the rest four: $\mathfrak{T}_c$, i_c, o_c, and b_c. By Theorem 2 we know $\mathfrak{T}_c = \varphi_{2,1}(c) = \{\langle\emptyset, \emptyset\rangle, \langle X, X\rangle\}$. By Theorem 3 we know $i_c = \varphi_{2,3}(c) : \mathfrak{S}(X, I) \longrightarrow \mathfrak{S}(X, I)$ is defined by $i_c(\langle X, X\rangle) = \langle X, X\rangle$ and $i_c(\langle F_1, F_2\rangle) = \langle\emptyset, \emptyset\rangle$ $(\forall\langle F_1, F_2\rangle \in (\mathfrak{S}(X, I) - \langle X, X\rangle))$. By Theorem 4 we know $o_c = \varphi_{2,4}(c) : \mathfrak{S}(X, I) \longrightarrow \mathfrak{S}(X, I)$ is defined by $o_c(\langle\emptyset, \emptyset\rangle) = \langle X, X\rangle$ and $o_c(\langle F_1, F_2\rangle) = \langle\emptyset, \emptyset\rangle$ $(\forall\langle F_1, F_2\rangle \in (\mathfrak{S}(X, I) - \langle\emptyset, \emptyset\rangle))$. By Theorem 5 we know $b_c = \varphi_{2,5}(c) : \mathfrak{S}(X, I) \longrightarrow \mathfrak{S}(X, I)$ is defined by $b_c(\langle\emptyset, \emptyset\rangle) = b_c(\langle X, X\rangle) = \langle\emptyset, \emptyset\rangle$ and $b_c(\langle F_1, F_2\rangle) = \langle X, X\rangle$ $(\forall\langle F_1, F_2\rangle \in (\mathfrak{S}(X, I) - \langle\emptyset, \emptyset\rangle - \langle X, X\rangle))$.

Example 3 Let $X = \{x, y\}$ be a two-element set, and $I = \{1, 2\}$. Then $\mathfrak{S}(X, I)$ has sixteen members (as in Example 1) and $i : \mathfrak{S}(X, I) \longrightarrow \mathfrak{S}(X, I)$, defined by $i(\langle F_1, F_2\rangle) = \langle\emptyset, \emptyset\rangle$ $(\forall\langle F_1, F_2\rangle \in \mathfrak{S}(X, I))$, is a soft weak interior operator on X indexed by I. Next we compute the rest four: $\mathfrak{T}_i$, c_i, o_i, and b_i. By Remark 4 we know $\mathfrak{T}_i = \varphi_{3,1}(i) = \{\langle\emptyset, \emptyset\rangle\}$. By Theorem 3 we know $c_i = \varphi_{3,2}(i) : \mathfrak{S}(X, I) \longrightarrow \mathfrak{S}(X, I)$ is defined by $c_i(\langle F_1, F_2\rangle) = \langle X, X\rangle$ $(\forall\langle F_1, F_2\rangle \in \mathfrak{S}(X, I))$. By Remark 5 we know $o_i = \varphi_{3,4}(i) : \mathfrak{S}(X, I) \longrightarrow \mathfrak{S}(X, I)$ is defined by $o_i(\langle F_1, F_2\rangle) = \langle\emptyset, \emptyset\rangle$ $(\forall\langle F_1, F_2\rangle \in \mathfrak{S}(X, I))$. By Remark 6 we know $b_i = \varphi_{3,5}(i) : \mathfrak{S}(X, I) \longrightarrow \mathfrak{S}(X, I)$ is defined by $b_i(\langle F_1, F_2\rangle) = \langle X, X\rangle$ $(\forall\langle F_1, F_2\rangle \in \mathfrak{S}(X, I))$.

Example 4 Let $X = \{x, y\}$ be a two-element set, and $I = \{1, 2\}$. Then $\mathfrak{S}(X, I)$ has sixteen members (as in Example 1) and $o : \mathfrak{S}(X, I) \longrightarrow \mathfrak{S}(X, I)$, defined by $o(\langle\emptyset, \emptyset\rangle) = \langle\emptyset, \{x\}\rangle$, $o(\langle\emptyset, \{x\}\rangle) = \langle\emptyset, \emptyset\rangle$, $o(\langle\emptyset, \{y\}\rangle) = \langle\emptyset, \{x\}\rangle$, $o(\langle\emptyset, X\rangle) = \langle\emptyset, \emptyset\rangle$, $o(\langle\{x\}, \emptyset\rangle) = \langle\emptyset, \{x\}\rangle$, $o(\langle\{x\}, \{x\}\rangle) = \langle\emptyset, \emptyset\rangle$, $o(\langle\{x\}, \{y\}\rangle) = \langle\emptyset, \{x\}\rangle$, $o(\langle\{x\}, X\rangle) = \langle\emptyset, \emptyset\rangle$, $o(\langle\{y\}, \emptyset\rangle) = \langle\emptyset, \{x\}\rangle$, $o(\langle\{y\}, \{x\}\rangle) = \langle\emptyset, \emptyset\rangle$, $o(\langle\{y\}, \{y\}\rangle) = \langle\emptyset, \{x\}\rangle$, $o(\langle\{y\}, X\rangle) = \langle\emptyset, \emptyset\rangle$, $o(\langle X, \emptyset\rangle) = \langle\emptyset, \{x\}\rangle$, $o(\langle X, \{x\}\rangle) = \langle\emptyset, \emptyset\rangle$, $o(\langle X, \{y\}\rangle) = \langle\emptyset, \{x\}\rangle$, $o(\langle X, X\rangle) = \langle\emptyset, \emptyset\rangle$, is a soft weak exterior operator on X indexed by I. Next we compute the rest four: $\mathfrak{T}_o$, c_o, i_o, and b_o.

By Remark 5 we know $\mathfrak{T}_o = \varphi_{4,1}(o) = \{\langle\emptyset, \emptyset\rangle, \langle\emptyset, \{x\}\rangle\}$. By Theorem 4 we know $c_o = \varphi_{4,2}(o) : \mathfrak{S}(X, I) \longrightarrow \mathfrak{S}(X, I)$ is defined by $c_o(\langle\emptyset, \emptyset\rangle) = \langle X, \{y\}\rangle$, $c_o(\langle\emptyset, \{x\}\rangle) = \langle X, X\rangle$, $c_o(\langle\emptyset, \{y\}\rangle) = \langle X, \{y\}\rangle$, $c_o(\langle\emptyset, X\rangle) = \langle X, X\rangle$, $c_o(\langle\{x\}, \emptyset\rangle) = \langle X, \{y\}\rangle$, $c_o(\langle\{x\}, \{x\}\rangle) = \langle X, X\rangle$, $c_o(\langle\{x\}, \{y\}\rangle) = \langle X, \{y\}\rangle$, $c_o(\langle\{x\}, X\rangle) = \langle X, X\rangle$, $c_o(\langle\{y\}, \emptyset\rangle) = \langle X, \{y\}\rangle$, $c_o(\langle\{y\}, \{x\}\rangle) = \langle X, X\rangle$, $c_o(\langle\{y\}, \{y\}\rangle) = \langle X, \{y\}\rangle$, $c_o(\langle\{y\}, X\rangle) = \langle X, X\rangle$, $c_o(\langle X, \emptyset\rangle) = \langle X, \{y\}\rangle$, $c_o(\langle X, \{x\}\rangle) = \langle X, X\rangle$, $c_o(\langle X, \{y\}\rangle) = \langle X, \{y\}\rangle$, $c_o(\langle X, X\rangle) = \langle X, X\rangle$.

By Remark 5 we know $i_o(\langle\emptyset, \emptyset\rangle) = \langle\emptyset, \emptyset\rangle$, $i_o(\langle\emptyset, \{x\}\rangle) = \langle\emptyset, \{x\}\rangle$, $i_o(\langle\emptyset, \{y\}\rangle) = \langle\emptyset, \emptyset\rangle$, $i_o(\langle\emptyset, X\rangle) = \langle\emptyset, \{x\}\rangle$, $i_o(\langle\{x\}, \emptyset\rangle) = \langle\emptyset, \emptyset\rangle$, $i_o(\langle\{x\}, \{x\}\rangle) = \langle\emptyset, \{x\}\rangle$, $i_o(\langle\{x\}, \{y\}\rangle) = \langle\emptyset, \emptyset\rangle$, $i_o(\langle\{x\}, X\rangle) = \langle\emptyset, \{x\}\rangle$, $i_o(\langle\{y\}, \emptyset\rangle) = \langle\emptyset, \emptyset\rangle$, $i_o(\langle\{y\}, \{x\}\rangle) = \langle\emptyset, \{x\}\rangle$, $i_o(\langle\{y\}, \{y\}\rangle) = \langle\emptyset, \emptyset\rangle$, $i_o(\langle\{y\}, X\rangle) = \langle\emptyset, \{x\}\rangle$, $i_o(\langle X, \emptyset\rangle) = \langle\emptyset, \emptyset\rangle$, $i_o(\langle X, \{x\}\rangle) = \langle\emptyset, \{x\}\rangle$, $i_o(\langle X, \{y\}\rangle) = \langle\emptyset, \emptyset\rangle$, $i_o(\langle X, X\rangle) = \langle\emptyset, \{x\}\rangle$.

By Remark 6 we know $b_o = \varphi_{4,5}(o) : \mathfrak{S}(X, I) \longrightarrow \mathfrak{S}(X, I)$ is defined by $b_o(\langle F_1, F_2\rangle) = \langle X, \{y\}\rangle \ \ (\forall \langle F_1, F_2\rangle \in \mathfrak{S}(X, I))$

Example 5 Let $X = \{x, y\}$ be a two-element set, and $I = \{1, 2\}$. Then $\mathfrak{S}(X, I)$ has sixteen members (as in Example 1) and $b : \mathfrak{S}(X, I) \longrightarrow \mathfrak{S}(X, I)$, defined by $b(\langle F_1, F_2\rangle) = \langle \{y\}, \{x\}\rangle \ \ (\forall \langle F_1, F_2\rangle \in \mathfrak{S}(X, I))$, is a soft weak boundary operator on X indexed by I. Next we compute the rest four: $\mathfrak{T}_b$, c_b, i_b, and o_b. By Remark 6 we know $\mathfrak{T}_b = \varphi_{5,1}(b) = \{\langle \emptyset, \emptyset\rangle, \langle \{x\}, \emptyset\rangle, \langle \emptyset, \{y\}\rangle, \langle \{x\}, \{y\}\rangle\}$.

By Theorem 5 we know $c_b = \varphi_{5,2}(b) : \mathfrak{S}(X, I) \longrightarrow \mathfrak{S}(X, I)$ is defined by $c_b(\langle \emptyset, \emptyset\rangle) = \langle \{y\}, \{x\}\rangle$, $c_b(\langle \emptyset, \{x\}\rangle) = \langle \{y\}, \{x\}\rangle$, $c_b(\langle \emptyset, \{y\}\rangle) = \langle \{y\}, X\rangle$, $c_b(\langle \emptyset, X\rangle) = \langle \{y\}, X\rangle$, $c_b(\langle \{x\}, \emptyset\rangle) = \langle X, \{x\}\rangle$, $c_b(\langle \{x\}, \{x\}\rangle) = \langle X, \{x\}\rangle$, $c_b(\langle \{x\}, \{y\}\rangle) = \langle X, X\rangle$, $c_b(\langle \{x\}, X\rangle) = \langle X, X\rangle$, $c_b(\langle \{y\}, \emptyset\rangle) = \langle \{y\}, \{x\}\rangle$, $c_b(\langle \{y\}, \{x\}\rangle) = \langle \{y\}, \{x\}\rangle$, $c_b(\langle \{y\}, \{y\}\rangle) = \langle \{y\}, X\rangle$, $c_b(\langle \{y\}, X\rangle) = \langle \{y\}, X\rangle$, $c_b(\langle X, \emptyset\rangle) = \langle X, \{x\}\rangle$, $c_b(\langle X, \{x\}\rangle) = \langle X, \{x\}\rangle$, $c_b(\langle X, \{y\}\rangle) = \langle X, X\rangle$, $c_b(\langle X, X\rangle) = \langle X, X\rangle$.

By Remark 6 we know $i_b = \varphi_{5,3}(b) : \mathfrak{S}(X, I) \longrightarrow \mathfrak{S}(X, I)$ is defined by $i_b(\langle \emptyset, \emptyset\rangle) = \langle \emptyset, \emptyset\rangle$, $i_b(\langle \emptyset, \{x\}\rangle) = \langle \emptyset, \emptyset\rangle$, $i_b(\langle \emptyset, \{y\}\rangle) = \langle \emptyset, \{y\}\rangle$, $i_b(\langle \emptyset, X\rangle) = \langle \emptyset, \{y\}\rangle$, $i_b(\langle \{x\}, \emptyset\rangle) = \langle \{x\}, \emptyset\rangle$, $i_b(\langle \{x\}, \{x\}\rangle) = \langle \{x\}, \emptyset\rangle$, $i_b(\langle \{x\}, \{y\}\rangle) = \langle \{x\}, \{y\}\rangle$, $i_b(\langle \{x\}, X\rangle) = \langle \{x\}, \{y\}\rangle$, $i_b(\langle \{y\}, \emptyset\rangle) = \langle \emptyset, \emptyset\rangle$, $i_b(\langle \{y\}, \{x\}\rangle) = \langle \emptyset, \emptyset\rangle$, $i_b(\langle \{y\}, \{y\}\rangle) = \langle \emptyset, \{y\}\rangle$, $i_b(\langle \{y\}, X\rangle) = \langle \emptyset, \{y\}\rangle$, $i_b(\langle X, \emptyset\rangle) = \langle \{x\}, \emptyset\rangle$, $i_b(\langle X, \{x\}\rangle) = \langle \{x\}, \emptyset\rangle$, $i_b(\langle X, \{y\}\rangle) = \langle \{x\}, \{y\}\rangle$, $i_b(\langle X, X\rangle) = \langle \{x\}, \{y\}\rangle$.

By Remark 6 we know $o_b = \varphi_{5,4}(b) : \mathfrak{S}(X, I) \longrightarrow \mathfrak{S}(X, I)$ is defined by $o_b(\langle \emptyset, \emptyset\rangle) = \langle \{x\}, \{y\}\rangle$, $o_b(\langle \emptyset, \{x\}\rangle) = \langle \{x\}, \{y\}\rangle$, $o_b(\langle \emptyset, \{y\}\rangle) = \langle \{x\}, \emptyset\rangle$, $o_b(\langle \emptyset, X\rangle) = \langle \{x\}, \emptyset\rangle$, $o_b(\langle \{x\}, \emptyset\rangle) = \langle \emptyset, \{y\}\rangle$, $o_b(\langle \{x\}, X\rangle) = \langle \emptyset, \emptyset\rangle$, $o_b(\langle \{x\}, \{x\}\rangle) = \langle \emptyset, \{y\}\rangle$, $o_b(\langle \{x\}, \{y\}\rangle) = \langle \emptyset, \emptyset\rangle$, $o_b(\langle \{y\}, \emptyset\rangle) = \langle \{x\}, \{y\}\rangle$, $o_b(\langle \{y\}, \{x\}\rangle) = \langle \{x\}, \{y\}\rangle$, $o_b(\langle \{y\}, \{y\}\rangle) = \langle \{x\}, \emptyset\rangle$, $o_b(\langle \{y\}, X\rangle) = \langle \{x\}, \emptyset\rangle$, $o_b(\langle X, \emptyset\rangle) = \langle \emptyset, \{y\}\rangle$, $o_b(\langle X, \{x\}\rangle) = \langle \emptyset, \{y\}\rangle$, $o_b(\langle X, \{y\}\rangle) = \langle \emptyset, \emptyset\rangle$, $o_b(\langle X, X\rangle) = \langle \emptyset, \emptyset\rangle$.

Acknowledgments This work is supported by the International Science and Technology Cooperation Foundation of China (2012DFA11270), and the National Natural Science Foundation of China (11501435).

References

1. Molodtsov, D.: Soft set theory – first results. Comput. Math. Appl. **37**, 19–31 (1999)
2. Akdag M., Ozkan A.: On soft β-open sets and soft β-continuous functions. Sci. World J. (2014). http://dx.doi.org/10.1155/2014/843456
3. Kharal A.: Soft approximations and uni-int decision making. Sci. World J. (2014). http://dx.doi.org/10.1155/2014/327408
4. Liu Z.C., Qin K.Y., Pei Z.: Similarity measure and entropy of fuzzy soft sets. Sci. World J. (2014). http://dx.doi.org/10.1155/2014/161607
5. Zhou M., Li S.G., Akram M.: Categorical properties of soft sets. Sci. World J. (2014). http://dx.doi.org/10.1155/2014/783056
6. Aktaş, H., Çağman, N.: Soft sets and soft groups. Inf. Sci. **177**, 2726–2735 (2007)
7. Jun, Y.B.: Soft BCK/BCI-algebras. Comput. Math. Appl. **56**, 1408–1413 (2008)
8. Feng, F., Jun, Y.B., Zhao, X.Z.: Soft semirings. Comput. Math. Appl. **56**, 2621–2628 (2008)

9. Zhan, J.M., Jun, Y.B.: Soft BL-algebras based on fuzzy sets. Comput. Math. Appl. **59**, 2037–2046 (2010)
10. Acar, U., Koyuncu, F., Tanay, B.: Soft sets and soft rings. Comput. Math. Appl. **59**, 3458–3463 (2010)
11. Atagün, A.O., Sezgin, A.: Soft substructures of rings, fields and modules. Comput. Math. Appl. **61**, 592–601 (2011)
12. Sezgin, A., Atagün, A.O.: Soft groups and normalistic soft groups. Comput. Math. Appl. **62**, 685–698 (2011)
13. Shabir, M., Naz, M.: On soft topological spaces. Comput. Math. Appl. **61**, 1786–1799 (2011)
14. Çağman, N., Karataş, S., Enginoglu, S.: Soft topology. Comput. Math. Appl. **62**, 351–358 (2011)
15. Maji, P.K., Biswas, R., Roy, A.R.: Soft set theory. Comput. Math. Appl. **45**, 555–562 (2003)
16. Adámek, J., Herrlich, H., Strecker, G.E.: Abstract and Concrete Categories. Wiley, NewYork (1990)
17. Liu, Y.M., Luo, M.K.: Fuzzy Topology. World Scientific Publishing, Singapore (1997)
18. Gierz, G., Hofmann, K.H., Keimel, K., Lawson, J.D., Mislowe, M., Scott, D.S.: Continuous Lattices and Domains. Encyclopedia of Mathematics and its Applications. Cambridge University Press, Cambridge (2003)

Čech Closure Molecular Lattices

Xue-Mei Dai, Hong-Xia Li, Sheng-Gang Li, Sheng-Quan Ma and Xiao-Fei Yang

Abstract Some results on connected fuzzy topological spaces are extend to the setting of Čech closure molecular lattices, a natural generalization of Mashhour and Ghanim's fuzzy closure spaces, Liu and Luo's quasi-subspaces, and Wang's topological molecular lattices (and also an analogous of knowledge space, implicational space, and learning space). It is proved that **CCML**, the category of Čech closure molecular lattices and generalized order-homomorphisms, has products and coproducts and that the product of a family of **CCML**-objects is connected if and only if each of these objects is connected.

Keywords Čech closure molecular lattice · Knowledge space · Implicational space · Learning space · Category · Connectedness.

X.-M. Dai · S.-G. Li
College of Mathematics and Information Science, Shaanxi Normal University, Xi'an 710062, People's Republic of China
e-mail: qhdaixuemei@126.com

S.-G. Li
e-mail: shengganglinew@126.com

X.-M. Dai
Department of Mathematics, Qinghai Normal University, Xining 810008, People's Republic of China

H.-X. Li
School of Mathematics and Statistics, Long Dong University, Qingyang 745000, People's Republic of China
e-mail: lhxia0929@163.com

S.-Q. Ma (✉)
College of Information Science and Technology, Hainan Normal University, Haikou 571158, People's Republic of China
e-mail: mashengquan@163.com

X.-F. Yang
School of Science, Xi'an Polytechnic University, Xi'an 710048, People's Republic of China
e-mail: yangxiaofei2002@163.com

T.-H. Fan et al. (eds.), *Quantitative Logic and Soft Computing 2016*, Advances in Intelligent Systems and Computing 510,
DOI 10.1007/978-3-319-46206-6_52

1 Introduction

Connectedness is a useful notion in topology, which has various generalizations in fuzzy topological spaces. One of such generalization is the fuzzy connectedness defined by Pu and Liu in [1] for an arbitrary fuzzy set, which has been developed and used by many authors. In this article, we will extend some results on Pu and Liu's connectedness for fuzzy topological spaces to the setting of Čech closure molecular lattices, a natural generalization of Mashhour and Ghanim's fuzzy closure space [2], Liu and Luo's quasi-subspace [3], and Wang's topological molecular lattice—TML [4] for short (and also an analogous of knowledge space, implicational space, and learning space, cf. [5–7]). In Sect. 1 we define the notions of Čech closure molecular lattice, continuous generalized order-homomorphism, and homeomorphic generalized order-homomorphism and investigate some fundamental properties. In Sect. 2 we study the categorical preparations of Čech closure molecular lattices, including existence and structures of **CCML**-products and **CCML**-coproducts, where **CCML** is the category of Čech closure molecular lattices and generalized order-homomorphisms. In the final section we define the notion of connected Čech closure molecular lattice and establish a series of properties of these closure molecular lattices. In particular, we show that the product of a family of **CCML**-objects is connected if and only if each of these objects is connected. We refer to [4] as a general reference on topological molecular lattice (or quasi-subspace).

2 Čech Closure Molecular Lattices and Continuous GOHs

Definition 1 Let L be a completely distributive complete lattice (CD lattice for short) with the smallest element 0_L and the largest element 1_L. A Čech closure operator on L is a mapping $\sim: L \longrightarrow L$ satisfying:
(1) $0_L^\sim = 0_L$;
(2) $a \leq a^\sim$ for every $a \in L$;
(3) $(a \vee b)^\sim = a^\sim \vee b^\sim$ for every $a, b \in L$.
When $\sim$ is a Čech closure operator, we call $(L, \sim)$ a Čech closure molecular lattice (CCML for short).

Remark 1 (1) Let L be a CD lattice, $\mathbb{C}_L^{\sim}$ the set of all Čech closure operators on L, $\mathbb{T}_L$ the set of all co-topologies [4] on L, and $\mathbb{C}_L$ the set of all all Čech closure operators on L satisfying the idempotent law (such kinds of Čech closure operators are called Kuratowski closure operators). For each $\sim \in \mathbb{C}_L$, let $\mathbb{F}(\sim) = \{a^\sim \mid a \in L\}$. Then we have a bijection $\mathbb{F} : \mathbb{C}_L \longrightarrow \mathbb{T}_L$, whose inverse $\mathbb{F}^{-1}$ maps a co-topology $\delta \in \mathbb{T}_L$ to a Kuratowski closure operator $\sim \in \mathbb{C}_L$ defined by $a^\sim = \bigwedge\{b \in \delta \mid a \leq b\} (\forall a \in L)$. For this reason, we will make no distinction between a Kuratowski closure molecular lattice (i.e. CCML $(L, \sim)$ with $\sim$ a Kuratowski closure operator) and the corresponding TML, and use $-$ to denote any Kuratowski closure operator on L in this paper. We note that $\mathbb{C}_L \subset \mathbb{C}_L^{\sim}$ for some CD lattice L (see Example 1).

Thus CCML is a natural and nontrivial generalization of fuzzy closure space [2], quasi-subspace [3], and TML [4].

(2) For any $\sim_1, \sim_2 \in \mathbb{C}_L^{\sim}$, we say $\sim_1$ is coarser than $\sim_2$ (or $\sim_2$ is finer than $\sim_1$), in symbols $\sim_1 \prec \sim_2$, iff $a^{\sim_2} \leq a^{\sim_1}$ $(\forall a \in L)$. One can verify that $(\mathbb{C}_L^{\sim}, \prec)$ is a complete lattice.

Example 1 (1) Suppose that L is a well-ordered set which has at least four elements. Let $0_L^{\sim} = 0_L$, $1_L^{\sim} = 1_L$, and $x^{\sim}$ the successor of x for every $x \in L - \{0_L, 1_L\}$. Then $\sim \in \mathbb{C}_L^{\sim} - \mathbb{C}_L$.

(2) Let R be the set of all real numbers, $L = 2^R$ the CD lattice consisting of all subsets of R and ordered by inclusion $\subseteq$. Let $\emptyset^{\sim} = \emptyset$ and $E^{\sim} = \{x \in R \mid |x - e| < 1$ for some $e \in E\}$ for every $E \in L - \{\emptyset\}$. Then $\sim \in \mathbb{C}_L^{\sim} - \mathbb{C}_L$.

Definition 2 Suppose that $(L_1, \sim)$ and $(L_2, \sim)$ are both CCMLs (for simplicity, we use the same symbol $\sim$ to denote the two different Čech closure operators on L_1 and L_2, respectively), and $f : L_1 \longrightarrow L_2$ is a generalized order-homomorphism [4] (GOH for short). f is said to be
(1) Continuous iff $f(a^{\sim}) \leq [f(a)]^{\sim}$ for every $a \in L_1$.
(2) Closed iff $f(a^{\sim}) = [f(a)]^{\sim}$ for every $a \in L_1$.
(3) Open iff for every $a \in L_1$ and $b \in L_2$ satisfying $f^*(b) \leq a^{\sim}$, there exists a $c \in L_2$ such that $b \leq c^{\sim}$ and $f^*(c^{\sim}) \leq a^{\sim}$, where $f^* : L_2 \longrightarrow L_1$ is the right adjoint of f.
(4) An homeomorphism (in this case, we say that $(L_1, \sim)$ is homeomorphic to $(L_2, \sim)$) iff it is a bijection, and both $f : (L_1, \sim) \longrightarrow (L_2, \sim)$ and $f^* : (L_2, \sim) \longrightarrow (L_1, \sim)$ are continuous.

A property P on CCMLs is said to be topological iff it is preserved under homeomorphic GOHs.

Theorem 1 *Using the same symbols as in Definition 2, we have*
(1) f is continuous if and only if $[f^*(b)]^{\sim} \leq f^*(b^{\sim})$ $(\forall b \in L_2)$.
(2) If f is a bijection. Then f is an homeomorphism if and only if both f and f^* are open (equivalently, both f and f^* are closed).

Proof We only show (2).

Suppose that f is a bijection. Then f^* is exactly the inverse mapping of f. First, assume that f is an homeomorphism, $a \in L_1$, $b \in L_2$ and $f^*(b) \leq a^{\sim}$. Then we take $c = f(a)$. As f and f^* are both continuous, it follows that $b \leq c^{\sim}$ and $f^*(c^{\sim}) \leq a^{\sim}$. Thus f is open. Similarly, f^* is also open. Next, suppose that both f and f^* are open. Then $[f(a)]^{\sim} \leq f(a^{\sim})$ and $f(a^{\sim}) \leq [f(a)]^{\sim}$ for every $a \in L_1$, i.e. f is a closed. Similarly, f^* is closed. Apparently, f is an homeomorphism if f and f^* are both closed and continuous. □

Theorem 2 *Let L be a CD lattice, $(X, \sim)$ a Čech closure space,*

$$A^{\sim_L} = \bigwedge \{[r] \vee \chi_{E^{\sim}} \mid A \leq [r] \vee \chi_{E^{\sim}}, r \in L, E \subseteq X\},$$

where $A \in L^X$ (the set of all L-subsets with pointwise order), $\chi_{E^\sim}$ is the characteristic function of $E^\sim$, $[r] \in L^X$ is the L-subset taking constant value r. Then
(1) $(X, \sim_L)$ is a fully stratified (i.e. it satisfies $[r]^{\sim_L} = [r]$ for all $r \in L$) L-Čech closure space (called induced L-Čech closure space).
(2) $(\chi_Y)^{\sim_L} = \chi_{Y^\sim}$ for every $Y \subseteq X$.

Proof We only show (1). It suffices to show that $(A \vee B)^{\sim_L} = A^{\sim_L} \vee B^{\sim_L}$ for all $A, B \in L^X$. Let

$$\mathfrak{A} = \{[r] \vee \chi_{E^\sim} \mid A \leq [r] \vee \chi_{E^\sim}\},$$

$$\mathfrak{B} = \{[r] \vee \chi_{\tilde{E}} \mid B \leq [r] \vee \chi_{E^\sim}\},$$

$$\mathfrak{C} = \{[r] \vee \chi_{E^\sim} \mid A \vee B \leq [r] \vee \chi_{E^\sim}\}.$$

Then $A^{\sim_L} = \bigwedge \mathfrak{A}$, $B^{\sim_L} = \bigwedge \mathfrak{B}$ and $(A \vee B)^{\sim_L} = \bigwedge \mathfrak{C}$. Obviously, $\mathfrak{C} \subseteq \mathfrak{A}$, thus $A^{\sim_L} \leq (A \vee B)^{\sim_L}$. Similarly, $B^{\sim_L} \leq (A \vee B)^{\sim_L}$. Therefore $A^{\sim_L} \vee B^{\sim_L} \leq (A \vee B)^{\sim_L}$. Next, we show $(A \vee B)^{\sim_L} \leq A^{\sim_L} \vee B^{\sim_L}$. It suffices to show $A \vee B \leq [r \vee s] \vee \chi_{(E \cup F)^\sim}$ whenever $A \leq [r] \vee \chi_{E^\sim}$ and $B \leq [r] \vee \chi_{F^\sim}$. Suppose that $e \in \mathrm{Copr}(L^X)$ satisfying $e \leq A \vee B$. Then $e \leq A$ or $e \leq B$. Without loss of generality we assume that $e \leq A$. Since $(X, \sim)$ is a Čech closure space, we have $E^\sim \cup F^\sim = (E \cup F)^\sim$, and thus

$$e \leq [r \vee s] \vee \chi_{(E^\sim \cup F^\sim)} = [r \vee s] \vee \chi_{(E \cup F)^{\sim_L}},$$

which implies $(A \vee B)^{\sim_L} = A^{\sim_L} \vee B^{\sim_L}$. □

3 The Category of Čech Closure Molecular Lattices

In the following, let **CD** (resp., **TML**, **CCML**) be the category of CD lattices (resp., TMLs, CCMLs) and GOHs (resp., continuous GOHs, continuous GOHs), and for any given CD lattice L, $\mathrm{Copr}(L)$ be the set of all nonzero co-prime elements of L.

Lemma 1 ([8]) *Let $\{L_t\}_{t \in T}$ be a family of CD lattices, $L_t^0 = L_t - \{0_{L_t}\}$, and $\bigotimes_{t \in T} L_t$ the collection of all $A \subseteq \prod_{t \in T} L_t^0$ satisfying the following conditions (where $\prod_{t \in T} L_t^0$ is direct product):*
(P1) A is a lower set, i.e. $A = \downarrow A$, where

$$\downarrow A = \left\{ \{y_t\}_{t \in T} \in \prod_{t \in T} L_t^0 \;\middle|\; \exists \{x_t\}_{t \in T} \in A,\ x_t \geq y_t\ (\forall t \in T) \right\}.$$

(P2) If $\emptyset \neq B_t \subseteq L_t^0$ and $\prod_{t \in T} B_t \subseteq A$, then $\{b_t\}_{t \in T} \in A$, where $b_t = \bigvee B_t$ $(t \in T)$.
Then the following statements hold:
(1) $(\bigotimes_{t \in T} L_t, \subseteq)$ is a CD lattice, the intersection and union of a family $\{A_s\}_{s \in S}$ of CD lattices is defined by $\bigwedge_{s \in S} A_s = \bigcap_{s \in S} A_s$ and

$$\bigvee_{s\in S} A_s = \left\{ \{b_t\}_{t\in T} \;\middle|\; \exists B_t \in 2^{L_t^0} - \{\emptyset\}(t \in T), \prod_{t\in T} B_t \subseteq \bigcup_{s\in S} A_s \; \bigvee B_t = b_t \right\}$$

respectively, where $\bigcap$ *and* $\bigcup$ *are the ordinary set theoretic intersection and union respectively.*
(2) $\{\downarrow\{a_t\}_{t\in T} \in \bigotimes_{t\in T} L_t \mid a_t \in Copr(L_t),\ \ t \in T\} \subseteq Copr(\bigotimes_{t\in T} L_t)$, *and it is a union-generating set of* $\bigotimes_{t\in T} L_t$.
(3) $\{\bigotimes_{t\in T} L_t,\ p_t \mid t \in T\}$, *breifly,* $\bigotimes_{t\in T} L_t$, *is the* **CD**-*product of* $\{L_t\}_{t\in T}$, *where* p_s, *defined by*

$$p_s(A) = \bigvee \{x_s \mid \{x_t\}_{t\in T} \in A\} \left(A \in \bigotimes_{t\in T} L_t,\ s \in T \right),$$

is called the projection from $\bigotimes_{t\in T} L_t$ *to* L_s.
(4) $\{(\bigotimes_{t\in T} L_t, \delta),\ p_t \mid t \in T\}$, *breifly,* $(\bigotimes_{t\in T} L_t, \delta)$, *is the* **TML**-*product of* $\{(L_t, \delta_t)\}_{t\in T}$, *where* δ *is the co-topology on* $\bigotimes_{t\in T} L_t$ *having* $\{p_t^*(Q) \mid Q \in \delta_t, t \in T\}$ *as a subbase.*

Theorem 3 **CCML** *has products and coproducts. Let* $\{(L_t, \sim)\}_{t\in T}$ *be a family of CCMLs,* $\{L, p_t\}_{t\in T}$ *and* $\{T, q_t\}_{t\in T}$ *the product and coproduct of* $\{L_t\}_{t\in T}$ *in* **CD** *respectively. For every* $a \in L$, *let* $a^\sim$ *be the union set of all* $e \in Copr(L)$ *satisfying following condition* $(*)$:
$(*)$ *If* $a = \bigvee_{k=1}^{b} a_k$, *then there exists a natural number* k *such that* $p_t(e) \le [p_t(a_k)]^\sim$ *for every* $t \in T$.

For each $b \in J = \prod_{t\in T} L_t$, *let* $b^\smile \in J$ *satisfying* $b^\smile(t) = [b(t)]^\sim$ *for every* $t \in T$. *Then*
(1) $\{(L, \sim), p_t\}_{t\in T}$ *and* $\{(J, \smile), q_t\}_{t\in T}$ *are the product and coproduct of* $\{(L_t, \sim)\}_{t\in T}$ *in* **CCML**, *respectively.*
(2) $\sim$ *is the coarsest Čech closure operator such that every* p_t *is continuous, and* $\smile$ *is the finest Čech closure operator such that every* q_t *is continuous.*
(3) *When* $\{(L_t, \sim)\}_{t\in T}$ *is a family of TMLs,* $\{(L, \sim), p_t\}_{t\in T}$ *and* $\{(J, \smile), q_t\}_{t\in T}$ *are the product and coproduct of* $\{(L_t, \sim)\}_{t\in T}$ *in category* **TML**, *respectively.*
(4) q_t *is a clopen (i.e. both closed and open) GOH for every* $t \in T$.

Proof We only show that $(L, \sim)$ is a CCML. Obviously, $0_L^\sim = 0_L$. Suppose that $e \in$Copr(L) satisfying $e \le a = \bigvee_{k=1}^{n} a_k \quad (a_k \in L)$. Then $e \le a_k$ for some $k \le n$ since $e \in$Copr(L), and thus $p_t(e) \le p_t(a_k) \le [p_t(a_k)]^\sim$ for every $t \in T$. It follows that $e \le a^\sim$ by the definition of $a^\sim$, i.e. $a \le a^\sim$. Next, let $a, b \in L$. Then $a^\sim \vee b^\sim \le (a \vee b)^\sim$ by the above definition of $\sim$. If $e \in$Copr(L) and $e \not\le a^\sim \vee b^\sim$, i.e. $e \not\le a^\sim$ and $e \not\le b^\sim$. Then there exist $t_1, t_2, \ldots, t_n \in T$ and $e_1, e_2, \ldots, e_m, e_{m+1}, \ldots, e_n \in L$ such that $a = \bigvee_{k=1}^{m} e_k, b = \bigvee_{k=m+1}^{n} e_k$ and $p_{t_k}(e) \not\le [p_{t_k}(e_{t_k})]^\sim \quad (k = 1, 2, \ldots, n)$, i.e. $e \not\le (a \vee b)^\sim$. Therefore $(L, \sim)$ is a CCML. □

Theorem 4 **TML** *is a coreflective subcategory of* **CCML**.

Proof Obviously, **TML** is a subcategory of **CCML**. For each **CCML**-object $A = (L, \sim)$, let $\delta = \{a^\sim \mid a \in L, a^{\sim\sim} = a^\sim\}$. Then $\delta \in \mathbb{T}_L$ and $- = \sigma^{-1}(\delta) : L \longrightarrow L$ is a Kuratowski closure operator on L. It can be easily seen that $r_A = id_L : A = (L, \sim) \longrightarrow (L, -)$ is a coreflection. □

Theorem 4, Lemma 2.3 in [9], and Proposition 27.9 (2) in [10] imply.

Theorem 5 **CCML** *is not a Cartesian closed category.*

4 Connectedness of Čech Closure Molecular Lattices

In this section, $(L, \sim)$ is always supposed to be a Čech closure moleculer lattice.

Definition 3 $a, b \in L$ is said to be separated iff $a^\sim \wedge b = a \wedge b^\sim = 0_L$. $c \in L$ are said to be connected in $(L, \sim)$ iff there exists no nonzero separated elements a and b such that $c = a \vee b$. $(L, \sim)$ is said to be separated iff 1_L is connected.

Theorem 6 *$e \in L$ is not connected in $(L, \sim)$ if and only if there exist $a, b \in L - \{0_L\}$ such that $a \vee b = e$, $a \wedge b = 0_L$, $a = a^\sim \wedge e$ and $b = b^\sim \wedge e$.*

Proof Assume that e is not connected in $(L, \sim)$, i.e. there exist $a, b \in L - \{0_L\}$ such that $a \vee b = e$ and $a \wedge b^\sim = a^\sim \wedge b = 0_L$. As $a \leq a^\sim$, we have $a \wedge b = 0_L$ and $a^\sim \wedge e = a^\sim \wedge (a \vee b) = (a^\sim \wedge a) \vee (a^\sim \wedge b) = a$. Similarly $b^\sim \wedge e = b$.

Conversely, assume the condition in Theorem 6. Then

$$a \wedge b^\sim = a^\sim \wedge (e \wedge e) \wedge b^\sim = [a^\sim \wedge (a \vee b)] \wedge [(a \vee b) \wedge b^\sim] = [a^\sim \wedge (a \wedge b) \wedge b^\sim] = a \wedge b = 0_L.$$

Similarly, $a^\sim \wedge b = 0_L$. This means that e is not connected in $(L, \sim)$. □

Corollary 1 *$(L, \sim)$ is connected if and only if there exists no $a, b \in L - \{0_L\}$ such that $a \vee b = 1_L$, $a \wedge b = 0_L$, $a = a^\sim$ and $b = b^\sim$.*

Analogous to the case of fuzzy topological spaces, we may show the following Theorems 7–10.

Theorem 7 *If $a \in L$ is connected in $(L, \sim)$. Then every $b \in L$ satisfying $a \leq b \leq a^\sim$ is also connected in $(L, \sim)$.*

Theorem 8 *Let $\{a_t\}_{t \in T}$ be a family of connected elements in $(L, \sim)$. If there exists an $s \in T$ such that a_t and a_s are not separated for every $t \in T$. Then $\bigvee_{t \in T} a_t$ is also connected in $(L, \sim)$.*

For each molecule $e \in L$, $[e] = \bigvee\{c \in L \mid c$ is connected in $(L, \sim)\}$ is a connected element in $(L, \sim)$, which is called the connected component containing e.

Theorem 9 *Let $\mathfrak{A}$ be the set of all connected components in $(L, \sim)$. Then*
(1) $\bigvee \mathfrak{A} = 1_L$;
(2) *If* $a, b \in \mathfrak{A}$ *and* $a \neq b$, *then* $a \wedge b = 0_L$;
(3) $a = a^{\sim}$ *for every* $a \in \mathfrak{A}$.

Theorem 10 *The image of a connected element under a continuous GOH is connected.*

Next, we consider the product of connected CCMLs.

Lemma 2 *Suppose that $(L, \sim)$ and $(J, \sim)$ are connected CCMLs and that the largest element 1_L of L is a molecule. Then the product $(L \otimes J, \sim)$ of $(L, \sim)$ and $(J, \sim)$ in* **CCML** *is connected.*

Proof Assume that $(L \otimes J, \sim)$ is not connected, i.e. there exist $A, B \in L \otimes J - \{0_{L\otimes J}\}$ such that $A \vee B = 1_{L\otimes J}$ and $A^{\sim} \wedge B = 0_{L\otimes J} = A \wedge B^{\sim}$, respectively. For every $y \in \text{Copr}\,(J)$, as $\downarrow(L_L, y) \in \text{Copr}\,(L \otimes J)$, we have $(1_L, y) \in A$ or $(1_L, y) \in B$. Let

$$A_2 = \bigvee\{y \mid y \in \text{Copr}(J), (1_L, y) \in A\},$$

$$B_2 = \bigvee\{y \mid y \in \text{Copr}(J), (1_L, y) \in B\}.$$

Then $A_2 \vee B_2 = 1_J$.

Suppose there exists a $y \in$Copr(J) such that $y \leq A_2^{\sim} \wedge B_2$. Then $\downarrow(1_L, y) \subseteq B$. Let $A = \bigvee_{k=1}^{n} C_k$. Then

$$p_2(\downarrow(1_L, y)) = y \leq A_2^{\sim} \leq [p_2(A)]^{\sim} = \left[\bigvee_{k=1}^{n} p_2(C_k)\right]^{\sim} = \bigvee_{k=1}^{n} [p_2(C_k)]^{\sim}.$$

Thus there exists a natural number k such that $p_2(\downarrow(1_L, y)) = y \leq [p_2(C_k)]^{\sim}$. By the definition of $A^{\sim}$, we have $\downarrow(1_L, y) \leq A^{\sim}$, i.e. $B \wedge A^{\sim} \neq 0_{L\otimes J}$. This is a contradiction. Therefore $B_2 \wedge A_2^{\sim} = 0_J$. similarly, $A_2 \wedge B_2^{\sim} = 0_J$, which means that $(J, \sim)$ is not connected, this is a contradiction. □

Theorem 11 *The product $(L, \sim)$ of a family $\{(L_t, \sim)\}_{t\in T}$ of connected Čech closure molecular lattices in* **CCML** *is connected if and only if $(L_t, \sim)$ is connected for all $t \in T$.*

Proof By Theorem 10, we only need to show the sufficient. Let $F(T)$ be the set of all nonempty fimite subsets of T. For each $t \in T$, take an $a_t \in$Copr(L_t), and for each $S \in F(T)$, let $C_S = \bigotimes_{t\in T} A_t$ be the product of the family $\{A_t\}_{t\in T}$ in **CD** where A_t is defined by

$$A_t = \begin{cases} \downarrow a_t, \ t \in S, \\ L_t, \ \ t \in T - S. \end{cases}$$

Similar to Lemma 2, we can show that C_S is connected in $(L, \sim)$ (notice that the largest element of C_S belongs to Copr(L)). Since

$$\{a_t\}_{t\in T} \in \bigcap\{C_s \mid S \in F(T)\},$$

$C = \bigvee\{C_S \mid S \in F(T)\}$ is connected in $(L, \sim)$ by Theorem 8. It is easy to verify that $C^{\sim} = 1_L$, and thus $(L, \sim)$ is connected by Theorem 7.

The notions of chain and layer compact lattice may be characterized in terms of layer compactness of L-topological spaces (see [11]). Similarly, the notion of anti-diamond lattice may be characterized in terms of connectedness of L-Čech closure spaces.

Theorem 12 *For a CD lattice L, the following conditions are equivalent:*
(1) L is an anti-diamond lattice (i.e. there exist no $a, b \in L - \{0_L\}$ *such that* $a \vee b = 1_L$ *and* $a \wedge b = 0_L$*);*
(2) All connected components in an L-Čech closure space are characteristic functions;
(3) A Čech closure space $(X, \sim)$ *is connected if and only if the L-Čech closure space* $(X, \sim_L)$*, induced by* $(X, \sim)$*, is connected (i.e. connectedness of L-Čech closure spaces is an L-extension of the connectedness of Čech closure spaces).*

Proof (1)$\Longrightarrow$(2): Let $(X, \sim)$ be an L-Čech closure space, and $A \in L^X$ a connected component in $(X, \sim)$. It suffices to show that $A(x) = 1_L$ whenever $A(x) \neq 0_L$. Suppose that $A(x) < 1_L$. We first show that $\lambda \wedge A(x) \neq 0_L$ for some molecule $\lambda \in L$ satisfying $\lambda \not\leq A(x)$. In fact, let $b = \bigvee\{\lambda \in$Copr$(L) \mid \lambda \leq A(x)\}$. Then $b \vee A(x) = 1_L$. Since L is an anti-diamond lattice, $b \wedge A(x) \neq 0_L$, and thus the above statement holds. Next, let B be the connected component such that $x_\lambda \leq B$ and $x_\lambda \in$Copr(L^X). As $\lambda \wedge A(x) \neq 0_L$, we may show that $A \vee B \neq B$ and $A \vee B$ is connected in $(X, \sim)$. This is a contradiction since B is a connected component.

(2)$\Longrightarrow$(3): Obviously $(X, \sim)$ is connected whenever $(X, \sim_L)$ is by Theorem 2 (2). Conversely, assume that $(X, \sim)$ is connected but $(X, \sim_L)$ is not connected. Then there exists a connected component A satisfying $A \notin \{0_{L^X}, 1_{L^X}\}$. By (2), $A = \chi_Y$ for some $Y \notin \{\emptyset, X\}$. By Theorem 2 (2), Y is connected in $(X, \sim)$. We will show that Y is a component, which implies (2)$\Longrightarrow$(3). Suppose that Z is a connected subset of X satisfying $Y \subseteq Z \neq Y$. Let $B = \chi_{Z-Y}$. Then $A, B \neq 0_{L^X}$, $A \vee B = \chi_Z$ and A and B are separated by (2) and Theorem 9 (3). It follows that Z is not connected in $(L, \sim)$. This is a contradiction.

(3) $\Longrightarrow$ (1): Assume that (1) does not holds, i.e. there exist $a, b \in L - \{0_L\}$ such that $a \vee b = 1_L$ and $a \wedge b = 0_L$. Take $X = \{x\}$. Then $(X, \sim)$ is a connected space, but the induced L-Čech closure space $(X, \sim_L)$ is not connected because $[a] = [a]^{\sim_L} \neq 0_{L^X}$, $[b] = [b]^{\sim_L} \neq 0_{L^X}$ and $[a] \vee [b] = 1_{L^X}$ which means that (3) does not hold either.

Acknowledgments This work is supported by the International Science and Technology Cooperation Foundation of China (2012DFA11270), the National Natural Science Foundation of China (11501435), and the Scientific Research Project of Gansu Province (2015A-144).

References

1. Pu, P.M., Liu, Y.M.: Fuzzy topology I: Neighborhood structure of a fuzzy point and Moore-Smith convergence. J. Math. Anal. Appl. **76**, 571–599 (1980)
2. Mashhour, A.S., Ghanim, M.H.: Fuzzy closure spaces. J. Math. Anal. Appl. **106**, 154–170 (1985)
3. Liu, Y.M., Luo, M.K.: Induced spases and fuzzy Stone-$\check{C}$ech compactifications. Scientia Sinica (Series A). **30**, 1034–1044 (1987)
4. Wang, G.J.: Theory of topological molecular lattices. Fuzzy Sets Syst. **47**, 351–376 (1992)
5. Doignon, J.P., Falmagne, J.C.: Knowledge Spaces. Springer, Berlin (1999)
6. Caspard, N., Monjardet, B.: The lattices of closure systems, closure operators, and implicational systems on a finite set: a survey. Discret. Appl. Math. **127**, 241–269 (2003)
7. Falmagne, J.C., Doignon, J.P.: Learning Spaces: Interdisciplinary Applied Mathematics. Springer, Berlin (2011)
8. Fan, T.H.: Product operations in the category of topological molecular lattices. Fuzzy Syst. Math. **2**, 32–40 (1988). (in Chinese)
9. Li, Y.M.: Exponentiable objects in the category of topological molecular lattices. Fuzzy Sets Syst. **104**, 407–414 (1999)
10. Adamek, J., Herrlich, H., Strecker, G.E.: Abstract and Concrete Categories. Wiley, New York (1990)
11. Li, S.G.: The layer compactness in L-fuzzy topological spaces. Fuzzy Sets Syst. **95**, 233–238 (1998)

Multi-L-soft Set and Its Application in Decision Making

Wen-Qing Fu and Yue Shen

Abstract In this paper, the concept of multi-L-soft set is proposed. It is a generalization of multi-fuzzy soft set. Then relations between multi-L-soft sets and operations on the multi-L-soft sets are defined, furthermore, properties of the operations are discussed. Finally, an illustrative example is given to show validity of the multi-interval-valued fuzzy soft set in decision making problem.

Keywords L-soft set · Multi-L-set · Multi-L-soft set · Multi-interval-valued fuzzy soft set · Decision making

1 Introduction and Preliminaries

Many subjects in academic studies, such as economics, engineering, environmental science, social science, medical science, et. al., are full of uncertainty, imprecision and vagueness. A range of existing theories such as probability theory, fuzzy set theory, rough set theory, vague set theory and interval mathematics are well known and are often useful to model vagueness. But in 1999, Molodtsov [1] pointed out that each of these theories has its inherent difficulties and he initiated soft set theory as a new mathematical tool for dealing with uncertainties which is free from difficulties affecting existing methods. And he also discussed the application of soft set theory in many fields, such as operations analysis, game theory, the smoothness of function, and so on [2].

In recent years, research on soft set theory has been rapidly developed, and great progress has been achieved, including works of theoretical soft set [2–5], soft set theory in abstract algebras [6–11], decision making, data analysis, information system, and so on [12–15].

W.-Q. Fu (✉) · Y. Shen
School of Science, Xi'an Technological University,
Xi'an 710032, People's Republic of China
e-mail: palace_2000@163.com

T.-H. Fan et al. (eds.), *Quantitative Logic and Soft Computing 2016*,
Advances in Intelligent Systems and Computing 510,
DOI 10.1007/978-3-319-46206-6_53

The "standard" soft set in [1] deals with a binary-valued information system. Maji et al. [16] first applied soft sets to solve decision making problem that is based on the concept of knowledge reduction in the theory of rough sets [17]. Chen et. al. [18] presented a new definition of soft set parametrization reduction to improve soft set based decision making in [16]. All the above mentioned studies in decision making problems were based on crisp soft sets. For a multi-valued information system, Herawan [19] introduced a concept of multi soft set, and they used the concept of multi-soft set and AND operation for finding reducts in a multi-valued information system. In 2001, P.K. Maji [20] presented the concept of fuzzy soft set which is based on a combination of fuzzy set and soft set models. Later, many researchers implied fuzzy soft set to decision making [12–14].

Sebastian [21] proposed the concept of multi-fuzzy set which is a more general fuzzy set using ordinary fuzzy sets as building blocks, its membership function is an ordered sequence of ordinary fuzzy membership functions. Then Yang et. al. [22] combined multi-fuzzy set and soft set, from which they obtained a new soft set model named multi-fuzzy soft set, and applied it to decision making.

There still a problem that in many fuzzy decision making applications, the related membership functions are extremely individual (dependent on experts' evaluation of alternatives) and thus cannot be lightly confirmed. It is more reasonable to give an interval-valued data to describe degree of membership; in other words, we can make use of interval-valued fuzzy sets which assign to each element an interval that approximates the "real" (but unknown) membership degree. In respond to this, Yang et al. [23, 24] defined a hybrid model called interval-valued fuzzy soft sets and investigated some of their basic properties. They also presented an algorithm to solve decision making problems based on interval-valued fuzzy soft sets. Feng [15] followed the line of exploration in [14] and gave deeper insights into interval-valued fuzzy soft set based decision making discussed in [24].

Since the set of all intervals in [0,1] forms a complete lattice under the pointwise partial order, we can generalize interval-valued fuzzy sets into L-sets, where L is a complete lattice, then we propose the concept of multi-L-soft set, which is a combination of multi-L-set and soft set. In this paper, we first review some background of soft sets, L-sets and soft sets in Sect. 2, and the concept of multi-L-soft set and some of its operations are also presented. In Sect. 3, as a special kind of multi-L-soft set, multi-interval-valued fuzzy soft set is used to analyze decision making problems and an algorithm is proposed.

2 Multi-L-soft Set

Throughout this paper, U refers to an initial universe set, E is a set of parameters, M is a nonempty set. L is a complete De Morgan algebra, that is, L is a complete lattice with a maximal element 1 and a minimal element 0, and is also equipped with an order reversing involution mapping $' : L \longrightarrow L$ (i.e., $b' \leq a'$ if $a \leq b$ and $(a')' = a$

for any $a, b \in L$). An L-set on U is a mapping $f : U \longrightarrow L$, let $\mathcal{L}(U)$ be the set of all L-set on U, that is $\mathcal{L}(U) = \{f : U \longrightarrow L\}$.

If L is the set of all intervals in [0,1], we denote it by $L^I = \{[a, b] \mid 0 \le a \le b \le 1\}$, and the order relation on L^I is given by $[a_1, b_1] \le [a_2, b_2] \Longleftrightarrow a_1 \le a_2, b_1 \le b_2$ $(\forall [a_1, b_1], [a_2, b_2] \in L^I)$, then an L^I-set is an interval-valued fuzzy set in [15]. The set of all interval-valued fuzzy set on U is denoted by $\mathcal{L}^I(U)$, that is $\mathcal{L}^I(U) =$
$\{f : U \longrightarrow L^I\}$.

Definition 1 ([1]) A soft set is a pair (F, A), where $A \subseteq E$, $F : A \longrightarrow \mathcal{P}(U)$ is a mapping, $\mathcal{P}(U)$ is the power set of U.

Definition 2 An L-soft set over U is a pair $(\tilde{F}, A)$, where $A \subseteq E$, and $\tilde{F}$ is a mapping given by $\tilde{F} : A \longrightarrow \mathcal{L}(U)$, that is for every $e \in A$, $\tilde{F}(e)$ is an L-set, we denote $\tilde{F}(e)(u)$ by $\tilde{F}(e, u)$ for short (for every $e \in A$ and $u \in U$).

An L^I-soft set is an interval-valued fuzzy soft set in [15].

Definition 3 A multi-L-set is a mapping $\mathcal{F} : M \longrightarrow \mathcal{L}(U)$. The set of all multi-$L$-sets on U is denoted by $M\mathcal{L}(U)$, That is $\mathcal{F}(m)$ is an L-set ($\forall m \in M$), we denote $\mathcal{F}(m)(u)$ by $\mathcal{F}(m, u)$ for short (for any $m \in M$ and $u \in U$).

Remark 1 (1) If M and U are countable set, then we can tabularize a multi-L-set $\mathcal{F}$ as follows:

$\mathcal{F}$	u_1	u_2	u_3	$\cdots$
m_1	$\mathcal{F}(m_1, u_1)$	$\mathcal{F}(m_1, u_2)$	$\mathcal{F}(m_1, u_3)$	$\cdots$
m_2	$\mathcal{F}(m_2, u_1)$	$\mathcal{F}(m_2, u_2)$	$\mathcal{F}(m_2, u_3)$	$\cdots$
$\vdots$		$\ddots$		

(2) Let $\mathcal{F} \in M\mathcal{L}(U)$. If $\mathcal{F}(m, u) = 0$ for all $m \in M$ and $u \in U$, then $\mathcal{F}$ is called the null multi-L-set, denoted by $\tilde{0}$. If $\mathcal{F}(m, u) = 1$ for all $m \in M$ and $u \in U$, then $\mathcal{F}$ is called the absolute multi-L-set, denoted by $\tilde{1}$.
(3) Let $\mathcal{F} \in M\mathcal{L}(U)$. Define $\mathcal{F}^c \in M\mathcal{L}(U)$ as follows: $\mathcal{F}^c(m, u) = (\mathcal{F}(m, u))'$ (for all $m \in M$ and $u \in U$), then we call $\mathcal{F}^c$ the complement of $\mathcal{F}$, where $' : L \longrightarrow L$ is the order reversing revolution on L.

Definition 4 Let $\mathcal{F}$ and $\mathcal{G}$ be two multi-L-sets on U, we define the following relations and operations:
(1) $\mathcal{F} \le \mathcal{G}$ iff $\mathcal{F}(m, u) \le \mathcal{G}(m, u)$ for every $m \in M$ and $u \in U$.
(2) $\mathcal{F} = \mathcal{G}$ iff $\mathcal{F}(m, u) = \mathcal{G}(m, u)$ for every $m \in M$ and $u \in U$.
(3) $\mathcal{F} \vee \mathcal{G}$ is a mapping $\mathcal{F} \vee \mathcal{G} : \mathcal{M} \longrightarrow \mathcal{L}(\mathcal{U})$ defined by $\mathcal{F} \vee \mathcal{G} = \mathcal{F}(m, u) \vee \mathcal{G}(m, u)$ for every $m \in M$ and $u \in U$.
(4) $\mathcal{F} \wedge \mathcal{G}$ is a mapping $\mathcal{F} \wedge \mathcal{G} : \mathcal{M} \longrightarrow \mathcal{L}(\mathcal{U})$ defined by $\mathcal{F} \wedge \mathcal{G} = \mathcal{F}(m, u) \wedge \mathcal{G}(m, u)$ for every $m \in M$ and $u \in U$.

Lemma 1 *Let $\mathcal{F}$ and $\mathcal{G}$ be two multi-L-sets on U, then*
(1) $(\mathcal{F} \vee \mathcal{G})^c = \mathcal{F}^c \wedge \mathcal{G}^c$;
(2) $(\mathcal{F} \wedge \mathcal{G})^c = \mathcal{F}^c \vee \mathcal{G}^c$.

Proof 1 (1) For any $m \in M$ and $u \in U$, $(\mathcal{F} \vee \mathcal{G})^c(m, u) = (\mathcal{F} \vee \mathcal{G})(m, u)' = (\mathcal{F}(m, u) \vee \mathcal{G}(m, u))' = \mathcal{F}(m, u)' \wedge \mathcal{G}(m, u)' = \mathcal{F}^c(m, u) \wedge \mathcal{G}^c(m, u)$, thus $(\mathcal{F} \vee \mathcal{G})^c = \mathcal{F}^c \wedge \mathcal{G}^c$.
(2) Similar to the proof of (1). □

Definition 5 A multi-L-soft set over U is a pair $(\mathfrak{F}, A)$, where $A \subseteq E$, amd $\mathfrak{F}$ is a mapping given by $\mathfrak{F} : A \longrightarrow M\mathcal{L}(U)$, that is for all $e \in A$, $\mathfrak{F}(e)$ is a multi-L-set, we denote $\mathfrak{F}(e)(m)(u)$ by $\mathfrak{F}(e, m, u)$ for short (for every $e \in A$, $m \in M$ and $u \in U$).

For a multi-L^I-soft set, we call it a multi-interval-valued fuzzy soft set.

Example 1 Now let us consider a set of houses $U = \{h_1, h_2, h_3\}$, A is the set of parameters of houses, that is $A = \{e_1, e_2, e_3\} = \{\text{beautiful, cheap, in good location}\}$. $M = \{m_1, m_2\}$ is a set of two observers. Every observer comments on the three houses, he may give an interval for every house's membership grade for every parameter, thus we get a multi-interval-valued fuzzy soft set $(\mathfrak{F}, A)$. It can be tabled as follows:

$\mathfrak{F}(e_1)$	h_1	h_2	h_3
m_1	[0.7, 0.9]	[0.6, 0.7]	[0.3, 0.4]
m_2	[0.4, 0.6]	[0.8, 1.0]	[0.5, 0.7]

$\mathfrak{F}(e_2)$	h_1	h_2	h_3
m_1	[0.8, 0.9]	[0.3, 0.5]	[0.7, 0.8]
m_2	[0.1, 0.3]	[0.5, 0.6]	[0.2, 0.5]

$\mathfrak{F}(e_3)$	h_1	h_2	h_3
m_1	[0.6, 0.8]	[0.3, 0.6]	[0.1, 0.4]
m_2	[0.2, 0.3]	[0.3, 0.4]	[0.6, 0.9]

Definition 6 Let $A, B \subseteq E$, $(\mathfrak{F}, A)$ and $(\mathfrak{G}, B)$ be two multi-L-soft sets. $(\mathfrak{F}, A)$ is said to be a multi-L-soft subset of $(\mathfrak{G}, B)$ if
(1) $A \subseteq B$;
(2) $\forall e \in A$, $\mathfrak{F}(e) \leq \mathfrak{G}(e)$.
In this case, we write $(\mathfrak{F}, A)\widetilde{\subseteq}(\mathfrak{G}, B)$.

Remark 2 Let $(\mathfrak{F}, A)$ be a multi-L-soft set, if $\mathfrak{F}(e) = \tilde{0}$ $(\forall e \in A)$, then it is called a null multi-L-soft set, and it is denoted by $\widetilde{0}_A$. If $\mathfrak{F}(e) = \tilde{1}$ $(\forall e \in A)$, then it is called a absolute multi-L-soft set, and it is denoted by $\widetilde{1}_A$. It is easy to see that for any multi-L-soft set $(\mathfrak{F}, A)$, $\widetilde{0}_A\widetilde{\subseteq}(\mathfrak{F}, A)\widetilde{\subseteq}\widetilde{1}_A$.

Definition 7 Let $A, B \subseteq E$, $(\mathfrak{F}, A)$ and $(\mathfrak{G}, B)$ be two multi-L-soft sets. If both $(\mathfrak{F}, A)\widetilde{\subseteq}(\mathfrak{G}, B)$ and $(\mathfrak{G}, B)\widetilde{\subseteq}(\mathfrak{F}, A)$, then $(\mathfrak{F}, A)$ and $(\mathfrak{G}, B)$ are called multi-L-soft equal. In this case, we write $(\mathfrak{F}, A)\widetilde{=}(\mathfrak{G}, B)$.

It is easy to see that, $(\mathfrak{F}, A)\widetilde{=}(\mathfrak{G}, B)$ iff $A = B$, and for every $e \in A = B$, $\mathfrak{F}(e) = \mathfrak{G}(e)$.

Definition 8 Let $(\mathfrak{F}, A)$ be a multi-L-soft set, the complement of $(\mathfrak{F}, A)$, which is denote by $(\mathfrak{F}, A)^c$, is defined by $(\mathfrak{F}, A)^c = (\mathfrak{F}^c, A)$, where $\mathfrak{F}^c(e) = (\mathfrak{F}(e))^c$ $(\forall e \in A)$.

Remark 3 (1) $((\mathfrak{F}, A)^c)^c = (\mathfrak{F}, A)$;
(2) $(\widetilde{0}_A)^c = \widetilde{1}_A$, $(\widetilde{1}_A)^c = \widetilde{0}_A$,

Definition 9 Let $(\mathfrak{F}, A)$ and $(\mathfrak{G}, B)$ be two multi-L-soft sets, $(\mathfrak{F}, A)$ AND $(\mathfrak{G}, B)$, which is denoted by $(\mathfrak{F}, A) \wedge (\mathfrak{G}, B)$, is a multi-$L$-soft set $(\mathfrak{H}, C)$, where $C = A \times B$, and $\mathfrak{H} : C \longrightarrow M\mathcal{L}(U)$ is as follows: $\mathfrak{H}(a, b) = \mathfrak{F}(a) \wedge \mathfrak{G}(b)$ (for all $a \in A$ and $b \in B$).

Definition 10 Let $(\mathfrak{F}, A)$ and $(\mathfrak{G}, B)$ be two multi-L-soft sets, $(\mathfrak{F}, A)$ OR $(\mathfrak{G}, B)$, which is denoted by $(\mathfrak{F}, A) \vee (\mathfrak{G}, B)$, is a multi-$L$-soft set $(\mathfrak{I}, C)$, where $C = A \times B$, and $\mathfrak{I} : C \longrightarrow M\mathcal{L}(U)$ is as follows: $\mathfrak{I}(a, b) = \mathfrak{F}(a) \vee \mathfrak{G}(b)$ (for all $a \in A$ and $b \in B$).

Theorem 1 *Let $(\mathfrak{F}, A)$ and $(\mathfrak{G}, B)$ be two multi-L-soft sets. Then*
(1) $[(\mathfrak{F}, A) \wedge (\mathfrak{G}, B)]^c = (\mathfrak{F}, A)^c \vee (\mathfrak{G}, B)^c$;
(2) $[(\mathfrak{F}, A) \vee (\mathfrak{G}, B)]^c = (\mathfrak{F}, A)^c \wedge (\mathfrak{G}, B)^c$.

Proof 2 (1) Suppose that $(\mathfrak{F}, A) \wedge (\mathfrak{G}, B) = (\mathfrak{H}, A \times B)$, therefore $[(\mathfrak{F}, A) \wedge (\mathfrak{G}, B)]^c = (\mathfrak{H}, A \times B)^c = (\mathfrak{H}^c, A \times B)$. For every $(a, b) \in A \times B$, $\mathfrak{H}^c(a, b) = (\mathfrak{H}(a, b))^c = (\mathfrak{H}(a, b))^c = (\mathfrak{F}(a) \wedge \mathfrak{G}(b))^c = \mathfrak{F}(a)^c \vee \mathfrak{G}(b)^c$. On the other hand, suppose $(\mathfrak{F}, A)^c \vee (\mathfrak{G}, B)^c = (\mathfrak{F}^c, A) \vee (\mathfrak{G}^c, B) = (\mathfrak{I}, A \times B)$, then for each $(a, b) \in A \times B$, $\mathfrak{I}(a, b) = \mathfrak{F}^c(a) \vee \mathfrak{G}^c(b) = \mathfrak{H}^c(a, b)$. Thus $(\mathfrak{H}, A \times B) = (\mathfrak{I}, A \times B)$, that is $[(\mathfrak{F}, A) \wedge (\mathfrak{G}, B)]^c = (\mathfrak{F}, A)^c \vee (\mathfrak{G}, B)^c$.

(2) Similar to the proof of (1). □

Definition 11 Let $(\mathfrak{F}, A)$ and $(\mathfrak{G}, B)$ be two multi-L-soft sets. The union of them is a multi-L-soft set $(\mathfrak{O}, C)$, where $C = A \cup B$, and for any $e \in C$,

$$\mathfrak{O}(e) = \begin{cases} \mathfrak{F}(e), & \text{if } e \in A - B; \\ \mathfrak{G}(e), & \text{if } e \in B - A; \\ \mathfrak{F}(e) \vee \mathfrak{G}(e), & \text{if } e \in A \cap B. \end{cases}$$

It is denoted by $(\mathfrak{F}, A)\widetilde{\cup}(\mathfrak{G}, B) = (\mathfrak{O}, C)$.

Definition 12 Let $(\mathfrak{F}, A)$ and $(\mathfrak{G}, B)$ be two multi-L-soft sets. The intersection of them is a multi-L-soft set $(\mathfrak{K}, D)$, where $D = A \cap B$, and for any $e \in D$, $\mathfrak{K}(e) = \mathfrak{F}(e) \wedge \mathfrak{G}(e)$. It is denoted by $(\mathfrak{F}, A)\widetilde{\cap}(\mathfrak{G}, B) = (\mathfrak{K}, D)$.

One can easily get the following results:

Theorem 2 *Let* $(\mathfrak{F}, A)$ *and* $(\mathfrak{G}, B)$ *be two multi-L-soft sets. Then*
(1) $(\mathfrak{F}, A)\tilde{\cup}(\mathfrak{F}, A) = (\mathfrak{F}, A)$, $(\mathfrak{F}, A)\tilde{\cap}(\mathfrak{F}, A) = (\mathfrak{F}, A)$
(2) $(\mathfrak{F}, A)\tilde{\cup}\tilde{0}_A = (\mathfrak{F}, A)$, $(\mathfrak{F}, A)\tilde{\cap}\tilde{0}_A = \tilde{0}_A$
(3) $(\mathfrak{F}, A)\tilde{\cup}\tilde{1}_A = \tilde{1}_A$, $(\mathfrak{F}, A)\tilde{\cap}\tilde{1}_A = (\mathfrak{F}, A)$
(4) $(\mathfrak{F}, A)\tilde{\cup}(\mathfrak{G}, B) = (\mathfrak{G}, B)\tilde{\cup}(\mathfrak{F}, A)$, $(\mathfrak{F}, A)\tilde{\cap}(\mathfrak{G}, B) = (\mathfrak{G}, B)\tilde{\cap}(\mathfrak{F}, A)$

Theorem 3 *Let* $(\mathfrak{F}, A)$ *and* $(\mathfrak{G}, B)$ *be two multi-L-soft sets. Then*
(1) $(\mathfrak{F}, A)^c\tilde{\cap}(\mathfrak{G}, B)^c\tilde{\subseteq}((\mathfrak{F}, A)\tilde{\cup}(\mathfrak{G}, B))^c\tilde{\subseteq}(\mathfrak{F}, A)^c\tilde{\cup}(\mathfrak{G}, B)^c$;
(2) $(\mathfrak{F}, A)^c\tilde{\cap}(\mathfrak{G}, B)^c\tilde{\subseteq}((\mathfrak{F}, A)\tilde{\cap}(\mathfrak{G}, B))^c\tilde{\subseteq}(\mathfrak{F}, A)^c\tilde{\cup}(\mathfrak{G}, B)^c$.

If the set of parameters of the two multi-L-soft sets $(\mathfrak{F}, A)$ and $(\mathfrak{G}, B)$ are equal, then we can get the following result:

Theorem 4 *Let* $(\mathfrak{F}, A)$ *and* $(\mathfrak{G}, A)$ *be two multi-L-soft sets. Then*
(1) $((\mathfrak{F}, A)\tilde{\cup}(\mathfrak{G}, A))^c = (\mathfrak{F}, A)^c\tilde{\cap}(\mathfrak{G}, A)^c$;
(2) $((\mathfrak{F}, A)\tilde{\cap}(\mathfrak{G}, A))^c = (\mathfrak{F}, A)^c\tilde{\cup}(\mathfrak{G}, A)^c$.

Proof 3 (1) Let $((\mathfrak{F}, A)\tilde{\cup}(\mathfrak{G}, A))^c = (\mathfrak{H}_1, C_1)$, that is $C_1 = A \cup A = A$, and for each $e \in C_1, \mathfrak{H}_1(e) = (\mathfrak{H}_1^c(e))^c = (\mathfrak{F}(e) \vee \mathfrak{G}(e))^c = \mathfrak{F}(e)^c \wedge \mathfrak{G}(e)^c$. Let $(\mathfrak{F}, A)^c\tilde{\cap}(\mathfrak{G}, A)^c = (\mathfrak{H}_2, C_2)$, that is $C_2 = A \cap A = A$, and for each $e \in C_2$, $\mathfrak{H}_2(e) = \mathfrak{F}^c(e) \wedge \mathfrak{G}^c(e)$. Thus, we get $C_1 = C_2$, and for any $e \in C_1$, $\mathfrak{H}_1(e) = \mathfrak{H}_2(e)$, so $((\mathfrak{F}, A)\tilde{\cup}(\mathfrak{G}, A))^c = (\mathfrak{F}, A)^c\tilde{\cap}(\mathfrak{G}, A)^c$.

(2) Let $((\mathfrak{F}, A)\tilde{\cap}(\mathfrak{G}, A))^c = (\mathfrak{H}_3, C_3)$, that is $C_3 = A \cap A = A$, and for every $e \in C_3$, $\mathfrak{H}_3(e) = (\mathfrak{H}_3^c(e))^c = (\mathfrak{F}(e) \wedge \mathfrak{G}(e))^c = \mathfrak{F}(e)^c \vee \mathfrak{G}(e)^c$. Let $(\mathfrak{F}, A)^c\tilde{\cup}(\mathfrak{G}, A)^c = (\mathfrak{H}_4, C_4)$, that is $C_4 = A \cap A = A$, and for each $e \in C_4$, $\mathfrak{H}_4(e) = \mathfrak{F}(e)^c \vee \mathfrak{G}(e)^c$. Thus, we get $C_3 = C_4$, and for each $e \in C_3$, $\mathfrak{H}_3(e) = \mathfrak{H}_4(e)$, so $((\mathfrak{F}, A)\tilde{\cap}(\mathfrak{G}, A))^c = (\mathfrak{F}, A)^c\tilde{\cup}(\mathfrak{G}, A)^c$. □

3 Application of Multi-L-soft Set in Decision Making

In this section, we present an application of multi-L-soft set in solving decision making problem, where L is the lattice consisted of all intervals of $[0, 1]$.

3.1 A Decision Making Problem

Let us consider the example of choosing houses. Let $U = \{h_1, h_2, h_3, h_4, h_5\}$ be the set of houses which one considers, $A = \{e_1, e_2, e_3, e_4\}$ = {beautiful, cheap, in good location, wooden} be the set of parameters, $M = \{m_1, m_2, m_3\}$ the set of observers, Every observer comment on the five houses, he may give an interval for the degree of house h_i having the parameter e_j ($\forall i = 1, 2, \ldots, 5$, $j = 1, 2, \ldots, 4$), thus we get a multi-interval-valued fuzzy soft set $(\mathfrak{F}, A)$. We table it as follows:

$\mathfrak{F}(e_1)$	h_1	h_2	h_3	h_4	h_5
m_1	[0.27, 0.33]	[0.27, 0.28]	[0.41, 0.45]	[0.35, 0.42]	[0.28, 0.29]
m_2	[0.46, 0.48]	[0.20, 0.37]	[0.19, 0.29]	[0.07, 0.22]	[0.18, 0.35]
m_3	[0.04, 0.13]	[0.18, 0.33]	[0.13, 0.21]	[0.18, 0.35]	[0.24, 0.33]

$\mathfrak{F}(e_1)$	h_1	h_2	h_3	h_4	h_5
m_1	[0.46, 0.48]	[0.04, 0.10]	[0.42, 0.60]	[0.21, 0.40]	[0.33, 0.36]
m_2	[0.09, 0.17]	[0.39, 0.52]	[0.18, 0.24]	[0.31, 0.44]	[0.21 0.28]
m_3	[0.25, 0.33]	[0.36, 0.37]	[0.09, 0.16]	[0.02, 0.14]	[0.25, 0.35]

$\mathfrak{F}(e_1)$	h_1	h_2	h_3	h_4	h_5
m_1	[0.35, 0.51]	[0.19, 0.26]	[0.35, 0.52]	[0.16, 0.22]	[0.32, 0.40]
m_2	[0.18, 0.21]	[0.34, 0.46]	[0.28, 0.33]	[0.31, 0.46]	[0.09, 0.23]
m_3	[0.24, 0.26]	[0.11, 0.22]	[0.02, 0.10]	[0.18, 0.28]	[0.20, 0.30]

$\mathfrak{F}(e_1)$	h_1	h_2	h_3	h_4	h_5
m_1	[0.27, 0.37]	[0.30, 0.36]	[0.03, 0.18]	[0.31, 0.44]	[0.31, 0.43]
m_2	[0.20, 0.29]	[0.28, 0.37]	[0.12, 0.20]	[0.05, 0.24]	[0.18, 0.30]
m_3	[0.27, 0.33]	[0.21, 0.26]	[0.33, 0.52]	[0.26, 0.32]	[0.05, 0.24]

One would like to choose a house referring to the data given above.

3.2 A Comparison Algorithm

In this section, we use the comparison algorithm to solve the decision making problem above. For the sake of universality of the algorithm, we consider more general case as follows.

Let $U = \{u_1, u_2, \ldots, u_n\}, M = \{m_1, m_2, \ldots, m_k\}, (\mathfrak{F}, A)$ a multi-interval-valued fuzzy soft set. For every $e \in A$, $\mathfrak{F}(e)$ can be expressed in the following matrix:

$$\mathfrak{F}(e) = \begin{pmatrix} [\mu_{11}^-, \mu_{11}^+] & [\mu_{12}^-, \mu_{12}^+] & \cdots & [\mu_{1n}^-, \mu_{1n}^+] \\ [\mu_{21}^-, \mu_{21}^+] & [\mu_{22}^-, \mu_{22}^+] & \cdots & [\mu_{2n}^-, \mu_{2n}^+] \\ \vdots & \vdots & \ddots & \vdots \\ [\mu_{k1}^-, \mu_{k1}^+] & [\mu_{k2}^-, \mu_{k2}^+] & \cdots & [\mu_{kn}^-, \mu_{kn}^+] \end{pmatrix}$$

where $[\mu_{ij}^-, \mu_{ij}^+] = \mathfrak{F}(e, m_i, u_j)$ is an interval contained in [0, 1].

Suppose that $\omega(e) = \{\omega_1, \omega_2, \ldots, \omega_k\}^T$ ($\sum_{i=1}^k \omega_i = 1$) is the relative weight correspond to $m_1, m_2, \ldots, m_k$ of e, we define an induced interval-valued fuzzy set $\mu_{\mathfrak{F}(e)}$ with respect to e as follows:

$$\mu_{\mathfrak{F}(e)} = \begin{pmatrix} [\sum_{i=1}^{k} \omega_i \mu_{i1}^-, \sum_{i=1}^{k} \omega_i \mu_{i1}^+] \\ [\sum_{i=1}^{k} \omega_i \mu_{i2}^-, \sum_{i=1}^{k} \omega_i \mu_{i2}^+] \\ \vdots \\ [\sum_{i=1}^{k} \omega_i \mu_{in}^-, \sum_{i=1}^{k} \omega_i \mu_{in}^+] \end{pmatrix} \triangleq \begin{pmatrix} [\mu_1^-, \mu_1^+] \\ [\mu_2^-, \mu_2^+] \\ \vdots \\ [\mu_n^-, \mu_n^+] \end{pmatrix}$$

Thus, by using this method, we change a multi-interval-valued fuzzy soft set to an interval-valued fuzzy soft set. Therefore, we can make a decision by the following algorithm.

1. Input the multi-interval-valued fuzzy soft set $(\mathfrak{F}, A)$. Input the relative weight $\omega(e_i)$ of every parameter $e_i \in A$.

2. Change $(\mathfrak{F}, A)$ into the normalized multi-interval-valued fuzzy soft set, we still denoted it by $(\mathfrak{F}, A)$, that is, if there exists some $u_j \in U$ such that $\sum_{i=1}^{k} \mu_{ij}^+ = l > 1$, then we change $([\mu_{1j}^-, \mu_{1j}^+], [\mu_{2j}^-, \mu_{2j}^+], \ldots, [\mu_{kj}^-, \mu_{kj}^+])$ to $\frac{1}{l}([\mu_{1j}^-, \mu_{1j}^+], [\mu_{2j}^-, \mu_{2j}^+], \ldots, [\mu_{kj}^-, \mu_{kj}^+])$.

3. Compute the induced interval-valued fuzzy soft set $(\mathcal{F}, A)$, where $\mathcal{F}(e) = \mu_{\mathfrak{F}(e)}$.

4. For every $u_i \in U$, compute value of the score function $S_i : A \longrightarrow [0, 1]$, where $S_i(e) = \frac{\mu_i^- + \mu_i^+}{2}$ $(\forall i = 1, 2, \ldots n)$.

5. Use t_{ij} to denote the number of the parameters $e \in A$ which satisfies $S_i(e) \geq S_j(e)$, for every $i, j = 1, 2, \ldots n$.

6. Compute $r_i = \sum_{j=1}^{n} t_{ij}$, $d_j = \sum_{i=1}^{n} t_{ij}$ for every pair of u_i and u_j in U, where $i, j = 1, 2, \ldots, n$.

7. Compute the final score $s_i = r_i - d_i$ of u_i, $i = 1, 2, \ldots, n$.

8. The optimal decision is to select u_j such that $s_j = \max_i s_i$.

9. If j in 8 are more than one, then any one of them can be chosen.

Use the algorithm above, we compute the decision making problem in Sect. 3.1 as follows.

We impose the following weights of the three observers for the parameters in A: for parameter "beautiful", $\omega(e_1) = \{0.5, 0.3, 0.2\}$, for parameter "cheap", $\omega(e_2) = \{0.4, 0.2, 0.4\}$, for parameter "in good location", $\omega(e_3) = \{0.1, 0.6, 0.3\}$, for parameter "wooden", $\omega(e_4) = \{0.2, 0.5, 0.3\}$. Thus we have an induced interval valued fuzzy soft set $(\mathcal{F}, A)$ which is as Table 1.

Table 1 $(\mathcal{F}, A)$

$\mathcal{F}$	e_1 $\omega(e_1)$	e_2 $\omega(e_2)$	e_3 $\omega(e_3)$	e_4 $\omega(e_4)$
h_1	[0.281, 0.335]	[0.302, 0.358]	[0.215, 0.255]	[0.235, 0.318]
h_2	[0.231, 0.317]	[0.238, 0.292]	[0.256, 0.368]	[0.263, 0.335]
h_3	[0.288, 0.354]	[0.240, 0.352]	[0.209, 0.280]	[0.165, 0.292]
h_4	[0.232, 0.346]	[0.154, 0.304]	[0.256, 0.382]	[0.165, 0.304]
h_5	[0.242, 0.316]	[0.274, 0.340]	[0.146, 0.268]	[0.167, 0.308]

Thus the sore $S_i(e_j)$ of every house u_i of every parameters $e_j \in A$ is as follows, and then we can get the following table of all t_{ij}s and all r_is, d_is and s_is:

$S_i(e_j)$	e_1	e_2	e_3	e_4
h_1	0.308	0.330	0.235	0.276
h_2	0.274	0.265	0.312	0.299
h_3	0.321	0.296	0.245	0.229
h_4	0.289	0.229	0.319	0.235
h_5	0.279	0.307	0.207	0.238

t_{ij}	h_1	h_2	h_3	h_4	h_5
h_1	4	2	2	3	4
h_2	2	4	2	2	2
h_3	2	2	4	2	2
h_4	1	2	2	4	2
h_5	0	2	2	2	4

	r_i	d_i	s_i
h_1	15	9	6
h_2	12	12	0
h_3	12	12	0
h_4	11	13	-2
h_5	10	14	-4

As can be seen in the table above, h_1 is the best choice.

3.3 A Threshold Value Algorithm

Feng et al. [14] presented an approach to fuzzy soft set based decision making problems by using level soft sets, and this new method can be successfully applied to some decision making problems that can not be solved by using the method in [12]. In [23], Yang et al. used Feng's algorithm to solve a decision making problem which is based on multi-fuzzy soft set. In the following, we will revise Yang's algorithm so that it can be used to solve some decision making problems based on multi-interval-valued fuzzy soft set.

Let $U = \{u_1, u_2, \ldots, u_n\}$, $M = \{m_1, m_2, \ldots, m_k\}$, $(\mathfrak{F}, A)$ is the same multi-interval-valued fuzzy soft set as in Sect. 3.2. By using the relative weight $\omega(e) = \{\omega_1, \omega_2, \ldots, \omega_k\}^T$ correspond to $m_1, m_2, \ldots, m_k$ of e $(\sum_{i=1}^{k} \omega_i = 1)$ we defined an induced interval-valued fuzzy set $\mu_{\mathfrak{F}(e)}$ with respect to e, then we can make a decision by the following algorithm.

1. Input the multi-interval-valued fuzzy soft set $(\mathfrak{F}, A)$. Input the relative weight $\omega(e_i)$ of every parameter $e_i \in A$.

2. Change $(\mathfrak{F}, A)$ into the normalized multi-interval-valued fuzzy soft set, we still denoted it by $(\mathfrak{F}, A)$, that is, if there exists some $u_j \in U$ such that $\sum_{i=1}^{k} \mu_{ij}^{+} = l > 1$,

then we change $([\mu_{1j}^-, \mu_{1j}^+], [\mu_{2j}^-, \mu_{2j}^+], \ldots, [\mu_{kj}^-, \mu_{kj}^+])$ to $\frac{1}{l}([\mu_{1j}^-, \mu_{1j}^+], [\mu_{2j}^-, \mu_{2j}^+], \ldots, [\mu_{kj}^-, \mu_{kj}^+])$.

3. Compute the induced interval-valued fuzzy soft set $(\mathcal{F}, A)$, where $\mathcal{F}(e) = \mu_{\mathfrak{F}(e)}$.

4. Input a threshold interval-valued fuzzy set $\lambda : A \longrightarrow L^I$ (or give a threshold interval $t = [t^-, t^+] \subseteq [0, 1]$; or choose the mid-level decision rule) for decision making.

5. Compute the level triple-valued fuzzy soft set $L((\mathfrak{F}, A), \lambda)$ of the $(\mathfrak{F}, A)$ with respect to the threshold interval-valued fuzzy set λ (or the t-level soft set $L((\mathfrak{F}, A), t)$; or the mid-level triple-valued soft set $L((\mathfrak{F}, A), \text{mid})$).

6. Present the level triple-valued fuzzy soft set $L((\mathfrak{F}, A), \lambda)$ (or $L((\mathfrak{F}, A), t)$; or $L((\mathfrak{F}, A), \text{mid})$; or $L((\mathfrak{F}, A), \max)$) in tabular form and compute the choice value c_i of u_i $(\forall i = 1, 2, \ldots, n)$.

7. The optimal decision is to select u_j such that $c_j = \max_i c_i$.

8. If j in 7 are more than one, then any one can be chosen.

Use the algorithm above, we compute the decision making problem in Sect. 3.1 as follows.

We also impose the same weights of the three observers for the parameters in A in Sect. 3.2: for parameter "beautiful", $\omega(e_1) = \{0.5, 0.3, 0.2\}$, for parameter "cheap", $\omega(e_2) = \{0.4, 0.2, 0.4\}$, for parameter "in good location", $\omega(e_3) = \{0.1, 0.6, 0.3\}$, for parameter "wooden", $\omega(e_4) = \{0.2, 0.5, 0.3\}$. Thus we have an induced interval valued fuzzy soft set $(\mathcal{F}, A)$ which is the same as with Table 1.

Then we choose the mid-level decision rule, the mid-threshold of $(\mathcal{F}, A)$ is an interval valued fuzzy set $(\text{mid}_{\mathcal{F}}, A)$

$\text{mid}_{(\mathcal{F},A)}$	e_1	e_2	e_3	e_4
	[0.2548, 0.3336]	[0.2416, 0.3292]	[0.2164, 0.3106]	[0.1990, 0.3114]

For every $i = 1, 2, \ldots, 4$, and every $j = 1, 2, \ldots, 5$, denote $\mathcal{F}(e_i, h_j) = [\mu_{ij}^-, \mu_{ij}^+]$, and $\text{mid}_{\mathcal{F}}(e_i) = [r_i^-, r_i^+]$, if both $\mu_{ij}^- \geq r_i^-$ and $\mu_{ij}^+ \geq r_i^+$, then h_j gets a "2", if only one of $\mu_{ij}^- \geq r_i^-$ and $\mu_{ij}^+ \geq r_i^+$ is valid, then h_j gets an "1", otherwise, h_j gets a "0". Then we can present the level triple-valued fuzzy soft set $L((\mathfrak{F}, A), \text{mid})$ of the induced interval-valued fuzzy soft set $(\mathcal{F}, A)$ as follows:

U	e_1	e_2	e_3	e_4	Choice value c_j
h_1	2	2	0	2	6
h_2	0	0	2	2	4
h_3	2	1	0	0	3
h_4	1	0	2	0	3
h_5	0	2	0	0	2

Thus one can choose h_1 as the best choice.

4 Concluding Remarks

In this paper, the concept of multi-fuzzy soft set is generated to multi-L-soft set, which is a combination of multi-L-set and soft set. Then some relations between multi-L-soft sets and some operations on multi-L-soft sets are defined, furthermore, some properties of the operations are discussed. Finally, an illustrative example is used to show the validity of multi-interval-valued fuzzy soft set in decision making problems.

Acknowledgments This research is supported by the Special Fund of the Shaanxi Provincial Education Department (grant Nr. 2013JK0567, Nr. 16JK1373), and Shaanxi Provincial Natural Science Foundation (grant Nr. 2014JM1018).

References

1. Molodtsov, D.: Soft set theory-first results. Comput. Math. Appl. **37**, 19–31 (1999)
2. Molodtsov, D.: The Theory of Soft Sets. URSS Publisher, Moscow (2004). (in Russian)
3. Maji, P.K., Bismas, R., Roy, A.R.: Soft set theory. Comput. Math. Appl. **45**, 555–562 (2003)
4. Pei, D., Miao, D.: From soft sets to information systems. In: IEEE International Conference on Granular Computing. pp. 617–621 (2005)
5. Ali, M.I., Feng, F., Liu, X.Y., Win, W.K., Shabir, M.: On some new operations in soft set theory. Comput. Math. Appl. **57**, 1547–1553 (2009)
6. Aktas, H., Cağman, N.: Soft sets and soft groups. Inf. Sci. **177**, 2726–2735 (2007)
7. Jun, Y.B.: Soft BCK/BCI-algrbras. Comput. Math. Appl. **56**, 1408–1413 (2008)
8. Jun, Y.B., Park, C.H.: Applications of soft sets in ideal theory of BCK/BCI-algrbras. Inf. Sci. **178**, 2466–2475 (2008)
9. Jun, Y.B., Lee, K.J., Park, C.H.: Soft set theory applied to ideals in d-algebras. Comput. Math. Appl. **57**, 367–378 (2009)
10. Park, C.H., Jun, Y.B., *Öztürk*, M.A.: Soft WS-algebras. Korean Math. Soc. Commun. **23**, 313–324 (2008)
11. Feng, F., Jun, Y.B., Zhao, X.: Soft semirings. Comput. Math. Appl. **56**, 2621–2628 (2008)
12. Roy, A.R., Maji, P.K.: A fuzzy soft set theoretic approach to decision making problems. J. Comput. Appl. Math. **203**, 412–418 (2007)
13. Kong, Z., Gao, L., Wang, L.: Comment on A fuzzy soft set theoretic approach to decision making problems. J. Comput. Appl. Math. **223**, 540–542 (2008)
14. Feng, F., Jun, Y.B.: An adjustable approach to fuzzy soft set based decision making. J. Comput. Appl. Math. **234**, 10–20 (2010)
15. Feng, F., Li, Y.M., Fotea, V.L.: Application of level soft sets in decision making based on interval-valued fuzzy soft set. Comput. Math. Appl. **60**, 1756–1767 (2010)
16. Maji, P.K., Roy, A.R., Bismas, R.: Soft set theory. Comput. Math. Appl. **44**, 1077–1083 (2002)
17. Pawlak, Z.: Rough Sets: Theoretical Aspects of Reasoning About Data. Kluwer Academic, Boston. MA (1991)
18. Chen, D., Tsang, E.C.C., Yeung, D.S., Wang, X.: The parametrization reduction of soft sets and its applications. Comput. Math. Appl. **49**, 757–763 (2005)
19. Herawan, T., Mustafa, M.D.: On Multi-soft Sets Construction in Information Systems. In: Huang, D.-S., Jo, K.-H., Lee, H.-H., Kang, H.J., Bevilacqua, V. (eds.) ICIC 2009. LNCS (LNAI), vol. 5755, pp. 101–110. Springer, Heidelberg (2009)
20. Maji, P.K., Biswas, R., Roy, A.R.: Fuzzy soft sets. Comput. Math. Appl. **9**, 589–602 (2001)

21. Sebastian, S., Ramakrishnan, T.V.: Multi-fuzzy sets: an extension of fuzzy sets. Comput. Math. Appl. **3**, 35–43 (2011)
22. Yang, Y., Tan, X., Meng, C.C.: The multi-fuzzy soft set and its application in decision making. Comput. Math. Appl. **37**, 4915–4923 (2013)
23. Yang, X.B., Lin, T.Y., Yang, J.Y., Li, Y., Yu, D.Y.: Combination of interval-valued fuzzy set and soft set. Comput. Math. Appl. **58**, 521–527 (2009)
24. Jun, Y.B., Yang, X.B.: A note on the paper Combination of interval-valued fuzzy set and soft set [Comput. Math. Appl. 58 (2009) 521–527]. Comput. Math. Appl. **61**, 1468–1470 (2011)

On Some New Generalizations of Yager's Implications

Feng-Xia Zhang and Xing-Fang Zhang

Abstract In this paper, Yager's implications are generalized, and two classes of implications, called generalized f- and g-implications, respectively, are introduced. Basic properties of these implications are discussed in detail.

Keywords Fuzzy implications · Additive generators · f-generators · g-generators

1 Introduction

In fuzzy logic, one of the main operators is fuzzy implications. The reason lies in two aspects: firstly, the management of fuzzy conditionals of the type "If p, then q" with p and q fuzzy statements is by these operators [17]; secondly, fuzzy implications are often used to perform inferences. The two main inference rules are modus ponens and modus tollens, used to perform forward and backward inferences, respectively. Thus fuzzy implication operators play an important role in fuzzy control and approximate reasoning [2, 7, 10, 11, 20, 22]. Furthermore, fuzzy implications are also very useful in many other fields such as fuzzy relational equations, fuzzy mathematical morphology, image processing, fuzzy DI-subsethood measures, data mining, computing with words and so on (see, for instance, [16, 17]).

To represent imprecise knowledge, many different models are proposed to perform fuzzy implications. By now, the most well-studied classes of fuzzy implications are those obtained from t-norms and t-conorms, viz., (S, N)-, R-, and QL-implications (see, for instance, [2, 4, 5, 21]). Moreover, these types have been extended because t-norms and t-conorms are special kinds of aggregation operators. Indeed, not only copulas, quasi-copulas, but also conjunctors in general [8], representable aggregation functions [6], uninorms and many other aggregation functions have been used for

F.-X. Zhang (✉) · X.-F. Zhang
School of Mathematics, Liaocheng University, Liaocheng 252059, Shandong, China
e-mail: fengxiazhang@163.com

X.-F. Zhang
e-mail: zhangxingfang2005@126.com

T.-H. Fan et al. (eds.), *Quantitative Logic and Soft Computing 2016*,
Advances in Intelligent Systems and Computing 510,
DOI 10.1007/978-3-319-46206-6_54

this purpose [1, 3, 9, 15, 18, 19]. Recently, Yager [23] has defined two classes of implications called f- and g-implications, respectively, by using the additive generators of continuous Archimedean t-norms and t-conorms (shortly, f- and g-generators) and has done an extensive analysis of the impact of these implications in approximate reasoning, by introducing concept like strictness of implications and sharpness of inferences, among others. By means of the additive generators of representable uninorms, Massanet and Torrens [16] extended Yager's f- and g-implications, and introduced h-implications. Also by means of g-generators and h-generators, and by introducing the concept of partial-inverse of additive generators, Liu [13, 14] has defined two classes of implications called $(g, \min)$-implications and $(h, \min)$-implications.

In this paper, by using f-generators and g-generators, we introduce two new classes of fuzzy implications, called generalized f- and g-implications which can be regarded as the generalizations of Yager's f-implications and g-implications, respectively. We study some properties of these implications such as left neutrality principle, exchange principle, identity principle, ordering property etc.

The paper is organized as follows. In Sect. 2, we present some notions concerning basic logic connectives employed in the sequel. In Sect. 3, we introduce a new class of fuzzy implications called generalized f-implications and discuss some of their properties. In Sect. 4, we introduce a new class of fuzzy implications called generalized g-implications and discuss their properties. The last section concludes the paper.

2 Preliminaries

In this section, we recall some of the concepts employed in the rest of the paper. Above all, definition of fuzzy implication is given.

Definition 1 ([2, 10, 23]) A function $I : [0, 1]^2 \to [0, 1]$ is called a fuzzy implication if it satisfies, for all $x, x_1, x_2, y, y_1, y_2 \in [0, 1]$, the following conditions:
(I1) If $x_1 \leq x_2$, then $I(x_1, y) \geq I(x_2, y)$;
(I2) If $y_1 \leq y_2$, then $I(x, y_1) \leq I(x, y_2)$;
(I3) $I(0, 0) = 1$;
(I4) $I(1, 1) = 1$;
(I5) $I(1, 0) = 0$.

Definition 2 ([2, 12]) A decreasing function $N : [0, 1] \to [0, 1]$ is called a fuzzy negation, if $N(0) = 1$, $N(1) = 0$. A fuzzy negation N is called
(i) strict, if it is strictly decreasing and continuous;
(ii) strong, if it is an involution, i.e., $N(N(x)) = x$ for all $x \in [0, 1]$.

Definition 3 (Definition 1.4.15 in [2]) Let $I : [0, 1]^2 \to [0, 1]$ be a fuzzy implication. The function $N_I : [0, 1] \to [0, 1]$, defined by $N_I(x) = I(x, 0)$ for any $x \in [0, 1]$, is said to be the natural negation of I.

Definition 4 ([2, 7, 10, 20, 22]) A fuzzy implication I is said to satisfy
(NP) the left neutrality property, if $I(1, y) = y$ for all $y \in [0, 1]$;
(EP) the exchange principle, if $I(x, I(y, z)) = I(y, I(x, z))$ for all $x, y, z \in [0, 1]$;
(IP) the identity principle, if $I(x, x) = 1$ for all $x \in [0, 1]$;
(OP) the order property, if $I(x, y) = 1 \Leftrightarrow x \leq y$ for all $x, y \in [0, 1]$;
(CP(N)) the law of contraposition with respect to a fuzzy negation N, if $I(x, y) = I(N(y), N(x))$ for all $x, y \in [0, 1]$.

Definition 5 ([23]) An f-generator is a function $f : [0, 1] \to [0, 1]\infty]$ that is a strictly decreasing and continuous function with $f(1) = 0$.

A g-generator is a function $g : [0, 1] \to [0, \infty]$ that is a strictly increasing and continuous function with $g(0) = 0$.

Definition 6 ([2, 23]) Let function f be an f-generator. The function $I_f : [0, 1] \to [0, 1]$ defined by

$$I_f(x, y) = f^{-1}(x \cdot f(y)), \quad x, y \in [0, 1],$$

with the understanding that $\infty \cdot 0 = 0$ is called an f-implication.

Definition 7 ([2, 23]) Let function g be a g-generator. The function $I_g : [0, 1] \to [0, 1]$ defined by

$$I_g(x, y) = g^{(-1)}\left(\frac{1}{x} \cdot g(y)\right), \quad x, y \in [0, 1],$$

with the understanding that $\frac{1}{0} = \infty$ and $\infty \cdot 0 = \infty$ is called a g-implication where $g^{(-1)} : [0, \infty] \to [0, 1]$ is the pseudo-inverse function of g given by

$$g^{(-1)}(y) = \begin{cases} 1 & if\, y \geq g(1), \\ g^{-1}(y) & if\, y < g(1). \end{cases}$$

The f-generator (g-generator) can be seen as a continuous additive generator of continuous Archimedean t-norm (t-conorm) [10, 12].

3 Generalization of Yager's f-Implications

In this section, we will give generalization of Yager's f-implications called generalized f-implications and discuss some properties such as left neutrality principle, exchange principle, identity principle, ordering property, the law of contraposition and the continuity of this class of implications.

3.1 Definition and Examples

Definition 8 Let f_1, f_2 be f-generators, the function $I_{f_1,f_2} : [0,1]^2 \to [0,1]$ defined by

$$I_{f_1,f_2}(x,y) = f_2^{(-1)}(x \cdot f_1(y)), \quad x, y \in [0,1] \tag{1}$$

with the understanding that $0 \cdot \infty = 0$, is called generalized f-operation generated from f_1, f_2, where $f_2^{(-1)} : [0,\infty] \to [0,1]$ is the pseudo-inverse function of f_2 given by

$$f_2^{(-1)}(y) = \begin{cases} 0 & if y \geq f_2(0), \\ f_2^{-1}(y) & if y < f_2(0). \end{cases}$$

f_1 is called the inner f-generator while f_2 is called the outer f-generator.

The following theorem gives necessary and sufficient conditions under which I_{f_1,f_2} is a fuzzy implication in the sense of Definition 1.

Theorem 1 *Let f_1, f_2 be f-generators, then I_{f_1,f_2} defined by* (1) *is a fuzzy implication if and only if $f_2(0) \leq f_1(0)$. In this case we call it generalized f-implication generated by f_1, f_2.*

Proof It is easy to see that I_{f_1,f_2} is decreasing in its first variable and increasing in its second one because of the monotonicity of f-generator. Moreover,
$I_{f_1,f_2}(0,0) = f_2^{(-1)}(0 \cdot f_1(0))) = f_2^{(-1)}(0) = 1,$
$I_{f_1,f_2}(1,1) = f_2^{(-1)}(1 \cdot f_1(1))) = f_2^{(-1)}(0) = 1,$
$I_{f_1,f_2}(1,0) = f_2^{(-1)}(1 \cdot f_1(0))) = 0 \Leftrightarrow f_2(0) \leq f_1(0).$ □

Remark 1 (i) If $f_1 = f_2$, then I_{f_1,f_2} defined by (1) is Yager's f-implication. So we can say that the class of generalized f-implications is a generalization of f-implications.

(ii) Given two f-generators f_1, f_2, there exists at least one generalized f-implication generated by f_1 and f_2. If $f_1(0) = f_2(0)$, we get two generalized f-implications, viz., I_{f_1,f_2} and I_{f_2,f_1}.

(iii) Formula (1) can also be written in the following form without explicitly using the pseudo-inverse of f_2:

$$I_{f_1,f_2}(x,y) = f_2^{-1}(\min(x \cdot f_1(y), f_2(0))), \quad x, y \in [0,1].$$

Example 1 (i) Let us consider the Frank's class of additive generators given by

$$f_1^s(x) = -\ln\left(\frac{s^x - 1}{s - 1}\right),$$

where $s > 0, s \neq 1$, as the inner f-generators, and the continuous additive generator of the product t-norm T_p given by $f_2(x) = -\ln x$ as the outer f-generator, then the corresponding generalized f-implication, for every s, is given by

$$I_{f_1^s, f_2}(x, y) = \left(\frac{s^y - 1}{s - 1}\right)^x, \quad x, y \in [0, 1].$$

(ii) If we take the Yager's class of additive generators, viz., $f_1^\lambda(x) = (1 - x)^\lambda$, where $\lambda \in (0, \infty)$, as the inner f-generators, and the additive generator of the Łukasieweicz t-norm T_L given by $f_2(x) = \frac{1}{2}(1 - x)$ as the outer f-generator, then the corresponding generalized f-implication, for every $\lambda \in (0, \infty)$, is given by

$$I_{f_1^\lambda, f_2}(x, y) = \begin{cases} 0 & \text{if } y \le 1 - \left(\frac{1}{2x}\right)^{\frac{1}{\lambda}}, \\ 1 - 2x(1 - y)^\lambda & \text{otherwise,} \end{cases} \quad x, y \in [0, 1].$$

3.2 Properties of Generalized f-Implications

It is well known that for Yager's f-implication I_f determined by $I_f(x, y) = f^{(-1)}(x \cdot f(y))$ for all $x, y \in [0, 1]$, the f-generators are unique up to a positive multiplicative constant (Theorem 3.1.4 in [2]). This implies that for an f-generator f with $f(0) < \infty$, we can define another f-generator f_0 by $f_0(x) = \frac{f(x)}{f(0)}$ such that $I_f = I_{f_0}$. In other words, it is enough to consider only decreasing generator for which $f(0) = \infty$ or $f(0) = 1$. As for generalized f-implications, we have the following result:

Theorem 2 *Let $f_{11}, f_{12}, f_{21}, f_{22}$ be any four f-generators with*

$$f_{11}(0) \ge f_{12}(0), f_{21}(0) \ge f_{22}(0),$$

then the following statements are equivalent:
(i) $I_{f_{11}, f_{12}} = I_{f_{21}, f_{22}}$*;*
(ii) *There exists a constant $c \in (0, \infty)$ such that $f_{21} = c \cdot f_{11}$, $f_{22} = c \cdot f_{12}$.*

Proposition 1 *Let f_1, f_2 be any two f-generators with $f_2(0) \le f_1(0)$.*
(i) *If $f_1(0) = \infty$, then the natural negation of I_{f_1, f_2} is the Gödel negation*

$$N_{D_1}(x) = \begin{cases} 1 & \text{if } x=0, \\ 0 & \text{if } x>0, \end{cases} \quad x \in [0, 1]$$

which is non-continuous;
(ii) *If $f_1(0) < \infty$, then the natural negation of I_{f_1, f_2} is a strict negation if and only if $f_1(0) = f_2(0)$. In this case,*

$$N_{I_{f_1, f_2}}(x) = f_2^{-1}(f_2(0) \cdot x).$$

(iii) *The natural negation of I_{f_1, f_2} is a strong negation if and only if the following holds:*
(a) $f_1(0) = f_2(0) < \infty$*;*

(b) *The function defined by*

$$f_{20}(x) = \frac{f_2(x)}{f_2(0)}, \quad x \in [0, 1]$$

is a strong negation.

Theorem 3 *Let f_1, f_2 be any two f-generators with $f_1(0) \geq f_2(0)$, then the generalized f-implication defined by* (1) *has the following properties:*
(i) *I_{f_1,f_2} satisfies (NP) if and only if $f_1 = f_2$;*
(ii) *If $f_2(0) = \infty$, then I_{f_1,f_2} satisfies (EP) if and only if there exists a constant $c \in (0, \infty)$ such that $f_2 = c \cdot f_1$;*
(iii) *If $f_1(0) = f_2(0) < \infty$, then I_{f_1,f_2} satisfies (EP) if and only if $f_1 = f_2$;*
(iv) *$I_{f_1,f_2}(x, x) = 1$ if and only if $x = 0$ or $x = 1$, i.e., I_{f_1,f_2} does not satisfy (IP);*
(v) *$I_{f_1,f_2}(x, y) = 1$ if and only if $x = 0$ or $y = 1$, i.e., I_{f_1,f_2} does not satisfy (OP);*
(vi) *If $f_1(0) = \infty$, then I_{f_1,f_2} does not satisfy (CP) with respect to any fuzzy negation N;*
(vii) *If $f_1(0) = f_2(0) < \infty$, then I_{f_1,f_2} satisfies (CP) with respect to N if and only if $N(x) = \frac{f_1(x)}{f_1(0)}$ is a strong negation.*

Remark 2 (i) Note that if $f_1(0) > f_2(0)$, the generalized f-implication I_{f_1,f_2} may not satisfy (EP). For instance, in Example 1 (ii), if we take $\lambda = 1$ and $x = 0.6$, $y = 0.3$, $z = 0.1$, then $I_{f_1^1,f_2}(0.6, I_{f_1^1,f_2}(0, 3, 0.1)) = 0.352$, while $I_{f_1^1,f_2}(0.3, I_{f_1^1,f_2}(0, 6, 0.1)) = 0.4$.

(ii) In the case $f_2(0) < f_1(0) < \infty$, if $N = f_{10}$ is a strong negation, then it is obvious that I_{f_1,f_2} satisfies (CP) with respect to N. But it should be noted that such a problem still remains: whether there exists a fuzzy negation N such that I_{f_1,f_2} satisfies (CP(N)) if f_{10} is not a strong negation?

From Proposition 1 and Theorem 3, the following result is immediate:

Corollary 1 *Let f_1, f_2 be any two f-generators with $f_1(0) = f_2(0) < \infty$, then the generalized f-implication I_{f_1,f_2} satisfies (CP) with respect to its natural negation $N_{I_{f_1,f_2}}$ if and only if $f_1 = f_2$ and the function f_{10} given by*

$$f_{10}(x) = \frac{f_1(x)}{f_1(0)}, \quad x \in [0, 1]$$

is a strong negation.

The following proposition discusses the continuity of generalized f-implications.

Proposition 2 *Let f_1, f_2 be f-generators that satisfy $f_2(0) \leq f_1(0)$, then*
(i) *I_{f_1,f_2} is continuous if and only if $f_1(0) < \infty$;*
(ii) *I_{f_1,f_2} is continuous except at the point $(0, 0)$ if and only if $f_1(0) = \infty$.*

4 Generalization of Yager's g-Implications

In this section, we will give a generalization of Yager's g-implications called generalized g-implications. Our discussion in this section will mirror the approach taken in the previous section.

4.1 Definition and Examples

Definition 9 Let g_1, g_2 be g-generators, the function $I_{g_1,g_2} : [0,1]^2 \to [0,1]$ defined by

$$I_{g_1,g_2}(x,y) = g_2^{(-1)}\left(\frac{1}{x} \cdot g_1(y)\right), \quad x, y \in [0,1] \tag{2}$$

with the understanding that $\frac{1}{0} = \infty$ and $0 \cdot \infty = \infty \cdot 0 = \infty$, is called a generalized g-operation generated by g_1, g_2, where $g_2^{(-1)} : [0,\infty] \to [0,1]$ is the pseudo-inverse function of g_2. g_1 is called the inner g-generator while g_2 is called the outer g-generator.

The following theorem gives the necessary and sufficient conditions under which I_{g_1,g_2} is a fuzzy implication in the sense of Definition 1.

Theorem 4 *Let g_1, g_2 be g-generators, then I_{g_1,g_2} defined by (2) is a fuzzy implication if and only if $g_2(1) \leq g_1(1)$. In this case we call it generalized g-implication generated from g_1, g_2.*

Proof It is easy to see that I_{g_1,g_2} is decreasing in its first variable and increasing in its second one because of the monotonicity of g-generator. Moreover,
$I_{g_1,g_2}(0,0) = g_2^{(-1)}(\infty \cdot g_1(0))) = g_2^{(-1)}(\infty) = 1,$
$I_{g_1,g_2}(1,1) = g_2^{(-1)}(1 \cdot g_1(1)) = 1 \Leftrightarrow g_2(1) \leq g_1(1),$
$I_{g_1,g_2}(1,0) = g_2^{(-1)}(1 \cdot g_1(0))) = 0.$ □

Remark 3 (i) If $g_1 = g_2$, then I_{g_1,g_2} defined by (2) is Yager's g-implication. So we can say that the class of generalized g-implications is a generalization of g-implications.

(ii) Given two g-generators g_1, g_2, there exists at least one generalized g-implication generated by g_1 and g_2. If $g_1(1) = g_2(1)$, we get two generalized g-implications: I_{g_1,g_2} and I_{g_2,g_1}.

(iii) Formula (2) can also be written in the following form without explicitly using the pseudo-inverse of g_2:

$$I_{g_1,g_2}(x,y) = g_2^{-1}\left(\min\left(\frac{1}{x} \cdot g_1(y), g_2(1)\right)\right), \quad x, y \in [0,1]$$

Example 2 (i) Let us consider the Frank's class of additive generators given by

$$g_1^s(x) = -\ln\left(\frac{s^{1-x}-1}{s-1}\right),$$

where $s > 0, s \neq 1$, as the inner g-generators, and the continuous additive generator of the Łukasiewicz t-conorm S_L given by $g_2(x) = x$ as the outer g-generator, then the corresponding generalized g-implication, for every s, is given by

$$I_{g_1^s,g_2}(x,y) = \begin{cases} 1 & \text{if } y \geq 1 - \log_s(1+(s-1)e^{-x}), \\ -\frac{1}{x}\cdot\ln\left(\frac{s^{1-y}-1}{s-1}\right) & \text{otherwise,} \end{cases} \quad x, y \in [0,1].$$

(ii) If we take the g-generators $g_1(x) = -\ln(1-x)$, and $g_2(x) = -\frac{1}{\ln x}$, then $g_1(1) = g_2(1) = \infty$, in this case, we get two generalized g-implications generated by g_1 and g_2.

$$I_{g_1,g_2}(x,y) = e^{\frac{x}{\ln(1-y)}}, \quad x, y \in [0,1].$$

$$I_{g_2,g_1}(x,y) = 1 - e^{\frac{1}{x\ln y}}, \quad x, y \in [0,1].$$

4.2 Properties of Generalized g-Implications

It is well known that for Yager's g-implication I_g determined by $I_g(x,y) = g^{(-1)}(\frac{1}{x}\cdot g(y))$ for all $x, y \in [0,1]$, the g-generators are unique up to a positive multiplicative constant (Theorem 3.2.5 in [2]). This implies that for a g-generator g with $g(1) < \infty$, we can define another g-generator g_0 by $g_0(x) = \frac{g(x)}{g(1)}$ such that $I_g = I_{g_0}$. In other words, it is enough to consider only decreasing generator for which $g(1) = \infty$ or $g(1) = 1$. As for generalized g-implications, we have the following result:

Theorem 5 *Let $g_{11}, g_{12}, g_{21}, g_{22}$ be any four g-generators with $g_{11}(1) \geq g_{12}(1)$ and $g_{21}(1) \geq g_{22}(1)$, then the following two statements are equivalent:*
(i) $I_{g_{11},g_{12}} = I_{g_{21},g_{22}}$;
(ii) There exists a constant $c \in (0,\infty)$ such that $g_{22} = c \cdot g_{12}$ and $g_{21}(x) = c \cdot g_{11}(x)$ for all $x \in [0, g_{11}^{-1} \circ g_{12}(1)]$.

Proposition 3 *Let g_1, g_2 be any two generators with $g_1(1) \geq g_2(1)$, then the natural negation of I_{g_1,g_2} is the Gödel negation $N_{D_1}(x)$, which is non-continuous.*

Theorem 6 *Let g_1, g_2 be any two g-generators with $g_1(1) \geq g_2(1)$, then the generalized g-implication defined by* (2) *has the following properties:*

(i) I_{g_1,g_2} satisfies (NP) if and only if $g_1 = g_2$;
(ii) If $g_2(1) = \infty$, then I_{g_1,g_2} satisfies (EP) if and only if there exists a constant $c \in (0,\infty)$ such that $g_2 = c \cdot g_1$;
(iii) If $g_2(1) < \infty$, and there exists a constant $c \in (0,\infty)$ such that $g_2 = c \cdot g_1$, then I_{g_1,g_2} satisfies (EP);

(iv) I_{g_1,g_2} satisfy (IP) if and only if $g_2(1) < \infty$ and $g_1(x) \geq g_2(1) \cdot x$ for all $x \in [0, 1]$;
(v) *If* $g_2(1) = \infty$, then $I_{g_1,g_2}(x, y) = 1$ if and only if $x = 0$ *or* $y = 1$, *i.e.*, I_{g_1,g_2} does not satisfy (OP) when $g_2(1) = \infty$;
(vi) I_{g_1,g_2} does not satisfy (CP) with respect to any fuzzy negation N.

It should be noted that such a problem still remains: In Theorem 6 (iii), does the converse implication holds? In other words, is there a constant $c \in (0, \infty)$ such that $g_2 = c \cdot g_1$ when I_{g_1,g_2} satisfies (EP)?

Theorem 7 *Let g_1, g_2 be any two g-generators with $g_1(1) \geq g_2(1)$, then the following statements are equivalent:*
(i) I_{g_1,g_2} satisfies (OP);
(ii) $g_2(1) < \infty$ and $g_1(x) = g_2(1) \cdot x$ for all $x \in [0, 1]$;
(iii) I_{g_1,g_2} has the form

$$I_{g_1,g_2}(x, y) = \begin{cases} 1 & \text{if } x \leq y, \\ g_2^{-1}\left(g_2(1) \cdot \frac{y}{x}\right) & \text{if } x > y, \end{cases} \quad x, \ y \in [0, 1]. \tag{3}$$

Proposition 4 *Let g_1, g_2 be any two generators with $g_1(1) \geq g_2(1)$, then I_{g_1,g_2} is continuous except at the point* $(0, 0)$.

5 Conclusions

In this paper, we have introduced two new generalizations of Yager's implications called generalized f-implications and g-implications, respectively. We showed that the inner f-generators and the outer f-generators of generalized f-implications are unique up to a positive multiplicative constant. However, for generalized g-implication, only the outer g-generators are unique up to a positive multiplicative constant and the inner g-generators are unique on a subinterval of $[0, 1]$ up to a positive multiplicative constant. We have discussed some properties such as left neutrality principle, exchange principle, identity principle, ordering property, law of contraposition, and given some conditions under which these properties hold. We have also discussed the continuity of f-implications and g-implications. This work will bring benefit for approximate reasoning, fuzzy control and other application areas.

Acknowledgments This work is supported by the National Natural Science Foundation of China (No.11471152).

References

1. Aguiló, I., Suñer, J., Torrens, J.: A characterization of residual implications derived from left-continuous uninorms. Inf. Sci. **180**, 3992–4005 (2010)
2. Baczyński, M., Jayaram, B.: Fuzzy Implications. Studies in Fuzziness and Soft Computing, vol. 231. Springer, Berlin (2008)
3. Baczyński, M., Jayaram, B.: (U, N)-implications and their characterizations. Fuzzy Sets Syst. **160**, 2049–2062 (2009)
4. Baczyński, M., Jayaram, B.: QL-implications: some properties and intersections. Fuzzy Sets Syst. **161**, 158–188 (2010)
5. Baczyński, M., Jayaram, B.: On the characterization of (S, N)-implications. Fuzzy Sets Syst. **158**, 1713–1727 (2007)
6. Carbonell, M., Torrens, J.: Continuous R-implications generated from representable aggregation functions. Fuzzy Sets Syst. **161**, 2276–2289 (2010)
7. Dubois, D., Prade, H.: Fuzzy sets in approximate reasoning, Part 1: Inference with possibility distributions. Fuzzy Sets Syst. **40**, 143–202 (1991)
8. Durante, F., Klement, E., Mesiar, R., Sempi, C.: Conjunctors and their residual implicators: characterizations and construction methods. Mediter. J. Math. **4**, 343–356 (2007)
9. De Baets, B., Fodor, J.C.: Residual operators of uninorms. Soft Comput. **3**, 89–100 (1999)
10. Fodor, J., Roubens, M.: Fuzzy Preference Modelling and Multicriteria Decision Support. Kluwer, Dordrecht (1994)
11. Gottwald, S.: A Treatise on Many-Valued Logics. Research Studies Press, Baldock (2001)
12. Klement, E.P., Mesiar, R., Pap, E.: Triangular Norms. Kluwer Academic Publishers, Dordrecht (2000)
13. Liu, H.W.: On a new class of implications: (g, min)-implications and several classical tautologies. Int. J. Uncertain. Fuzziness Knowl. Syst. **20**(1), 1–20 (2012)
14. Liu, H.W.: A new class of fuzzy implications derived from generalized h-generators. Fuzzy Sets Syst. **224**, 63–92 (2013)
15. Mas, M., Monserrat, M., Torrens, J.: Two types of implications derived from uninorms. Fuzzy Sets Syst. **158**, 2612–2626 (2007)
16. Massanet, S., Torrens, J.: On a new class of fuzzy implications: h-Implications and generalizations. Inf. Sci. **181**, 2111–2127 (2011)
17. Mas, M., Monserrat, M., Torrens, J., Trillas, E.: A survey on fuzzy implication functions. IEEE Trans. Fuzzy Syst. **15**(6), 1107–1121 (2007)
18. Ruiz-Aguilera, D., Torrens, J.: S- and R-implications from uninorms continuous in $(0, 1)^2$ and their distributivity over uninorms. Fuzzy Sets Syst. **160**, 832–852 (2009)
19. Ruiz-Aguilera, D., Torrens, J.: Distributivity of residual implications over conjunctive and disjunctive uninorms. Fuzzy Sets Syst. **158**, 23–37 (2007)
20. Smets, P., Magrez, P.: Implication in fuzzy logic. Int. J. Approx. Reason. **1**, 327–347 (1987)
21. Shi, Y., Van Gasse, B., Ruan, D., Kerre, E.E.: On the first place antitonicity in QL-implications. Fuzzy Sets Syst. **159**, 2988–3013 (2008)
22. Trillas, E., Valverde, L.: On implication and indistinguishability in the setting of fuzzy logic. Manag. Decis. Support Syst. Using Fuzzy Sets Possibility Theory, pp. 198–212. Verlag TUV-Rhineland, Cologne (1985)
23. Yager, R.R.: On some new classes of implication operators and their role in approximate reasoning. Inf. Sci. **167**, 193–216 (2004)

Part VII
Artificial Intelligence and Soft Computing

An Online Mall CRM Model Based on Data Mining

Dao-Lei Liang and Hai-Bo Chen

Abstract For the past few years, some e-commerce enterprises such as Taobao, Jingdong had experienced rapid development, and some electronic malls had accumulated a large number of customer information and transaction data. In order to find the value customers and to retain customers, it was very necessary to use data mining technology for client segmentation. Combined with RFM and Analytic Hierarchy Process, a data mining model is used for Customer Relation Management in this paper. Under the guidance of domain experts, the disadvantage of data excessive fitting is overcome in the model for an online mall. The result is put into practice and the commercial effect is tangible.

Keywords Data mining · CRM · Online Mall

1 Introduction

Since the beginning of twenty-first Century, as China has vigorously promoted informationization, e-commerce applications have developed rapidly. Many companies and individuals are scrambling to set up e-malls and e-stores, the development trend is more and more obvious. Data mining [1] technology applied in customer relationship management [2] can help decision-maker to understand the customers' needs, analyze customer behavior and evaluate customer value, and then targeted market strategies can be made to carry out accurate marketing, raise enterprise's profit and improve the core competence.

D.-L. Liang (✉) · H.-B. Chen
School of Science, ZheJiang Sci-Tech University, HangZhou 310018, China
e-mail: iangdaoleil@sina.com

T.-H. Fan et al. (eds.), *Quantitative Logic and Soft Computing 2016*,
Advances in Intelligent Systems and Computing 510,
DOI 10.1007/978-3-319-46206-6_55

1.1 Related Work

Many scholars had done the relevant research in the area of data mining and customer relationship management for market precision. Barry Keating [3] pointed out that in market data mining can help to make decisions, and help people to identify the market objects, realize cross-selling and sequence-selling, etc. Candace Gunnarsson, Mary M. Walker, Kenneth Swann [4] used the case of data mining of customer churn in newspaper industry to illustrate that people should share the experiences and lessons in the process of mining, and only in this way can promote the wide application of data-driven decisions. Coskun Samli, Terrance L. Pohlen1 and Nenad Bozovic [5] stated that Data mining can provide strong support for market segmentation decisions. And based on the results of data mining, people can evaluate products which can be adjusted to meet the needs of market. In 2005, by using business intelligence, and using the existed customer behavior to predict the stock customers' needs and purchasing behavior, NTT DoCoMo of Japan let its customer churn rate decrease from 1.01 % to 0.77 %, and can save about $84 million in revenue for the company annually.

1.2 Main Contributions of this Paper

In this paper, we study electronic store customer relationship management based on data mining, which can help to understand customer needs, analyze customer behavior and assess customer value, and then make marketing strategies pertinently to carry out accurate marketing. The CRM can greatly enhance the relevance and effectiveness of marketing, and save marketing costs and increase profits as well. This is one development trend of modern enterprise management.

2 Theory and Methodology

RFM [6] is used to maximize revenue of the existing customers. Customers can be segmented by three indexes named Recently, Frequency and Monetary. RFM is a smart and useful model widely used to customer. The Analytical Hierarchy Process [7] was a systematic, hierarchical analysis method proposed by American Thomas L. Saaty, and is an effective method to integrate data and information factors of a complex decision-making system. APH, simulating the process of decision making, can analyze multi-factors system especially social system according to a combination of qualitative and quantitative methods. Taking an online mall as an example, a customer segmentation model is proposed, and the effectiveness is proved.

2.1 Business Background and Business Understanding

The company has several stores from Tianmao, Jingdong, Dangdang, and the goods are shirts, T-shirts, suits, jackets, pants. All store sales was nearly one million per day with a membership of nearly 200,000. We hoped to cluster customers and find the most valuable customers by data mining.

Combined with characters of sale life cycle: early, development, stable and recession stage in customer relationship management and customer segmentation framework, customers would be classified as low-value customers, general customers, important development customers, important keeping customers, important retaining customers.

2.2 Data Understanding and Data Preparation

We vacuumed up data from all stores in 2015, which contains product databases, customer database, browse databases and transaction databases. The fields were sale goods name, sale time, unit price, total price, discounts, goods customers had explored, explored time, unit price and discounts and so on. We cleaned up the data, discarded the missing value and rectified the wrong name, unified renamed and integrated the data in different sale platforms. We transformed some data and added some new properties, such as browsing time, amount of consumption, times of consumption, consumer items, lever of items and so on, get a total of 32000 records and 18 properties.

2.3 The Model of Customer Value Segmentation

Model steps: Firstly we create RFM table based on different values of customers in different life-cycle stage, and use K-means algorithm to get a preliminary cluster scheme, then conduct a customer value segmentation model by AHP and finally adjust the former cluster scheme. The steps of customer value segmentation are shown in Fig. 1.

There were twelve property fields in the model. we select those variables which can truly reflect customer behaviors as a core of grouping. Different from the traditional RFM model, we took L, R, F, M and C as customer segmentation parameters. L, R, F, M and C are respectively recent time of consumption, total consumption, times of consumption, goods, and lever of goods. After transforming, the first five records of data are shown in Table 2.

We analyze five parameters that are L, R, F, M, C based on Pearsons correlation analysis, and the results are shown in Table 2 named index parameter correlation matrix. There are negative correlations both between R and L, and also between F

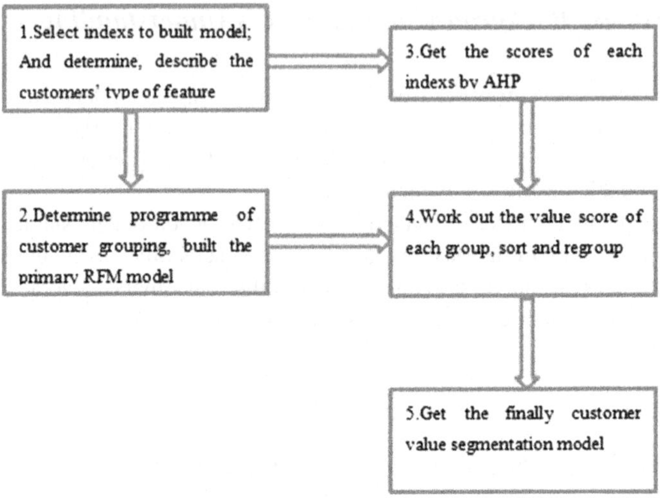

Fig. 1 Steps of customer value segmentation

Table 1 The LRFMC source data

Member_No	L	R	F	M	C
000001	96.0	6.6	3.0	18770	0.658
000002	96.0	3.8	24.0	35087	0.616
000003	95.8	6.6	9.0	20660	0.522
000004	91.6	1.0	12.0	23071	0.511
000005	73.8	3.17	3.0	2897	0.954

Table 2 Index parameter correlation matrix

	L	R	F	M	C
L	1.000	−0.120	0.190	0.173	0.080
R	−0.120	1.000	−0.405	−0.370	−0.021
F	0.190	−0.405	1.000	0.850	0.139
M	0.173	−0.370	0.850	1.000	0.108
C	0.080	−0.021	0.139	0.108	1.000

and M, which is in accordance with the actual situation. In terms of values of the correlation, only correlation coefficient of F and M is 0.85, so their correlation is the lager. However, in the model, It is observed that the frequency and the amount of consumption presented the customer loyalty to the store. At the same time, segmentation parameters of the model are more persuasive than just a consumer gear or a single index separately. As a result, according to practical significance, we selected F and M.

Table 3 Customer-type characteristics

The type of customers	L	F or M	C	R
DIC	Lower	Lower	Higher	Lower
RIC	Higher	Higher	Higher	Lower
IDC	Higher		Higher	Higher
LVC and GC	Uncertainty	Lower	Lower	Uncertainty

Table 4 Customer-type under experts guidance

Category	L	R	F	M	C	Type
1	1	1	1	1	1	IDC
2	1	1	1	0	1	IDC
3	1	1	1	1	0	IDC
4	1	1	0	1	1	DIC
5	1	0	1	1	1	RIC
6	0	1	1	1	1	DIC, RIC
...						

According to the characteristics of the market, we select five parameters that are L, R, F, M, C, and used K_mean clustering algorithm to divide customers into five groups, which are Low Value Customers, General Customers, Development Important Customers, Retain Important Customers, Important Detention Customers. We got the characters of the five groups of customers, as shown in Table 3 named customer-type characteristics.

2.4 *The Domain Experts Guidance*

The model on the basis of statistical graph, comparing every client basic L, R, F, M, C with the total L, R, F, M, C average value, there are two conditions: greater than(equal to) or less than the average, altogether 32 categories. If the average is greater than or equal to the total average, we mark the value as 1, otherwise, we marked it 0. So the 32 categories of customers can be clustered to 5 types. With the guidance of experts in the field, we give particular client type of each category. Under experts guidance, the customer are clustered and the types are shown in Table 4.

Using AHP (Analytic Hierarchy Process), we get the weight of each target. Combined with the average value of each group of customers, we can calculate the value of each group, and then sort the customers by the points of customers value, thus quantitatively compare the differences of value between each group of customers, and adjust the customer type. represent the customer value of the customers in group j, the computational method is as following

$$C_I^j = W_L C_L^j + W_R C_R^j + W_F C_F^j + W_M C_M^j + W_C C_C^j \tag{1}$$

In formula (1) C_L^j, C_R^j, C_F^j, C_M^j, C_C^j represents respectively the value of L, R, F, M, C(after being standardized) of customers in group j, j = 1,. . .,n represents the category number after segmentation. W_L, W_R, W_F, W_M, W_C respectively represents the weight of L, R, F, M, C. About 13 persons are involved such as department managers in shops, key members in customer department, industry experts, long-term customers. The weight number is made with 9 scales by pair comparison with the importance of such five variables, and the judgment matrix were made, the result is $W_L = 0.04$, $W_R = 0.07$, $W_F = 0.27$, $W_M = 0.23$, $W_C = 0.39$.

2.5 The Data Model Conclusion

Under the guidance of field experts, the former cluster scheme is corrected. From the final chart, the general and low value customers occupies a great proportion, accounting for 57 %. The important customers accounts for 20 percent, which is in accordance with the two-eight market rule. The important development customer increased to 15 percent, which had a good result. The data indicate the importance of customer segmentation in customer relationship management. Excessive costs can be effectively avoided on the low value customers as making marketing plans. With customer segmentation, people can not only save costs but also improve the response rate of market, thus more profits would be brought for the enterprises. The proportion was shown in Fig. 2.

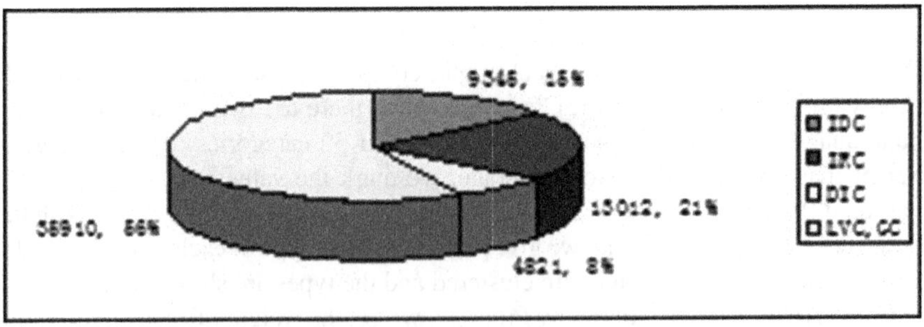

Fig. 2 The proportion of customer value segmentation

2.6 *Marketing Strategies*

After the customer segmentation, a membership system would be put into practice and all customers would be divided into five parts. Customers who had not yet reached a better consumption line would be given a reminder or the promotion in customer service, so that lower-level members became members of a higher level, and enjoy a better discount or promotion service. As to the important customers to develop, the store should encourage this type of customers to increase their consumption in stores and try to strengthen their satisfaction. The important customers to keep were the major contributor customer base, so they should be given preferential access to resources and get the best services. As the changes of the important retain customers are uncertain, these customers should be visited and contacted in particular, in order to extend their life cycle.

3 Generalization and Achievement

By building the customer segmentation model, each member was classified under the membership system and different sales strategies were implemented to different types of customers. Half a year later, without the increase of company's marketing costs, the store turnover increased by 12 %. Nearly 8 % of important customers to develop became important customers to keep. The proportion of important customers to develop and to keep had an increase from 35 % to 47 %. What's more, the customer churn rate decreased by 3 %. The expected results of this model had been achieved.

4 Conclusion

There are many factors affecting customer segmentation. Taking into account the principles of practicality, ease of operation and so on, and combined with advices from some relevant experts, this model considers customer segmentation problem from the perspective of customer behavior and customer value. The model uses LRFMC index, which is improved from the traditional RFM index, as a parameter of customer segmentation, a query model of customer segmentation table is established in the paper. AHP is used to quantify the weight of indices, to correct the customer segmentation table and to provide reliable basis for segmentation. By way of practice, this model has great practical value and research significance.

Acknowledgments This work was supported by the Natural Science Foundation of China (No. 11171308).

References

1. Kantardzic, M.: Data Mining: Concept, Models, Methods, and Algorithms. Wiley-IEEE press, Piscataway (2003)
2. Berry, M.J.A., Linoff, G.: Data Mining Techniques: For Marketing, Sales, and Customer Relationship Management. Wiley Computer Publishing, New York (2004)
3. Keating, B.: Data mining: What is it and how is it used? J. Bus. Forecast. **9**, 33–35 (2008)
4. Candace, L., Mary, M., Walatka, V.: A case study using data mining in the newspaper industry. J. Database Mark. Cust. Strategy Manag. **4**, 271–280 (2007)
5. Coskun, S.A., Terrance, L., Nenad, B.: A review of data mining techniques as they apply to marketing. Generating strategic information to develop market segments. Mark. Rev. **3**, 211–227 (2002)
6. Miglautsch, J.R.: Thoughts on RFM scoring. Database Mark. **1**, 67–72 (1995)
7. Saaty, T.L.: The Analytic Hierarchy Process. McGraw-Hill, New York (1980)

Color Image Segmentation Based on Superpixel and Improved Nyström Algorithm

Jing Zhao, Han-Qiang Liu and Feng Zhao

Abstract Image segmentation methods based on spectral clustering overcome some drawbacks of the so-called central-grouping. Nyström is one of them, which uses only a partial smaller set of samples to replace the whole image pixels. In order to utilize the region information and select the sample set of the image, an image segmentation algorithm based on superpixel and improved Nyström algorithm is proposed in this paper. Firstly, region information is obtained by the superpixel method. Then the similarity measure for regions is constructed. Finally, an interval sample strategy is designed in Nyström and the regions are clustered to create the image segmentation result. With this method, the instability of random sampling is overcome, and the time complexity of color image segmentation is reduced. This method is applied to some images selected from Berkeley and VOC segmentation images. Experimental results show that our method has more advantages than FCM and Nyström algorithm in segmenting images.

Keywords Image segmentation · Nyström algorithm · Superpixel · Interval sampling · Region information

1 Introduction

Image segmentation is an important technique in image, video and computer vision applications. Its purpose is to divide an image into a number of non-overlapping regions [1]. In the past years, many algorithms have been applied to image segmentation, such as clustering methods [2, 3], threshold methods [4], region growing

J. Zhao (✉) · H.-Q. Liu
School of Computer Science, Shaanxi Normal University, Xi'an, People's Republic of China
e-mail: zxh_jzhao@163.com

F. Zhao
School of Telecommunication and Information Engineering, Xi'an University of Posts and Telecommunication, Xi'an, People's Republic of China

T.-H. Fan et al. (eds.), *Quantitative Logic and Soft Computing 2016*, Advances in Intelligent Systems and Computing 510,
DOI 10.1007/978-3-319-46206-6_56

methods [5], and model based methods [6]. The fuzzy clustering method, especially the fuzzy c-means (FCM) algorithm [7], is one of the most popular data clustering algorithms for image segmentation. However, FCM and other clustering algorithms are easily to fall into the drawback of the so-called central-grouping. In order to solve the problem, spectral clustering is proposed, which turns the clustering problem into a graph partitioning problem. But, traditional spectral clustering method faces both the challenges of time and resource. For this reason, Fowlkes et al. [8] proposed a technique based on sampling which solves the problem for a considerably smaller subset of points and later extrapolates to the full set: the method of Nyström approximation [9].

Although Nyström has been achieved some good results in image segmentation, there are still some problems. Firstly, the sampling points are randomly selected in Nyström, so the final segmentation result is unstable. Secondly, compared with the traditional spectral clustering algorithm, the complexity of time has been reduced. However, the image processing is still based on the pixel level, it is still too long to perform the algorithm. Finally, Nyström does not consider the region information of the image, it makes the object in segmentation result incomplete. To overcome instability and reduce the time complexity of Nyström, an improved Nyström algorithm using superpixel, interval sampling and regional similarity is proposed in this paper. With our method, stability of the final result of segmentation is increased, and time complexity is reduced. In the experiments, FCM, Nyström, FCM with superpixel and Nyström with superpixel are chosen as the comparison methods.

2 Background of Nystrom

Generally, spectral clustering algorithms are carried out in three steps: constructing the similarity matrix based on input, calculating the eigenvectors and eigenvalues of the matrix and clustering the first k eigenvectors using the clustering algorithm such as K-means and so on. The Nyström is one of the classical spectral clustering methods. The algorithm is shown as follows:

1. Enter dataset with N samples. Set the number m of random samples. The number of the others remaining sample points is n (n = N − M).
2. Construct matrix:

$$W = \begin{pmatrix} A & B \\ B^T & C \end{pmatrix} \tag{1}$$

In Eq. 1, A represents the similarity among random samples, B represents the similarity from random samples to the rest of samples and C represents the similarity of among the rest of the samples. Because of the number of samples is very small, then C would be huge to calculate. The Nyström method approximates C using [8]

$$\overline{W} = \begin{pmatrix} A & B \\ B^T & B^T A^{-1} B \end{pmatrix} \tag{2}$$

3. Diagonalize the matrix to obtain the eigenvalues and eigenvectors.
4. Get the final image segmentation by clustering the eigenvectors.

3 Color Image Segmentation Based on Superpixel and Improved Nyström Algorithm

In order to utilize the region information and overcome the randomness and higher time complexity of Nyström, an improved Nyström algorithm using superpixel, interval sampling and regional similarity is proposed in this paper.

3.1 Superpixel Algorithm

Superpixel algorithm is used to group pixels into perceptually meaning atomic regions which can be used to replace the rigid structure of the pixel grid [10]. The superpixel is usually constructed by grouping similar pixels, and the methods for superpixel extraction can be broadly classified into two groups: graph-based [11, 12] and gradient-based solutions [10, 13]. They have become key building blocks of many computer vision algorithms, such as top scoring multiclass object segmentation entries to the PASCAL VOC challenge [14], depth estimation, segmentation and so on. The SLIC superpixel method clusters pixels based on their color similarity and proximity in the image plane [10]. The method is used and understood simply, because the only parameter of the algorithm is k, which is the number of superpixel. For color images in the CIELAB color space, two feature vectors, $C_i = [L_i, a_i, b_i]^T$ and $S_i = [x_i, y_i]^T$ are defined to represent the color values and 2D positions of the ith pixel. The procedure can be summarized as follows:

1. Initialize the seeds: distribute the seeds evenly in the image according to the number of superpixel. The SLIC algorithm takes as input as a desired number of approximately equally-sized superpixel K, then for an image with N pixels, the approximate size of each superpixel is N/K. For roughly equally sized superpixels, there would be a superpixel center at every grid interval $S = \sqrt{N/K}$.

2. Re-select the seeds in the 3*3 neighborhood: calculate the gradient values of all pixels in the neighborhood. In order to avoid the seed points to fall into the noise, move them to the neighborhood where gradient is the smallest. The gradient equation is as follows:

$$G(x, y) = [v(x+1, y) - v(x-1, y)]^2 + [v(x, y+1) - v(x, y-1)]^2 \tag{3}$$

3. Redistribute the pixels in 2S*2S neighborhood of each seed. Therefore, instead of using a simple Euclidean norm in the 5D space, we use a distance measure D_{ij} between ith and jth defined as follows:

$$d_{lab} = \sqrt{(l_i - l_j)^2 + (a_i - a_j)^2 + (b_i - b_j)^2} \tag{4}$$

$$d_{xy} = \sqrt{(x_i - x_j)^2 + (y_i - y_j)^2} \tag{5}$$

$$D_{ij} = d_{lab} + \frac{m}{s} d_{xy} \tag{6}$$

where D_{ij} is the sum of the lab distance and the xy plane distance normalized by the grid interval S. A variable m is introduced in D_{ij} allowing us to control the compactness of a superpixel. The greater the value of m, the more spatial proximity is emphasized and the more compact the cluster. This value can be in the range 1~20. This roughly matches the empirical maximum perceptually meaningful CIELAB distance and offers a good balance between color similarity and spatial proximity.

4. Move the cluster center: iteration continues until to each of the clusters are un-changing.

5. Enhance the connectivity, reallocate the smaller and unconnected regions.

The clustering and updating processes are repeated until a predefined number of iteration is achieved. The SLIC algorithm can generate compact and nearly uniform superpixel in the case of very low amount of calculation. Superpixel can capture region information in the image, and greatly reduce the complexity of subsequent image processing tasks.

3.2 Construct the Similarity of Image Region

In the traditional spectral clustering, similarity of structure is based on the pixel level. In order to extract the characteristics of each super-pixel block, a similarity measure method between regions is proposed. The main idea is selecting a window in each region to represent the region. The window can be chosen by any shapes regularly, such as: rectangular, circular, triangular, rhombus and so on. The similarity between two super-pixel blocks can be considered as the similarity between the windows. The advantage of this approach is that: the central region of pixels instead of blocks can avoid the misclassification as much as possible.

In this paper, we select a rectangular window of 3*3 in each of the super blocks. The similarity between the blocks is as in Fig. 1.

As the picture shown, the first step process the is superpixel image by SLIC algorithm. Then, select a rectangular window in each of the superpixel blocks, extract the RGB values of the nine pixels, finally, calculate the similarity. We must ensure

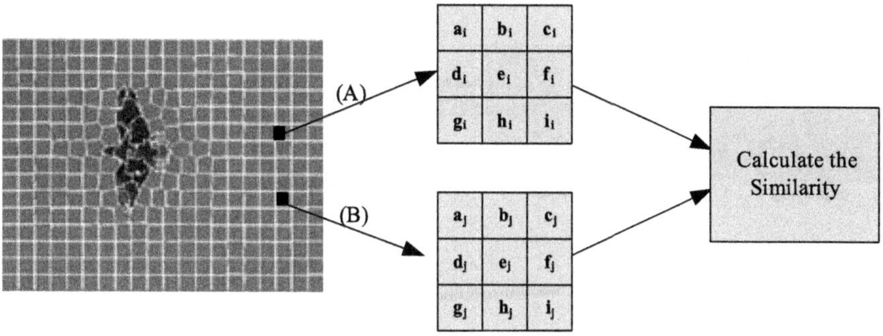

Fig. 1 Construct the similarity of regions, select a rectangular window of 3*3 in each of the superpixel blocks (*A*) and (*B*), extract the feature in the window and calculate the similarity

that the windows fall in the blocks. Only in this way, can we use the windows value to represent the blocks. The similarity between any two superpixels can be measured by the following equation:

$$S_{ij} = e^{-\frac{(a_i - a_j)^2 + (b_i - b_j)^2 + \ldots (i_i - i_j)^2}{2 * \sigma^2}} \tag{7}$$

3.3 *Interval Sampling for Nyström*

In the algorithm of Nyström, the most critical step is selection of the sample points. If the points are selected too unscientifically (for example, all in the background or all in the target), the final result is undesirable. However, a large number of samples is impossible [8]. On the one hand: there is not a standard for non-stop sampling, on the other hand, with the increasing of samplings, the complexity of the algorithm will increase definitely. In the traditional Nyström, sampling points are selected randomly, but the strategy make the algorithm unstable. In order to disperse sampling points in the background and target, proportionating is preferably, and we propose a method of selecting the sampling by interval.

In the superpixel image, all blocks are labeled, we can select points by these labels. In order to get the best result, when the target and background are of similar size, the interval (m) can be select larger (value about 10), otherwise, the value should be chosen as small as possible ($m \geq 1$).

From the previous description, the general procedure can be summarized as follows: taking an image as input, the SLIC algorithm is run to make the input to superpixel image; selecting the sampling points in the blocks by the number of interval; choosing a window of equal size in the center of each superpixel blocks and extracting the RGB value; resegmented using the Nyström method. Image segmentation is then obtained simply by extrapolating the cluster of each mode to the ultra-pixels it represents. A visual representation of the method is presented in Fig. 2.

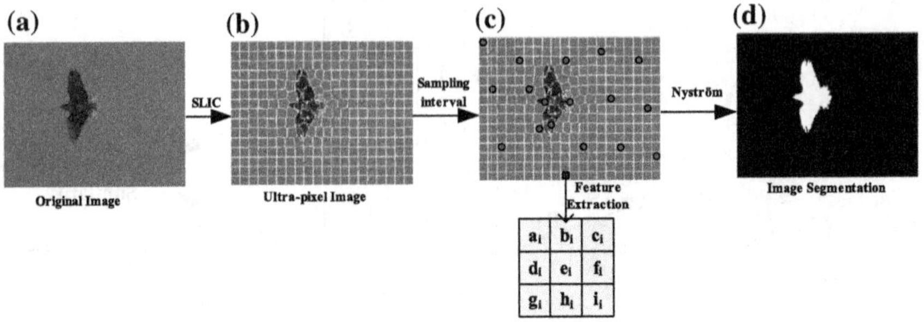

Fig. 2 Diagram showing the key steps for the proposed algorithm: **a** Taking an image as input, **b** The SLIC algorithm is run to make the input to superpixel image. **c** Selecting the sampling points and choosing a window in each of the superpixel blocks. **d** Using the Nyström approximation to get the segmentation image

4 Experimental Result and Analysis

In order to demonstrate the effectiveness of our method, we perform the segmentation experiments on two images (#2009_001466 and #2009_005130) selected from VOC Segmentation Dataset, and one image (238011) from Berkeley Segmentation Dataset. In the experimental section, fuzzy c-means(FCM), Nyström, FCM with super-pixel, Nyström with ultra-pixel are used as the comparison methods. In the Nyström algorithm, we all select 0.1 % points as the sampling randomly in the image. For each image, the initial segmentation is the same as that used by SLIC. In the Nyström with superpixel and our method, the number of sampling points are the same and the size of window in our method are all rectangular of 3*3.

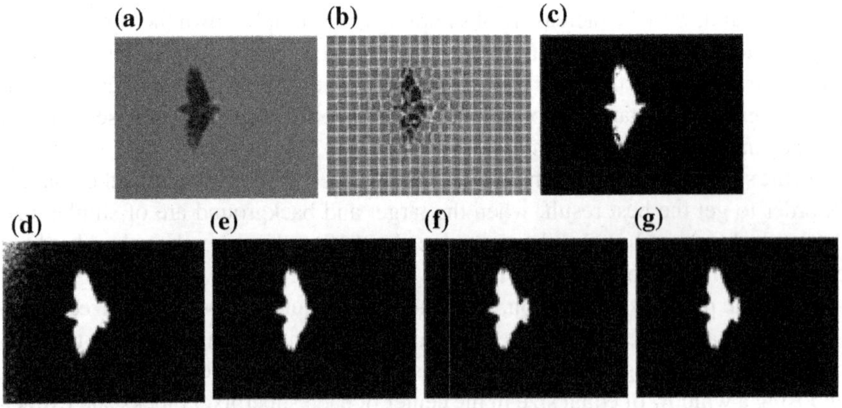

Fig. 3 Segmentation result on the image #2009_001466. In this experiment, the number of *clusters* is 2 and the number of *superpixels* is 300. **a** Taking an image as input. **b** The SLIC algorithm is run to make the input to superpixel image. **c** Selecting the sampling points and choosing a window in each of the superpixel blocks. **d** Using the Nyström approximation to get the segmentation image

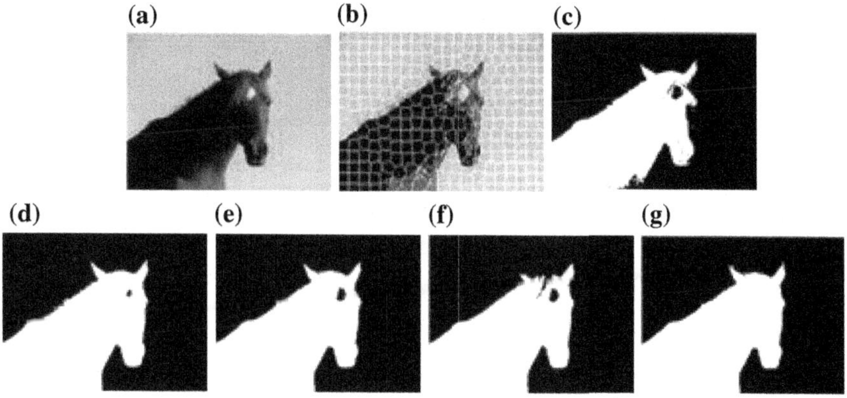

Fig. 4 Segmentation result on the image #2009_005130. In this experiment, the number of *cluster* is 2 and the number of *superpixels* is 305. **a** Taking an image as input. **b** The SLIC algorithm is run to make the input to superpixel image. **c** Selecting the sampling points and choosing a window in each of the superpixel blocks. **d** Using the Nyström approximation to get the segmentation image

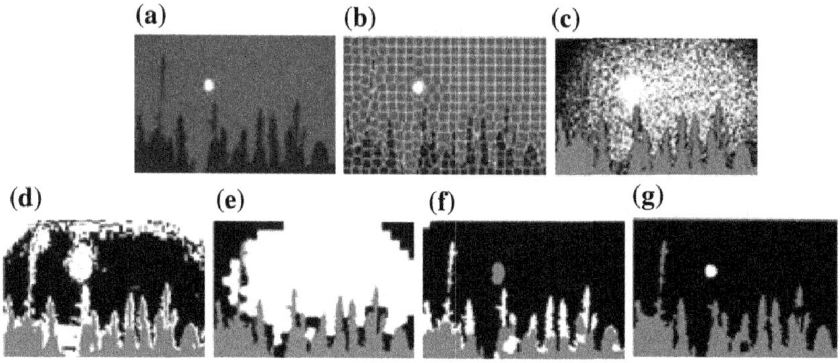

Fig. 5 Segmentation result on the image238011. In this experiment, the number of the *cluster* is 3 and the number of *superpixels* is 365. **a** Taking an image as input. **b** The SLIC algorithm is run to make the input to superpixel image. **c** Selecting the sampling points and choosing a window in each of the superpixel blocks. **d** Using the Nyström approximation to get the segmentation image

The experimental results show that the effect is significant, and our method obtains the best result among the comparison methods. In FCM and Nyström, there always have some points which are misclassified. Because the Nyström is unstable, when introducing of superpixel (f) one has to perform more times. (d) and (f) are the best result of the experiment repeating 10 times (Figs. 3, 4 and 5).

Since the introduction of SLIC, the complexity was reduced, so that, the time of computation was decreased eventually. The computation time for experiments used the algorithm of Nyström, Nyström with SLIC and our method, are shown in Fig. 6.

In Fig. 6, the unit of time is second. In the algorithm of Nyström and Nyström with SLIC, if we want to obtain the desired effect experiment, we have to repeat the

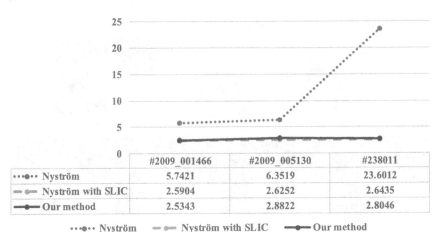

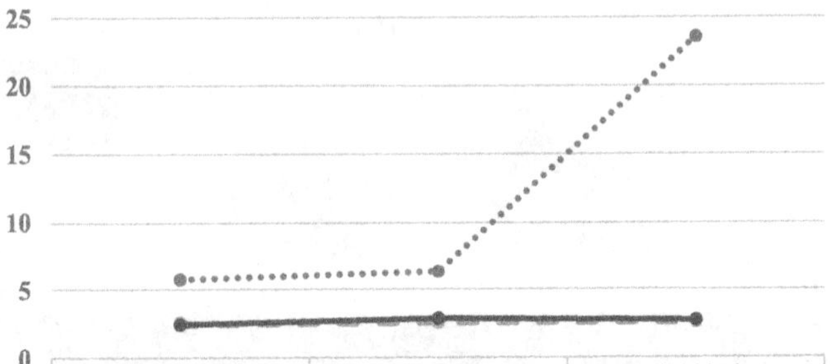

	#2009_001466	#2009_005130	#238011
Nyström	5.7421	6.3519	23.6012
Nyström with SLIC	2.5904	2.6252	2.6435
Our method	2.5343	2.8822	2.8046

Nyström Nyström with SLIC Our method

Fig. 6 Comparison of the time(s) segmented by these three methods on the above-mentioned images

experiment many times. The results of these algorithms are the average of 10 times. From the experiment, the result is obvious that, our method not only can get better results than the other four algorithms but also that the time complexity are the lowest.

5 Conclusions

In the article, a color image segmentation based on superpixel and improved Nyström algorithm is presented. In the proposed method, for color image we can obtain the desired segmentation not only satisfactorily but we also reduce the complexity of time. The improved method overcome the drawback of the Nyström which is unstable caused by sampling points randomly. The experimental results show that our method outperforms FCM, Nyström, FCM with SLIC, and Nyström with SLIC.

In this method, the numbers of sampling points and cluster are set manually. How to automatically set them is our future research.

Acknowledgments This work is supported by the National Natural Science Foundation of China (Grant Nos. 61102095, 61202153 and 61571361), the Science and Technology Plan in Shaanxi Province of China (Grant No. 2014KJXX-72), the Fundamental Research Funds for the Central Universities (Grant No. GK201503063).

References

1. Gonzalez, R.C., Woods, R.E.: Digital Image Processing. Addison-Wesley, Massachusetts (1992)
2. Chen, S.C., Zhang, D.Q.: Robust image segmentation using FCM with spatial constraints based on new kernel-induced distance measure. IEEE Trans., Syst., Man, Cybern., Part B: (Cybernetics) **34**(4), 1907–1916 (2004)
3. Pal, N.R., Pal, S.K.: A review on image segmentation to techniques. Pattern Recognit. **26**(9), 1277–1294 (1993)
4. Cheriet, M., Said, J.N., Suen, C.Y.: A recursive thresholding technique for image segmentation. IEEE Trans. on Image Process. **7**(6), 918–921 (1998)
5. Frank, Y.S., Cheng, S.: Automatic seeded region growing for color image segmentation. Image Vis. Comput. **23**(10), 877–886 (2005)
6. Yang, X., Krishnan, S.M.: Image segmentation using finite mixtures and spatial information. Image Vis. Comput. **22**(9), 735–745 (2004)
7. Bezdek, J.C.: Pattern Recognition with Fuzzy Objective Function Algorithms. Plenum Press, New York (1981)
8. Fowlkes, C., Belongie, S., Chung, F., Malik, J.: Spectral grouping using the Nyström. IEEE Trans. on Pattern Anal. Mach. Intell. **26**(2), 214–225 (2004)
9. Garcia, J.F.G., Andraca, S.E.V.: Region-based approach for the spectra clustering Nyström approximation with an application to burn depth assessment. Mach. Vis. Appl. **26**(2), 353–368 (2015)
10. Achanta, R., Kevin, S.A.S.: SLIC superpixels compared to state-of-the-art superpixel methods. IEEE Trans. on Pattern Anal. Mach. Intell. **34**(11), 2274–2281 (2012)
11. Ayvaci, A., Soatto, S.: Motion Segmentation with Occlusions on the Superpixel Graph. In: 12th International Conference Computer Vision, pp. 727-734. Kyoto (2009)
12. Veksler, O., Bpykov, Y., Mehrani, P.: Superpixels and Supervoxels in an Energy Optimization Framework. In: 11th European Conference on Computer Vision, pp:211-224. Springer, Berlin, Heidelberg (2010)
13. Comanciu, D., Meer, P.: Mean shift: A robust approach toward feature space analysis. IEEE Trans. on Pattern Anal. Mach. Intell. **24**(5), 603–619 (2002)
14. Yang, Y., Hallman, S., Ramanan, D. and Fawlkes, C.: Layered Object Detection for Multi-class Segmentation. In: Computer Vision and Pattern Recognition (CVPR), pp. 3113–3120. (2010)

A Vague Sets Based Vertical Handoff Algorithm in Heterogeneous Networks

Ming-Di Hu and Ming-Ming Tan

Abstract The integration of WLAN and cellular network is the development trend of the next generation mobile communications. With the movement of mobile devices, in order to keep the best quality of service, mobile devices are required to carry seamlessly handoff. The vertical handoff is a key point to solve the seamlessly handoff of heterogeneous networks. A vague based algorithm for vertical handoff in heterogeneous networks is proposed. The algorithm maps parameters to area vague sets, and a new score function on vague set is given, which improves the intelligence of vertical handoff. Our simulation results have proved that the algorithm reduces unnecessary handoff frequencies, and improves the efficiency of vertical handoff.

Keywords VAGUE sets · Heterogeneous networks · Vertical handoff

1 Introduction

With the rapid development of wireless technology and wireless network, today's wireless network is no longer a single technology network, it is a integration based on a variety of different access technologies, named heterogeneous network [1]. The purpose of network convergence is to combine the advantages of different networks and optimize the performance of network, which consequently demands a seamless connection between networks and enable users make real-time handoff [2–5]. The core technology is the seamless connection by handoff between networks, which plays the vital role in the upgrading of wireless communication. The procedure of vertical handoff can be divided into three parts: finding networks, deciding handoff and conducting handoff [3]. The algorithm of determining a handoff is based on a series of parameters, such as available bandwidth, latency, vibration, access cost, error

M.-D. Hu (✉) · M.-M. Tan
Xi'an University of Posts and Telecommunications, Xi'an 710021, China
e-mail: Mendy2013@163.com

M.-M. Tan
e-mail: 444709718@qq.com

T.-H. Fan et al. (eds.), *Quantitative Logic and Soft Computing 2016*,
Advances in Intelligent Systems and Computing 510,
DOI 10.1007/978-3-319-46206-6_57

rate, transmission power, power supply and the preference of users, in other words, the parameters of QoS(Quality of Service) and handoff measurement should be taken into account. The deciding vertical handoff is the most important, namely, mobile terminal determines which network can be connected. Considering the necessity of considering multiple factors, the best option is to adopt the MCDM(—Multiple Criteria Decision Making) [6] method.

Nowdays more and more researches concerning vertical handoff decision continue to boom, from the research on the algorithm of multi-attribute decision-making [6], which is based on simple weighting, ordinal preference and analytic hierarchy process, to the algorithm of artificial intelligence decision based on fuzzy algorithm [7, 8], neural network, genetic algorithm and the composite algorithm of the above mentioned algorithm [3, 9–12]. But in these literatures on the algorithm of artificial intelligence and fuzzy inference, the value of fuzzy subjection degree in fuzzy inference machine is three-valued (low, medium, hight), being far from enough to explain the complexity of the recognition of human brain. Gau et al. presented the concepts of vague sets. Vague set is a generalization of fuzzy set [14]. They used a truth-membership function and false-membership function to characterize the lower bounds on element of membership function. The bounds, one reflection of the elements must have s information, the other is the upper bound of the change of information turbulence. The degree of membership of elements of upper and lower bounds is considered, which make our grasp the information more practical, thus making vague set in information processing more flexible and expressive, more fit the fact that variability and complexity. Therefore, vague set for fuzzy decision rule of multiple objective is endless [15–18], and was used in many aspects [19–21], but vertical handoff algorithm for heterogeneous networks based on the vague set has not yet been seen.

According to the variety and complexity of wireless network parameter data, we gives a new study on vertical handoff algorithm for heterogeneous network based on vague set. First, the parameters is transformed into vague set interval. Secondly, a new scoring function is constructed and its validity is proved, the score function is introduced into the fuzzy inference machine, vague set based algorithm for vertical handoff in heterogeneous networks is given, and in the heterogeneous network with three parameters, namely, signal strength, bandwidth, price three aspects as the impact of switching factors, the simulation results, the simulation of the vertical switching frequency. For commonly used WLAN and UMTS, seamless intelligent switching is concerned, and there have been some mobile terminal businesses using an unidirectional intelligent switching technology. In this paper, the switching of these two heterogeneous networks is simulated. It can be seen in simulated test that the vertical handover algorithm based on vague sets method makes the inference result more intelligentized and less affected by ping-pong effect.

2 Basics

2.1 Vague Sets and Operation

Definition 1 Let X be a space of points (objects), with a generic element of X denoted by x. A vague set V in X is characterized by a truth-membership function t_V and a false-membership function f_V. $t_V(x)$ is a lower bound on the grade of membership of x derived from the evidence for x, and $f_V(x)$ is a lower bound on the negation of x derived from the evidence against x. Both $t_V(x)$ and $f_V(x)$ associate a real number in the interval [0, 1] with each point in X, where $0 \leq t_V(x) + f_V(x) \leq 1$. This approach bounds the grade of membership of x to a subinterval $[t_V(x), 1 - f_V(x)]$ of [0, 1].

When X is continuous, a vague set V can be written as

$$V = \int_X [t_V(x), 1 - f_V(x)] / x, x \in X \tag{1}$$

When X is discrete, a vague set V can be written as

$$V = \sum [t_V(x), 1 - f_V(x)] / x, x \in X \tag{2}$$

For example, assume that $Z = \{1, 2, \ldots, 10\}$. *small* is a vague set of Z defined by

$$small = [1, 1]/1 + [0.9, 1]/2 + [0.6, 0.8]/3 + [0.3, 0.5]/4 + [0.1, 0.2]/5.$$

Definition 2 The complement of a vague set V is denoted by V' and is denoted by

$$\forall x \in V\prime$$

$$t_{V\prime}(x) = f_V(x), \tag{3}$$

$$1 - f_{V\prime}(x) = 1 - t_V(x), \tag{4}$$

$$V' = \{(x, f_V(x), 1 - t_V(x)), x \in Z\}. \tag{5}$$

Definition 3 The intersection of two vague sets $A = [t_A, 1 - f_A]$ and $B = [t_B, 1 - f_B]$ is a vague set C, written as $C = A \bigcap B$, whose truth-membership and false-membership functions are related to those of A and B by

$$t_C = \min\{t_A, t_B\} \tag{6}$$

$$1 - f_C = \min\{1 - f_A, 1 - f_B\} = 1 - \max\{f_A, f_B\} \tag{7}$$

$$C = [\min\{t_A, t_B\}, \min\{1 - f_A, 1 - f_B\}] \tag{8}$$

Definition 4 The union of two vague sets $A = [t_A, 1 - f_A]$ and $B = [t_B, 1 - f_B]$ is a vague set C, written as $C = A \bigcup B$, whose truth-membership and false-membership functions are related to those of A and B by

$$t_C = \max\{t_A, t_B\} \tag{9}$$

$$1 - f_C = \max\{1 - f_A, 1 - f_B\} = 1 - \min\{f_A, f_B\} \tag{10}$$

$$C = [\max\{t_A, t_B\}, \max\{1 - f_A, 1 - f_B\}] \tag{11}$$

2.2 The Method of Handling Multicriteria Decision-Making Problems Based on Vague Set Theory

Multicriteria fuzzy decision-making problems: Let $A = \{A_1, A_2, \ldots A_m\}$ be a set of alternatives and let $C = \{C_1, C_2, \ldots C_n\}$ be a set of criteria, Assume that the characteristics of the alternative A_i is represented by the vague set shown as follows:

$$A_i = \{(C_1, [t_{i1}, 1 - f_{i1}]), (C_2, [t_{i2}, 1 - f_{i2}]), \ldots (C_n, [t_{in}, 1 - f_{in}])\} \tag{12}$$

where t_{ij} indicates the degree that the alternative A_i satisfies criteria $C_j(j = 1, 2, 3 \ldots n)$, f_{ij} indicates the degree that the alternative A_i does not satisfy criteria $C_j(j = 1, 2, 3 \ldots n)$. Assume that there is a decision-maker who wants to choose an alternative which satisfies the criteria $c_j, c_k \ldots c_p$ or which satisfies the criteria c_s, then we can represent the decision-maker's requirement by the following expression: c_j *and* c_k *and* ... *and* c_p*or*c_s, the degrees that the alternative A_i satisfies and not satisfies the decision-maker's requirement can be measured by the evaluation function E:

$$\begin{aligned} E(A_i) &= (([t_{ij}, 1 - f_{ij}]) \bigcap ([t_{ik}, 1 - f_{ik}]) \bigcap \cdots ([t_{ip}, 1 - f_{ip}])) \bigcup ([t_{is}, 1 - f_{is}]) \\ &= [t_{A_i}, 1 - f_{A_i}] \end{aligned}$$

where, $t_{A_i} = \max\{\min\{t_{ij}, t_{ik}, \ldots t_{ip}\}, t_{is}\}$, $1 - f_{A_i} = \max\{\min\{1 - f_{ij}, 1 - f_{ik}, \ldots 1 - f_{ip}\}, 1 - f_{is}\}$. Multi-criteria decision-making based on Vague Sets question is how to choose the best option to meet the requirements of decision-makers and to meet from property index level candidates represented by Vague set. At present, the scoring function is used to indicate the degree of decision-makers plan to meet the requirements, the greater the scoring function value, the more satisfied the requirements of decision-makers.

2.3 Existing Integral Function

(1) Chen and Tan [22] score function:

$$S(E(A_i)) = t_{A_i} - f_{A_i} \tag{13}$$

The larger the value of $S(E(A_i))$,the more the suitability that the alternative A_i satisfies the decision-maker's requirement.

Example 1 $E(A_1) = [0.6, 0.7]$, $E(A_2) = [0.5, 0.6]$, By applying (14), we can get $S(E(A_1)) = 0.3$, $S(E(A_2)) = 0.1$. Therefore, we can see that the alternative A_1 is his best choice, but sometimes there will be indistinguishable cases, as shown in Example 2.

Example 2 $E(A_3) = [0.2, 0.9]$, $E(A_4) = [0.5, 0.6]$, By applying (14), we can get $S(E(A_3)) = S(E(A_4)) = 0.1$. We can not judge the merits of the program.
(2) After Hong and Choi [23] analyzed the deficiency of formula (14), the exact function is added:

$$H(E(A_i)) = t_{A_i} + f_{A_i} \tag{14}$$

By applying (15), we can get $H(E(A_3)) = 0.3$, $H(E(A_4)) = 0.9$. Therefore, we can see that the alternative A_4 is his best choice, but sometimes there will have indistinguishable cases, as shown in Example 3.

Example 3 $E(A_5) = [0.4, 0.1]$, $E(A_6) = [0.3, 0.9]$, By applying (15), we can get $H(E(A_5)) = H(E(A_6)) = 0.4$. We can not judge the merits of the program.

3 Two New Score Functions

3.1 The New Score Function

By analyzing the above example, We can see that the Chen, Hong [22, 23] score function is not able to make decisions on the target well. In this paper, we use the idea stepwise, Considering the three dimensional nature of the vague set, which are $t_v(x)$, $f_v(x)$, $\pi_v(x)$, where $\pi_v(x) = 1 - t_v(x) - f_v(x)$, $\pi_v(x)$ is called the neutral degrees of element x in V. We construct a scoring function:

$$G_1 = t_v(x) - f_v(x) - 0.5 * \pi_v(x) \tag{15}$$

The greater the value of G_1, the better the strategy is better. When we construct the scoring function, we should also consider to reduce the switching times in heterogeneous network, so we make half of the neutral degree as it may not support network switch part and it needs remedy. Because there will be G_1 equal situations, then construct the second scoring function:

$$G_2 = t_v(x) - f_v(x) \tag{16}$$

The greater the value of G_2, the better the strategy is. We prove that, with G_1, G_2 we can only determine the merits of the two vague values. Suppose there are two different vague sets $A = [t_A, 1 - f_A]$ and $B = [t_B, 1 - f_B]$. Use the G_1, G_2 to calculate, we still cannot determine size of A and B. In other words, A calculation using G_2, it would be that

$$G_2(A) = G_2(B)$$

Namely

$$t_A - f_A = t_B - f_B$$

$$t_A - t_B = f_A - f_B \tag{17}$$

For calculation of G_2, then G_1 is sure to be calculated, we can see that the G_1 values are equal:

$$G_1(A) = G_1(B)$$

Namely

$$t_A - f_A - 0.5 * (1 - t_A - f_A) = t_B - f_B - 0.5 * (1 - t_B - f_B)$$

Simplification:

$$3 * (t_A - t_B) = (f_A - f_B) \tag{18}$$

Combine (18) and (19) we have

$$t_A - t_B = f_A - f_B, 3 * (t_A - t_B) = (f_A - f_B)$$

$$t_A = t_B, f_A = f_B$$

That is to say the vague values of $A = [t_A, 1 - f_A]$ and $B = [t_B, 1 - f_B]$ are equal, which contradicts the hypothesis, so that after formulas G_1, G_2, we can uniquely determine the merits of two different vague values.

3.2 The Weighting Function

In fact, the importance of constraints is different. Assume that the degree of importance of the criteria $c_j, c_k \dots c_p$ entered by the decision-maker are $w_j, w_k, \dots w_p$, respectively, where $w_j, w_k, \ldots w_p \in [0, 1]$, and $w_j + w_k + \cdots + w_p = 1$, then the weighting function $W(A_i)$ is the following:

$$W(A_i) = \max\{G_1([t_{ij}, 1 - f_{ij}]) * w_j + G_1([t_{ik}, 1 - f_{ik}]) * w_k) + \cdots + G_1([t_{ip}, 1 - f_{ip}])w_p, G_1([t_{is}, 1 - f_{is}])\}$$

If the case occurs when the value is equal to the result of G_1, we can continue to compare the G_2 values and get the optimal.

4 Vertical Handoff Algorithm Based on Vague Set

In this paper, three parameters are considered: the received signal strength (RSS), the bandwidth (B) and the price (C). First, we turn on the switch indicator parameters vague fuzzy, and then use them in multi-attribute decision algorithm.

4.1 *Parameters of Vague*

For convenience, we use the vague parameter fuzzy method of Ref. [24], we take p = 2: Let $X = \{x_1, x_2, \ldots x_n\}$ be a set of criteria, and let $A_i (i = 1, 2, 3, \ldots, m)$ be a set of alternatives. x_{ij} is an indicator of program A_i, $x_{ij} \geq 0$. $x_{jmin} = \min x_{1j}, x_{2j}, \ldots x_{mj}$, Then

$$A_i(x_j) = [t_{ij}, 1 - f_{ij}] = \left[\frac{x_{ij}^2 - x_{jmin}^2}{x_{jmax}^2 - x_{jmin}^2}, 1 - \frac{x_{jmax} - x_{ij}}{x_{jmax} - x_{jmin}}\right] \tag{19}$$

where, $x_{jmin} = \min x_{1j}, x_{2j}, \ldots x_{mj}$, $x_{jmax} = \max x_{1j}, x_{2j}, \ldots x_{mj}$.
In our construction of UMTS and WLAN networks, x_{ij} is the primitive parameter switching index of j, $[t_{ij}, 1 - f_{ij}]$ is the value of vagueness. Where t_{ij} indicates the degree that the network i satisfies criteria j, f_{ij} indicates the degree that the network i does not satisfy criteria j, x_{jmin} indicates min of criteria j, x_{jmax} indicates max of criteria j. By applying (21), we can get

$$A_i(x_j) = [t_{ij}, 1 - f_{ij}] = \left[\frac{x_{ij}^2 - x_{jmin}^2}{x_{jmax}^2 - x_{jmin}^2}, 1 - \frac{x_{jmax} - x_{ij}}{x_{jmax} - x_{jmin}}\right] \tag{20}$$

Where $i = WorU$, $j = R, B, C$.

4.2 Rule of Handover Decision

By applying formula G, we can get $G(E(A_i(x_j)))$ and use it for decision. Due to the fact that different parameters have different impact on performance of network and the importance of the handover decision, the weighting may be selected dynamically based on network conditions. However, we consider the computational complexity and use fixed weighting as R and C have different importance on the handover decision: $w_R = 0.8$, $w_C = 0.2$, $w_R + w_C = 1$. By applying (20), we can get the comprehensive evaluation value of different networks(PEV):

$$PEV_i = \max G[A_i(x_R)] * w_R + G[A_i(x_C)] * w_C, G[A_i(x_B)] \tag{21}$$

Where $i = WorU$.

There are two steps in making a decision to switch, one is vague fuzzy control, the other is handover decision. The basic process of vague fuzzy control is to input the vague fuzzy parameters, get the vague language variables, and then according to the rules of vague fuzzy to make inference, the final result is vague score. The vague score is the realization of defuzzification, converted into classical values, and finally the output value is used to determine whether or not to switch. The above vague control system is shown in Fig. 1

Now, we give the vertical handoff decision process

(a)When the mobile terminal switches from UMTS to WLAN, if $PEV_U \geqslant PEV_W$, then do not switch; if $PEV_U < PEV_W$, then do witch;

(b)When the mobile terminal to switch from WLAN to UMTS, if $PEV_U > PEV_W$, then do switch; if $PEV_U \leq PEV_W$, then do not witch;

Vertical handoff algorithm for heterogeneous networks based on VAGUE set is shown in Fig. 2:

Example 4 $UW = \{U_1, W_1, U_2, W_2, U_3, W_3, U_4, W_4, U_5, W_5\}$represents five different locations for UMTS and WLAN networks, there is only one mobile terminal in the network. $Z = \{R, C, B\}$ is the attribute set, where R represents the network signal strength, C represents the network cost, B represents the network bandwidth. The vague set for each attribute criterion is expressed as the network at different time:

$$U_1 = \{(R, [0.0405, 0.1373]), (C, [0, 0]), (B, [1, 1])\}$$

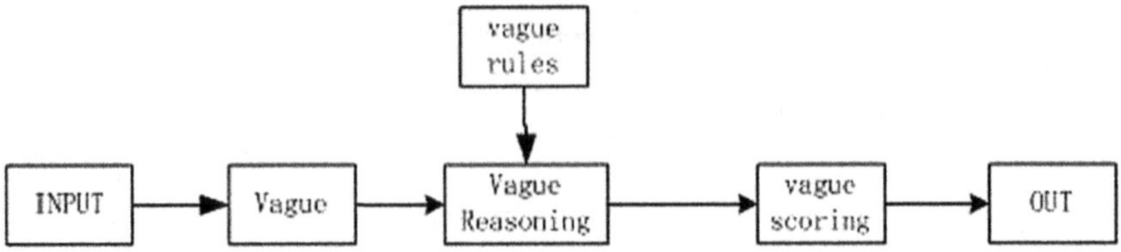

Fig. 1 Vague control system

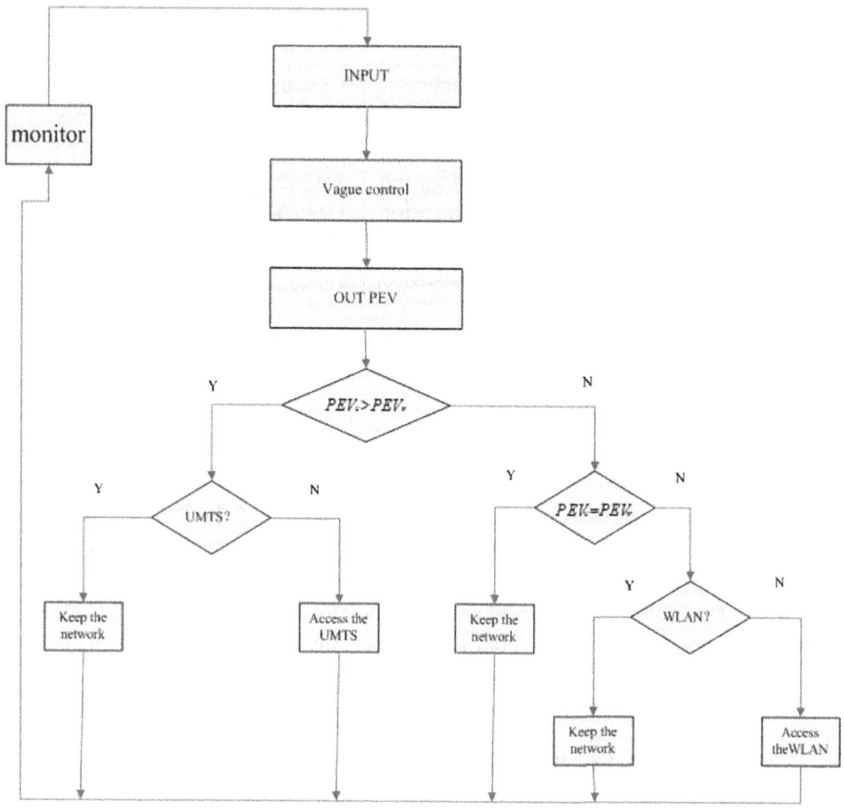

Fig. 2 Vertical handoff algorithm based on vague set

$$W_1 = \{(R, [0.0050, 0.0686]), (C, [1, 1]), (B, [0, 0])\}$$

$$U_2 = \{(R, [0.0503, 0.1603]), (C, [0, 0]), (B, [0, 0])\}$$

$$W_2 = \{(R, [0.1390, 0.3714]), (C, [1, 1]), (B, [1, 1])\}$$

$$U_3 = \{(R, [0.0409, 0.1383]), (C, [0, 0]), (B, [0, 0])\}$$

$$W_3 = \{(R, [0.5349, 0.7307]), (C, [1, 1]), (B, [1, 1])\}$$

$$U_4 = \{(R, [0.0688, 0.1991]), (C, [0, 0]), (B, [1, 1])\}$$

$$W_4 = \{(R, [0.0731, 0.2686]), (C, [1, 1]), (B, [0, 0])\}$$

$$U_5 = \{(R, [0.0994, 0.2545]), (C, [0, 0]), (B, [0, 0])\}$$

$$W_5 = \{(R, [0.1819, 0.3251]), (C, [1, 1]), (B, [1, 1])\}$$

By applying (23), we can get Comprehensive performance evaluation value(PEV):
$PEV_{U1} = 1$, $PEV_{W1} = -0.5665$, In position 1, UMTS is better than WLAN.
$PEV_{U2} = -0.8755$, $PEV_{W2} = 1$, In position 2, WLAN is better than UMTS.
$PEV_{U3} = -0.8956$, $PEV_{W3} = 1$, In position 3, WLAN is better than UMTS.
$PEV_{U4} = 1$, $PEV_{W4} = -0.4049$, In position 4, UMTS is better than WLAN.
$PEV_{U5} = -0.7789$, $PEV_{W5} = 1$, In position 5, WLAN is better than UMTS.

5 Simulation

5.1 System Model

Users in multiple networks may go through multiple WLAN. When entering the WLAN, WLAN can be chosen to provide services, otherwise, one user may not choose it, and instead he may use the original UMTS network. If one user plans to leave the WLAN and then use UMTS, that is to say, user must switch to UMTS in order to obtain services. At this time, we can build the system model shown in Fig. 3.

5.2 Simulation Calculation and Analysis

This paper considers the 3G-WLAN system model [8] as shown in Fig. 3. There are two UMTS base stations and two WLAN access points. The coverage radius of UMTS is 1000m, the bandwidth of UMTS is 3Mbps. The coverage radius of WLAN is 200 m, the bandwidth of WLAN is 54Mbps. According to Ref. [8], two same simulation scenarios are also setted down as follows: (i) there is a walk along the

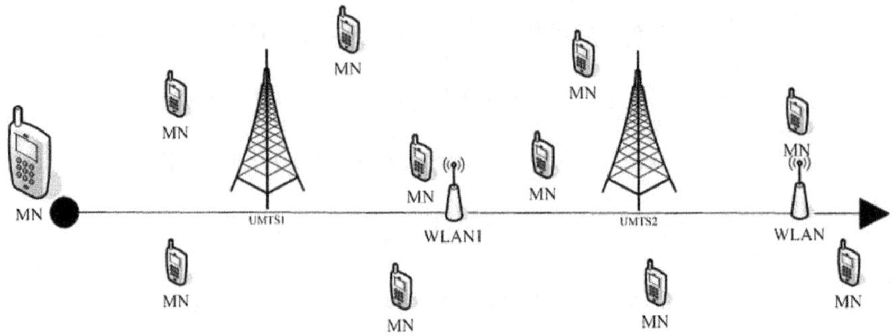

Fig. 3 System model

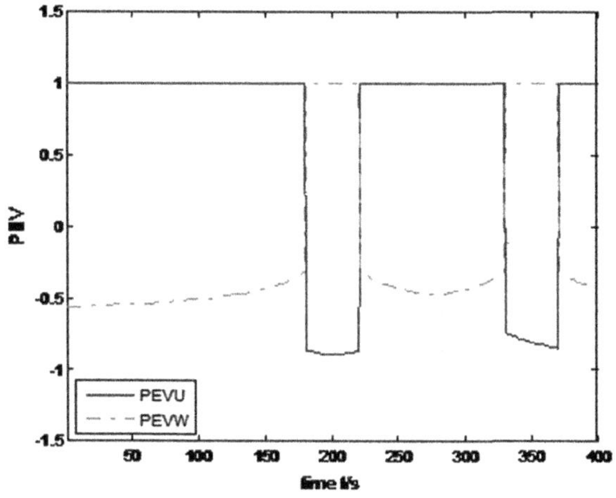

Fig. 4 MNs Handoff Decisions

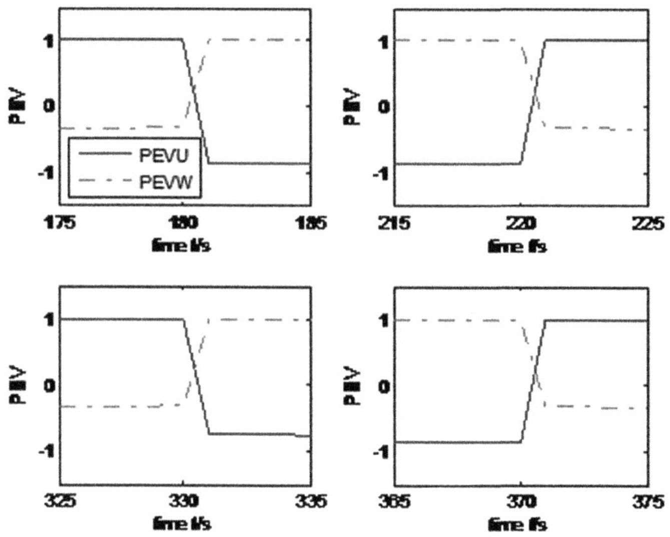

Fig. 5 The detailed MNs Handoff Decisions

path of the mobile station MN with solid line distance of 4000m, in such a sequence: $UMTS1 \rightarrow WLAN1 \rightarrow UMTS2 \rightarrow WLAN2 \rightarrow UMTS2$. (ii) Fig. 3 covers 10 MN(Figure smaller MN). Also evenly distributed, and moves freely. The algorithms mentioned herein and Ref. [8] compares the total number of handoff in 10 MN, a total of 20 times simulation. The results are shown in Fig. 6.

As shown in Figs. 4 and 5, a total of four times handover occurred. The algorithms described in this article is more accurate and feasible. The average handoff shown in

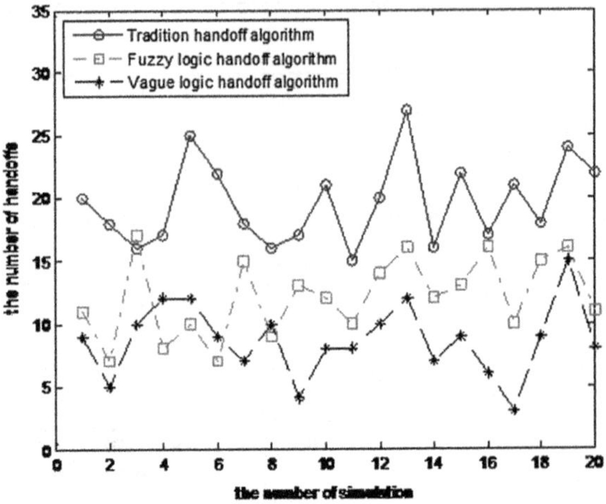

Fig. 6 The comparative results between three kinds of algorithms

Fig. 6 is less than that in Ref [8], the results illustrate that the proposed algorithm is better than that in Ref. [8] in reducing the number of handoffs is.

6 Conclusions

In this paper, the membership function of the switching parameters of heterogeneous network is extended by considering two aspects of true membership degree and false membership degree, and a new vague score function is constructed. Based on three parameters (signal strength,bandwidth,cost) as the impact of switching factors, by vague reasoning, and calculating the scoring results, the final vertical handoff decision is made. The simulation scenario is constructed and carried out. According to the result obtained, the vague sets based vertical handoff algorithm reduces the number of handoffs as compared with the other counterparts.

Acknowledgments Project supported by National Natural Science Foundation of China Research Grant (61502386); Department of education of shanxi province (2013JK1074).

References

1. Lee, S.K., Sriram, K., Kim, K., et al.: Vertical handoff decision algorithms for providing optimized performance in heterogeneous wireless networks. IEEE Trans. Veh. Technol. **58**(2), 865–881 (2009)
2. So, J.W.: Vertical handoff in integrated CDMA and WLAN systems. AEU-Int. J. of Electron. Commun. **62**(6), 78–482 (2008)
3. Calhan, A., Ceken, C.: An optimum vertical handoff decision algorithm based on adaptive fuzzy logic and genetic algorithm. Wirel. Pers. Commun. **64**, 647–664 (2012)

4. Wang, Y.H., Hsu, C.P., Huang, K.F., Huang, W.C.: Handoff Decision Scheme with Guaranteed QoS in Heterogeneous Network. In: First IEEE International Conference on Ubi-Media Computing, pp. 138-143. (2008)
5. Dai, Z., Fracchia, R., Gosteau, J et al.: Vertical handover criteria and algorithm in IEEE 802.11 and 802.16 hybrid networks. In: IEEE International Conference on Communications, pp. 2480-2484 (2008)
6. He, X., Li, B.: Switching Technologies of Heterogeneous Wireless Networks. Beijing University of Posts and Telecommunications Press, Beijing (2008)
7. Xie, S.D., Wu, M.: A fuzzy control based decision algorithm for vertical handoff. J. Nanjing Univ. Posts Telecommun. (Nat. Sci.) **27**(5), 6–10 (2007)
8. He, Q., Chen, G.: A fuzzy control based algorithm for vertical handoff in 3G-WLAN hybrid wireless environment. Bull. Sci. Technol. **26**(2), 261–264 (2010)
9. Onel, T., Ersoy, C., Cayirci, E., Par, G.: A multicriteria handoff decision schema for the next generation tactical communications systems. Int. J. Comput. Telecommun. Netw. **46**(5), 695–708 (2004)
10. Ling, Y., Yi, B., Zhu, Q.: An improved vertical handoff decision algorithm for heterogeneous wireless networks. Wirel. Commun. Netw. Mob. Comput. WiCOM **08**, 1–3 (2008)
11. Nkansah-Gyekye,Y., Agbinya,J.I.: A vertical handoff decision algorithm for next genera- tion wireless networks. In: Third international conference on broadband communications, information technology biomedical applications. pp. 358-364 (2008)
12. Alkhawlani, M., Ayesh, A.: Access network selection based on fuzzy logic and genetic algorithms. Adv. Artic. Intell. **8**(1), 1–12 (2008)
13. Gau, W.L., Buehrer, D.J.: Vague sets. J. IEEE Trans. Syst. Man Cybern. **23**(2), 610–614 (1993)
14. Wang, G.J.: Computational Intelligence. Higher Education Press, Beijing (2005)
15. Li, F., Lu, A., Cai, L.J.: Multicriteria decision making based on vague set. J. Huazhong Univ. Sci. Technol. **29**(7), 1–3 (2001)
16. Liu, H.W.: Vague set methods of multi-criteria fuzzy decision making. Syst. Eng. Theory Pract. **24**(5), 103–109 (2004)
17. Zhou, Z., Wu, Q.Z.: Multicriteria fuzzy decision making method based on vague set. J. Chinese Comput. Syst. **26**(8), 1350–1353 (2005)
18. Zhou, X.G., Tan, C.Q., Zhang, Q.: Decision Theories Methods based on Vague Set. Science Press, Beijing (2009)
19. Chen, J.L., Hu, Z.G., Liu, Q.: Research on fuzzy multi objective decision making of diversion scheme of hydropower project based on Vague set. China Water Transport. **03**, 334–336 (2015)
20. Luo, J.G., Xie, J.C.: Multicriteria decision making and application based on vague set. Math. Pract. Theory **38**(20), 114–122 (2008)
21. Xu, T.J.: Optimization of oil field development scheme based on fuzzy multi objective decision making method based on Vague set, vol. 07, pp. 9–10. http://www.yqtdmgc.com (2013)
22. Chen, S.M., Tan, J.M.: Handling multi-criteria fuzzy decision-making problems based on vague set theory. Fuzzy Sets Syst. **67**(2), 163–172 (1994)
23. Hong, D.H., Choi, C.H.: Multi-criteria fuzzy decision-ma king problems based on vague set theory. Fuzzy Sets Syst. **114**, 103–113 (2000)
24. Wang, H.X.: Definition and transforming formulas from single valued data to Vague valued data. Comput. Eng. Appl. **46**(24), 42–44 (2010)

Image Thresholding by Maximizing the Similarity Degree Based on Intuitionistic Fuzzy Sets

Rong Lan, Jiu-Lun Fan, Ying Liu and Feng Zhao

Abstract In this paper a new image thresholding method is proposed by using a similarity measure on intuitionistic fuzzy sets. Based on the 'vote model' of intuitionistic fuzzy sets, an image is mapped to an intuitionistic fuzzy set which is constructed from a fuzzy set. The corresponding fuzzy set's membership degree is calculated by Gamma distribution. The proposed technique maximizes the similarity degree to select the best threshold.

Keywords Intuitionistic fuzzy set · Similarity measure · Image segmentation · Thresholding

1 Introduction

In 1986, Atanassov [1] introduced the concept of intuitionistic fuzzy sets (IFSs), which is an extension of the concept of fuzzy sets (FSs, proposed by Zadeh [2]). IFSs use two characteristic functions to express the membership degree and the non-membership degree of elements in the universe belonging to an intuitionistic fuzzy set (IFS), respectively. Therefore, the idea of using positive and (independently) negative information becomes the core of IFSs. Since then, the theory of IFSs has been

R. Lan (✉) · J.-L. Fan · Y. Liu · F. Zhao
School of Communication and Information Engineering, Xian University of Posts and Telecommunications, Xian 710121, China
e-mail: ronglanlogic@163.com

J.-L. Fan
e-mail: jiulunf@163.com

Y. Liu
e-mail: ly_yolanda@sina.com

F. Zhao
e-mail: fengz1119@163.com

T.-H. Fan et al. (eds.), *Quantitative Logic and Soft Computing 2016*, Advances in Intelligent Systems and Computing 510,
DOI 10.1007/978-3-319-46206-6_58

widely and deeply discussed [3–7]. Recently, IFSs have been successfully applied to many areas, such as pattern recognition [8–13], decision analysis [14], approximate reasoning [15] and image processing [16–21].

It is well known that image segmentation plays important role in image processing analyses. As a popular and effective tool, threshold method is often used to segment objects from background in an image. In real cases, there are always more or less fuzziness in images because of the limitations and characteristics of equipment, uneven illumination and unavoidable noises. Therefore, it is a hot issue to combine threshold method with FS or IFS theory. Reference [22] provided an idea which maps an image to a FS. Huang and Wang [23] proposed two measures of fuzziness and selected the threshold by minimizing the measures of fuzziness. Chaira and Ray [24] used fuzzy divergence, proposed by means of fuzzy exponential entropy in [25], to obtain the optimal threshold. These fuzzy techniques can be naturally extended to intuitionistic fuzzy cases for dealing with image segmentation. Vlachos and Sergiadis [16] proposed symmetric discrimination information measure for IFSs, which was obtained by means of the logarithmic cross-entropy measure defined by Kullback [26, 27], and then formed an objective function for image segmentation by discrimination information measure. In [19], the exponential fuzzy divergence [25] was extended to IFSs and an image edge detection method was proposed by using the exponential information fuzzy divergence. The exponential divergence on IFSs was used to medical image segmentation later [18]. In [21], an algorithm for constructing interval type-2 fuzzy sets (a particular case of type-2 fuzzy sets and is equivalent to IFS) model for images was provided, and then segmentation threshold was given by calculating the entropy of interval type-2 fuzzy set. Ananthi, Balasubramaniam and Lim [17] used the hesitation degree to construct an IFS from several FSs on application to image segmentation by minimizing the entropy on IFSs.

According to the references mentioned above, it is important for image segmentation based on IFSs to transfer an image to an IFS. Most of these existent methods have parameter that should be taken in advance. Unlike the existent methods above, a new method without parameter is presented to obtain intuitionistic fuzzy membership degree and non-membership degree function from fuzzy membership degree function. The idea of proposed method is based on the 'vote model' of intuitionistic fuzzy value (IFV). So the actual explanation of the proposed method can be given.

In order to search for an optimal threshold to segment an image, one needs to select a proper measure to construct objective function. In this paper, we will employ similarity measure on IFSs to segment an image. The similarity measure, proposed in [28], considers not only membership degree and non-membership degree but also the relationship between them. The new proposed method searches for optimal threshold by maximizing the similarity degree. Comparing with several fuzzy and intuitionistic fuzzy methods, the proposed method works well and shows better performance in image segmentation.

The remainder of this paper is organized as follows. In Sect. 2, the notion of IFS and a similarity measure on IFSs are introduced. In Sect. 3 we describe briefly the elements of image segmentation and develops a novel thresholding method for image segmentation based on IFS. In Sect. 4, we compare the proposed method with 3 methods and show the results of these methods. Finally, conclusion is drawn in Sect. 5.

2 Intuitionistic Fuzzy Set and Its Similarity Measure

In this section we present the basic elements of IFSs theory, which will be needed in the following analysis.

2.1 Intuitionistic Fuzzy Set

Definition 1 ([2]) Let X be a non-empty finite universe and $Card(X) = n$. A fuzzy set A defined on a universe X is given by:

$$A = \{(x, \mu_A(x))|x \in X\}$$

where the mapping $\mu_A : X \to [0, 1]$ is called the membership function.

Definition 2 ([1]) Let X be an non-empty finite universe and $Card(X) = n$. An intuitionistic fuzzy set $\widetilde{A}$ defined on the universe X is given by:

$$\widetilde{A} = \{(x, \mu_{\widetilde{A}}(x), \nu_{\widetilde{A}}(x))|x \in X\} \quad (1)$$

where $\mu_{\widetilde{A}} : X \to [0, 1]$ and $\nu_{\widetilde{A}} : X \to [0, 1]$ with the condition $0 \leq \mu_{\widetilde{A}}(x) + \nu_{\widetilde{A}}(x) \leq 1$ for all $x \in X$. The numbers $\mu_{\widetilde{A}}(x)$ and $\nu_{\widetilde{A}}(x)$ denote the degree of membership and the degree of non-membership of x belonging to $\widetilde{A}$, respectively. We will denote the set of all the intuitionistic fuzzy sets on X by $IFSs(X)$.

In particular, if there is only one element in a universe X, i.e. $X = \{x\}$, an intuitionistic fuzzy set $\widetilde{A} = \{(x, \mu_{\widetilde{A}}(x), \nu_{\widetilde{A}}(x))\}$ defined on X is called an intuitionistic fuzzy value, and denoted by $x = (\mu_x, \nu_x)$, where $\mu_x = \mu_{\widetilde{A}}(x), \nu_x = \nu_{\widetilde{A}}(x)$.

It can be easily observed that $\widetilde{A} = \{(x, \mu_{\widetilde{A}}(x), \nu_{\widetilde{A}}(x))|x \in X\} = \{(x, \mu_{\widetilde{A}}(x), 1 - \mu_{\widetilde{A}}(x))|x \in X\}$ if $\mu_{\widetilde{A}}(x) + \nu_{\widetilde{A}}(x) = 1$, i.e., fuzzy set is a particular case of intuitionistic fuzzy set.

Definition 3 ([6]) Let $\widetilde{A} = \{(x, \mu_{\widetilde{A}}(x), \nu_{\widetilde{A}}(x))|x \in X\} \in IFSs(X)$. For all $x \in X$, we call

$$\pi_{\widetilde{A}}(x) = 1 - \mu_{\widetilde{A}}(x) - \nu_{\widetilde{A}}(x) \quad (2)$$

an intuitionistic fuzzy index (or a hesitation margin) of x to $\widetilde{A}$.

Obviously, $0 \le \pi_{\tilde{A}}(x) \le 1$ for all $x \in X$, and it expresses a lack of knowledge of whether x belongs to $\tilde{A}$ or not.

In the following, we show some basic operations on IFSs which will be needed in the following discussion.

Definition 4 ([6]) Let $\tilde{A}, \tilde{B} \in IFSs(X)$, then
(1) $\tilde{A} \le \tilde{B}$ if $\forall x \in X$, $\mu_{\tilde{A}}(x) \le \mu_{\tilde{B}}(x)$ and $\nu_{\tilde{A}}(x) \ge \nu_{\tilde{B}}(x)$;
(2) $\tilde{A} = \tilde{B}$ if $\forall x \in X$, $\mu_{\tilde{A}}(x) = \mu_{\tilde{B}}(x)$ and $\nu_{\tilde{A}}(x) = \nu_{\tilde{B}}(x)$;
(3) $\tilde{A}^c = \{(x, \nu_{\tilde{A}}(x), \mu_{\tilde{A}}(x)) | x \in X\}$, and $\tilde{A}^c$ is called the complement of $\tilde{A}$.

2.2 Similarity Measure Between Intuitionistic Fuzzy Sets

In [28], a similarity measure with parameters was presented and applied to pattern recognition. It is defined as follows:

Definition 5 ([28]) Let $X = \{x_1, x_2, \ldots, x_n\}$ be a finite universe of discourse. And let $\tilde{A}, \tilde{B} \in IFSs(X)$, $\tilde{A} = \{(x_i, \mu_{\tilde{A}}(x_i), \nu_{\tilde{A}}(x_i)) | i = 1, 2, \ldots, n\}$ and $\tilde{B} = \{(x_i, \mu_{\tilde{B}}(x_i), \nu_{\tilde{B}}(x_i)) | i = 1, 2, \ldots, n\}$. The degree of similarity between $\tilde{A}$ and $\tilde{B}$ is defined by:

$$S(\tilde{A}, \tilde{B}) = \frac{1}{n} \sum\nolimits_{i=1}^{n} (1 - \lambda_1 |\mu_{\tilde{A}}(x_i) - \mu_{\tilde{B}}(x_i)| - \lambda_2 |\nu_{\tilde{A}}(x_i) - \nu_{\tilde{B}}(x_i)| - \lambda_3 |\varphi_{\tilde{A}}(x_i) - \varphi_{\tilde{B}}(x_i)|) \tag{3}$$

where $\varphi_{\tilde{A}}(x_i) = \frac{\mu_{\tilde{A}}(x_i) + (1 - \nu_{\tilde{A}}(x_i))}{2}$ and $\varphi_{\tilde{B}}(x_i) = \frac{\mu_{\tilde{B}}(x_i) + (1 - \nu_{\tilde{B}}(x_i))}{2}$, $\lambda_1 \ge 0, \lambda_2 \ge 0, \lambda_3 \ge 0$, and $\lambda_1 + \lambda_2 + \lambda_3 = 1$. Furthermore, there are at least two non-zero values in the three parameters λ_1, λ_2 and λ_3.

The similarity measure mentioned above has the following properties.

Proposition 1 ([28]) *Let $\tilde{A}, \tilde{B} \in IFSs(X)$, $\tilde{A} = \{(x_i, \mu_{\tilde{A}}(x_i), \nu_{\tilde{A}}(x_i)) | i = 1, \ldots, n\}$ and $\tilde{B} = \{(x_i, \mu_{\tilde{B}}(x_i), \nu_{\tilde{B}}(x_i)) | i = 1, \ldots, n\}$. Then $S(\tilde{A}, \tilde{B})$ defined in Definition 5 has the following properties.*
(1) $0 \le S(\tilde{A}, \tilde{B}) \le 1$;
(2) $S(\tilde{A}, \tilde{B}) = S(\tilde{B}, \tilde{A})$;
(3) $S(\tilde{A}, \tilde{B}) = 1$ *if and only if* $\tilde{A} = \tilde{B}$;
(4) $S(\tilde{A}, \tilde{B}) = S(\tilde{A}^c, \tilde{B}^c)$;
(5) $S(\tilde{A}, \tilde{B}) = 0$ *if and only if* $\tilde{A} = \{(x_i, \mu_{\tilde{A}}(x_i) = 1, \nu_{\tilde{A}}(x_i) = 0) | i = 1, \ldots, n\}$ *and* $\tilde{B} = \{(x_i, \mu_{\tilde{B}}(x_i) = 0, \nu_{\tilde{B}}(x_i) = 1) | i = 1, \ldots, n\}$ *or* $\tilde{A} = \{(x_i, \mu_{\tilde{A}}(x_i) = 0, \nu_{\tilde{A}}(x_i) = 1) | i = 1, \ldots, n\}$ *and* $\tilde{B} = \{(x_i, \mu_{\tilde{B}}(x_i) = 1, \nu_{\tilde{B}}(x_i) = 0) | i = 1, \ldots, n\}$;
(6) $\forall \tilde{A}, \tilde{B}, \tilde{C} \in IFSs(X)$, $S(\tilde{A}, \tilde{B}) \ge S(\tilde{A}, \tilde{C})$ *and* $S(\tilde{B}, \tilde{C}) \ge S(\tilde{A}, \tilde{C})$ *if* $\tilde{A} \le \tilde{B} \le \tilde{C}$.

Remark 1 In fact, it is reasonable to regard the importance of the membership degree $\mu_{\tilde{A}}(x)$ and the non-membership degree $\nu_{\tilde{A}}(x)$ the same due to lack of enough knowledge. Therefore, the parameters λ_1 and λ_2 should be equal in formula (3), i.e.,

$\lambda_1 = \lambda_2 = \lambda$, so we have $\lambda_3 = 1 - 2\lambda$. Then, the similarity measure (3) reduces to a simple formula as follows

$$S^*(\widetilde{A}, \widetilde{B}) = \frac{1}{n}\sum_{i=1}^{n}(1 - \lambda|\mu_{\widetilde{A}}(x_i) - \mu_{\widetilde{B}}(x_i)| - \lambda|\nu_{\widetilde{A}}(x_i) - \nu_{\widetilde{B}}(x_i)| \\ -(1-2\lambda)|\varphi_{\widetilde{A}}(x_i) - \varphi_{\widetilde{B}}(x_i)|). \quad (4)$$

3 Image Thresholding Method Based on Intuitionistic Fuzzy Set

In this section, we study image segmentation by using the above-mentioned similarity measure on IFSs. And a novel image segmentation technique using IFS will be proposed.

Let us consider an image I of size $M \times N$ pixels, having L gray levels g ranging from 0 to $L - 1$. According to Pal and King [22], any image can be considered as an array of fuzzy singletons. Each element of the array denotes the membership degree of the gray level g_{ij}, corresponding to the $(i, j)th$ pixel, with respect to an image property. Therefore, the image I can be expressed by the following FS.

$$I = \{(g_{ij}, \mu_I(g_{ij}))|g_{ij} \in \{0, 1, \ldots, L-1\}\} \quad (5)$$

where $i \in \{1, \ldots, M\}$ and $j \in \{1, \ldots, N\}$.

Given a certain threshold T that separates the foreground (object) from the background, the average gray levels of the background and the foreground are given by

$$m_B = \frac{\sum_{g=0}^{T} g h_I(g)}{\sum_{g=0}^{T} h_I(g)}, \; m_F = \frac{\sum_{g=T+1}^{L-1} g h_I(g)}{\sum_{g=T+1}^{L-1} h_I(g)} \quad (6)$$

where h_I is the histogram of image I. Chaira and Ray [24] calculated the membership degree of each pixel of the image by using the Gamma distribution function as follows.

$$\mu_I(g_{ij}, T) = \begin{cases} exp(-c|g_{ij} - m_B|), \; if \; g_{ij} \le T \; for \; background, \\ exp(-c|g_{ij} - m_F|), \; if \; g_{ij} > T \; for \; foreground \end{cases} \quad (7)$$

where T is any chosen threshold, c is the constant $c = \frac{1}{g_{max} - g_{min}}$ and g_{max} and g_{min} are the maximum and minimum gray levels of the image respectively.

3.1 Images Intuitionistic Fuzzy Model Based on 'The Vote Model'

The intuitionistic fuzzy thresholding method starts from the transferring of an image to an intuitionistic fuzzy model, i.e. an IFS. Since an IFS is a generalization of a FS, common method to get an IFS is to generate it from an existent FS. Several existing constructive methods need to fix the value of parameters, such that in [16, 18, 19]. Apparently, the selection of parameter mainly depends on the results of image segmentation. So it takes time to do many tests then.

In this paper, we try to explore another possibility of constructing an IFS by means of a FS. In fact, an IFV can be explained by 'voting model'. Let $x = (\mu_x, \nu_x)$ be an IFV, $\mu_x = 0.3$, $\nu_x = 0.6$. The IFV $x = (0.3, 0.6)$ can be interpreted as 'the vote for resolution is 3 persons in favor, 6 persons against, and 1 abstention'.

Let $A = \{(x, \mu_A(x))|x \in X\}$ be a FS on X. $\mu_A(x)$ represents $x's$ approval rating. According to the concept of FSs, $1 - \mu_A(x)$ is $x's$ against rating. In real world, when one person lives in a society or a community, he cannot discount people around him. That is to say, people always are more or less influenced by others, especially in voting. Though the viewpoints of two sides, supporter and anti, are different, they may affect each other. On the one hand, the anti can affect the supporter, we have $(1 - \mu_A(x)) \cdot \mu_A(x)$; on the other hand, the supporter can also affect the anti, so we have $\mu_A(x) \cdot (1 - \mu_A(x))$. Therefore, a hesitation margin of IFS $\widetilde{A}$ is

$$\pi_{\widetilde{A}}(x) = (1 - \mu_A(x)) \cdot \mu_A(x) + \mu_A(x) \cdot (1 - \mu_A(x)) = 2\mu_A(x) \cdot (1 - \mu_A(x)).$$

At the same time, we can get that a new membership function is

$$\mu_{\widetilde{A}}(x) = \mu_A(x) - (1 - \mu_A(x)) \cdot \mu_A(x) = \mu_A^2(x).$$

And the corresponding non-membership function is

$$\nu_{\widetilde{A}}(x) = (1 - \mu_A(x)) - \mu_A(x) \cdot (1 - \mu_A(x)) = (1 - \mu_A(x))^2$$

In this way, we obtain a new IFS on X as follows

$$\widetilde{A} = \{(x, \mu_{\widetilde{A}}(x), \nu_{\widetilde{A}}(x))|x \in X\} = \{(x, \mu_A^2(x), (1 - \mu_A(x))^2)|x \in X\} \quad (8)$$

Based on the membership function defined by formula (7), an IFS can be constructed by means of (8). The corresponding membership and non-membership function are defined by

$$\mu_{\widetilde{I}}(g_{ij}, T) = \mu_I^2(g_{ij}, T), \nu_{\widetilde{I}}(g_{ij}, T) = (1 - \mu_I(g_{ij}, T))^2.$$

So we can map image I to the following IFS,

$$\begin{aligned}\widetilde{I} &= \{(g_{ij}, \mu_{\widetilde{I}}(g_{ij}, T), \nu_{\widetilde{I}}(g_{ij}, T))|g_{ij} \in \{0, 1, \ldots, L-1\}\} \\ &= \{(g_{ij}, \mu_{\widetilde{I}}^2(g_{ij}, T), (1-\mu_I(g_{ij}, T))^2)|g_{ij} \in \{0, 1, \ldots, L-1\}\}.\end{aligned} \tag{9}$$

3.2 *Intuitionistic Fuzzy Objective Function*

The idea of the proposed method is the maximization of the similarity degree between the actual and the ideally thresholded image. Owing to lack of prior knowledge, it is a reasonable assumption that the membership degree is the same importance as the non-membership degree. So we calculate the similarity degree between an IFS of image and that of an ideally segmented image by using formula (4).

For an ideally threshold image, the membership degree $\mu_{\widetilde{B}}(g_{ij}) = 1$ and the non-membership degree $\nu_{\widetilde{I}}(g_{ij}) = 0$ for all $g_{ij} \in \{0, 1, \ldots, L-1\}$. Therefore, we have

$$S^*(\widetilde{I}, \widetilde{B}, T) = \frac{1}{L-1}\sum\nolimits_{g_{ij}=1}^{L-1}(1 - \lambda|\mu_{\widetilde{I}}(g_{ij}) - \mu_{\widetilde{B}}(g_{ij})| - \lambda|\nu_{\widetilde{I}}(g_{ij}) - \nu_{\widetilde{B}}(g_{ij})|$$
$$-(1-2\lambda)|\varphi_{\widetilde{I}}(g_{ij}) - \varphi_{\widetilde{B}}(g_{ij})|)$$

$$\begin{aligned}&= \frac{1}{L-1}\sum\nolimits_{g_{ij}=1}^{L-1}\left(1 - \lambda(1-\mu_{\widetilde{I}}(g_{ij})) - \lambda\nu_{\widetilde{I}}(g_{ij}) - (1-2\lambda)\left(\frac{1+\nu_{\widetilde{I}}(g_{ij}) - \mu_{\widetilde{I}}(g_{ij})}{2}\right)\right) \\ &= \frac{1}{L-1}\sum\nolimits_{g_{ij}=1}^{L-1}\frac{1}{2}(1+\mu_{\widetilde{I}}(g_{ij}) - \nu_{\widetilde{I}}(g_{ij}))\end{aligned}$$

The optimization criterion is the following:

$$T_{opt} = arg\max_{T}\{S^*(\widetilde{I}, \widetilde{B}, T)\}, \tag{10}$$

where T_{opt} is the optimal threshold.

4 Experimental Results

In order to show the performance of the proposed method, the exponential fuzzy divergence method (EFDM) [24], the symmetric intuitionistic fuzzy discrimination information method (SIFDIM) (Let $\lambda = 0.2$ in the experimentation. Simulations have shown that setting parameter $\lambda = 0.2$ yields the overall best result [16].) and the intuitionistic fuzzy similarity measure method with parameter (IFSMMP) are compared with it. All 3 methods use the gamma-distribution to map an image to a

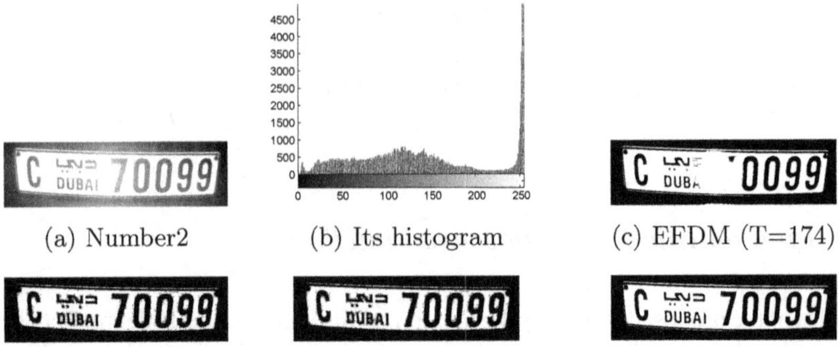

(a) Number2 (b) Its histogram (c) EFDM (T=174)

(d) SIFDIM (T=250) (e) IFSMMP (T=250) (f) The proposed method (T=203)

Fig. 1 The results of the first image

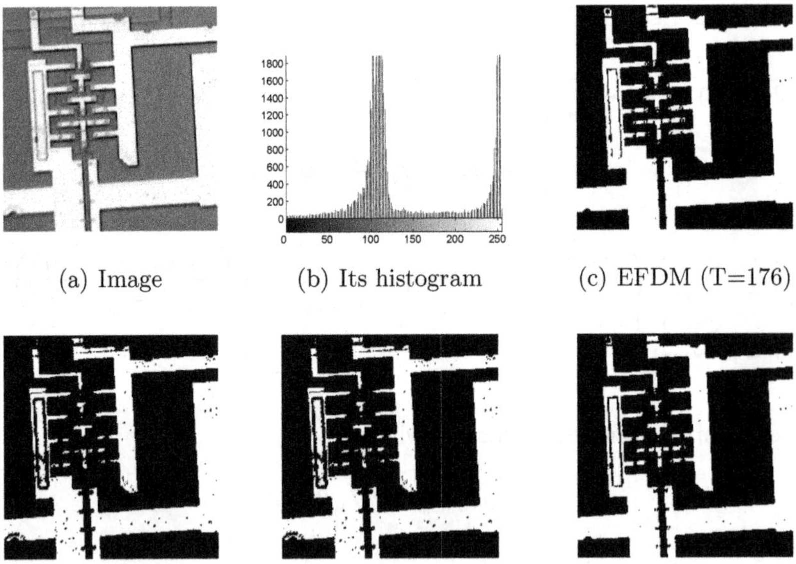

(a) Image (b) Its histogram (c) EFDM (T=176)

(d) SIFDIM (T=235) (e) IFSMMP (T=235) (f) The proposed method (T=206)

Fig. 2 The results of the second image

FS. Among them, the EFDM is a fuzzy thresholding algorithm and the SIFDIM is an intuitionistic fuzzy thresholding algorithm. The IFSMMP uses the same objective function to opt the best threshold as the proposed method and the same image's IFS model as the SIFDIM (Let $\lambda = 0.2$ in the experimentation, it makes IFSMMP to yield the overall best result.) in order to show the effectiveness of the image IFS model without parameter based on 'the vote model'. In Figs. 1 and 2 we show the thresholds of the 4 methods.

5 Conclusions

It is well known that the notion of IFS is an extension of FS, and the IFSs have been applied to many areas. In this paper, we proposed a new thresholding method to segment image based on a similarity measure between two IFSs. The method starts from the representation of an image by using an IFS, which is generated by an IFS's own actual meaning but not using any parameter. Therefore, it does not require a great number of tests for searching the proper value of parameter. The segmentation method selects the best threshold associated with the maximum of the similarity measure between an IFS of image and that of an ideally segmented image, and yields the better result than the other 3 methods, even including the intuitionistic fuzzy similarity measure method with parameter.

Acknowledgments This work is supported by the National Natural Science Foundation of China under Grant No. 61571361, the Scientific Research Program Funded by Shaanxi Provincial Education Department (Program No. 16JK1709), the National Natural Science Youth Foundation of China under Grant No. 61202183, the National Natural Science Youth Foundation of China under Grant No. 61102095, and the Science and Technology Plan in Shaanxi Province of China (No. 2014KJXX-72).

References

1. Atanassov, K.T.: Intuitionistic fuzzy sets. Fuzzy Set Syst. **20**, 87–96 (1986)
2. Zadeh, L.A.: Fuzzy sets. Inf. Control **8**, 338–356 (1965)
3. Gau, W.L., Buehrer, D.J.: Vague sets. IEEE Trans. Syst. Man Cybern. **23**, 610–614 (1993)
4. Burillo, P., Bustine, H.: Vague sets are intuitionistic fuzzy sets. Fuzzy Set Syst. **79**, 403–405 (1996)
5. Burillo, P., Bustince, H.: Construction theorems for intuitionistic fuzzy sets. Fuzzy Set Syst. **84**, 271–281 (1996)
6. Atanassov, K.T.: Intuitionistic Fuzzy Sets: Theory and Application. Physica-Verlag, Heidelberg (1999)
7. Deschrijver, G., Kerre, E.E.: On the relationship between some extensions of fuzzy set theory. Fuzzy Set Syst. **133**, 227–235 (2003)
8. Li, D.F., Cheng, C.T.: New similarity measures of intuitionistic fuzzy sets and applications to pattern recognitions. Pattern Recogn. Lett. **23**, 221–225 (2002)
9. Mitchell, H.B.: On the Dengfeng-Chuntian similarity measure and its application to pattern recognition. Pattern Recogn. Lett. **24**, 3101–3104 (2003)
10. Liang, Z.Z., Shi, P.F.: Similarity measures on intuitionistic fuzzy sets. Pattern Recogn. Lett. **24**, 2687–2693 (2003)
11. Hung, W.L., Yang, M.S.: Similarity measures of intuitionistic fuzzy sets based on the Hausdorff distance. Pattern Recogn. Lett. **25**, 1603–1611 (2004)
12. Hung, W.L., Yang, M.S.: On the J-divergence of intuitionistic fuzzy sets with its application to pattern recognition. Inf. Sci. **178**, 1641–1650 (2008)
13. Hung, W.L., Yang, M.S.: Similarity measures of intuitionistic fuzzy sets based on Lp metric. Int. J. Approx. Reason. **46**, 120–136 (2007)
14. Guo, K.H.: Amount of information and attitudinal-based method for ranking Atanassov's intuitionistic fuzzy values. IEEE Trans. Fuzzy Syst. **22**, 177–188 (2014)

15. Szmidt, E., Kacprzyk, J.: Distance between intuitionistic fuzzy sets and their applications in reasoning. Stud. Comput. Intell. **2**, 101–116 (2005)
16. Vlachos, I.K., Sergiadis, G.D.: Intuitionistic fuzzy information - application to pattern recognition. Pattern Recogn. Lett. **28**, 197–206 (2007)
17. Ananthi, V.P., Balasubramaniam, P., Lim, C.P.: Segmentation of gray scale image based on intuitionistic fuzzy sets constructed from several membership functions. Pattern Recogn. **47**, 3870–3880 (2014)
18. Chaira, T.: Intuitionistic fuzzy segmentation of medical images. IEEE Trans. Bio-Med. Eng. **57**, 1430–1436 (2010)
19. Chaira, T., Ray, A.K.: A new measure using fuzzy set theory and its application to edge detection. Appl. Soft. Comput. **8**, 919–927 (2008)
20. Chaira, T., Chaira, T.: Intuitionistic fuzzy set: application to medical image segmentation. Stud. Comput. Intell. **85**, 51–68 (2008)
21. Pagola, M., Lopez-Molina, C., Fernandez, J., Barrenechea, E., Bustine, H.: Interval type-2 fuzzy sets constructed from several membership functions: application to the fuzzy thresholding algorithm. IEEE Trans. Fuzzy Syst. **21**, 230–244 (2013)
22. Pal, S.K., King, R.A.: Image enhancement using smoothing with fuzzy sets. IEEE Trans. Syst. Man Cyb. **11**, 495–501 (1981)
23. Huang, L.K., Wang, M.J.: Image thresholding by minimizing the measures of fuzziness. Pattern Recogn. Lett. **28**, 41–51 (1995)
24. Chaira, T., Ray, A.K.: Segmentation using fuzzy divergence. Pattern Recogn. Lett. **24**, 1837–1844 (2003)
25. Fan, J.L., Ma, Y.L., Xie, W.X.: On some properties of distance measure. Fuzzy Set Syst. **117**, 355–361 (2001)
26. Kullback, S.: Information Theory and Statistics, 2nd edition. Dover Publication, New York (1968)
27. Kullback, S., Leibler, R.A.: On information and sufficiency. Ann. Math. Stat. **22**, 79–86 (1951)
28. Lan, R., Fan, J.L.: Similarity measures on vague values and three-parameter vague values. Pattern Recogn. Artif. Intell. **23**, 341–348 (2010). (in Chinese)

Image Super Resolution Using Expansion Move Algorithm

Dong-Xiao Zhang, Guo-Rong Cai, Zong-Qi Liang and Huan Huang

Abstract In multi-frame image super resolution (SR), graph-cut is an effective algorithm to minimize the energy function for SR. As a kind of graph-cut algorithm, α-expansion move algorithm can effectively minimize energy functions such as class $\mathcal{F}^2$. However, the energy functions for SR established in Markov random field usually don't fall into this class and need some approximations, which may lead to poor results. In this paper, we propose a new method, with which we make the energy function for SR a form of class $\mathcal{F}^2$ without approximation. Experimental results show that our motivation is valid and the proposed method is effective for not only synthetic low-resolution images but also real images.

Keywords Super resolution · Expanded neighborhood · α-expansion move · Graph-cut

1 Introduction

Image super resolution (SR) aims to reconstruct a high-resolution image from a single low-resolution image, which is called single-frame image SR. If the reconstruction is based on a serial of low-resolution images, the SR process is extended and called multi-frame image SR. Single-frame image SR usually utilizes the codebook pre-trained based on local self-similarity [1–3], image characters [4], sparse

D.-X. Zhang (✉) · Z.-Q. Liang · H. Huang
School of Science, Jimei University, Xiamen 361021, Fujian, China
e-mail: dx.z@foxmail.com

Z.-Q. Liang
e-mail: liangzq2719@163.com

H. Huang
e-mail: hhuangjy@126.com

G.-R. Cai
Computer Engineering College, Jimei University, Xiamen 361021, Fujian, China
e-mail: kiser@jmu.edu.cn

T.-H. Fan et al. (eds.), *Quantitative Logic and Soft Computing 2016*,
Advances in Intelligent Systems and Computing 510,
DOI 10.1007/978-3-319-46206-6_59

representation [5, 6] and so on. Multi-frame image SR takes advantage of the redundant information between low-resolution images. Single-frame SR tries to maintain the details of the low-resolution image in the process of image magnification, while multi-frame SR attempts to reconstruct some information which may be lost in the process of imaging low-resolution images, to identify the details unrecognized in low-resolution images.

In this paper, we mainly consider multi-frame SR because we devote to improving the ability of identifying image details. There are many methods of Multi-frame SR, such as regularization [7, 8], wavelet [9], maximum a posteriori (MAP) [10–12], and sub-pixel based method [13], which are respectively discussed under different assumptions. Anyway, it is necessary to make the inverse problem of SR well-conditioned [14]. To solve this problem, Mudenagudi, etc. [10] introduced a tool called the zone of influence on low-resolution pixels. Based on this technique, they proposed an energy function for SR using a maximum a posteriori estimate in Markov random field and chose α-expansion move algorithm to minimize the energy function. The experiments show that this method is very effective on some images.

As a kind of graph-cut algorithm, α-expansion move algorithms [15] can effectively minimize energy functions such as class $\mathcal{F}^2$ [16]. However, to make the energy function for SR belong to the class $\mathcal{F}^2$, Mudenagudi, etc. [10] crudely remove the triple prior term from the energy function, which may bring the errors of reconstruction information to some extent. Otherwise, there is no doubt that the more accurate energy function can produce the better reconstruction results. Therefore, it is crucial how to treat this triple prior term.

In this paper, we attempt to retain the triple prior term to present the more accurate energy function for SR. We propose an expanded neighborhood and convert the triple prior term the binary one such that the energy function belongs to the class $\mathcal{F}^2$ without approximation. Experimental results show that our motivation is valid and the proposed method is effective for both synthetic low-resolution images and real images.

The rest of the paper is organized as follows. The energy function for SR is introduced in Sect. 2. In Sect. 3, we propose the expanded neighborhood and present a new way of converting triple prior term into a binary one. Experimental results and conclusion are given in Sects. 4 and 5, respectively.

2 Energy Function

2.1 Image Degradation Model

In accord with [10], $g_1, g_2, \ldots, g_n$ denote low-resolution images, which maybe come from imaging systems. And f is a high-resolution image to be reconstructed. The process of degradation from high-resolution image f to low-resolution image g_k (See Fig. 1) can be modeled as

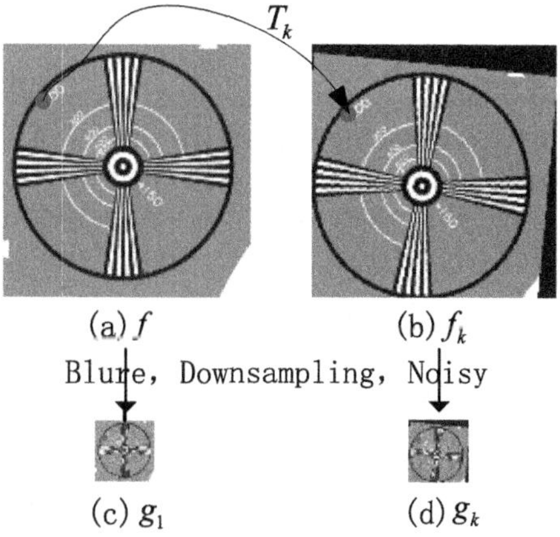

Fig. 1 The process of image degradation. **a** High-resolution image to be reconstructed. **b** High-resolution image after the transformation T_k is applied to **a**. **c** First low-resolution image treated as a reference frame. **d** kth low-resolution image

$$g_k = DH_kT_kf + \eta_k,\ k = 1, 2, \ldots, n, \tag{1}$$

where T_k is a geometric transformation that models camera shake and scene movement, H_k is a blur kernel that models the point spread function (PSF) of camera, D is a decimation operator, and η_k is noise signal. In the process of reconstruction, we can obtain T_k by registration algorithm such as [17] and model H_k as a parametric or a nonparametric PSF just like [10]. In the case of image registration, the first low-resolution image is always treated as a reference frame. In this paper, we consider reconstruction with 4 × 4 times magnification, thus D is carried out with four down-sampling factors in two directions.

2.2 Energy Function for SR in MRF-MAP Framework

The energy function for SR posed in the maximum a posteriori-Markov random field (MAP-MRF) framework by Mudenagudi, etc. [10] is as following:

$$E(f|g) = \sum_{p\in S}\sum_{k=1}^{n}\alpha_k(p,p')(h * f(p) - g_k(p'))^2 + \lambda\sum_{p,q\in N}V_{p,q}(f(p), f(q)), \tag{2}$$

where p is a pixel in high-resolution image space S, just like the red point shown in Fig. 1a. p' is a pixel in low-resolution image corresponding to p, just like the red point shown in Fig. 1d. $\alpha_k(p, p')$ is a switching function defined as follows:

$$\alpha_k\left(p,p'\right) = \begin{cases} 1, & d(p',p'') < \theta, \\ 0, & \text{otherwise.} \end{cases} \tag{3}$$

where θ is a threshold, $d(\cdot)$ is the Euclidean distance and $p'' = DT_kp$ is the projection of p onto the kth low-resolution image. It is worthwhile to note that p'' need not be an integer pixel. Equation (3) is used to determine whether p is within the zone of influence of p' (corresponding to $\alpha_k(p, p') = 1$) or not (corresponding to $\alpha_k(p, p') = 0$). Otherwise, it is necessary to normalize α_k such that $\sum_k \alpha_k = 1$. The symbol h in (2) is a discrete form of PSF and $h * f$ is the convolution of h and f. The smooth prior term $V_{p,q}(f(p), f(q))$ is defined as $V_{p,q}(f(p), f(q)) = min(\Theta, |f(p) - f(q)|)$ where Θ is a threshold. λ is a parameter of regularization used to adjust weight of prior term. $f(p)$ is the gray value of pixel p and is written as f_p in the sequel.

2.3 Expansion Move Algorithm

One of the most effective algorithms for minimizing discontinuity-preserving energy functions like (2) is α-expansion move [15], which can only minimize the following function:

$$E(f) = \sum_{p} D_p(f_p) + \sum_{p,q \in N} V_{p,q}(f_p, f_q), \tag{4}$$

where $\sum_{p} D_p(f_p)$ is the data term that measures the total cost of assigning label f_p to pixel p, $\sum_{p,q \in N} V_{p,q}(f_p, f_q)$ is the smooth prior term that measures the total cost of assigning label f_p and f_q to adjacent pixels p and q.

Otherwise, Kolmogorov and Zabih [16] propose two classes of energy functions, called class $\mathcal{F}^2$ and $\mathcal{F}^3$, which are respectively denoted by (5) and (6), where $x_1, x_2, \ldots, x_n \in \{0, 1\}$.

$$E(x_1, \ldots, x_n) = \sum_{i} E^i(x_i) + \sum_{i<j} E^{i,j}(x_i, x_j), \tag{5}$$

$$E(x_1, \ldots, x_n) = \sum_{i} E^i(x_i) + \sum_{i<j} E^{i,j}(x_i, x_j) + \sum_{i<j<k} E^{i,j,k}(x_i, x_j, x_k). \tag{6}$$

It should be noted that the energy of class $\mathcal{F}^2$ is more efficient and need less time to be minimized than the one of class $\mathcal{F}^3$.

As pointed out in [16], class $\mathcal{F}^2$ of (5) can be minimized using α-expansion if and only if $E^{i,j}$ satisfies the regularity condition

$$E^{i,j}(0, 0) + E^{i,j}(1, 1) \leq E^{i,j}(0, 1) + E^{i,j}(1, 0). \tag{7}$$

Now we are obliged to give method to make general energy function like (4) a binary function like (5). An α-expansion determines whose pixels' grey level should be changed to α such that the energy function is minimum in a single expansion move. Assignation of variable x_i depends on whether the gray value of ith pixel is changed to α. If $x_i = 1$ signs change and $x_i = 0$ signs no change, the energy function like (4) will become a binary one in the form of (5).

2.4 Rewriting the Energy Function

Obviously, the first term of (2) is not of the form of the data term shown in (4), because $h * f(p)$ shown in (2) is relative to the neighbors of the pixel p. Therefore, we should rewrite it. Let N_p be the neighborhood of the pixel p, then the convolution of $h * f$ can be calculated as follows

$$h * f(p) = \omega_{pp} f_p + \sum_{q \in N_p} \omega_{pq} f_q,$$

where ω_{pp} and ω_{pq} are PSF weights, which are determined by h. Then the first term of (2) is given by

$$\begin{aligned}
&\sum_{p \in S} \sum_{k=1}^{n} \alpha_k (h * f(p) - g_k(p'))^2 \\
&= \sum_{p \in S} \sum_{k=1}^{n} \alpha_k \left(\omega_{pp} f_p - g_k(p')\right)^2 + \sum_{p \in S} \sum_{k=1}^{n} \alpha_k \left(\sum_{q \in N_p} \omega_{pq} f_q\right)^2 \\
&\quad + \sum_{p \in S} \sum_{k=1}^{n} 2\alpha_k \left(\omega_{pp} f_p - g_k(p')\right) \sum_{q \in N_p} \omega_{pq} f_q
\end{aligned}$$

Considering $\sum_{k=1}^{n} \alpha_k = 1$, we have

$$\begin{aligned}
\sum_{p \in S} \sum_{k=1}^{n} \alpha_k \left(\sum_{q \in N_p} \omega_{pq} f_q\right)^2 &= \sum_{p \in S} \left(\sum_{q \in N_p} \omega_{pq} f_q\right)^2 \\
&= \sum_{p \in S} \sum_{q \in N_p} \left(\omega_{pq} f_q\right)^2 + \sum_{p \in S} \sum_{q,r \in N_p} \left(\omega_{pq} \omega_{pr}\right) \left(f_q f_r\right)
\end{aligned}$$

Then, the energy function of (2) is rewritten as

$$\begin{aligned}
E(f|g) = &\sum_{p \in S} \sum_{k=1}^{n} \alpha_k (\omega_{pp} f_p - g_k(p'))^2 + \sum_{p \in S} \sum_{q \in N_p} \sum_{k=1}^{n} 2\alpha_k \left(\omega_{pp} f_p - g_k(p')\right) \omega_{pq} f_q \\
&+ \sum_{p \in S} \sum_{q \in N_p} \left(\omega_{pq} f_q\right)^2 + \lambda \sum_{p,q \in N} V_{p,q}(f_p, f_q) + \sum_{p \in S} \sum_{q,r \in N_p} \left(\omega_{pq} \omega_{pr}\right) \left(f_q f_r\right)
\end{aligned} \tag{8}$$

In Eq. (8), the first term isn't concerned with the neighbors of pixel p and is the data term. The second and third terms depend on the neighbors of pixel p and are called the binary prior terms while the fourth one is called the smooth prior term. The last one is concerned with three mutually adjacent pixels and we call it a triple prior term. This triple prior term is discarded in [10], which may introduce errors in SR reconstruction to some extent. In the following section, we try to conquer this problem and retain the triple prior term.

3 Energy Minimization

3.1 *How to Deal with the Energy Function*

The energy function of (8) is in the form of class $\mathcal{F}^3$ shown in (6), but not yet in the form of class $\mathcal{F}^2$ shown in (5) due to the triple prior term. There are two immediate ways to deal with the triple prior term: it is discarded such that the energy function belongs to class $\mathcal{F}^2$ just as that in [10] yet loses accuracy; it is retained such that the energy function belongs to class $\mathcal{F}^3$ yet need more time to be minimized. We choose neither of them and try to convert the triple prior term to a binary one such that the energy function of (8) belongs to class $\mathcal{F}^2$.

3.2 *From Triple Prior Term to Binary One*

In order to convert the triple prior term shown in (8) into a binary one, we suppose without loss of generality that the discrete PSF is a 3×3 filter shown in Fig. 2a. The case for other PSF of different size can be considered similarly.

Considering the fact that $p \in N_q$ if and only if $q \in N_p$, we can interchange the order of summation in the triple prior term as shown in (9).

$$\sum_{p \in S} \sum_{q,r \in N_p} \left(\omega_{pq}\omega_{pr}\right)\left(f_q f_r\right) = \sum_{q \in S} \sum_{p \in N_q, r \in N_p} \left(\omega_{pq}\omega_{pr}\right)\left(f_q f_r\right). \quad (9)$$

For any pixel q, there are twenty-four sites whose gray value f_r may be multiplied by f_q. We use consecutive numbers to identify these sites as shown in Fig. 2b and

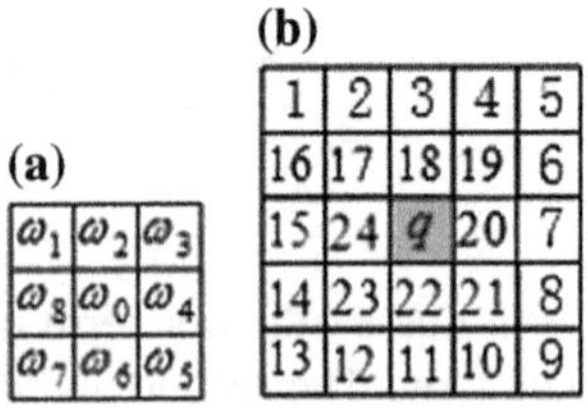

Fig. 2 **a** 3×3 filter used to model PSF. **b** Expanded neighborhood of q, $N_q^{\exp}$, including all possible site multiplied with f_q

denote the set of all these sites by $N_q^{\exp}$ and call it the expanded neighborhood of q. Then $\{r|p \in N_q, r \in N_p\} = \{r|r \in N_q^{\exp}, p \in N_r \cap N_p\}$. Thus we can write the triple prior term of (9) as

$$\sum_{p \in S} \sum_{q,r \in N_p} (\omega_{pq}\omega_{pr}) (f_q f_r) = \sum_{q \in S} \sum_{r \in N_q^{\exp}, p \in N_r \cap N_p} (\omega_{pq}\omega_{pr}) (f_q f_r) . \quad (10)$$

Considering that fact that some pixels are in S, we will calculate all coefficients of $f_q f_r$ for every site $r \in N_q^{\exp}$. Taking the site 1 shown in Fig. 2b into account, we can find out only one case in which some pixel p is located in the neighborhood of both site 1 and q as shown in Fig. 3a. Corresponding to Fig. 2a, the coefficient of $f_q f_1$ is $\omega_1\omega_5$. Then we write this coefficient at the site 1, top left of the 5×5 mask shown in Fig. 4a. Similarly, we can find out three cases about site 3 as shown in Fig. 3b–d. And their coefficients are $\omega_3\omega_5$, $\omega_2\omega_6$ and $\omega_1\omega_7$, respectively. Thus the ultimate coefficient of $f_q f_3$ is $\omega_3\omega_5 + \omega_2\omega_6 + \omega_1\omega_7 = \omega_1\omega_7 + \omega_2\omega_6 + \omega_3\omega_5$, which is written at the site 3 as shown in Fig. 4a. For the site 24, there are four cases as shown in Fig. 3e–h. Their coefficients are $\omega_6\omega_7$, $\omega_1\omega_2$, $\omega_2\omega_3$ and $\omega_5\omega_6$, respectively. Thus the ultimate coefficient of $f_q f_{24}$ is $\omega_1\omega_2 + \omega_2\omega_3 + \omega_5\omega_6 + \omega_6\omega_7$, which is written at the site 24 as shown in Fig. 4a. Analogous to the preceding cases, we can calculate all coefficients about twenty-four sites shown in Fig. 2b. Finally, we write all these coefficients in the 5×5 mask shown in Fig. 4a.

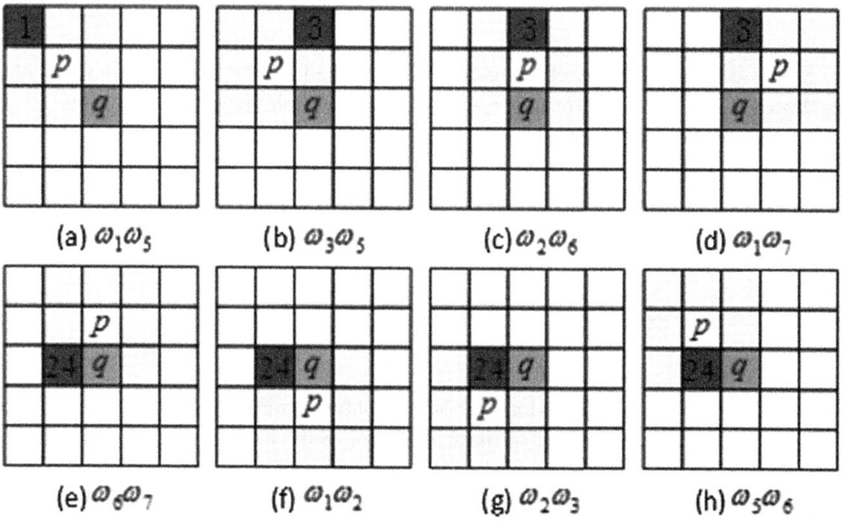

Fig. 3 All possible cases that both r and q are in the neighborhood of some p for a given $r \in N_q^{\exp}$. **a** The case of $r = 1$. **b–d** The cases of $r = 3$. **e–h** The cases of $r = 24$

We denote these coefficients shown in Fig. 4a by c_i whose subscripts correspond to the pixel sites shown in Fig. 2b. Then we can write the triple prior term of (10) as

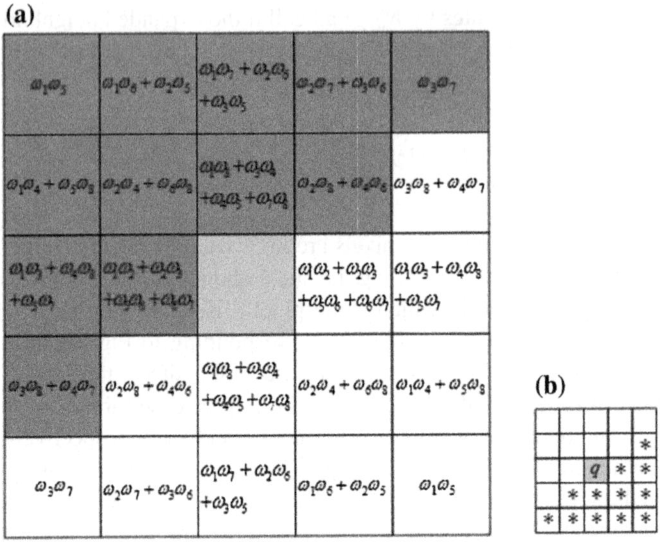

Fig. 4 **a** Coefficients of $f_q f_r$ for all $r \in N_q^{\exp}$. The value at site r is just right the coefficient of $f_q f_r$. **b** New expanded neighborhood of q. Only these sites of *asterisk* are involved in the calculations

$$\sum_{p \in S} \sum_{q,r \in N_p} \left(\omega_{pq}\omega_{pr}\right)\left(f_q f_r\right) = \sum_{q \in S} \sum_{r \in N_q^{\exp}} c_r f_q f_r. \tag{11}$$

Suppose that q is the pixel being considered at hand and it is located in the center as shown in Fig. 2b, then for any $r \in N_q^{\exp}$, we need calculate the product of f_q, f_r and their coefficient c_r. Otherwise, when pixel r is located in the center, we also need calculate the product of f_r, f_q and their coefficient c_q. Considering the central symmetry of $N_q^{\exp}$ shown in Fig. 4a, we can conclude that $c_q = c_r$, which guarantees that we need only calculate the half amount of product as shown in the following equation

$$\sum_{p \in S} \sum_{q,r \in N_p} \left(\omega_{pq}\omega_{pr}\right)\left(f_q f_r\right) = 2\sum_{q \in S} \sum_{r \in N_q^{\exp^*}} c_r f_q f_r, \tag{12}$$

where $N_q^{\exp^*}$ is a new expanded neighborhood shown in Fig. 4b.

In addition, the summation of the triple prior term shown in (12) is not relative to p. However, we should take p as a center in the triple prior term in order to make energy function of (8) consistent. Considering the fact that p, q and r are only the symbols to indicate some pixels, we can swap symbol q and r for p and q respectively as shown in following equation

$$\sum_{q \in S} \sum_{r \in N_q^{\exp^*}} c_r f_q f_r = \sum_{p \in S} \sum_{q \in N_p^{\exp^*}} c_q f_p f_q. \tag{13}$$

Therefore, the ultimate energy function is

$$E(f|g) = \sum_{p \in S} D_p(f_p) + \sum_{p,q \in N} \left(\phi_{pq}(f_p, f_q) + \lambda V_{p,q}(f_p, f_q)\right) + 2\sum_{p \in S} \sum_{q \in N_p^{\exp *}} c_q f_p f_q, \tag{14}$$

where the data term is

$$D_p(f_p) = \sum_{k=1}^{n} \alpha_k \left(\omega_{pp} f_p - g_k(p')\right)^2. \tag{15}$$

The binary prior term is

$$\phi_{pq}(f_p, f_q) = \sum_{k=1}^{n} 2\alpha_k \left(\omega_{pp} f_p - g_k(p')\right) \omega_{pq} f_q + \left(\omega_{pq} f_q\right)^2. \tag{16}$$

3.3 A New Neighborhood System

Obviously, the subscript $p, q \in N$ of a summation symbol $\sum_{p,q \in N}$ can be interpreted as both $p \in N_q$ and $q \in N_p$. This fact is contradictory to the requirement of $i < j$ in the subscript of $\sum_{i<j} E^{i,j}(x_i, x_j)$ shown in (5). In order to conquer this problem, we introduce a new neighborhood system shown in Fig. 5.

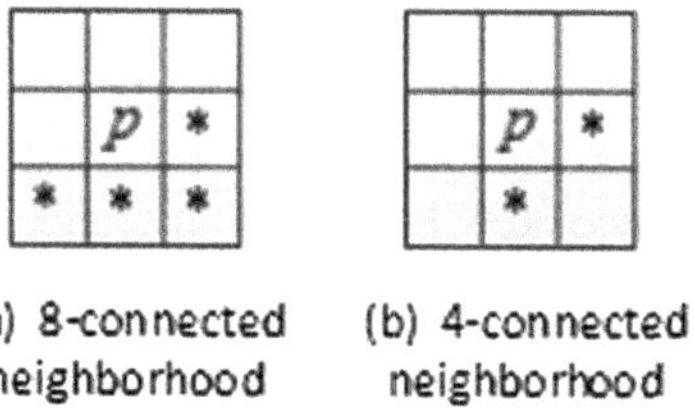

Fig. 5 **a** A new neighborhood systems N^*. **a** and **b** are new neighborhood of p with regard to 8-connected and 4-connected neighborhood, respectively. Only the pixels of *asterisk* are connected with p

Obviously, $p \in N_q^*$ and $q \in N_p^*$ are mutually exclusive. Considering the fact that $V_{p,q}(f_p, f_q) = V_{q,p}(f_q, f_p)$ and $N_p^* \subset N_p^{\exp^*}$, we write the energy function of (14) as follows,

$$E(f|g) = \underbrace{\sum_{p \in S} D_p^*(f_p)}_{\text{Data Term}} + \underbrace{\sum_{p \in S} \sum_{q \in N_p^{\exp^*}} \left\{\beta(q)\left[\phi_{pq}^*(f_p, f_q) + \lambda' V_{p,q}(f_p, f_q)\right] + 2c_q f_p f_q\right\}}_{\text{Binary Prior Term}}, \tag{17}$$

where $\lambda' = 2\lambda$, $\phi_{pq}^*(f_p, f_q) = \phi_{pq}(f_p, f_q) + \phi_{qp}(f_q, f_p)$ and $\beta(q)$ is a switching function whose value is 1 if $q \in N_p^*$ and 0 otherwise. Suppose that

$$\begin{aligned}
E^i(0) &= D_p^*(f_p), \quad E^i(1) = D_p^*(\alpha) \\
E^{ij}(0,0) &= \beta(q)\left[\phi_{pq}^*(f_p, f_q) + \lambda' V_{p,q}(f_p, f_q)\right] + 2c_q f_p f_q, \\
E^{ij}(0,1) &= \beta(q)\left[\phi_{pq}^*(f_p, \alpha) + \lambda' V_{p,q}(f_p, \alpha)\right] + 2c_q f_p \alpha, \\
E^{ij}(1,0) &= \beta(q)\left[\phi_{pq}^*(\alpha, f_q) + \lambda' V_{p,q}(\alpha, f_q)\right] + 2c_q \alpha f_q, \\
E^{ij}(1,1) &= \beta(q)\left[\phi_{pq}^*(\alpha, \alpha) + \lambda' V_{p,q}(\alpha, \alpha)\right] + 2c_q \alpha\alpha
\end{aligned}$$

for all $x_i \in \{0, 1\}$, then the energy function of (17) is in the form of class $\mathcal{F}^2$ shown in (5).

3.4 Satisfying Regularity Condition

The energy function of class $\mathcal{F}^2$ must satisfy the regularity condition as shown in (7). So we should guarantee that energy function of (17) satisfies this condition. Obviously, the following inequality holds,

$$\lambda' V_{p,q}(f_p, f_q) + \lambda' V_{p,q}(\alpha, \alpha) \le \lambda' V_{p,q}(f_p, \alpha) + \lambda' V_{p,q}(\alpha, f_q).$$

To satisfy the regularity of the binary prior term $E^{i,j}$, it is only necessary to make $\phi_{pq}^*(f_p, f_q)$ and $2c_q f_p f_q$ meet the following inequality,

$$\begin{aligned}
\phi_{pq}^*(f_p, f_q) + \phi_{pq}^*(\alpha, \alpha) &\le \phi_{pq}^*(f_p, \alpha) + \phi_{pq}^*(\alpha, f_q), \\
2c_q f_p f_q + 2c_q \alpha\alpha &\le 2c_q f_p \alpha + 2c_q \alpha f_q.
\end{aligned}$$

We consider the regularity of $\phi^*_{pq}(f_p, f_q)$ firstly. Noting that $\sum_{k=1}^{n} \alpha_k = 1$, we can conclude that

$$\begin{aligned}&\phi^*_{pq}(f_p, f_q) + \phi^*_{pq}(\alpha, \alpha) - \phi^*_{pq}(f_p, \alpha) - \phi^*_{pq}(\alpha, f_q)\\&\quad = 2\left(\omega_{pp}\omega_{pq} + \omega_{qq}\omega_{qp}\right)(\alpha - f_q)(\alpha - f_p).\end{aligned} \tag{18}$$

It is not possible to guarantee that $(\alpha - f_q)(\alpha - f_p) \leq 0$ always holds as both f_p and f_q may be assigned any gray value. Thus, we need approximate ϕ^*_{pq}.

During the derivation of (18), it is not difficult to find out that f_p and f_q of (18) are inherited from $\omega_{pp} f_p - g_k(p')$ and $\omega_{qq} f_q - g_k(q')$, respectively. Therefore, an approximation of $\omega_{pp} f_p - g_k(p')$ by $\omega_{pp}\alpha - g_k(p')$ can guarantee the regularity of ϕ^*_{pq} in an α-expansion move. Similarly, we can guarantee the regularity of $2c_q f_p f_q$ by approximating it by $2c_q \alpha f_q$.

4 Experimental Results

In this section, we demonstrate the SR reconstruction results for both synthetic and real images. In all cases, the magnification factor is 4 in each direction. We compare mainly our proposed method with the one by Mudenagudi et al. [10] and always keep their parameters the same. In addition, we also compare it with the wavelet algorithm by Ji and Fermller [9], whose parameters are as follows: $\mu_1 = 0.05$ and $\mu_2 = 0.01$.

With regard to the synthetic images, we simulate the degradation model shown in (1) and randomly generate thirty-two low-resolution images with a rotation range of $-\frac{\pi}{32}$ to $\frac{\pi}{32}$, a translation range of −16–16 pixels with no noise.

For real images, we obtain the low-resolution images with camera of Canon EOS 600d. The lens is Canon EF-S18-55mmISII, whose parameters are set as shown in Table 1.

In our implementation of SR reconstruction, we have used the graph-cut library that can be found on Kolmogorov's personal web page at http://pub.ist.ac.at/~vnk/.

Table 1 Camera parameters set when obtaining low-resolution images

Focal length	Aperture	ISO	Shutter speed	Flash	Picture style	Quality setting
18 mm	f/3.5	200	1/100 s	Off	Monochrome	S3 (480 × 720)

4.1 Validity of Our Motivation

Indeed, our method in Sect. 3 is proposed on the premise of the fact that the triple prior term shown in (8) is necessary for SR reconstruction. In this experiment, we try to verify our motivation by showing the effect of the triple prior term on SR. Therefore, we use only the triple prior term and do not use the binary one shown in (16) for this experiment.

In this experiment, we generate thirty-two low-resolution images (of size 64×64) with ground truth images (of size 256×256) shown in Fig. 6. In addition, we use the combination of all possible cases of following parameters:

$$\lambda = 0.01, 0.06, 0.1, 0.6; \quad \Theta = 8, 10, 20; \quad \theta = 0.3, 0.35, 0.4, 0.45.$$

To show the effect of the triple prior term quantitatively, we calculate peak signal-to-noise ratio (PSNR) between ground truth images and reconstruction results and then show the statistical results of PSNR about all cases of parameters in Fig. 7.

Fig. 6 Ground truth images (256×256)

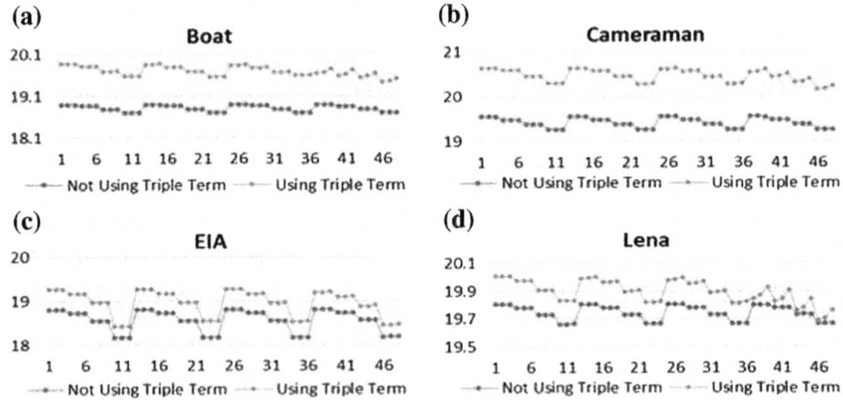

Fig. 7 Statistics of PSNR with triple term

According to the comparison results shown in Fig. 7, we can conclude that the triple prior term does improve the reconstruction results, especially about the images of boat and cameraman, whose PSNR increase about 1 dB.

4.2 Synthetic Images

To compare the proposed method with the algorithm of [10] quantitatively, we calculate the PSNR between ground truth images shown in Fig. 6 and the results of both our proposed method and [10]. Both low-resolution images and parameters are the same as the above experiment. The ultimate statistical results of the PSNR are shown in Fig. 8. Obviously, all PSNR of our proposed method are larger than the ones of [10]. Therefore, it is illustrated that our proposed method outperforms the method of [10] in terms of PSNR.

Some other comparison results are presented in Fig. 10, where the ground truth image and input low-resolution images are shown in Fig. 9a, b, respectively. The parameters used for this experiment are as follows: $\lambda = 0.01$, $\Theta = 10$ and $\theta = 0.3$. As shown in Fig. 10a, plane logos are so blurred that we can't distinguish them at all, which illustrates that the input images lose a great deal of definition. As shown in Fig. 10b, the backgrounds, especially clouds, are so real that they are almost identical to the ground truth image shown in Fig. 9a, although the plane logos are also blurred, which illustrates that the wavelet algorithm [9] is not so good at reconstructing details. As shown in Fig. 10c, d, their plane logos, even serial number, are clear and almost distinguishable, however the whole images are so bright that their backgrounds are not good enough. Therefore, we apply histogram equalization to them and show the

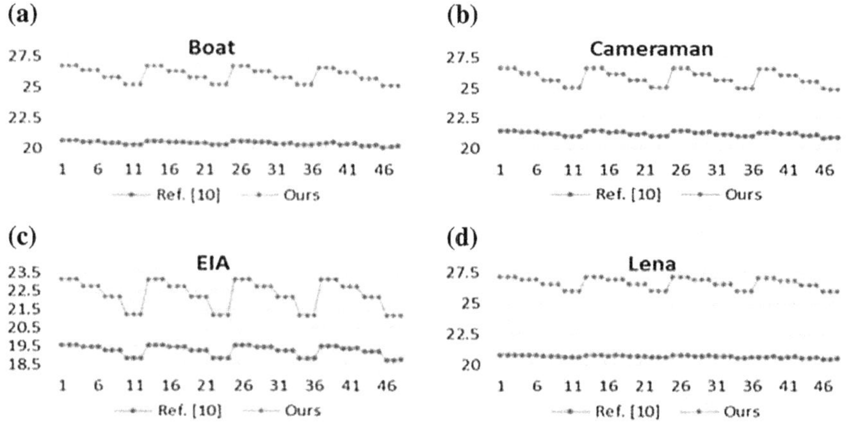

Fig. 8 Statistics of PSNR of our method and Ref. [10]

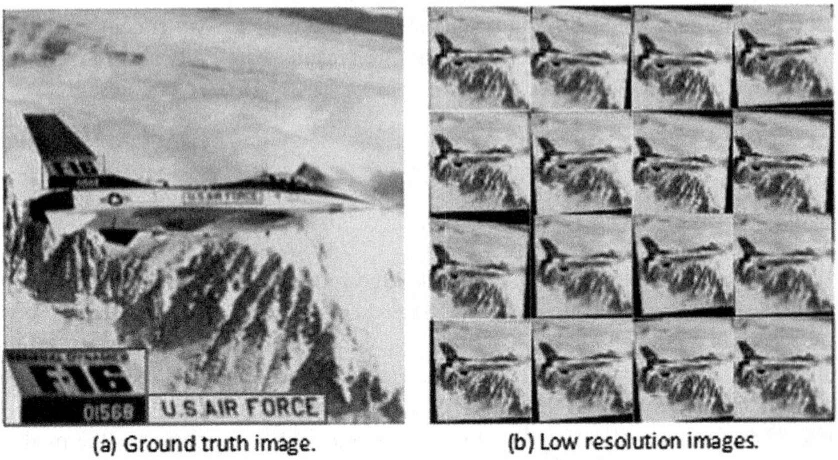

Fig. 9 **a** Ground truth image (512 × 512). Close-ups of the plane logos appear in the *lower-left corner*. **b** Low-resolution images randomly chosen from all inputs (32, each of size 128 × 128). The first one (*upper-left*) is a reference image

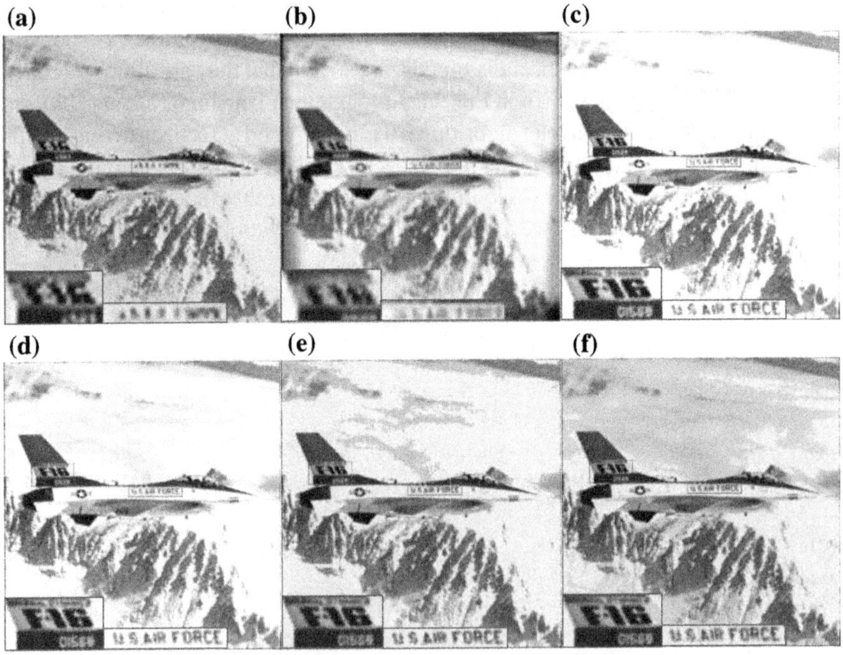

Fig. 10 Super resolution reconstruction using thirty-two input images half shown in Fig. 9b, each of size 128 × 128, and super resolved images of size 512 × 512. Close-ups of plane logos appear in the *lower-left corner* of each super resolved image. **a** Bicubic interpolation of the reference image. **b** Wavelet algorithm [9]. **c** Algorithm of [10]. **d** The proposed algorithm. **e** and **f** are the results of applying histogram equalization to **c** and **d**, respectively

results in Fig. 10e, f, which illustrate that our result has more real clouds than the result by Mudenagudi et al. [10].

4.3 Real Images

In this experiment, we demonstrate the performance of our proposed method for real images. We firstly take pictures of the scene shown in Fig. 11a with camera parameters

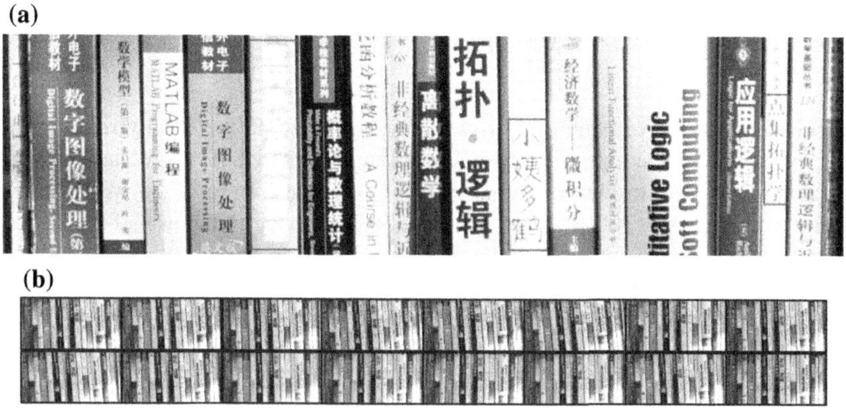

Fig. 11 **a** Real-world scene. **b** Input images obtained with camera parameters shown in Table 1

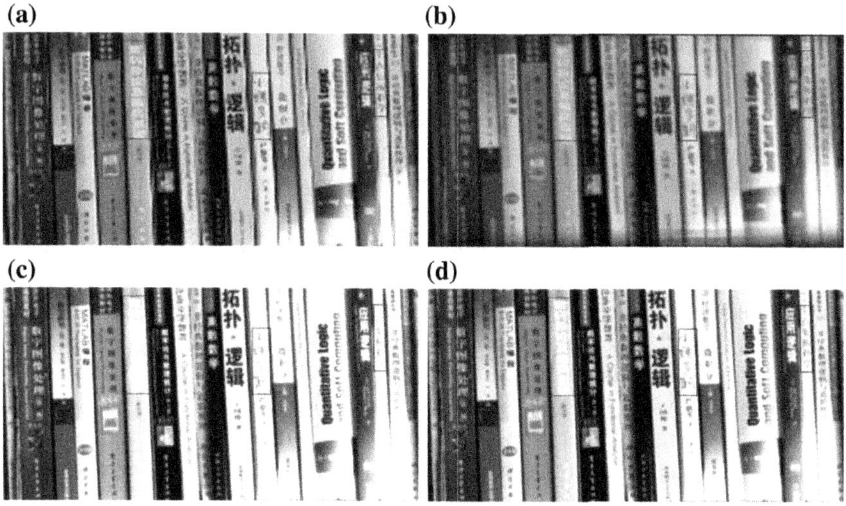

Fig. 12 Super resolution reconstruction using thirty-two input images partly shown in Fig. 11b. **a** Bicubic interpolation. **b** Wavelet algorithm [9]. **c** Algorithm of [10]. **d** The proposed algorithm

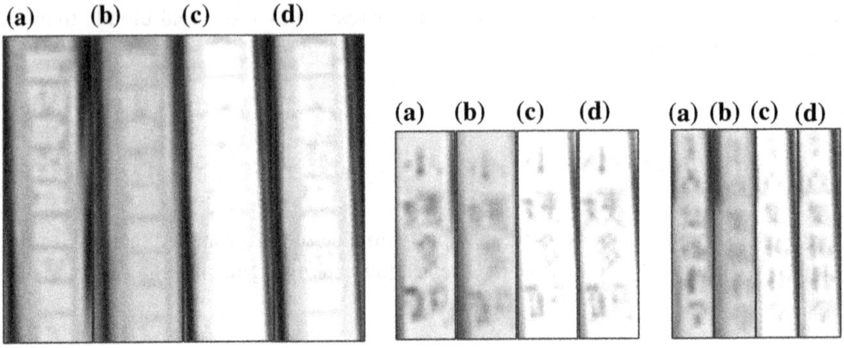

Fig. 13 Close-ups of super resolution reconstruction shown in Fig. 12

shown in Table 1. We obtain thirty-two images of size 480×720 by rotating the camera about optical axis with the maximal angle of about 10 degree and moving it from top to bottom with maximal distance of about 20 mm. Then we cut out a definite region of size 100×200 from every obtained image and show half in Fig. 11b. Finally, we take these images as inputs and compare the proposed method with others. The results and close-ups are shown in Figs. 12 and 13, respectively. As shown in Fig. 13d, our results have more Chinese character strokes and can distinguish almost all word. In contrast, as shown in Fig. 13b, c, the results by wavelet algorithm [9] are so blurred that many words can't be recognized and the results by Mudenagudi et al. [10] have few strokes.

5 Conclusions

Graph-cut is effective in minimizing a kind of special energy functions such as class $\mathcal{F}^2$. In order to use this algorithm, Mudenagudi, etc. [10] removed the triple prior term from the energy function for SR. Actually, the experimental result in Sect. 4.1 show that this triple prior term is necessary for SR.

In this paper, we propose an expanded neighborhood, with which we retain the triple prior term and make the energy function for SR a form that can be minimized using graph-cut. In order to verify this method, we obtain the low-resolution images by two ways: simulating image degradation and taking photos. Reconstruction results increased 4 times in each direction show that the proposed method is valid for not only synthetic low-resolution images but also real images. Otherwise, the proposed method also applies to other problems, such as image denoising and segmentation, as long as their energy functions include such formulas as the triple prior term of (9). This will be discussed in future studies.

Acknowledgments This work was supported by the NSF of Fujian Province of China under Grant 2016J01310 and 2016J01022, the Key Program of Fujian Province of China under Grant 2014H0034, the Huang Huizhen Discipline Construction Fund of Jimei University of China under Grant ZC2014010, and by NSF of China under Grant 61103052.

References

1. Suetake, N., Sakano, M., Uchino, E.: Image super-resolution based on local self-similarity. Opt. Rev. **1**, 26–30 (2008)
2. Glasner, D., Bagon, S., Irani, M.: Super-resolution from a single image. In: 12th IEEE International Conference on Computer Vision, pp. 349–356. IEEE Press, Kyoto (2009)
3. Freedman, G., Fattal, R.: Image and video upscaling from local self-examples. ACM Trans. Graph. **2**, 1–11 (2011)
4. Kim, C., Choi, K., Ra, J.B.: Example-based super-resolution via structure analysis of patches. IEEE Signal Process. Lett. **4**, 407–410 (2013)
5. Yang, J., Wright, J., Huang, T.S., Ma, Y.: Image super-resolution via sparse representation. IEEE Trans. Image Process. **11**, 2861–2873 (2010)
6. Ren, J., Liu, J., Guo, Z.: Context-aware sparse decomposition for image denoising and super-resolution. IEEE Trans. Image Process. **4**, 1456–1469 (2013)
7. Farsiu, S., Robinson, M.D.: Fast and robust multiframe super resolution. IEEE Trans. Image Process. **10**, 1327–1344 (2013)
8. Li, X., Hu, Y., Gao, X., Tao, D., Ning, B.: A multi-frame image super-resolution method. Signal Process. **2**, 405–414 (2010)
9. Ji, H., Fermller, C.: Robust wavelet-based super-resolution reconstruction: theory and algorithm. IEEE Trans. Pattern Anal. Mach. Intell. **4**, 649–660 (2009)
10. Mudenagudi, U., Banerjee, S., Kalra, P.K.: Space-time super-resolution using graph-cut optimization. IEEE Trans. Pattern Anal. Mach. Intell. **5**, 995–1008 (2011)
11. Zhang, D., Jodoin, P., Li, C., Wu, Y., Cai, G.: Novel graph cuts method for multi-frame super-resolution. IEEE Signal Process. Lett. **12**, 2279–2283 (2015)
12. Faramarzi, E., Rajan, D., Christensen, M.P.: A unified blind method for multi-image super-resolution and single/multi-image blur deconvolution. IEEE Trans. Image Process. **6**, 2101–2114 (2013)
13. Zhang, D., Lu, L., Li, C., Jin, T.: Ssuper-resolution image reconstruction algorithm based on sub-pixel shift (in Chinese). acta autom. sin. **12**, 2851–2861 (2014)
14. Baker, S., Kanade, T.: Limits on super-resolution and how to break them. IEEE Trans. Pattern Anal. Mach. Intell. **9**, 1167C–1183 (2002)
15. Boykov, Y., Veksler, O., Zabih, R.: Fast approximate energy minimization via graph cuts. IEEE Trans. Pattern Anal. Mach. Intell. **11**, 1222–1239 (2001)
16. Kolmogorov, V., Zabih, R.: What energy functions can be minimized via graph cuts. IEEE Trans. Pattern Anal. Mach. Intell. **2**, 147–159 (2004)
17. Bergen J. R., Anandan P., Hanna K. J., Hingorani R.: Hierarchical model-based motion estimation. In: 2nd European Conference on Computer Vision, pp. 237–252. Springer, London (1992)

The Minimum Spectral Radius of Strongly Connected Bipartite Digraphs with Complete Bipartite Subdigraph

Shu-Ting Chen, Shui-Li Chen and Wei-Quan Liu

Abstract Let $\mathfrak{D}_{n,p,q}$ be the set of strongly connected bipartite digraphs on n vertices with complete bipartite digraph, where p, q, n are positive integers and $p + q \leq n$. In this paper, the minimum spectral radius of $\mathfrak{D}_{n,p,q}$ is studied and extremal graphs with the minimum spectral radius are characterized.

Keywords Strongly connected bipartite digraph · Spectral radius · Eigenvalue · Eigenvectors · Adjacency matrix

1 Introduction

In the theory of spectra of graph, the largest eigenvalue or the spectral radius is one of the most important and intensively studied spectra. The problem of spectral radius actually originated in chemical theoretical study in the 1930s. In the theory of H*ü*ckel molecular orbital of quantum chemistry, the adjacency spectral radius of graph represents the lowest π electron energies of molecular electronics orbital. Of course, the mathematical meaning of the spectral radius of a matrix or graph is very important. In 1986, R.A. Brualdi and E.S. Solheid presented the following problems in the study of spectral radius:

Question. For a class of given graphs, we give the upper bound of the maximum eigenvalue and the lower bound of the minimum eigenvalue, and also characterize the extremal graphs which attain the upper and lower bounds, respectively.

The study of the problem caught many scholars's attention and interest. The problem of upper bounds and lower bounds of adjacent spectral radius, Laplace

S.-T. Chen · S.-L. Chen (✉) · W.-Q. Liu
Chengyi University College, Jimei University, Fujian 361021, Xiamen, China
e-mail: sgzx@jmu.edu.cn

S.-T. Chen
e-mail: stchen0725@163.com

W.-Q. Liu
e-mail: wqliu1026@163.com

T.-H. Fan et al. (eds.), *Quantitative Logic and Soft Computing 2016*,
Advances in Intelligent Systems and Computing 510,
DOI 10.1007/978-3-319-46206-6_60

spectral radius and signless Laplacian spectral radius, and the characterization of extremal undirected graphs are well treated in the literature, see [2, 4–8] and so on, but there is no much known results about digraphs. In 2010, R.A. Brualdi wrote a stimulating survey on the spectra of digraph [1]. Furthermore, some sharp upper or lower bounds on the spectral radius or the signless Laplacian spectral radius were obtained for digraphs with given graph parameters, such as girth, clique number and vertex connectivity, and the corresponding extremal graphs are characterized [3, 9]. In this paper, we study the spectral radius of strongly connected bipartite digraphs which contain a complete bipartite subdigraph $\overleftrightarrow{K}_{p,q}$ $(p \geq q > 1)$, and give characterization of the extremal graph with the least spectral radius.

2 Preliminaries

Let $D = (V(D), A(D))$ be a digraph, where $V(D)$ and $E(D)$ are the vertex set and arc set of D, respectively. For an arc $a = (i, j) \in E(D)$, where $a = (i, j)$ is the arc from i to j, that is to say i is the initial vertex of a, j is the terminal vertex of a and vertex i is a tail of vertex j. Let $D^{'} = (V(D^{'}), A(D^{'}))$ be a subdigraph of D, if $V(D^{'}) \subseteq V(D)$, $A(D^{'}) \subseteq A(D)$. A simple digraph is one which has neither loops nor multiple arcs. Let $\overrightarrow{P_n}$ and $\overrightarrow{C_n}$ denote the directed path and the directed cycle on n vertices respectively.

For a digraph $D = (V(D), A(D))$, if there is a non-empty sequence $P = v_1 a_1 v_2 a_2 \ldots a_{k-1} v_k$ $(k \geq 1)$, where $v_i \in V(D)$ $(i = 1, \ldots, k)$, $a_i \in A(D)$ $(i = 1, \ldots, k-1)$, then P is called a directed path of D.

We call the vertex i arrivable to vertex j in D, if there exists a directed path (i, j) in D. For any $i, j \in V(D)$, if there exists a directed path from i to j or a directed path from j to i, then D is called a connected digraph. We call D a strongly connected digraph if for any $i, j \in V(D)$, there exists a directed path from i to j and a directed path from j to i.

Let $G = (W, A)$ be a digraph, if $W = V \cup U$, $V \cap U = \emptyset$ and for any arc $(i, j) \in A$, $i \in V$ and $j \in U$ or $j \in V$ and $i \in U$, then the digraph $G = (W, A)$ is called a bipartite digraph. Let $\overleftrightarrow{K}_{p,q}$ be a complete bipartite digraph whose vertices can be partitioned into two subsets V_p and V_q such that no arc has both endpoints in the same subset, and every possible arc that could connect vertices in different subsets is part of the digraph. In this paper, we consider finite, simple strongly connected bipartite digraphs. For a digraph D of order n, let the adjacency matrix $A(D) = (a_{ij})$ be a $n \times n$ nonnegative matrix whose entry a_{ij} is defined as the number of arcs (i, j). Otherwise, and let $\rho(D)$ denote its spectral radius, the largest modulus of an eigenvalue of $A(D)$. Let X^T be the transpose of a vector X. Let $X = (x_1, x_2, \ldots, x_n)$ be the nonnegative unit eigenvector corresponding to $\rho(D)$, that is $AX = \rho(D)X$, then we can call $X = (x_1, x_2, \ldots, x_n)$ the Perron vector of digraph D.

In the rest of this section, let $x = (x_1, x_2, \ldots, x_n)^T$ be the unqiue positive unit eigenvector corresponding to $\rho(D)$, which corresponds to the vertex i.

Lemma 1 ([9]) *Let $D = (V(D), E(D))$ be a simple digraph on n vertices, u, v, w distinct vertices of $V(D)$, and $(u, v) \in E(D)$. Let $H = D - \{(u, v)\} + \{(u, w)\}$. If $x_w \geq x_v$, then $\rho(H) \geq \rho(D)$. Furthermore, if H is strongly connected and $x_w > x_v$, then $\rho(H) > \rho(D)$.*

Let $D = (V(D), E(D))$ be a digraph with $(u, v) \in E(D)$ and $w \notin V(D)$, $D^w = (V(D^w), E(D^w))$ with $V(D^w) = V(D) \cup \{w\}$, $E(D^w) = E(D) - \{(u, v)\} + \{(u, w), (w, v)\}$.

Lemma 2 ([9]) *Let $D(\neq \overrightarrow{C_n})$ be a strongly connected digraph, $w \notin V(D)$ and defined as before. Then $\rho(D) > \rho(D^w)$.*

Lemma 3 ([9]) *Let $D(\neq \overrightarrow{C_n})$ be a strongly connected digraph with $V(D) = \{u_1, u_2, \ldots, u_n\}$, and $\overrightarrow{P} = u_1 u_2 \ldots u_k$ $(k \geq 3)$ be a directed path of D with $d_D^+(u_i) = 1$ $(i = 2, 3, \ldots, k-1)$. Then we have $x_2 < x_3 < \cdots < x_{k-1} < x_k$.*

Lemma 4 ([9]) *Let D be a digraph and $D^1, \ldots, D^s$ be the strongly connected components of D. Then $\rho(D) = max\{\rho(D^1), \ldots, \rho(D^s)\}$.*

Lemma 5 ([9]) *Let $D_{p,q}$ be a digraph and $H_{k,l}$ be a subdigraph of $D_{p,q}$. Then $\rho(H_{k,l}) \leq \rho(D_{p,q})$. If $D_{p,q}$ is strongly connected, and $H_{k,l}$ is a proper subdigraph of $D_{p,q}$, then $\rho(H_{k,l}) < \rho(D_{p,q})$.*

3 The Spectral Radius of Strongly Connected Bipartite Digraphs Which Contain a Complete Bipartite Subdigraph

In this section, we will show that if $n \not\equiv p + q(mod\ 2)$ then $B^1_{n,p,q}$ is the unique bipartite digraph with the minimum spectral radius among all bipartite digraphs which have the complete bipartite subdigraph $\overleftrightarrow{K}_{p,q}$ $(p \geq q > 1)$, otherwise, if $n \equiv p + q(mod\ 2)$ then $D \cong B^5_{n,p,q}$ or $D \cong B^6_{n,p,q}$ is the unique bipartite digraph with the least spectral radius among all bipartite digraphs which have the complete bipartite subdigraph $\overleftrightarrow{K}_{p,q}$ $(p \geq q > 1)$.

Let $\overleftrightarrow{K}_{p,q}$ be a complete bipartite digraph with $V(\overleftrightarrow{K}_{p,q}) = V_p \cup V_q$, $A(\overleftrightarrow{K}_{p,q}) = \{(u, v), (v, u)\}$, $(u \in U, v \in V)$ and $|V_p| = p$, $|V_q| = q$. Let $\mathfrak{D}_{n,p,q}$ denote the set of strongly connected bipartite digraphs on n vertices with complete bipartite digraph $\overleftrightarrow{K}_{p,q}$. As we know, if $p + q = n$, then $\mathfrak{D}_{n,p,q} = \{\overleftrightarrow{K}_{p,q}\}$ and $\rho(\overleftrightarrow{K}_{p,q}) = \sqrt{pq}$. Thus we only discuss the cases when $p + q \leq n - 1$ and $p \geq q \geq 1$. In the rest of this section, we just discuss under this assumption.

Let $B^1_{n,p,q} = (V(B_{n,p,q}), E(B_{n,p,q}))$ be a digraph obtained by adding a directed path $\overrightarrow{P_{n-p-q}} = u_1 u_{p+q+1} u_{p+q+2} \ldots u_n u_p$ to a complete bipartite digraph $\overleftrightarrow{K}_{p,q}$ such that $V(K_{p,q}) \cap V(\overrightarrow{P_{n-p-q}}) = \{u_1, u_p\}$ (as shown in Fig. 2a), where $V(B^1_{n,p,q}) = \{u_1, u_2, \ldots, u_n\}$. Clearly, $B^1_{n,p,q} \in \mathfrak{D}_{n,p,q}$.

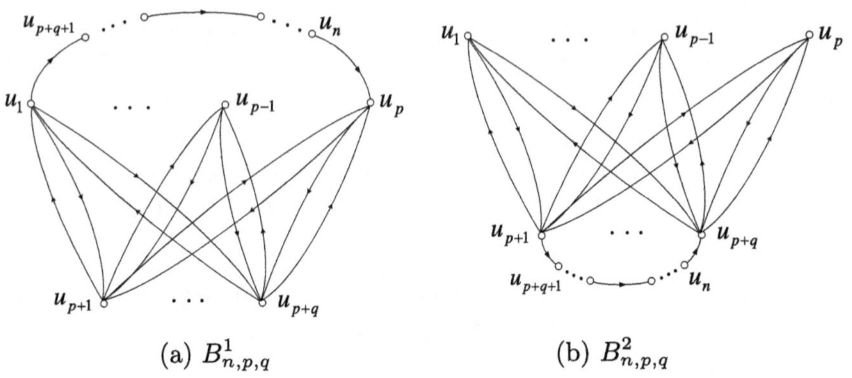

Fig. 1 $B^1_{n,p,q}$ and $B^2_{n,p,q}$

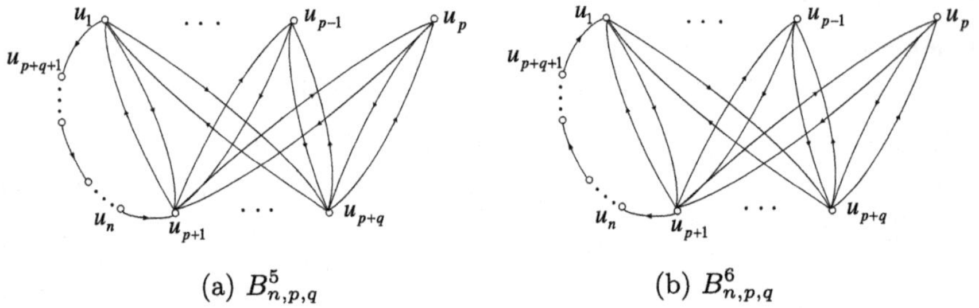

Fig. 2 $B^5_{n,p,q}$ and $B^6_{n,p,q}$

Let $B^2_{n,p,q} = (V(B_{n,p,q}), E(B_{n,p,q}))$ be a digraph obtained by adding a directed path $\overrightarrow{P_{n-p-q}} = u_{p+1}u_{p+q+1}u_{p+q+2}\dots u_n u_{p+q}$ such that $V(K_{p,q}) \cap V(\overrightarrow{P_{n-p-q}}) = \{u_{p+1}, u_{p+q}\}$ (as shown in Fig. 1b), where $V(B^2_{n,p,q}) = \{u_1, u_2, \dots, u_n\}$. Clearly, $B^2_{n,p,q} \in \mathfrak{D}_{n,p,q}$.

Let $B^3_{n,p,q} = (V(B_{n,p,q}), E(B_{n,p,q}))$ be a digraph obtained by adding a directed cycle $\overrightarrow{C_{n-p-q}} = u_1 u_{p+q+1} u_{p+q+2} \dots u_n u_p u_1$ to a complete bipartite digraph $\overleftrightarrow{K_{p,q}}$ such that $V(\overleftrightarrow{K_{p,q}}) \cap V(\overrightarrow{C_{n-p-q}}) = \{u_1\}$ (as shown in Fig. 2a), where $V(B^3_{n,p,q}) = \{u_1, u_2, \dots, u_n\}$. Clearly, $B^3_{n,p,q} \in \mathfrak{D}_{n,p,q}$.

Let $B^5_{n,p,q} = (V(B_{n,p,q}), E(B_{n,p,q}))$ be a digraph obtained by adding a directed path $\overrightarrow{P_{n-p-q}} = u_1 u_{p+q+1} u_{p+q+2} \dots u_n u_{p+1}$ to a complete bipartite digraph $\overleftrightarrow{K_{p,q}}$, such that $V(K_{p,q}) \cap V(\overrightarrow{P_{n-p-q}}) = \{u_1, u_{p+1}\}$, where $V(B^5_{n,p,q}) = \{u_1, u_2, \dots, u_n\}$ (as shown in Fig. 2a). Clearly, $B^5_{n,p,q} \in \mathfrak{D}_{n,p,q}$.

Lemma 6 *Let $x = (x_1, x_2, \dots, x_n)^T$ be the Perron eigenvector corresponding to $\rho = \rho(B^1_{n,p,q})$, where x_i corresponds to the vertex u_i, then we have:*

(i) $x_1 > x_p$;

(ii) $x_{p+1} > x_p$.

Proof (i) Since

$$(A(B^1_{n,p,q})x)_1 = \rho x_1 = \sum_{u \in N^+_{(u_1)}} x_u = \sum_{u \in V_q} x_u + x_{p+q+1}, \tag{1}$$

$$(A(B^1_{n,p,q})x)_p = \rho x_p = \sum_{u \in N^+_{(u_p)}} x_u = \sum_{u \in V_q} x_u. \tag{2}$$

From (1)–(2), then we have

$$\rho(x_1 - x_p) = x_{p+q+1} > 0. \tag{3}$$

Thus $x_1 > x_p$ by $\rho > 0$.

(ii) Since

$$(A(B^1_{n,p,q})x)_i = \rho x_i = \sum_{u \in N^+_{(u_i)}} x_u = \sum_{u \in V_q} x_u \ (i = 2, 3, \ldots, p),$$

then $\rho x_2 = \rho x_3 = \cdots = \rho x_p$. Noting that $B^1_{n,p,q}$ is strongly connected and by using Perron–Frobenius Theorem, we have $\rho > 0$. Thus

$$x_2 = x_3 = \cdots = x_p \triangleq x_p. \tag{4}$$

Since

$$(A(B^1_{n,p,q})x)_j = \rho x_j = \sum_{u \in N^+_{(u_j)}} x_u = \sum_{u \in V_p} x_u \ (j = p+1, p+2, ..., p+q),$$

then similarly, we have

$$x_{p+1} = x_{p+2} = \cdots = x_{p+q} \triangleq x_{p+1}, \tag{5}$$

From (3)–(5), we have

$$(A(B^1_{n,p,q})x)_p = \rho x_p = \sum_{u \in N^+_{(u_p)}} x_u = \sum_{u \in V_q} x_u = q x_{p+1}, \tag{6}$$

$$(A(B^1_{n,p,q})x)_{p+1} = \rho x_{p+1} = \sum_{u \in V_p} x_u + x_1 = (p-1)x_p + x_1 > p x_p. \tag{7}$$

From (6) and (7), we can briefly have

$$\begin{cases} \rho = \frac{qx_{p+1}}{x_p} \\ \rho > \frac{px_p}{x_{p+1}}. \end{cases}$$

Noting that $p \geq q \geq 1, x_p > 0, x_{p+1} > 0$, then

$$\frac{x_p}{x_{p+1}} < \sqrt{\frac{q}{p}} \leq 1.$$

Thus $x_{p+1} > x_p$. □

Theorem 1 *Let $B^4_{n,p,q} = B^2_{n,p,q} - \{(u_n, u_{p+q})\} + \{(u_n, u_{p+1})\}$, then*

$$\rho(B^2_{n,p,q}) < \rho(B^4_{n,p,q})$$

Proof Clearly, $B^4_{n,p,q}$ is strongly connected. Let $x = (x_1, x_2, ..., x_n)^T$ be the Perron eigenvector corresponding to $\rho = \rho(B^2_{n,p,q})$, where x_i corresponds to the vertex u_i. By Lemma 1, we only need to show $x_{p+1} > x_{p+q}$.

Since

$$(A(B^2_{n,p,q})x)_{p+1} = \rho x_{p+1} = \sum_{u \in N^+_{(u_{p+1})}} x_u = \sum_{u \in V_p} x_u + x_{p+q+1} \quad (8)$$

and

$$(A(B^2_{n,p,q})x)_{p+q} = \rho x_{p+q} = \sum_{u \in N^+_{(u_{p+q})}} x_u = \sum_{u \in V_p} x_u \quad (9)$$

From (8)–(9) we have

$$\rho(x_{p+1} - x_{p+q}) = x_{p+q+1} > 0.$$

Thus, $x_{p+1} > x_{p+q}$, by $\rho > \sqrt{pq} > 0$. □

Theorem 2 *Let $B^3_{n,p,q} = B^1_{n,p,q} - \{(u_n, u_p)\} + \{(u_n, u_1)\}$, then*

$$\rho(B^1_{n,p,q}) < \rho(B^3_{n,p,q})$$

Proof Clearly, $B^3_{n,p,q}$ is strongly connected. Let $x = (x_1, x_2, ..., x_n)^T$ be the Perron eigenvector corresponding to $\rho = \rho(B^1_{n,p,q})$, where x_i corresponds to the vertex u_i. Thus $\rho(B^1_{n,p,q}) < \rho(B^3_{n,p,q})$ by Lemmas 1 and 6. □

Let $\rho(D)$ be the spectral radius of strongly connected digraph D. According to the Perron–Frobenius Theorem, we have

$$\rho(D) = \max_{||Y||=1} Y^T A(D) Y = X^T A(D) X,$$

where X is the Perron vector corresponds to $\rho(D)$.

Theorem 3 $\rho(B^5_{n,p,q}) = \rho(B^6_{n,p,q})$.

Proof Let $x = (x_1, x_2, ..., x_n)^T$ be the Perron eigenvector corresponding to $\rho(B^5_{n,p,q})$, where x_i corresponds to the vertex u_i. Let $Y = (y_1, y_2, ..., y_n)^T$ be the Perron eigenvector corresponding to $\rho(B^6_{n,p,q})$, where y_i corresponding to the vertex u_i. By Perron–Frobenius Theorem, we have

$$\rho(B^5_{n,p,q}) = X^T A(B^5_{n,p,q}) X = Y^T A(B^6_{n,p,q}) Y = \rho(B^6_{n,p,q})$$

Thus $\rho(B^5_{n,p,q}) = \rho(B^6_{n,p,q})$. □

Theorem 4 *Let* $B^5_{n,p,q} = B^1_{n,p,q} - \{(u_n, u_p)\} + \{(u_n, u_{p+1})\}$, *then*

$$\rho(B^1_{n,p,q}) < \rho(B^5_{n,p,q}),$$

Thus $\rho(B^1_{n,p,q}) < \rho(B^5_{n-1,p,q})$.

Proof Clearly, $B_{n,p,q}$ is strongly connected. Let $x = (x_1, x_2, ..., x_n)^T$ be the Perron eigenvector corresponding to $\rho = \rho(B^1_{n,p,q})$, where x_i corresponds to the vertex u_i. By Lemmas 1 and 6, we have $\rho(B^5_{n,p,q}) > \rho(B^1_{n,p,q})$. Then $\rho(B^5_{n-1,p,q}) > \rho(B^5_{n,p,q})$ by Lemma 2, thus $\rho(B^5_{n-1,p,q}) > \rho(B^1_{n,p,q})$. □

Theorem 5 $\rho(B^5_{n,p,q}) \leq \rho(B^1_{n-1,p,q})$.

Proof Let $B^{1*}_{n,p,q} = B^5_{n,p,q} - \{(u_{n-1}, u_n)\} + \{(u_{n-1}, u_p)\}$ and $x = (x_1, x_2, ..., x_n)^T$ be the Perron eigenvector corresponding to $\rho = \rho(B^5_{n,p,q})$, where x_i corresponds to the vertex u_i. By Lemma 1, we only need to show $x_p > x_n$.

Since $(A(B^5_{n,p,q})x)_p = \rho x_p = \sum_{u \in N^+_{(u_p)}} x_u = \sum_{u \in V_q} x_u = q x_{p+1}$,

$$(A(B^5_{n,p,q})x)_n = \rho x_n = \sum_{u \in N^+_{(u_n)}} x_u = x_{p+1}.$$

Then $x_p \geq x_n$, by $q \geq 1$. Thus $\rho(B^{1*}_{n,p,q}) \geq \rho(B^5_{n,p,q})$. Since $B^{1*}_{n,p,q} = B^1_{n-1,p,q} \cup \{u_n\}$, we have $\rho(B^{1*}_{n,p,q}) = max\{\rho(B^1_{n-1,p,q}), \rho(u_n)\}$ by Lemma 4, thus $\rho(B^1_{n-1,p,q}) = \rho(B^{1*}_{n,p,q}) \geq \rho(B^5_{n,p,q})$. □

Theorem 6 $\rho(B^1_{n,p,q}) = \sqrt[n-p-q+1]{\frac{pq+\sqrt{(pq)^2+4q}}{2}}$

Proof Let $x = (x_1, x_2, ..., x_n)^T$ be the Perron eigenvector corresponding to $\rho(B^1_{n,p,q})$, where x_i corresponds to the vertex u_i. By Lemma 6, we have

$$(A(B^1_{n,p,q})x)_1 = \rho x_1 = \sum_{u \in N^+_{(u_1)}} x_u = \sum_{u \in V_q} x_u + x_{p+q+1} = q x_{p+1} + x_{p+q+1} \tag{10}$$

$$(A(B^1_{n,p,q})x)_{p+1} = \rho x_{p+1} = \sum_{u \in V_p} x_u + x_1 = x_1 + (p-1)x_p \tag{11}$$

$$(A(B^1_{n,p,q})x)_p = \rho x_p = \sum_{u \in N^+_{(u_p)}} x_u = \sum_{u \in V_q} x_u = q x_{p+1} \tag{12}$$

$$\rho^{n-p-q} x_{p+q+1} = x_p \tag{13}$$

From (10)–(13), we have

$$\begin{aligned} \rho * (11) = \rho^2 x_{p+1} &= \rho x_1 + (p-1)\rho x_p \\ &= q x_{p+1} + x_{p+q+1} + (p-1) q x_{p+1} \\ &= q x_{p+1} + \frac{q}{\rho^{n-p-q+1}} x_{p+1} + (p-1) q x_{p+1}, \end{aligned}$$

multiply both sides of the equation by $\rho^{n-p-q+1}$, we have

$$\rho^{n-p-q+3} - pq\rho^{n-p-q+1} - q = 0.$$

Let $t = \rho^{n-p-q+1}(B^1_{n,p,q})$, then we have $t^2 - pqt - q = 0$, since $\rho > 0$, we have

$$\rho^{n-p-q+1}(B^1_{n,p,q}) = t = \frac{pq + \sqrt{(pq)^2 + 4q}}{2},$$

and thus

$$\rho(B^1_{n,p,q}) = \sqrt[n-p-q+1]{\frac{pq + \sqrt{(pq)^2 + 4q}}{2}}$$

□

Corollary 1 $\rho(B^1_{n,p,q}) \leq \rho(B^2_{n,p,q})$.

Proof We have

$$\rho(B^2_{n,p,q}) = \sqrt[n-p-q+1]{\frac{qp + \sqrt{(qp)^2 + 4p}}{2}},$$

since $B^2_{n,p,q} \cong B^1_{n,q,p}$,
We get

$$\rho(B^1_{n,p,q}) = \sqrt[n-p-q+1]{\frac{pq + \sqrt{(pq)^2 + 4q}}{2}} \leq \sqrt[n-p-q+1]{\frac{pq + \sqrt{(pq)^2 + 4p}}{2}} = \rho(B^2_{n,p,q}),$$

since $p \geq q$, thus $\rho(B^1_{n,p,q}) \leq \rho(B^2_{n,p,q})$. □

Theorem 7 *Let $p \geq q > 1$, $p+q \leq n-1$, $n \not\equiv p+q (mod\ 2)$ and $D \in \mathfrak{D}_{n,p,q}$ be a bipartite digraph, then $\rho(D) \geq \rho(B^1_{n,p,q})$ and the equality holds if and only if $D \cong B^1_{n,p,q}$.*

Proof Clearly, $\overleftrightarrow{K}_{p,q}$ is a proper subdigraph of D since $D \in \mathfrak{D}_{n,p,q}$. Since D is strongly connected, it is possible to obtain a digraph H from D by deleting vertices and arcs in such a way that one has a subdigraph $\overleftrightarrow{K}_{p,q}$. Therefore:

(i) $H \cong B^1_{p+q+l,p,q}$ $(l \equiv 1\ (mod 2), l \geq 2)$ or
(ii) $H \cong B^2_{p+q+l,p,q}$ $(l \equiv 1\ (mod 2), l \geq 2)$ or
(iii) $H \cong B^3_{p+q+l,p,q}$ $(l \equiv 1\ (mod 2), l \geq 2)$ or
(iv) $H \cong B^4_{p+q+l,p,q}$ $(l \equiv 1\ (mod 2), l \geq 2)$ or
(v) $H \cong B^5_{p+q+k,p,q}$ $(k \equiv 0\ (mod 2), k \geq 2)$ or
(vi) $H \cong B^6_{p+q+k,p,q}$ $(k \equiv 0\ (mod 2), k \geq 2)$

By Lemma 5, $\rho(H) \leq \rho(D)$, the equality holds if and only if $H \cong D$.

Case (i). $H \cong B^1_{p+q+l,p,q}$ $(l \equiv 1\ (mod 2), l \geq 2)$.

For $n \not\equiv p+q(mod\ 2)$, insert $n-p-q-l$ vertices into the directed path $\overrightarrow{P_l}$ such that the resulting bipartite digraph is $B^1_{n,p,q}$. Then $\rho(B^1_{n,p,q}) < \rho(H)$ by using Lemma 2 repeatedly $n-p-q-l$ times.

Case (ii). $H \cong B^2_{p+q+l,p,q}$ $(l \equiv 1\ (mod 2), l \geq 2)$.

Insert $n-p-q-l$ vertices into the directed path $\overrightarrow{P_l}$ such that the resulting bipartite digraph is $B^2_{n,p,q}$. Then $\rho(B^2_{n,p,q}) < \rho(H)$ by using Lemma 2 repeatedly $n-p-q-l$ times, and thus $\rho(B^1_{n,p,q}) \leq \rho(B^2_{n,p,q}) < \rho(H)$ by Corollary 1.

Case (iii). $H \cong B^3_{p+q+l,p,q}$ $(l \equiv 1\ (mod 2), l \geq 2)$.

Insert $n-p-q-l$ vertices into the directed cycle $\overrightarrow{C_l}$ such that the resulting bipartite digraph is $B^3_{n,p,q}$. Then $\rho(B^3_{n,p,q}) < \rho(H)$ by using Lemma 2 repeatedly $n-p-q-l$ times, and thus $\rho(B^1_{n,p,q}) < \rho(B^3_{n,p,q}) < \rho(H)$ by Theorem 6.

Case (iv). $H \cong B^4_{p+q+l,p,q}$ $(l \equiv 1\ (mod 2), l \geq 2)$.

Insert $n-p-q-l$ vertices into the directed cycle $\overrightarrow{C_l}$ such that the resulting digraph is $B^4_{n,p,q}$. Then $\rho(B^4_{n,p,q}) < \rho(H)$ by using Lemma 2 repeatedly $n-p-q-l$ times, and thus $\rho(B^1_{n,p,q}) \leq \rho(B^2_{n,p,q}) < \rho(B^4_{n,p,q}) < \rho(H)$ by Corollary 1 and Theorem 1.

Case (v). $H \cong B^5_{p+q+k,p,q}$ $(k \equiv 0\ (mod 2), l \geq 2)$.

Insert $n-p-q-k-1$ vertices into the directed path $\overrightarrow{P_k}$ such that the resulting bipartite digraph is $B^5_{n-1,p,q}$. Then $\rho(B^5_{n-1,p,q}) < \rho(H)$ by using Lemma 2 repeatedly $n-p-q-k-l$ times and thus $\rho(B^1_{n,p,q}) < \rho(B^5_{n-1,p,q}) < \rho(H)$ by Theorem 4.

Case (vi). $H \cong B^6_{p+q+k,p,q}$ $(k \equiv 0\ (mod 2), l \geq 2)$.

Insert $n-p-q-k-1$ vertices into the directed path $\overrightarrow{P_k}$, such that the resulting bipartite digraph is $B^6_{n-1,p,q}$. Then $\rho(B^6_{n-1,p,q}) < \rho(H)$ by using Lemma 2 repeatedly $n-p-q-k-1$ times, and thus $\rho(B^1_{n,p,q}) < \rho(B^5_{n-1,p,q}) = \rho(B^6_{n-1,p,q}) < \rho(H)$, by Theorems 3 and 4.

Combining the above six cases, we have $\rho(D) \geq \rho(B^1_{n,p,q})$, the equality holds if and only if $D \cong B^1_{n,p,q}$, where $n \not\equiv p+q(mod\ 2)$ and $p > q \geq 2$. □

Corollary 2 *Let $p \geq q \geq 1$, $p+q \leq n-1$, $n \not\equiv p+q(mod\ 2)$ and $D \in \mathfrak{D}_{n,p,q}$ be a bipartite digraph, then $\rho(D) \geq \sqrt[n-p-q+1]{\frac{pq+\sqrt{(pq)^2+4q}}{2}}$.* □

Theorem 8 *Let $p \geq q \geq 1$, $p+q \leq n-1$, $n \equiv p+q(mod\ 2)$ and $D \in \mathfrak{D}_{n,p,q}$ be a bipartite digraph, then $\rho(D) \geq \rho(B^5_{n,p,q}) = \rho(B^6_{n,p,q})$ and the equality holds if and only if $D \cong B^5_{n,p,q}$ or $D \cong B^6_{n,p,q}$.*

Proof Clearly, $\overleftrightarrow{K}_{p,q}$ is a proper subdigraph of D since $D \in \mathfrak{D}_{n,p,q}$. Since D is strongly connected, it is possible to obtain a digraph H from D by deleting vertices and arcs in such a way that one has subdigraph $\overleftrightarrow{K}_{p,q}$. Therefore:

(1) $H \cong B^5_{p+q+k,p,q}$ $(k \equiv 1\ (mod 2), k \geq 2)$ or
(2) $H \cong B^6_{p+q+k,p,q}$ $(k \equiv 1\ (mod 2), k \geq 2)$or
(3) $H \cong B^1_{p+q+l,p,q}$ $(l \equiv 1\ (mod 2), l \geq 2)$ or
(4) $H \cong B^2_{p+q+l,p,q}$ $(l \equiv 1\ (mod 2), l \geq 2)$ or
(5) $H \cong B^3_{p+q+l,p,q}$ $(l \equiv 0\ (mod 2), l \geq 2)$ or
(6) $H \cong B^4_{p+q+l,p,q}$ $(l \equiv 0\ (mod 2), l \geq 2)$

By Lemma 5, $\rho(H) \leq \rho(D)$, the equality holds if and only if $H \cong D$.

Case (i). $H \cong B^5_{p+q+k,p,q}$ $(k \equiv 1\ (mod 2), l \geq 2)$

Insert $n-p-q-k$ vertices into the directed path $\overrightarrow{P_l}$ such that the resulting bipartite digraph is $B^5_{n,p,q}$. Then $\rho(B^5_{n,p,q}) = \rho(B^6_{n,p,q}) < \rho(H)$ by using Lemma 2 repeatedly $n-p-q-k$ times and Theorem 3.

Case (ii). $H \cong B^6_{p+q+k,p,q}$ $(k \equiv 1\ (mod 2), l \geq 2)$

Insert $n-p-q-k$ vertices into the directed cycle $\overrightarrow{P_k}$ such that the resulting bipartite digraph is $B^6_{n,p,q}$. Then $\rho(B^6_{n,p,q}) < \rho(H)$ by using Lemma 2 repeatedly $n-p-q-k$ times, and thus $\rho(B^5_{n,p,q}) = \rho(B^6_{n,p,q}) < \rho(H)$ by Theorem 3.

Case (iii). $H \cong B^1_{p+q+l,p,q}$ $(l \equiv 1\ (mod 2), l \geq 2)$

Insert $n-p-q-l-1$ vertices into the directed path $\overrightarrow{P_l}$ such that the resulting bipartite digraph is $B^1_{n-1,p,q}$, then $\rho(B^1_{n-1,p,q}) < \rho(H)$ by using Lemma 2 repeatedly

$n - p - q - l - 1$ times, and thus $\rho(B^6_{n,p,q}) = \rho(B^5_{n,p,q}) \leq \rho(B^1_{n-1,p,q}) < \rho(H)$ by Theorems 3 and 5.

Case (iv). $H \cong B^2_{p+q+l,p,q} \quad (l \equiv 1\ (mod 2), l \geq 2)$

Insert $n - p - q - l - 1$ vertices into the directed path $\overrightarrow{P_l}$ such that the resulting digraph is $B^2_{n-1,p,q}$, then $\rho(B^2_{n-1,p,q}) < \rho(H)$ by using Lemma 2 repeatedly $n - p - q - l - 1$ times, and thus $\rho(B^6_{n,p,q}) = \rho(B^5_{n,p,q}) \leq \rho(B^1_{n-1,p,q}) \leq \rho(B^2_{n-1,p,q}) < \rho(H)$ by Theorems 3 and 5, Corollary 1.

Case (v). $H \cong B^3_{p+q+l,p,q} \quad (l \equiv 0\ (mod 2), l \geq 2)$

Insert $n - p - q - l - 1$ vertices into the directed cycle $\overrightarrow{C_l}$ such that the resulting digraph is $B^3_{n-1,p,q}$, then $\rho(B^3_{n-1,p,q}) < \rho(H)$ by using Lemma 2 repeatedly $n - p - q - l - 1$ times, and thus $\rho(B^6_{n,p,q}) = \rho(B^5_{n,p,q}) \leq \rho(B^1_{n-1,p,q}) < \rho(B^3_{n-1,p+q}) < \rho(H)$ by Theorems 3, 5 and 6.

Case (vi). $H \cong B^4_{p+q+l,p,q} \quad (l \equiv 0\ (mod 2), l \geq 2)$

Insert $n - p - q - l - 1$ vertices into the directed cycle $\overrightarrow{C_l}$ such that the resulting digraph is $B^4_{n-1,p,q}$, then $\rho(B^4_{n-1,p,q}) < \rho(H)$, by using Lemma 2 repeatedly $n - p - q - l - 1$ times, and thus $\rho(B^6_{n,p,q}) = \rho(B^5_{n,p,q}) \leq \rho(B^1_{n-1,p,q}) \leq \rho(B^2_{n-1,p,q}) < \rho(B^4_{n-1,p,q}) < \rho(H)$ by Theorems 1, 3 and 5, Corollary 1.

Combining the above six cases, we have $\rho(D) \geq \rho(B^5_{n,p,q}) = \rho(B^6_{n,p,q})$, the equality holds if and only if either $D \cong B^5_{n,p,q}$ or $D \cong B^6_{n,p,q}$, where $n \equiv p + q (mod\ 2)$ and $p \geq q \geq 1$. □

Acknowledgments This work was supported by the Natural Science Foundation of Fujian Province of China (No. 2016J01309, No. 2016J01310) and the key projects of Science and Technology Agency of Fujian Province of China (No. 2014H0034).

References

1. Brualdi, R.: Spectra of digraphs. Linear Algebra Appl. **432**, 2181–2213 (2010)
2. Drury, S.W., Lin, H.Q.: Extremal digraphs with given clique number. Linear Algebra Appl. **439**, 328–345 (2013)
3. Hong, W., You, L.H.: Spectral radius and signless Laplacian spectral radius of strongly connected digraphs. Linear Algebra Appl. **457**, 93–113 (2014)
4. Jin, Y.L., Zhang, X.D.: On the spectral radius of simple digraphs with prescribed number of arcs. Discrete Math. **338**, 1555–1564 (2015)
5. Lin, H.Q., Drury, S.W.: The maximum Perron roots of digraphs with some given parameters. Discrete Math. **313**, 2607–2613 (2013)
6. Lin, H.Q., Shu, J.L.: Spectral radius of digraphs with given dichromatic number. Linear Algebra Appl. **434**, 2462–2467 (2011)
7. Lin, H.Q., Shu, J.L.: The distance spectral radius of digraphs. Discrete Appl. Math. **161**, 2537–2543 (2013)
8. Lin, H.Q., Yang, W.H., Zhang, H.L., Shu, J.L.: Distance spectral radius of digraphs with given con-nectivity. Discrete Math. **312**, 1849–1856 (2012)
9. Lin, H.Q., Shu, J.L., Wu, Y.R., Yu, L.G.: Spectral radius of strongly connected digraphs. Discrete Math. **312**, 3663–3669 (2012)

A Fuzzy Support Vector Machine Algorithm and Its Application in Telemarketing

Ming Liu, Ya-Mei Yan and Yu-De He

Abstract Telemarketing applications in various industries are increasingly popular in modern society. Telemarketing is one of the important means to contact with customers in insurance companies, banks and other financial systems. For example, if we accurately predict telemarketing successfully, we can appropriately reduce the cost and the scope of marketing in banking, which has very important significance. In this paper, the authors present a fuzzy support vector machine (SVM) algorithm, based on the data of customer information obtained in telemarketing campaigns in a financial institution of Portugal, using Weka and Matlab software, predict the success of telemarketing. Experimental results show that the fuzzy SVM algorithm outperforms the traditional SVM with 92.89 % predicting accuracy rate.

Keywords Telemarketing · Fuzzy algorithm · Support vector machine · Cluster analysis

1 Introduction

With the progress of society and the development of science and technology, competitions in the business community are more intense in the 21st century. Now, in the fierce global competition, the marketing model of business community is constantly developing and changing. Direct Marketing is a model of providing information, by e-mail or telephone or other means to get a direct response from consumers for the purpose of business activity. Now this marketing model has been applied by an increasing number of corporations, especially in the financial service industries, banks and insurance companies, whose core is the study of how to identify potential customers. Telemarketing is a direct marketing method which takes a more scientific approach to select customers who are more likely to get a feedback. Telemarketing is one of the important methods for banks to build customer relationships.

M. Liu (✉) · Y.-M. Yan · Y.-D. He
College of Basic Sciences, Changchun University of Technology, Changchun 130012, China
e-mail: jlcclm@163.com

T.-H. Fan et al. (eds.), *Quantitative Logic and Soft Computing 2016*,
Advances in Intelligent Systems and Computing 510,
DOI 10.1007/978-3-319-46206-6_61

By predicting the success of telemarketing result, banks can reduce the cost and the scope of marketing appropriately.

In many scientific and engineering fields, classification or prediction from the obtained data is a key issue. In recent years, along with the advances of data mining technology, many foreign scholars have used data mining methods to identify potential customers of direct mail marketing, that is, customer acquisition forecast. Ling and Li [1] used Naive Bayes and decision tree C4.5 algorithm to predict the three different data sets, which is an effective solution to the insufficiency of only using the predicting accuracy rate as evaluation criteria when the data sample distribution is extremely uneven (response rate is only 1 %), and solves the problem of how to select the appropriate algorithm when training set is too big. Research shows that using data mining as an effective tool of direct mail marketing can get more profits for banks, insurance companies and retails. Putten made a brief introduction of data mining in direct mail marketing application in 1998, including data and algorithm selection, evaluation results, and the use of BP neural network algorithm for the relevant empirical research [2]. Bawsens and Vianene and others used Bayesian neural network for forecasting, and got a good result [3].

Support vector machine (SVM), as an important branch of statistical learning theory, has many applications in the classification field. In Jacaheris study, using a SVM classification algorithm as a predictive algorithm of potential customers and verifies empirical research in Parisian bank, enhances the effect of forecast three times compared to the original algorithm [4]. But the best hyperplane of the traditional SVM classifier depends on part of the trained data point, it may be sensitive to noise or outliers in the training set. To overcome this drawback, many scholars have done lot of research work. Shigeo Abe put forward a fuzzy SVM to solve the multi-classification problem [5]. Zhousuo Zhang and others proposed a fuzzy SVM based on the differences of data importance to solve the mechanical malfunction diagnose problems [6]. Juang and Hsieh proposed TSFS-SVR algorithm, and the estimation of the weight adopted fuzzy clustering and linear SVR [7]. Hao and Chiang used symmetrical membership functions and typical combination of SVM kernel function, to provide a new method for fuzzy set theory and SVM parameters combination [8].

In this paper, we propose a forecasting model based on a fuzzy SVM algorithm to predict the acquisition of the potential customers of banks. The forecasting model uses a symmetric triangular membership function, first fuzzes input variables, then uses the fuzzy Multilayer Perception (FMLP) neural network as the fuzzy SVM kernel function, and then adopts FMLP-SVM algorithm to predict fuzzify input variables. This method uses data in direct marketing activities of Portugal financial institution, to predict whether customers will subscribe to deposits so as to get the predicting result of telemarketing.

2 Methods

2.1 Fuzzy SVM Algorithm

For this paper, we propose a method for the prediction of the potential customers of telemarketing based on an optimization model as follows:

$$minimize \frac{1}{2}\|w\|^2 + F\sum_{i=1}^{N}(\xi_i + \xi_i^*) \tag{1}$$

$$s.t. \begin{cases} Y_i - (w^T \cdot \phi(x_i) + a) \le \varepsilon - \xi_i \\ (w^T \cdot \phi(x_i) + a) - Y_i \le \varepsilon - \xi_i^* \\ \xi_i, \xi_i^* \ge 0, \forall i = 1, 2, \ldots, N \end{cases} \tag{2}$$

Here, w is an N-dimensional vector, F is the regularization parameter, a is a scalar, ϕ is the kernel function, ξ_i, ξ_i^* are the Slack variables and ε is the maximum error limit.

In fuzzy SVM, the input values are expressed as fuzzy vector $\widetilde{Y}_i = (Y_i^C, Y_i^S)$. Using symmetric triangular membership functions $u_{\widetilde{Y}_i}$, Y_i^C and Y_i^S is the center and extension of $\widetilde{Y}_i$ respectively. Membership function is in the following form:

$$u_{\widetilde{Y}_i} = 1 - \frac{|Y - Y_i^C|}{K_i Y_i^S} \tag{3}$$

Here, $X = [x_1, x_2, \ldots, x_N]^T$ is the tilt factor of $\widetilde{Y}_i$. For symmetric membership functions, $K_i = 1$. Each of the non-fuzzy input vectors $X = [x_1, x_2, \ldots, x_N]^T$ is transformed by the kernel function $\phi(x) = [\varphi(x_1), \varphi(x_1), \ldots, \varphi(x_1)]^T$. In the fuzzy weight $w^* = (w, v)$, w and v represent the central vector and the extended vector of w^* respectively, and N represents the dimension of the Eigen functions space. In the formula (3), $\widetilde{Y}_i$ is replaced by its approximation $Y_i^* = w^*\phi(x_i) + A^*$, in which $A^* = (a, b)$ is the bias term, a and b, respectively, indicates the center and the extension. Then $u_{\widetilde{Y}_i}$ can be expressed as the following form:

$$u_{\widetilde{Y}_i} = \begin{cases} 1 - \frac{|Y - (w^T\phi(x_i) + a)|}{K_i(v^T\,|\phi(x_i)| + b)} & x_i \ne 0, \forall Y \\ 1 - \frac{1}{K_i} & x_i = 0, Y = 0 \\ 0 & x_i = 0, Y \ne 0 \end{cases} \tag{4}$$

The fitting degree of the fuzzy model can be expressed as $H = \min(h_i), i = 1, 2, \ldots, M$. Here, h_i is the fitting degree, H is the fuzzy model, and M is the size of the training data set. The semantic definition of fuzzy output is a linear combination of the expansion of the model parameters. For an acceptable regression model, the semantics should be minimized. Therefore, the authors use $\sum_{j=1}^{n} v_j + b$ to replace the objective function in Formula 1.

Similar to the traditional SVM method, the fuzzy model requires minimum control capability term $\|w\|^2$. The constraint condition of SVM is rewritten, so the fitting degree of the training data set is greater than that of the fuzzy model. Thus, SVM is rewritten in the following form:

$$\begin{aligned} \min \quad & \tfrac{1}{2}\|w\|^2 + F\sum_{j=1}^{M} v_j + b \\ s.t. \quad & h_i \geq H, i = 1, 2, \ldots, M \end{aligned} \tag{5}$$

According to Tanaka's viewpoint, the constraint condition $h_i \geq H$ can be expanded into the following formula:

$$\begin{cases} (w^T\phi(x_i) + a) + (1 - H)(v^T\,|\phi(x_i)| + a) \geq Y_i \\ -(w^T\phi(x_i) + a) + (1 - H)(v^T\,|\phi(x_i)| + a) \geq -Y_i \end{cases} \tag{6}$$

The main difference between the SVM model and the corresponding fuzzy model is that the former is to find a function which makes the maximum ε criterion satisfied in the observed Y_i; the latter is trying to find a fuzzy function containing the fuzzy kernel parameters, and the expected target is considered.

2.2 FMLP Kernel Function

Choosing the appropriate kernel function can make the SVM to separate the data in the Eigen functions space effectively, even though they are inseparable in the original space. Comparison of different kernel functions, the radial basis function (RBF) kernel function has less parameter than linear and polynomial kernel function.

In this paper, the FMLP neural network function is used as kernel function of fuzzy SVM (formula 6). It will show that the fuzzy SVM model with application of FMLP kernel function has advantage in classification and identification.

In this paper, we only introduce a FMLP with an S type of hidden layer and linear output to approximate the nonlinear function, so FMLP can be written as

$$\widetilde{Y}_i = f^{(2)}(a_k^2 + \sum_{j=1}^{m} w_{kj}^{(2)} \times f^{(1)}(a_j^2 + \sum_{i=1}^{l} w_{ij}^{(1)} + x_i)) \tag{7}$$

where, x_i is the value of the ith node of the input layer, $\widetilde{Y}_i$ is the estimate of the corresponding x_i, and a_j is the deviation of the jth node of the hidden layer. a_k is the fuzzy offset of the kth output value, and w_{ij} is the fuzzy connection weights of the ith node of the input layer and the jth node of the hidden layer. L is the number of input neurons, M is the number of hidden layer neurons, $f^{(1)}$ and $f^{(2)}$ are the activation functions of the hidden layer and the output layer respectively.

3 Experimental Analysis

3.1 *Telemarketing Data of Banks*

In this paper, the data comes from the UCI data sets, which is about telemarketing belonging to direct marketing activities of a financial institution in Portugal [9]. Typically, in order to estimate if the product, such as bank deposits, is signed, it needs to contact the same customer at least once. Based on the sample data analysis conclusion, we can analyze customers consumption characteristics, risk preferences to determine whether its recent consumer behavior changes significantly, thus sums up the characteristics of the customers needs and anticipates changes in demand, then guide the potential customers actively and push the product successfully to market.

The data used in this paper is randomly selected 10 % from all samples [10]. In order to test the effect of the model, divide the sample data into the test set and the training set, for instance, 75 % of the groups data is the training set, and the rest is a test data set. Using the training set to develop prediction model based on FMLP-SVM method, predict the data with the developed model. For each set of data, it contains 17 variables. Among them, the former 16 variables as independent variables, shows some basic information of customers, namely, age, job, marital, education, default, balance, housing, loan, contact, day, month, duration, campaign, pdays, previous,poutcome. In these 16 independent variables, the first eight variables are some basic information of bank customers, the following four variables are the last contact between the client and the bank related to current marketing activities, and the other four variables are some other attribute data. The 17th variable is the dependent variable for each data set, it is a binary variable, "Yes" and "No" represent clients subscribe and unsubscribe bank deposits respectively, related to the telemarketing success. For each set of data, specific information and the related interpretation of some variables are shown in Table 1.

3.2 *Forecast and Analysis*

For the data in this paper, we hope to achieve the prediction of the telemarketing success rate, and then carry on targeted telemarketing based on the results that is predicted to improve efficiency and save cost. In this paper, we take if customers subscribe to deposits as the goal, and as the output variable, use additional information about the customer as an input variable, and develop a prediction model based on FMLP-SVM method so as to carry on data mining and predictive analysis.

Table 1 The information of variables

Data		Interpretation	Type	Remarks
Bank client data	Default	Has credit in default?	String	Binary: "yes", "no"
	Balance	Average yearly balance	Numeric	In euro
	Housing	Has housing loan?	String	Binary: "yes", "no"
	Loan	Has personal loan?	String	Binary: "yes", "no"
Related with the last contact of the current campaign	Contact	Contact communication type	String	Categorical: "unknown", "telephone", "cellular"
	Duration	Last contact duration, In seconds	Numeric	
Other attributes	Campaign	Number of contacts performed during this campaign and for this client	Numeric	Includes last contact
	Pdays	Number of days that passed by after the client was last contacted from a previous campaign	Numeric	−1 means client was not previously contacted
	Previous	Number of contacts performed before this campaign and for this client	Numeric	
	Poutcome	Outcome of the previous marketing campaign	String	Categorical: "unknown", "other", "failure", "success"
Output variable (desired target)	y	Has the client subscribed a term deposit?	String	Binary: "yes", "no"

To demonstrate the predictive accuracy of the model, we will compare and contrast the FMLP-SVM with the other three methods. We use the Bayesian networks, decision trees, and SVM for data to predict the corresponding prediction accuracy and make the comparison and contrast. In this paper, Weka software is implemented for forecasting of the four methods, and Matlab software is implemented to draw receiver operating characteristic (ROC) curve according to the predicting results of

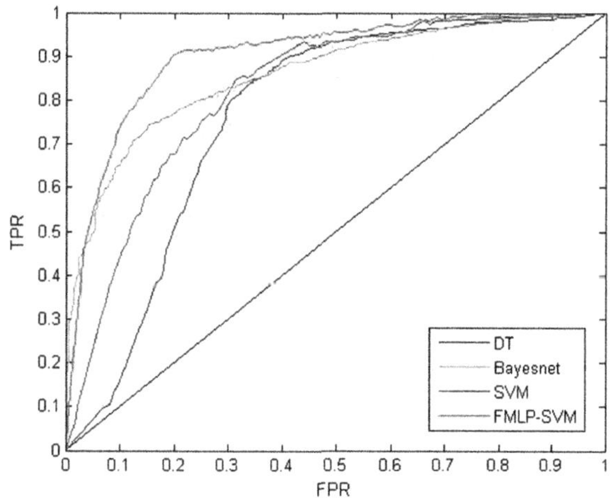

Fig. 1 ROC curve of the predict results of the four models

the four models and calculate the value of the corresponding area under the ROC curve (AUC). ROC curve is shown in Fig. 1.

ROC curve is based on a different set of dichotomous way (cut-off value or decision threshold), we use the true positive rate (sensitivity) as the vertical axis, the false positive rate (1-specificity) as the abscissa of the curve. The closer ROC curve is to the upper left corner, the higher the accuracy of the test is. From Fig. 1, we can see that, predictions of FMLP-SVM method is superior to other methods.

The size of AUC is between 1.0 and 0.5. In the case of AUC>0.5, The closer to 1 the AUC, the better the diagnosis. For instance, if it is between 0.5 and 0.7, it has less accuracy. AUC has some accuracy from 0.7 to 0.9. If AUC is above 0.9, it has high accuracy. What's more, if AUC $= 0.5$, it indicates that the diagnosis method is completely ineffective with no diagnostic value. If AUC<0.5, it does not comply with the real situation, this rarely occurs in practice. In this paper, we calculate the AUC values of the various methods, the Bayesian Network AUC value is 0.8608, the decision tree AUC is 0.7621, the traditional SVM AUC value is 0.8245, while the AUC FMLP-SVM value is 0.9048. Experimental data shows that the prediction of the first three methods has some accuracy but FMLP-SVM prediction we proposed has much higher accuracy.

Meanwhile, we select samples of different sizes to predict, and results of samples of different sizes corresponding to various methods are shown in Table 2. From Table 2, it can be seen that, for each sample size, the prediction accuracy of FMLP-SVM method is higher than other methods.

In summary, FMLP-SVM method used here to predict the success rate of telemarketing has good accuracy, and is significantly better than other predictive methods.

Table 2 Forecast accuracy comparison

Samplesize (%)	Bayes net (%)	DT (%)	SVM (%)	FMLP-SVM (%)
5	88.38	88.10	88.48	89.52
10	88.50	87.81	88.57	89.76
20	89.30	89.66	88.78	90.92
30	88.97	89.29	88.75	90.55
40	88.54	89.20	88.54	91.02
50	87.74	89.56	88.36	92.57
60	87.78	88.94	88.22	92.31
75	87.83	89.16	88.35	92.89

4 Conclusions

In this paper, the authors study the predicting success of commercial banks telemarketing, analyze the bank customers relevant information that has impact on telemarketing success, and finally propose the prediction model FMLP-SVM Method. The prediction model using fuzzy algorithm first fuzzifies input variables and uses fuzzified MLP neural network as a function of the fuzzy SVM kernel function, and then uses fuzzy SVM algorithm to predict the fuzzified data in order to get the forecast results of telemarketing success. Experimental data show that as for prediction of telemarketing success issues, this prediction results of FMLP-SVM method is superior to the other three forecasting methods with 92.89 % prediction accuracy rate, thus can help the bank better solve the forecasting problems of telemarketing success.

Acknowledgments We thank the editor, associate editor, and reviewers for thoughtful and constructive comments. This research is supported by the Natural Science Foundation of China under grant 11226335.

References

1. Ling, C.X., Li, C.: Data Mining for Direct Marketing: Problems and Solutions. In: KDD. vol. 98, pp. 73–79 (1998)
2. Van Der Putten, P.: Complexity and Management: A Collection of Essays. Data mining in direct marketing databases. World Scientific Publishers, Singapore (1999)
3. Baesens, B., Viaene, S., Van den Poel, D., Vanthienen, J., Dedene, G.: Bayesian neural network learning for repeat purchase modelling in direct marketing. Eur. J. Oper. Res. **138**, 191–211 (2002)
4. Javaheri, S.H., Sepehri, M.M., Teimourpour, B.: Response Modeling in Direct Marketing CA Data Mining Based Approach for Target Selection. Data Mining Applications with R. pp. 153–178 (2007)

5. Inoue, T., Abe, S.: Fuzzy support vector machines for pattern classification. In: Proceedings. IJCNN'01. International Joint Conference on IEEE Neural Networks, vol. 2, pp. 1449–1454 (2001)
6. Zhang, Z., Hou, Z., Sun, C., He, Z.: Hybrid intelligent diagnosis technology based on granular computing. J. Xi'an Jiaotong Univ. **1**, 011 (2011)
7. Tsang, E.C., Yeung, D.S., Chan, P.P.: Fuzzy support vector machines for solving two-class problems. In: Machine Learning and Cybernetics. vol. 2, pp. 1080–1083. IEEE Press (2003)
8. Vazifeh, M.R., Hao, P., Abbasi, F.: Fault diagnosis based on multikernel classification and information fusion decision. Comput. Technol. Appl. **4**, 404–409 (2013)
9. Moro, S., Cortez, P., Rita, P.: A data-driven approach to predict the success of bank telemarketing. Decis. Support Syst. **62**, 22–31 (2014)
10. Liu, M., He, Y., Wang, J., Lee, H.P., Liang, Y.: Hybrid intelligent algorithm and its application in geological hazard risk assessment. Neurocomputing **149**, 847–853 (2015)

Printed in the United States
By Bookmasters